# Chapter 5

# Chapter 6

# Chapter 7

# Chapter 8

# Chapter 9

# Chapter 10

# Electrical Engineering

## Principles and Applications

# Electrical Engineering
## Principles and Applications
**SEVENTH EDITION**

## Allan R. Hambley

Department of Electrical and Computer Engineering
Michigan Technological University
arhamble@mtu.edu

330 Hudson Street, NY NY 10013

Vice President and Editorial Director, ECS: Marcia J. Horton
Editor-in-Chief: Julian Partridge
Portfolio Manager: Julie Bai
Portfolio Manager Assistant: Michelle Bayman
Executive Marketing Manager: Tim Galligan
Director of Marketing: Christy Lesko
Field Marketing Manager: Demetrius Hall
Marketing Assistant: Jon Bryant
Managing Content Producer: Scott Disanno
Global HE Director of Vendor Sourcing and Procurement: Diane Hynes

Director of Operations: Nick Sklitsis
Operations Specialist: Maura Zaldivar-Garcia
Creative Director: Blair Brown
Cover Design: Black Horse Designs
Cover Image: Chesky/Shutterstock
Manager, Rights and Permissions: Ben Ferrini
Cover Printer: Phoenix Color/Hagerstown
Composition/Full-Service Project Management: SPi Global/ Shylaja Gattupalli

MATLAB is a registered trademark of The MathWorks, Inc. Company and product names mentioned herein are the trademarks or registered trademarks of their respective owners.

The author and publisher of this book have used their best efforts in preparing this book. These efforts include the development, research, and testing of the theories and programs to determine their effectiveness. The author and publisher make no warranty of any kind, expressed or implied, with regard to these programs or the documentation contained in this book. The author and publisher shall not be liable in any event for incidental or consequential damages in connection with, or arising out of, the furnishing, performance, or use of these programs.

**Library of Congress Cataloging-in-Publication Data**

Names: Hambley, Allan R., author.
Title: Electrical engineering : principles & applications / Allan R. Hambley.
Description: Seventh edition. | Hoboken : Pearson Education, 2016. | Includes index.
Identifiers: LCCN 2016020630| ISBN 9780134484143 | ISBN 0134484142
Subjects: LCSH: Electrical engineering—Textbooks.
Classification: LCC TK146 .H22 2016 | DDC 621.3—dc23
LC record available at https://lccn.loc.gov/2016020630

2   17

ISBN-10:    0-13-448414-2
ISBN-13: 978-0-13-448414-3

To my family Judy, Tony, Pam, and Mason
and to my special friend, Carol

# Practical Applications
# of Electrical Engineering Principles

# Contents

# Preface

As in the previous editions, my guiding philosophy in writing this book has three elements. The first element is my belief that in the long run students are best served by learning basic concepts in a general setting. Second, I believe that students need to be motivated by seeing how the principles apply to specific and interesting problems in their own fields. The third element of my philosophy is to take every opportunity to make learning free of frustration for the student.

This book covers circuit analysis, digital systems, electronics, and electromechanics at a level appropriate for either electrical-engineering students in an introductory course or nonmajors in a survey course. The only essential prerequisites are basic physics and single-variable calculus. Teaching a course using this book offers opportunities to develop theoretical and experimental skills and experiences in the following areas:

- Basic circuit analysis and measurement
- First- and second-order transients
- Steady-state ac circuits
- Resonance and frequency response
- Digital logic circuits
- Microcontrollers
- Computer-based instrumentation
- Diode circuits
- Electronic amplifiers
- Field-effect and bipolar junction transistors
- Operational amplifiers
- Transformers
- Ac and dc machines
- Computer-aided circuit analysis using MATLAB

While the emphasis of this book is on basic concepts, a key feature is the inclusion of short articles scattered throughout showing how electrical-engineering concepts are applied in other fields. The subjects of these articles include anti-knock signal processing for internal combustion engines, a cardiac pacemaker, active noise control, and the use of RFID tags in fisheries research, among others.

I welcome comments from users of this book. Information on how the book could be improved is especially valuable and will be taken to heart in future revisions. My e-mail address is `arhamble@mtu.edu`

# your work...

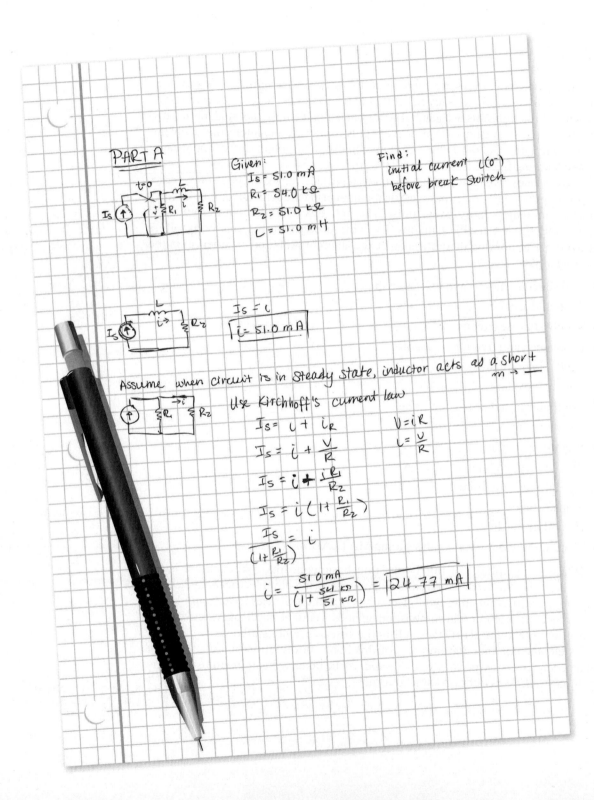

PART A

Given:
$I_S = 51.0 \text{ mA}$
$R_1 = 54.0 \text{ k}\Omega$
$R_2 = 51.0 \text{ k}\Omega$
$L = 51.0 \text{ mH}$

Find:
initial current $i(0^-)$
before break switch

$I_S = i$

$\boxed{i = 51.0 \text{ mA}}$

Assume when circuit is in steady state, inductor acts as a short $m \to -$

Use Kirchhoff's current law

$I_S = i + i_R$          $V = iR$

$I_S = i + \dfrac{V}{R}$          $i = \dfrac{V}{R}$

$I_S = i + \dfrac{iR_1}{R_2}$

$I_S = i\left(1 + \dfrac{R_1}{R_2}\right)$

$\dfrac{I_S}{\left(1 + \dfrac{R_1}{R_2}\right)} = i$

$i = \dfrac{51.0 \text{ mA}}{\left(1 + \dfrac{54 \text{ k}\Omega}{51 \text{ k}\Omega}\right)} = \boxed{24.77 \text{ mA}}$

# your answer specific feedback

**Express your answer to three significant figures and include the appropriate units.**

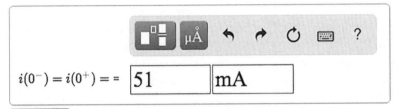

$i(0^-) = i(0^+) = =$ | `51` | `mA`

**Submit**   Hints   My Answers   Give Up   Review Part

**Incorrect; Try Again; 5 attempts remaining**

Note that elements in series have the same current but the inductor is not in series with the current source. Use Kirchhoff's current law or the current divider to find the initial inductor current.

**Express your answer to three significant figures and include the appropriate units.**

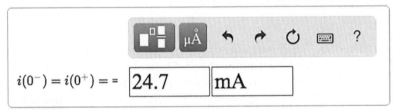

$i(0^-) = i(0^+) = =$ | `24.7` | `mA`

**Submit**   Hints   My Answers   Give Up   Review Part

**Incorrect; Try Again; 4 attempts remaining**

It appears you have found the current through the resistor, $R_1$. Find the current through the resistor in series with the inductor.

# www.MasteringEngineering®.com

## ON-LINE STUDENT RESOURCES

- **MasteringEngineering.** Tutorial homework problems emulate the instructor's office-hour environment, guiding students through engineering concepts with self-paced individualized coaching. These in-depth tutorial homework problems are designed to coach students with feedback specific to their errors and optional hints that break problems down into simpler steps. Video Solutions and coaching activities also provide complete, step-by-step solution walkthroughs of representative homework problems from each chapter. Access can be purchased bundled with the textbook or online at www.masteringengineering.com.

- Pearson eText, which is a complete on-line version of the book that includes highlighting, note-taking, and search capabilities is also available through MasteringEngineering.

- **Resource Website.** An open access website is available at www.pearsonhighered. com/engineering-resources. Resources include:

  - A Student Solutions Manual. A PDF file for each chapter includes full solutions for the in-chapter exercises, answers for the end-of-chapter problems that are marked with asterisks, and full solutions for the Practice Tests.

  - A MATLAB folder that contains the m-files discussed in the book.

## INSTRUCTOR RESOURCES

Resources for instructors include:

- **MasteringEngineering.** This online Tutorial Homework program allows you to integrate dynamic homework with automatic grading and personalized feedback. MasteringEngineering allows you to easily track the performance of your entire class on an assignment-by-assignment basis, or the detailed work of an individual student.

- A complete Instructor's Solutions Manual

- PowerPoint slides with all the figures from the book

Instructor Resources are available for download by adopters of this book at the Pearson Higher Education website: www.pearsonhighered.com. If you are in need of a login and password, please contact your local Pearson representative.

## WHAT'S NEW IN THIS EDITION

- We have continued and added items to the popular Practice Tests that students can use in preparing for course exams at the end of each chapter. Answers for the Practice Tests appear in Appendix D and complete solutions are included in the on-line Student Solutions Manual files.

- New examples have been added in Chapters 1 through 7.

- Approximately half of the end-of-chapter problems have been replaced or modified.

- Coverage of computers, microcontrollers and computer-based instrumentation has been merged from two chapters into Chapter 8 for this edition.

- Appendix C has been modified to keep up with new developments in the Fundamentals of Engineering Exam.
- We have updated the coverage of MATLAB and the Symbolic Toolbox for network analysis in Chapters 2 through 6.
- Relatively minor corrections and improvements appear throughout the book.

## PREREQUISITES

The essential prerequisites for a course from this book are basic physics and single-variable calculus. A prior differential equations course would be helpful but is not essential. Differential equations are encountered in Chapter 4 on transient analysis, but the skills needed are developed from basic calculus.

## PEDAGOGICAL FEATURES

The book includes various pedagogical features designed with the goal of stimulating student interest, eliminating frustration, and engendering an awareness of the relevance of the material to their chosen profession. These features are:

- Statements of learning objectives open each chapter.
- Comments in the margins emphasize and summarize important points or indicate common pitfalls that students need to avoid.
- Short boxed articles demonstrate how electrical-engineering principles are applied in other fields of engineering. For example, see the articles on active noise cancellation (page 296) and electronic pacemakers (starting on page 394).
- Step-by-step problem solving procedures. For example, see the step-by-step summary of node-voltage analysis (on pages 76–80) or the summary of Thévenin equivalents (on page 252).
- A Practice Test at the end of each chapter gives students a chance to test their knowledge. Answers appear in Appendix D.
- Complete solutions to the in-chapter exercises and Practice Tests, included as PDF files on-line, build student confidence and indicate where additional study is needed.
- Summaries of important points at the end of each chapter provide references for students.
- Key equations are highlighted in the book to draw attention to important results.

## MEETING ABET-DIRECTED OUTCOMES

Courses based on this book provide excellent opportunities to meet many of the directed outcomes for accreditation. The Criteria for Accrediting Engineering Programs require that graduates of accredited programs have "an ability to apply knowledge of mathematics, science, and engineering" and "an ability to identify, formulate, and solve engineering problems." This book, in its entirety, is aimed at developing these abilities.

Furthermore, the criteria require "an ability to function on multi-disciplinary teams" and "an ability to communicate effectively." Courses based on this book contribute to these abilities by giving nonmajors the knowledge and vocabulary to communicate effectively with electrical engineers. The book also helps to inform electrical engineers about applications in other fields of engineering. To aid in communication skills, end-of-chapter problems that ask students to explain electrical-engineering concepts in their own words are included.

## CONTENT AND ORGANIZATION

### Basic Circuit Analysis

Chapter 1 defines current, voltage, power, and energy. Kirchhoff's laws are introduced. Voltage sources, current sources, and resistance are defined.

Chapter 2 treats resistive circuits. Analysis by network reduction, node voltages, and mesh currents is covered. Thévenin equivalents, superposition, and the Wheatstone bridge are treated.

Capacitance, inductance, and mutual inductance are treated in Chapter 3.

Transients in electrical circuits are discussed in Chapter 4. First-order $RL$ and $RC$ circuits and time constants are covered, followed by a discussion of second-order circuits.

Chapter 5 considers sinusoidal steady-state circuit behavior. (A review of complex arithmetic is included in Appendix A.) Power calculations, ac Thévenin and Norton equivalents, and balanced three-phase circuits are treated.

Chapter 6 covers frequency response, Bode plots, resonance, filters, and digital signal processing. The basic concept of Fourier theory (that signals are composed of sinusoidal components having various amplitudes, phases, and frequencies) is qualitatively discussed.

### Digital Systems

Chapter 7 introduces logic gates and the representation of numerical data in binary form. It then proceeds to discuss combinatorial and sequential logic. Boolean algebra, De Morgan's laws, truth tables, Karnaugh maps, coders, decoders, flip-flops, and registers are discussed.

Chapter 8 treats microcomputers with emphasis on embedded systems using the Freescale Semiconductor HCS12/9S12 as the primary example. Computer organization and memory types are discussed. Digital process control using microcontrollers is described in general terms. Selected instructions and addressing modes for the CPU12 are described. Assembly language programming is treated very briefly. Finally, computer-based instrumentation systems including measurement concepts, sensors, signal conditioning, and analog-to-digital conversion are discussed.

### Electronic Devices and Circuits

Chapter 9 presents the diode, its various models, load-line analysis, and diode circuits, such as rectifiers, Zener-diode regulators, and wave shapers.

In Chapter 10, the specifications and imperfections of amplifiers that need to be considered in applications are discussed from a users perspective. These include gain, input impedance, output impedance, loading effects, frequency response, pulse response, nonlinear distortion, common-mode rejection, and dc offsets.

Chapter 11 covers the MOS field-effect transistor, its characteristic curves, loadline analysis, large-signal and small-signal models, bias circuits, the common-source amplifier, and the source follower.

Chapter 12 gives a similar treatment for bipolar transistors. If desired, the order of Chapters 11 and 12 can be reversed. Another possibility is to skip most of both chapters so more time can be devoted to other topics.

Chapter 13 treats the operational amplifier and many of its applications. Nonmajors can learn enough from this chapter to design and use op-amp circuits for instrumentation applications in their own fields.

## Electromechanics

Chapter 14 reviews basic magnetic field theory, analyzes magnetic circuits, and presents transformers.

DC machines and ac machines are treated in Chapters 15 and 16, respectively. The emphasis is on motors rather than generators because the nonelectrical engineer applies motors much more often than generators. In Chapter 15, an overall view of motors in general is presented before considering DC machines, their equivalent circuits, and performance calculations. The universal motor and its applications are discussed.

Chapter 16 deals with AC motors, starting with the three-phase induction motor. Synchronous motors and their advantages with respect to power-factor correction are analyzed. Small motors including single-phase induction motors are also discussed. A section on stepper motors and brushless dc motors ends the chapter.

# ACKNOWLEDGMENTS

I wish to thank my colleagues in the Electrical and Computer Engineering Department at Michigan Technological University, all of whom have given me help and encouragement at one time or another in writing this book and in my other projects.

I have received much excellent advice from professors at other institutions who reviewed the manuscript in various stages over the years. This advice has improved the final result a great deal, and I am grateful for their help.

Current and past reviewers include:

Ibrahim Abdel-Motaled, Northwestern University
William Best, Lehigh University
Steven Bibyk, Ohio State University
D. B. Brumm, Michigan Technological University
Karen Butler-Purry, Texas A&M University
Robert Collin, Case Western University
Joseph A. Coppola, Syracuse University
Norman R. Cox, University of Missouri at Rolla
W. T. Easter, North Carolina State University
Zoran Gajic, Rutgers University
Edwin L. Gerber, Drexel University
Victor Gerez, Montana State University
Walter Green, University of Tennessee
Elmer Grubbs, New Mexico Highlands University
Jasmine Henry, University of Western Australia
Ian Hutchinson, MIT
David Klemer, University of Wisconsin, Milwaukee
Richard S. Marleau, University of Wisconsin

Sunanda Mitra, Texas Tech University
Phil Noe, Texas A&M University
Edgar A. O'Hair, Texas Tech University
John Pavlat, Iowa State University
Clifford Pollock, Cornell University
Michael Reed, Carnegie Mellon University
Gerald F. Reid, Virginia Polytechnic Institute
Selahattin Sayil, Lamar University
William Sayle II, Georgia Institute of Technology
Len Trombetta, University of Houston
John Tyler, Texas A&M University
Belinda B. Wang, University of Toronto
Carl Wells, Washington State University
Al Wicks, Virginia Tech
Edward Yang, Columbia University
Subbaraya Yuvarajan, North Dakota State University
Rodger E. Ziemer, University of Colorado, Colorado Springs

Over the years, many students and faculty using my books at Michigan Technological University and elsewhere have made many excellent suggestions for improving the books and correcting errors. I thank them very much.

I am indebted to Julie Bai, my present and past editors at Pearson, for keeping me pointed in the right direction and for many excellent suggestions that have improved my books a great deal. A very special thank you, also, to Scott Disanno for a great job of managing the production of this and past editions of this book.

Also, I want to thank Tony, Pam, and Mason for their continuing encouragement and valuable insights. I thank Judy, my late wife, for many good things much too extensive to list.

ALLAN R. HAMBLEY

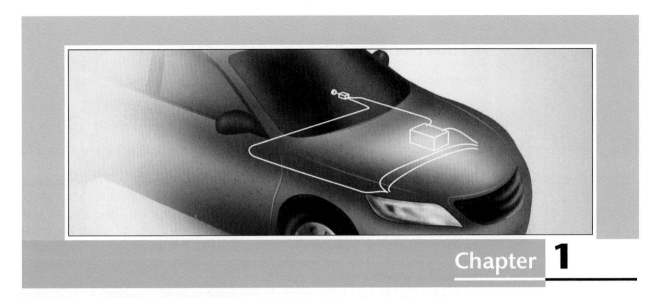

# Chapter 1

# Introduction

## Study of this chapter will enable you to:

- Recognize interrelationships between electrical engineering and other fields of science and engineering.

- List the major subfields of electrical engineering.

- List several important reasons for studying electrical engineering.

- Define current, voltage, and power, including their units.

- Calculate power and energy and determine whether energy is supplied or absorbed by a circuit element.

- State and apply Kirchhoff's current and voltage laws.

- Recognize series and parallel connections.

- Identify and describe the characteristics of voltage and current sources.

- State and apply Ohm's law.

- Solve for currents, voltages, and powers in simple circuits.

## Introduction to this chapter:

In this chapter, we introduce electrical engineering, define circuit variables (current, voltage, power, and energy), study the laws that these circuit variables obey, and meet several circuit elements (current sources, voltage sources, and resistors).

## 1.1  OVERVIEW OF ELECTRICAL ENGINEERING

Electrical engineers design systems that have two main objectives:

**1.** To gather, store, process, transport, and present *information*.

**2.** To distribute, store, and convert *energy* between various forms.

In many electrical systems, the manipulation of energy and the manipulation of information are interdependent.

For example, numerous aspects of electrical engineering relating to information are applied in weather prediction. Data about cloud cover, precipitation, wind speed, and so on are gathered electronically by weather satellites, by land-based radar stations, and by sensors at numerous weather stations. (Sensors are devices that convert physical measurements to electrical signals.) This information is transported by electronic communication systems and processed by computers to yield forecasts that are disseminated and displayed electronically.

In electrical power plants, energy is converted from various sources to electrical form. Electrical distribution systems transport the energy to virtually every factory, home, and business in the world, where it is converted to a multitude of useful forms, such as mechanical energy, heat, and light.

No doubt you can list scores of electrical engineering applications in your daily life. Increasingly, electrical and electronic features are integrated into new products. Automobiles and trucks provide just one example of this trend. The electronic content of the average automobile is growing rapidly in value. Self-driving vehicles are in rapid development and will eventually become the norm. Auto designers realize that electronic technology is a good way to provide increased functionality at lower cost. Table 1.1 shows some of the applications of electrical engineering in automobiles.

As another example, we note that many common household appliances contain keypads or touch screens for operator control, sensors, electronic displays, and computer chips, as well as more conventional switches, heating elements, and motors. Electronics have become so intimately integrated with mechanical systems that the name **mechatronics** is used for the combination.

> You may find it interesting to search the web for sites related to "mechatronics."

### Subdivisions of Electrical Engineering

Next, we give you an overall picture of electrical engineering by listing and briefly discussing eight of its major areas.

**1. Communication systems** transport information in electrical form. Cellular phone, radio, satellite television, and the Internet are examples of communication systems. It is possible for virtually any two people (or computers) on the globe to communicate almost instantaneously. A climber on a mountaintop in Nepal can call or send e-mail to friends whether they are hiking in Alaska or sitting in a New York City office. This kind of connectivity affects the way we live, the way we conduct business, and the design of everything we use. For example, communication systems will change the design of highways because traffic and road-condition information collected by roadside sensors can be transmitted to central locations and used to route traffic. When an accident occurs, an electrical signal can be emitted automatically when the airbags deploy, giving the exact location of the vehicle, summoning help, and notifying traffic-control computers.

**2. Computer** process and store information in digital form. No doubt you have already encountered computer applications in your own field. Besides the

> Computers that are part of products such as appliances and automobiles are called *embedded computers*.

**Table 1.1**  Current and Emerging Electronic/Electrical Applications in Automobiles and Trucks

Safety
    Antiskid brakes
    Inflatable restraints
    Collision warning and avoidance
    Blind-zone vehicle detection (especially for large trucks)
    Infrared night vision systems
    Heads-up displays
    Automatic accident notification
    Rear-view cameras

Communications and entertainment
    AM/FM radio
    Digital audio broadcasting
    CD/DVD player
    Cellular phone
    Computer/e-mail
    Satellite radio

Convenience
    Electronic GPS navigation
    Personalized seat/mirror/radio settings
    Electronic door locks

Emissions, performance, and fuel economy
    Vehicle instrumentation
    Electronic ignition
    Tire inflation sensors
    Computerized performance evaluation and maintenance scheduling
    Adaptable suspension systems

Alternative propulsion systems
    Electric vehicles
    Advanced batteries
    Hybrid vehicles

computers of which you are aware, there are many in unobvious places, such as household appliances and automobiles. A typical modern automobile contains several dozen special-purpose computers. Chemical processes and railroad switching yards are routinely controlled through computers.

3. **Control systems** gather information with sensors and use electrical energy to control a physical process. A relatively simple control system is the heating/cooling system in a residence. A sensor (thermostat) compares the temperature with the desired value. Control circuits operate the furnace or air conditioner to achieve the desired temperature. In rolling sheet steel, an electrical control system is used to obtain the desired sheet thickness. If the sheet is too thick (or thin), more (or less) force is applied to the rollers. The temperatures and flow rates in chemical processes are controlled in a similar manner. Control systems have even been installed in tall buildings to reduce their movement due to wind.

4. **Electromagnetics** is the study and application of electric and magnetic fields. The device (known as a magnetron) used to produce microwave energy in an oven is one application. Similar devices, but with much higher power levels,

are employed in manufacturing sheets of plywood. Electromagnetic fields heat the glue between layers of wood so that it will set quickly. Cellular phone and television antennas are also examples of electromagnetic devices.

5. **Electronics** is the study and application of materials, devices, and circuits used in amplifying and switching electrical signals. The most important electronic devices are transistors of various kinds. They are used in nearly all places where electrical information or energy is employed. For example, the cardiac pacemaker is an electronic circuit that senses heart beats, and if a beat does not occur when it should, applies a minute electrical stimulus to the heart, forcing a beat. Electronic instrumentation and electrical sensors are found in every field of science and engineering. Many of the aspects of electronic amplifiers studied later in this book have direct application to the instrumentation used in your field of engineering.

Electronic devices are based on controlling electrons. Photonic devices perform similar functions by controlling photons.

6. **Photonics** is an exciting new field of science and engineering that promises to replace conventional computing, signal-processing, sensing, and communication devices based on manipulating electrons with greatly improved products based on manipulating photons. Photonics includes light generation by lasers and light-emitting diodes, transmission of light through optical components, as well as switching, modulation, amplification, detection, and steering light by electrical, acoustical, and photon-based devices. Current applications include readers for DVD disks, holograms, optical signal processors, and fiber-optic communication systems. Future applications include optical computers, holographic memories, and medical devices. Photonics offers tremendous opportunities for nearly all scientists and engineers.

7. **Power systems** convert energy to and from electrical form and transmit energy over long distances. These systems are composed of generators, transformers, distribution lines, motors, and other elements. Mechanical engineers often utilize electrical motors to empower their designs. The selection of a motor having the proper torque speed characteristic for a given mechanical application is another example of how you can apply the information in this book.

8. **Signal processing** is concerned with information-bearing electrical signals. Often, the objective is to extract useful information from electrical signals derived from sensors. An application is machine vision for robots in manufacturing. Another application of signal processing is in controlling ignition systems of internal combustion engines. The timing of the ignition spark is critical in achieving good performance and low levels of pollutants. The optimum ignition point relative to crankshaft rotation depends on fuel quality, air temperature, throttle setting, engine speed, and other factors.

If the ignition point is advanced slightly beyond the point of best performance, *engine knock* occurs. Knock can be heard as a sharp metallic noise that is caused by rapid pressure fluctuations during the spontaneous release of chemical energy in the combustion chamber. A combustion-chamber pressure pulse displaying knock is shown in Figure 1.1. At high levels, knock will destroy an engine in a very short time. Prior to the advent of practical signal-processing electronics for this application, engine timing needed to be adjusted for distinctly suboptimum performance to avoid knock under varying combinations of operating conditions.

By connecting a sensor through a tube to the combustion chamber, an electrical signal proportional to pressure is obtained. Electronic circuits process this signal to determine whether the rapid pressure fluctuations characteristic of knock are present. Then electronic circuits continuously adjust ignition timing for optimum performance while avoiding knock.

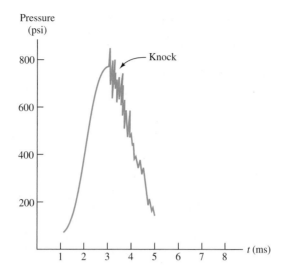

**Figure 1.1** Pressure versus time for an internal combustion engine experiencing knock. Sensors convert pressure to an electrical signal that is processed to adjust ignition timing for minimum pollution and good performance.

## Why You Need to Study Electrical Engineering

As a reader of this book, you may be majoring in another field of engineering or science and taking a required course in electrical engineering. Your immediate objective is probably to meet the course requirements for a degree in your chosen field. However, there are several other good reasons to learn and retain some basic knowledge of electrical engineering:

1. *To pass the Fundamentals of Engineering (FE) Examination as a first step in becoming a Registered Professional Engineer.* In the United States, before performing engineering services for the public, you will need to become registered as a Professional Engineer (PE). This book gives you the knowledge to answer questions relating to electrical engineering on the registration examinations. Save this book and course notes to review for the FE examination. (See Appendix C for more on the FE exam.)

   Save this book and course notes to review for the FE exam.

2. *To have a broad enough knowledge base so that you can lead design projects in your own field.* Increasingly, electrical engineering is interwoven with nearly all scientific experiments and design projects in other fields of engineering. Industry has repeatedly called for engineers who can see the big picture and work effectively in teams. Engineers or scientists who narrow their focus strictly to their own field are destined to be directed by others. (Electrical engineers are somewhat fortunate in this respect because the basics of structures, mechanisms, and chemical processes are familiar from everyday life. On the other hand, electrical engineering concepts are somewhat more abstract and hidden from the casual observer.)

3. *To be able to operate and maintain electrical systems, such as those found in control systems for manufacturing processes.* The vast majority of electrical-circuit malfunctions can be readily solved by the application of basic electrical-engineering principles. You will be a much more versatile and valuable engineer or scientist if you can apply electrical-engineering principles in practical situations.

4. *To be able to communicate with electrical-engineering consultants.* Very likely, you will often need to work closely with electrical engineers in your career. This book will give you the basic knowledge needed to communicate effectively.

### Content of This Book

Electrical engineering is too vast to cover in one or two courses. Our objective is to introduce the underlying concepts that you are most likely to need. Circuit theory is the electrical engineer's fundamental tool. That is why the first six chapters of this book are devoted to circuits.

Embedded computers, sensors, and electronic circuits will be an increasingly important part of the products you design and the instrumentation you use as an engineer or scientist. Chapters 7 and 8 treat digital systems with emphasis on embedded computers and instrumentation. Chapters 9 through 13 deal with electronic devices and circuits.

As a mechanical, chemical, civil, industrial, or other engineer, you will very likely need to employ energy-conversion devices. The last three chapters relate to electrical energy systems treating transformers, generators, and motors.

Because this book covers many basic concepts, it is also sometimes used in introductory courses for electrical engineers. Just as it is important for other engineers and scientists to see how electrical engineering can be applied to their fields, it is equally important for electrical engineers to be familiar with these applications.

## 1.2    CIRCUITS, CURRENTS, AND VOLTAGES

### Overview of an Electrical Circuit

Before we carefully define the terminology of electrical circuits, let us gain some basic understanding by considering a simple example: the headlight circuit of an automobile. This circuit consists of a battery, a switch, the headlamps, and wires connecting them in a closed path, as illustrated in Figure 1.2.

The battery voltage is a measure of the energy gained by a unit of charge as it moves through the battery.

Chemical forces in the battery cause electrical charge (electrons) to flow through the circuit. The charge gains energy from the chemicals in the battery and delivers energy to the headlamps. The battery voltage (nominally, 12 volts) is a measure of the energy gained by a unit of charge as it moves through the battery.

Electrons readily move through copper but not through plastic insulation.

The wires are made of an excellent electrical conductor (copper) and are insulated from one another (and from the metal auto body) by electrical insulation (plastic) coating the wires. Electrons readily move through copper but not through the plastic insulation. Thus, the charge flow (electrical current) is confined to the wires until it reaches the headlamps. Air is also an insulator.

The switch is used to control the flow of current. When the conducting metallic parts of the switch make contact, we say that the switch is **closed** switch and current flows through the circuit. On the other hand, when the conducting parts of the switch do not make contact, we say that the switch is **open** and current does not flow.

Electrons experience collisions with the atoms of the tungsten wires, resulting in heating of the tungsten.

The headlamps contain special tungsten wires that can withstand high temperatures. Tungsten is not as good an electrical conductor as copper, and the electrons experience collisions with the atoms of the tungsten wires, resulting in heating of the tungsten. We say that the tungsten wires have electrical resistance. Thus, energy is transferred by the chemical action in the battery to the electrons and then to the tungsten, where it appears as heat. The tungsten becomes hot enough so that copious light is emitted. We will see that the power transferred is equal to the product of current (rate of flow of charge) and the voltage (also called electrical potential) applied by the battery.

Energy is transferred by the chemical action in the battery to the electrons and then to the tungsten.

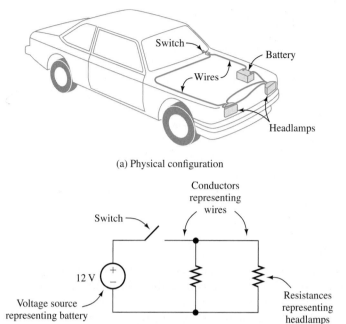

(a) Physical configuration

(b) Circuit diagram

**Figure 1.2**  The headlight circuit. (a) The actual physical layout of the circuit. (b) The circuit diagram.

(Actually, the simple description of the headlight circuit we have given is most appropriate for older cars. In more modern automobiles, light emitting diodes (LEDs) are used in place of the tungsten filaments. Furthermore, sensors provide information to an embedded computer about the ambient light level, whether or not the ignition is energized, and whether the transmission is in park or drive. The dashboard switch merely inputs a logic level to the computer, indicating the intention of the operator with regard to the headlights. Depending on these inputs, the computer controls the state of an electronic switch in the headlight circuit. When the ignition is turned off and if it is dark, the computer keeps the lights on for a few minutes so the passengers can see to exit and then turns them off to conserve energy in the battery. This is typical of the trend to use highly sophisticated electronic and computer technology to enhance the capabilities of new designs in all fields of engineering.)

## Fluid-Flow Analogy

Electrical circuits are analogous to fluid-flow systems. The battery is analogous to a pump, and charge is analogous to the fluid. Conductors (usually copper wires) correspond to frictionless pipes through which the fluid flows. Electrical current is the counterpart of the flow rate of the fluid. Voltage corresponds to the pressure differential between points in the fluid circuit. Switches are analogous to valves. Finally, the electrical resistance of a tungsten headlamp is analogous to a constriction in a fluid system that results in turbulence and conversion of energy to heat. Notice that current is a measure of the flow of charge *through* the cross section of a circuit element, whereas voltage is measured *across* the ends of a circuit element or *between* any other two points in a circuit.

The fluid-flow analogy can be very helpful initially in understanding electrical circuits.

Now that we have gained a basic understanding of a simple electrical circuit, we will define the concepts and terminology more carefully.

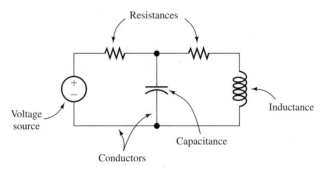

**Figure 1.3** An electrical circuit consists of circuit elements, such as voltage sources, resistances, inductances, and capacitances, connected in closed paths by conductors.

## Electrical Circuits

An electrical circuit consists of various types of circuit elements connected in closed paths by conductors.

An **electrical circuit** consists of various types of circuit elements connected in closed paths by conductors. An example is illustrated in Figure 1.3. The circuit elements can be resistances, inductances, capacitances, and voltage sources, among others. The symbols for some of these elements are illustrated in the figure. Eventually, we will carefully discuss the characteristics of each type of element.

Charge flows easily through conductors.

Charge flows easily through conductors, which are represented by lines connecting circuit elements. Conductors correspond to connecting wires in physical circuits. Voltage sources create forces that cause charge to flow through the conductors and other circuit elements. As a result, energy is transferred between the circuit elements, resulting in a useful function.

## Electrical Current

Current is the time rate of flow of electrical charge. Its units are amperes (A), which are equivalent to coulombs per second (C/s).

**Electrical current** is the time rate of flow of electrical charge through a conductor or circuit element. The units are amperes (A), which are equivalent to coulombs per second (C/s). (The charge on an electron is $-1.602 \times 10^{-19}$ C.)

Conceptually, to find the current for a given circuit element, we first select a cross section of the circuit element roughly perpendicular to the flow of current. Then, we select a **reference direction** along the direction of flow. Thus, the reference direction points from one side of the cross section to the other. This is illustrated in Figure 1.4.

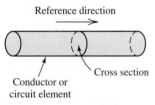

**Figure 1.4** Current is the time rate of charge flow through a cross section of a conductor or circuit element.

Next, suppose that we keep a record of the net charge flow through the cross section. Positive charge crossing in the reference direction is counted as a positive contribution to net charge. Positive charge crossing opposite to the reference is counted as a negative contribution. Furthermore, negative charge crossing in the reference direction is counted as a negative contribution, and negative charge against the reference direction is a positive contribution to charge.

Thus, in concept, we obtain a record of the net charge in coulombs as a function of time in seconds denoted as $q(t)$. The electrical current flowing through the element in the reference direction is given by

Colored shading is used to indicate key equations throughout this book.

$$i(t) = \frac{dq(t)}{dt} \tag{1.1}$$

A constant current of one ampere means that one coulomb of charge passes through the cross section each second.

To find charge given current, we must integrate. Thus, we have

$$q(t) = \int_{t_0}^{t} i(t)\, dt + q(t_0) \qquad (1.2)$$

in which $t_0$ is some initial time at which the charge is known. (Throughout this book, we assume that time $t$ is in seconds unless stated otherwise.)

Current flow is the same for all cross sections of a circuit element. (We reexamine this statement when we introduce the capacitor in Chapter 3.) The current that enters one end flows through the element and exits through the other end.

**Example 1.1   Determining Current Given Charge**

Suppose that charge versus time for a given circuit element is given by

$$q(t) = 0 \qquad \text{for } t < 0$$

and

$$q(t) = 2 - 2e^{-100t}\ \text{C} \qquad \text{for } t > 0$$

Sketch $q(t)$ and $i(t)$ to scale versus time.

**Solution**   First we use Equation 1.1 to find an expression for the current:

$$i(t) = \frac{dq(t)}{dt}$$
$$= 0 \qquad \text{for } t < 0$$
$$= 200e^{-100t}\ \text{A} \qquad \text{for } t > 0$$

Plots of $q(t)$ and $i(t)$ are shown in Figure 1.5. ∎

### Reference Directions

In analyzing electrical circuits, we may not initially know the *actual direction* of current flow in a particular circuit element. Therefore, we start by assigning current

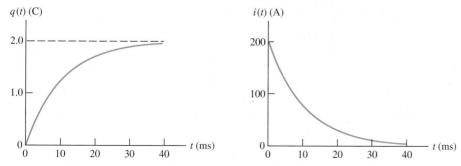

**Figure 1.5** Plots of charge and current versus time for Example 1.1. *Note:* The time scale is in milliseconds (ms). One millisecond is equivalent to $10^{-3}$ seconds.

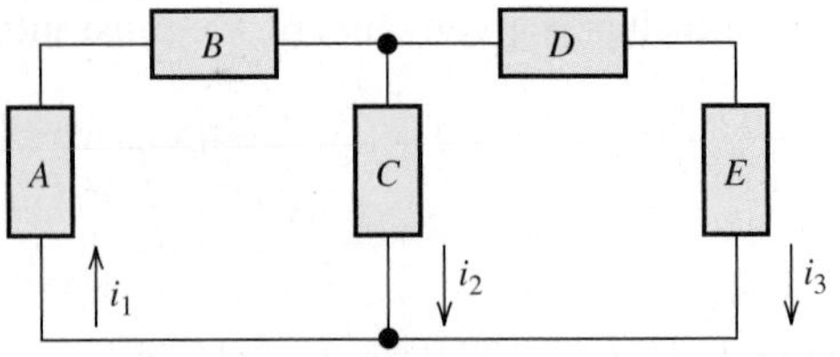

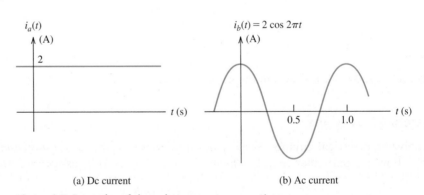

**Figure 1.6** In analyzing circuits, we frequently start by assigning current variables $i_1$, $i_2$, $i_3$, and so forth.

variables and arbitrarily selecting a *reference direction* for each current of interest. It is customary to use the letter $i$ for currents and subscripts to distinguish different currents. This is illustrated by the example in Figure 1.6, in which the boxes labeled $A$, $B$, and so on represent circuit elements. After we solve for the current values, we may find that some currents have negative values. For example, suppose that $i_1 = -2$ A in the circuit of Figure 1.6. Because $i_1$ has a negative value, we know that current actually flows in the direction opposite to the reference initially selected for $i_1$. Thus, the actual current is 2 A flowing downward through element $A$.

### Direct Current and Alternating Current

Dc currents are constant with respect to time, whereas ac currents vary with time.

When a current is constant with time, we say that we have **direct current**, abbreviated as dc. On the other hand, a current that varies with time, reversing direction periodically, is called **alternating current**, abbreviated as ac. Figure 1.7 shows the values of a dc current and a sinusoidal ac current versus time. When $i_b(t)$ takes a negative value, the actual current direction is opposite to the reference direction for $i_b(t)$. The designation ac is used for other types of time-varying currents, such as the triangular and square waveforms shown in Figure 1.8.

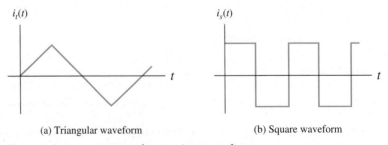

(a) Dc current    (b) Ac current

**Figure 1.7** Examples of dc and ac currents versus time.

(a) Triangular waveform    (b) Square waveform

**Figure 1.8** Ac currents can have various waveforms.

## Double-Subscript Notation for Currents

So far we have used arrows alongside circuit elements or conductors to indicate reference directions for currents. Another way to indicate the current and reference direction for a circuit element is to label the ends of the element and use double subscripts to define the reference direction for the current. For example, consider the resistance of Figure 1.9. The current denoted by $i_{ab}$ is the current through the element with its reference direction pointing from $a$ to $b$. Similarly, $i_{ba}$ is the current with its reference directed from $b$ to $a$. Of course, $i_{ab}$ and $i_{ba}$ are the same in magnitude and opposite in sign, because they denote the same current but with opposite reference directions. Thus, we have

$$i_{ab} = -i_{ba}$$

**Figure 1.9** Reference directions can be indicated by labeling the ends of circuit elements and using double subscripts on current variables. The reference direction for $i_{ab}$ points from $a$ to $b$. On the other hand, the reference direction for $i_{ba}$ points from $b$ to $a$.

**Exercise 1.1**   A constant current of 2 A flows through a circuit element. In 10 seconds (s), how much net charge passes through the element?
**Answer**   20 C.                                                                                     ☐

**Exercise 1.2**   The charge that passes through a circuit element is given by $q(t) = 0.01 \sin(200t)$ C, in which the angle is in radians. Find the current as a function of time.
**Answer**   $i(t) = 2 \cos(200t)$ A.                                                                  ☐

**Exercise 1.3**   In Figure 1.6, suppose that $i_2 = 1$ A and $i_3 = -3$ A. Assuming that the current consists of positive charge, in which direction (upward or downward) is charge moving in element $C$? In element $E$?
**Answer**   Downward in element $C$ and upward in element $E$.                                        ☐

## Voltages

When charge moves through circuit elements, energy can be transferred. In the case of automobile headlights, stored chemical energy is supplied by the battery and absorbed by the headlights where it appears as heat and light. The **voltage** associated with a circuit element is the energy transferred per unit of charge that flows through the element. The units of voltage are volts (V), which are equivalent to joules per coulomb (J/C).

Voltage is a measure of the energy transferred per unit of charge when charge moves from one point in an electrical circuit to a second point.

For example, consider the storage battery in an automobile. The voltage across its terminals is (nominally) 12 V. This means that 12 J are transferred to or from the battery for each coulomb that flows through it. When charge flows in one direction, energy is supplied by the battery, appearing elsewhere in the circuit as heat or light or perhaps as mechanical energy at the starter motor. If charge moves through the battery in the opposite direction, energy is absorbed by the battery, where it appears as stored chemical energy.

Notice that voltage is measured across the ends of a circuit element, whereas current is a measure of charge flow through the element.

Voltages are assigned polarities that indicate the direction of energy flow. If positive charge moves from the positive polarity through the element toward the negative polarity, the element absorbs energy that appears as heat, mechanical energy, stored chemical energy, or as some other form. On the other hand, if positive charge moves from the negative polarity toward the positive polarity, the element supplies energy. This is illustrated in Figure 1.10. For negative charge, the direction of energy transfer is reversed.

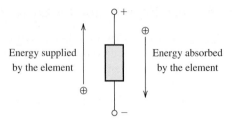

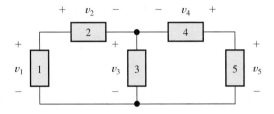

Figure 1.10  Energy is transferred when charge flows through an element having a voltage across it.

Figure 1.11  If we do not know the voltage values and polarities in a circuit, we can start by assigning voltage variables choosing the reference polarities arbitrarily. (The boxes represent unspecified circuit elements.)

### Reference Polarities

When we begin to analyze a circuit, we often do not know the actual polarities of some of the voltages of interest in the circuit. Then, we simply assign voltage variables choosing *reference* polarities arbitrarily. (Of course, the *actual* polarities are not arbitrary.) This is illustrated in Figure 1.11. Next, we apply circuit principles (discussed later), obtaining equations that are solved for the voltages. If a given voltage has an actual polarity opposite to our arbitrary choice for the reference polarity, we obtain a negative value for the voltage. For example, if we find that $v_3 = -5$ V in Figure 1.11, we know that the voltage across element 3 is 5 V in magnitude and its actual polarity is opposite to that shown in the figure (i.e., the actual polarity is positive at the bottom end of element 3 and negative at the top).

In circuit analysis, we frequently assign reference polarities for voltages arbitrarily. If we find at the end of the analysis that the value of a voltage is negative, then we know that the true polarity is opposite of the polarity selected initially.

We usually do not put much effort into trying to assign "correct" references for current directions or voltage polarities. If we have doubt about them, we make arbitrary choices and use circuit analysis to determine true directions and polarities (as well as the magnitudes of the currents and voltages).

Voltages can be constant with time or they can vary. Constant voltages are called **dc voltages**. On the other hand, voltages that change in magnitude and alternate in polarity with time are said to be **ac voltages**. For example,

$$v_1(t) = 10 \text{ V}$$

is a dc voltage. It has the same magnitude and polarity for all time. On the other hand,

$$v_2(t) = 10 \cos(200\pi t) \text{ V}$$

is an ac voltage that varies in magnitude and polarity. When $v_2(t)$ assumes a negative value, the actual polarity is opposite the reference polarity. (We study sinusoidal ac currents and voltages in Chapter 5.)

### Double-Subscript Notation for Voltages

Figure 1.12  The voltage $v_{ab}$ has a reference polarity that is positive at point *a* and negative at point *b*.

Another way to indicate the reference polarity of a voltage is to use double subscripts on the voltage variable. We use letters or numbers to label the terminals between which the voltage appears, as illustrated in Figure 1.12. For the resistance shown in the figure, $v_{ab}$ represents the voltage between points $a$ and $b$ with the positive reference at point $a$. The two subscripts identify the points between which the voltage appears, and the first subscript is the positive reference. Similarly, $v_{ba}$ is the voltage between $a$ and $b$ with the positive reference at point $b$. Thus, we can write

$$v_{ab} = -v_{ba} \qquad (1.3)$$

because $v_{ba}$ has the same magnitude as $v_{ab}$ but has opposite polarity.

Still another way to indicate a voltage and its reference polarity is to use an arrow, as shown in Figure 1.13. The positive reference corresponds to the head of the arrow.

## Switches

Switches control the currents in circuits. When an ideal switch is open, the current through it is zero and the voltage across it is determined by the remainder of the circuit. When an ideal switch is closed, the voltage across it is zero and the current through it is determined by the remainder of the circuit.

**Figure 1.13** The positive reference for $v$ is at the head of the arrow.

---

**Exercise 1.4**   The voltage across a given circuit element is $v_{ab} = 20$ V. A positive charge of 2 C moves through the circuit element from terminal $b$ to terminal $a$. How much energy is transferred? Is the energy supplied by the circuit element or absorbed by it?

**Answer**   40 J are supplied by the circuit element.   ☐

---

## 1.3   POWER AND ENERGY

Consider the circuit element shown in Figure 1.14. Because the current $i$ is the rate of flow of charge and the voltage $v$ is a measure of the energy transferred per unit of charge, the product of the current and the voltage is the rate of energy transfer. In other words, the product of current and voltage is power:

$$p = vi \qquad (1.4)$$

The physical units of the quantities on the right-hand side of this equation are

$$\text{volts} \times \text{amperes} =$$

$$\text{(joules/coulomb)} \times \text{(coulombs/second)} =$$

$$\text{joules/second} =$$

$$\text{watts}$$

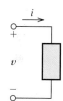

**Figure 1.14** When current flows through an element and voltage appears across the element, energy is transferred. The rate of energy transfer is $p = vi$.

## Passive Reference Configuration

Now we may ask whether the power calculated by Equation 1.4 represents energy supplied by or absorbed by the element. Refer to Figure 1.14 and notice that the current reference enters the positive polarity of the voltage. We call this arrangement the **passive reference configuration**. Provided that the references are picked in this manner, a positive result for the power calculation implies that energy is being absorbed by the element. On the other hand, a negative result means that the element is supplying energy to other parts of the circuit.

If the current reference enters the negative end of the reference polarity, we compute the power as

$$p = -vi \tag{1.5}$$

Then, as before, a positive value for $p$ indicates that energy is absorbed by the element, and a negative value shows that energy is supplied by the element.

If the circuit element happens to be an electrochemical battery, positive power means that the battery is being charged. In other words, the energy absorbed by the battery is being stored as chemical energy. On the other hand, negative power indicates that the battery is being discharged. Then the energy supplied by the battery is delivered to some other element in the circuit.

Sometimes currents, voltages, and powers are functions of time. To emphasize this fact, we can write Equation 1.4 as

$$p(t) = v(t)i(t) \tag{1.6}$$

### Example 1.2    Power Calculations

Consider the circuit elements shown in Figure 1.15. Calculate the power for each element. If each element is a battery, is it being charged or discharged?

**Solution**    In element $A$, the current reference enters the positive reference polarity. This is the passive reference configuration. Thus, power is computed as

$$p_a = v_a i_a = 12 \text{ V} \times 2 \text{ A} = 24 \text{ W}$$

Because the power is positive, energy is absorbed by the device. If it is a battery, it is being charged.

In element $B$, the current reference enters the negative reference polarity. (Recall that the current that enters one end of a circuit element must exit from the other end, and vice versa.) This is opposite to the passive reference configuration. Hence, power is computed as

$$p_b = -v_b i_b = -(12 \text{ V}) \times 1 \text{ A} = -12 \text{ W}$$

Since the power is negative, energy is supplied by the device. If it is a battery, it is being discharged.

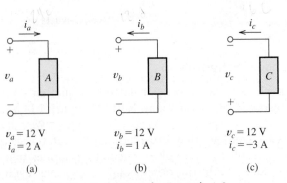

$$
\begin{array}{ccc}
v_a = 12 \text{ V} & v_b = 12 \text{ V} & v_c = 12 \text{ V} \\
i_a = 2 \text{ A} & i_b = 1 \text{ A} & i_c = -3 \text{ A} \\
\text{(a)} & \text{(b)} & \text{(c)}
\end{array}
$$

**Figure 1.15** Circuit elements for Example 1.2.

In element $C$, the current reference enters the positive reference polarity. This is the passive reference configuration. Thus, we compute power as

$$p_c = v_c i_c = 12 \text{ V} \times (-3 \text{ A}) = -36 \text{ W}$$

Since the result is negative, energy is supplied by the element. If it is a battery, it is being discharged. (Notice that since $i_c$ takes a negative value, current actually flows downward through element $C$.) ∎

## Energy Calculations

To calculate the energy $w$ delivered to a circuit element between time instants $t_1$ and $t_2$, we integrate power:

$$w = \int_{t_1}^{t_2} p(t)\, dt \tag{1.7}$$

Here we have explicitly indicated that power can be a function of time by using the notation $p(t)$.

---

**Example 1.3    Energy Calculation**

Find an expression for the power for the voltage source shown in Figure 1.16. Compute the energy for the interval from $t_1 = 0$ to $t_2 = \infty$.

**Solution**   The current reference enters the positive reference polarity. Thus, we compute power as

$$p(t) = v(t)i(t)$$
$$= 12 \times 2e^{-t}$$
$$= 24e^{-t} \text{ W}$$

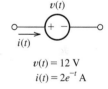

$v(t) = 12 \text{ V}$
$i(t) = 2e^{-t} \text{ A}$

**Figure 1.16** Circuit element for Example 1.3.

Subsequently, the energy transferred is given by

$$w = \int_0^\infty p(t)\, dt$$
$$= \int_0^\infty 24e^{-t}\, dt$$
$$= [-24e^{-t}]_0^\infty = -24e^{-\infty} - (-24e^0) = 24 \text{ J}$$

Because the energy is positive, it is absorbed by the source. ∎

## Prefixes

In electrical engineering, we encounter a tremendous range of values for currents, voltages, powers, and other quantities. We use the prefixes shown in Table 1.2 when working with very large or small quantities. For example, 1 milliampere (1 mA) is equivalent to $10^{-3}$ A, 1 kilovolt (1 kV) is equivalent to 1000 V, and so on.

**Table 1.2** Prefixes Used for Large or Small Physical Quantities

| Prefix | Abbreviation | Scale Factor |
|---|---|---|
| giga- | G | $10^9$ |
| meg- or mega- | M | $10^6$ |
| kilo- | k | $10^3$ |
| milli- | m | $10^{-3}$ |
| micro- | $\mu$ | $10^{-6}$ |
| nano- | n | $10^{-9}$ |
| pico- | p | $10^{-12}$ |
| femto- | f | $10^{-15}$ |

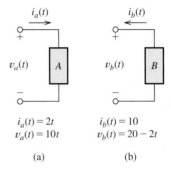

$$i_a(t) = 2t$$
$$v_a(t) = 10t$$

$$i_b(t) = 10$$
$$v_b(t) = 20 - 2t$$

(a)        (b)

**Figure 1.17**  See Exercise 1.6.

**Exercise 1.5**   The ends of a circuit element are labeled $a$ and $b$, respectively. Are the references for $i_{ab}$ and $v_{ab}$ related by the passive reference configuration? Explain.
**Answer**   The reference direction for $i_{ab}$ enters terminal $a$, which is also the positive reference for $v_{ab}$. Therefore, the current reference direction enters the positive reference polarity, so we have the passive reference configuration.  □

**Exercise 1.6**   Compute the power as a function of time for each of the elements shown in Figure 1.17. Find the energy transferred between $t_1 = 0$ and $t_2 = 10$ s. In each case is energy supplied or absorbed by the element?
**Answer**   **a.** $p_a(t) = 20t^2$ W, $w_a = 6667$ J; since $w_a$ is positive, energy is absorbed by element $A$. **b.** $p_b(t) = 20t - 200$ W, $w_b = -1000$ J; since $w_b$ is negative, energy is supplied by element $B$.  □

## 1.4  KIRCHHOFF'S CURRENT LAW

Kirchhoff's current law states that the net current entering a node is zero.

A **node** in an electrical circuit is a point at which two or more circuit elements are joined together. Examples of nodes are shown in Figure 1.18.

An important principle of electrical circuits is **Kirchhoff's current law**: *The net current entering a node is zero.* To compute the *net* current entering a node, we add the currents entering and subtract the currents leaving. For illustration, consider the nodes of Figure 1.18. Then, we can write:

$$\text{Node } a: \quad i_1 + i_2 - i_3 = 0$$
$$\text{Node } b: \quad i_3 - i_4 = 0$$
$$\text{Node } c: \quad i_5 + i_6 + i_7 = 0$$

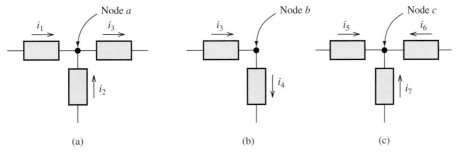

**Figure 1.18** Partial circuits showing one node each to illustrate Kirchhoff's current law.

Notice that for node $b$, Kirchhoff's current law requires that $i_3 = i_4$. In general, if only two circuit elements are connected at a node, their currents must be equal. The current flows into the node through one element and out through the other. Usually, we will recognize this fact and assign a single current variable for both circuit elements.

For node $c$, either all of the currents are zero or some are positive while others are negative.

We abbreviate Kirchhoff's current law as KCL. There are two other equivalent ways to state KCL. One way is: *The net current leaving a node is zero.* To compute the net current leaving a node, we add the currents leaving and subtract the currents entering. For the nodes of Figure 1.18, this yields the following:

$$\text{Node } a: \quad -i_1 - i_2 + i_3 = 0$$
$$\text{Node } b: \quad -i_3 + i_4 = 0$$
$$\text{Node } c: \quad -i_5 - i_6 - i_7 = 0$$

Of course, these equations are equivalent to those obtained earlier.

Another way to state KCL is: *The sum of the currents entering a node equals the sum of the currents leaving a node.* Applying this statement to Figure 1.18, we obtain the following set of equations:

$$\text{Node } a: \quad i_1 + i_2 = i_3$$
$$\text{Node } b: \quad i_3 = i_4$$
$$\text{Node } c: \quad i_5 + i_6 + i_7 = 0$$

An alternative way to state Kirchhoff's current law is that the sum of the currents entering a node is equal to the sum of the currents leaving a node.

Again, these equations are equivalent to those obtained earlier.

## Physical Basis for Kirchhoff's Current Law

An appreciation of why KCL is true can be obtained by considering what would happen if it were violated. Suppose that we could have the situation shown in Figure 1.18(a), with $i_1 = 3$ A, $i_2 = 2$ A, and $i_3 = 4$ A. Then, the net current entering the node would be

$$i_1 + i_2 - i_3 = 1 \text{ A} = 1 \text{ C/s}$$

In this case, 1 C of charge would accumulate at the node during each second. After 1 s, we would have $+1$ C of charge at the node, and $-1$ C of charge somewhere else in the circuit.

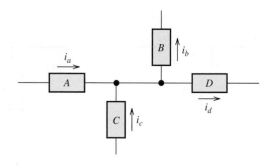

Figure 1.19 Elements *A, B, C,* and *D* can be considered to be connected to a common node, because all points in a circuit that are connected directly by conductors are electrically equivalent to a single point.

Suppose that these charges are separated by a distance of one meter (m). Recall that unlike charges experience a force of attraction. The resulting force turns out to be approximately $8.99 \times 10^9$ newtons (N) (equivalent to $2.02 \times 10^9$ pounds). Very large forces are generated when charges of this magnitude are separated by moderate distances. In effect, KCL states that such forces prevent charge from accumulating at the nodes of a circuit.

All points in a circuit that are connected directly by conductors can be considered to be a single node. For example, in Figure 1.19, elements $A, B, C,$ and $D$ are connected to a common node. Applying KCL, we can write

$$i_a + i_c = i_b + i_d$$

All points in a circuit that are connected directly by conductors can be considered to be a single node.

## Series Circuits

We make frequent use of KCL in analyzing circuits. For example, consider the elements $A$, $B$, and $C$ shown in Figure 1.20. When elements are connected end to end, we say that they are connected in **series**. *In order for elements A and B to be in series, no other path for current can be connected to the node joining A and B. Thus, all elements in a series circuit have identical currents.* For example, writing Kirchhoff's current law at node 1 for the circuit of Figure 1.20, we have

$$i_a = i_b$$

At node 2, we have

$$i_b = i_c$$

Thus, we have

$$i_a = i_b = i_c$$

The current that enters a series circuit must flow through each element in the circuit.

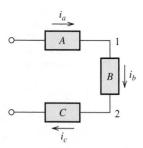

Figure 1.20 Elements *A, B,* and C are connected in series.

---

| Example 1.4 | Kirchhoff's Current Law |

Consider the circuit shown in Figure 1.21.

**a.** Which elements are in series?

**b.** What is the relationship between $i_d$ and $i_c$?

**c.** Given that $i_a = 6$ A and $i_c = -2$ A, determine the values of $i_b$ and $i_d$.

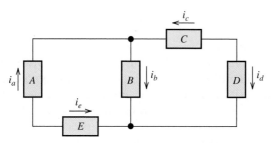

Figure 1.21  Circuit for Example 1.4.

## Solution

**a.** Elements $A$ and $E$ are in series, and elements $C$ and $D$ are in series.

**b.** Because elements $C$ and $D$ are in series, the currents are equal in magnitude. However, because the reference directions are opposite, the algebraic signs of the current values are opposite. Thus, we have $i_c = -i_d$.

**c.** At the node joining elements $A$, $B$, and $C$, we can write the KCL equation $i_b = i_a + i_c = 6 - 2 = 4$ A. Also, we found earlier that $i_d = -i_c = 2$ A.   ∎

**Exercise 1.7**   Use KCL to determine the values of the unknown currents shown in Figure 1.22.
**Answer**   $i_a = 4$ A, $i_b = -2$ A, $i_c = -8$ A.   □

**Exercise 1.8**   Consider the circuit of Figure 1.23. Identify the groups of circuit elements that are connected in series.
**Answer**   Elements $A$ and $B$ are in series; elements $E$, $F$, and $G$ form another series combination.   □

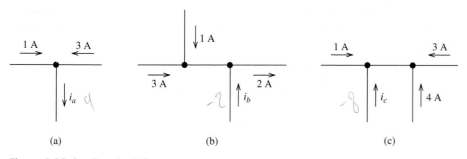

(a)                                (b)                                (c)

Figure 1.22  See Exercise 1.7.

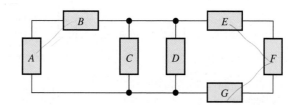

Figure 1.23  Circuit for Exercise 1.8.

## 1.5  KIRCHHOFF'S VOLTAGE LAW

Kirchhoff's voltage law (KVL) states that the algebraic sum of the voltages equals zero for any closed path (loop) in an electrical circuit.

A **loop** in an electrical circuit is a closed path starting at a node and proceeding through circuit elements, eventually returning to the starting node. Frequently, several loops can be identified for a given circuit. For example, in Figure 1.23, one loop consists of the path starting at the top end of element *A* and proceeding clockwise through elements *B* and *C*, returning through *A* to the starting point. Another loop starts at the top of element *D* and proceeds clockwise through *E*, *F*, and *G*, returning to the start through *D*. Still another loop exists through elements *A*, *B*, *E*, *F*, and *G* around the periphery of the circuit.

**Kirchhoff's voltage law** (KVL) states: *The algebraic sum of the voltages equals zero for any closed path (loop) in an electrical circuit.* In traveling around a loop, we encounter various voltages, some of which carry a positive sign while others carry a negative sign in the algebraic sum. A convenient convention is to use the first polarity mark encountered for each voltage to decide if it should be added or subtracted in the algebraic sum. If we go through the voltage from the positive polarity reference to the negative reference, it carries a plus sign. If the polarity marks are encountered in the opposite direction (minus to plus), the voltage carries a negative sign. This is illustrated in Figure 1.24.

For the circuit of Figure 1.25, we obtain the following equations:

$$\text{Loop 1:}\quad -v_a + v_b + v_c = 0$$
$$\text{Loop 2:}\quad -v_c - v_d + v_e = 0$$
$$\text{Loop 3:}\quad v_a - v_b + v_d - v_e = 0$$

Notice that $v_a$ is subtracted for loop 1, but it is added for loop 3, because the direction of travel is different for the two loops. Similarly, $v_c$ is added for loop 1 and subtracted for loop 2.

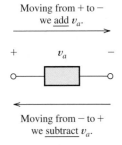

Moving from + to − we <u>add</u> $v_a$.

+    $v_a$    −

Moving from − to + we <u>subtract</u> $v_a$.

**Figure 1.24** In applying KVL to a loop, voltages are added or subtracted depending on their reference polarities relative to the direction of travel around the loop.

### Kirchhoff's Voltage Law Related to Conservation of Energy

KVL is a consequence of the law of energy conservation. Consider the circuit shown in Figure 1.26. This circuit consists of three elements connected in series. Thus, the same current *i* flows through all three elements. The power for each of the elements is given by

$$\text{Element } A:\quad p_a = v_a i$$
$$\text{Element } B:\quad p_b = -v_b i$$
$$\text{Element } C:\quad p_c = v_c i$$

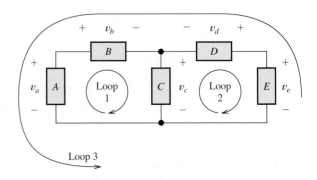

**Figure 1.25** Circuit used for illustration of Kirchhoff's voltage law.

Notice that the current and voltage references have the passive configuration (the current reference enters the plus polarity mark) for elements $A$ and $C$. For element $B$, the relationship is opposite to the passive reference configuration. That is why we have a negative sign in the calculation of $p_b$.

At a given instant, the sum of the powers for all of the elements in a circuit must be zero. Otherwise, for an increment of time taken at that instant, more energy would be absorbed than is supplied by the circuit elements (or vice versa):

$$p_a + p_b + p_c = 0$$

Substituting for the powers, we have

$$v_a i - v_b i + v_c i = 0$$

Canceling the current $i$, we obtain

$$v_a - v_b + v_c = 0$$

This is exactly the same equation that is obtained by adding the voltages around the loop and setting the sum to zero for a clockwise loop in the circuit of Figure 1.26.

One way to check our results after solving for the currents and voltages in a circuit is the check to see that the power adds to zero for all of the elements.

## Parallel Circuits

We say that two circuit elements are connected in **parallel** if both ends of one element are connected directly (i.e., by conductors) to corresponding ends of the other. For example, in Figure 1.27, elements $A$ and $B$ are in parallel. Similarly, we say that the three circuit elements $D, E$, and $F$ are in parallel. Element $B$ is *not* in parallel with $D$ because the top end of $B$ is not *directly* connected to the top end of $D$.

The voltages across parallel elements are equal in magnitude and have the same polarity. For illustration, consider the partial circuit shown in Figure 1.28. Here elements $A, B$, and $C$ are connected in parallel. Consider a loop from the bottom end of $A$ upward and then down through element $B$ back to the bottom of $A$. For this clockwise loop, we have $-v_a + v_b = 0$. Thus, KVL requires that

$$v_a = v_b$$

Next, consider a clockwise loop through elements $A$ and $C$. For this loop, KVL requires that

$$-v_a - v_c = 0$$

This implies that $v_a = -v_c$. In other words, $v_a$ and $v_c$ have opposite algebraic signs. Furthermore, one or the other of the two voltages must be negative (unless both are zero). Therefore, one of the voltages has an actual polarity opposite to the reference

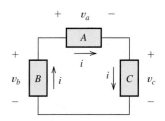

**Figure 1.26** In this circuit, conservation of energy requires that $v_b = v_a + v_c$.

Two circuit elements are connected in parallel if both ends of one element are connected directly (i.e., by conductors) to corresponding ends of the other.

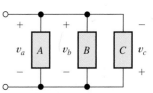

**Figure 1.28** For this circuit, we can show that $v_a = v_b = -v_c$. Thus, the magnitudes and *actual* polarities of all three voltages are the same.

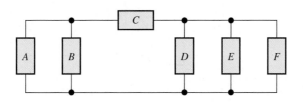

**Figure 1.27** In this circuit, elements $A$ and $B$ are in parallel. Elements $D, E$, and $F$ form another parallel combination.

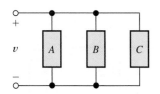

**Figure 1.29** Analysis is simplified by using the same voltage variable and reference polarity for elements that are in parallel.

polarity shown in the figure. Thus, the actual polarities of the voltages are the same (either both are positive at the top of the circuit or both are positive at the bottom).

Usually, when we have a parallel circuit, we simply use the same voltage variable for all of the elements as illustrated in Figure 1.29.

---

**Example 1.5    Kirchhoff's Voltage Law**

Consider the circuit shown in Figure 1.30.

**a.** Which elements are in parallel?

**b.** Which elements are in series?

**c.** What is the relationship between $v_d$ and $v_f$?

**d.** Given that $v_a = 10$ V, $v_c = 15$ V, and $v_e = 20$ V, determine the values of $v_b$ and $v_f$.

**Solution**

**a.** Elements $D$ and $F$ are in parallel.

**b.** Elements $A$ and $E$ are in series.

**c.** Because elements $D$ and $F$ are in parallel, $v_d$ and $v_f$ are equal in magnitude. However, because the reference directions are opposite, the algebraic signs of their values are opposite. Thus, we have $v_d = -v_f$.

**d.** Applying KVL to the loop formed by elements $A$, $B$, and $E$, we have:

$$v_a + v_b - v_e = 0$$

Solving for $v_b$ and substituting values, we find that $v_b = 10$ V.

Applying KVL to the loop around the outer perimeter of the circuit, we have:

$$v_a - v_c + v_f = 0$$

Solving for $v_f$ and substituting values, we find that $v_f = 5$ V. ∎

---

**Exercise 1.9**    Use repeated application of KVL to find the values of $v_c$ and $v_e$ for the circuit of Figure 1.31.
**Answer**    $v_c = 8$ V, $v_e = -2$ V. ☐

---

**Exercise 1.10**    Identify elements that are in parallel in Figure 1.31. Identify elements in series.
**Answer**    Elements $E$ and $F$ are in parallel; elements $A$ and $B$ are in series. ☐

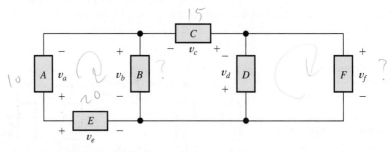

**Figure 1.30** Circuit for Example 1.5.

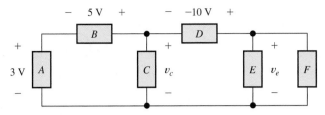

**Figure 1.31**  Circuit for Exercises 1.9 and 1.10.

# 1.6   INTRODUCTION TO CIRCUIT ELEMENTS

In this section, we carefully define several types of ideal circuit elements:

Conductors

Voltage sources

Current sources

Resistors

Later in the book, we will encounter additional elements, including inductors and capacitors. Eventually, we will be able to use these idealized circuit elements to describe (model) complex real-world electrical devices.

## Conductors

We have already encountered conductors. Ideal conductors are represented in circuit diagrams by unbroken lines between the ends of other circuit elements. We define ideal circuit elements in terms of the relationship between the voltage across the element and the current through it.

*The voltage between the ends of an ideal conductor is zero regardless of the current flowing through the conductor.* When two points in a circuit are connected together by an ideal conductor, we say that the points are **shorted** together. Another term for an ideal conductor is **short circuit**. All points in a circuit that are connected by ideal conductors can be considered as a single node.

If no conductors or other circuit elements are connected between two parts of a circuit, we say that an **open circuit** exists between the two parts of the circuit. No current can flow through an ideal open circuit.

The voltage between the ends of an ideal conductor is zero regardless of the current flowing through the conductor.

All points in a circuit that are connected by ideal conductors can be considered as a single node.

## Independent Voltage Sources

An **ideal independent voltage source** maintains a specified voltage across its terminals. The voltage across the source is independent of other elements that are connected to it and of the current flowing through it. We use a circle enclosing the reference polarity marks to represent independent voltage sources. The value of the voltage is indicated alongside the symbol. The voltage can be constant or it can be a function of time. Several voltage sources are shown in Figure 1.32.

In Figure 1.32(a), the voltage across the source is constant. Thus, we have a dc voltage source. On the other hand, the source shown in Figure 1.32(b) is an ac voltage source having a sinusoidal variation with time. We say that these are *independent* sources because the voltages across their terminals are independent of all other voltages and currents in the circuit.

An ideal independent voltage source maintains a specified voltage across its terminals.

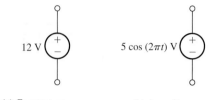

(a) Constant or
dc voltage source

(b) Ac voltage
source

Figure 1.32 Independent voltage sources.

## Ideal Circuit Elements versus Reality

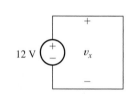

**Figure 1.33** We avoid self-contradictory circuit diagrams such as this one.

Here we are giving definitions of *ideal* circuit elements. It is possible to draw ideal circuits in which the definitions of various circuit elements conflict. For example, Figure 1.33 shows a 12-V voltage source with a conductor connected across its terminals. In this case, the definition of the voltage source requires that $v_x = 12$ V. On the other hand, the definition of an ideal conductor requires that $v_x = 0$. In our study of ideal circuits, we avoid such conflicts.

In the real world, an automobile battery is nearly an ideal 12-V voltage source, and a short piece of heavy-gauge copper wire is nearly an ideal conductor. If we place the wire across the terminals of the battery, a very large current flows through the wire, stored chemical energy is converted to heat in the wire at a very high rate, and the wire will probably melt or the battery be destroyed.

When we encounter a contradictory idealized circuit model, we often have an undesirable situation (such as a fire or destroyed components) in the real-world counterpart to the model. In any case, a contradictory circuit model implies that we have not been sufficiently careful in choosing circuit models for the real circuit elements. For example, an automobile battery is not exactly modeled as an ideal voltage source. We will see that a better model (particularly if the currents are very large) is an ideal voltage source in series with a resistance. (We will discuss resistance very soon.) A short piece of copper wire is not modeled well as an ideal conductor, in this case. Instead, we will see that it is modeled better as a small resistance. If we have done a good job at picking circuit models for real-world circuits, we will not encounter contradictory circuits, and the results we calculate using the model will match reality very well.

## Dependent Voltage Sources

A **dependent** or **controlled voltage source** is similar to an independent source except that the voltage across the source terminals is a function of other voltages or currents in the circuit. Instead of a circle, it is customary to use a diamond to represent controlled sources in circuit diagrams. Two examples of dependent sources are shown in Figure 1.34.

A **voltage-controlled voltage source** is a voltage source having a voltage equal to a constant times the voltage across a pair of terminals elsewhere in the network. An example is shown in Figure 1.34(a). The dependent voltage source is the diamond symbol. The reference polarity of the source is indicated by the marks inside the diamond. The voltage $v_x$ determines the value of the voltage produced by the source. For example, if it should turn out that $v_x = 3$ V, the source voltage is $2v_x = 6$ V.

A voltage-controlled voltage source maintains a voltage across its terminals equal to a constant times a voltage elsewhere in the circuit.

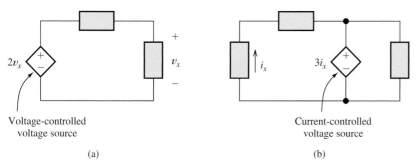

**Figure 1.34** Dependent voltage sources (also known as controlled voltage sources) are represented by diamond-shaped symbols. The voltage across a controlled voltage source depends on a current or voltage that appears elsewhere in the circuit.

If $v_x$ should equal $-7$ V, the source produces $2v_x = -14$ V (in which case, the actual positive polarity of the source is at the bottom end).

A **current-controlled voltage source** is a voltage source having a voltage equal to a constant times the current through some other element in the circuit. An example is shown in Figure 1.34(b). In this case, the source voltage is three times the value of the current $i_x$. The factor multiplying the current is called the **gain parameter**. We assume that the voltage has units of volts and the current is in amperes. Thus, the gain parameter [which is 3 in Figure 1.34(b)] has units of volts per ampere (V/A). (Shortly, we will see that the units V/A are the units of resistance and are called ohms.)

Returning our attention to the voltage-controlled voltage source in Figure 1.34(a), we note that the gain parameter is 2 and is unitless (or we could say that the units are V/V).

Later in the book, we will see that controlled sources are very useful in modeling transistors, amplifiers, and electrical generators, among other things.

> A current-controlled voltage source maintains a voltage across its terminals equal to a constant times a current flowing through some other element in the circuit.

### Independent Current Sources

An ideal **independent current source** forces a specified current to flow through itself. The symbol for an independent current source is a circle enclosing an arrow that gives the reference direction for the current. The current through an independent current source is independent of the elements connected to it and of the voltage across it. Figure 1.35 shows the symbols for a dc current source and for an ac current source.

If an open circuit exists across the terminals of a current source, we have a contradictory circuit. For example, consider the 2-A dc current source shown in Figure 1.35(a). This current source is shown with an open circuit across its terminals. By definition, the current flowing into the top node of the source is 2 A. Also by definition, no current can flow through the open circuit. Thus, KCL is not satisfied at this node. In good models for actual circuits, this situation does not occur. Thus, we will avoid current sources with open-circuited terminals in our discussion of ideal networks.

A battery is a good example of a voltage source, but an equally familiar example does not exist for a current source. However, current sources are useful in constructing theoretical models. Later, we will see that a good approximation to an ideal current source can be achieved with electronic amplifiers.

> An ideal independent current source forces a specified current to flow through itself.

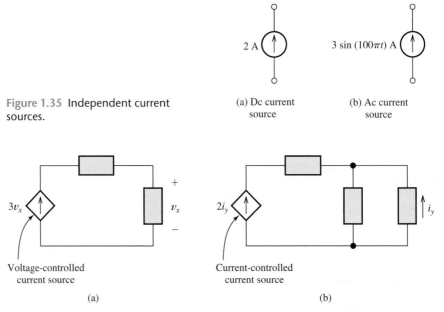

**Figure 1.35** Independent current sources.

(a) Dc current source

(b) Ac current source

Voltage-controlled current source

(a)

Current-controlled current source

(b)

**Figure 1.36** Dependent current sources. The current through a dependent current source depends on a current or voltage that appears elsewhere in the circuit.

### Dependent Current Sources

The current flowing through a dependent current source is determined by a current or voltage elsewhere in the circuit.

The current flowing through a **dependent current source** is determined by a current or voltage elsewhere in the circuit. The symbol is a diamond enclosing an arrow that indicates the reference direction. Two types of controlled current sources are shown in Figure 1.36.

In Figure 1.36(a), we have a **voltage-controlled current source**. The current through the source is three times the voltage $v_x$. The gain parameter of the source (3 in this case) has units of A/V (which we will soon see are equivalent to siemens or inverse ohms). If it turns out that $v_x$ has a value of 5 V, the current through the controlled current source is $3v_x = 15$ A.

Figure 1.36(b) illustrates a **current-controlled current source**. In this case, the current through the source is twice the value of $i_y$. The gain parameter, which has a value of 2 in this case, has units of A/A (i.e., it is unitless).

Like controlled voltage sources, controlled current sources are useful in constructing circuit models for many types of real-world devices, such as electronic amplifiers, transistors, transformers, and electrical machines. If a controlled source is needed for some application, it can be implemented by using electronic amplifiers. In sum, these are the four kinds of controlled sources:

1. Voltage-controlled voltage sources
2. Current-controlled voltage sources
3. Voltage-controlled current sources
4. Current-controlled current sources

### Resistors and Ohm's Law

The voltage $v$ across an ideal **resistor** is proportional to the current $i$ through the resistor. The constant of proportionality is the resistance $R$. The symbol used for

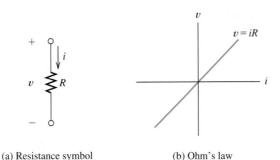

**Figure 1.37** Voltage is proportional to current in an ideal resistor. Notice that the references for v and i conform to the passive reference configuration.

(a) Resistance symbol          (b) Ohm's law

a resistor is shown in Figure 1.37(a). Notice that the current reference and voltage polarity reference conform to the passive reference configuration. In other words, the reference direction for the current is into the positive polarity mark and out of the negative polarity mark. In equation form, the voltage and current are related by **Ohm's law**:

$$v = iR$$

The units of resistance are V/A, which are called ohms. The uppercase Greek letter omega ($\Omega$) represents ohms. In practical circuits, we encounter resistances ranging from milliohms (m$\Omega$) to megohms (M$\Omega$).

Except for rather unusual situations, the resistance $R$ assumes positive values. (In certain types of electronic circuits, we can encounter negative resistance, but for now we assume that $R$ is positive.) In situations for which the current reference direction enters the *negative* reference of the voltage, Ohm's law becomes

$$v = -iR$$

This is illustrated in Figure 1.38.

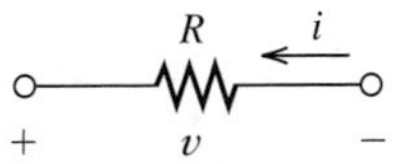

The relationship between current direction and voltage polarity can be neatly included in the equation for Ohm's law if double-subscript notation is used. (Recall that to use double subscripts, we label the ends of the element under consideration, which is a resistance in this case.) If the order of the subscripts is the same for the current as for the voltage ($i_{ab}$ and $v_{ab}$, for example), the current reference direction enters the first terminal and the positive voltage reference is at the first terminal. Thus, we can write

**Figure 1.38** If the references for v and i are opposite to the passive configuration, we have v = −Ri.

$$v_{ab} = i_{ab}R$$

On the other hand, if the order of the subscripts is not the same, we have

$$v_{ab} = -i_{ba}R$$

## Conductance

Solving Ohm's law for current, we have

$$i = \frac{1}{R}v$$

We call the quantity $1/R$ a **conductance**. It is customary to denote conductances with the letter $G$:

$$G = \frac{1}{R} \tag{1.8}$$

Conductances have the units of inverse ohms ($\Omega^{-1}$), which are called siemens (abbreviated S). Thus, we can write Ohm's law as

$$i = Gv \tag{1.9}$$

## Resistors

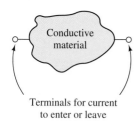

**Figure 1.39** We construct resistors by attaching terminals to a piece of conductive material.

It turns out that we can construct nearly ideal resistors by attaching terminals to many types of conductive materials. This is illustrated in Figure 1.39. Conductive materials that can be used to construct resistors include most metals, their alloys, and carbon.

On a microscopic level, current in metals consists of electrons moving through the material. (On the other hand, in solutions of ionic compounds, current is carried partly by positive ions.) The applied voltage creates an electric field that accelerates the electrons. The electrons repeatedly collide with the atoms of the material and lose their forward momentum. Then they are accelerated again. The net effect is a constant average velocity for the electrons. At the macroscopic level, we observe a current that is proportional to the applied voltage.

## Resistance Related to Physical Parameters

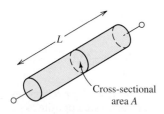

**Figure 1.40** Resistors often take the form of a long cylinder (or bar) in which current enters one end and flows along the length.

The dimensions and geometry of the resistor as well as the particular material used to construct a resistor influence its resistance. We consider only resistors that take the form of a long cylinder or bar with terminals attached at the ends, as illustrated in Figure 1.40. The cross-sectional area $A$ is constant along the length of the cylinder or bar. If the length $L$ of the resistor is much greater than the dimensions of its cross section, the resistance is approximately given by

$$R = \frac{\rho L}{A} \tag{1.10}$$

in which $\rho$ is the *resistivity* of the material used to construct the resistor. The units of resistivity are ohm meters ($\Omega$m).

Materials can be classified as conductors, semiconductors, or insulators, depending on their resistivity. **Conductors** have the lowest resistivity and easily conduct electrical current. **Insulators** have very high resistivity and conduct very little current (at least for moderate voltages). **Semiconductors** fall between conductors and insulators. We will see in Chapters 9, 11, and 12 that certain semiconductors are very useful in constructing electronic devices Table 1.3 gives approximate values of resistivity for several materials.

**Table 1.3** Resistivity Values ($\Omega$m) for Selected Materials at 300 K

| | |
|---|---|
| Conductors | |
| Aluminum | $2.73 \times 10^{-8}$ |
| Carbon (amorphous) | $3.5 \times 10^{-5}$ |
| Copper | $1.72 \times 10^{-8}$ |
| Gold | $2.27 \times 10^{-8}$ |
| Nichrome | $1.12 \times 10^{-6}$ |
| Silver | $1.63 \times 10^{-8}$ |
| Tungsten | $5.44 \times 10^{-8}$ |
| Semiconductors | |
| Silicon (device grade) depends on impurity concentration | $10^{-5}$ to 1 |
| Insulators | |
| Fused quartz | $> 10^{21}$ |
| Glass (typical) | $1 \times 10^{12}$ |
| Teflon | $1 \times 10^{19}$ |

**Example 1.6   Resistance Calculation**

Compute the resistance of a copper wire having a diameter of 2.05 mm and a length of 10 m.

**Solution**   First, we compute the cross-sectional area of the wire:

$$A = \frac{\pi d^2}{4} = \frac{\pi (2.05 \times 10^{-3})^2}{4} = 3.3 \times 10^{-6} \, \text{m}^2$$

Then, the resistance is given by

$$R = \frac{\rho L}{A} = \frac{1.72 \times 10^{-8} \times 10}{3.3 \times 10^{-6}} = 0.052 \, \Omega$$

These are the approximate dimensions of a piece of 12-gauge copper wire that we might find connecting an electrical outlet to the distribution box in a residence. Of course, two wires are needed for a complete circuit.   ∎

## Power Calculations for Resistances

Recall that we compute power for a circuit element as the product of the current and voltage:

$$p = vi \tag{1.11}$$

If $v$ and $i$ have the passive reference configuration, a positive sign for power means that energy is being absorbed by the device. Furthermore, a negative sign means that energy is being supplied by the device.

If we use Ohm's law to substitute for $v$ in Equation 1.11, we obtain

$$p = Ri^2 \qquad (1.12)$$

On the other hand, if we solve Ohm's law for $i$ and substitute into Equation 1.11, we obtain

$$p = \frac{v^2}{R} \qquad (1.13)$$

Notice that power for a resistance is positive regardless of the sign of $v$ or $i$ (assuming that $R$ is positive, which is ordinarily the case). Thus, power is absorbed by resistances. If the resistance results from collisions of electrons with the atoms of the material composing a resistor, this power shows up as heat.

Some applications for conversion of electrical power into heat are heating elements for ovens, water heaters, cooktops, and space heaters. In a typical space heater, the heating element consists of a nichrome wire that becomes red hot in operation. (Nichrome is an alloy of nickel, chromium, and iron.) To fit the required length of wire in a small space, it is coiled rather like a spring.

## PRACTICAL APPLICATION    1.1

### Using Resistance to Measure Strain

Civil and mechanical engineers routinely employ the dependence of resistance on physical dimensions of a conductor to measure strain. These measurements are important in experimental stress strain analysis of mechanisms and structures. (Strain is defined as fractional change in length, given by $\epsilon = \Delta L/L$.)

A typical resistive strain gauge consists of nickel–copper alloy foil that is photoetched to obtain multiple conductors aligned with the direction of the strain to be measured. This is illustrated in Figure PA1.1. Typically, the conductors are bonded to a thin polyimide (a tough flexible plastic) backing, which in turn is attached to the structure under test by a suitable adhesive, such as cyanoacrylate cement.

The resistance of a conductor is given by

$$R = \frac{\rho L}{A}$$

As strain is applied, the length and area change, resulting in changes in resistance. The strain and the change in resistance are related by the gauge factor:

$$G = \frac{\Delta R/R_0}{\epsilon}$$

in which $R_0$ is the resistance of the gauge before strain. A typical gauge has $R_0 = 350\ \Omega$ and $G = 2.0$. Thus, for a strain of 1%, the change in resistance is $\Delta R = 7\ \Omega$. Usually, a Wheatstone bridge (discussed in Chapter 2) is used to measure the small changes in resistance associated with accurate strain determination.

Sensors for force, torque, and pressure are constructed by using resistive strain gauges.

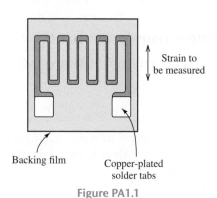

Backing film    Copper-plated solder tabs

**Figure PA1.1**

## Resistors versus Resistances

As an aside, we mention that resistance is often useful in modeling devices in which electrical power is converted into forms other than heat. For example, a loudspeaker appears to have a resistance of 8 Ω. Part of the power delivered to the loudspeaker is converted to acoustic power. Another example is a transmitting antenna having a resistance of 50 Ω. The power delivered to an antenna is radiated, traveling away as an electromagnetic wave.

There is a slight distinction between the terms *resistor* and *resistance*. A resistor is a two-terminal device composed of a conductive material. Resistance is a circuit property for which voltage is proportional to current. Thus, resistors have the property of resistance. However, resistance is also useful in modeling antennas and loudspeakers, which are quite different from resistors. Often, we are not careful about this distinction in using these terms.

---

**Example 1.7**   Determining Resistance for Given Power and Voltage Ratings

A certain electrical heater is rated for 1500 W when operated from 120 V. Find the resistance of the heater element and the operating current. (Resistance depends on temperature, and we will find the resistance at the operating temperature of the heater.)

**Solution**   Solving Equation 1.13 for resistance, we obtain

$$R = \frac{v^2}{p} = \frac{120^2}{1500} = 9.6 \ \Omega$$

Then, we use Ohm's law to find the current:

$$i = \frac{v}{R} = \frac{120}{9.6} = 12.5 \ \text{A} \qquad \blacksquare$$

---

**Exercise 1.11**   The 9.6-Ω resistance of Example 1.7 is in the form of a nichrome wire having a diameter of 1.6 mm. Find the length of the wire. (*Hint:* The resistivity of nichrome is given in Table 1.3.)
**Answer**   $L = 17.2$ m.   □

---

**Exercise 1.12**   Suppose we have a typical incandescent electric light bulb that is rated for 100 W and 120 V. Find its resistance (at operating temperature) and operating current.
**Answer**   $R = 144 \ \Omega$, $i = 0.833$ A.   □

---

**Exercise 1.13**   A 1-kΩ resistor used in a television receiver is rated for a maximum power of 1/4 W. Find the current and voltage when the resistor is operated at maximum power.
**Answer**   $v_{max} = 15.8$ V, $i_{max} = 15.8$ mA.   □

---

## 1.7   INTRODUCTION TO CIRCUITS

In this chapter, we have defined electrical current and voltage, discussed Kirchhoff's laws, and introduced several ideal circuit elements: voltage sources, current sources, and resistances. Now we illustrate these concepts by considering a few relatively

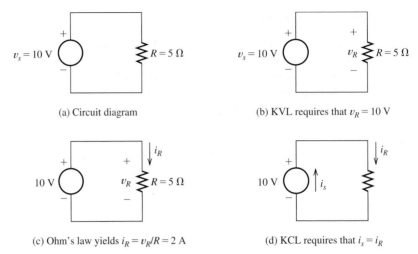

(a) Circuit diagram

(b) KVL requires that $v_R = 10$ V

(c) Ohm's law yields $i_R = v_R/R = 2$ A

(d) KCL requires that $i_s = i_R$

**Figure 1.41** A circuit consisting of a voltage source and a resistance.

simple circuits. In the next chapter, we consider more complex circuits and analysis techniques.

Consider the circuit shown in Figure 1.41(a). Suppose that we want to know the current, voltage, and power for each element. To obtain these results, we apply the basic principles introduced in this chapter. At first, we proceed in small, methodical steps. Furthermore, for ease of understanding, we initially select reference polarities and directions that agree with the actual polarities and current directions.

KVL requires that the sum of the voltages around the circuit shown in Figure 1.41 must equal zero. Thus, traveling around the circuit clockwise, we have $v_R - v_s = 0$. Consequently, $v_R = v_s$, and the voltage across the resistor $v_R$ must have an actual polarity that is positive at the top end and a magnitude of 10 V.

An alternative way of looking at the voltages in this circuit is to notice that the voltage source and the resistance are in parallel. (The top ends of the voltage source and the resistance are connected, and the bottom ends are also connected.) Recall that when elements are in parallel, the voltage magnitude and polarity are the same for all elements.

Now consider Ohm's law. Because 10 V appears across the 5-$\Omega$ resistance, the current is $i_R = 10/5 = 2$ A. This current flows through the resistance from the positive polarity to the negative polarity. Thus, $i_R = 2$ A flows downward through the resistance, as shown in Figure 1.41(c).

According to KCL, the sum of the currents entering a given node must equal the sum of the currents leaving. There are two nodes for the circuit of Figure 1.41: one at the top and one at the bottom. The current $i_R$ leaves the top node through the resistance. Thus, an equal current must enter the top node through the voltage source. The actual direction of current flow is upward through the voltage source, as shown in Figure 1.41(d).

Another way to see that the currents $i_s$ and $i_R$ are equal is to notice that the voltage source and the resistance are in series. In a series circuit, the current that flows in one element must continue through the other element. (Notice that for this circuit the voltage source and the resistance are in parallel and they are also in series. A two-element circuit is the only case for which this occurs. If more than two elements are interconnected, a pair of elements that are in parallel cannot also be in series, and vice versa.)

Notice that in Figure 1.41, the current in the voltage source flows from the negative polarity toward the positive polarity. It is only for resistances that the current is required to flow from plus to minus. For a voltage source, the current can flow in either direction, depending on the circuit to which the source is connected.

Now let us calculate the power for each element. For the resistance, we have several ways to compute power:

$$p_R = v_R i_R = 10 \times 2 = 20 \text{ W}$$

$$p_R = i_R^2 R = 2^2 \times 5 = 20 \text{ W}$$

$$p_R = \frac{v_R^2}{R} = \frac{10^2}{5} = 20 \text{ W}$$

Of course, all the calculations yield the same result. Energy is delivered to the resistance at the rate of 20 J/s.

To find the power for the voltage source, we have

$$p_s = -v_s i_s$$

where the minus sign is used because the reference direction for the current enters the negative voltage reference (opposite to the passive reference configuration). Substituting values, we obtain

$$p_s = -v_s i_s = -10 \times 2 = -20 \text{ W}$$

Because $p_s$ is negative, we understand that energy is being delivered by the voltage source.

As a check, if we add the powers for all the elements in the circuit, the result should be zero, because energy is neither created nor destroyed in an electrical circuit. Instead, it is transported and changed in form. Thus, we can write

$$p_s + p_R = -20 + 20 = 0$$

## Using Arbitrary References

In the previous discussion, we selected references that agree with actual polarities and current directions. This is not always possible at the start of the analysis of more complex circuits. Fortunately, it is not necessary. We can pick the references in an arbitrary manner. Application of circuit laws will tell us not only the magnitudes of the currents and voltages but the true polarities and current directions as well.

### Example 1.8   Circuit Analysis Using Arbitrary References

Analyze the circuit of Figure 1.41 using the current and voltage references shown in Figure 1.42. Verify that the results are in agreement with those found earlier.

**Solution**   Traveling clockwise and applying KVL, we have

$$-v_s - v_x = 0$$

This yields $v_x = -v_s = -10$ V. Since $v_x$ assumes a negative value, the actual polarity is opposite to the reference. Thus, as before, we conclude that the voltage across the resistance is actually positive at the top end.

It is only for resistances that the current is required to flow from plus to minus. Current may flow in either direction for a voltage source depending on the other elements in the circuit.

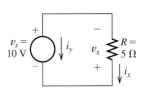

**Figure 1.42**  Circuit for Example 1.8.

According to Ohm's law,

$$i_x = -\frac{v_x}{R}$$

where the minus sign appears because $v_x$ and $i_x$ have references opposite to the passive reference configuration. Substituting values, we get

$$i_x = -\frac{-10}{5} = 2 \text{ A}$$

Since $i_x$ assumes a positive value, the actual current direction is downward through the resistance.

Next, applying KCL at the bottom node of the circuit, we have

$$\text{total current entering} = \text{total current leaving } i_y + i_x = 0$$

Thus, $i_y = -i_x = -2$ A, and we conclude that a current of 2 A actually flows upward through the voltage source.

The power for the voltage source is

$$p_s = v_s i_y = 10 \times (-2) = -20 \text{ W}$$

Finally, the power for the resistance is given by

$$p_R = -v_x i_x$$

where the minus sign appears because the references for $v_x$ and $i_x$ are opposite to the passive reference configuration. Substituting, we find that $p_R = -(-10) \times (2) = 20$ W. Because $p_R$ has a positive value, we conclude that energy is delivered to the resistance. ■

Sometimes circuits can be solved by repeated application of Kirchhoff's laws and Ohm's law. We illustrate with an example.

### Example 1.9    Using KVL, KCL, and Ohm's Law to Solve a Circuit

Solve for the source voltage in the circuit of Figure 1.43 in which we have a current-controlled current source and we are given that the voltage across the 5-$\Omega$ resistance is 15 V.

**Solution**    First, we use Ohm's Law to determine the value of $i_y$:

$$i_y = \frac{15 \text{ V}}{5 \text{ }\Omega} = 3 \text{ A}$$

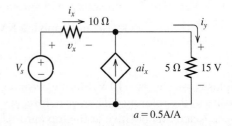

**Figure 1.43** Circuit for Example 1.9.

Next, we apply KCL at the top end of the controlled source:

$$i_x + 0.5i_x = i_y$$

Substituting the value found for $i_y$ and solving, we determine that $i_x = 2$ A. Then Ohm's law yields $v_x = 10i_x = 20$ V. Applying KCL around the periphery of the circuit gives

$$V_s = v_x + 15$$

Finally, substituting the value found for $v_x$ yields $V_s = 35$ V.  ■

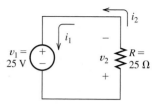

**Figure 1.44** Circuit for Exercise 1.14.

**Exercise 1.14**   Analyze the circuit shown in Figure 1.44 to find the values of $i_1$, $i_2$, and $v_2$. Use the values found to compute the power for each element.
**Answer**   $i_1 = i_2 = -1$ A, $v_2 = -25$ V, $p_R = 25$ W, $p_s = -25$ W.   □

**Exercise 1.15**   Figure 1.45 shows an independent current source connected across a resistance. Analyze to find the values of $i_R$, $v_R$, $v_s$, and the power for each element.
**Answer**   $i_R = 2$ A, $v_s = v_R = 80$ V, $p_s = -160$ W, $p_R = 160$ W.   □

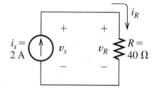

**Figure 1.45** Circuit for Exercise 1.15.

## Summary

1. Electrical and electronic features are increasingly integrated into the products and systems designed by engineers in other fields. Furthermore, instrumentation in all fields of engineering and science is based on the use of electrical sensors, electronics, and computers.

2. Some of the main areas of electrical engineering are communication systems, computer systems, control systems, electromagnetics, photonics, electronics, power systems, and signal processing.

3. Some important reasons to learn basic electrical engineering principles are to pass the Fundamentals of Engineering Examination, to have a broad enough knowledge base to lead design projects in your own field, to be able to identify and correct simple malfunctions in electrical systems, and to be able to communicate efficiently with electrical engineering consultants.

4. Current is the time rate of flow of electrical charge. Its units are amperes (A), which are equivalent to coulombs per second (C/s)

5. The voltage associated with a circuit element is the energy transferred per unit of charge that flows through the element. The units of voltages are volts (V), which are equivalent to joules per

coulomb (J/C). If positive charge moves from the positive reference to the negative reference, energy is absorbed by the circuit element. If the charge moves in the opposite direction, energy is delivered by the element.

6. In the passive reference configuration, the current reference direction enters the positive reference polarity.

7. If the references have the passive configuration, power for a circuit element is computed as the product of the current through the element and the voltage across it:

$$p = vi$$

If the references are opposite to the passive configuration, we have

$$p = -vi$$

In either case, if $p$ is positive, energy is being absorbed by the element.

8. A node in an electrical circuit is a point at which two or more circuit elements are joined together. All points joined by ideal conductors are electrically equivalent and constitute a single node.

9. Kirchhoff's current law (KCL) states that the sum of the currents entering a node equals the sum of the currents leaving.

10. Elements connected end to end are said to be in series. For two elements to be in series, no other current path can be connected to their common node. The current is identical for all elements in a series connection.

11. A loop in an electrical circuit is a closed path starting at a node and proceeding through circuit elements eventually returning to the starting point.

12. Kirchhoff's voltage law (KVL) states that the algebraic sum of the voltages in a loop must equal zero. If the positive polarity of a voltage is encountered first in going around the loop, the voltage carries a plus sign in the sum. On the other hand, if the negative polarity is encountered first, the voltage carries a minus sign.

13. Two elements are in parallel if both ends of one element are directly connected to corresponding ends of the other element. The voltages of parallel elements are identical.

14. The voltage between the ends of an ideal conductor is zero regardless of the current flowing through the conductor. All points in a circuit that are connected by ideal conductors can be considered as a single point.

15. An ideal independent voltage source maintains a specified voltage across its terminals independent of other elements that are connected to it and of the current flowing through it.

16. For a controlled voltage source, the voltage across the source terminals depends on other voltages or currents in the circuit. A voltage-controlled voltage source is a voltage source having a voltage equal to a constant times the voltage across a pair of terminals elsewhere in the network. A current-controlled voltage source is a voltage source having a voltage equal to a constant times the current through some other element in the circuit.

17. An ideal independent current source forces a specified current to flow through itself, independent of other elements that are connected to it and of the voltage across it.

18. For a controlled current source, the current depends on other voltages or currents in the circuit. A voltage-controlled current source produces a current equal to a constant times the voltage across a pair of terminals elsewhere in the network. A current-controlled current source produces a current equal to a constant times the current through some other element in the circuit.

19. For constant resistances, voltage is proportional to current. If the current and voltage references have the passive configuration, Ohm's law states that $v = Ri$. For references opposite to the passive configuration, $v = -Ri$.

## Problems

### Section 1.1: Overview of Electrical Engineering

**P1.1.** Broadly speaking, what are the two main objectives of electrical systems?

**P1.2.** List four reasons why other engineering students need to learn the fundamentals of electrical engineering.

**P1.3.** List eight subdivisions of electrical engineering.

**P1.4.** Write a few paragraphs describing an interesting application of electrical engineering in your field. Consult engineering journals and trade magazines such as the *IEEE Spectrum,*
*Automotive Engineering, Chemical Engineering,* or *Civil Engineering* for ideas.

### Section 1.2: Circuits, Currents, and Voltages

**P1.5.** Carefully define or explain the following terms in your own words (give units where appropriate): **a.** Electrical current. **b.** Voltage. **c.** An open switch. **d.** A closed switch. **e.** Direct current. **f.** Alternating current.

**P1.6.** In the fluid-flow analogy for electrical circuits, what is analogous to **a.** a conductor; **b.** an open switch; **c.** a resistance; **d.** a battery?

**P1.7.** The charge of an electron is $-1.60 \times 10^{-19}$ C. A current of 1 A flows in a wire carried by electrons. How many electrons pass through a cross section of the wire each second?

**\*P1.8.** The ends of a length of wire are labeled $a$ and $b$. If the current in the wire is $i_{ab} = -5$ A, are electrons moving toward $a$ or $b$? How much charge passes through a cross section of the wire in 3 seconds?

**P1.9.** The circuit element shown in Figure P1.9 has $v = 12$ V and $i_{ba} = -2$ A. What is the value of $v_{ba}$? Be sure to give the correct algebraic sign. What is the value of $i$? Is energy delivered to the element or taken from it?

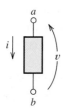

Figure P1.9

**P1.10.** To stop current from flowing through the headlight circuit of Figure 1.2 on page 7, should the switch be open or closed? In the fluid-flow analogy for the circuit, would the valve corresponding to the switch be open or closed? What state for a valve, open or closed, is analogous to an open switch?

**\*P1.11.** The net charge through a cross section of a circuit element is given by $q(t) = 2 + 3t$ C. Find the current through the element.

**P1.12.** The current through a particular circuit element is given by $i(t) = 10 \sin(200\pi t)$ A in which the angle is in radians. **a.** Sketch $i(t)$ to scale versus time. **b.** Determine the net charge that passes through the element between $t = 0$ and $t = 5$ ms. **c.** Repeat for the interval from $t = 0$ to $t = 10$ ms.

**\*P1.13.** The current through a given circuit element is given by

$$i(t) = 2e^{-t} \text{ A}$$

Find the net charge that passes through the element in the interval from $t = 0$ to $t = \infty$.

[*Hint:* Current is the rate of flow of charge. Thus, to find charge, we must integrate current with respect to time.]

**P1.14.** The net charge through a cross section of a certain circuit element is given by

$$q(t) = 3 - 3e^{-2t} \text{ C}$$

Determine the current through the element.

**P1.15.** A copper wire has a diameter of 2.05 mm and carries a current of 15 A due solely to electrons. (These values are common in residential wiring.) Each electron has a charge of $-1.60 \times 10^{-19}$ C. Assume that the free-electron (these are the electrons capable of moving through the copper) concentration in copper is $10^{29}$ electrons/m$^3$. Find the average velocity of the electrons in the wire.

**\*P1.16.** A certain lead acid storage battery has a mass of 30 kg. Starting from a fully charged state, it can supply 5 amperes for 24 hours with a terminal voltage of 12 V before it is totally discharged. **a.** If the energy stored in the fully charged battery is used to lift the battery with 100-percent efficiency, what height is attained? Assume that the acceleration due to gravity is 9.8 m/s$^2$ and is constant with height. **b.** If the energy stored is used to accelerate the battery with 100-percent efficiency, what velocity is attained? **c.** Gasoline contains about $4.5 \times 10^7$ J/kg. Compare this with the energy content per unit mass for the fully charged battery.

**P1.17.** A circuit element having terminals $a$ and $b$ has $v_{ab} = 10$ V and $i_{ba} = 2$ A. Over a period of 20 seconds, how much charge moves through the element? If electrons carry the charge, which terminal do they enter? How much energy is transferred? Is it delivered to the element or taken from it?

**P1.18.** An electron moves through a voltage of 9 V from the positive polarity to the negative polarity. How much energy is transferred? Does the electron gain or lose energy? Each electron has a change of $-1.60 \times 10^{-19}$ C.

**\*P1.19.** A typical "deep-cycle" battery (used for electric trolling motors for fishing boats)

\*Denotes that answers are contained in the Student Solutions files. See Appendix E for more information about accessing the Student Solutions.

is capable of delivering 12 V and 5 A for a period of 10 hours. How much charge flows through the battery in this interval? How much energy is delivered by the battery?

### Section 1.3: Power and Energy

**P1.20.** Define the term *passive reference configuration*. When do we have this configuration when using double subscript notation?

**\*P1.21.** Compute the power for each element shown in Figure P1.21. For each element, state whether energy is being absorbed by the element or supplied by it.

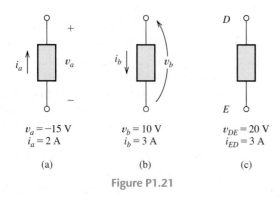

$$v_a = -15 \text{ V}$$
$$i_a = 2 \text{ A}$$

$$v_b = 10 \text{ V}$$
$$i_b = 3 \text{ A}$$

$$v_{DE} = 20 \text{ V}$$
$$i_{ED} = 3 \text{ A}$$

(a)          (b)          (c)

**Figure P1.21**

**P1.22.** The terminals of an electrical device are labeled $a$ and $b$. If $v_{ab} = -10\text{V}$, how much energy is exchanged when a charge of 3 C moves through the device from $a$ to $b$? Is the energy delivered to the device or taken from the device?

**\*P1.23.** The terminals of a certain battery are labeled $a$ and $b$. The battery voltage is $v_{ab} = 12$ V. To increase the chemical energy stored in the battery by 600 J, how much charge must move through the battery? Should electrons move from $a$ to $b$ or from $b$ to $a$?

**P1.24.** The element shown in Figure P1.24 has $v(t) = 10$ V and $i(t) = 2e^{-t}$ A. Compute the power for the circuit element. Find the energy transferred between $t = 0$ and $t = \infty$. Is this

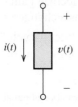

**Figure P1.24**

energy absorbed by the element or supplied by it?

**P1.25.** The current and voltage of an electrical device are $i_{ab}(t) = 5$ A and $v_{ab}(t) = 10 \sin(200\pi t)$ V in which the angle is in radians. **a.** Find the power delivered to the device and sketch it to scale versus time. **b.** Determine the energy delivered to the device for the interval from $t = 0$ to $t = 5$ ms. **c.** Repeat for the interval from $t = 0$ to $t = 10$ ms.

**\*P1.26.** Suppose that the cost of electrical energy is \$0.12 per kilowatt hour and that your electrical bill for 30 days is \$60. Assume that the power delivered is constant over the entire 30 days. What is the power in watts? If this power is supplied by a voltage of 120 V, what current flows? Part of your electrical load is a 60 W light that is on continuously. By what percentage can your energy consumption be reduced by turning this light off?

**P1.27.** Figure P1.27 shows an ammeter (AM) and voltmeter (VM) connected to measure the current and voltage, respectively, for circuit element $A$. When current actually enters the + terminal of the ammeter, the reading is positive, and when current leaves the + terminal, the reading is negative. If the actual voltage polarity is positive at the + terminal of the VM, the reading is positive; otherwise, it is negative. (Actually, for the connection shown, the ammeter reads the sum of the current in element $A$ and the very small current taken by the voltmeter. For purposes of this problem, assume that the current taken by the voltmeter is negligible.) Find the power for element $A$ and state whether energy is being delivered to element $A$ or taken from it if **a.** the ammeter reading is +2 A and the voltmeter reading is +30 V; **b.** the ammeter reading is −2 A and the voltmeter reading is +30 V; **c.** the ammeter reading is −2 A and the voltmeter reading is −30 V.

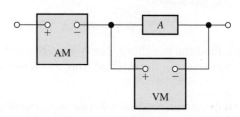

**Figure P1.27**

*P1.28. Repeat Problem P1.27 with the meters connected as shown in Figure P1.28.

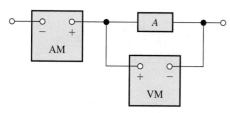

Figure P1.28

P1.29. A certain type of D-cell battery that costs $0.50 is capable of producing 1.2 V and a current of 0.1 A for a period of 75 hours. Determine the cost of the energy delivered by this battery per kilowatt hour. (For comparison, the approximate cost of energy purchased from electric utilities in the United States is $0.12 per kilowatt hour.)

P1.30. The electronics aboard a certain sailboat consume 50 W when operated from a 12.6-V source. If a certain fully charged deep-cycle lead acid storage battery is rated for 12.6 V and 100 ampere hours, for how many hours can the electronics be operated from the battery without recharging? (The ampere-hour rating of the battery is the operating time to discharge the battery multiplied by the current.) How much energy in kilowatt hours is initially stored in the battery? If the battery costs $75 and has a life of 300 charge-discharge cycles, what is the cost of the energy in dollars per kilowatt hour? Neglect the cost of recharging the battery.

## Section 1.4: Kirchhoff's Current Law

P1.31. What is a *node* in an electrical circuit? Identify the nodes in the circuit of Figure P1.31. Keep

in mind that all points connected by ideal conductors are considered to be a single node in electrical circuits.

P1.32. State Kirchhoff's current law.

P1.33. Two electrical elements are connected in series. What can you say about the currents through the elements?

P1.34. Suppose that in the fluid-flow analogy for an electrical circuit the analog of electrical current is volumetric flow rate with units of $cm^3$/s. For a proper analogy to electrical circuits, must the fluid be compressible or incompressible? Must the walls of the pipes be elastic or inelastic? Explain your answers.

*P1.35. Identify elements that are in series in the circuit of Figure P1.31.

P1.36. Consider the circuit shown in Figure P1.36. **a.** Which elements are in series? **b.** What is the relationship between $i_d$ and $i_c$? **c.** Given that $i_a = 3$ A and $i_c = 1$ A, determine the values of $i_b$ and $i_d$.

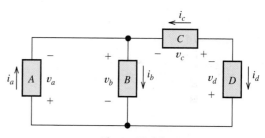

Figure P1.36

*P1.37. Use KCL to find the values of $i_a$, $i_c$, and $i_d$ for the circuit of Figure P1.37. Which elements are connected in series in this circuit?

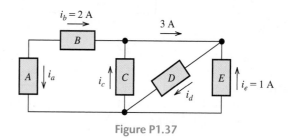

Figure P1.37

*P1.38. Find the values of the other currents in Figure P1.38 if $i_a = 2$ A, $i_b = 3$ A, $i_d = -5$ A, and $i_h = 4$ A.

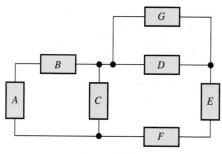

Figure P1.31

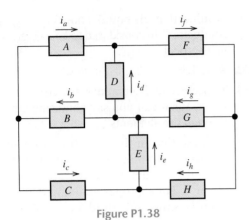

**Figure P1.38**

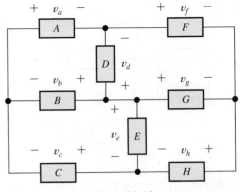

**Figure P1.43**

**P1.39.** Find the values of the other currents in Figure P1.38 if $i_a = -1$ A, $i_c = 3$ A, $i_g = 5$ A, and $i_h = 1$ A.

## Section 1.5: Kirchhoff's Voltage Law

**P1.40.** State Kirchhoff's voltage law.

**P1.41.** Consider the circuit shown in Figure P1.36. **a.** Which elements are in parallel? **b.** What is the relationship between $v_a$ and $v_b$? **c.** Given that $v_a = 2$ V and $v_d = -5$ V, determine the values of $v_b$ and $v_c$.

**\*P1.42.** Use KVL to solve for the voltages $v_a$, $v_b$, and $v_c$ in Figure P1.42.

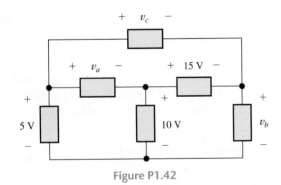

**Figure P1.42**

**P1.43.** Solve for the other voltages shown in Figure P1.43 given that $v_a = 5$ V, $v_b = 7$ V, $v_f = -10$ V, and $v_h = 6$ V.

**\*P1.44.** Use KVL and KCL to solve for the labeled currents and voltages in Figure P1.44. Compute the power for each element and show that power is conserved (i.e., the algebraic sum of the powers is zero).

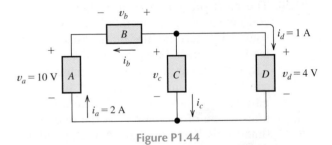

**Figure P1.44**

**P1.45.** Identify elements that are in parallel **a.** in Figure P1.37, **b.** in Figure P1.43, **c.** in Figure P1.44.

**P1.46.** Points $a, b, c$, and $d$ appear in a certain circuit. We know that $v_{ab} = 5$ V, $v_{cb} = 15$ V, and $v_{da} = -10$ V. Determine the values of $v_{ac}$ and $v_{cd}$.

## Section 1.6: Introduction to Circuit Elements

**P1.47.** In your own words, define **a.** an ideal conductor; **b.** an ideal voltage source; **c.** an ideal current source.

**P1.48.** Name four types of dependent sources and give the units for the gain parameter for each type.

**P1.49.** State Ohm's law, including references.

*P1.50. Draw a circuit that contains a 5-Ω resistance, a 10-V independent voltage source, and a 2-A independent current source. Connect all three elements in series. Because the polarity of the voltage source and reference direction for the current source are not specified, several correct answers are possible.

P1.51. Repeat Problem P1.50, placing all three elements in parallel.

P1.52. The resistance of a certain copper wire is 0.5 Ω. Determine the resistance of a tungsten wire having the same dimensions as the copper wire.

P1.53. Draw a circuit that contains a 5-Ω resistor, a 10-V voltage source, and a voltage-controlled voltage source having a gain constant of 0.5. Assume that the voltage across the resistor is the control voltage for the controlled source. Place all three elements in series.

P1.54. Draw a circuit that contains a 5-Ω resistor, a 10-V voltage source, and a current-controlled voltage source having a gain constant of 2 Ω. Assume that the current through the resistor is the control current for the controlled source. Place all three elements in series.

*P1.55. A power of 100 W is delivered to a certain resistor when the applied voltage is 100 V. Find the resistance. Suppose that the voltage is reduced by 10 percent (to 90 V). By what percentage is the power reduced? Assume that the resistance remains constant.

P1.56. The voltage across a 10-Ω resistor is given by $v(t) = 5e^{-2t}$ V. Determine the energy delivered to the resistor between $t = 0$ and $t = \infty$.

P1.57. The voltage across a 10-Ω resistor is given by $v(t) = 5\sin(2\pi t)$ V. Determine the energy delivered to the resistor between $t = 0$ and $t = 10$ s.

P1.58. A certain wire has a resistance of 0.5 Ω. Find the new resistance a. if the length of the wire is doubled, b. if the diameter of the wire is doubled.

### Section 1.7: Introduction to Circuits

P1.59. Plot $i$ versus $v$ to scale for each of the parts of Figure P1.59.

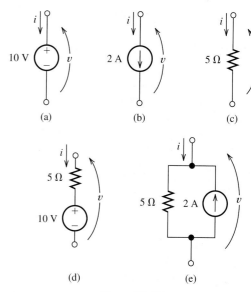

(a)          (b)          (c)

(d)          (e)

Figure P1.59

*P1.60. Which of the following are self-contradictory combinations of circuit elements? a. A 12-V voltage source in parallel with a 2-A current source. b. A 2-A current source in series with a 3-A current source. c. A 2-A current source in parallel with a short circuit. d. A 2-A current source in series with an open circuit. e. A 5-V voltage source in parallel with a short circuit.

P1.61. Consider the circuit shown in Figure P1.61. Find the power for the voltage source and for the current source. Which source is absorbing power?

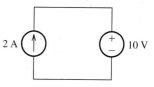

2 A          10 V

Figure P1.61

*P1.62. Consider the circuit shown in Figure P1.62. Find the current $i_R$ flowing through the resistor.

Find the power for each element in the circuit. Which elements are absorbing power?

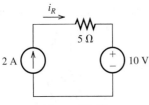

Figure P1.62

**P1.63.** Consider the circuit shown in Figure P1.63. Find the current $i_R$ flowing through the resistor. Find the power for each element in the circuit. Which elements are receiving power?

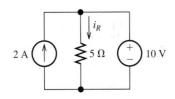

Figure P1.63

**\*P1.64.** Consider the circuit shown in Figure P1.64. Use Ohm's law, KVL, and KCL to find $V_x$.

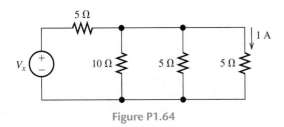

Figure P1.64

**P1.65.** Determine the value of $I_x$ in the circuit shown in Figure P1.65.

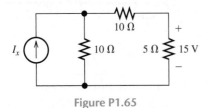

Figure P1.65

**P1.66.** Consider the circuit shown in Figure P1.66. **a.** Which elements are in series? **b.** Which

elements are in parallel? **c.** Apply Ohm's and Kirchhoff's laws to solve for $V_x$.

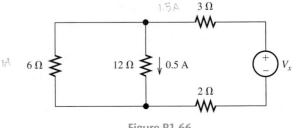

Figure P1.66

**P1.67.** The circuit shown in Figure P1.67 is the electrical model for an electronic megaphone, in which the 8-$\Omega$ resistance models a loudspeaker, the source $V_x$ and the 5-k$\Omega$ resistance represent a microphone, and the remaining elements model an amplifier. Given that the power delivered to the 8-$\Omega$ resistance is 8 W, determine the current circulating in the right-hand loop of the circuit. Also, determine the value of the microphone voltage $V_x$.

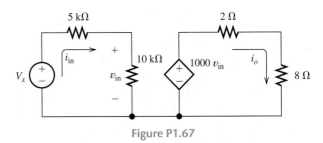

Figure P1.67

**P1.68.** Consider the circuit shown in Figure P1.68. **a.** Which elements are in series? **b.** Which elements are in parallel? **c.** Apply Ohm's and Kirchhoff's laws to solve for $R_x$.

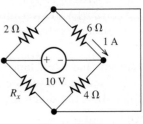

Figure P1.68

**P1.69.** Solve for the currents shown in Figure P1.69.

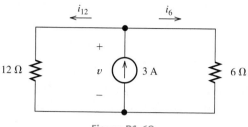

Figure P1.69

*P1.70. The circuit shown in Figure P1.70 contains a voltage-controlled voltage source. **a.** Use KVL to write an equation relating the voltages and solve for $v_x$. **b.** Use Ohm's law to find the current $i_x$. **c.** Find the power for each element in the circuit and verify that power is conserved.

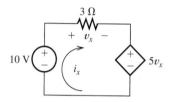

Figure P1.70

**P1.71.** Determine the value of $v_x$ and $i_y$ in the circuit shown in Figure P1.71.

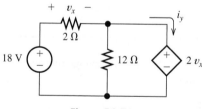

Figure P1.71

**P1.72.** A 10-V independent voltage source is in series with a 2-A independent current source. What single source is equivalent to this series combination? Give the type and value of the equivalent source.

**P1.73.** A 10-V independent voltage source is in parallel with a 2-A independent current source. What single source is equivalent to this parallel combination? Give the type and value of the equivalent source.

**P1.74.** Consider the circuit shown in Figure P1.74. **a.** Use KVL to write an equation relating the voltages. **b.** Use Ohm's law to write equations relating $v_1$ and $v_2$ to the current $i$. **c.** Substitute the equations from part (b) into the equation from part (a) and solve for $i$. **d.** Find the power for each element in the circuit and verify that power is conserved.

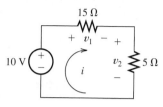

Figure P1.74

*P1.75. The circuit shown in Figure P1.75 contains a voltage-controlled current source. Solve for $v_s$.

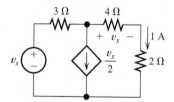

Figure P1.75

**P1.76.** For the circuit shown in Figure P1.76, solve for $i_s$. What types of sources are present in this circuit?

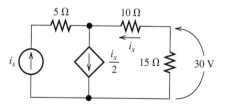

Figure P1.76

**P1.77.** For the circuit shown in Figure P1.77, solve for the current $i_x$. What types of sources are present in this circuit?

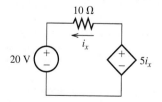

Figure P1.77

## Practice Test

Here is a practice test you can use to check your comprehension of the most important concepts in this chapter. Answers can be found in Appendix D and complete solutions are included in the Student Solutions files.

See Appendix E for more information about the Student Solutions.

**T1.1.** Match each entry in Table T1.1(a) with the best choice from the list given in Table T1.1(b).

### Table T1.1

| Item | Best Match |
|------|-----------|
| **(a)** | |
| **a.** Node | |
| **b.** Loop | |
| **c.** KVL | |
| **d.** KCL | |
| **e.** Ohm's law | |
| **f.** Passive reference configuration | |
| **g.** Ideal conductor | |
| **h.** Open circuit | |
| **i.** Current source | |
| **j.** Parallel connected elements | |
| **k.** Controlled source | |
| **l.** Units for voltage | |
| **m.** Units for current | |
| **n.** Units for resistance | |
| **o.** Series connected elements | |

**(b)**

1. $v_{ab} = Ri_{ab}$
2. The current reference for an element enters the positive voltage reference
3. A path through which no current can flow
4. Points connected by ideal conductors
5. An element that carries a specified current
6. An element whose current or voltage depends on a current or voltage elsewhere in the circuit
7. A path starting at a node and proceeding from node to node back to the starting node
8. An element for which the voltage is zero
9. A/V
10. V/A
11. J/C
12. C/V
13. C/s
14. Elements connected so their currents must be equal
15. Elements connected so their voltages must be equal
16. The algebraic sum of voltages for a closed loop is zero
17. The algebraic sum of the voltages for elements connected to a node is zero
18. The sum of the currents entering a node equals the sum of those leaving

[Items in Table T1.1(b) may be used more than once or not at all.]

**T1.2.** Consider the circuit of Figure T1.2 with $I_s = 3$ A, $R = 2\ \Omega$, and $V_s = 10$ V. **a.** Determine the value of $v_R$. **b.** Determine the magnitude of the power for the voltage source and state whether the voltage source is absorbing energy or delivering it. **c.** How many nodes does this circuit have? **d.** Determine the magnitude of the power for the current source and state whether the current source is absorbing energy or delivering it.

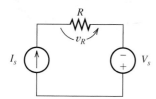

Figure T1.2

**T1.3.** The circuit of Figure T1.3 has $I_1 = 3$ A, $I_2 = 1$ A, $R_1 = 12\ \Omega$, and $R_2 = 6\ \Omega$. **a.** Determine the value of $v_{ab}$. **b.** Determine the power for each current source and state whether it is absorbing energy or delivering it. **c.** Compute the power absorbed by $R_1$ and by $R_2$.

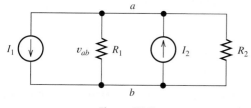

Figure T1.3

**T1.4.** The circuit shown in Figure T1.4 has $V_s = 12$ V, $v_2 = 4$ V, and $R_1 = 4\ \Omega$. **a.** Find the values of: **a.** $v_1$; **b.** $i$; **c.** $R_2$.

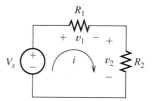

Figure T1.4

**T1.5.** We are given $V_s = 15$ V, $R = 10\ \Omega$, and $a = 0.3$ S for the circuit of Figure T1.5. Find the value of the current $i_{sc}$ flowing through the short circuit.

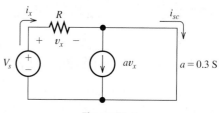

Figure T1.5

**T1.6.** We are given $i_4 = 2$ A for the circuit of Figure T1.6. Use Ohm's law, KCL, and KVL to find the values of $i_1$, $i_2$, $i_3$ and $v_s$.

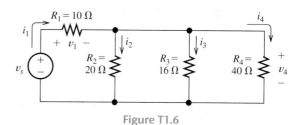

Figure T1.6

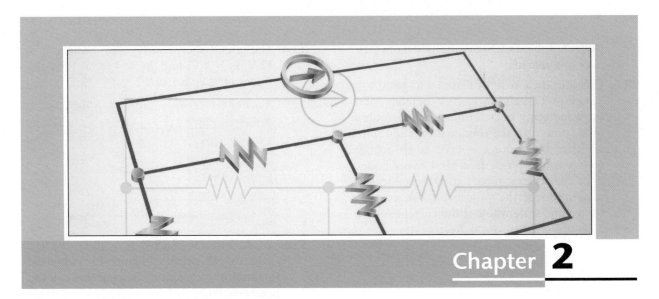

# Chapter 2

# Resistive Circuits

## Study of this chapter will enable you to:

- Solve circuits (i.e., find currents and voltages of interest) by combining resistances in series and parallel.

- Apply the voltage-division and current-division principles.

- Solve circuits by the node-voltage technique.

- Solve circuits by the mesh-current technique.

- Find Thévenin and Norton equivalents and apply source transformations.

- Use MATLAB® to solve circuit equations numerically and symbolically.

- Understand and apply the superposition principle.

- Draw the circuit diagram and state the principles of operation for the Wheatstone bridge.

## Introduction to this chapter:

In applications of electrical engineering, we often face circuit-analysis problems for which the structure of a circuit, including element values, is known and the currents, voltages, and powers need to be found. In this chapter, we examine techniques for analyzing circuits composed of resistances, voltage sources, and current sources. Later, we extend many of these concepts to circuits containing inductance and capacitance.

Over the years, you will meet many applications of electrical engineering in your field of engineering or science. This chapter will give you the skills needed to work effectively with the electronic instrumentation and other circuits that you will encounter. The material in this book will help you to answer questions on the Fundamentals of Engineering Examination and become a Registered Professional Engineer.

## 2.1   RESISTANCES IN SERIES AND PARALLEL

In this section, we show how to replace series or parallel combinations of resistances by equivalent resistances. Then, we demonstrate how to use this knowledge in solving circuits.

### Series Resistances

Consider the series combination of three resistances shown in Figure 2.1(a). Recall that in a series circuit the elements are connected end to end and that the same current flows through all of the elements. By Ohm's law, we can write

$$v_1 = R_1 i \tag{2.1}$$

$$v_2 = R_2 i \tag{2.2}$$

and

$$v_3 = R_3 i \tag{2.3}$$

Using KVL, we can write

$$v = v_1 + v_2 + v_3 \tag{2.4}$$

Substituting Equations 2.1, 2.2, and 2.3 into Equation 2.4, we obtain

$$v = R_1 i + R_2 i + R_3 i \tag{2.5}$$

Factoring out the current $i$, we have

$$v = (R_1 + R_2 + R_3)i \tag{2.6}$$

Now, we define the equivalent resistance $R_{eq}$ to be the sum of the resistances in series:

$$R_{eq} = R_1 + R_2 + R_3 \tag{2.7}$$

Using this to substitute into Equation 2.6, we have

$$v = R_{eq} i \tag{2.8}$$

Thus, we conclude that the three resistances in series can be replaced by the equivalent resistance $R_{eq}$ shown in Figure 2.1(b) with no change in the relationship

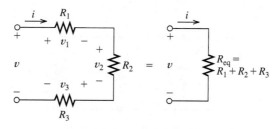

**Figure 2.1**  Series resistances can be combined into an equivalent resistance.

(a) Three resistances in series

(b) Equivalent resistance

between the voltage $v$ and current $i$. If the three resistances are part of a larger circuit, replacing them by a single equivalent resistance would make no changes in the currents or voltages in other parts of the circuit.

This analysis can be applied to any number of resistances. For example, two resistances in series can be replaced by a single resistance equal to the sum of the original two. To summarize, *a series combination of resistances has an equivalent resistance equal to the sum of the original resistances.*

> A series combination of resistances has an equivalent resistance equal to the sum of the original resistances.

### Parallel Resistances

Figure 2.2(a) shows three resistances in parallel. In a parallel circuit, the voltage across each element is the same. Applying Ohm's law in Figure 2.2(a), we can write

$$i_1 = \frac{v}{R_1} \tag{2.9}$$

$$i_2 = \frac{v}{R_2} \tag{2.10}$$

$$i_3 = \frac{v}{R_3} \tag{2.11}$$

The top ends of the resistors in Figure 2.2(a) are connected to a single node. (Recall that all points in a circuit that are connected by conductors constitute a node.) Thus, we can apply KCL to the top node of the circuit and obtain

$$i = i_1 + i_2 + i_3 \tag{2.12}$$

Now using Equations 2.9, 2.10, and 2.11 to substitute into Equation 2.12, we have

$$i = \frac{v}{R_1} + \frac{v}{R_2} + \frac{v}{R_3} \tag{2.13}$$

Factoring out the voltage, we obtain

$$i = \left( \frac{1}{R_1} + \frac{1}{R_2} + \frac{1}{R_3} \right) v \tag{2.14}$$

Now, we define the equivalent resistance as

$$R_{eq} = \frac{1}{1/R_1 + 1/R_2 + 1/R_3} \tag{2.15}$$

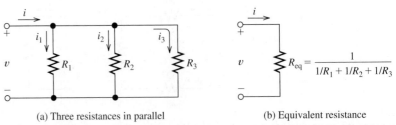

(a) Three resistances in parallel                    (b) Equivalent resistance

**Figure 2.2** Parallel resistances can be combined into an equivalent resistance.

In terms of the equivalent resistance, Equation 2.14 becomes

$$i = \frac{1}{R_{eq}}v \qquad (2.16)$$

Comparing Equations 2.14 and 2.16, we see that $i$ and $v$ are related in the same way by both equations provided that $R_{eq}$ is given by Equation 2.15. Therefore, a parallel combination of resistances can be replaced by its equivalent resistance without changing the currents and voltages in other parts of the circuit. The equivalence is illustrated in Figure 2.2(b).

A parallel combination of resistances can be replaced by its equivalent resistance without changing the currents and voltages in other parts of the circuit.

This analysis can be applied to any number of resistances in parallel. For example, if four resistances are in parallel, the equivalent resistance is

$$R_{eq} = \frac{1}{1/R_1 + 1/R_2 + 1/R_3 + 1/R_4} \qquad (2.17)$$

Similarly, for two resistances, we have

$$R_{eq} = \frac{1}{1/R_1 + 1/R_2} \qquad (2.18)$$

This can be put into the form

$$R_{eq} = \frac{R_1 R_2}{R_1 + R_2} \qquad (2.19)$$

(Notice that Equation 2.19 applies only for two resistances. The product over the sum does not apply for more than two resistances.)

The product over the sum does not apply for more than two resistances.

Sometimes, resistive circuits can be reduced to a single equivalent resistance by repeatedly combining resistances that are in series or parallel.

---

**Example 2.1    Combining Resistances in Series and Parallel**

Find a single equivalent resistance for the network shown in Figure 2.3(a).

**Solution**    First, we look for a combination of resistances that is in series or in parallel. In Figure 2.3(a), $R_3$ and $R_4$ are in series. (In fact, as it stands, no other two resistances in this network are either in series or in parallel.) Thus, our first step is to combine $R_3$ and $R_4$, replacing them by their equivalent resistance. Recall that for a series combination, the equivalent resistance is the sum of the resistances in series:

1. Find a series or parallel combination of resistances.

$$R_{eq1} = R_3 + R_4 = 5 + 15 = 20 \ \Omega$$

2. Combine them.

Figure 2.3(b) shows the network after replacing $R_3$ and $R_4$ by their equivalent resistance. Now we see that $R_2$ and $R_{eq1}$ are in parallel. The equivalent resistance for this combination is

3. Repeat until the network is reduced to a single resistance (if possible).

$$R_{eq2} = \frac{1}{1/R_{eq1} + 1/R_2} = \frac{1}{1/20 + 1/20} = 10 \ \Omega$$

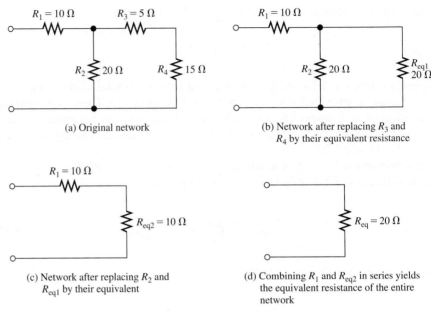

(a) Original network

(b) Network after replacing $R_3$ and $R_4$ by their equivalent resistance

(c) Network after replacing $R_2$ and $R_{eq1}$ by their equivalent

(d) Combining $R_1$ and $R_{eq2}$ in series yields the equivalent resistance of the entire network

**Figure 2.3** Resistive network for Example 2.1.

Making this replacement gives the equivalent network shown in Figure 2.3(c).

Finally, we see that $R_1$ and $R_{eq2}$ are in series. Thus, the equivalent resistance for the entire network is

$$R_{eq} = R_1 + R_{eq2} = 10 + 10 = 20 \ \Omega$$

■

**Exercise 2.1**    Find the equivalent resistance for each of the networks shown in Figure 2.4. [*Hint for part (b): $R_3$ and $R_4$ are in parallel.*]
**Answer**    **a.** 3 Ω; **b.** 5 Ω; **c.** 52.1 Ω; **d.** 1.5 kΩ.    ◻

## Conductances in Series and Parallel

Recall that conductance is the reciprocal of resistance. Using this fact to change resistances to conductances for a series combination of $n$ elements, we readily obtain:

Combine conductances in series as you would resistances in parallel. Combine conductances in parallel as you would resistances in series.

$$G_{eq} = \frac{1}{1/G_1 + 1/G_2 + \cdots + 1/G_n} \tag{2.20}$$

Thus, we see that conductances in series combine as do resistances in parallel. For two conductances in series, we have:

$$G_{eq} = \frac{G_1 G_2}{G_1 + G_2}$$

For $n$ conductances in parallel, we can show that

$$G_{eq} = G_1 + G_2 + \cdots + G_n \tag{2.21}$$

Conductances in parallel combine as do resistances in series.

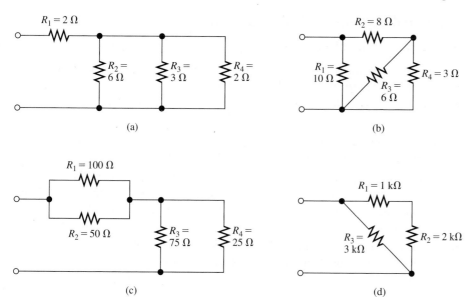

**Figure 2.4** Resistive networks for Exercise 2.1.

## Series versus Parallel Circuits

An element such as a toaster or light bulb that absorbs power is called a **load**. When we want to distribute power from a single voltage source to various loads, we usually place the loads in parallel. A switch in series with each load can break the flow of current to that load without affecting the voltage supplied to the other loads.

Sometimes, to save wire, strings of Christmas lights consist of bulbs connected in series. The bulbs tend to fail or "burn out" by becoming open circuits. Then the entire string is dark and the defective bulb can be found only by trying each in turn. If several bulbs are burned out, it can be very tedious to locate the failed units. In a parallel connection, only the failed bulbs are dark.

*When we want to distribute power from a single voltage source to various loads, we usually place the loads in parallel.*

## 2.2  NETWORK ANALYSIS BY USING SERIES AND PARALLEL EQUIVALENTS

An electrical **network** (or electrical circuit) consists of circuit elements, such as resistances, voltage sources, and current sources, connected together to form closed paths. **Network analysis** is the process of determining the current, voltage, and power for each element, given the circuit diagram and the element values. In this and the sections that follow, we study several useful techniques for network analysis.

Sometimes, we can determine the currents and voltages for each element in a resistive circuit by repeatedly replacing series and parallel combinations of resistances by their equivalent resistances. Eventually, this may reduce the circuit sufficiently that the equivalent circuit can be solved easily. The information gained from the simplified circuit is transferred to the previous steps in the chain of equivalent circuits. In the end, we gain enough information about the original circuit to determine all the currents and voltages.

*An electrical network consists of circuit elements such as resistances, voltage sources, and current sources, connected together to form closed paths.*

## Circuit Analysis Using Series/Parallel Equivalents

Here are the steps in solving circuits using series/parallel equivalents:

Some good advice for beginners: Don't try to combine steps. Be very methodical and do one step at a time. Take the time to redraw each equivalent carefully and label unknown currents and voltages consistently in the various circuits. The slow methodical approach will be faster and more accurate when you are learning. Walk now—later you will be able to run.

1. Begin by locating a combination of resistances that are in series or parallel. Often the place to start is farthest from the source.
2. Redraw the circuit with the equivalent resistance for the combination found in step 1.
3. Repeat steps 1 and 2 until the circuit is reduced as far as possible. Often (but not always) we end up with a single source and a single resistance.
4. Solve for the currents and voltages in the final equivalent circuit. Then, transfer results back one step and solve for additional unknown currents and voltages. Again transfer the results back one step and solve. Repeat until all of the currents and voltages are known in the original circuit.
5. Check your results to make sure that KCL is satisfied at each node, KVL is satisfied for each loop, and the powers add to zero.

---

**Example 2.2**    Circuit Analysis Using Series/Parallel Equivalents

Find the current, voltage, and power for each element of the circuit shown in Figure 2.5(a).

Steps 1, 2, and 3.

**Solution**    First, we combine resistances in series and parallel. For example, in the original circuit, $R_2$ and $R_3$ are in parallel. Replacing $R_2$ and $R_3$ by their parallel equivalent, we obtain the circuit shown in Figure 2.5(b). Next, we see that $R_1$ and $R_{eq1}$ are in series. Replacing these resistances by their sum, we obtain the circuit shown in Figure 2.5(c).

After we have reduced a network to an equivalent resistance connected across the source, we solve the simplified network. Then, we transfer results back through

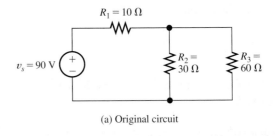

(a) Original circuit

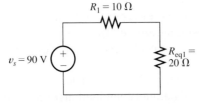

(b) Circuit after replacing $R_2$ and $R_3$ by their equivalent

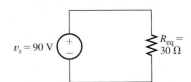

(c) Circuit after replacing $R_1$ and $R_{eq1}$ by their equivalent

**Figure 2.5**  A circuit and its simplified versions. See Example 2.2.

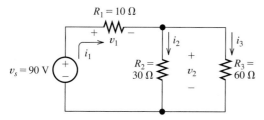

(a) Third, we use known values of $i_1$ and $v_2$
to solve for the remaining currents and voltages

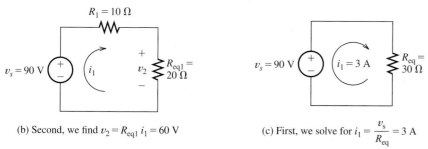

(b) Second, we find $v_2 = R_{eq1} i_1 = 60$ V          (c) First, we solve for $i_1 = \dfrac{v_s}{R_{eq}} = 3$ A

**Figure 2.6** After reducing the circuit to a source and an equivalent resistance, we solve the simplified circuit. Then, we transfer results back to the original circuit. Notice that the logical flow in solving for currents and voltages starts from the simplified circuit in (c).

the chain of equivalent circuits. We illustrate this process in Figure 2.6. (Figure 2.6 is identical to Figure 2.5, except for the currents and voltages shown in Figure 2.6. Usually, in solving a network by this technique, we first draw the chain of equivalent networks and then write results on the same drawings. However, this might be confusing in our first example.)

First, we solve the simplified network shown in Figure 2.6(c). Because $R_{eq}$ is in parallel with the 90-V voltage source, the voltage across $R_{eq}$ must be 90 V, with its positive polarity at the top end. Thus, the current flowing through $R_{eq}$ is given by

Step 4.

$$ i_1 = \frac{v_s}{R_{eq}} = \frac{90 \text{ V}}{30 \text{ } \Omega} = 3 \text{ A} $$

We know that this current flows downward (from plus to minus) through $R_{eq}$. Since $v_s$ and $R_{eq}$ are in series in Figure 2.6(c), the current must also flow upward through $v_s$. Thus, $i_1 = 3$ A flows clockwise around the circuit, as shown in Figure 2.6(c).

Because $R_{eq}$ is the equivalent resistance seen by the source in all three parts of Figure 2.6, the current through $v_s$ must be $i_1 = 3$ A, flowing upward in all three equivalent circuits. In Figure 2.6(b), we see that $i_1$ flows clockwise through $v_s$, $R_1$, and $R_{eq1}$. The voltage across $R_{eq1}$ is given by

$$ v_2 = R_{eq1}i_1 = 20 \text{ } \Omega \times 3 \text{ A} = 60 \text{ V} $$

Because $R_{eq1}$ is the equivalent resistance for the parallel combination of $R_2$ and $R_3$, the voltage $v_2$ also appears across $R_2$ and $R_3$ in the original network.

At this point, we have found that the current through $v_s$ and $R_1$ is $i_1 = 3$ A. Furthermore, the voltage across $R_2$ and $R_3$ is 60 V. This information is shown in Figure 2.6(a). Now, we can compute the remaining values desired:

$$i_2 = \frac{v_2}{R_2} = \frac{60 \text{ V}}{30 \text{ }\Omega} = 2 \text{ A}$$

$$i_3 = \frac{v_2}{R_3} = \frac{60 \text{ V}}{60 \text{ }\Omega} = 1 \text{ A}$$

(As a check, we can use KCL to verify that $i_1 = i_2 + i_3$.)

Next, we can use Ohm's law to compute the value of $v_1$:

$$v_1 = R_1 i_1 = 10 \text{ }\Omega \times 3 \text{ A} = 30 \text{ V}$$

Step 5.

(As a check, we use KVL to verify that $v_s = v_1 + v_2$.)

Now, we compute the power for each element. For the voltage source, we have

$$p_s = -v_s i_1$$

We have included the minus sign because the references for $v_s$ and $i_1$ are opposite to the passive configuration. Substituting values, we have

$$p_s = -(90 \text{ V}) \times 3 \text{ A} = -270 \text{ W}$$

Because the power for the source is negative, we know that the source is supplying energy to the other elements in the circuit.

The powers for the resistances are

$$p_1 = R_1 i_1^2 = 10 \text{ }\Omega \times (3 \text{ A})^2 = 90 \text{ W}$$

$$p_2 = \frac{v_2^2}{R_2} = \frac{(60 \text{ V})^2}{30 \text{ }\Omega} = 120 \text{ W}$$

$$p_3 = \frac{v_2^2}{R_3} = \frac{(60 \text{ V})^2}{60 \text{ }\Omega} = 60 \text{ W}$$

(As a check, we verify that $p_s + p_1 + p_2 + p_3 = 0$, showing that power is conserved.)    ∎

## Power Control by Using Heating Elements in Series or Parallel

Resistances are commonly used as heating elements for the reaction chamber of chemical processes. For example, the catalytic converter of an automobile is not effective until its operating temperature is achieved. Thus, during engine warm-up, large amounts of pollutants are emitted. Automotive engineers have proposed and studied the use of electrical heating elements to heat the converter more quickly, thereby reducing pollution. By using several heating elements that can be operated individually, in series, or in parallel, several power levels can be achieved. This is useful in controlling the temperature of a chemical process.

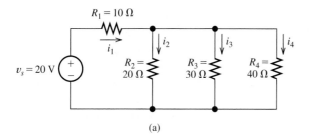

(a)

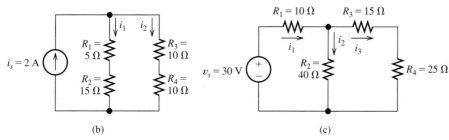

(b)                                         (c)

**Figure 2.7** Circuits for Exercise 2.2.

**Exercise 2.2**   Find the currents labeled in Figure 2.7 by combining resistances in series and parallel.

**Answer**   **a.** $i_1 = 1.04$ A, $i_2 = 0.480$ A, $i_3 = 0.320$ A, $i_4 = 0.240$ A;  **b.** $i_1 = 1$ A, $i_2 = 1$ A; **c.** $i_1 = 1$ A, $i_2 = 0.5$ A, $i_3 = 0.5$ A.   □

## 2.3  VOLTAGE-DIVIDER AND CURRENT-DIVIDER CIRCUITS

### Voltage Division

When a voltage is applied to a series combination of resistances, a fraction of the voltage appears across each of the resistances. Consider the circuit shown in Figure 2.8. The equivalent resistance seen by the voltage source is

$$R_{eq} = R_1 + R_2 + R_3 \qquad (2.22)$$

The current is the total voltage divided by the equivalent resistance:

$$i = \frac{v_{total}}{R_{eq}} = \frac{v_{total}}{R_1 + R_2 + R_3} \qquad (2.23)$$

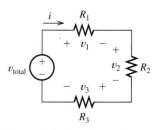

**Figure 2.8** Circuit used to derive the voltage-division principle.

Furthermore, the voltage across $R_1$ is

$$v_1 = R_1 i = \frac{R_1}{R_1 + R_2 + R_3} v_{total} \qquad (2.24)$$

Similarly, we have

$$v_2 = R_2 i = \frac{R_2}{R_1 + R_2 + R_3} v_{total} \qquad (2.25)$$

and

$$v_3 = R_3 i = \frac{R_3}{R_1 + R_2 + R_3} v_{\text{total}} \tag{2.26}$$

Of the total voltage, the fraction that appears across a given resistance in a series circuit is the ratio of the given resistance to the total series resistance.

We can summarize these results by the statement: *Of the total voltage, the fraction that appears across a given resistance in a series circuit is the ratio of the given resistance to the total series resistance.* This is known as the **voltage-division principle**.

We have derived the voltage-division principle for three resistances in series, but it applies for any number of resistances as long as they are connected in series.

---

**Example 2.3**    Application of the Voltage-Division Principle

Find the voltages $v_1$ and $v_4$ in Figure 2.9.

**Solution**    Using the voltage-division principle, we find that $v_1$ is the total voltage times the ratio of $R_1$ to the total resistance:

$$v_1 = \frac{R_1}{R_1 + R_2 + R_3 + R_4} v_{\text{total}}$$

$$= \frac{1000}{1000 + 1000 + 2000 + 6000} \times 15 = 1.5 \text{ V}$$

Similarly,

$$v_4 = \frac{R_4}{R_1 + R_2 + R_3 + R_4} v_{\text{total}}$$

$$= \frac{6000}{1000 + 1000 + 2000 + 6000} \times 15 = 9 \text{ V}$$

Notice that the largest voltage appears across the largest resistance in a series circuit.    ∎

## Current Division

The total current flowing into a parallel combination of resistances divides, and a fraction of the total current flows through each resistance. Consider the circuit shown in Figure 2.10. The equivalent resistance is given by

$$R_{\text{eq}} = \frac{R_1 R_2}{R_1 + R_2} \tag{2.27}$$

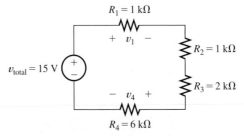

Figure 2.9  Circuit for Example 2.3.

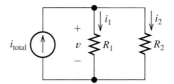

**Figure 2.10** Circuit used to derive the current-division principle.

The voltage across the resistances is given by

$$v = R_{eq}i_{total} = \frac{R_1 R_2}{R_1 + R_2} i_{total} \qquad (2.28)$$

Now, we can find the current in each resistance:

$$i_1 = \frac{v}{R_1} = \frac{R_2}{R_1 + R_2} i_{total} \qquad (2.29)$$

and

$$i_2 = \frac{v}{R_2} = \frac{R_1}{R_1 + R_2} i_{total} \qquad (2.30)$$

We can summarize these results by stating the **current-division principle**: *For two resistances in parallel, the fraction of the total current flowing in a resistance is the ratio of the other resistance to the sum of the two resistances.* Notice that this principle applies only for two resistances. If we have more than two resistances in parallel, we should combine resistances so we only have two before applying the current-division principle.

For two resistances in parallel, the fraction of the total current flowing in a resistance is the ratio of the other resistance to the sum of the two resistances.

An alternative approach is to work with conductances. For *n* conductances in parallel, it can be shown that

$$i_1 = \frac{G_1}{G_1 + G_2 + \cdots + G_n} i_{total}$$

Current division using conductances uses a formula with the same form as the formula for voltage division using resistances.

$$i_2 = \frac{G_2}{G_1 + G_2 + \cdots + G_n} i_{total}$$

and so forth. In other words, current division using conductances uses a formula with the same form as the formula for voltage division using resistances.

| **Example 2.4** | **Applying the Current- and Voltage-Division Principles** |

Use the voltage-division principle to find the voltage $v_x$ in Figure 2.11(a). Then find the source current $i_s$ and use the current-division principle to compute the current $i_3$.

**Solution**   The voltage-division principle applies only for resistances in series. Therefore, we first must combine $R_2$ and $R_3$. The equivalent resistance for the parallel combination of $R_2$ and $R_3$ is

$$R_x = \frac{R_2 R_3}{R_2 + R_3} = \frac{30 \times 60}{30 + 60} = 20 \ \Omega$$

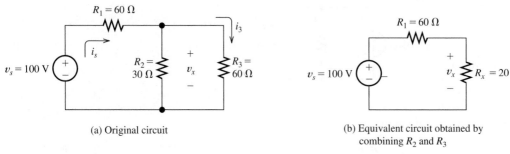

(a) Original circuit

(b) Equivalent circuit obtained by combining $R_2$ and $R_3$

**Figure 2.11**  Circuit for Example 2.4.

The equivalent network is shown in Figure 2.11(b).

Now, we can apply the voltage-division principle to find $v_x$. The voltage $v_x$ is equal to the total voltage times $R_x$ divided by the total series resistance:

$$v_x = \frac{R_x}{R_1 + R_x} v_s = \frac{20}{60 + 20} \times 100 = 25 \text{ V}$$

The source current $i_s$ is given by

$$i_s = \frac{v_s}{R_1 + R_x} = \frac{100}{60 + 20} = 1.25 \text{ A}$$

Now, we can use the current-division principle to find $i_3$. The fraction of the source current $i_s$ that flows through $R_3$ is $R_2/(R_2 + R_3)$. Thus, we have

$$i_3 = \frac{R_2}{R_2 + R_3} i_s = \frac{30}{30 + 60} \times 1.25 = 0.417 \text{ A}$$

As a check, we can also compute $i_3$ another way:

$$i_3 = \frac{v_x}{R_3} = \frac{25}{60} = 0.417 \text{ A}$$ ∎

## Example 2.5    Application of the Current-Division Principle

Use the current-division principle to find the current $i_1$ in Figure 2.12(a).

**Solution**   The current-division principle applies for two resistances in parallel. Therefore, our first step is to combine $R_2$ and $R_3$:

*The current-division principle applies for two resistances in parallel. Therefore, our first step is to combine $R_2$ and $R_3$.*

$$R_{eq} = \frac{R_2 R_3}{R_2 + R_3} = \frac{30 \times 60}{30 + 60} = 20 \text{ }\Omega$$

The resulting equivalent circuit is shown in Figure 2.12(b). Applying the current-division principle, we have

$$i_1 = \frac{R_{eq}}{R_1 + R_{eq}} i_s = \frac{20}{10 + 20} 15 = 10 \text{ A}$$

Reworking the calculations using conductances, we have

$$G_1 = \frac{1}{R_1} = 100 \text{ mS}, \quad G_2 = \frac{1}{R_2} = 33.33 \text{ mS}, \quad \text{and} \quad G_3 = \frac{1}{R_3} = 16.67 \text{ mS}$$

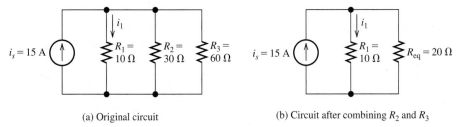

(a) Original circuit          (b) Circuit after combining $R_2$ and $R_3$

**Figure 2.12**  Circuit for Example 2.5.

Then, we compute the current

$$i_1 = \frac{G_1}{G_1 + G_2 + G_3}i_s = \frac{100}{100 + 33.33 + 16.67}15 = 10 \text{ A}$$

which is the same value that we obtained working with resistances.     ■

## Position Transducers Based on the Voltage-Division Principle

Transducers are used to produce a voltage (or sometimes a current) that is proportional to a physical quantity of interest, such as distance, pressure, or temperature. For example, Figure 2.13 shows how a voltage that is proportional to the rudder angle of a boat or aircraft can be obtained. As the rudder turns, a sliding contact moves along a resistance such that $R_2$ is proportional to the rudder angle $\theta$. The total resistance $R_1 + R_2$ is fixed. Thus, the output voltage is

$$v_o = v_s\frac{R_2}{R_1 + R_2} = K\theta$$

where $K$ is a constant of proportionality that depends on the source voltage $v_s$ and the construction details of the transducer. Many examples of transducers such as this are employed in all areas of science and engineering.

**Exercise 2.3**  Use the voltage-division principle to find the voltages labeled in Figure 2.14.
**Answer   a.**   $v_1 = 10$ V, $v_2 = 20$ V, $v_3 = 30$ V, $v_4 = 60$ V;   **b.**   $v_1 = 6.05$ V, $v_2 = 5.88$ V, $v_4 = 8.07$ V.     □

**Exercise 2.4**  Use the current-division principle to find the currents labeled in Figure 2.15
**Answer   a.** $i_1 = 1$ A, $i_3 = 2$ A; **b.** $i_1 = i_2 = i_3 = 1$ A.     □

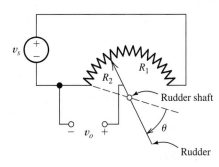

**Figure 2.13**  The voltage-division principle forms the basis for some position sensors. This figure shows a transducer that produces an output voltage $v_o$ proportional to the rudder angle $\theta$.

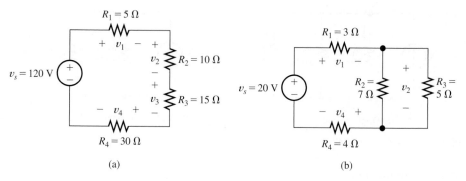

(a)

(b)

**Figure 2.14**  Circuits for Exercise 2.3.

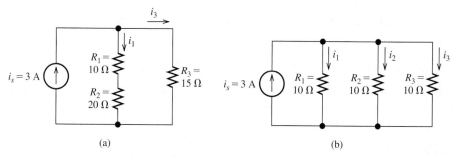

(a)

(b)

**Figure 2.15**  Circuits for Exercise 2.4.

## 2.4  NODE-VOLTAGE ANALYSIS

Although they are very important concepts, series/parallel equivalents and the current/voltage division principles are not sufficient to solve all circuits.

The network analysis methods that we have studied so far are useful, but they do not apply to all networks. For example, consider the circuit shown in Figure 2.16. We cannot solve this circuit by combining resistances in series and parallel because no series or parallel combination of resistances exists in the circuit. Furthermore, the voltage-division and current-division principles cannot be applied to this circuit. In this section, we learn **node-voltage analysis**, which is a general technique that can be applied to any circuit.

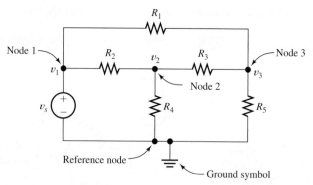

**Figure 2.16**  The first step in node analysis is to select a reference node and label the voltages at each of the other nodes.

## Selecting the Reference Node

A **node** is a point at which two or more circuit elements are joined together. In node-voltage analysis, we first select one of the nodes as the **reference node**. In principle, any node can be picked to be the reference node. However, the solution is usually facilitated by selecting one end of a voltage source as the reference node. We will see why this is true as we proceed.

For example, the circuit shown in Figure 2.16 has four nodes. Let us select the bottom node as the reference node. We mark the reference node by the **ground symbol**, as shown in the figure.

## Assigning Node Voltages

Next, we label the voltages at each of the other nodes. For example, the voltages at the three nodes are labeled $v_1$, $v_2$, and $v_3$ in Figure 2.16. The voltage $v_1$ is the voltage between node 1 and the reference node. The reference polarity for $v_1$ is positive at node 1 and negative at the reference node. Similarly, $v_2$ is the voltage between node 2 and the reference node. The reference polarity for $v_2$ is positive at node 2 and negative at the reference node. *In fact, the negative reference polarity for each of the node voltages is at the reference node.* We say that $v_1$ is the voltage at node 1 with respect to the reference node.

*The negative reference polarity for each of the node voltages is at the reference node.*

## Finding Element Voltages in Terms of the Node Voltages

In node-voltage analysis, we write equations and eventually solve for the node voltages. Once the node voltages have been found, it is relatively easy to find the current, voltage, and power for each element in the circuit.

For example, suppose that we know the values of the node voltages and we want to find the voltage across $R_3$ with its positive reference on the left-hand side. To avoid additional labels in Figure 2.16, we have made a second drawing of the circuit, which is shown in Figure 2.17. The node voltages and the voltage $v_x$ across $R_3$ are shown in Figure 2.17. Notice that $v_2$, $v_x$, and $v_3$ are the voltages encountered in traveling around the closed path through $R_4$, $R_3$, and $R_5$. Thus, these voltages must obey Kirchhoff's voltage law. Traveling around the loop clockwise and summing voltages, we have

*Once the node voltages have been determined, it is relatively easy to determine other voltages and currents in the circuit.*

$$-v_2 + v_x + v_3 = 0$$

Solving for $v_x$, we obtain

$$v_x = v_2 - v_3$$

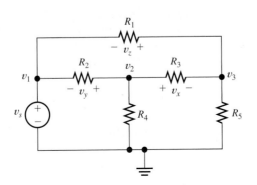

**Figure 2.17** Assuming that we can determine the node voltages $v_1$, $v_2$, and $v_3$, we can use KVL to determine $v_x$, $v_y$, and $v_z$. Then using Ohm's law, we can find the current in each of the resistances. Thus, the key problem is in determining the node voltages.

This is the same circuit shown in Figure 2.16. We have redrawn it simply to avoid cluttering the original diagram with the voltages $v_x$, $v_y$, and $v_z$ that are not involved in the final node equations.

Thus, we can find the voltage across any element in the network as the difference between node voltages. (If one end of an element is connected to the reference node, the voltage across the element is a node voltage.)

After the voltages are found, Ohm's law and KCL can be used to find the current in each element. Then, power can be computed by taking the product of the voltage and current for each element.

**Exercise 2.5**   In the circuit of Figure 2.17, find expressions for $v_y$ and $v_z$ in terms of the node voltages $v_1$, $v_2$, and $v_3$.

**Answer**   $v_y = v_2 - v_1, v_z = v_3 - v_1$.                                     □

## Writing KCL Equations in Terms of the Node Voltages

After choosing the reference node and assigning the voltage variables, we write equations that can be solved for the node voltages.

After choosing the reference node and assigning the voltage variables, we write equations that can be solved for the node voltages. We demonstrate by continuing with the circuit of Figure 2.16.

In Figure 2.16, the voltage $v_1$ is the same as the source voltage $v_s$:

$$v_1 = v_s$$

(In this case, one of the node voltages is known without any effort. This is the advantage in selecting the reference node at one end of an independent voltage source.)

Therefore, we need to determine the values of $v_2$ and $v_3$, and we must write two independent equations. We usually start by trying to write current equations at each of the nodes corresponding to an unknown node voltage. For example, at node 2 in Figure 2.16, the current leaving through $R_4$ is given by

$$\frac{v_2}{R_4}$$

This is true because $v_2$ is the voltage across $R_4$ with its positive reference at node 2. Thus, the current $v_2/R_4$ flows from node 2 toward the reference node, which is away from node 2.

Next, referring to Figure 2.17, we see that the current flowing out of node 2 through $R_3$ is given by $v_x/R_3$. However, we found earlier that $v_x = v_2 - v_3$. Thus, the current flowing out of node 2 through $R_3$ is given by

$$\frac{v_2 - v_3}{R_3}$$

To find the current flowing out of node $n$ through a resistance toward node $k$, we subtract the voltage at node $k$ from the voltage at node $n$ and divide the difference by the resistance.

At this point, we pause in our analysis to make a useful observation. *To find the current flowing out of node n through a resistance toward node k, we subtract the voltage at node k from the voltage at node n and divide the difference by the resistance.* Thus, if $v_n$ and $v_k$ are the node voltages and $R$ is the resistance connected between the nodes, the current flowing from node $n$ toward node $k$ is given by

$$\frac{v_n - v_k}{R}$$

Applying this observation in Figure 2.16 to find the current flowing out of node 2 through $R_2$, we have

$$\frac{v_2 - v_1}{R_2}$$

[In Exercise 2.5, we found that $v_y = v_2 - v_1$ (see Figure 2.17). The current flowing to the left through $R_2$ is $v_y/R_2$. Substitution yields the aforementioned expression.]

Of course, if the resistance is connected between node $n$ and the reference node, the current away from node $n$ toward the reference node is simply the node voltage $v_n$ divided by the resistance. For example, as we noted previously, the current leaving node 2 through $R_4$ is given by $v_2/R_4$.

Now we apply KCL, adding all of the expressions for the currents leaving node 2 and setting the sum to zero. Thus, we obtain

$$\frac{v_2 - v_1}{R_2} + \frac{v_2}{R_4} + \frac{v_2 - v_3}{R_3} = 0$$

Writing the current equation at node 3 is similar. We try to follow the same pattern in writing each equation. Then, the equations take a familiar form, and mistakes are less frequent. We usually write expressions for the currents leaving the node under consideration and set the sum to zero. Applying this approach at node 3 of Figure 2.16, we have

$$\frac{v_3 - v_1}{R_1} + \frac{v_3}{R_5} + \frac{v_3 - v_2}{R_3} = 0$$

In many networks, we can obtain all of the equations needed to solve for the node voltages by applying KCL to the nodes at which the unknown voltages appear.

### Example 2.6   Node-Voltage Analysis

Write equations that can be solved for the node voltages $v_1$, $v_2$, and $v_3$ shown in Figure 2.18.

**Solution**   We use KCL to write an equation at node 1:

$$\frac{v_1}{R_1} + \frac{v_1 - v_2}{R_2} + i_s = 0$$

Each term on the left-hand side of this equation represents a current leaving node 1. Summing the currents leaving node 2, we have

$$\frac{v_2 - v_1}{R_2} + \frac{v_2}{R_3} + \frac{v_2 - v_3}{R_4} = 0$$

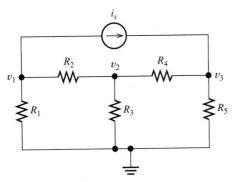

Figure 2.18  Circuit for Example 2.6.

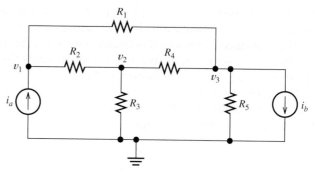

Figure 2.19 Circuit for Exercise 2.6.

Similarly, at node 3, we get

$$\frac{v_3}{R_5} + \frac{v_3 - v_2}{R_4} = i_s$$

Here, the currents leaving node 3 are on the left-hand side and the current entering is on the right-hand side. ■

**Exercise 2.6**  Use KCL to write equations at each node (except the reference node) for the circuit shown in Figure 2.19.
**Answer**

$$\text{Node 1:} \quad \frac{v_1 - v_3}{R_1} + \frac{v_1 - v_2}{R_2} = i_a$$

$$\text{Node 2:} \quad \frac{v_2 - v_1}{R_2} + \frac{v_2}{R_3} + \frac{v_2 - v_3}{R_4} = 0$$

$$\text{Node 3:} \quad \frac{v_3}{R_5} + \frac{v_3 - v_2}{R_4} + \frac{v_3 - v_1}{R_1} + i_b = 0$$   ▢

### Circuit Equations in Standard Form

Once we have written the equations needed to solve for the node voltages, we put the equations into standard form. We group the node-voltage variables on the left-hand sides of the equations and place terms that do not involve the node voltages on the right-hand sides. For two node voltages, this eventually puts the node-voltage equations into the following form:

$$g_{11}v_1 + g_{12}v_2 = i_1 \tag{2.31}$$
$$g_{21}v_1 + g_{22}v_2 = i_2 \tag{2.32}$$

If we have three unknown node voltages, the equations can be put into the form

$$g_{11}v_1 + g_{12}v_2 + g_{13}v_3 = i_1 \tag{2.33}$$
$$g_{21}v_1 + g_{22}v_2 + g_{23}v_3 = i_2 \tag{2.34}$$
$$g_{31}v_1 + g_{32}v_2 + g_{33}v_3 = i_3 \tag{2.35}$$

We have chosen the letter $g$ for the node-voltage coefficients because they are often (but not always) conductances with units of siemens. Similarly, we have used $i$ for the terms on the right-hand sides of the equations because they are often currents.

In matrix form, the equations can be written as

$$\mathbf{GV} = \mathbf{I}$$

in which we have

$$\mathbf{G} = \begin{bmatrix} g_{11} & g_{12} \\ g_{21} & g_{22} \end{bmatrix} \quad \text{or} \quad \mathbf{G} = \begin{bmatrix} g_{11} & g_{12} & g_{13} \\ g_{21} & g_{22} & g_{23} \\ g_{31} & g_{32} & g_{33} \end{bmatrix}$$

depending on whether we have two or three unknown node voltages. Also, **V** and **I** are column vectors:

$$\mathbf{V} = \begin{bmatrix} v_1 \\ v_2 \end{bmatrix} \quad \text{or} \quad \mathbf{V} = \begin{bmatrix} v_1 \\ v_2 \\ v_3 \end{bmatrix} \quad \text{and} \quad \mathbf{I} = \begin{bmatrix} i_1 \\ i_2 \end{bmatrix} \quad \text{or} \quad \mathbf{I} = \begin{bmatrix} i_1 \\ i_2 \\ i_3 \end{bmatrix}$$

As the number of nodes and node voltages increases, the dimensions of the matrices increase.

One way to solve for the node voltages is to find the inverse of **G** and then compute the solution vector as:

$$\mathbf{V} = \mathbf{G}^{-1}\mathbf{I}$$

## A Shortcut to Writing the Matrix Equations

If we put the node equations for the circuit of Exercise 2.6 (Figure 2.19) into matrix form, we obtain

$$\begin{bmatrix} \dfrac{1}{R_1} + \dfrac{1}{R_2} & -\dfrac{1}{R_2} & -\dfrac{1}{R_1} \\[2ex] -\dfrac{1}{R_2} & \dfrac{1}{R_2} + \dfrac{1}{R_3} + \dfrac{1}{R_4} & -\dfrac{1}{R_4} \\[2ex] -\dfrac{1}{R_1} & -\dfrac{1}{R_4} & \dfrac{1}{R_1} + \dfrac{1}{R_4} + \dfrac{1}{R_5} \end{bmatrix} \begin{bmatrix} v_1 \\ v_2 \\ v_3 \end{bmatrix} = \begin{bmatrix} i_a \\ 0 \\ -i_b \end{bmatrix}$$

Let us take a moment to compare the circuit in Figure 2.19 with the elements in this equation. First, look at the elements on the diagonal of the **G** matrix, which are

$$g_{11} = \frac{1}{R_1} + \frac{1}{R_2} \quad g_{22} = \frac{1}{R_2} + \frac{1}{R_3} + \frac{1}{R_4} \quad \text{and} \quad g_{33} = \frac{1}{R_1} + \frac{1}{R_4} + \frac{1}{R_5}$$

We see that the diagonal elements of **G** are equal to the sums of the conductances connected to the corresponding nodes. Next, notice the off diagonal terms:

$$g_{12} = -\frac{1}{R_2} \quad g_{13} = -\frac{1}{R_1} \quad g_{21} = -\frac{1}{R_2} \quad g_{23} = -\frac{1}{R_4} \quad g_{31} = -\frac{1}{R_1} \quad g_{32} = -\frac{1}{R_4}$$

In each case, $g_{jk}$ is equal to the negative of the conductance connected between node $j$ and $k$. The terms in the **I** matrix are the currents pushed into the corresponding nodes by the current sources. These observations hold whenever the network consists of resistances and independent current sources, assuming that we follow our usual pattern in writing the equations.

Thus, if a circuit consists of resistances and independent current sources, we can use the following steps to rapidly write the node equations directly in matrix form.

1. Make sure that the circuit contains only resistances and independent current sources.

2. The diagonal terms of **G** are the sums of the conductances connected to the corresponding nodes.

3. The off diagonal terms of **G** are the negatives of the conductances connected between the corresponding nodes.

4. The elements of **I** are the currents pushed into the corresponding nodes by the current sources.

*This is a shortcut way to write the node equations in matrix form, provided that the circuit contains only resistances and independent current sources.*

Keep in mind that if the network contains voltage sources or controlled sources this pattern does not hold.

**Exercise 2.7**   Working directly from Figure 2.18 on page 63, write its node-voltage equations in matrix form
**Answer**

$$
\begin{bmatrix}
\dfrac{1}{R_1} + \dfrac{1}{R_2} & -\dfrac{1}{R_2} & 0 \\[2mm]
-\dfrac{1}{R_2} & \dfrac{1}{R_2} + \dfrac{1}{R_3} + \dfrac{1}{R_4} & -\dfrac{1}{R_4} \\[2mm]
0 & -\dfrac{1}{R_4} & \dfrac{1}{R_4} + \dfrac{1}{R_5}
\end{bmatrix}
\begin{bmatrix} v_1 \\ v_2 \\ v_3 \end{bmatrix}
=
\begin{bmatrix} -i_s \\ 0 \\ i_s \end{bmatrix}
$$

$\square$

### Example 2.7   Node-Voltage Analysis

Write the node-voltage equations in matrix form for the circuit of Figure 2.20.

**Solution**   Writing KCL at each node, we have

$$\frac{v_1}{5} + \frac{v_1 - v_2}{4} + 3.5 = 0$$

$$\frac{v_2 - v_1}{4} + \frac{v_2}{2.5} + \frac{v_2 - v_3}{5} = 3.5$$

$$\frac{v_3 - v_2}{5} + \frac{v_3}{10} = 2$$

Manipulating the equations into standard form, we have

$$0.45v_1 - 0.25v_2 = -3.5$$

$$-0.25v_1 + 0.85v_2 - 0.2v_3 = 3.5$$

$$-0.2v_2 + 0.35v_3 = 2$$

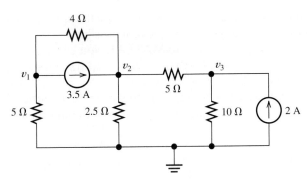

Figure 2.20  Circuit for Example 2.7.

Then, in matrix form, we obtain

$$\begin{bmatrix} 0.45 & -0.25 & 0 \\ -0.25 & 0.85 & -0.20 \\ 0 & -0.20 & 0.30 \end{bmatrix} \begin{bmatrix} v_1 \\ v_2 \\ v_3 \end{bmatrix} = \begin{bmatrix} -3.5 \\ 3.5 \\ 2 \end{bmatrix} \qquad (2.36)$$

Because the circuit contains no voltage sources or controlled sources, we could have used the shortcut method to write the matrix form directly. For example, $g_{11} = 0.45$ is the sum of the conductances connected to node 1, $g_{12} = -0.25$ is the negative of the conductance connected between nodes 1 and 2, $i_3 = 2$ is the current pushed into node 3 by the 2-A current source, and so forth. ∎

## Solving the Network Equations

After we have obtained the equations in standard form, we can solve them by a variety of methods, including substitution, Gaussian elimination, and determinants. As an engineering student, you may own a powerful calculator such as the TI-84 or TI-89 that has the ability to solve systems of linear equations. You should learn to do this by practicing on the exercises and the problems at the end of this chapter.

In some situations, you may not be allowed to use one of the more advanced calculators or a notebook computer. For example, only fairly simple scientific calculators are allowed on the Fundamentals of Engineering (FE) Examination, which is the first step in becoming a registered professional engineer in the United States. The calculator policy for the professional engineering examinations can be found at http://ncees.org/. Thus, even if you own an advanced calculator, you may wish to practice with one of those allowed in the FE Examination.

**Exercise 2.8**   Use your calculator to solve Equation 2.36.
**Answer**   $v_1 = -5$ V, $v_2 = 5$ V, $v_3 = 10$ V. ☐

## Using MATLAB to Solve Network Equations

When you have access to a computer and MATLAB software, you have a very powerful system for engineering and scientific calculations. This software is available to students at many engineering schools and is very likely to be encountered in some of your other courses.

In this and the next several chapters, we illustrate the application of MATLAB to various aspects of circuit analysis, but we cannot possibly cover all of its many useful features in this book. If you are new to MATLAB, you can gain access to a variety of online interactive tutorials at `http://www.mathworks.com/academia/student_center/tutorials/`. If you have already used the program, the MATLAB commands we present may be familiar to you. In either case, you should be able to easily modify the examples we present to work out similar circuit problems.

Next, we illustrate the solution for Equation 2.36 using MATLAB. Instead of using $\mathbf{V} = \mathbf{G}^{-1}\mathbf{I}$ to compute node voltages, MATLAB documentation recommends using the command $\mathbf{V} = \mathbf{G}\backslash\mathbf{I}$ which invokes a more accurate algorithm for computing solutions to systems of linear equations.

The comments following the % sign are ignored by MATLAB. For improved clarity, we use a **bold** font for the input commands, a regular font for comments, and a color font for the responses from MATLAB, otherwise the following has the appearance of the MATLAB command screen for this problem. ($>>$ is the MATLAB command prompt.)

```
>> clear % First we clear the work space.
>> % Then, we enter the coefficient matrix of Equation 2.36 with
>> % spaces between elements in each row and semicolons between rows.
>> G = [0.45 -0.25 0; -0.25 0.85 -0.2; 0 -0.2 0.30]
G =
     0.4500   -0.2500        0
    -0.2500    0.8500   -0.2000
         0   -0.2000    0.3000
>> % Next, we enter the column vector for the right-hand side.
>> I = [-3.5; 3.5; 2]
I =
    -3.5000
     3.5000
     2.0000
>> % The MATLAB documentation recommends computing the node
>> % voltages using V = G\I instead of using V = inv(G)*I.
>> V = G\I
V =
    -5.0000
     5.0000
    10.0000
```

Thus, we have $v_1 = -5$ V, $v_2 = 5$ V, and $v_3 = 10$ V, as you found when working Exercise 2.8 with your calculator.

**Note:** You can download m-files for some of the exercises and examples in this book that use MATLAB. See Appendix E for information on how to do this.

---

Example 2.8    Node-Voltage Analysis

Solve for the node voltages shown in Figure 2.21 and determine the value of the current $i_x$.

**Solution**    Our first step in solving a circuit is to select the reference node and assign the node voltages. This has already been done, as shown in Figure 2.21.

Next, we write equations. In this case, we can write a current equation at each node. This yields

$$\text{Node 1:} \quad \frac{v_1}{10} + \frac{v_1 - v_2}{5} + \frac{v_1 - v_3}{20} = 0$$

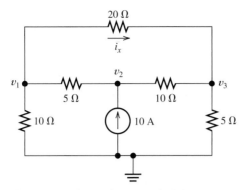

Figure 2.21   Circuit for Example 2.8.

Node 2:  $\dfrac{v_2 - v_1}{5} + \dfrac{v_2 - v_3}{10} = 10$

Node 3:  $\dfrac{v_3}{5} + \dfrac{v_3 - v_2}{10} + \dfrac{v_3 - v_1}{20} = 0$

Next, we place these equations into standard form:

$$0.35v_1 - 0.2v_2 - 0.05v_3 = 0$$

$$-0.2v_1 + 0.3v_2 - 0.10v_3 = 10$$

$$-0.05v_1 - 0.10v_2 + 0.35v_3 = 0$$

In matrix form, the equations are

$$\begin{bmatrix} 0.35 & -0.2 & -0.05 \\ -0.2 & 0.3 & -0.1 \\ -0.05 & -0.1 & 0.35 \end{bmatrix} \begin{bmatrix} v_1 \\ v_2 \\ v_3 \end{bmatrix} = \begin{bmatrix} 0 \\ 10 \\ 0 \end{bmatrix}$$

or $\mathbf{GV} = \mathbf{I}$ in which $\mathbf{G}$ represents the coefficient matrix of conductances, $\mathbf{V}$ is the column vector of node voltages, and $\mathbf{I}$ is the column vector of currents on the right-hand side.

Here again, we could write the equations directly in standard or matrix form using the short cut method because the circuit contains only resistances and independent current sources.

The MATLAB solution is:

```
>> clear
>> G = [0.35 -0.2 -0.05; -0.2 0.3 -0.1; -0.05 -0.1 0.35];
>> % A semicolon at the end of a command suppresses the
>> % MATLAB response.
>> I = [0; 10; 0];
>> V = G\I
V =
    45.4545
    72.7273
    27.2727
>> % Finally, we calculate the current.
>> Ix = (V(1) - V(3))/20
Ix =
     0.9091
```

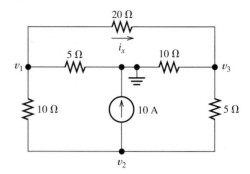

**Figure 2.22** Circuit of Example 2.8 with a different choice for the reference node. See Exercise 2.9.

**Exercise 2.9**   Repeat the analysis of the circuit of Example 2.8, using the reference node and node voltages shown in Figure 2.22. **a.** First write the network equations. **b.** Put the network equations into standard form. **c.** Solve for $v_1$, $v_2$, and $v_3$. (The values will be different than those we found in Example 2.8 because $v_1$, $v_2$, and $v_3$ are not the same voltages in the two figures.) **d.** Find $i_x$. (Of course, $i_x$ is the same in both figures, so it should have the same value.)

**Answer**

**a.**

$$\frac{v_1 - v_3}{20} + \frac{v_1}{5} + \frac{v_1 - v_2}{10} = 0$$

$$\frac{v_2 - v_1}{10} + 10 + \frac{v_2 - v_3}{5} = 0$$

$$\frac{v_3 - v_1}{20} + \frac{v_3}{10} + \frac{v_3 - v_2}{5} = 0$$

**b.**

$$0.35v_1 - 0.10v_2 - 0.05v_3 = \quad 0$$

$$-0.10v_1 + 0.30v_2 - 0.20v_3 = -10$$

$$-0.05v_1 - 0.20v_2 + 0.35v_3 = \quad 0$$

**c.** $v_1 = -27.27$, $v_2 = -72.73$, $v_3 = -45.45$

**d.** $i_x = 0.909$ A                                                                                           ◻

## Circuits with Voltage Sources

When a circuit contains a single voltage source, we can often pick the reference node at one end of the source, and then we have one less unknown node voltage for which to solve.

---

**Example 2.9**    Node-Voltage Analysis

Write the equations for the network shown in Figure 2.23 and put them into standard form.

**Solution**    Notice that we have selected the reference node at the bottom end of the voltage source. Thus, the voltage at node 3 is known to be 10 V, and we do not need to assign a variable for that node.

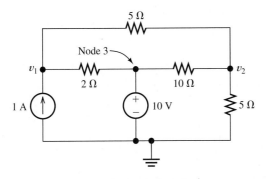

**Figure 2.23** Circuit for Example 2.9.

Writing current equations at nodes 1 and 2, we obtain

$$\frac{v_1 - v_2}{5} + \frac{v_1 - 10}{2} = 1$$

$$\frac{v_2}{5} + \frac{v_2 - 10}{10} + \frac{v_2 - v_1}{5} = 0$$

Now if we group terms and place the constants on the right-hand sides of the equations, we have

$$0.7v_1 - 0.2v_2 = 6$$
$$-0.2v_1 + 0.5v_2 = 1$$

Thus, we have obtained the equations needed to solve for $v_1$ and $v_2$ in standard form. ∎

**Exercise 2.10**   Solve the equations of Example 2.9 for $v_1$ and $v_2$.
**Answer**   $v_1 = 10.32$ V, $v_2 = 6.129$ V. ☐

**Exercise 2.11**   Solve for the node voltages $v_1$ and $v_2$ in the circuit of Figure 2.24.
**Answer**   $v_1 = 6.77$ V, $v_2 = 4.19$ V. ☐

Sometimes, the pattern for writing node-voltage equations that we have illustrated so far must be modified. For example, consider the network and node voltages shown in Figure 2.25. Notice that $v_3 = -15$ V because of the 15-V source connected between node 3 and the reference node. Therefore, we need two equations relating the unknowns $v_1$ and $v_2$.

If we try to write a current equation at node 1, we must include a term for the current through the 10-V source. We could assign an unknown for this current, but then we would have a higher-order system of equations to solve. Especially if we are solving the equations manually, we want to minimize the number of unknowns. For this circuit, it is not possible to write a current equation in terms of the node voltages for any single node (even the reference node) because a voltage source is connected to each node.

Another way to obtain a current equation is to form a **supernode**. This is done by drawing a dashed line around several nodes, including the elements connected between them. This is shown in Figure 2.25. Two supernodes are indicated, one enclosing each of the voltage sources.

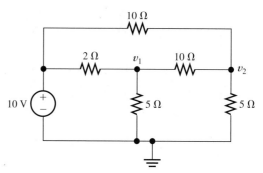

Figure 2.24  Circuit for Exercise 2.11.

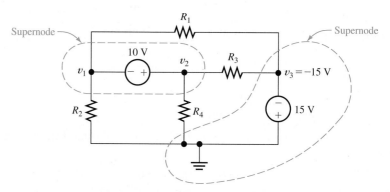

Figure 2.25  A supernode is formed by drawing a dashed line enclosing several nodes and any elements connected between them.

Another way to state Kirchhoff's current law is that the net current flowing through any closed surface must equal zero.

We can state Kirchhoff's current law in a slightly more general form than we have previously: *The net current flowing through any closed surface must equal zero.* Thus, we can apply KCL to a supernode. For example, for the supernode enclosing the 10-V source, we sum currents leaving and obtain

$$\frac{v_1}{R_2} + \frac{v_1 - (-15)}{R_1} + \frac{v_2}{R_4} + \frac{v_2 - (-15)}{R_3} = 0 \qquad (2.37)$$

Each term on the left-hand side of this equation represents a current leaving the supernode through one of the resistors. Thus, by enclosing the 10-V source within the supernode, we have obtained a current equation without introducing a new variable for the current in the source.

We obtain dependent equations if we use all of the nodes in a network to write KCL equations.

Next, we might be tempted to write another current equation for the other supernode. However, we would find that the equation is equivalent to the one already written. *In general, we obtain dependent equations if we use all of the nodes in writing current equations.* Nodes 1 and 2 were part of the first supernode, while node 3 and the reference node are part of the second supernode. Thus, in writing equations for both supernodes, we would have used all four nodes in the network.

If we tried to solve for the node voltages by using substitution, at some point all of the terms would drop out of the equations and we would not be able to solve for those voltages. In MATLAB, you will receive a warning that the G matrix is singular, in other words, its determinant is zero. If this happens, we know that we should return to writing equations and find another equation to use in the solution. This will not happen if we avoid using all of the nodes in writing current equations.

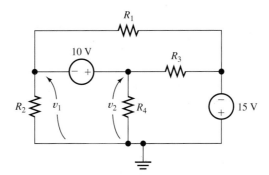

**Figure 2.26** Node voltages $v_1$ and $v_2$ and the 10-V source form a closed loop to which KVL can be applied. (This is the same circuit as that of Figure 2.25.)

There is a way to obtain an independent equation for the network under consideration. We can use KVL because $v_1$, the 10-V source, and $v_2$ form a closed loop. This is illustrated in Figure 2.26, where we have used arrows to indicate the polarities of $v_1$ and $v_2$. Traveling clockwise and summing the voltages around the loop, we obtain

$$-v_1 - 10 + v_2 = 0 \qquad (2.38)$$

Equations 2.37 and 2.38 form an independent set that can be used to solve for $v_1$ and $v_2$ (assuming that the resistance values are known).

**Exercise 2.12**   Write the current equation for the supernode that encloses the 15-V source in Figure 2.25. Show that your equation is equivalent to Equation 2.37.   □

**Exercise 2.13**   Write a set of independent equations for the node voltages shown in Figure 2.27.
**Answer**

KVL:

$$-v_1 + 10 + v_2 = 0$$

KCL for the supernode enclosing the 10-V source:

$$\frac{v_1}{R_1} + \frac{v_1 - v_3}{R_2} + \frac{v_2 - v_3}{R_3} = 1$$

KCL for node 3:

$$\frac{v_3 - v_1}{R_2} + \frac{v_3 - v_2}{R_3} + \frac{v_3}{R_4} = 0$$

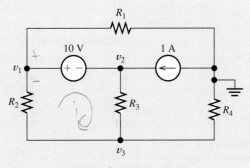

**Figure 2.27** Circuit for Exercise 2.13.

When a voltage source is connected between nodes so that current equations cannot be written at the individual nodes, first write a KVL equation, including the voltage source, and then enclose the voltage source in a supernode and write a KCL equation for the supernode.

KCL at the reference node:

$$\frac{v_1}{R_1} + \frac{v_3}{R_4} = 1$$

For independence, the set must include the KVL equation. Any two of the three KCL equations can be used to complete the three-equation set. (The three KCL equations use all of the network nodes and, therefore, do not form an independent set.)    □

## Circuits with Controlled Sources

Controlled sources present a slight additional complication of the node-voltage technique. (Recall that the value of a controlled source depends on a current or voltage elsewhere in the network.) In applying node-voltage analysis, first we write equations exactly as we have done for networks with independent sources. Then, we express the controlling variable in terms of the node-voltage variables and substitute into the network equations. We illustrate with two examples.

### Example 2.10    Node-Voltage Analysis with a Dependent Source

Write an independent set of equations for the node voltages shown in Figure 2.28.

**Solution**    First, we write KCL equations at each node, including the current of the controlled source just as if it were an ordinary current source:

$$\frac{v_1 - v_2}{R_1} = i_s + 2i_x \tag{2.39}$$

$$\frac{v_2 - v_1}{R_1} + \frac{v_2}{R_2} + \frac{v_2 - v_3}{R_3} = 0 \tag{2.40}$$

$$\frac{v_3 - v_2}{R_3} + \frac{v_3}{R_4} + 2i_x = 0 \tag{2.41}$$

Next, we find an expression for the controlling variable $i_x$ in terms of the node voltages. Notice that $i_x$ is the current flowing away from node 3 through $R_3$. Thus, we can write

$$i_x = \frac{v_3 - v_2}{R_3} \tag{2.42}$$

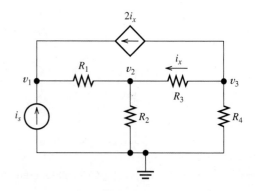

Figure 2.28  Circuit containing a current-controlled current source. See Example 2.10.

Finally, we use Equation 2.42 to substitute into Equations 2.39, 2.40, and 2.41. Thus, we obtain the required equation set:

$$\frac{v_1 - v_2}{R_1} = i_s + 2\frac{v_3 - v_2}{R_3} \tag{2.43}$$

$$\frac{v_2 - v_1}{R_1} + \frac{v_2}{R_2} + \frac{v_2 - v_3}{R_3} = 0 \tag{2.44}$$

$$\frac{v_3 - v_2}{R_3} + \frac{v_3}{R_4} + 2\frac{v_3 - v_2}{R_3} = 0 \tag{2.45}$$

Assuming that the value of $i_s$ and the resistances are known, we could put this set of equations into standard form and solve for $v_1$, $v_2$, and $v_3$.   ■

### Example 2.11   Node-Voltage Analysis with a Dependent Source

Write an independent set of equations for the node voltages shown in Figure 2.29.

**Solution**   First, we ignore the fact that the voltage source is a dependent source and write equations just as we would for a circuit with independent sources. We cannot write a current equation at either node 1 or node 2, because of the voltage source connected between them. However, we can write a KVL equation:

$$-v_1 + 0.5v_x + v_2 = 0 \tag{2.46}$$

Then, we use KCL to write current equations. For a supernode enclosing the controlled voltage source,

$$\frac{v_1}{R_2} + \frac{v_1 - v_3}{R_1} + \frac{v_2 - v_3}{R_3} = i_s$$

For node 3,

$$\frac{v_3}{R_4} + \frac{v_3 - v_2}{R_3} + \frac{v_3 - v_1}{R_1} = 0 \tag{2.47}$$

For the reference node,

$$\frac{v_1}{R_2} + \frac{v_3}{R_4} = i_s \tag{2.48}$$

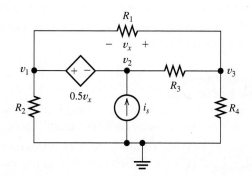

**Figure 2.29**  Circuit containing a voltage-controlled voltage source. See Example 2.11.

Of course, these current equations are dependent because we have used all four nodes in writing them. We must use Equation 2.46 and two of the KCL equations to form an independent set. However, Equation 2.46 contains the controlling variable $v_x$, which must be eliminated before we have equations in terms of the node voltages.

Thus, our next step is to write an expression for the controlling variable $v_x$ in terms of the node voltages. Notice that $v_1$, $v_x$, and $v_3$ form a closed loop. Traveling clockwise and summing voltages, we have

$$-v_1 - v_x + v_3 = 0$$

Solving for $v_x$, we obtain

$$v_x = v_3 - v_1$$

Now if we substitute into Equation 2.46, we get

$$v_1 = 0.5(v_3 - v_1) + v_2 \qquad (2.49)$$

Equation 2.49 along with any two of the KCL equations forms an independent set that can be solved for the node voltages.    ■

Using the principles we have discussed in this section, we can write node-voltage equations for any network consisting of sources and resistances. Thus, given a computer or calculator to help in solving the equations, we can compute the currents and voltages for any network.

## Step-by-Step Node-Voltage Analysis

Next, we summarize the steps in analyzing circuits by the node-voltage technique:

1. First, combine any series resistances to reduce the number of nodes. Then, select a reference node and assign variables for the unknown node voltages. If the reference node is chosen at one end of an independent voltage source, one node voltage is known at the start, and fewer need to be computed.

2. Write network equations. First, use KCL to write current equations for nodes and supernodes. Write as many current equations as you can without using all of the nodes, including those within supernodes. Then if you do not have enough equations because of voltage sources connected between nodes, use KVL to write additional equations.

3. If the circuit contains dependent sources, find expressions for the controlling variables in terms of the node voltages. Substitute into the network equations, and obtain equations having only the node voltages as unknowns.

4. Put the equations into standard form and solve for the node voltages.

5. Use the values found for the node voltages to calculate any other currents or voltages of interest.

---

**Example 2.12**  Node Voltage Analysis

Use node voltages to solve for the value of $i_x$ in the circuit of Figure 2.30(a). (This rather complex circuit has been contrived mainly to display all of the steps listed above.)

**Solution**    First, we combine the 1 Ω, 2 Ω, and 3 Ω resistances in series to eliminate nodes $A$ and $G$. Then, we select node $C$ at one end of the 20-V source as the reference

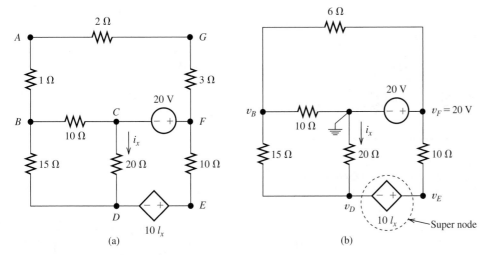

**Figure 2.30** Circuit of Example 2.12.

node. Thus, we know that the voltage at node $F$ is 20 V. (Of course, any node could be chosen for the reference node, but if we chose node $B$, for example, we would have one more variable in the equations.) The resulting circuit is shown in Figure 2.30(b).

We cannot write KCL equations at any single node, except node $B$, because each    Step 2 of the other nodes has a voltage source connected. The KCL equation at node $B$ is

$$\frac{v_B - 20}{6} + \frac{v_B}{10} + \frac{v_B - v_D}{15} = 0$$

Multiplying all terms by 30 and rearranging, we have

$$10v_B - 2v_D = 100$$

Next, we form a super node enclosing the controlled voltage source as indicated in Figure 2.30(b). This results in

$$\frac{v_E - 20}{10} + \frac{v_D}{20} + \frac{v_D - v_B}{15} = 0$$

(Another option would have been a super node enclosing the 20 V source.)

Multiplying all terms by 60 and rearranging, we have

$$-4v_B + 7v_D + 6v_E = 120$$

No options for another KCL equation exist without using all of the circuit nodes and producing dependent equations.

Thus, we write a KVL equation starting from the reference node to one end of the controlled voltage source, through the source, and back to the reference node. This results in $v_E = 10i_x + v_D$.

Next, we note that $i_x$ is the current through and $v_D$ is the voltage across the 20-$\Omega$    Step 3 resistance. The current reference enters the negative end of the voltage, so we have $v_D = -20\,i_x$. Combining these two equations eventually results in

$$v_D - 2v_E = 0$$

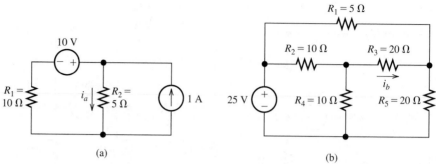

(a)

(b)

**Figure 2.31** Circuits for Exercise 2.14.

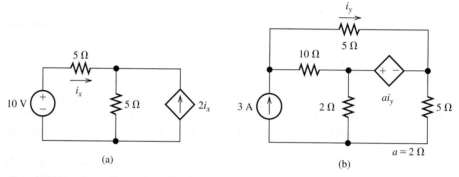

(a)

(b)

**Figure 2.32** Circuits for Exercise 2.15.

Step 4

Thus, we have these three equations to solve for the node voltages:

$$10v_B - 2v_D = 100$$

$$-4v_B + 7v_D + 6v_E = 120$$

$$v_D - 2v_E = 0$$

Solving these three equations results in $v_D = 17.3913$ V.
    Then, we have $i_x = -v_D/20 = -0.8696$ A.    ∎

---

**Exercise 2.14**   Use the node-voltage technique to solve for the currents labeled in the circuits shown in Figure 2.31.
**Answer   a.** $i_a = 1.33$ A; **b.** $i_b = -0.259$ A.    □

---

**Exercise 2.15**   Use the node-voltage technique to solve for the values of $i_x$ and $i_y$ in Figure 2.32.
**Answer**   $i_x = 0.5$ A, $i_y = 2.31$ A.    □

---

## Using the MATLAB Symbolic Toolbox to Obtain Symbolic Solutions

If the Symbolic Toolbox is included with your version of MATLAB, you can use it to solve node voltage and other equations symbolically. We illustrate by solving Equations 2.43, 2.44, and 2.45 from Example 2.10 on page 74.

```
>> % First we clear the work space.
>> clear all
>> % Next, we identify the symbols used in the
```

```
>> % equations to be solved.
>> syms V1 V2 V3 R1 R2 R3 R4 Is
>> % Then, we enter the equations into the solve command
>> % followed by the variables for which we wish to solve.
>> [V1, V2, V3] = solve((V1 - V2)/R1 == Is + 2*(V3 - V2)/R3, ...
                     (V2 - V1)/R1 + V2/R2 + (V2 - V3)/R3 == 0, ...
                     (V3 - V2)/R3 + V3/R4 + 2*(V3 - V2)/R3 == 0, ...
                     V1, V2, V3)
V1 =
(Is*(R1*R2 + R1*R3 + 3*R1*R4 + R2*R3 + 3*R2*R4))/(3*R2 + R3 + 3*R4)
V2 =
(Is*R2*(R3 + 3*R4))/(3*R2 + R3 + 3*R4)
V3 = (3*Is*R2*R4)/(3*R2 + R3 +3*R4)
>> % The solve command gives the answers, but in a form that is
>> % somewhat difficult to read.
>> % A more readable version of the answers is obtained using the
>> % pretty command. We combine the three commands on one line
>> % by placing commas between them.
>> pretty(V1), pretty(V2), pretty(V3)
   Is R1 R2 + Is R1 R3 + 3 Is R1 R4 + Is R2 R3 + 3 Is R2 R4
   --------------------------------------------------------
                    3 R2 + R3 + 3 R4

   Is R2 R3 + 3 Is R2 R4
   ---------------------
     3 R2 + R3 + 3 R4

        3 Is R2 R4
   -------------------
   3 R2 + R3 + 3 R4
```

(Here we have shown the results obtained using a particular version of MATLAB; other versions may give results different in appearance but equivalent mathematically.) In more standard mathematical format, the results are:

$$v_1 = \frac{i_s R_1 R_2 + i_s R_1 R_3 + 3i_s R_1 R_4 + i_s R_2 R_3 + 3i_s R_2 R_4}{3R_2 + R_3 + 3R_4}$$

$$v_2 = \frac{i_s R_2 R_3 + 3i_s R_2 R_4}{3R_2 + R_3 + 3R_4}$$

$$\text{and } v_3 = \frac{3i_s R_2 R_4}{3R_2 + R_3 + 3R_4}$$

## Checking Answers

As usual, it is a good idea to apply some checks to the answers. First of all, make sure that the answers have proper units, which are volts in this case. If the units don't check, look to see if any of the numerical values entered in the equations have units. Referring to the circuit (Figure 2.28 on page 74), we see that the only numerical parameter entered into the equations was the gain of the current-controlled current source, which has no units.

Again referring to the circuit diagram, we can see that we should have $v_2 = v_3$ for $R_3 = 0$, and we check the results to see that this is the case. Another check is obtained by observing that we should have $v_3 = 0$ for $R_4 = 0$. Still another check of the results comes from observing that, in the limit as $R_3$ approaches infinity, we should have $i_x = 0$, (so the controlled current source becomes an open circuit), $v_3 = 0$, $v_1 = i_s(R_1 + R_2)$, and $v_2 = i_s R_2$. Various other checks of a similar nature can be applied. This type of checking may not guarantee correct results, but it can find a lot of errors.

**Exercise 2.16**   Use the symbolic math features of MATLAB to solve Equations 2.47, 2.48, and 2.49 for the node voltages in symbolic form.
**Answer**

$$v_1 = \frac{2i_s R_1 R_2 R_3 + 3i_s R_1 R_2 R_4 + 2i_s R_2 R_3 R_4}{3 R_1 R_2 + 2 R_1 R_3 + 3 R_1 R_4 + 2 R_2 R_3 + 2 R_3 R_4}$$

$$v_2 = \frac{3i_s R_1 R_2 R_3 + 3i_s R_1 R_2 R_4 + 2i_s R_2 R_3 R_4}{3 R_1 R_2 + 2 R_1 R_3 + 3 R_1 R_4 + 2 R_2 R_3 + 2 R_3 R_4}$$

$$v_3 = \frac{3i_s R_1 R_2 R_4 + 2i_s R_2 R_3 R_4}{3 R_1 R_2 + 2 R_1 R_3 + 3 R_1 R_4 + 2 R_2 R_3 + 2 R_3 R_4}$$

Depending on the version of MATLAB and the Symbolic Toolbox that you use, your answers may have a different appearance but should be algebraically equivalent to these.

## 2.5 MESH-CURRENT ANALYSIS

In this section, we show how to analyze networks by using another general technique, known as mesh-current analysis. Networks that can be drawn on a plane without having one element (or conductor) crossing over another are called **planar networks**. On the other hand, circuits that must be drawn with one or more elements crossing others are said to be **nonplanar**. We consider only planar networks.

Let us start by considering the planar network shown in Figure 2.33(a). Suppose that the source voltages and resistances are known and that we wish to solve for the currents. We first write equations for the currents shown in Figure 2.33(a), which are called branch currents because a separate current is defined in each branch of the network. However, we will eventually see that using the mesh currents illustrated in Figure 2.33(b) makes the solution easier.

Three independent equations are needed to solve for the three branch currents shown in Figure 2.33(a). In general, the number of independent KVL equations that can be written for a planar network is equal to the number of open areas defined by the network layout. For example, the circuit of Figure 2.33(a) has two open areas: one defined by $v_A$, $R_1$, and $R_3$, while the other is defined by $R_3$, $R_2$, and $v_B$. Thus, for this network, we can write only two independent KVL equations. We must employ KCL to obtain the third equation.

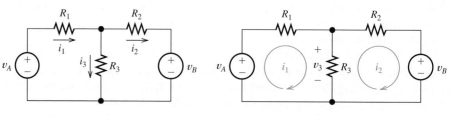

(a) Circuit with branch currents                    (b) Circuit with mesh currents

**Figure 2.33** Circuit for illustrating the mesh-current method of circuit analysis.

Application of KVL to the loop consisting of $v_A$, $R_1$, and $R_3$ yields

$$R_1 i_1 + R_3 i_3 = v_A \qquad (2.50)$$

Similarly, for the loop consisting of $R_3$, $R_2$, and $v_B$, we get

$$-R_3 i_3 + R_2 i_2 = -v_B \qquad (2.51)$$

Applying KCL to the node at the top end of $R_3$, we have

$$i_1 = i_2 + i_3 \qquad (2.52)$$

Next, we solve Equation 2.52 for $i_3$ and substitute into Equations 2.50 and 2.51. This yields the following two equations:

$$R_1 i_1 + R_3 (i_1 - i_2) = v_A \qquad (2.53)$$

$$-R_3 (i_1 - i_2) + R_2 i_2 = -v_B \qquad (2.54)$$

Thus, we have used the KCL equation to reduce the KVL equations to two equations in two unknowns.

Now, consider the mesh currents $i_1$ and $i_2$ shown in Figure 2.33(b). As indicated in the figure, mesh currents are considered to flow around closed paths. Hence, mesh currents automatically satisfy KCL. *When several mesh currents flow through one element, we consider the current in that element to be the algebraic sum of the mesh currents.* Thus, assuming a reference direction pointing downward, the current in $R_3$ is $(i_1 - i_2)$. Thus, $v_3 = R_3 (i_1 - i_2)$. Now if we follow $i_1$ around its loop and apply KVL, we get Equation 2.53 directly. Similarly, following $i_2$, we obtain Equation 2.54 directly.

When several mesh currents flow through one element, we consider the current in that element to be the algebraic sum of the mesh currents.

Because mesh currents automatically satisfy KCL, some work is saved in writing and solving the network equations. The circuit of Figure 2.33 is fairly simple, and the advantage of mesh currents is not great. However, for more complex networks, the advantage can be quite significant.

### Choosing the Mesh Currents

For a planar circuit, we can choose the current variables to flow through the elements around the periphery of each of the open areas of the circuit diagram. For consistency, we usually define the mesh currents to flow clockwise.

Two networks and suitable choices for the mesh currents are shown in Figure 2.34. When a network is drawn with no crossing elements, it resembles a window, with each open area corresponding to a pane of glass. Sometimes it is said that the mesh currents are defined by "soaping the window panes."

We usually choose the current variables to flow clockwise around the periphery of each of the open areas of the circuit diagram.

Keep in mind that, if two mesh currents flow through a circuit element, we consider the current in that element to be the algebraic sum of the mesh currents. For example, in Figure 2.34(a), the current in $R_2$ referenced to the left is $i_3 - i_1$. Furthermore, the current referenced upward in $R_3$ is $i_2 - i_1$.

**Exercise 2.17**   Consider the circuit shown in Figure 2.34(b). In terms of the mesh currents, find the current in **a.** $R_2$ referenced upward; **b.** $R_4$ referenced to the right; **c.** $R_8$ referenced downward; **d.** $R_8$ referenced upward.
**Answer   a.** $i_4 - i_1$; **b.** $i_2 - i_1$; **c.** $i_3 - i_4$; **d.** $i_4 - i_3$. [Notice that the answer for part (d) is the negative of the answer for part (c).]   □

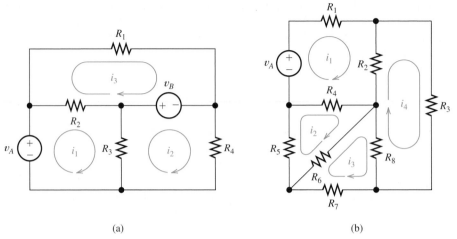

(a)                                         (b)

**Figure 2.34** Two circuits and their mesh-current variables.

## Writing Equations to Solve for Mesh Currents

If a network contains only resistances and independent voltage sources, we can write the required equations by following each current around its mesh and applying KVL. (We do not need to apply KCL because the mesh currents flow out of each node that they flow into.)

| Example 2.13 | Mesh-Current Analysis |

Write the equations needed to solve for the mesh currents in Figure 2.34(a).

**Solution**    Using a pattern in solving networks by the mesh-current method helps to avoid errors. Part of the pattern that we use is to select the mesh currents to flow clockwise. Then, we write a KVL equation for each mesh, going around the meshes clockwise. As usual, we add a voltage if its positive reference is encountered first in traveling around the mesh, and we subtract the voltage if the negative reference is encountered first. Our pattern is always to take the first end of each resistor encountered as the positive reference for its voltage. Thus, we are always adding the resistor voltages.

> If a network contains only resistances and independent voltage sources, we can write the required equations by following each current around its mesh and applying KVL.

For example, in mesh 1 of Figure 2.34(a), we first encounter the left-hand end of $R_2$. The voltage across $R_2$ referenced positive on its left-hand end is $R_2(i_1 - i_3)$. Similarly, we encounter the top end of $R_3$ first, and the voltage across $R_3$ referenced positive at the top end is $R_3(i_1 - i_2)$. By using this pattern, we add a term for each resistor in the KVL equation, consisting of the resistance times the current in the mesh under consideration minus the current in the adjacent mesh (if any). Using this pattern for mesh 1 of Figure 2.34(a), we have

$$R_2(i_1 - i_3) + R_3(i_1 - i_2) - v_A = 0$$

Similarly, for mesh 2, we obtain

$$R_3(i_2 - i_1) + R_4 i_2 + v_B = 0$$

Finally, for mesh 3, we have

$$R_2(i_3 - i_1) + R_1 i_3 - v_B = 0$$

Notice that we have taken the positive reference for the voltage across $R_3$ at the top in writing the equation for mesh 1 and at the bottom for mesh 3. This is not an error because the terms for $R_3$ in the two equations are opposite in sign.

In standard form, the equations become:

$$(R_2 + R_3)i_1 - R_3 i_2 - R_2 i_3 = v_A$$
$$-R_3 i_1 + (R_3 + R_4)i_2 = -v_B$$
$$-R_2 i_1 + (R_1 + R_2)i_3 = v_B$$

In matrix form, we have

$$
\begin{bmatrix}
(R_2 + R_3) & -R_3 & -R_2 \\
-R_3 & (R_3 + R_4) & 0 \\
-R_2 & 0 & (R_1 + R_2)
\end{bmatrix}
\begin{bmatrix}
i_1 \\
i_2 \\
i_3
\end{bmatrix}
=
\begin{bmatrix}
v_A \\
-v_B \\
v_B
\end{bmatrix}
$$

Often, we use $\mathbf{R}$ to represent the coefficient matrix, $\mathbf{I}$ to represent the column vector of mesh currents, and $\mathbf{V}$ to represent the column vector of the terms on the right-hand sides of the equations in standard form. Then, the mesh-current equations are represented as:

$$\mathbf{RI} = \mathbf{V}$$

We refer to the element of the $i$th row and $j$th column of $\mathbf{R}$ as $r_{ij}$. ■

**Exercise 2.18**   Write the equations for the mesh currents in Figure 2.34(b) and put them into matrix form.

**Answer**   Following each mesh current in turn, we obtain

$$R_1 i_1 + R_2(i_1 - i_4) + R_4(i_1 - i_2) - v_A = 0$$
$$R_5 i_2 + R_4(i_2 - i_1) + R_6(i_2 - i_3) = 0$$
$$R_7 i_3 + R_6(i_3 - i_2) + R_8(i_3 - i_4) = 0$$
$$R_3 i_4 + R_2(i_4 - i_1) + R_8(i_4 - i_3) = 0$$   □

$$
\begin{bmatrix}
(R_1 + R_2 + R_4) & -R_4 & 0 & -R_2 \\
-R_4 & (R_4 + R_5 + R_6) & -R_6 & 0 \\
0 & -R_6 & (R_6 + R_7 + R_8) & -R_8 \\
-R_2 & 0 & -R_8 & (R_2 + R_3 + R_8)
\end{bmatrix}
\begin{bmatrix}
i_1 \\
i_2 \\
i_3 \\
i_4
\end{bmatrix}
=
\begin{bmatrix}
v_A \\
0 \\
0 \\
0
\end{bmatrix}
$$

$$(2.55)$$

## Solving Mesh Equations

After we write the mesh-current equations, we can solve them by using the methods that we discussed in Section 2.4 for the node-voltage approach. We illustrate with a simple example.

**Example 2.14**   **Mesh-Current Analysis**

Solve for the current in each element of the circuit shown in Figure 2.35.

**Solution**   First, we select the mesh currents. Following our standard pattern, we define the mesh currents to flow clockwise around each mesh of the circuit. Then, we write a KVL equation around mesh 1:

$$20(i_1 - i_3) + 10(i_1 - i_2) - 70 = 0 \tag{2.56}$$

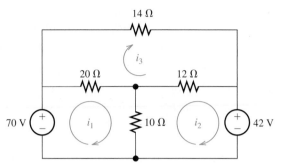

**Figure 2.35**  Circuit of Example 2.14.

For meshes 2 and 3, we have:

$$10(i_2 - i_1) + 12(i_2 - i_3) + 42 = 0 \qquad (2.57)$$

$$20(i_3 - i_1) + 14i_3 + 12(i_3 - i_2) = 0 \qquad (2.58)$$

Putting the equations into standard form, we have:

$$30i_1 - 10i_2 - 20i_3 = 70 \qquad (2.59)$$

$$-10i_1 + 22i_2 - 12i_3 = -42 \qquad (2.60)$$

$$-20i_1 - 12i_2 + 46i_3 = 0 \qquad (2.61)$$

In matrix form, the equations become:

$$
\begin{bmatrix} 30 & -10 & -20 \\ -10 & 22 & -12 \\ -20 & -12 & 46 \end{bmatrix}
\begin{bmatrix} i_1 \\ i_2 \\ i_3 \end{bmatrix} =
\begin{bmatrix} 70 \\ -42 \\ 0 \end{bmatrix}
$$

These equations can be solved in a variety of ways. We will demonstrate using MATLAB. We use **R** for the coefficient matrix, because the coefficients often are resistances. Similarly, we use **V** for the column vector for the right-hand side of the equations and **I** for the column vector of the mesh currents. The commands and results are:

```
>> R = [30 -10 -20; -10 22 -12; -20 -12 46];
>> V = [70; -42; 0];
>> I = R\V % Try to avoid using i, which represents the square root of
>> % -1 in MATLAB.
I =
    4.0000
    1.0000
    2.0000
```

Thus, the values of the mesh currents are $i_1 = 4$ A, $i_2 = 1$ A, and $i_3 = 2$ A. Next, we can find the current in any element. For example, the current flowing downward in the 10-$\Omega$ resistance is $i_1 - i_2 = 3$ A.    ∎

**Exercise 2.19**  Use mesh currents to solve for the current flowing through the 10-$\Omega$ resistance in Figure 2.36. Check your answer by combining resistances in series and parallel to solve the circuit. Check a second time by using node voltages.
**Answer**  The current through the 10-$\Omega$ resistance is 5 A.    □

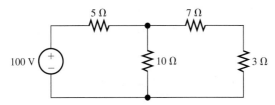

Figure 2.36  Circuit of Exercise 2.19.

**Exercise 2.20**   Use mesh currents to solve for the current flowing through the 2-$\Omega$ resistance in Figure 2.24 on page 72.
**Answer**   The current is 1.613 A directed toward the right.  ☐

## Writing Mesh Equations Directly in Matrix Form

If a circuit contains only resistances and independent voltage sources, and if we select the mesh currents flowing clockwise, the mesh equations can be obtained directly in matrix form using these steps:

1. Make sure that the circuit contains only resistances and independent voltage sources. Select all of the mesh currents to flow in the clockwise direction.

2. Write the sum of the resistances contained in each mesh as the corresponding element on the main diagonal of **R**. In other words, $r_{jj}$ equals the sum of the resistances encountered in going around mesh $j$.

3. Insert the negatives of the resistances common to the corresponding meshes as the off diagonal terms of **R**. Thus, for $i \neq j$, the elements $r_{ij}$ and $r_{ji}$ are the same and are equal to negative of the sum of the resistances common to meshes $i$ and $j$.

4. For each element of the **V** matrix, go around the corresponding mesh clockwise, *subtracting* the values of voltage sources for which we encounter the positive reference first and *adding* the values of voltage sources for which we encounter the negative reference first. (We have reversed the rules for adding or subtracting the voltage source values from what we used when writing KVL equations because the elements of **V** correspond to terms on the opposite side of the KVL equations.)

*This is a shortcut way to write the mesh equations in matrix form, provided that the circuit contains only resistances and independent voltage sources.*

Keep in mind that this procedure does not apply to circuits having current sources or controlled sources.

### Example 2.15   Writing Mesh Equations Directly in Matrix Form

Write the mesh equations directly in matrix form for the circuit of Figure 2.37.

**Solution**   The matrix equation is:

$$\begin{bmatrix} (R_2 + R_4 + R_5) & -R_2 & -R_5 \\ -R_2 & (R_1 + R_2 + R_3) & -R_3 \\ -R_5 & -R_3 & (R_3 + R_5 + R_6) \end{bmatrix} \begin{bmatrix} i_1 \\ i_2 \\ i_3 \end{bmatrix} = \begin{bmatrix} -v_A + v_B \\ v_A \\ -v_B \end{bmatrix}$$

Notice that mesh 1 includes $R_2$, $R_4$, and $R_5$, so the $r_{11}$ element of **R** is the sum of these resistances. Similarly, mesh 2 contains $R_1$, $R_2$, and $R_3$, so $r_{22}$ is the sum of these resistances. Because $R_2$ is common to meshes 1 and 2, we have $r_{12} = r_{21} = -R_2$. Similar observations can be made for the other elements of **R**.

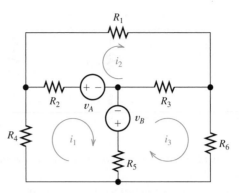

Figure 2.37 Circuit of Example 2.15.

As we go around mesh 1 clockwise, we encounter the positive reference for $v_A$ first and the negative reference for $v_B$ first, so we have $v_1 = -v_A + v_B$, and so forth. ∎

**Exercise 2.21**    Examine the circuit of Figure 2.34(a) on page 82, and write its mesh equations directly in matrix form.
**Answer**

$$
\begin{bmatrix}
(R_2 + R_3) & -R_3 & -R_2 \\
-R_3 & (R_3 + R_4) & 0 \\
-R_2 & 0 & (R_1 + R_2)
\end{bmatrix}
\begin{bmatrix}
i_1 \\
i_2 \\
i_3
\end{bmatrix}
=
\begin{bmatrix}
v_A \\
-v_B \\
v_B
\end{bmatrix}
$$                              ☐

## Mesh Currents in Circuits Containing Current Sources

Recall that a current source forces a specified current to flow through its terminals, but the voltage across its terminals is not predetermined. Instead, the voltage across a current source depends on the circuit to which the source is connected. Often, it is not easy to write an expression for the voltage across a current source. *A common mistake made by beginning students is to assume that the voltages across current sources are zero.*

> A common mistake made by beginning students is to assume that the voltages across current sources are zero.

Consequently, when a circuit contains a current source, we must depart from the pattern that we use for circuits consisting of voltage sources and resistances. First, consider the circuit of Figure 2.38. As usual, we have defined the mesh currents flowing clockwise. If we were to try to write a KVL equation for mesh 1, we would need to include an unknown for the voltage across the current source. Because we do not wish to increase the number of unknowns in our equations, we avoid writing KVL equations for loops that include current sources. In the circuit in Figure 2.38,

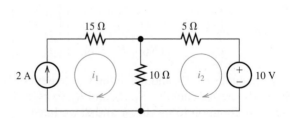

Figure 2.38 In this circuit, we have $i_1 = 2$ A.

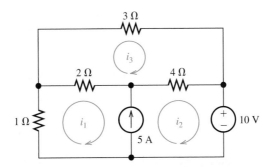

**Figure 2.39** A circuit with a current source common to two meshes.

we have defined the current in the current source as $i_1$. However, we know that this current is 2 A. Thus, we can write

$$i_1 = 2 \text{ A} \tag{2.62}$$

The second equation needed can be obtained by applying KVL to mesh 2, which yields

$$10(i_2 - i_1) + 5i_2 + 10 = 0 \tag{2.63}$$

Equations 2.62 and 2.63 can readily be solved for $i_2$. Notice that in this case the presence of a current source facilitates the solution.

Now let us consider the somewhat more complex situation shown in Figure 2.39. As usual, we have defined the mesh currents flowing clockwise. We cannot write a KVL equation around mesh 1 because the voltage across the 5-A current source is unknown (and we do not want to increase the number of unknowns in our equations). A solution is to combine meshes 1 and 2 into a **supermesh**. In other words, we write a KVL equation around the periphery of meshes 1 and 2 combined. This yields

$$i_1 + 2(i_1 - i_3) + 4(i_2 - i_3) + 10 = 0 \tag{2.64}$$

Next, we can write a KVL equation for mesh 3:

$$3i_3 + 4(i_3 - i_2) + 2(i_3 - i_1) = 0 \tag{2.65}$$

Finally, we recognize that we have defined the current in the current source referenced upward as $i_2 - i_1$. However, we know that the current flowing upward through the current source is 5 A. Thus, we have

$$i_2 - i_1 = 5 \tag{2.66}$$

It is important to realize that Equation 2.66 is not a KCL equation. Instead, it simply states that we have defined the current referenced upward through the current source in terms of the mesh currents as $i_2 - i_1$, but this current is known to be 5 A. Equations 2.64, 2.65, and 2.66 can be solved for the mesh currents.

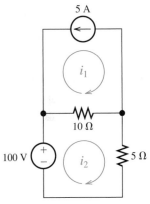

**Figure 2.40** The circuit for Exercise 2.22.

It is important to realize that Equation 2.66 is not a KCL equation.

**Exercise 2.22** Write the equations needed to solve for the mesh currents in Figure 2.40.
**Answer**

$$i_1 = -5 \text{ A}$$
$$10(i_2 - i_1) + 5i_2 - 100 = 0$$

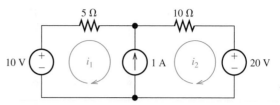

Figure 2.41  The circuit for Exercise 2.23.

**Exercise 2.23**   Write the equations needed to solve for the mesh currents in Figure 2.41. Then solve for the currents.
**Answer**   The equations are $i_2 - i_1 = 1$ and $5i_1 + 10i_2 + 20 - 10 = 0$. Solving, we have $i_1 = -4/3$ A and $i_2 = -1/3$ A.                                               □

## Circuits with Controlled Sources

Controlled sources present a slight additional complication to the mesh-current technique. First, we write equations exactly as we have done for networks with independent sources. Then, we express the controlling variables in terms of the mesh-current variables and substitute into the network equations. We illustrate with an example.

---

**Example 2.16**   **Mesh-Current Analysis with Controlled Sources**

Solve for the currents in the circuit of Figure 2.42(a), which contains a voltage-controlled current source common to the two meshes.

**Solution**   First, we write equations for the mesh currents as we have done for independent sources. Since there is a current source common to mesh 1 and mesh 2, we start by combining the meshes to form a supermesh and write a voltage equation:

$$-20 + 4i_1 + 6i_2 + 2i_2 = 0 \tag{2.67}$$

Then, we write an expression for the source current in terms of the mesh currents:

$$av_x = 0.25v_x = i_2 - i_1 \tag{2.68}$$

Next, we see that the controlling voltage is

$$v_x = 2i_2 \tag{2.69}$$

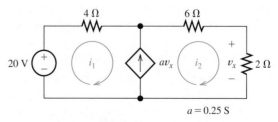

Figure 2.42  A circuit with a voltage-controlled current source. See Example 2.16.

Using Equation 2.58 to substitute for $v_x$ in Equation 2.57, we have

$$\frac{i_2}{2} = i_2 - i_1 \tag{2.70}$$

Finally, we put Equations 2.67 and 2.70 into standard form, resulting in

$$4i_1 + 8i_2 = 20 \tag{2.71}$$

$$i_1 - \frac{i_2}{2} = 0 \tag{2.72}$$

Solving these equations yields $i_1 = 1$ A and $i_2 = 2$ A.     ∎

Using the principles we have discussed in this section, we can write mesh-current equations for any planar network consisting of sources and resistances.

## Step-by-Step Mesh-Current Analysis

Next, we summarize the steps in analyzing planar circuits by the mesh-current technique:

1. If necessary, redraw the network without crossing conductors or elements. Consider combining resistances in parallel to reduce circuit complexity. Then, define the mesh currents flowing around each of the open areas defined by the network. For consistency, we usually select a clockwise direction for each of the mesh currents, but this is not a requirement.

2. Write network equations, stopping after the number of equations is equal to the number of mesh currents. First, use KVL to write voltage equations for meshes that do not contain current sources. Next, if any current sources are present, write expressions for their currents in terms of the mesh currents. Finally, if a current source is common to two meshes, write a KVL equation for the supermesh.

3. If the circuit contains dependent sources, find expressions for the controlling variables in terms of the mesh currents. Substitute into the network equations, and obtain equations having only the mesh currents as unknowns.

4. Put the equations into standard form. Solve for the mesh currents by use of determinants or other means.

5. Use the values found for the mesh currents to calculate any other currents or voltages of interest.

*Here is a convenient step-by-step guide to mesh-current analysis.*

| Example 2.17 | Mesh Current Analysis |

Use mesh currents to solve for the value of $v_x$ in the circuit of Figure 2.43(a). (This rather complex circuit has been contrived mainly to illustrate the steps listed above.)

**Solution**   First, we combine the 15-$\Omega$ resistances in parallel to eliminate two meshes. The resulting circuit is shown in Figure 2.30(b). As usual, we select the mesh currents flowing clockwise around the open areas.

*Step 1*

We cannot write KVL equations for meshes 1 or 2 because we do not know the voltage across the 22-A current source, and we do not want to introduce another unknown. Thus, we write a KVL equation for mesh 3:

*Step 2*

$$10i_3 + 2v_x + 6(i_3 - i_1) = 0$$

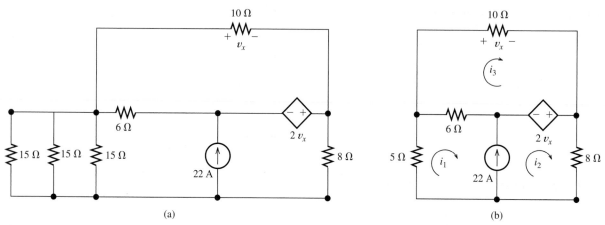

**Figure 2.43** Circuit of Example 2.17.

Next, in terms of the mesh currents the current flowing upward through the current source is $i_2 - i_1$. However, we know that this current is 22 A. Thus, we have:

$$i_2 - i_1 = 22$$

Next, we write a KVL equation for the super mesh formed by combining meshes 1 and 2:

$$5i_1 + 6(i_1 - i_3) - 2v_x + 8i_2 = 0$$

Next, Ohm's law gives

$$v_x = 10i_3$$

Substituting this into the previous equations and putting them into a standard form produces:

$$-6i_1 + 36i_3 = 0$$
$$-i_1 + i_2 = 22$$
$$11i_1 + 8i_2 - 26i_3 = 0$$

Solving these equations produces $i_1 = -12$ A, $i_2 = 10$ A, and $i_3 = -2$ A. Then, we have $v_x = 10i_3 = -20$ V. ∎

**Exercise 2.24**   Use the mesh-current technique to solve for the currents labeled in the circuits shown in Figure 2.31 on page 78.
**Answer   a.** $i_a = 1.33$ A; **b.** $i_b = -0.259$ A.  ☐

**Exercise 2.25**   Use the mesh-current technique to solve for the values of $i_x$ and $i_y$ in Figure 2.32 on page 78.
**Answer**   $i_x = 0.5$ A, $i_y = 2.31$ A.  ☐

## 2.6  THÉVENIN AND NORTON EQUIVALENT CIRCUITS

In this section, we learn how to replace two-terminal circuits containing resistances and sources by simple equivalent circuits. By a two-terminal circuit, we mean that the original circuit has only two points that can be connected to other circuits. The original circuit can be any complex interconnection of resistances and sources. However, a restriction is that the controlling variables for any controlled sources must appear inside the original circuit.

## Thévenin Equivalent Circuits

One type of equivalent circuit is the **Thévenin equivalent**, which consists of an independent voltage source in series with a resistance. This is illustrated in Figure 2.44.

Consider the Thévenin equivalent with open-circuited terminals as shown in Figure 2.45. By definition, no current can flow through an open circuit. Therefore, no current flows through the Thévenin resistance, and the voltage across the resistance is zero. Applying KVL, we conclude that

$$V_t = v_{oc}$$

Both the original circuit and the equivalent circuit are required to have the same open-circuit voltage. *Thus, the Thévenin source voltage $V_t$ is equal to the open-circuit voltage of the original network.*

Now, consider the Thévenin equivalent with a short circuit connected across its terminals as shown in Figure 2.46. The current flowing in this circuit is

$$i_{sc} = \frac{V_t}{R_t}$$

The short-circuit current $i_{sc}$ is the same for the original circuit as for the Thévenin equivalent. Solving for the Thévenin resistance, we have

$$R_t = \frac{V_t}{i_{sc}} \tag{2.73}$$

The Thévenin equivalent circuit consists of an independent voltage source in series with a resistance.

The Thévenin voltage $v_t$ is equal to the open-circuit voltage of the original network.

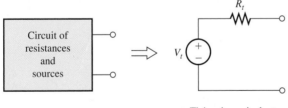

Thévenin equivalent circuit

**Figure 2.44**  A two-terminal circuit consisting of resistances and sources can be replaced by a Thévenin equivalent circuit.

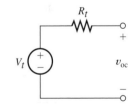

**Figure 2.45**  Thévenin equivalent circuit with open-circuited terminals. The open-circuit voltage $v_{oc}$ is equal to the Thévenin voltage $V_t$.

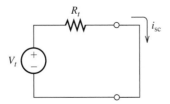

**Figure 2.46**  Thévenin equivalent circuit with short-circuited terminals. The short-circuit current is $i_{sc} = V_t/R_t$.

Using the fact that the Thévenin voltage is equal to the open-circuit voltage of the network, we have

$$R_t = \frac{v_{oc}}{i_{sc}} \tag{2.74}$$

The Thévenin resistance is equal to the open-circuit voltage divided by the short-circuit current.

Thus, to determine the Thévenin equivalent circuit, we can start by analyzing the original network for its open-circuit voltage and its short-circuit current. The Thévenin voltage equals the open-circuit voltage, and the Thévenin resistance is given by Equation 2.74.

---

**Example 2.18**    **Determining the Thévenin Equivalent Circuit**

Find the Thévenin equivalent for the circuit shown in Figure 2.47(a).

**Solution**    First, we analyze the circuit with open-circuited terminals. This is shown in Figure 2.47(b). The resistances $R_1$ and $R_2$ are in series and have an equivalent resistance of $R_1 + R_2$. Therefore, the current circulating is

$$i_1 = \frac{v_s}{R_1 + R_2} = \frac{15}{100 + 50} = 0.10 \text{ A}$$

The open-circuit voltage is the voltage across $R_2$:

$$v_{oc} = R_2 i_1 = 50 \times 0.10 = 5 \text{ V}$$

Thus, the Thévenin voltage is $V_t = 5$ V.

Now, we consider the circuit with a short circuit connected across its terminals as shown in Figure 2.47(c). By definition, the voltage across a short circuit is zero. Hence, the voltage across $R_2$ is zero, and the current through it is zero, as shown in

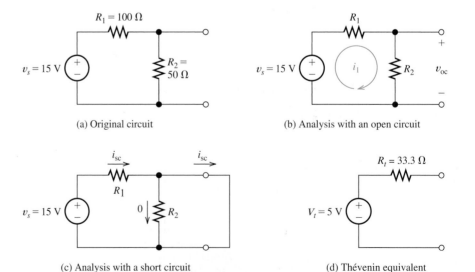

(a) Original circuit

(b) Analysis with an open circuit

(c) Analysis with a short circuit

(d) Thévenin equivalent

**Figure 2.47**  Circuit for Example 2.18.

the figure. Therefore, the short-circuit current $i_{sc}$ flows through $R_1$. The source voltage $v_s$ appears across $R_1$, so we can write

$$i_{sc} = \frac{v_s}{R_1} = \frac{15}{100} = 0.15 \text{ A}$$

Now, we can use Equation 2.74 to determine the Thévenin resistance:

$$R_t = \frac{v_{oc}}{i_{sc}} = \frac{5 \text{ V}}{0.15 \text{ A}} = 33.3 \text{ }\Omega$$

The Thévenin equivalent circuit is shown in Figure 2.47(d).   ■

**Exercise 2.26**   Find the Thévenin equivalent circuit for the circuit shown in Figure 2.48.
**Answer**   $V_t = 50 \text{ V}, R_t = 50 \text{ }\Omega.$   □

**Finding the Thévenin Resistance Directly.**   If a network contains no dependent sources, there is an alternative way to find the Thévenin resistance. First, we *zero* the sources in the network. In zeroing a voltage source, we reduce its voltage to zero. A voltage source with zero voltage is equivalent to a short circuit.

In zeroing a current source, we reduce its current to zero. By definition, an element that always carries zero current is an open circuit. *Thus, to zero the independent sources, we replace voltage sources with short circuits and replace current sources with open circuits.*

Figure 2.49 shows a Thévenin equivalent before and after zeroing its voltage source. Looking back into the terminals after the source is zeroed, we see the Thévenin resistance. *Thus, we can find the Thévenin resistance by zeroing the sources in the original network and then computing the resistance between the terminals.*

> When zeroing a current source, it becomes an open circuit. When zeroing a voltage source, it becomes a short circuit.

> We can find the Thévenin resistance by zeroing the sources in the original network and then computing the resistance between the terminals.

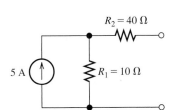

**Figure 2.48**  Circuit for Exercise 2.26.

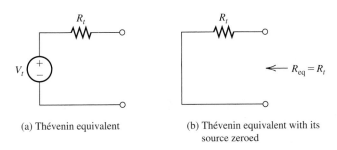

(a) Thévenin equivalent

(b) Thévenin equivalent with its source zeroed

**Figure 2.49**  When the source is zeroed, the resistance seen from the circuit terminals is equal to the Thévenin resistance.

### Example 2.19   Zeroing Sources to Find Thévenin Resistance

Find the Thévenin resistance for the circuit shown in Figure 2.50(a) by zeroing the sources. Then, find the short-circuit current and the Thévenin equivalent circuit.

**Solution**   To zero the sources, we replace the voltage source by a short circuit and replace the current source by an open circuit. The resulting circuit is shown in Figure 2.50(b).

The Thévenin resistance is the equivalent resistance between the terminals. This is the parallel combination of $R_1$ and $R_2$, which is given by

$$R_t = R_{eq} = \frac{1}{1/R_1 + 1/R_2} = \frac{1}{1/5 + 1/20} = 4 \ \Omega$$

Next, we find the short-circuit current for the circuit. The circuit is shown in Figure 2.50(c). In this circuit, the voltage across $R_2$ is zero because of the short circuit. Thus, the current through $R_2$ is zero:

$$i_2 = 0$$

Furthermore, the voltage across $R_1$ is equal to 20 V. Thus, the current is

$$i_1 = \frac{v_s}{R_1} = \frac{20}{5} = 4 \text{ A}$$

Finally, we write a current equation for the node joining the top ends of $R_2$ and the 2-A source. Setting the sum of the currents entering equal to the sum of the currents leaving, we have

$$i_1 + 2 = i_2 + i_{sc}$$

This yields $i_{sc} = 6$ A.

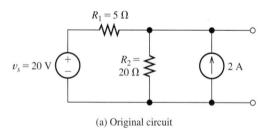

(a) Original circuit

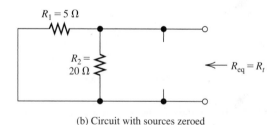

(b) Circuit with sources zeroed

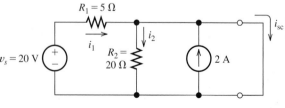

(c) Circuit with a short circuit

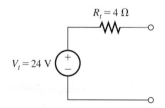

(d) Thévenin equivalent circuit

**Figure 2.50** Circuit for Example 2.19.

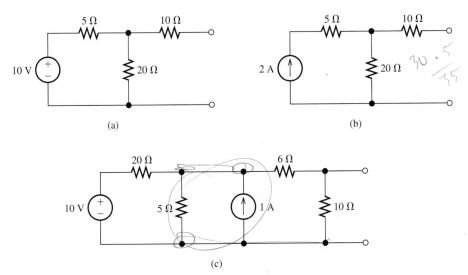

Figure 2.51  Circuits for Exercise 2.28.

Now, the Thévenin voltage can be found. Applying Equation 2.74, we get

$$V_t = R_t i_{sc} = 4 \times 6 = 24 \text{ V}$$

The Thévenin equivalent circuit is shown in Figure 2.50(d).   ■

**Exercise 2.27**   Use node-voltage analysis of the circuit shown in Figure 2.50(a) to show that the open-circuit voltage is equal to the Thévenin voltage found in Example 2.19.

**Exercise 2.28**   Find the Thévenin resistance for each of the circuits shown in Figure 2.51 by zeroing the sources.
**Answer   a.** $R_t = 14 \ \Omega$; **b.** $R_t = 30 \ \Omega$; **c.** $R_t = 5 \ \Omega$.   ☐

We complete our discussion of Thévenin equivalent circuits with one more example.

**Example 2.20   Thévenin Equivalent of a Circuit with a Dependent Source**

Find the Thévenin equivalent for the circuit shown in Figure 2.52(a).

**Solution**   Because this circuit contains a dependent source, we cannot find the Thévenin resistance by zeroing the sources and combining resistances in series and parallel. Thus, we must analyze the circuit to find the open-circuit voltage and the short-circuit current.

We start with the open-circuit voltage. Consider Figure 2.52(b). We use node-voltage analysis, picking the reference node at the bottom of the circuit. Then, $v_{oc}$ is the unknown node-voltage variable. First, we write a current equation at node 1.

*If a circuit contains a dependent source, we cannot find the Thévenin resistance by zeroing the sources and combining resistances in series and parallel.*

$$i_x + 2i_x = \frac{v_{oc}}{10} \qquad (2.75)$$

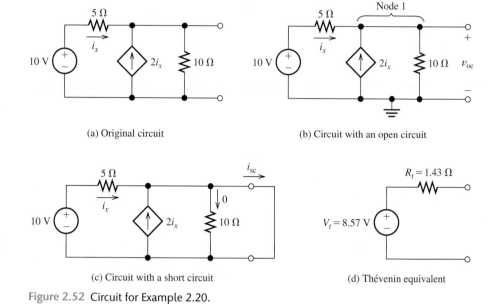

(a) Original circuit                              (b) Circuit with an open circuit

(c) Circuit with a short circuit                 (d) Thévenin equivalent

**Figure 2.52** Circuit for Example 2.20.

Next, we write an expression for the controlling variable $i_x$ in terms of the node voltage $v_{oc}$:

$$i_x = \frac{10 - v_{oc}}{5}$$

Substituting this into Equation 2.75, we have

$$3\frac{10 - v_{oc}}{5} = \frac{v_{oc}}{10}$$

Solving, we find that $v_{oc} = 8.57$ V.

Now, we consider short-circuit conditions as shown in Figure 2.52(c). In this case, the current through the 10-$\Omega$ resistance is zero. Furthermore, we get

$$i_x = \frac{10\ \text{V}}{5\ \Omega} = 2\ \text{A}$$

and

$$i_{sc} = 3i_x = 6\ \text{A}$$

Next, we use Equation 2.74 to compute the Thévenin resistance:

$$R_t = \frac{v_{oc}}{i_{sc}} = \frac{8.57\ \text{V}}{6\ \text{A}} = 1.43\ \Omega$$

Finally, the Thévenin equivalent circuit is shown in Figure 2.52(d). ∎

### Norton Equivalent Circuit

Another type of equivalent, known as the **Norton equivalent circuit**, is shown in Figure 2.53. It consists of an independent current source $I_n$ in parallel with the Thévenin resistance. Notice that if we zero the Norton current source, replacing

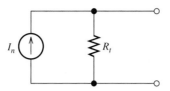

Figure 2.53 The Norton equivalent circuit consists of an independent current source $I_n$ in parallel with the Thévenin resistance $R_t$.

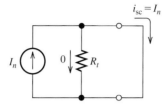

Figure 2.54 The Norton equivalent circuit with a short circuit across its terminals.

it by an open circuit, the Norton equivalent becomes a resistance of $R_t$. This also happens if we zero the voltage source in the Thévenin equivalent by replacing the voltage source by a short circuit. Thus, the resistance in the Norton equivalent is the same as the Thévenin resistance.

Consider placing a short circuit across the Norton equivalent as shown in Figure 2.54. In this case, the current through $R_t$ is zero. *Therefore, the Norton current is equal to the short-circuit current:*

$$I_n = i_{sc}$$

We can find the Norton equivalent by using the same techniques as we used for the Thévenin equivalent.

### Step-by-Step Thévenin/Norton-Equivalent-Circuit Analysis

1. Perform two of these:
   a. Determine the open-circuit voltage $V_t = v_{oc}$.
   b. Determine the short-circuit current $I_n = i_{sc}$.
   c. Zero the independent sources and find the Thévenin resistance $R_t$ looking back into the terminals. Do not zero dependent sources.
2. Use the equation $V_t = R_t I_n$ to compute the remaining value.
3. The Thévenin equivalent consists of a voltage source $V_t$ in series with $R_t$.
4. The Norton equivalent consists of a current source $I_n$ in parallel with $R_t$.

---

**Example 2.21**   **Norton Equivalent Circuit**

Find the Norton equivalent for the circuit shown in Figure 2.55(a).

Solution   Because the circuit contains a controlled source, we cannot zero the sources and combine resistances to find the Thévenin resistance. First, we consider the circuit with an open circuit as shown in Figure 2.53(a). We treat $v_{oc}$ as a node-voltage variable. Writing a current equation at the top of the circuit, we have

$$\frac{v_x}{4} + \frac{v_{oc} - 15}{R_1} + \frac{v_{oc}}{R_2 + R_3} = 0 \qquad (2.76)$$

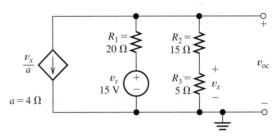

(a) Original circuit under open-circuit conditions

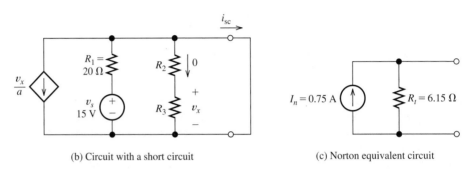

(b) Circuit with a short circuit          (c) Norton equivalent circuit

Figure 2.55  Circuit of Example 2.21.

Next, we use the voltage-divider principle to write an expression for $v_x$ in terms of resistances and $v_{oc}$:

$$v_x = \frac{R_3}{R_2 + R_3} v_{oc} = 0.25 v_{oc}$$

Substituting into Equation 2.76, we find that

$$\frac{0.25 v_{oc}}{4} + \frac{v_{oc} - 15}{R_1} + \frac{v_{oc}}{R_2 + R_3} = 0$$

Substituting resistance values and solving, we observe that $v_{oc} = 4.62$ V.

Next, we consider short-circuit conditions as shown in Figure 2.55(b). In this case, the current through $R_2$ and $R_3$ is zero. Thus, $v_x = 0$, and the controlled current source appears as an open circuit. The short-circuit current is given by

$$i_{sc} = \frac{v_s}{R_1} = \frac{15 \text{ V}}{20 \text{ }\Omega} = 0.75 \text{ A}$$

Now, we can find the Thévenin resistance:

$$R_t = \frac{v_{oc}}{i_{sc}} = \frac{4.62}{0.75} = 6.15 \text{ }\Omega$$

The Norton equivalent circuit is shown in Figure 2.55(c).    ■

**Exercise 2.29**   Find the Norton equivalent for each of the circuits shown in Figure 2.56.
**Answer**   **a.** $I_n = 1.67$ A, $R_t = 9.375$ $\Omega$; **b.** $I_n = 2A$, $R_t = 15$ $\Omega$.    □

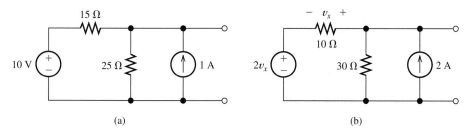

(a)                                    (b)

**Figure 2.56** Circuits for Exercise 2.29.

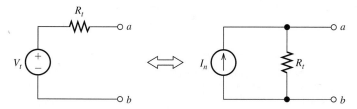

**Figure 2.57** A voltage source in series with a resistance is externally equivalent to a current source in parallel with the resistance, provided that $I_n = V_t/R_t$.

## Source Transformations

We can replace a voltage source in series with a resistance by a Norton equivalent circuit, which consists of a current source in parallel with the resistance. This is called a **source transformation** and is illustrated in Figure 2.57. The two circuits are identical in terms of their external behavior. In other words, the voltages and currents at terminals $a$ and $b$ remain the same after the transformation is made. However, in general, the current flowing through $R_t$ is different for the two circuits. For example, suppose that the two circuits shown in Figure 2.57 are open circuited. Then no current flows through the resistor in series with the voltage source, but the current $I_n$ flows through the resistance in parallel with the current source.

In making source transformations, it is very important to maintain the proper relationship between the reference direction for the current source and the polarity of the voltage source. If the positive polarity is closest to terminal $a$, the current reference must point toward terminal $a$, as shown in Figure 2.57.

Sometimes, we can simplify the solution of a circuit by source transformations. This is similar to solving circuits by combining resistances in series or parallel. We illustrate with an example.

Here is a "trick" question that you might have some fun with: Suppose that the circuits of Figure 2.57 are placed in identical black boxes with the terminals accessible from outside the box. How could you determine which box contains the Norton equivalent? An answer can be found at the end of the chapter summary on the top of page 111.

| **Example 2.22** | **Using Source Transformations** |

Use source transformations to aid in solving for the currents $i_1$ and $i_2$ shown in Figure 2.58(a).

**Solution**   Several approaches are possible. One is to transform the 1-A current source and $R_2$ into a voltage source in series with $R_2$. This is shown in Figure 2.58(b). Notice that the positive polarity of the 10-V source is at the top, because the 1-A source reference points upward. The single-loop circuit of Figure 2.58(b) can be

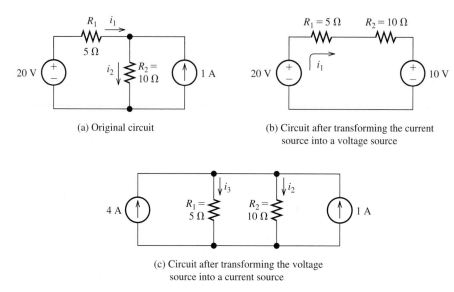

(a) Original circuit

(b) Circuit after transforming the current source into a voltage source

(c) Circuit after transforming the voltage source into a current source

**Figure 2.58**  Circuit for Example 2.22.

solved by writing a KVL equation. Traveling clockwise and summing voltages, we have

$$R_1 i_1 + R_2 i_1 + 10 - 20 = 0$$

Solving and substituting values, we get

$$i_1 = \frac{10}{R_1 + R_2} = 0.667 \text{ A}$$

Then in the original circuit, we can write a current equation at the top node and solve for $i_2$:

$$i_2 = i_1 + 1 = 1.667 \text{ A}$$

Another approach is to transform the voltage source and $R_1$ into a current source in parallel with $R_1$. Making this change to the original circuit yields the circuit shown in Figure 2.58(c). Notice that we have labeled the current through $R_1$ as $i_3$ rather than $i_1$. This is because the current in the resistance of the transformed source is not the same as in the original circuit. Now, in Figure 2.58(c), we see that a total current of 5 A flows into the parallel combination of $R_1$ and $R_2$. Using the current-division principle, we find the current through $R_2$:

$$i_2 = \frac{R_1}{R_1 + R_2} i_{\text{total}} = \frac{5}{5 + 10}(5) = 1.667 \text{ A}$$

This agrees with our previous result.  ∎

**Exercise 2.30**   Use two different approaches employing source transformations to solve for the values of $i_1$ and $i_2$ in Figure 2.59.

In the first approach, transform the current source and $R_1$ into a voltage source in series with $R_1$. (Make sure in making the transformation that the polarity of the

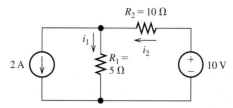

**Figure 2.59** Circuit for Exercise 2.30.

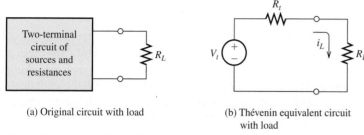

(a) Original circuit with load

(b) Thévenin equivalent circuit with load

**Figure 2.60** Circuits for analysis of maximum power transfer.

voltage source bears the correct relationship to the current reference direction.) Then solve the transformed circuit and determine the values of $i_1$ and $i_2$.

In the second approach, starting with the original circuit, transform the 10-V source and $R_2$ into a current source in parallel with $R_2$. Then solve the transformed circuit and determine the values of $i_1$ and $i_2$. Of course, the answers should be the same for both approaches.

**Answer**   $i_1 = -0.667$ A, $i_2 = 1.333$ A.

## Maximum Power Transfer

Suppose that we have a two-terminal circuit and we want to connect a load resistance $R_L$ such that the maximum possible power is delivered to the load. This is illustrated in Figure 2.60(a). To analyze this problem, we replace the original circuit by its Thévenin equivalent as shown in Figure 2.60(b). The current flowing through the load resistance is given by

$$i_L = \frac{V_t}{R_t + R_L}$$

The power delivered to the load is

$$p_L = i_L^2 R_L$$

Substituting for the current, we have

$$p_L = \frac{V_t^2 R_L}{(R_t + R_L)^2} \tag{2.77}$$

To find the value of the load resistance that maximizes the power delivered to the load, we set the derivative of $p_L$ with respect to $R_L$ equal to zero:

$$\frac{dp_L}{dR_L} = \frac{V_t^2(R_t + R_L)^2 - 2V_t^2 R_L(R_t + R_L)}{(R_t + R_L)^4} = 0$$

Solving for the load resistance, we have

$$R_L = R_t$$

*Thus, the load resistance that absorbs the maximum power from a two-terminal circuit is equal to the Thévenin resistance.* The maximum power is found by substituting $R_L = R_t$ into Equation 2.77. The result is

$$P_{L\text{ max}} = \frac{V_t^2}{4R_t} \tag{2.78}$$

**An All-Too-Common Example.**   You may have had difficulty in starting your car on a frigid morning. The battery in your car can be represented by a Thévenin equivalent circuit. It turns out that the Thévenin voltage of the battery does not change greatly with temperature. However, when the battery is very cold, the chemical reactions occur much more slowly and its Thévenin resistance is much higher. Thus, the power that the battery can deliver to the starter motor is greatly reduced.

---

### Example 2.23   Determining Maximum Power Transfer

Find the load resistance for maximum power transfer from the circuit shown in Figure 2.61. Also, find the maximum power.

**Solution**   First, we must find the Thévenin equivalent circuit. Zeroing the voltage source, we find that the resistances $R_1$ and $R_2$ are in parallel. Thus, the Thévenin resistance is

$$R_t = \frac{1}{1/R_1 + 1/R_2} = \frac{1}{1/20 + 1/5} = 4\ \Omega$$

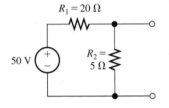

$R_1 = 20\ \Omega$

$R_2 = 5\ \Omega$

50 V

**Figure 2.61** Circuit for Example 2.23.

The Thévenin voltage is equal to the open-circuit voltage. Using the voltage-division principle, we find that

$$V_t = v_{oc} = \frac{R_2}{R_1 + R_2}(50) = \frac{5}{5 + 20}(50) = 10\ \text{V}$$

Hence, the load resistance that receives maximum power is

$$R_L = R_t = 4\ \Omega$$

and the maximum power is given by Equation 2.78:

$$P_{L\text{ max}} = \frac{V_t^2}{4R_t} = \frac{10^2}{4 \times 4} = 6.25\ \text{W} \quad\blacksquare$$

---

### PRACTICAL APPLICATION   2.1

#### An Important Engineering Problem: Energy-Storage Systems for Electric Vehicles

Imagine pollution-free electric vehicles with exciting performance and 500-mile range. They do not exist, but they are the target of an ongoing large-scale engineering effort to which you may contribute. Such electric vehicles (EVs) are a worthwhile goal because they can be very efficient in their use of energy, particularly in stop-and-go traffic. Kinetic energy can be recovered during braking and saved for later use during acceleration. Furthermore, EVs emit little pollution into crowded urban environments.

So far, EV range and performance remains less than ideal. The availability of suitable energy-storage devices is the key stumbling block in achieving better EVs (and a multitude of other highly desirable devices, such as smart phones that do not need recharging for a week).

In Chapter 3, we will see that capacitors and inductors are capable of storing electrical energy. However, it turns out that their energy content per unit volume is too small to make them a practical solution for EVs. The energy content of modern rechargeable batteries is better but still not on a par with the energy content of gasoline, which is approximately 10,000 watt-hours/liter (Wh/L). In contrast, the energy content of nickel-metal hydride batteries used in current EVs is about 175 Wh/L. Lithium-ion batteries under current development are expected to increase this to about 300 Wh/L. Thus, even allowing for the relative inefficiency of the internal combustion engine in converting chemical energy to mechanical energy, much more usable energy can be obtained from gasoline than from current batteries of comparable volume.

Although EVs do not emit pollutants at the point of use, the mining, refining, and disposal of metals pose grave environmental dangers. We must always consider the entire environmental (as well as economic) impact of the systems we design. As an engineer, you can do a great service to humanity by accepting the challenge to develop safe, clean systems for storing energy in forms that are readily converted to and from electrical form.

Naturally, one possibility currently under intense development is improved electrochemical batteries based on nontoxic chemicals. Another option is a mechanical flywheel system that would be coupled through an electrical generator to electric drive motors. Still another solution is a hybrid vehicle that uses a small internal combustion engine, an electrical generator, an energy-storage system, and electrical drive motors. The engine achieves low pollution levels by being optimized to run at a constant load while charging a relatively small energy-storage system. When the storage capacity becomes full, the engine shuts down automatically and the vehicle runs on stored energy. The engine is just large enough to keep up with energy demands under high-speed highway conditions.

Whatever form the ultimate solution to vehicle pollution may take, we can anticipate that it will include elements from mechanical, chemical, manufacturing, and civil engineering in close combination with electrical-engineering principles.

**Application of Maximum Power Transfer.**    When a load resistance equals the internal Thévenin resistance of the source, half of the power is dissipated in the source resistance and half is delivered to the load. In higher power applications for which efficiency is important, we do not usually design for maximum power transfer. For example, in designing an electric vehicle, we would want to deliver the energy stored in the batteries mainly to the drive motors and minimize the power loss in the resistance of the battery and wiring. This system would approach maximum power transfer rarely when maximum acceleration is needed.

On the other hand, when small amounts of power are involved, we would design for maximum power transfer. For example, we would design a radio receiver to extract the maximum signal power from the receiving antenna. In this application, the power is very small, typically much less than one microwatt, and efficiency is not a consideration.

## 2.7   SUPERPOSITION PRINCIPLE

Suppose that we have a circuit composed of resistances, linear dependent sources, and $n$ independent sources. (We will explain the term *linear* dependent source shortly.) The current flowing through a given element (or the voltage across it) is called a **response**, because the currents and voltages appear in response to the independent sources.

Recall that we zeroed the independent sources as a method for finding the Thévenin resistance of a two-terminal circuit. To zero a source, we reduce its value to zero. Then, current sources become open circuits, and voltage sources become short circuits.

Now, consider zeroing all of the independent sources except the first, observe a particular response (a current or voltage), and denote the value of that response as $r_1$. (We use the symbol $r$ rather than $i$ or $v$ because the response could be either a current or a voltage.) Similarly, with only source 2 activated, the response is denoted as $r_2$, and so on. The response with all the sources activated is called the total response, denoted as $r_T$. The **superposition principle** states that the total response is the sum of the responses to each of the independent sources acting individually. In equation form, this is

> The superposition principle states that any response in a linear circuit is the sum of the responses for each independent source acting alone with the other independent sources zeroed. When zeroed, current sources become open circuits and voltage sources become short circuits.

$$r_T = r_1 + r_2 + \cdots + r_n \tag{2.79}$$

Next, we illustrate the validity of superposition for the example circuit shown in Figure 2.62. In this circuit, there are two independent sources: the first is the voltage source $v_{s1}$, and the second is the current source $i_{s2}$. Suppose that the response of interest is the voltage across the resistance $R_2$.

First, we solve for the total response $v_T$ by solving the circuit with both sources in place. Writing a current equation at the top node, we obtain

$$\frac{v_T - v_{s1}}{R_1} + \frac{v_T}{R_2} + K i_x = i_{s2} \tag{2.80}$$

The control variable $i_x$ is given by

$$i_x = \frac{v_T}{R_2} \tag{2.81}$$

Substituting Equation 2.81 into Equation 2.80 and solving for the total response, we get

$$v_T = \frac{R_2}{R_1 + R_2 + K R_1} v_{s1} + \frac{R_1 R_2}{R_1 + R_2 + K R_1} i_{s2} \tag{2.82}$$

If we set $i_{s2}$ to zero, we obtain the response to $v_{s1}$ acting alone:

$$v_1 = \frac{R_2}{R_1 + R_2 + K R_1} v_{s1} \tag{2.83}$$

Similarly, if we set $v_{s1}$ equal to zero in Equation 2.82, the response due to $i_{s2}$ is given by

$$v_2 = \frac{R_1 R_2}{R_1 + R_2 + K R_1} i_{s2} \tag{2.84}$$

Comparing Equations 2.82, 2.83, and 2.84, we see that

$$v_T = v_1 + v_2$$

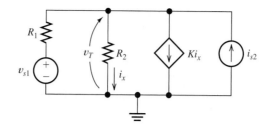

**Figure 2.62** Circuit used to illustrate the superposition principle.

Thus, as expected from the superposition principle, the total response is equal to the sum of the responses for each of the independent sources acting individually.

Notice that if we zero both of the independent sources ($v_{s1} = 0$ and $i_{s2} = 0$), the response becomes zero. Hence, the dependent source does not contribute to the total response. However, the dependent source affects the contributions of the two independent sources. This is evident because the gain parameter $K$ of the dependent source appears in the expressions for both $v_1$ and $v_2$. *In general, dependent sources do not contribute a separate term to the total response, and we must not zero dependent sources in applying superposition.*

Dependent sources do not contribute a separate term to the total response, and we must not zero dependent sources in applying superposition.

## Linearity

If we plot voltage versus current for a resistance, we have a straight line. This is illustrated in Figure 2.63. Thus, we say that Ohm's law is a **linear equation**. Similarly, the current in the controlled source shown in Figure 2.62 is given by $i_{cs} = Ki_x$, which is also a linear equation. In this book, the term **linear controlled source** means a source whose value is a constant times a control variable that is a current or a voltage appearing in the network.

Some examples of nonlinear equations are

$$v = 10i^2$$

$$i_{cs} = K\cos(i_x)$$

and

$$i = e^v$$

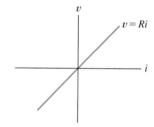

**Figure 2.63** A resistance that obeys Ohm's law is linear.

*The superposition principle does not apply to any circuit that has element(s) described by nonlinear equation(s).* We will encounter nonlinear elements later in our study of electronic circuits.

Furthermore, superposition does not apply for power in resistances, because $P = v^2/R$ and $P = i^2R$ are nonlinear equations.

The superposition principle does not apply to any circuit that has element(s) described by nonlinear equation(s).

## Using Superposition to Solve Circuits

We can apply superposition in circuit analysis by analyzing the circuit for each source separately. Then, we add the individual responses to find the total response. Sometimes, the analysis of a circuit is simplified by considering each independent source separately. We illustrate with an example.

### Example 2.24    Circuit Analysis Using Superposition

Use superposition in solving the circuit shown in Figure 2.64(a) for the voltage $v_T$.

**Solution**    We analyze the circuit with only one source activated at a time and add the responses. Figure 2.64(b) shows the circuit with only the voltage source active. The response can be found by applying the voltage-division principle:

$$v_1 = \frac{R_2}{R_1 + R_2}v_s = \frac{5}{5 + 10}(15) = 5 \text{ V}$$

Next, we analyze the circuit with only the current source active. The circuit is shown in Figure 2.64(c). In this case, the resistances $R_1$ and $R_2$ are in parallel, and the equivalent resistance is

$$R_{eq} = \frac{1}{1/R_1 + 1/R_2} = \frac{1}{1/10 + 1/5} = 3.33 \text{ }\Omega$$

The voltage due to the current source is given by

$$v_2 = i_s R_{eq} = 2 \times 3.33 = 6.66 \text{ V}$$

Finally, we obtain the total response by adding the individual responses:

$$v_T = v_1 + v_2 = 5 + 6.66 = 11.66 \text{ V}$$    ∎

**Exercise 2.31**    Find the responses $i_1$, $i_2$, and $i_T$ for the circuit of Figure 2.64.
**Answer**    $i_1 = 1$ A, $i_2 = -0.667$ A, $i_T = 0.333$ A.    □

**Exercise 2.32**    Use superposition to find the responses $v_T$ and $i_T$ for the circuit shown in Figure 2.65.
**Answer**    $v_1 = 5.45$ V, $v_2 = 1.82$ V, $v_T = 7.27$ V, $i_1 = 1.45$ A, $i_2 = -0.181$ A, $i_T = 1.27$ A.    □

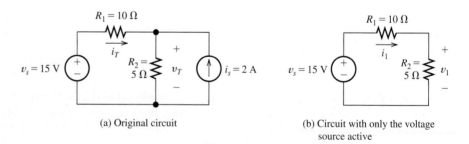

(a) Original circuit

(b) Circuit with only the voltage
source active

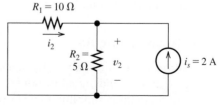

(c) Circuit with only the current
source active

**Figure 2.64**  Circuit for Example 2.24 and Exercise 2.31.

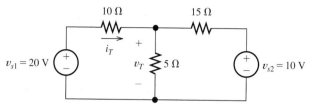

Figure 2.65  Circuit for Exercise 2.32.

## 2.8  WHEATSTONE BRIDGE

The **Wheatstone bridge** is a circuit used to measure unknown resistances. For example, it is used by mechanical and civil engineers to measure the resistances of strain gauges in experimental stress studies of machines and buildings. The circuit is shown in Figure 2.66. The circuit consists of a dc voltage source $v_s$, a detector, the unknown resistance to be measured $R_x$, and three precision resistors, $R_1$, $R_2$, and $R_3$. Usually, $R_2$ and $R_3$ are adjustable resistances, which is indicated in the figure by the arrow drawn through the resistance symbols.

The Wheatstone bridge is used by mechanical and civil engineers to measure the resistances of strain gauges in experimental stress studies of machines and buildings.

The detector is capable of responding to very small currents (less than one microampere). However, it is not necessary for the detector to be calibrated. It is only necessary for the detector to indicate whether or not current is flowing through it. Often, the detector has a pointer that deflects one way or the other, depending on the direction of the current through it.

In operation, the resistors $R_2$ and $R_3$ are adjusted in value until the detector indicates zero current. In this condition, we say that the bridge is **balanced**. Then, the current $i_g$ and the voltage across the detector $v_{ab}$ are zero.

Applying KCL at node $a$ (Figure 2.66) and using the fact that $i_g = 0$, we have

$$i_1 = i_3 \qquad (2.85)$$

Similarly, at node $b$, we get

$$i_2 = i_4 \qquad (2.86)$$

Writing a KVL equation around the loop formed by $R_1$, $R_2$, and the detector, we obtain

$$R_1 i_1 + v_{ab} = R_2 i_2 \qquad (2.87)$$

However, when the bridge is balanced, $v_{ab} = 0$, so that

$$R_1 i_1 = R_2 i_2 \qquad (2.88)$$

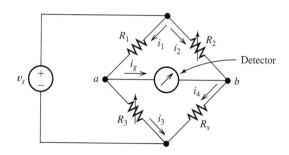

Figure 2.66  The Wheatstone bridge. When the Wheatstone bridge is balanced, $i_g = 0$ and $v_{ab} = 0$.

Similarly, for the loop consisting of $R_3$, $R_4$, and the detector, we have

$$R_3 i_3 = R_x i_4 \qquad (2.89)$$

Using Equations 2.85 and 2.86 to substitute into Equation 2.89, we obtain

$$R_3 i_1 = R_x i_2 \qquad (2.90)$$

Dividing each side of Equation 2.90 by the respective side of Equation 2.88, we find that

$$\frac{R_3}{R_1} = \frac{R_x}{R_2}$$

Finally, solving for the unknown resistance, we have

$$R_x = \frac{R_2}{R_1} R_3 \qquad (2.91)$$

Often, in commercial bridges, a multiposition switch selects an order-of-magnitude scale factor $R_2/R_1$ by changing the value of $R_2$. Then, $R_3$ is adjusted by means of calibrated switches until balance is achieved. Finally, the unknown resistance $R_x$ is the scale factor times the value of $R_3$.

---

### Example 2.25   Using a Wheatstone Bridge to Measure Resistance

In a certain commercial Wheatstone bridge, $R_1$ is a fixed 1-k$\Omega$ resistor, $R_3$ can be adjusted in 1-$\Omega$ steps from 0 to 1100 $\Omega$, and $R_2$ can be selected to be 1 k$\Omega$, 10 k$\Omega$, 100 k$\Omega$, or 1 M$\Omega$. **a.** Suppose that the bridge is balanced with $R_3 = 732\ \Omega$ and $R_2 = 10$ k$\Omega$. What is the value of $R_x$? **b.** What is the largest value of $R_x$ for which the bridge can be balanced? **c.** Suppose that $R_2 = 1$ M$\Omega$. What is the increment between values of $R_x$ for which the bridge can be precisely balanced?

**Solution**

**a.** From Equation 2.91, we have

$$R_x = \frac{R_2}{R_1} R_3 = \frac{10\ \text{k}\Omega}{1\ \text{k}\Omega} \times 732\ \Omega = 7320\ \Omega$$

Notice that $R_2/R_1$ is a scale factor that can be set at 1, 10, 100, or 1000, depending on the value selected for $R_2$. The unknown resistance is the scale factor times the value of $R_3$ needed to balance the bridge.

**b.** The maximum resistance for which the bridge can be balanced is determined by the largest values available for $R_2$ and $R_3$. Thus,

$$R_{x\,\text{max}} = \frac{R_{2\,\text{max}}}{R_1} R_{3\,\text{max}} = \frac{1\ \text{M}\Omega}{1\ \text{k}\Omega} \times 1100\ \Omega = 1.1\ \text{M}\Omega$$

**c.** The increment between values of $R_x$ for which the bridge can be precisely balanced is the scale factor times the increment in $R_3$:

$$R_{x\text{inc}} = \frac{R_2}{R_1} R_{3\text{inc}} = \frac{1\ \text{M}\Omega}{1\ \text{k}\Omega} \times 1\ \Omega = 1\ \text{k}\Omega$$

■

## Strain Measurements

The Wheatstone bridge circuit configuration is often employed with strain gauges in measuring strains of beams and other mechanical structures. (See the Practical Application on page 30 for more information about strain gauges.)

For example, consider the cantilevered beam subject to a downward load force at its outer end as shown in Figure 2.67(a). Two strain gauges are attached to the top of the beam where they are stretched, increasing their resistance by $\Delta R$ when the load is applied. The change in resistance is given by

$$\Delta R = R_0 G \frac{\Delta L}{L} \qquad (2.92)$$

in which $\Delta L/L$ is the strain for the surface of the beam to which the gauge is attached, $R_0$ is the gauge resistance before strain is applied, and $G$ is the **gauge factor** which is typically about 2. Similarly, two gauges on the bottom of the beam are compressed, reducing their resistance by $\Delta R$ with load. (For simplicity, we have assumed that the strain magnitude is the same for all four gauges.)

The four gauges are connected in a Wheatstone bridge as shown in Figure 2.67(b). The resistances labeled $R_0 + \Delta R$ are the gauges on the top of the beam and are being stretched, and those labeled $R_0 - \Delta R$ are those on the bottom and are being compressed. Before the load is applied, all four resistances have a value of $R_0$, the Wheatstone bridge is balanced, and the output voltage $v_o$ is zero.

It can be shown that the output voltage $v_o$ from the bridge is given by

$$v_o = V_s \frac{\Delta R}{R_0} = V_s G \frac{\Delta L}{L} \qquad (2.93)$$

Thus, the output voltage is proportional to the strain of the beam.

In principle, the resistance of one of the gauges could be measured and the strain determined from the resistance measurements. However, the changes in resistance are very small, and the measurements would need to be very precise. Furthermore, gauge resistance changes slightly with temperature. In the bridge arrangement with the gauges attached to the beam, the temperature changes tend to track very closely and have very little effect on $v_o$.

Usually, $v_o$ is amplified by an instrumentation-quality differential amplifier such as that discussed in Section 13.8 which starts on page 676. The amplified voltage can be converted to digital form and input to a computer or relayed wirelessly to a remote location for monitoring.

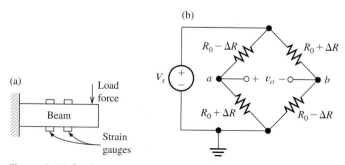

**Figure 2.67** Strain measurements using the Wheatstone bridge.

## Summary

1. Series resistances have an equivalent resistance equal to their sum. For $n$ resistances in series, we have

$$R_{eq} = R_1 + R_2 + \cdots + R_n$$

2. Parallel resistances have an equivalent resistance equal to the reciprocal of the sum of their reciprocals. For $n$ resistances in parallel, we get

$$R_{eq} = \frac{1}{1/R_1 + 1/R_2 + \cdots + 1/R_n}$$

3. Some resistive networks can be solved by repeatedly combining resistances in series or parallel. The simplified network is solved, and results are transferred back through the chain of equivalent circuits. Eventually, the currents and voltages of interest in the original circuit are found.

4. The voltage-division principle applies when a voltage is applied to several resistances in series. A fraction of the total voltage appears across each resistance. The fraction that appears across a given resistance is the ratio of the given resistance to the total series resistance.

5. The current-division principle applies when current flows through two resistances in parallel. A fraction of the total current flows through each resistance. The fraction of the total current flowing through $R_1$ is equal to $R_2/(R_1 + R_2)$.

6. The node-voltage method can be used to solve for the voltages in any resistive network. A step-by-step summary of the method is given starting on page 76.

7. A step-by-step procedure to write the node-voltage equations directly in matrix form for circuits consisting of resistances and independent current sources appears on page 66.

8. The mesh-current method can be used to solve for the currents in any planar resistive network. A step-by-step summary of the method is given on page 89.

9. A step-by-step procedure to write the mesh-current equations directly in matrix form for circuits consisting of resistances and independent voltage sources appears on page 85. For this method to apply, all of the mesh currents must flow in the clockwise direction.

10. A two-terminal network of resistances and sources has a Thévenin equivalent that consists of a voltage source in series with a resistance. The Thévenin voltage is equal to the open-circuit voltage of the original network. The Thévenin resistance is the open-circuit voltage divided by the short-circuit current of the original network. Sometimes, the Thévenin resistance can be found by zeroing the independent sources in the original network and combining resistances in series and parallel. When independent voltage sources are zeroed, they are replaced by short circuits. Independent current sources are replaced by open circuits. Dependent sources must not be zeroed.

11. A two-terminal network of resistances and sources has a Norton equivalent that consists of a current source in parallel with a resistance. The Norton current is equal to the short-circuit current of the original network. The Norton resistance is the same as the Thévenin resistance. A step-by-step procedure for determining Thévenin and Norton equivalent circuits is given on page 97.

12. Sometimes source transformations (i.e., replacing a Thévenin equivalent with a Norton equivalent or vice versa) are useful in solving networks.

13. For maximum power from a two-terminal network, the load resistance should equal the Thévenin resistance.

14. The superposition principle states that the total response in a resistive circuit is the sum of the responses to each of the independent sources acting individually. The superposition principle does not apply to any circuit that has element(s) described by nonlinear equation(s).

15. The Wheatstone bridge is a circuit used to measure unknown resistances. The circuit consists of a voltage source, a detector, three precision calibrated resistors, of which two are adjustable, and the unknown resistance. The resistors are adjusted until the bridge is balanced, and then the unknown resistance is given in terms of the three known resistances.

Here's the answer to the trick question on page 97: Suppose that we open circuit the terminals. Then, no current flows through the Thévenin equivalent, but a current $I_n$ circulates in the Norton equivalent. Thus, the box containing the Norton equivalent will become warm because of power dissipation in the resistance. The point of this question is that the circuits are equivalent in terms of their terminal voltage and current, not in terms of their internal behavior.

## Problems

### Section 2.1: Resistances in Series and Parallel

*P2.1. Reduce each of the networks shown in Figure P2.1 to a single equivalent resistance by combining resistances in series and parallel.

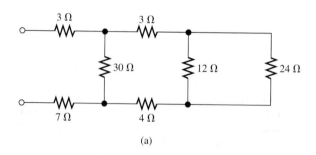

(a)

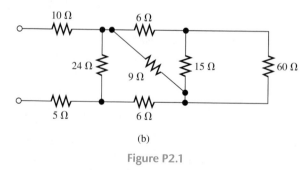

(b)

Figure P2.1

*P2.2. A 4-$\Omega$ resistance is in series with the parallel combination of a 20-$\Omega$ resistance and an unknown resistance $R_x$. The equivalent resistance for the network is 8 $\Omega$. Determine the value of $R_x$.

*P2.3. Find the equivalent resistance looking into terminals $a$ and $b$ in Figure P2.3.

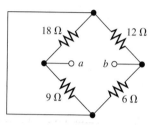

Figure P2.3

*P2.4. Suppose that we need a resistance of 1.5 k$\Omega$ and you have a box of 1-k$\Omega$ resistors. Devise a network of 1-k$\Omega$ resistors so the equivalent resistance is 1.5 k$\Omega$. Repeat for an equivalent resistance of 2.2 k$\Omega$.

*P2.5. Find the equivalent resistance between terminals $a$ and $b$ in Figure P2.5.

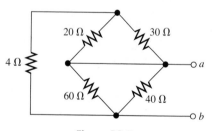

Figure P2.5

P2.6. Find the equivalent resistance between terminals $a$ and $b$ for each of the networks shown in Figure P2.6.

---

*Denotes that answers are contained in the Student Solutions files. See Appendix E for more information about accessing the Student Solutions.

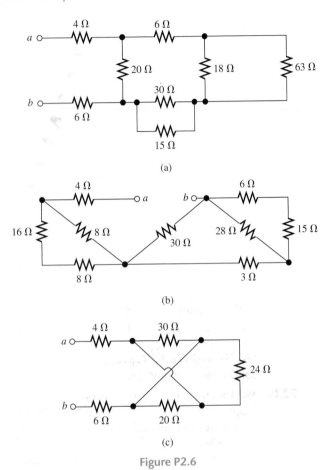

(a)

(b)

(c)

Figure P2.6

**P2.7.** What resistance in parallel with 120 $\Omega$ results in an equivalent resistance of 48 $\Omega$?

**P2.8. a.** Determine the resistance between terminals $a$ and $b$ for the network shown in Figure P2.8. **b.** Repeat after connecting $c$ and $d$ with a short circuit.

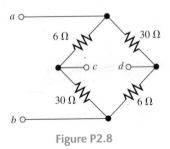

Figure P2.8

**P2.9.** Two resistances having values of $R$ and $2R$ are in parallel. $R$ and the equivalent resistance are both positive integers. What are the possible values for $R$?

**P2.10.** A network connected between terminals $a$ and $b$ consists of two parallel combinations that are in series. The first parallel combination is composed of a 16-$\Omega$ resistor and a 48-$\Omega$ resistor. The second parallel combination is composed of a 12-$\Omega$ resistor and a 24-$\Omega$ resistor. Draw the network and determine its equivalent resistance.

**P2.11.** Two resistances $R_1$ and $R_2$ are connected in parallel. We know that $R_1 = 90$ $\Omega$ and that the current through $R_2$ is three times the value of the current through $R_1$. Determine the value of $R_2$.

**P2.12.** Find the equivalent resistance for the infinite network shown in Figure P2.12(a). Because of its form, this network is called a semi-infinite ladder. [*Hint:* If another section is added to the ladder as shown in Figure P2.12(b), the equivalent resistance is the same. Thus, working from Figure P2.12(b), we can write an expression for $R_{eq}$ in terms of $R_{eq}$. Then, we can solve for $R_{eq}$.]

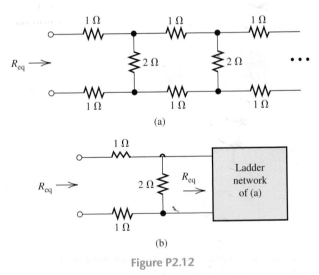

(a)

(b)

Figure P2.12

**P2.13.** If we connect $n$ 1000-$\Omega$ resistances in parallel, what value is the equivalent resistance?

**P2.14.** The heating element of an electric cook top has two resistive elements, $R_1 = 57.6$ $\Omega$ and $R_2 = 115.2$ $\Omega$, that can be operated separately, in series, or in parallel from voltages of either 120 V or 240 V. For the lowest power, $R_1$ is in series with $R_2$, and the combination is operated from 120 V. What is the lowest power? For the highest power, how should the elements be operated? What power results? List three more modes of operation and the resulting power for each.

**P2.15.** We are designing an electric space heater to operate from 120 V. Two heating elements with resistances $R_1$ and $R_2$ are to be used that can be operated in parallel, separately, or in series. The highest power is to be 1280 W, and the lowest power is to be 240 W. What values are needed for $R_1$ and $R_2$? What intermediate power settings are available?

**P2.16.** Sometimes, we can use symmetry considerations to find the resistance of a circuit that cannot be reduced by series or parallel combinations. A classic problem of this type is illustrated in Figure P2.16. Twelve 1-$\Omega$ resistors are arranged on the edges of a cube, and terminals $a$ and $b$ are connected to diagonally opposite corners of the cube. The problem is to find the resistance between the terminals. Approach the problem this way: Assume that 1 A of current enters terminal $a$ and exits through terminal $b$. Then, the voltage between terminals $a$ and $b$ is equal to the unknown resistance. By symmetry considerations, we can find the current in each resistor. Then, using KVL, we can find the voltage between $a$ and $b$.

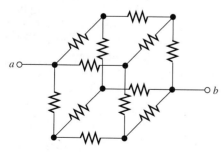

Figure P2.16 Each resistor has a value of 1 $\Omega$.

**P2.17.** The equivalent resistance between terminals $a$ and $b$ in Figure P2.17 is $R_{ab} = 23\ \Omega$. Determine the value of $R$.

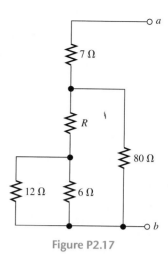

Figure P2.17

**P2.18. a.** Three conductances $G_1$, $G_2$, and $G_3$ are in series. Write an expression for the equivalent conductance $G_{eq} = 1/R_{eq}$ in terms of $G_1$, $G_2$, and $G_3$. **b.** Repeat part (a) with the conductances in parallel.

**P2.19.** Most sources of electrical power behave as (approximately) ideal voltage sources. In this case, if we have several loads that we want to operate independently, we place the loads in parallel with a switch in series with each load. Thereupon, we can switch each load on or off without affecting the power delivered to the other loads.

How would we connect the loads and switches if the source is an ideal independent current source? Draw the diagram of the current source and three loads with on–off switches such that each load can be switched on or off without affecting the power supplied to the other loads. To turn a load off, should the corresponding switch be opened or closed? Explain.

**P2.20.** The resistance for the network shown in Figure P2.20 between terminals $a$ and $b$ with $c$ open circuited is $R_{ab} = 50\ \Omega$. Similarly, the resistance between terminals $b$ and $c$ with $a$ open is $R_{bc} = 100\ \Omega$, and between $c$ and $a$ with $b$ open is $R_{ca} = 70\ \Omega$. Now, suppose that a short circuit is connected from terminal $b$ to terminal $c$, and determine the resistance between terminal $a$ and the shorted terminals $b$–$c$.

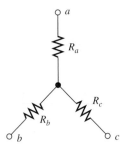

Figure P2.20

**P2.21.** Often, we encounter delta-connected loads, such as that illustrated in Figure P2.21, in three-phase power distribution systems (which are treated in Section 5.7). If we only have access to the three terminals, a method for determining the resistances is to repeatedly short two terminals together and measure the resistance between the shorted terminals and the third terminal. Then, the resistances can be calculated from the three measurements. Suppose that the measurements are $R_{as} = 12\ \Omega$, $R_{bs} = 20\ \Omega$, and $R_{cs} = 15\ \Omega$. Where $R_{as}$ is the resistance between terminal $a$ and the short between $b$ and $c$, etc. Determine the values of $R_a$, $R_b$, and $R_c$. (*Hint:* You may find the equations easier to deal with if you work in terms of conductances rather than resistances. Once the conductances are known, you can easily invert their values to find the resistances.)

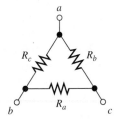

Figure P2.21

### Section 2.2: Network Analysis by Using Series and Parallel Equivalents

**P2.22.** What are the steps in solving a circuit by network reduction (series/parallel combinations)? Does this method always provide the solution? Explain.

**\*P2.23.** Find the values of $i_1$ and $i_2$ in Figure P2.23.

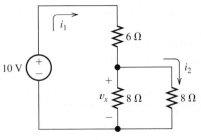

Figure P2.23

**\*P2.24.** Find the voltages $v_1$ and $v_2$ for the circuit shown in Figure P2.24 by combining resistances in series and parallel.

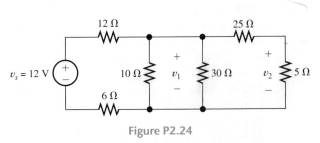

Figure P2.24

**\*P2.25.** Find the values of $v$ and $i$ in Figure P2.25.

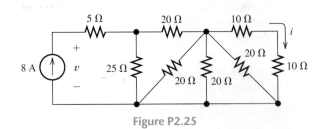

Figure P2.25

**P2.26.** Consider the circuit shown in Figure P2.24. Suppose that the value of $v_s$ is adjusted until $v_2 = 5$ V. Determine the new value of $v_s$. [*Hint:* Start at the right-hand side of the circuit and compute currents and voltages, moving to the left until you reach the source.]

**P2.27.** Find the voltage $v$ and the currents $i_1$ and $i_2$ for the circuit shown in Figure P2.27.

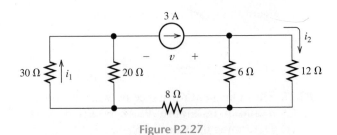

Figure P2.27

**P2.28.** Find the values of $v_s$, $v_1$, and $i_2$ in Figure P2.28.

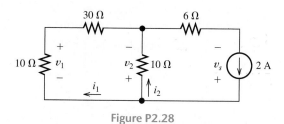

Figure P2.28

**P2.29.** Find the values of $i_1$ and $i_2$ in Figure P2.29.

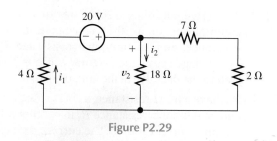

Figure P2.29

**P2.30.** Consider the circuit shown in Figure P2.30. Find the values of $v_1$, $v_2$, and $v_{ab}$.

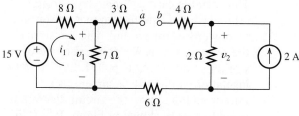

Figure P2.30

**P2.31.** Solve for the values of $i_1$, $i_2$, and the powers for the sources in Figure P2.31. Is the current source absorbing energy or delivering energy? Is the voltage source absorbing energy or delivering it?

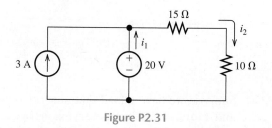

Figure P2.31

**P2.32.** The 12-V source in Figure P2.32 is delivering 36 mW of power. All four resistors have the same value $R$. Find the value of $R$.

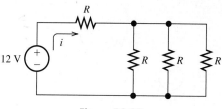

Figure P2.32

**P2.33.** Refer to the circuit shown in Figure P2.33. With the switch open, we have $v_2 = 8$ V. On the other hand, with the switch closed, we have $v_2 = 6$ V. Determine the values of $R_2$ and $R_L$.

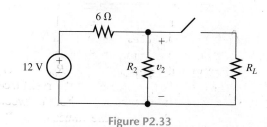

Figure P2.33

**\*P2.34.** Find the values of $i_1$ and $i_2$ in Figure P2.34. Find the power for each element in the circuit, and state whether each is absorbing or delivering energy. Verify that the total power absorbed equals the total power delivered.

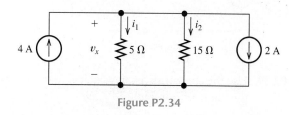

Figure P2.34

**\*P2.35.** Find the values of $i_1$ and $i_2$ in Figure P2.35.

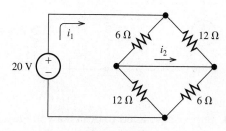

Figure P2.35

**Section2.3: Voltage-Divider and Current-Divider Circuits**

*P2.36. Use the voltage-division principle to calculate $v_1$, $v_2$, and $v_3$ in Figure P2.36.

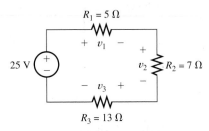

Figure P2.36

*P2.37. Use the current-division principle to calculate $i_1$ and $i_2$ in Figure P2.37.

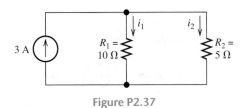

Figure P2.37

*P2.38. Use the voltage-division principle to calculate $v$ in Figure P2.38.

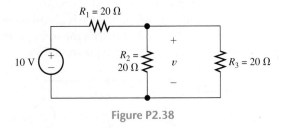

Figure P2.38

P2.39. Use the current-division principle to calculate the value of $i_3$ in Figure P2.39.

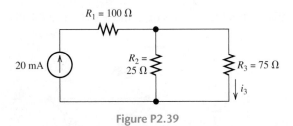

Figure P2.39

P2.40. Suppose we need to design a voltage-divider circuit to provide an output voltage $v_o = 5$ V from a 15-V source as shown in Figure P2.40.

The current taken from the 15-V source is to be 200 mA. **a.** Find the values of $R_1$ and $R_2$. **b.** Now suppose that a load resistance of 200 $\Omega$ is connected across the output terminals (i.e., in parallel with $R_2$). Find the value of $v_o$.

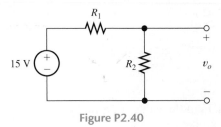

Figure P2.40

P2.41. A source supplies 120 V to the series combination of a 10-$\Omega$ resistance, a 5-$\Omega$ resistance, and an unknown resistance $R_x$. The voltage across the 5-$\Omega$ resistance is 20 V. Determine the value of the unknown resistance.

P2.42. We have a 60-$\Omega$ resistance, a 20-$\Omega$ resistance, and an unknown resistance $R_x$ in parallel with a 15 mA current source. The current through the unknown resistance is 10 mA. Determine the value of $R_x$.

*P2.43. A worker is standing on a wet concrete floor, holding an electric drill having a metallic case. The metallic case is connected through the ground wire of a three-terminal power outlet to power-system ground. The resistance of the ground wire is $R_g$. The resistance of the worker's body is $R_w = 500$ $\Omega$. Due to faulty insulation in the drill, a current of 2 A flows into its metallic case. The circuit diagram for this situation is shown in Figure P2.43. Find the maximum value of $R_g$ so that the current through the worker does not exceed 0.1 mA.

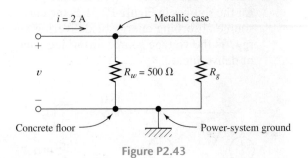

Figure P2.43

P2.44. Suppose we have a load that absorbs power and requires a current varying between 0 and 50 mA. The voltage across the load must remain between 4.7 and 5.0 V. A 15-V source is available. Design a voltage-divider network to supply the load. You may assume that resistors of any value desired are available. Also, determine the maximum power for each resistor.

**P2.45.** We have a load resistance of 50 Ω that we wish to supply with 5 V. A 12.6-V voltage source and resistors of any value needed are available. Draw a suitable circuit consisting of the voltage source, the load, and one additional resistor. Specify the value of the resistor.

**P2.46.** We have a load resistance of 1 kΩ that we wish to supply with 25 mW. A 20-mA current source and resistors of any value needed are available. Draw a suitable circuit consisting of the current source, the load, and one additional resistor. Specify the value of the resistor.

**P2.47.** The circuit of Figure P2.47 is similar to networks used in digital-to-analog converters. For this problem, assume that the circuit continues indefinitely to the right. Find the values of $i_1$, $i_2$, $i_3$, and $i_4$. How is $i_{n+2}$ related to $i_n$? What is the value of $i_{18}$? (*Hint:* See Problem P2.12.)

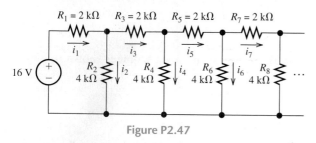

Figure P2.47

**Section 2.4:** Node-Voltage Analysis

*P2.48. Write equations and solve for the node voltages shown in Figure P2.48. Then, find the value of $i_1$.

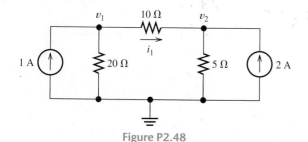

Figure P2.48

*P2.49. Solve for the node voltages shown in Figure P2.49. Then, find the value of $i_s$.

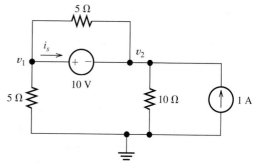

Figure P2.49

**P2.50.** Solve for the node voltages shown in Figure P2.50. What are the new values of the node voltages after the direction of the current source is reversed? How are the values related?

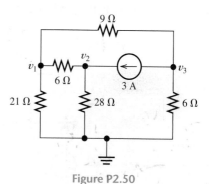

Figure P2.50

**P2.51.** Given $R_1 = 4\ \Omega$, $R_2 = 5\ \Omega$, $R_3 = 8\ \Omega$, $R_4 = 10\ \Omega$, $R_5 = 2\ \Omega$, and $I_s = 2$ A, solve for the node voltages shown in Figure P2.51.

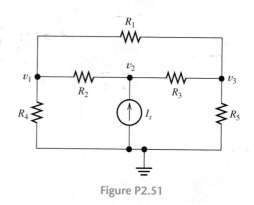

Figure P2.51

**P2.52.** Determine the value of $i_1$ in Figure P2.52 using node voltages to solve the circuit. Select the location of the reference node to minimize the number of unknown node voltages. What effect does the 20-$\Omega$ resistance have on the answer? Explain.

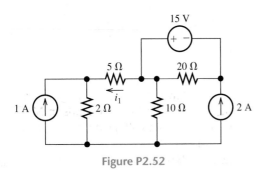

Figure P2.52

**P2.53.** Given $R_1 = 15\ \Omega$, $R_2 = 5\ \Omega$, $R_3 = 20\ \Omega$, $R_4 = 10\ \Omega$, $R_5 = 8\ \Omega$, $R_6 = 4\ \Omega$, and $I_s = 5$ A, solve for the node voltages shown in Figure P2.53.

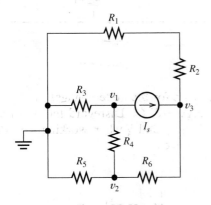

Figure P2.53

**P2.54.** In solving a network, what rule must you observe when writing KCL equations? Why?

**P2.55.** Use the symbolic features of MATLAB to find an expression for the equivalent resistance for the network shown in Figure P2.55. [*Hint:* First, connect a 1-A current source across terminals $a$ and $b$. Then, solve the network by the node-voltage technique. The voltage across the current source is equal in value to the equivalent resistance.] Finally, use the subs command to evaluate for $R_1 = 15\ \Omega$, $R_2 = 5\ \Omega$, $R_3 = 20\ \Omega$, $R_4 = 10\ \Omega$, and $R_5 = 8\ \Omega$.

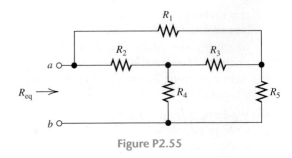

Figure P2.55

**\*P2.56.** Solve for the values of the node voltages shown in Figure P2.56. Then, find the value of $i_x$.

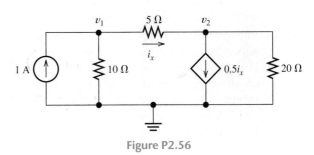

Figure P2.56

**\*P2.57.** Solve for the node voltages shown in Figure P2.57.

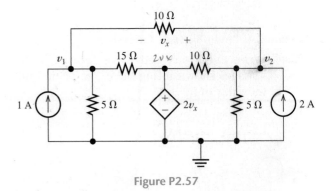

Figure P2.57

**P2.58** Solve for the power delivered to the 8-$\Omega$ resistance and for the node voltages shown in Figure P2.58.

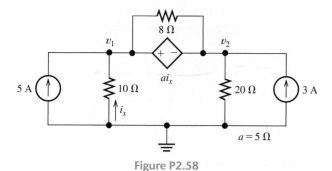

Figure P2.58

**P2.59.** Solve for the node voltages shown in Figure P2.59.

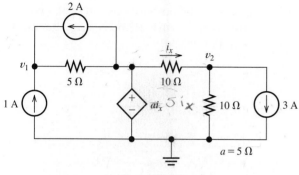

Figure P2.59

**P2.60.** Find the equivalent resistance looking into terminals for the network shown in Figure P2.60. [*Hint:* First, connect a 1-A current source across terminals $a$ and $b$. Then, solve the network by the node-voltage technique. The voltage across the current source is equal in value to the equivalent resistance.]

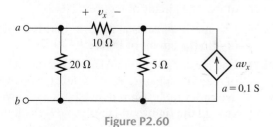

Figure P2.60

**P2.61.** Find the equivalent resistance looking into terminals for the network shown in Figure P2.61. [*Hint:* First, connect a 1-A current source across terminals $a$ and $b$. Then, solve the network by the node-voltage technique. The voltage across the current source is equal in value to the equivalent resistance.]

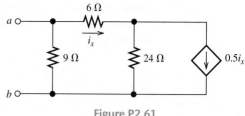

Figure P2.61

**P2.62.** Figure P2.62 shows an unusual voltage-divider circuit. Use node-voltage analysis and the symbolic math commands in MATLAB to solve for the voltage division ratio $V_{out}/V_{in}$ in terms of the resistances. Notice that the node-voltage variables are $V_1$, $V_2$, and $V_{out}$.

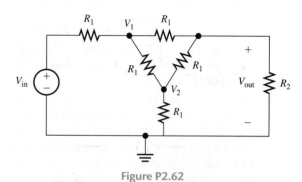

Figure P2.62

**P2.63.** Solve for the node voltages in the circuit of Figure P2.63. Disregard the mesh currents, $i_1, i_2, i_3,$ and $i_4$ when working with the node voltages.

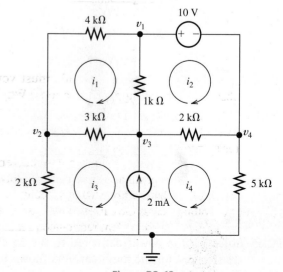

Figure P2.63

**P2.64.** We have a cube with 1-Ω resistances along each edge as illustrated in Figure P2.64 in which we are looking into the front face which has corners at nodes 1, 2, 7, and the reference node. Nodes 3, 4, 5, and 6 are the corners on the rear face of the cube. (Alternatively, you can consider it to be a planar network.) We want to find the resistance between adjacent nodes, such as node 1 and the reference node. We do this by connecting a 1-A current source as shown and solving for $v_1$, which is equal in value to the resistance between any two adjacent nodes. **a.** Use MATLAB to solve the matrix equation $\mathbf{GV} = \mathbf{I}$ for the node voltages and determine the resistance. **b.** Modify your work to determine the resistance between nodes at the ends of a diagonal across a face, such as node 2 and the reference node. **c.** Finally, find the resistance between opposite corners of the cube. [*Comment:* Part (c) is the same as Problem 2.16 in which we suggested using symmetry to solve for the resistance. Parts (a) and (b) can also be solved by use of symmetry and the fact that nodes having the same value of voltage can be connected by short circuits without changing the currents and voltages. With the shorts in place, the resistances can be combined in series and parallel to obtain the answers. Of course, if the resistors have arbitrary values, the MATLAB approach will still work, but considerations of symmetry will not.]

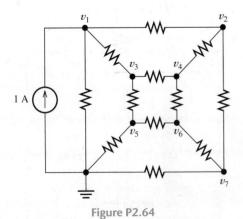

Figure P2.64

**Section 2.5: Mesh-Current Analysis**

**\*P2.65.** Solve for the power delivered to the 15-Ω resistor and for the mesh currents shown in Figure P2.65.

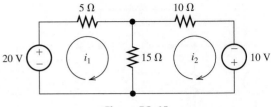

Figure P2.65

**\*P2.66.** Determine the value of $v_2$ and the power delivered by the source in the circuit of Figure P2.24 by using mesh-current analysis.

**\*P2.67.** Use mesh-current analysis to find the value of $i_1$ in the circuit of Figure P2.48.

**P2.68.** Solve for the power delivered by the voltage source in Figure P2.68, using the mesh-current method.

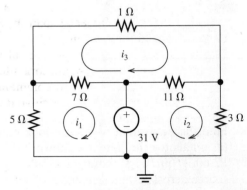

Figure P2.68

**P2.69.** Use mesh-current analysis to find the value of $v$ in the circuit of Figure P2.38.

**P2.70.** Use mesh-current analysis to find the value of $i_3$ in the circuit of Figure P2.39.

**P2.71.** Use mesh-current analysis to find the values of $i_1$ and $i_2$ in Figure P2.27. Select $i_1$ clockwise around the left-hand mesh, $i_2$ clockwise around the right-hand mesh, and $i_3$ clockwise around the center mesh.

**P2.72.** Find the power delivered by the source and the values of $i_1$ and $i_2$ in the circuit of Figure P2.23, using mesh-current analysis.

**P2.73.** Use mesh-current analysis to find the values of $i_1$ and $i_2$ in Figure P2.29. First, select $i_A$ clockwise around the left-hand mesh and $i_B$ clockwise around the right-hand mesh.

After solving for the mesh currents, $i_A$ and $i_B$, determine the values of $i_1$ and $i_2$.

**P2.74.** Use mesh-current analysis to find the values of $i_1$ and $i_2$ in Figure P2.28. First, select $i_A$ clockwise around the left-hand mesh and $i_B$ clockwise around the right-hand mesh. After solving for the mesh currents, $i_A$ and $i_B$, determine the values of $i_1$ and $i_2$.

**P2.75.** The circuit shown in Figure P2.75 is the dc equivalent of a simple residential power distribution system. Each of the resistances labeled $R_1$ and $R_2$ represents various parallel-connected loads, such as lights or devices plugged into outlets that nominally operate at 120 V, while $R_3$ represents a load, such as the heating element in an oven that nominally operates at 240 V. The resistances labeled $R_w$ represent the resistances of wires. $R_n$ represents the "neutral" wire. **a.** Use mesh-current analysis to determine the voltage magnitude for each load. **b.** Now suppose that due to a fault in the wiring at the distribution panel, the neutral wire becomes an open circuit. Again compute the voltages across the loads and comment on the probable outcome for a sensitive device such as a computer or plasma television that is part of the 15-$\Omega$ load

**P2.77.** Connect a 1-V voltage source across terminals $a$ and $b$ of the network shown in Figure P2.55. Then, solve the network by the mesh-current technique to find the current through the source. Finally, divide the source voltage by the current to determine the equivalent resistance looking into terminals $a$ and $b$. The resistance values are $R_1 = 6\ \Omega$, $R_2 = 5\ \Omega$, $R_3 = 4\ \Omega$, $R_4 = 8\ \Omega$, and $R_5 = 2\ \Omega$.

**P2.78.** Connect a 1-V voltage source across the terminals of the network shown in Figure P2.1(a). Then, solve the network by the mesh-current technique to find the current through the source. Finally, divide the source voltage by the current to determine the equivalent resistance looking into the terminals. Check your answer by combining resistances in series and parallel.

**P2.79.** Use MATLAB to solve for the mesh currents in Figure P2.63.

**Section 2.6:** Thévenin and Norton Equivalent Circuits

**\*P2.80.** Find the Thévenin and Norton equivalent circuits for the two-terminal circuit shown in Figure P2.80.

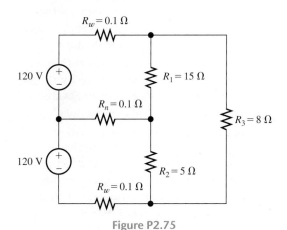

**Figure P2.75**

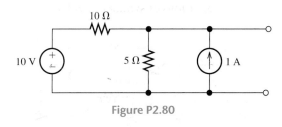

**Figure P2.80**

**P2.76.** Use MATLAB and mesh-current analysis to determine the value of $v_3$ in the circuit of Figure P2.51. The component values are $R_1 = 4\ \Omega$, $R_2 = 5\ \Omega$, $R_3 = 8\ \Omega$, $R_4 = 10\ \Omega$, $R_5 = 2\ \Omega$, and $I_s = 2$ A.

**\*P2.81.** We can model a certain battery as a voltage source in series with a resistance. The open-circuit voltage of the battery is 9 V. When a 100-$\Omega$ resistor is placed across the terminals of the battery, the voltage drops to 6 V. Determine the internal resistance (Thévenin resistance) of the battery.

**P2.82.** Find the Thévenin and Norton equivalent circuits for the circuit shown in Figure P2.82.

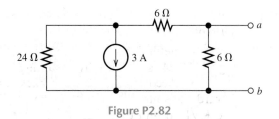

Figure P2.82

**P2.83.** Find the Thévenin and Norton equivalent circuits for the two-terminal circuit shown in Figure P2.83.

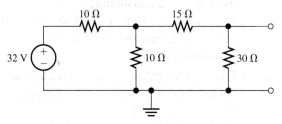

Figure P2.83

**P2.84.** Find the Thévenin and Norton equivalent circuits for the circuit shown in Figure P2.84. Take care that you orient the polarity of the voltage source and the direction of the current source correctly relative to terminals *a* and *b*. What effect does the 7-Ω resistor have on the equivalent circuits? Explain your answer.

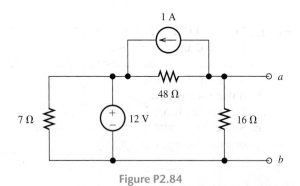

Figure P2.84

**P2.85.** An automotive battery has an open-circuit voltage of 12.6 V and supplies 100 A when a 0.1-Ω resistance is connected across the battery terminals. Draw the Thévenin and Norton equivalent circuits, including values for the circuit parameters. What current can this battery deliver to a short circuit? Considering that the energy stored in the

battery remains constant under open-circuit conditions, which of these equivalent circuits seems more realistic? Explain.

**P2.86.** A certain two-terminal circuit has an open-circuit voltage of 15 V. When a 2-kΩ load is attached, the voltage across the load is 10 V. Determine the Thévenin resistance for the circuit.

**P2.87.** If we measure the voltage at the terminals of a two-terminal network with two known (and different) resistive loads attached, we can determine the Thévenin and Norton equivalent circuits.

When a 2.2-kΩ load is attached to a two-terminal circuit, the load voltage is 4.4 V. When the load is increased to 10 kΩ, the load voltage becomes 5 V. Find the Thévenin voltage and resistance for this circuit.

**P2.88.** Find the Thévenin and Norton equivalent circuits for the circuit shown in Figure P2.88.

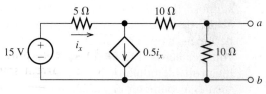

Figure P2.88

**P2.89.** Find the maximum power that can be delivered to a resistive load by the circuit shown in Figure P2.80. For what value of load resistance is the power maximum?

**P2.90.** Find the maximum power that can be delivered to a resistive load by the circuit shown in Figure P2.82. For what value of load resistance is the power maximum?

**\*P2.91.** Figure P2.91 shows a resistive load $R_L$ connected to a Thévenin equivalent circuit. For what value of Thévenin resistance is the power delivered to the load maximized? Find the maximum power delivered to the load. [*Hint:* Be careful; this is a trick question if you don't stop to think about it.]

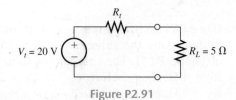

Figure P2.91

**P2.92.** Starting from the Norton equivalent circuit with a resistive load $R_L$ attached, find an expression for the power delivered to the load in terms of $I_n$, $R_t$, and $R_L$. Assuming that $I_n$ and $R_t$ are fixed values and that $R_L$ is variable, show that maximum power is delivered for $R_L = R_t$. Find an expression for maximum power delivered to the load in terms of $I_n$ and $R_t$.

**P2.93.** A battery can be modeled by a voltage source $V_t$ in series with a resistance $R_t$. Assuming that the load resistance is selected to maximize the power delivered, what percentage of the power taken from the voltage source $V_t$ is actually delivered to the load? Suppose that $R_L = 9R_t$; what percentage of the power taken from $V_t$ is delivered to the load? Usually, we want to design battery-operated systems so that nearly all of the energy stored in the battery is delivered to the load. Should we design for maximum power transfer?

## Section 2.7: Superposition Principle

**\*P2.94.** Use superposition to find the current $i$ in Figure P2.94. First, zero the current source and find the value $i_v$ caused by the voltage source alone. Then, zero the voltage source and find the value $i_c$ caused by the current source alone. Finally, add the results algebraically.

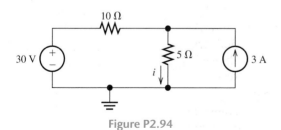

Figure P2.94

**\*P2.95.** Solve for $i_s$ in Figure P2.49 by using superposition.

**P2.96.** Solve the circuit shown in Figure P2.48 by using superposition. First, zero the 1-A source and find the value of $i_1$ with only the 2-A source activated. Then, zero the 2-A source and find the value of $i_1$ with only the 1-A source activated. Finally, find the total value of $i_1$ with both sources activated by algebraically adding the previous results.

**P2.97.** Solve for $i_1$ in Figure P2.34 by using superposition.

**P2.98.** Another method of solving the circuit of Figure P2.24 is to start by assuming that $v_2 = 1$ V. Accordingly, we work backward toward the source, using Ohm's law, KCL, and KVL to find the value of $v_s$. Since we know that $v_2$ is proportional to the value of $v_s$, and since we have found the value of $v_s$ that produces $v_2 = 1$ V, we can calculate the value of $v_2$ that results when $v_s = 12$ V. Solve for $v_2$ by using this method.

**P2.99.** Use the method of Problem P2.98 for the circuit of Figure P2.23, starting with the assumption that $i_2 = 1$ A.

**P2.100.** Solve for the actual value of $i_6$ for the circuit of Figure P2.100, starting with the assumption that $i_6 = 1$ A. Work back through the circuit to find the value of $I_s$ that results in $i_6 = 1$ A. Then, use proportionality to determine the value of $i_6$ that results for $I_s = 10$ A.

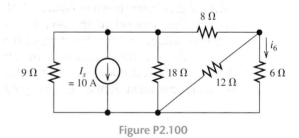

Figure P2.100

**P2.101.** Device $A$ shown in Figure P2.101 has $v = 3i^2$ for $i \geq 0$ and $v = 0$ for $i < 0$.
   **a.** Solve for $v$ with the 2-A source active and the 1-A source zeroed.
   **b.** Solve for $v$ with the 1-A source active and the 2-A source zeroed.
   **c.** Solve for $v$ with both sources active. Why doesn't superposition apply?

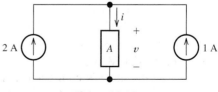

Figure P2.101

## Section 2.8: Wheatstone Bridge

**P2.102.  a.** The Wheatstone bridge shown in Figure 2.66 is balanced with $R_1 = 10$ k$\Omega$, $R_3 = 3419$ $\Omega$, and $R_2 = 1\ k\Omega$. Find $R_x$. **b.** Repeat if $R_2$ is $100\ k\Omega$ and the other values are unchanged.

**\*P2.103.** The Wheatstone bridge shown in Figure 2.66 has $v_s = 10\ V$, $R_1 = 10\ k\Omega$, $R_2 = 10\ k\Omega$, and $R_x = 5932\ \Omega$. The detector can be modeled as a 5-k$\Omega$ resistance. **a.** What value of $R_3$ is required to balance the bridge? **b.** Suppose that $R_3$ is 1 $\Omega$ higher than the value found in part (a). Find the current through the detector. [*Hint:* Find the Thévenin equivalent for the circuit with the detector removed. Then, place the detector across the Thévenin equivalent and solve for the current.] Comment.

**P2.104.** In theory, any values can be used for $R_1$ and $R_3$ in the Wheatstone bridge of Figure 2.66. For the bridge to balance, it is only the *ratio* $R_3/R_1$ that is important. What practical problems might occur if the values are very small? What practical problems might occur if the values are very large?

**P2.105.** Derive expressions for the Thévenin voltage and resistance "seen" by the detector in the Wheatstone bridge in Figure 2.66. (In other words, remove the detector from the circuit and determine the Thévenin resistance for the remaining two-terminal circuit.) What is the value of the Thévenin voltage when the bridge is balanced?

**P2.106.** Derive Equation 2.93 for the bridge circuit of Figure 2.67 on page 109.

**P2.107.** Consider a strain gauge in the form of a long thin wire having a length $L$ and a cross-sectional area $A$ before strain is applied. After the strain is applied, the length increases slightly to $L + \Delta L$ and the area is reduced so the volume occupied by the wire is constant. Assume that $\Delta L/L \ll 1$ and that the resistivity $\rho$ of the wire material is constant. Determine the gauge factor

$$G = \frac{\Delta R/R_0}{\Delta L/L}$$

[*Hint:* Make use of Equation 1.10 on page 28.]

**P2.108.** Explain what would happen if, in wiring the bridge circuit of Figure 2.67 on page 109, the gauges in tension (i.e., those labeled $R + \Delta R$) were both placed on the top of the bridge circuit diagram, shown in part (b) of the figure, and those in compression were both placed at the bottom of the bridge circuit diagram.

---

Here is a practice test you can use to check your comprehension of the most important concepts in this chapter. Answers can be found in Appendix D and complete solutions are included in the Student Solutions files.

See Appendix E for more information about the Student Solutions.

**T2.1.** Match each entry in Table T2.1(a) with the best choice from the list given in Table

### Table T2.1

| Item | Best Match |
|---|---|
| **(a)** | |
| **a.** The equivalent resistance of parallel-connected resistances... | |
| **b.** Resistances in parallel combine as do... | |
| **c.** Loads in power distribution systems are most often connected... | |
| **d.** Solving a circuit by series/parallel combinations applies to... | |
| **e.** The voltage-division principle applies to... | |
| **f.** The current-division principle applies to... | |
| **g.** The superposition principle applies to... | |

| Item | Best Match |
|------|-----------|

**h.** Node-voltage analysis can be applied to...

**i.** In this book, mesh-current analysis is applied to...

**j.** The Thévenin resistance of a two-terminal circuit equals...

**k.** The Norton current source value of a two-terminal circuit equals...

**l.** A voltage source in parallel with a resistance is equivalent to...

**(b)**

1. conductances in parallel
2. in parallel
3. all circuits
4. resistances or conductances in parallel
5. is obtained by summing the resistances
6. is the reciprocal of the sum of the reciprocals of the resistances
7. some circuits
8. planar circuits
9. a current source in series with a resistance
10. conductances in series
11. circuits composed of linear elements
12. in series
13. resistances or conductances in series
14. a voltage source
15. the open-circuit voltage divided by the short-circuit current
16. a current source
17. the short-circuit current

T2.1(b) for circuits composed of sources and resistances. [Items in Table T2.1(b) may be used more than once or not at all.]

**T2.2.** Consider the circuit of Figure T2.2 with $v_s = 96$ V, $R_1 = 6\ \Omega$, $R_2 = 48\ \Omega$, $R_3 = 16\ \Omega$, and $R_4 = 60\ \Omega$. Determine the values of $i_s$ and $i_4$.

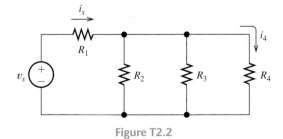

Figure T2.2

**T2.3.** Write MATLAB code to solve for the node voltages for the circuit of Figure T2.3.

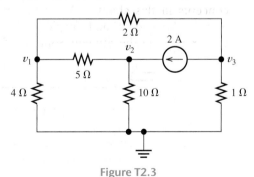

Figure T2.3

**T2.4.** Write a set of equations that can be used to solve for the mesh currents of Figure T2.4. Be sure to indicate which of the equations you write form the set.

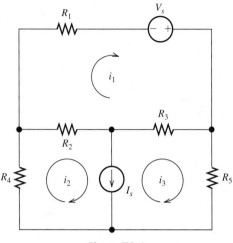

Figure T2.4

**T2.5.** Determine the Thévenin and Norton equivalent circuits for the circuit of Figure T2.5. Draw the equivalent circuits labeling the terminals to correspond with the original circuit.

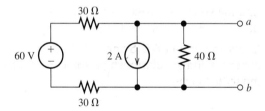

Figure T2.5

**T2.6.** According to the superposition principle, what percentage of the total current flowing through the 5-$\Omega$ resistance in the circuit of Figure T2.6 results from the 5-V source? What percentage of the power supplied to the 5-$\Omega$ resistance is supplied by the 5-V source? Assume that both sources are active when answering both questions.

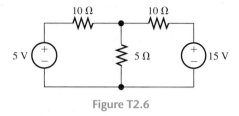

Figure T2.6

**T2.7.** Determine the equivalent resistance between terminals $a$ and $b$ in Figure T2.7.

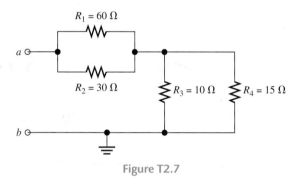

Figure T2.7

**T2.8.** Transform the 2-A current source and 6-$\Omega$ resistance in Figure T2.8 into an equivalent series combination. Then, combine the series voltage sources and resistances. Draw the circuit after each step.

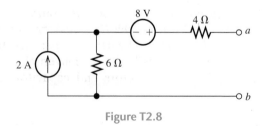

Figure T2.8

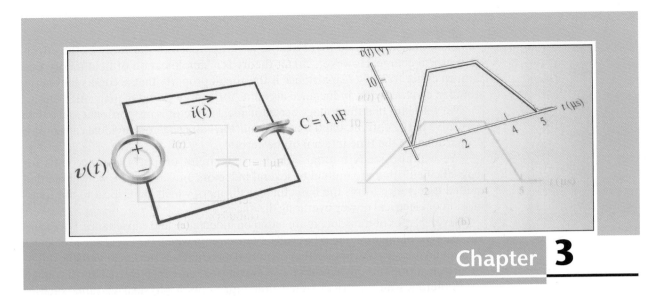

# Chapter 3

# Inductance and Capacitance

## Study of this chapter will enable you to:

- Find the current (voltage) for a capacitance or inductance given the voltage (current) as a function of time.

- Compute the capacitances of parallel-plate capacitors.

- Compute the energies stored in capacitances or inductances.

- Describe typical physical construction of capacitors and inductors and identify parasitic effects.

- Find the voltages across mutually coupled inductances in terms of the currents.

- Apply the MATLAB Symbolic Toolbox to the current–voltage relationships for capacitances and inductances.

## Introduction to this chapter:

Previously, we studied circuits composed of resistances and sources. In this chapter, we discuss two additional circuit elements: inductors and capacitors. Whereas resistors convert electrical energy into heat, inductors and capacitors are **energy-storage elements**. They can store energy and later return it to the circuit. Capacitors and inductors do not generate energy—only the energy that has been put into these elements can be extracted. Thus, like resistors, they are said to be **passive** elements.

Electromagnetic field theory is the basic approach to the study of the effects of electrical charge. However, circuit theory is a simplification of field theory that is much easier to apply. Capacitance is the circuit property that accounts for energy stored in electric fields. Inductance accounts for energy stored in magnetic fields.

We will learn that the voltage across an ideal inductor is proportional to the time derivative of the current. On the other hand, the voltage across an ideal capacitor is proportional to the time integral of the current.

We will also study mutual inductance, a circuit property that accounts for magnetic fields that are mutual to several inductors. In Chapter 14, we will see that mutual inductance forms the basis for transformers, which are critical to the transmission of electrical power over long distances.

Several types of transducers are based on inductance and capacitance. For example, one type of microphone is basically a capacitor in which the capacitance changes with sound pressure. An application of mutual inductance is the linear variable differential transformer in which position of a moving iron core is converted into a voltage.

Sometimes an electrical signal that represents a physical variable such as displacement is noisy. For example, in an active (electronically controlled) suspension for an automobile, the position sensors are affected by road roughness as well as by the loading of the vehicle. To obtain an electrical signal representing the displacement of each wheel, the rapid fluctuations due to road roughness must be eliminated. Later, we will see that this can be accomplished using inductance and capacitance in circuits known as filters.

After studying this chapter, we will be ready to extend the basic circuit-analysis techniques learned in Chapter 2 to circuits having inductance and capacitance.

## 3.1 CAPACITANCE

Capacitors are constructed by separating two conducting plates, which are usually metallic, by a thin layer of insulating material.

Capacitors are constructed by separating two sheets of conductor, which is usually metallic, by a thin layer of insulating material. In a parallel-plate capacitor, the sheets are flat and parallel as shown in Figure 3.1. The insulating material between the plates, called a **dielectric**, can be air, Mylar®, polyester, polypropylene, mica, or a variety of other materials.

Let us consider what happens as current flows through a capacitor. Suppose that current flows downward, as shown in Figure 3.2(a). In most metals, current consists of electrons moving, and conventional current flowing downward represents electrons actually moving upward. As electrons move upward, they collect on the lower plate of the capacitor. Thus, the lower plate accumulates a net negative charge that produces an electric field in the dielectric. This electric field forces electrons to leave the upper plate at the same rate that they accumulate on the lower plate. Therefore, current appears to flow through the capacitor. As the charge builds up, voltage appears across the capacitor.

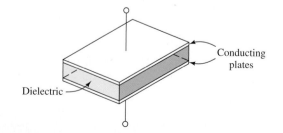

Figure 3.1 A parallel-plate capacitor consists of two conductive plates separated by a dielectric layer.

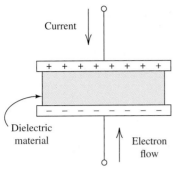
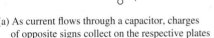
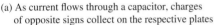

Current

+ + + + + + + +

− − − − − − − −

Dielectric
material

Electron
flow

(a) As current flows through a capacitor, charges
of opposite signs collect on the respective plates

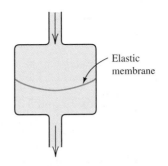

Elastic
membrane

(b) Fluid-flow analogy for capacitance

**Figure 3.2** A capacitor and its fluid-flow analogy.

We say that the charge accumulated on one plate is stored in the capacitor. However, the total charge on both plates is always zero, because positive charge on one plate is balanced by negative charge of equal magnitude on the other plate.

Positive charge on one plate is balanced by negative charge of equal magnitude on the other plate.

## Fluid-Flow Analogy

In terms of the fluid-flow analogy, a capacitor represents a reservoir with an elastic membrane separating the inlet and outlet as shown in Figure 3.2(b). As the fluid flows into the inlet, the membrane is stretched, creating a force (analogous to capacitor voltage) that opposes further flow. The displaced fluid volume starting from the unstretched membrane position is analogous to the charge stored on one plate of the capacitor.

In terms of the fluid-flow analogy, a capacitor represents a reservoir with an elastic membrane separating the inlet and outlet.

## Stored Charge in Terms of Voltage

In an ideal capacitor, the stored charge $q$ is proportional to the voltage between the plates:

In an ideal capacitor, the stored charge $q$ is proportional to the voltage between the plates.

$$q = Cv \tag{3.1}$$

The constant of proportionality is the capacitance $C$, which has units of farads (F). Farads are equivalent to coulombs per volt.

To be more precise, the charge $q$ is the net charge on the plate corresponding to the positive reference for $v$. Thus, if $v$ is positive, there is positive charge on the plate corresponding to the positive reference for $v$. On the other hand, if $v$ is negative, there is negative charge on the plate corresponding to the positive reference.

A farad is a very large amount of capacitance. In most applications, we deal with capacitances in the range from a few picofarads (1 pF $= 10^{-12}$ F) up to perhaps 0.01 F. Capacitances in the femtofarad (1 fF $= 10^{-15}$ F) range are responsible for limiting the performance of computer chips.

In most applications, we deal with capacitances in the range from a few picofarads up to perhaps 0.01 F.

## Current in Terms of Voltage

Recall that current is the time rate of flow of charge. Taking the derivative of each side of Equation 3.1 with respect to time, we have

$$i = \frac{dq}{dt} = \frac{d}{dt}(Cv) \tag{3.2}$$

Ordinarily, capacitance is not a function of time. (An exception is the capacitor microphone mentioned earlier.) Thus, the relationship between current and voltage becomes

$$i = C\frac{dv}{dt} \tag{3.3}$$

Equations 3.1 and 3.3 show that as voltage increases, current flows through the capacitance and charge accumulates on each plate. If the voltage remains constant, the charge is constant and the current is zero. Thus, a capacitor appears to be an open circuit for a steady dc voltage.

**Capacitors act as open circuits for steady dc voltages.**

The circuit symbol for capacitance and the references for $v$ and $i$ are shown in Figure 3.3. Notice that the references for the voltage and current have the passive configuration. In other words, the current reference direction points into the positive reference polarity. If the references were opposite to the passive configuration, Equation 3.3 would have a minus sign:

$$i = -C\frac{dv}{dt} \tag{3.4}$$

**Figure 3.3** The circuit symbol for capacitance, including references for the current $i(t)$ and voltage $v(t)$.

Sometimes, we emphasize the fact that in general the voltage and current are functions of time by denoting them as $v(t)$ and $i(t)$.

### Example 3.1    Determining Current for a Capacitance Given Voltage

Suppose that the voltage $v(t)$ shown in Figure 3.4(b) is applied to a 1-$\mu$F capacitance. Plot the stored charge and the current through the capacitance versus time.

**Solution**    The charge stored on the top plate of the capacitor is given by Equation 3.1. [We know that $q(t)$ represents the charge on the top plate because that is the plate corresponding to the positive reference for $v(t)$.] Thus,

$$q(t) = Cv(t) = 10^{-6}v(t)$$

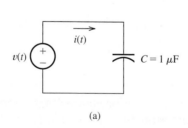

(a)

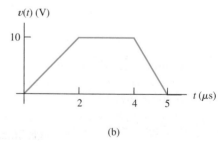

(b)

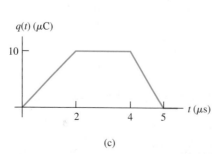

(c)

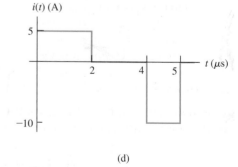

(d)

**Figure 3.4** Circuit and waveforms for Example 3.1.

This is shown in Figure 3.4(c).

The current flowing through the capacitor is given by Equation 3.3:

$$i(t) = C\frac{dv(t)}{dt} = 10^{-6}\frac{dv(t)}{dt}$$

Of course, the derivative of the voltage is the slope of the voltage versus time plot. Hence, for $t$ between 0 and 2 $\mu$s, we have

$$\frac{dv(t)}{dt} = \frac{10\text{ V}}{2 \times 10^{-6}\text{ s}} = 5 \times 10^6 \text{ V/s}$$

and

$$i(t) = C\frac{dv(t)}{dt} = 10^{-6} \times 5 \times 10^6 = 5\text{ A}$$

Between $t = 2$ and 4 $\mu$s, the voltage is constant ($dv/dt = 0$) and the current is zero. Finally, between $t = 4$ and 5 $\mu$s, we get

$$\frac{dv(t)}{dt} = \frac{-10\text{ V}}{10^{-6}\text{ s}} = -10^7 \text{ V/s}$$

and

$$i(t) = C\frac{dv(t)}{dt} = 10^{-6} \times (-10^7) = -10\text{ A}$$

A plot of $i(t)$ is shown in Figure 3.4(d).

Notice that as the voltage increases, current flows through the capacitor and charges accumulate on the plates. For constant voltage, the current is zero and the charge is constant. When the voltage decreases, the direction of the current reverses, and the stored charge is removed from the capacitor. ∎

**Exercise 3.1**   The charge on a 2-$\mu$F capacitor is given by

$$q(t) = 10^{-6} \sin(10^5 t) \text{ C}$$

Find expressions for the voltage and for the current. (The angle is in radians.)
**Answer** $v(t) = 0.5 \sin(10^5 t)$ V, $i(t) = 0.1 \cos(10^5 t)$ A.   ☐

## Voltage in Terms of Current

Suppose that we know the current $i(t)$ flowing through a capacitance $C$ and we want to compute the charge and voltage. Since current is the time rate of charge flow, we must integrate the current to compute charge. Often in circuit analysis problems, action starts at some initial time $t_0$, and the initial charge $q(t_0)$ is known. Then, charge as a function of time is given by

$$q(t) = \int_{t_0}^{t} i(t)\, dt + q(t_0) \tag{3.5}$$

Setting the right-hand sides of Equations 3.1 and 3.5 equal to each other and solving for the voltage $v(t)$, we have

$$v(t) = \frac{1}{C}\int_{t_0}^{t} i(t)\, dt + \frac{q(t_0)}{C} \tag{3.6}$$

However, the initial voltage across the capacitance is given by

$$v(t_0) = \frac{q(t_0)}{C} \tag{3.7}$$

Substituting this into Equation 3.6, we have

$$v(t) = \frac{1}{C} \int_{t_0}^{t} i(t) \, dt + v(t_0) \tag{3.8}$$

Usually, we take the initial time to be $t_0 = 0$.

---

**Example 3.2**    **Determining Voltage for a Capacitance Given Current**

After $t_0 = 0$, the current in a 0.1-$\mu$F capacitor is given by

$$i(t) = 0.5 \sin(10^4 t) \text{ A}$$

(The argument of the sin function is in radians.) The initial charge on the capacitor is $q(0) = 0$. Plot $i(t)$, $q(t)$, and $v(t)$ to scale versus time.

**Solution**    First, we use Equation 3.5 to find an expression for the charge:

$$
\begin{aligned}
q(t) &= \int_0^t i(t) \, dt + q(0) \\
&= \int_0^t 0.5 \sin(10^4 t) \, dt \\
&= -0.5 \times 10^{-4} \cos(10^4 t)\big|_0^t \\
&= 0.5 \times 10^{-4}[1 - \cos(10^4 t)] \text{ C}
\end{aligned}
$$

Solving Equation 3.1 for voltage, we have

$$
\begin{aligned}
v(t) &= \frac{q(t)}{C} = \frac{q(t)}{10^{-7}} \\
&= 500[1 - \cos(10^4 t)] \text{ V}
\end{aligned}
$$

Plots of $i(t)$, $q(t)$, and $v(t)$ are shown in Figure 3.5. Immediately after $t = 0$, the current is positive and $q(t)$ increases. After the first half-cycle, $i(t)$ becomes negative and $q(t)$ decreases. At the completion of one cycle, the charge and voltage have returned to zero. ∎

## Stored Energy

The power delivered to a circuit element is the product of the current and the voltage (provided that the references have the passive configuration):

$$p(t) = v(t)i(t) \tag{3.9}$$

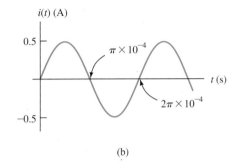

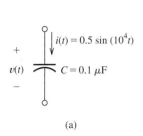

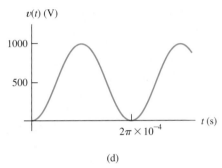

(a)

(b)

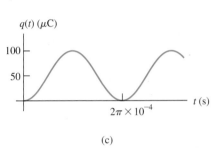

(c)

(d)

**Figure 3.5** Waveforms for Example 3.2.

Using Equation 3.3 to substitute for the current, we have

$$p(t) = Cv\frac{dv}{dt} \tag{3.10}$$

Suppose we have a capacitor that initially has $v(t_0) = 0$. Then the initial stored electrical energy is zero, and we say that the capacitor is uncharged. Furthermore, suppose that between time $t_0$ and some later time $t$ the voltage changes from 0 to $v(t)$ volts. As the voltage magnitude increases, energy is delivered to the capacitor, where it is stored in the electric field between the plates.

If we integrate the power delivered from $t_0$ to $t$, we find the energy delivered:

$$w(t) = \int_{t_0}^{t} p(t)\, dt \tag{3.11}$$

Using Equation 3.10 to substitute for power, we find that

$$w(t) = \int_{t_0}^{t} Cv\frac{dv}{dt}\, dt \tag{3.12}$$

Canceling differential time and changing the limits to the corresponding voltages, we have

$$w(t) = \int_{0}^{v(t)} Cv\, dv \tag{3.13}$$

Integrating and evaluating, we get

$$w(t) = \frac{1}{2}Cv^2(t) \tag{3.14}$$

This represents energy stored in the capacitance that can be returned to the circuit.

Solving Equation 3.1 for $v(t)$ and substituting into Equation 3.14, we can obtain two alternative expressions for the stored energy:

$$w(t) = \frac{1}{2}v(t)q(t) \tag{3.15}$$

$$w(t) = \frac{q^2(t)}{2C} \tag{3.16}$$

**Example 3.3    Current, Power, and Energy for a Capacitance**

Suppose that the voltage waveform shown in Figure 3.6(a) is applied to a 10-$\mu$F capacitance. Find and plot the current, the power delivered, and the energy stored for time between 0 and 5 s.

**Solution**    First, we write expressions for the voltage as a function of time:

$$v(t) = \begin{cases} 1000t \text{ V} & \text{for } 0 < t < 1 \\ 1000 \text{ V} & \text{for } 1 < t < 3 \\ 500(5 - t) \text{ V} & \text{for } 3 < t < 5 \end{cases}$$

Using Equation 3.3, we obtain expressions for the current:

$$i(t) = C\frac{dv(t)}{dt}$$

$$i(t) = \begin{cases} 10 \times 10^{-3} \text{ A} & \text{for } 0 < t < 1 \\ 0 \text{ A} & \text{for } 1 < t < 3 \\ -5 \times 10^{-3} \text{ A} & \text{for } 3 < t < 5 \end{cases}$$

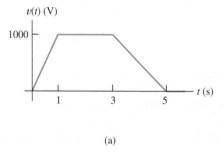

(a)

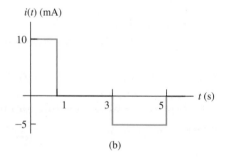

(b)

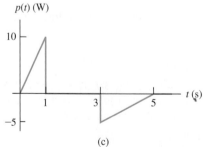

(c)

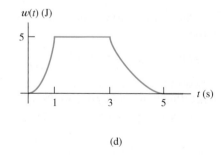

(d)

**Figure 3.6** Waveforms for Example 3.3.

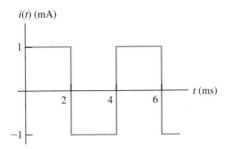

$i(t)$ (mA)

$t$ (ms)

Figure 3.7  Square-wave current for Exercise 3.2.

The plot of $i(t)$ is shown in Figure 3.6(b).

Next, we find expressions for power by multiplying the voltage by the current:

$$p(t) = v(t)i(t)$$

$$p(t) = \begin{cases} 10t \text{ W} & \text{for } 0 < t < 1 \\ 0 \text{ W} & \text{for } 1 < t < 3 \\ 2.5(t-5) \text{ W} & \text{for } 3 < t < 5 \end{cases}$$

The plot of $p(t)$ is shown in Figure 3.6(c). Notice that between $t = 0$ and $t = 1$ power is positive, showing that energy is being delivered to the capacitance. Between $t = 3$ and $t = 5$, energy flows out of the capacitance back into the rest of the circuit.

Next, we use Equation 3.14 to find expressions for the stored energy:

$$w(t) = \frac{1}{2}Cv^2(t)$$

$$w(t) = \begin{cases} 5t^2 \text{ J} & \text{for } 0 < t < 1 \\ 5 \text{ J} & \text{for } 1 < t < 3 \\ 1.25(5-t)^2 \text{ J} & \text{for } 3 < t < 5 \end{cases}$$

The plot of $w(t)$ is shown in Figure 3.6(d).  ■

**Exercise 3.2**  The current through a 0.1-$\mu$F capacitor is shown in Figure 3.7. At $t_0 = 0$, the voltage across the capacitor is zero. Find the charge, voltage, power, and stored energy as functions of time and plot them to scale versus time.
**Answer**  The plots are shown in Figure 3.8.  ☐

## 3.2  CAPACITANCES IN SERIES AND PARALLEL

### Capacitances in Parallel

Suppose that we have three capacitances in parallel as shown in Figure 3.9. Of course, the same voltage appears across each of the elements in a parallel circuit. The currents are related to the voltage by Equation 3.3. Thus, we can write

$$i_1 = C_1\frac{dv}{dt} \qquad (3.17)$$

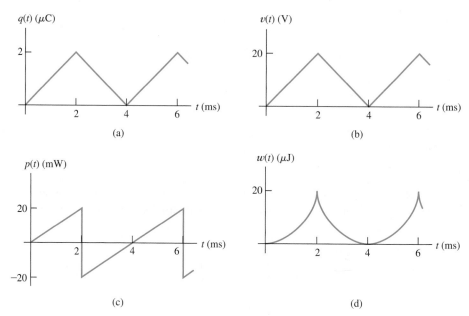

(a)

(b)

(c)

(d)

Figure 3.8 Answers for Exercise 3.2.

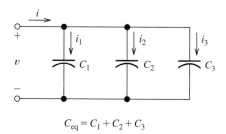

Figure 3.9 Three capacitances in parallel.

$$C_{eq} = C_1 + C_2 + C_3$$

$$i_2 = C_2 \frac{dv}{dt} \tag{3.18}$$

$$i_3 = C_3 \frac{dv}{dt} \tag{3.19}$$

Applying KCL at the top node of the circuit, we have

$$i = i_1 + i_2 + i_3 \tag{3.20}$$

Using Equations 3.17, 3.18, and 3.19 to substitute into Equation 3.20, we obtain

$$i = C_1 \frac{dv}{dt} + C_2 \frac{dv}{dt} + C_3 \frac{dv}{dt} \tag{3.21}$$

This can be written as

$$i = (C_1 + C_2 + C_3) \frac{dv}{dt} \tag{3.22}$$

Now, we define the equivalent capacitance as the sum of the capacitances in parallel:

$$C_{eq} = C_1 + C_2 + C_3 \tag{3.23}$$

We add parallel capacitances to find the equivalent capacitance.

Using this definition in Equation 3.22, we find that

$$i = C_{eq} \frac{dv}{dt} \tag{3.24}$$

Thus, the current in the equivalent capacitance is the same as the total current flowing through the parallel circuit.

In sum, we add parallel capacitances to find the equivalent capacitance. Recall that for resistances, the resistances are added if they are in *series* rather than parallel. Thus, we say that capacitances in parallel are combined like resistances in series.

Capacitances in parallel are combined like resistances in series.

## Capacitances in Series

By a similar development, it can be shown that the equivalent capacitance for three series capacitances is

$$C_{eq} = \frac{1}{1/C_1 + 1/C_2 + 1/C_3} \tag{3.25}$$

Capacitances in series are combined like resistances in parallel.

We conclude that capacitances in series are combined like resistances in parallel.

A technique for obtaining high voltages from low-voltage sources is to charge $n$ capacitors in parallel with the source, and then to switch them to a series combination. The resulting voltage across the series combination is $n$ times the source voltage. For example, in some cardiac pacemakers, a 2.5-V battery is used, but 5 V need to be applied to the heart muscle to initiate a beat. This is accomplished by charging two capacitors from the 2.5-V battery. The capacitors are then connected in series to deliver a brief 5-V pulse to the heart.

---

**Example 3.4    Capacitances in Series and Parallel**

Determine the equivalent capacitance between terminals $a$ and $b$ in Figure 3.10(a).

**Solution**    First, notice that the 12-$\mu$F and 24-$\mu$F capacitances are in series. Thus, their equivalent capacitance is:

$$\frac{1}{1/12 + 1/24} = 8 \; \mu\text{F}$$

The resulting equivalent is shown in Figure 3.10(b).

Then, the 8-$\mu$F and 4-$\mu$F capacitances are in parallel. Their equivalent is 12-$\mu$F as shown in Figure 3.10(c).

Finally we combine the 6-$\mu$F and 12-$\mu$F capacitances in series resulting in 4-$\mu$F as shown in Figure 3.10(d). ∎

---

**Exercise 3.3**    Derive Equation 3.25 for the three capacitances shown in Figure 3.11.

**Exercise 3.4    a.** Two capacitances of 2 $\mu$F and 1 $\mu$F are in series. Find the equivalent capacitance. **b.** Repeat if the capacitances are in parallel.
**Answer    a.** 2/3 $\mu$F; **b.** 3 $\mu$F. ☐

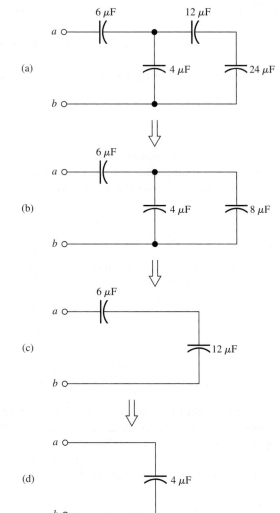

Figure 3.10  Circuit of Example 3.4.

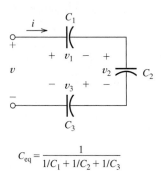

$$C_{eq} = \frac{1}{1/C_1 + 1/C_2 + 1/C_3}$$

**Figure 3.11**  Three capacitances in series.

## 3.3  PHYSICAL CHARACTERISTICS OF CAPACITORS

### Capacitance of the Parallel-Plate Capacitor

A parallel-plate capacitor is shown in Figure 3.12, including dimensions. The area of each plate is denoted as $A$. (Actually, $A$ is the area of one side of the plate.) The rectangular plate shown has a width $W$, length $L$, and area $A = W \times L$. The plates are parallel, and the distance between them is denoted as $d$.

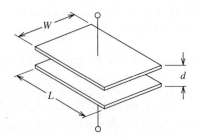

**Figure 3.12**  A parallel-plate capacitor, including dimensions.

**Table 3.1** Relative Dielectric Constants for Selected Materials

| | |
|---|---|
| Air | 1.0 |
| Diamond | 5.5 |
| Mica | 7.0 |
| Polyester | 3.4 |
| Quartz | 4.3 |
| Silicon dioxide | 3.9 |
| Water | 78.5 |

If the distance $d$ between the plates is much smaller than both the width and the length of the plates, the capacitance is approximately given by

$$C = \frac{\epsilon A}{d} \qquad (3.26)$$

in which $\epsilon$ is the **dielectric constant** of the material between the plates. For vacuum, the dielectric constant is

$$\epsilon = \epsilon_0 \cong 8.85 \times 10^{-12} \text{ F/m}$$

Dielectric constant of vacuum.

For other materials, the dielectric constant is

$$\epsilon = \epsilon_r \epsilon_0 \qquad (3.27)$$

where $\epsilon_r$ is the **relative dielectric constant** which has no physical units. Values of the relative dielectric constant for selected materials are given in Table 3.1.

---

**Example 3.5**   **Calculating Capacitance Given Physical Parameters**

Compute the capacitance of a parallel-plate capacitor having rectangular plates 10 cm by 20 cm separated by a distance of 0.1 mm. The dielectric is air. Repeat if the dielectric is mica.

**Solution**   First, we compute the area of a plate:

$$A = L \times W = (10 \times 10^{-2}) \times (20 \times 10^{-2}) = 0.02 \text{ m}^2$$

From Table 3.1, we see that the relative dielectric constant of air is 1.00. Thus, the dielectric constant is

$$\epsilon = \epsilon_r \epsilon_0 = 1.00 \times 8.85 \times 10^{-12} \text{ F/m}$$

Then, the capacitance is

$$C = \frac{\epsilon A}{d} = \frac{8.85 \times 10^{-12} \times 0.02}{10^{-4}} = 1770 \times 10^{-12} \text{ F}$$

For a mica dielectric, the relative dielectric constant is 7.0. Thus, the capacitance is seven times larger than for air or vacuum:

$$C = 12{,}390 \times 10^{-12} \text{ F}$$

■

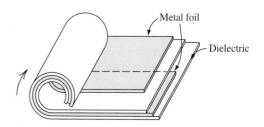

**Figure 3.13** Practical capacitors can be constructed by interleaving the plates with two dielectric layers and rolling them up. By staggering the plates, connection can be made to one plate at each end of the roll.

**Exercise 3.5** We want to design a 1-$\mu$F capacitor. Compute the length required for rectangular plates of 2-cm width if the dielectric is polyester of 15-$\mu$m thickness.
**Answer** $L = 24.93$ m. □

## Practical Capacitors

To achieve capacitances on the order of a microfarad, the dimensions of parallel-plate capacitors are too large for compact electronic circuits such as portable computers or cellular telephones. Frequently, capacitors are constructed by alternating the plates with two layers of dielectric, which are then rolled to fit in a smaller area. By staggering the plates before rolling, electrical contact can be made with the plates from the ends of the roll. This type of construction is illustrated in Figure 3.13.

To achieve small-volume capacitors, a very thin dielectric having a high dielectric constant is desirable. However, dielectric materials break down and become conductors when the electric field intensity (volts per meter) is too high. Thus, real capacitors have maximum voltage ratings. For a given voltage, the electric field intensity becomes higher as the dielectric layer becomes thinner. Clearly, an engineering trade-off exists between compact size and voltage rating.

*Real capacitors have maximum voltage ratings.*

*An engineering trade-off exists between compact size and high voltage rating.*

## Electrolytic Capacitors

In **electrolytic capacitors**, one of the plates is metallic aluminum or tantalum, the dielectric is an oxide layer on the surface of the metal, and the other "plate" is an electrolytic solution. The oxide-coated metallic plate is immersed in the electrolytic solution.

*Only voltages of the proper polarity should be applied to electrolytic capacitors.*

This type of construction results in high capacitance per unit volume. However, only one polarity of voltage should be applied to electrolytic capacitors. For the opposite polarity, the dielectric layer is chemically attacked, and a conductive path appears between the plates. (Usually, the allowed polarity is marked on the outer case.) On the other hand, capacitors constructed with polyethylene, Mylar®, and so on can be used in applications where the voltage polarity reverses. When the application results in voltages of only one polarity and a large-value capacitance is required, designers frequently use electrolytic capacitors.

## Parasitic Effects

Real capacitors are not always well modeled simply as a capacitance. A more complete circuit model for a capacitor is shown in Figure 3.14. In addition to the capacitance $C$, series resistance $R_s$ appears because of the resistivity of the material composing the plates. A series inductance $L_s$ (we discuss inductance later in this chapter) occurs because the current flowing through the capacitor creates a magnetic field. Finally, no practical material is a perfect insulator, and the resistance $R_p$ represents conduction through the dielectric.

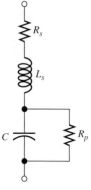

**Figure 3.14** The circuit model for a capacitor, including the parasitic elements $R_s$, $L_s$, and $R_p$.

We call $R_s$, $L_s$, and $R_p$ **parasitic elements**. We design capacitors to minimize the effects of parasitic circuit elements consistent with other requirements such as physical size and voltage rating. However, parasitics are always present to some degree. In designing circuits, care must be used to select components for which the parasitic effects do not prevent proper operation of the circuit.

---

**Example 3.6**   **What Happened to the Missing Energy?**

Consider the situation shown in Figure 3.15. Prior to $t = 0$, the capacitor $C_1$ is charged to a voltage of $v_1 = 100$ V and the other capacitor has no charge (i.e., $v_2 = 0$). At $t = 0$, the switch closes. Compute the total energy stored by both capacitors before and after the switch closes.

**Solution**   The initial stored energy for each capacitor is

$$w_1 = \frac{1}{2}C_1 v_1^2 = \frac{1}{2}(10^{-6})(100)^2 = 5 \text{ mJ}$$

$$w_2 = 0$$

and the total energy is

$$w_{\text{total}} = w_1 + w_2 = 5 \text{ mJ}$$

To find the voltage and stored energy after the switch closes, we make use of the fact that the total charge on the top plates cannot change when the switch closes. This is true because there is no path for electrons to enter or leave the upper part of the circuit.

The charge stored on the top plate of $C_1$ prior to $t = 0$ is given by

$$q_1 = C_1 v_1 = 1 \times 10^{-6} \times 100 = 100 \ \mu\text{C}$$

Furthermore, the initial charge on $C_2$ is zero:

$$q_2 = 0$$

Thus, after the switch closes, the charge on the equivalent capacitance is

$$q_{\text{eq}} = q_1 + q_2 = 100 \ \mu\text{C}$$

Also, notice that after the switch is closed, the capacitors are in parallel and have an equivalent capacitance of

$$C_{\text{eq}} = C_1 + C_2 = 2 \ \mu\text{F}$$

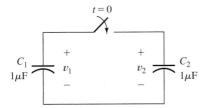

**Figure 3.15** See Example 3.6.

The voltage across the equivalent capacitance is

$$v_{eq} = \frac{q_{eq}}{C_{eq}} = \frac{100\ \mu C}{2\ \mu F} = 50\ V$$

Of course, after the switch is closed, $v_1 = v_2 = v_{eq}$.

Now, we compute the stored energy with the switch closed:

$$w_1 = \frac{1}{2} C_1 v_{eq}^2 = \frac{1}{2}(10^{-6})(50)^2 = 1.25\ mJ$$

$$w_2 = \frac{1}{2} C_2 v_{eq}^2 = \frac{1}{2}(10^{-6})(50)^2 = 1.25\ mJ$$

The total stored energy with the switch closed is

$$w_{total} = w_1 + w_2 = 2.5\ mJ$$

Thus, we see that the stored energy after the switch is closed is half of the value before the switch is closed. What happened to the missing energy?

Usually, the answer to this question is that it is absorbed in the parasitic resistances. It is impossible to construct capacitors that do not have some parasitic effects. Even if we use superconductors for the wires and capacitor plates, there would be parasitic inductance. If we included the parasitic inductance in the circuit model, we would not have missing energy. (We study circuits with time-varying voltages and currents in Chapter 4.)

To put it another way, a physical circuit that is modeled exactly by Figure 3.15 does not exist. Invariably, if we use a realistic model for an actual circuit, we can account for all of the energy.  ■

**Usually, the missing energy is absorbed in the parasitic resistances.**

**A physical circuit that is modeled exactly by Figure 3.15 does not exist.**

## 3.4  INDUCTANCE

**Inductors are usually constructed by coiling wire around a form.**

An inductor is usually constructed by coiling a wire around some type of form. Several examples of practical construction are illustrated in Figure 3.16. Current flowing through the coil creates a magnetic field or flux that links the coil. Frequently, the coil form is composed of a magnetic material such as iron or iron oxides that increases the magnetic flux for a given current. (Iron cores are often composed of

(a) Toroidal inductor

(b) Coil with an iron-oxide slug that can be screwed in or out to adjust the inductance

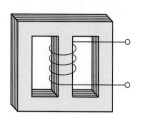

(c) Inductor with a laminated iron core

**Figure 3.16** An inductor is constructed by coiling a wire around some type of form.

thin sheets called **laminations**. We discuss the reason for this construction technique in Chapter 14.)

When the current changes in value, the resulting magnetic flux changes. According to Faraday's law of electromagnetic induction, time-varying magnetic flux linking a coil induces voltage across the coil. For an ideal inductor, the voltage is proportional to the time rate of change of the current. Furthermore, the polarity of the voltage is such as to oppose the change in current. The constant of proportionality is called inductance, usually denoted by the letter $L$.

The circuit symbol for inductance is shown in Figure 3.17. In equation form, the voltage and current are related by

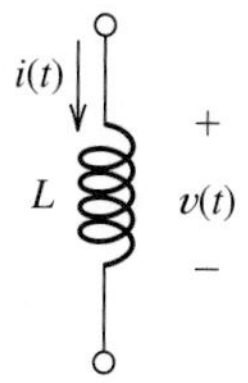

$$v(t) = L\frac{di}{dt}$$

**Figure 3.17** Circuit symbol and the $v - i$ relationship for inductance.

$$v(t) = L\frac{di}{dt} \tag{3.28}$$

As usual, we have assumed the passive reference configuration. In case the references are opposite to the passive configuration, Equation 3.28 becomes

$$v(t) = -L\frac{di}{dt} \tag{3.29}$$

Inductance has units of henries (H), which are equivalent to volt seconds per ampere. Typically, we deal with inductances ranging from a fraction of a microhenry ($\mu$H) to several tens of henries.

Inductance has units of henries (H), which are equivalent to volt seconds per ampere.

## Fluid-Flow Analogy

The fluid-flow analogy for inductance is the inertia of the fluid flowing through a *frictionless* pipe of constant diameter. The pressure differential between the ends of the pipe is analogous to voltage, and the flow rate or velocity is analogous to current. Thus, the acceleration of the fluid is analogous to rate of change of current. A pressure differential exists between the ends of the pipe only when the flow rate is increasing or decreasing.

The fluid-flow analogy for inductance is the inertia of the fluid flowing through a frictionless pipe of constant diameter.

One place where the inertia of flowing fluid is encountered is when a valve (typically operated by an electrical solenoid) closes suddenly, cutting off the flow. For example, in a washing machine, the sudden change in velocity of the water flow can cause high pressure, resulting in a bang and vibration of the plumbing. This is similar to electrical effects that occur when current in an inductor is suddenly interrupted. An application for the high voltage that appears when current is suddenly interrupted is in the ignition system for a gasoline-powered internal combustion engine.

## Current in Terms of Voltage

Suppose that we know the initial current $i(t_0)$ and the voltage $v(t)$ across an inductance. Furthermore, suppose that we need to compute the current for $t > t_0$. Rearranging Equation 3.28, we have

$$di = \frac{1}{L}v(t)\,dt \tag{3.30}$$

Integrating both sides, we find that

$$\int_{i(t_0)}^{i(t)} di = \frac{1}{L}\int_{t_0}^{t} v(t)\,dt \tag{3.31}$$

Notice that the integral on the right-hand side of Equation 3.31 is with respect to time. Furthermore, the limits are the initial time $t_0$ and the time variable $t$. The integral on the left-hand side is with respect to current with limits that correspond to the time limits on the right-hand side. Integrating, evaluating, and rearranging, we have

$$i(t) = \frac{1}{L} \int_{t_0}^{t} v(t)\, dt + i(t_0) \tag{3.32}$$

Notice that as long as $v(t)$ is finite, $i(t)$ can change only by an incremental amount in a time increment. Thus, $i(t)$ must be continuous with no instantaneous jumps in value (i.e., discontinuities). (Later, we encounter idealized circuits in which infinite voltages appear briefly, and then the current in an inductance can change instantaneously.)

### Stored Energy

Assuming that the references have the passive configuration, we compute the power delivered to a circuit element by taking the product of the current and the voltage:

$$p(t) = v(t)i(t) \tag{3.33}$$

Using Equation 3.28 to substitute for the voltage, we obtain

$$p(t) = Li(t)\frac{di}{dt} \tag{3.34}$$

Consider an inductor having an initial current $i(t_0) = 0$. Then, the initial electrical energy stored is zero. Furthermore, assume that between time $t_0$ and some later time $t$, the current changes from 0 to $i(t)$. As the current magnitude increases, energy is delivered to the inductor, where it is stored in the magnetic field.

Integrating the power from $t_0$ to $t$, we find the energy delivered:

$$w(t) = \int_{t_0}^{t} p(t)\, dt \tag{3.35}$$

Using Equation 3.34 to substitute for power, we have

$$w(t) = \int_{t_0}^{t} Li\frac{di}{dt}\, dt \tag{3.36}$$

Canceling differential time and changing the limits to the corresponding currents, we get

$$w(t) = \int_{0}^{i(t)} Li\, di \tag{3.37}$$

Integrating and evaluating, we obtain

$$w(t) = \frac{1}{2}Li^2(t) \tag{3.38}$$

This represents energy stored in the inductance that is returned to the circuit if the current changes back to zero.

### Example 3.7    Voltage, Power, and Energy for an Inductance

The current through a 5-H inductance is shown in Figure 3.18(a). Plot the voltage, power, and stored energy to scale versus time for $t$ between 0 and 5 s.

**Solution**    We use Equation 3.28 to compute voltages:

$$v(t) = L\frac{di}{dt}$$

The time derivative of the current is the slope (rise over run) of the current versus time plot. For $t$ between 0 and 2 s, we have $di/dt = 1.5$ A/s and thus, $v = 7.5$ V. For $t$ between 2 and 4 s, $di/dt = 0$, and therefore, $v = 0$. Finally, between 4 and 5 s, $di/dt = -3$ A/s and $v = -15$ V. A plot of the voltage versus time is shown in Figure 3.18(b).

Next, we obtain power by taking the product of current and voltage at each point in time. The resulting plot is shown in Figure 3.18(c).

Finally, we use Equation 3.38 to compute the stored energy as a function of time:

$$w(t) = \frac{1}{2}Li^2(t)$$

The resulting plot is shown in Figure 3.18(d).

Notice in Figure 3.18 that as current magnitude increases, power is positive and stored energy accumulates. When the current is constant, the voltage is zero, the

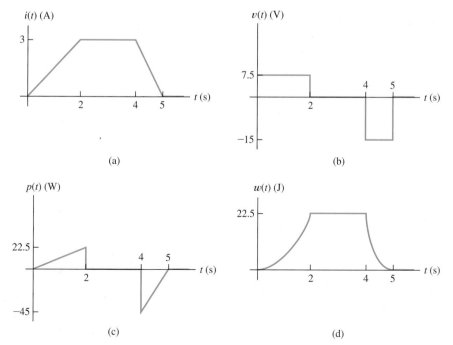

Figure 3.18 Waveforms for Example 3.7.

power is zero, and the stored energy is constant. When the current magnitude falls toward zero, the power is negative, showing that energy is being returned to the other parts of the circuit.    ■

### Example 3.8    Inductor Current with Constant Applied Voltage

Consider the circuit shown in Figure 3.19(a). In this circuit, we have a switch that closes at $t = 0$, connecting a 10-V source to a 2-H inductance. Find the current as a function of time.

**Solution**    Notice that because the voltage applied to the inductance is finite, the current must be continuous. Prior to $t = 0$, the current must be zero. (Current cannot flow through an open switch.) Thus, the current must also be zero immediately after $t = 0$.

The voltage across the inductance is shown in Figure 3.19(b). To find the current, we employ Equation 3.32:

$$i(t) = \frac{1}{L} \int_{t_0}^{t} v(t)\, dt + i(t_0)$$

In this case, we take $t_0 = 0$, and we have $i(t_0) = i(0) = 0$. Substituting values, we get

$$i(t) = \frac{1}{2} \int_{0}^{t} 10\, dt$$

where we have assumed that $t$ is greater than zero. Integrating and evaluating, we obtain

$$i(t) = 5t \text{ A} \qquad \text{for } t > 0$$

A plot of the current is shown in Figure 3.19(c).

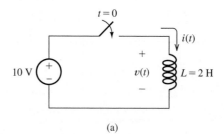

(a)

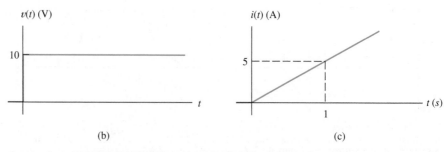

(b)    (c)

Figure 3.19 Circuit and waveforms for Example 3.8.

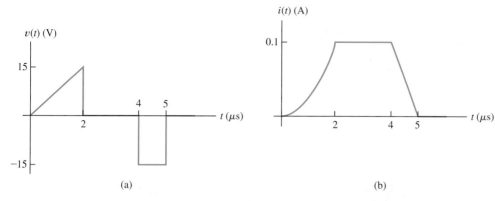

Figure 3.20  See Exercise 3.7.

Notice that the current in the inductor gradually increases after the switch is closed. Because a constant voltage is applied after $t = 0$, the current increases at a steady rate as predicted by Equation 3.28, which is repeated here for convenience:

$$v(t) = L\frac{di}{dt}$$

If $v(t)$ is constant, the rate of change of the current $di/dt$ is constant.   ∎

Suppose that at $t = 1$ s, we open the switch in the circuit of Figure 3.19. Ideally, current cannot flow through an open switch. Hence, we expect the current to fall abruptly to zero at $t = 1$ s. However, the voltage across the inductor is proportional to the time rate of change of the current. For an abrupt change in current, this principle predicts infinite voltage across the inductor. This infinite voltage would last for only the instant at which the current falls. Later, we introduce the concept of an impulse function to describe this situation (and similar ones). For now, we simply point out that very large voltages can appear when we switch circuits that contain inductances.

If we set up a real circuit corresponding to Figure 3.19(a) and open the switch at $t = 1$ s, we will probably find that the high voltage causes an arc across the switch contacts. The arc persists until the energy in the inductor is used up. If this is repeated, the switch will soon be destroyed.

**Exercise 3.6**   The current through a 10-mH inductance is $i(t) = 0.1 \cos(10^4 t)$ A. Find the voltage and stored energy as functions of time. Assume that the references for $v(t)$ and $i(t)$ have the passive configuration. (The angle is in radians.)
**Answer**   $v(t) = -10 \sin(10^4 t)$ V, $w(t) = 50 \cos^2(10^4 t)$ μJ.   ☐

**Exercise 3.7**   The voltage across a 150-μH inductance is shown in Figure 3.20(a). The initial current is $i(0) = 0$. Find and plot the current $i(t)$ to scale versus time. Assume that the references for $v(t)$ and $i(t)$ have the passive configuration.
**Answer**   The current is shown in Figure 3.20(b).   ☐

## 3.5  INDUCTANCES IN SERIES AND PARALLEL

It can be shown that the equivalent inductance for a series circuit is equal to the sum of the inductances connected in series. On the other hand, for inductances in parallel, we find the equivalent inductance by taking the reciprocal of the sum

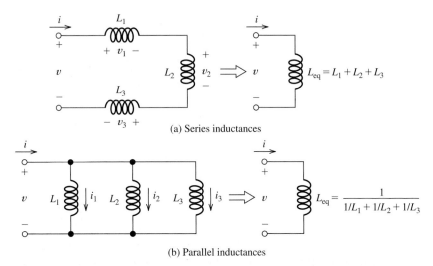

(a) Series inductances

(b) Parallel inductances

**Figure 3.21** Inductances in series and parallel are combined in the same manner as resistances.

Inductances in series and parallel are combined by using the same rules as for resistances: series inductances are added; parallel inductances are combined by taking the reciprocal of the sum of the reciprocals of the individual inductances.

of the reciprocals of the parallel inductances. Series and parallel equivalents for inductances are illustrated in Figure 3.21. Notice that inductances are combined in exactly the same way as are resistances. These facts can be proven by following the pattern used earlier in this chapter to derive the equivalents for series capacitances.

---

**Example 3.9    Inductances in Series and Parallel**

Determine the equivalent inductance between terminals $a$ and $b$ in Figure 3.22(a).

**Solution**    First, notice that the 3-H, 6-H, and 2-H inductances are in parallel. Thus, their equivalent inductance is:

$$\frac{1}{1/3 \ + \ 1/6 \ + \ 1/2} = 1 \text{ H}$$

The resulting equivalent is shown in Figure 3.22(b).
    Finally, we combine the 4-H and 1-H inductances in series resulting in 5 H as shown in Figure 3.22(c).    ∎

**Exercise 3.8**    Prove that inductances in series are added to find the equivalent inductance.

**Exercise 3.9**    Prove that inductances in parallel are combined according to the formula given in Figure 3.21(b).

**Exercise 3.10**    Find the equivalent inductance for each of the circuits shown in Figure 3.23.
**Answer    a.** 3.5 H; **b.** 8.54 H.    □

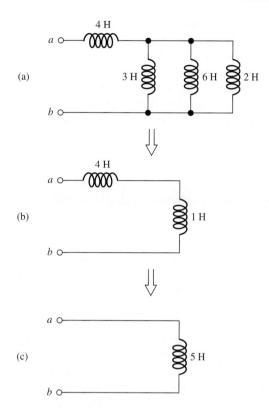

**Figure 3.22** Circuit of Example 3.9.

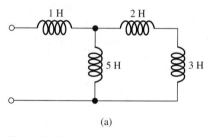

(a)

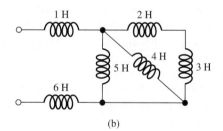

(b)

**Figure 3.23** See Exercise 3.10.

## 3.6 PRACTICAL INDUCTORS

Real inductors take a variety of appearances, depending on their inductance and the application. (Some examples were shown earlier in Figure 3.16.) For example, a 1-$\mu$H inductor could consist of 25 turns of fine (say, number 28) wire wound on an iron oxide toroidal (doughnut-shaped) core having an outside diameter of 1/2 cm. On the other hand, a typical 5-H inductor consists of several hundred turns of number 18 wire on an iron form having a mass of 1 kg.

Usually, metallic iron forms, also called *cores*, are made of thin sheets called *laminations*. [See Figure 3.16(c) for an example.] This is necessary because voltages are induced in the core by the changing magnetic field. These voltages cause **eddy currents** to flow in the core, dissipating energy. Usually, this **core loss** is undesirable. Using laminations that are insulated from one another helps to reduce eddy-current loss. The laminations are arranged perpendicular to the expected current direction.

Another way to defeat eddy currents is to use a core composed of **ferrites**, which are oxides of iron that are electrical insulators. Still another approach is to combine powdered iron with an insulating binder.

---

## PRACTICAL APPLICATION    3.1

### Electronic Photo Flash

Figure PA3.1 shows the electrical circuit of an electronic photo flash such as you may have seen on a camera. The objective of the unit is to produce a bright flash of light by passing a high current through the flash tube while the camera shutter is open. As much as 1000 W is supplied to the flash tube during the flash, which lasts for less than a millisecond. Although the power level is quite high, the total energy delivered is not great because of the short duration of the flash. (The energy is on the order of a joule.)

It is not possible to deliver the power directly from the battery to the flash tube for several reasons. First, practical batteries supply a few tens of volts at most, while several hundred volts are needed to operate the flash tube. Second, applying the principle of maximum power transfer, the maximum power available from the battery is limited to 1 W by its internal Thévenin resistance. (See Equation 2.78 and the related discussion.) This does not nearly meet the needs of the flash tube. Instead, energy is delivered by the battery over a period of several

seconds and stored in the capacitor. The stored energy can be quickly extracted from the capacitor because the parasitic resistance in series with the capacitor is very low.

The electronic switch alternates between open and closed approximately 10,000 times per second. (In some units, you can hear a high-pitched whistle resulting from incidental conversion of some of the energy to acoustic form.) While the electronic switch is closed, the battery causes the current in the inductor to build up. Then when the switch opens, the inductor forces current to flow through the diode, charging the capacitor. (Recall that the current in an inductor cannot change instantaneously.) Current can flow through the diode only in the direction of the arrow. Thus, the diode allows charge to flow into the capacitor when the electronic switch is open and prevents charge from flowing off the capacitor when the electronic switch is closed. Thus, the charge stored on the capacitor increases each time the electronic switch opens. Eventually, the voltage on the capacitor reaches several hundred volts. When

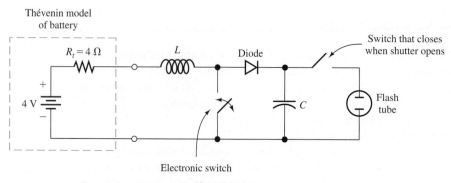

Figure PA3.1

the camera shutter is opened, another switch is closed, allowing the capacitor to discharge through the flash tube.

A friend of the author has a remote cabin on the north shore of Lake Superior that has an unusual water system (illustrated in Figure PA3.2) analogous to the electronic flash circuit. Water flows through a large pipe immersed in the river. Periodically, a valve on the bottom end of the pipe

suddenly closes, stopping the flow. The inertia of the flowing water creates a pulse of high pressure when the valve closes. This high pressure forces water through a one-way ball valve into a storage tank. Air trapped in the storage tank is compressed and forces water to flow to the cabin as needed.

Can you identify the features in Figure PA3.2 that are analogous to each of the circuit elements in Figure PA3.1?

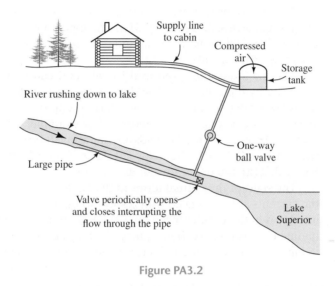

Figure PA3.2

## Parasitic Effects for Real Inductors

Real inductors have parasitic effects in addition to the desired inductance. A circuit model for a real inductor is shown in Figure 3.24. The series resistance $R_s$ is caused by the resistivity of the material composing the wire. (This parasitic effect can be avoided by using wire composed of a superconducting material, which has zero resistivity.) The parallel capacitance is associated with the electric field in the dielectric (insulation) between the coils of wire. It is called **interwinding capacitance**. The parallel resistance $R_p$ represents core loss due, in part, to eddy currents in the core.

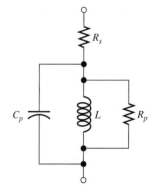

Figure 3.24 Circuit model for real inductors including several parasitic elements.

Actually, the circuit model for a real inductor shown in Figure 3.24 is an approximation. The series resistance is distributed along the length of the wire, as is the interwinding capacitance. A more accurate model for a real inductor would break each of the parasitic effects into many segments (possibly, an infinite number). Ultimately, we could abandon circuit models altogether and use electromagnetic field theory directly.

Rarely is this degree of detail necessary. Usually, modeling a real inductor as an inductance, including at most a few parasitic effects, is sufficiently accurate. Of course, computer-aided circuit analysis allows us to use more complex models and achieve more accurate results than traditional mathematical analysis.

## 3.7 MUTUAL INDUCTANCE

Sometimes, several coils are wound on the same form so that magnetic flux produced by one coil links the others. Then a time-varying current flowing through one coil induces voltages in the other coils. The circuit symbols for two mutually coupled inductances are shown in Figure 3.25. The **self inductances** of the two coils are denoted as $L_1$ and $L_2$, respectively. The **mutual inductance** is denoted as $M$, which also has units of henries. Notice that we have selected the passive reference configuration for each coil in Figure 3.25.

The equations relating the voltages to the currents are also shown in Figure 3.25. The mutual terms, $M\, di_1/dt$ and $M\, di_2/dt$, appear because of the mutual coupling of the coils. The self terms, $L_1\, di_1/dt$ and $L_2\, di_2/dt$, are the voltages induced in each coil due to its own current.

> The magnetic flux produced by one coil can either aid or oppose the flux produced by the other coil.

The magnetic flux produced by one coil can either aid or oppose the flux produced by the other coil. The dots on the ends of the coils indicate whether the fields are aiding or opposing. If one current enters a dotted terminal and the other leaves, the fields oppose one another. For example, if both $i_1$ and $i_2$ have positive values in Figure 3.25(b), the fields are opposing. If both currents enter the respective dots (or if both leave), the fields aid. Thus, if both $i_1$ and $i_2$ have positive values in Figure 3.25(a), the fields are aiding.

The signs of the mutual terms in the equations for the voltages depend on how the currents are referenced with respect to the dots. If both currents are referenced into (or if both are referenced out of) the dotted terminals, as in Figure 3.25(a), the mutual term is positive. If one current is referenced into a dot and the other out, as in Figure 3.25(b), the mutual term carries a negative sign.

### Linear Variable Differential Transformer

An application of mutual inductance can be found in a position transducer known as the linear variable differential transformer (LVDT), illustrated in Figure 3.26. An ac source connected to the center coil sets up a magnetic field that links both halves of the secondary coil. When the iron core is centered in the coils, the voltages induced in the two halves of the secondary cancel so that $v_o(t) = 0$. (Notice that the two halves

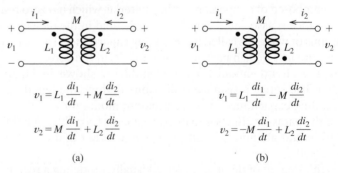

(a)                              (b)

**Figure 3.25** Circuit symbols and $v - i$ relationships for mutually coupled inductances.

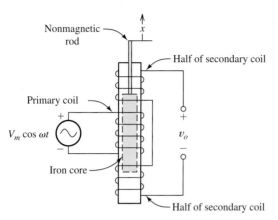

**Figure 3.26** A linear variable differential transformer used as a position transducer.

of the secondary winding are wound in opposite directions.) As the core moves up or down, the couplings between the primary and the halves of the secondary change. The voltage across one half of the coil becomes smaller, and the voltage across the other half becomes greater. Ideally, the output voltage is given by

$$v_o(t) = Kx \cos(\omega t)$$

where $x$ is the displacement of the core. LVDTs are used in applications such as automated manufacturing operations to measure displacements.

## 3.8   SYMBOLIC INTEGRATION AND DIFFERENTIATION USING MATLAB

As we have seen, finding the current given the voltage (or vice versa) for an energy storage element involves integration or differentiation. Thus, we may sometimes need to find symbolic answers for integrals or derivatives of complex functions, which can be very difficult by traditional methods. Then, we can resort to using symbolic mathematics software. Several programs are available including Maple™ from Maplesoft Corporation, Mathematica™ from Wolfram Research, and the Symbolic Toolbox which is an optional part of MATLAB from Mathsoft. Each of these programs has its strengths and weaknesses, and when a difficult problem warrants the effort, all of them should be tried. Because MATLAB is widely used in Electrical Engineering, we confine our brief discussion to the Symbolic Toolbox.

**One note of caution:** We have checked the examples, exercises, and problems using MATLAB version R2015b. Keep in mind that if you use versions other than R2015b, you may not be able to reproduce our results. Try running our example m-files before sinking a lot of time into solving the problems. Hopefully, your instructor can give you some guidance on what to expect with the MATLAB versions available to you.

In the following, we assume that you have some familiarity with MATLAB. A variety of online interactive tutorials are available at `https://www.mathworks.com/`. However, you may find it easier to write MATLAB instructions for the exercises and problems in this chapter by modeling your solutions after the code in our examples.

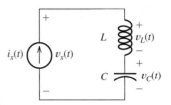

Figure 3.27  Circuit of Example 3.10.

---

**Example 3.10**    **Integration and Differentiation Using the MATLAB Symbolic Toolbox**

Use MATLAB to find expressions for the three voltages shown in Figure 3.27 given $v_C(0) = 0$ and

$$i_x(t) = kt^2 \exp(-at) \sin(\omega t) \text{ for } t \geq 0$$

$$= 0 \text{ for } t < 0 \tag{3.39}$$

Also, plot the current and the voltages for $k = 3$, $a = 2$, $\omega = 1$, $L = 0.5 \text{ H}$, $C = 1 \text{ F}$, and $t \geq 0$. (These values have been chosen mainly to facilitate the demonstration of MATLAB capabilities.) The currents are in amperes, voltages are in volts, $\omega t$ is in radians, and time $t$ is in seconds.

**Solution**    At first, we use symbols to represent the various parameters ($k$, $a$, $\omega$, $L$, and $C$), denoting the current and the voltages as ix, vx, vL, and vC. Then, we substitute the numerical values for the symbols and denote the results as ixn, vxn, vLn, and vCn. (The letter "n" is selected to suggest that the "numerical" values of the parameters have been substituted into the expressions.)

We show the commands in **boldface**, comments in regular font, and MATLAB responses in color. Comments (starting with the % sign) are ignored by MATLAB. We present the work as if we were entering the commands and comments one at a time in the MATLAB command window, however, it is usually more convenient to place all of the commands in an m-file and execute them as a group.

To start, we define the various symbols as symbolic objects in MATLAB, define the current ix, and substitute the numerical values of the parameters to obtain ixn.

```
>> clear all % Clear work area of previous work.
>> syms vx ix vC vL vxn ixn vCn vLn k a w t L C
>> % Names for symbolic objects must start with a letter and
>> % contain only alpha-numeric characters.
>> % Next, we define ix.
>> ix=k*t^2*exp(-a*t)*sin(w*t)
      ix =
      (k*t^2*sin(w*t))/exp(a*t)
>> % Next, we substitute k=3, a=2, and w=1
>> % into ix and denote the result as ixn.
>> ixn = subs(ix,[k a w],[3 2 1])
      ixn =
      (3*t^2*sin(t))/exp(2*t)
```

Next, we want to plot the current versus time. We need to consider what range of $t$ should be used for the plot. In standard mathematical typesetting, the expression we need to plot is

$$i_x(t) = 3t^2 \exp(-2t) \sin(t) \text{ for } t \geq 0$$

$$= 0 \text{ for } t < 0$$

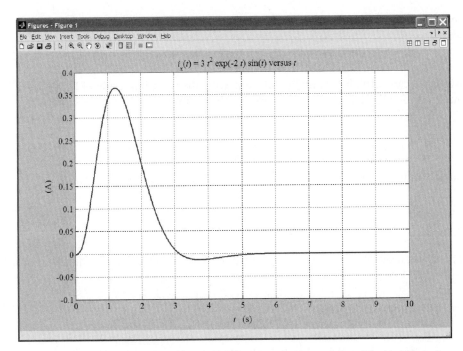

**Figure 3.28**  Plot of $i_x(t)$ produced by MATLAB. Reprinted with permission of The MathWorks, Inc.

Thoughtful examination of this expression (perhaps supplemented with a little work with a calculator) reveals that the current is zero at $t = 0$, builds up quickly after $t = 0$ because of the $t^2$ term, and decays to relatively small values after about $t = 10$ s because of the exponential term. Thus, we select the range from $t = 0$ to $t = 10$ s for the plot. Continuing in MATLAB, we have

```
>> % Next, we plot ixn for t ranging from 0 to 10 s.
>> ezplot(ixn,[0,10])
```

This opens a window with a plot of the current versus time as shown in Figure 3.28. As expected, the current increases rapidly after $t = 0$ and decays to insignificant values by $t = 10$ s. (We have used various Edit menu commands to improve the appearance of the plot for inclusion in this book.)

Next, we determine the inductance voltage, which is given by

$$v_L(t) = L\frac{di_x(t)}{dt}$$

in which the parameters, $a$, $k$, and $\omega$ are treated as constants. The corresponding MATLAB command and the result are:

```
>> vL=L*diff(ix,t)   % L × the derivative of ix with respect to t.
   vL =
   L*((2*k*t*sin(t*w))/exp(a*t) - (a*k*t^2*sin(t*w))/exp(a*t) +
   (k*t^2*w*cos(t*w))/exp(a*t))
>> % A nicer display for vL is produced with the command:
>> pretty(vL)
   L k t (2 sin(t w) - a t sin(t w) + t w cos(t w))
   ------------------------------------------------
                    exp(a t)
```

In more standard mathematical typesetting, this becomes

$$v_L(t) = Lkt \exp(-at)[2\sin(\omega t) - at\sin(\omega t) + \omega t\cos(\omega t)]$$

which we can verify by manually differentiating the right-hand side of Equation 3.39 and multiplying by $L$. Next, we determine the voltage across the capacitance.

$$v_C(t) = \frac{1}{C}\int_0^t i_x(t)dt + v_C(0) \text{ for } t \geq 0$$

Substituting the expressions for the current and initial voltage we obtain,

$$v_C(t) = \frac{1}{C}\int_0^t kt^2 \exp(-at)\sin(\omega t)dt \text{ for } t \geq 0$$

This is not a simple integration to perform by hand, but we can accomplish it easily with MATLAB:

```
>> % Integrate ix with respect to t with limits from 0 to t.
>> vC=(1/C)*int(ix,t,0,t);
>> % We included the semicolon to suppress the output, which is
>> % much too complex for easy interpretation.
>> % Next, we find the total voltage vx.
>> vx = vC + vL;
>> % Now we substitute numerical values for the parameters.
>> vLn=subs(vL,[k a w L C],[3 2 1 0.5 1]);
>> vCn=subs(vC,[k a w L C],[3 2 1 0.5 1]);
>> vxn=subs(vx,[k a w L C],[3 2 1 0.5 1]);
>> % Finally, we plot all three voltages in the same window.
>> figure % Open a new figure for this plot.
>> ezplot(vLn,[0,10])
>> hold on % Hold so the following two plots are on the same axes.
>> ezplot(vCn,[0,10])
>> ezplot(vxn,[0,10])
```

The resulting plot is shown in Figure 3.29. (Here again, we have used various items on the Edit menu to change the scale of the vertical axis and dress up the plot for inclusion in this book.)

The commands for this example are included as an m-file named Example_3_10 in the MATLAB files. (See Appendix E for information about accessing these MATLAB files.) If you copy the file and place it in a folder in the MATLAB path for your computer, you can run the file and experiment with it. For example, after running the m-file, if you enter the command

```
>> vC
```

you will see the rather complicated symbolic mathematical expression for the voltage across the capacitance.    ■

**Exercise 3.11**    Use MATLAB to work Example 3.2 on page 132 resulting in plots like those in Figure 3.5.

**Answer**    The MATLAB commands including some explanatory comments are:

```
clear   % Clear the work area.
% We avoid using i alone as a symbol for current because
% we reserve i for the square root of -1 in MATLAB. Thus, we
```

```
% will use iC for the capacitor current.
syms t iC qC vC % Define t, iC, qC and vC as symbolic objects.
iC = 0.5*sin((1e4)*t);
ezplot(iC, [0 3*pi*1e-4])
qC=int(iC,t,0,t);  % qC equals the integral of iC.
figure  % Plot the charge in a new window.
ezplot(qC, [0 3*pi*1e-4])
vC = 1e7*qC;
figure % Plot the voltage in a new window.
ezplot(vC, [0 3*pi*1e-4])
```

The plots are very similar to those of Figure 3.5 on page 133. An m-file (named Exercise_3_11) can be found in the MATLAB folder. □

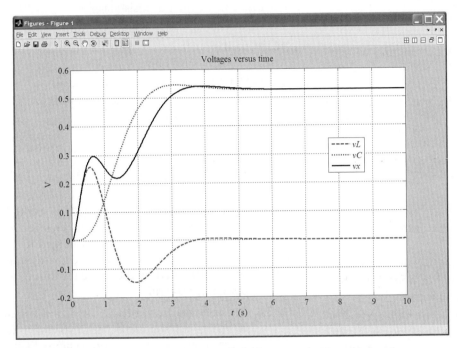

**Figure 3.29** Plots of the voltages for Example 3.10. Reprinted with permission of The MathWorks, Inc.

## Summary

1. Capacitance is the circuit property that accounts for electric-field effects. The units of capacitance are farads (F), which are equivalent to coulombs per volt.

2. The charge stored by a capacitance is given by $q = Cv$.

3. The relationships between current and voltage for a capacitance are

$$i = C\frac{dv}{dt}$$

and

$$v(t) = \frac{1}{C}\int_{t_0}^{t} i(t)\, dt + v(t_0)$$

4. The energy stored by a capacitance is given by

$$w(t) = \frac{1}{2}Cv^2(t)$$

5. Capacitances in series are combined in the same manner as resistances in parallel.

6. Capacitances in parallel are combined in the same manner as resistances in series.

7. The capacitance of a parallel-plate capacitor is given by

$$C = \frac{\epsilon A}{d}$$

For vacuum, the dielectric constant is $\epsilon = \epsilon_0 \cong 8.85 \times 10^{-12}$ F/m. For other materials, the dielectric constant is $\epsilon = \epsilon_r \epsilon_0$, where $\epsilon_r$ is the relative dielectric constant.

8. Real capacitors have several parasitic effects.

9. Inductance accounts for magnetic-field effects. The units of inductance are henries (H).

10. The relationships between current and voltage for an inductance are

$$v(t) = L\frac{di}{dt}$$

and

$$i(t) = \frac{1}{L}\int_{t_0}^{t} v(t)\,dt + i(t_0)$$

11. The energy stored in an inductance is given by

$$w(t) = \frac{1}{2}Li^2(t)$$

12. Inductances in series or parallel are combined in the same manner as resistances.

13. Real inductors have several parasitic effects.

14. Mutual inductance accounts for mutual coupling of magnetic fields between coils.

15. MATLAB is a powerful tool for symbolic integration, differentiation, and plotting of functions.

## Problems

### Section 3.1: Capacitance

**P3.1.** What is a dielectric material? Give two examples.

**P3.2.** Briefly discuss how current can flow "through" a capacitor even though a nonconducting layer separates the metallic parts.

**P3.3.** What current flows through an ideal capacitor if the voltage across the capacitor is constant with time? To what circuit element is an ideal capacitor equivalent in circuits for which the currents and voltages are constant with time?

**P3.4.** Describe the internal construction of capacitors.

**P3.5.** A voltage of 50 V appears across a 10-$\mu$F capacitor. Determine the magnitude of the net charge stored on each plate and the total net charge on both plates.

**\*P3.6.** A 2000-$\mu$F capacitor, initially charged to 100 V, is discharged by a steady current of 100 $\mu$A. How long does it take to discharge the capacitor to 0 V?

**P3.7.** A 5-$\mu$F capacitor is charged to 1000 V. Determine the initial stored charge and energy. If this capacitor is discharged to 0 V in a time interval of 1 $\mu$s, find the average power delivered by the capacitor during the discharge interval.

**\*P3.8.** The voltage across a 10-$\mu$F capacitor is given by $v(t) = 100\sin(1000t)$. Find expressions for the current, power, and stored energy. Sketch the waveforms to scale versus time.

**P3.9.** The voltage across a 1-$\mu$F capacitor is given by $v(t) = 100e^{-100t}$. Find expressions for the current, power, and stored energy. Sketch the waveforms to scale versus time.

**P3.10.** Prior to $t = 0$, a 100-$\mu$F capacitance is uncharged. Starting at $t = 0$, the voltage across the capacitor is increased linearly with time to 100 V in 2 s. Then, the voltage remains constant at 100 V. Sketch the voltage, current, power, and stored energy to scale versus time.

**P3.11.** The current through a 0.5-$\mu$F capacitor is shown in Figure P3.11. At $t = 0$, the voltage is zero. Sketch the voltage, power, and stored energy to scale versus time.

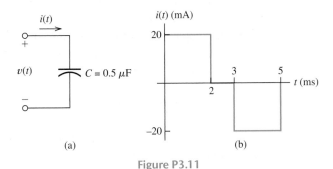

(a)

(b)

Figure P3.11

**P3.12.** Determine the capacitor voltage, power, and stored energy at $t = 20$ ms in the circuit of Figure P3.12.

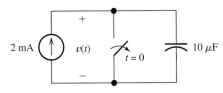

Figure P3.12

**P3.13.** A current given by $i(t) = I_m \cos(\omega t)$ flows through a capacitance $C$. The voltage is zero at $t = 0$. Suppose that $\omega$ is very large, ideally, approaching infinity. For this current does the capacitance approximate either an open or a short circuit? Explain.

**P3.14.** The current through a 3-$\mu$F capacitor is shown in Figure P3.14. At $t = 0$, the voltage is $v(0) = 10$ V. Sketch the voltage, power, and stored energy to scale versus time.

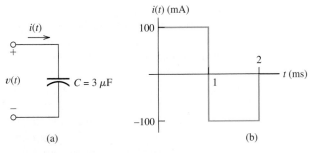

(a)

(b)

Figure P3.14

**\*P3.15.** A constant (dc) current $i(t) = 3$ mA flows into a 50-$\mu$F capacitor. The voltage at $t = 0$ is $v(0) = -20$ V. The references for $v(t)$ and $i(t)$ have the passive configuration. Find the power at $t = 0$ and state whether the power

flow is into or out of the capacitor. Repeat for $t = 1$ s.

**P3.16.** The energy stored in a 20-$\mu$F capacitor is 200 J and is increasing at 500 J/s at $t = 3$ s. Determine the voltage magnitude and current magnitude at $t = 3$ s. Does the current enter or leave the positive terminal of the capacitor?

**P3.17.** At $t = t_0$ the voltage across a certain capacitance is zero. A pulse of current flows through the capacitance between $t_0$ and $t_0 + \Delta t$, and the voltage across the capacitance increases to $V_f$. What can you say about the peak amplitude $I_m$ and area under the pulse waveform (i.e., current versus time)? What are the units and physical significance of the area under the pulse? What must happen to the peak amplitude and area under the pulse as $\Delta t$ approaches zero, assuming that $V_f$ remains the same?

**P3.18.** An unusual capacitor has a capacitance that is a function of time given by

$$C = 2 + \cos(2000t)\ \mu F$$

in which the argument of the cosine function is in radians. A constant voltage of 50 V is applied to this capacitor. Determine the current as a function of time.

**P3.19.** For a resistor, what resistance corresponds to a short circuit? For an uncharged capacitor, what value of capacitance corresponds to a short circuit? Explain your answers. Repeat for an open circuit.

**P3.20.** Suppose we have a very large capacitance (ideally, infinite) charged to 10 V. What other circuit element has the same current–voltage relationship? Explain your answer.

**\*P3.21.** We want to store sufficient energy in a 0.01-F capacitor to supply 5 horsepower (hp) for 1 hour. To what voltage must the capacitor be charged? (*Note:* One horsepower is equivalent to 745.7 watts.) Does this seem to be a practical method for storing this amount of energy? Do you think that an electric automobile design based on capacitive energy storage is feasible?

**P3.22.** A 100-$\mu$F capacitor has a voltage given by $v(t) = 10 - 10 \exp(-2t)$ V. Find the power at $t = 0$ and state whether the power flow is into or out of the capacitor. Repeat for $t_2 = 0.5$ s.

## Section 3.2: Capacitances in Series and Parallel

**P3.23.** How are capacitances combined in series and in parallel? Compare with how resistances are combined.

**\*P3.24.** Find the equivalent capacitance for each of the circuits shown in Figure P3.24.

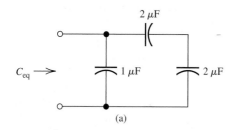

(a)

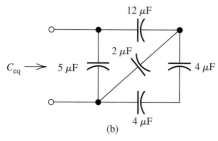

(b)

Figure P3.24

**P3.25.** Find the equivalent capacitance between terminals $x$ and $y$ for each of the circuits shown in Figure P3.25.

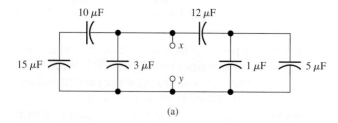

(a)

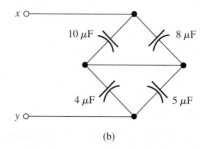

(b)

Figure P3.25

**P3.26.** A network has a 5-$\mu$F capacitance in series with the parallel combination of a 12-$\mu$F capacitance and an 8-$\mu$F capacitance. Sketch the circuit diagram and determine the equivalent capacitance of the combination.

**P3.27.** What are the minimum and maximum values of capacitance that can be obtained by connecting four 2-$\mu$F capacitors in series and/or parallel? How should the capacitors be connected?

**P3.28.** Two initially uncharged capacitors $C_1 = 15\ \mu$F and $C_2 = 10\ \mu$F are connected in series. Then, a 10-V source is connected to the series combination, as shown in Figure P3.28. Find the voltages $v_1$ and $v_2$ after the source is applied. [*Hint:* The charges stored on the two capacitors must be equal, because the current is the same for both capacitors.]

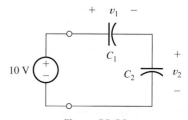

Figure P3.28

**\*P3.29.** Suppose that we are designing a cardiac pacemaker circuit. The circuit is required to deliver pulses of 1-ms duration to the heart, which can be modeled as a 500-$\Omega$ resistance. The peak amplitude of the pulses is required to be 5 V. However, the battery delivers only 2.5 V. Therefore, we decide to charge two equal-value capacitors in parallel from the 2.5-V battery and then switch the capacitors in series with the heart during the 1-ms pulse. What is the minimum value of the capacitances required so the output pulse amplitude remains between 4.9 V and 5.0 V throughout its 1-ms duration? If the pulses occur once every second, what is the average current drain from the battery? Use approximate calculations, assuming constant current during the output pulse. Find the ampere-hour rating of the battery so it lasts for five years.

**P3.30.** Suppose that we have two 100-$\mu$F capacitors. One is charged to an initial voltage of 50 V, and the other is charged to 100 V. If they are placed in series with the positive terminal of the first connected to the negative terminal of the second, determine the equivalent capacitance and its initial voltage. Now compute the total energy stored in the two capacitors. Compute the energy stored in the equivalent capacitance. Why is it less than the total energy stored in the original capacitors?

### Section 3.3: Physical Characteristics of Capacitors

*P3.31. Determine the capacitance of a parallel-plate capacitor having plates 10 cm by 30 cm separated by 0.01 mm. The dielectric has $\epsilon_r = 15$.

P3.32. A 100-pF capacitor is constructed of parallel plates of metal, each having a width $W$ and a length $L$. The plates are separated by air with a distance $d$. Assume that $L$ and $W$ are both much larger than $d$. What is the new capacitance if **a.** both $L$ and $W$ are doubled and the other parameters are unchanged? **b.** the separation $d$ is doubled and the other parameters are unchanged from their initial values? **c.** the air dielectric is replaced with oil having a relative dielectric constant of 25 and the other parameters are unchanged from their initial values?

P3.33. We have a parallel-plate capacitor with plates of metal each having a width $W$ and a length $L$. The plates are separated by the distance $d$. Assume that $L$ and $W$ are both much larger than $d$. The maximum voltage that can be applied is limited to $V_{max} = Kd$, in which $K$ is called the breakdown strength of the dielectric. Derive an expression for the maximum energy that can be stored in the capacitor in terms of $K$ and the volume of the dielectric. If we want to store the maximum energy per unit volume, does it matter what values are chosen for $L$, $W$, and $d$? What parameters are important?

*P3.34. Suppose that we have a 1000-pF parallel-plate capacitor with air dielectric charged to 1000 V. The capacitor terminals are open circuited. Find the stored energy. If the plates are moved farther apart so that $d$ is doubled, determine the new voltage on the capacitor and the new stored energy. Where did the extra energy come from?

P3.35. Two 1-$\mu$F capacitors have an initial voltage of 100 V (before the switch is closed), as shown in Figure P3.35. Find the total stored energy before the switch is closed. Find the voltage across each capacitor and the total stored energy after the switch is closed. What could have happened to the energy?

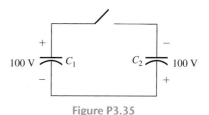

**Figure P3.35**

P3.36. A liquid-level transducer consists of two parallel plates of conductor immersed in an insulating liquid, as illustrated in Figure P3.36. When the tank is empty (i.e., $x = 0$), the capacitance of the plates is 200 pF. The relative dielectric constant of the liquid is 25. Determine an expression for the capacitance $C$ as a function of the height $x$ of the liquid.

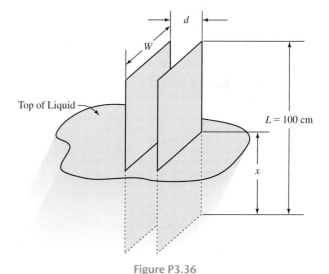

**Figure P3.36**

P3.37. A parallel-plate capacitor like that shown in Figure P3.36 has a capacitance of 2000 pF when the tank is full so the plates are totally immersed in the insulating liquid. (The dielectric constant of the fluid is different for this problem than for Problem P3.36.) The capacitance is 200 pF when the tank is empty and the space between the plates is filled with air. Suppose that the tank is full and the capacitance is charged to 1000 V. Then, the capacitance is open circuited so the charge on the plates cannot change, and the tank is drained. Compute the electrical energy stored in the capacitor before and after the tank is drained. With the plates open circuited, there is no electrical source for the extra energy. Where could it have come from?

**P3.38.** A parallel-plate capacitor is used as a vibration sensor. The plates have an area of 100 cm$^2$, the dielectric is air, and the distance between the plates is a function of time given by

$$d(t) = 1 + 0.01 \sin(200t) \text{ mm}$$

A constant voltage of 200 V is applied to the sensor. Determine the current through the sensor as a function of time by using the approximation $1/(1 + x) \cong 1 - x$ for $x \ll 1$. (The argument of the sinusoid is in radians.)

**P3.39.** A 0.1-$\mu$F capacitor has a parasitic series resistance of 10 $\Omega$, as shown in Figure P3.39. Suppose that the voltage across the capacitance is $v_c(t) = 10 \cos(100t)$; find the voltage across the resistance. In this situation, to find the total voltage $v(t) = v_r(t) + v_c(t)$ to within 1 percent accuracy, is it necessary to include the parasitic resistance? Repeat if $v_c(t) = 0.1 \cos(10^7 t)$.

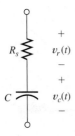

**Figure P3.39**

**\*P3.40.** Suppose that a parallel-plate capacitor has a dielectric that breaks down if the electric field exceeds $K$ V/m. Thus, the maximum voltage rating of the capacitor is $V_{max} = Kd$, where $d$ is the separation between the plates. In working Problem P3.33, we find that the maximum energy

that can be stored is $w_{max} = \frac{1}{2}\epsilon_r\epsilon_0 K^2$ (Vol) in which Vol is the volume of the dielectric. Given that $K = 32 \times 10^5$ V/m and that $\epsilon_r = 1$ (the approximate values for air), find the dimensions of a parallel-plate capacitor having square plates if it is desired to store 1 mJ at a voltage of 1000 V in the least possible volume.

**Section 3.4: Inductance**

**P3.41.** Briefly discuss how inductors are constructed.

**P3.42.** The current flowing through an inductor is increasing in magnitude. Is energy flowing into or out of the inductor?

**P3.43.** If the current through an ideal inductor is constant with time, what is the value of the voltage across the inductor? Comment. To what circuit element is an ideal inductor equivalent for circuits with constant currents and voltages?

**P3.44.** Briefly discuss the fluid-flow analogy for an inductor.

**\*P3.45.** The current flowing through a 2-H inductance is shown in Figure P3.45. Sketch the voltage, power, and stored energy versus time.

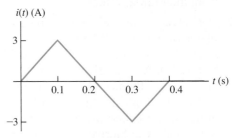

**Figure P3.45**

**P3.46.** The current flowing through a 100-mH inductance is given by 0.5 sin (1000t) A, in which the angle is in radians. Find expressions and sketch the waveforms to scale for the voltage, power, and stored energy.

**P3.47.** The current flowing through a 2-H inductance is given by 5 exp(−20t) A. Find expressions for the voltage, power, and stored energy. Sketch the waveforms to scale for $0 < t < 100$ ms.

**P3.48.** The voltage across a 2-H inductance is shown in Figure P3.48. The initial current in the inductance is $i(0) = 0$. Sketch the current, power, and stored energy to scale versus time.

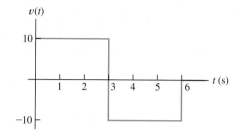

**Figure P3.48**

**P3.49.** The voltage across a 10-$\mu$ H inductance is given by $v(t) = 5\sin(10^6 t)$ V. The initial current is $i(0) = -0.5$ A. Find expressions for the current, power, and stored energy for $t > 0$. Sketch the waveforms to scale versus time.

**P3.50.** A 2-H inductance has $i(0) = 0$ and $v(t) = t\exp(-t)$ for $0 \le t$. Find an expression for $i(t)$. Then, using the computer program of your choice, plot $v(t)$ and $i(t)$ for $0 \le t \le 10$ s.

**\*P3.51.** A constant voltage of 10 V is applied to a 50-$\mu$H inductance, as shown in Figure P3.51. The current in the inductance at $t = 0$ is $-100$ mA. At what time $t_x$ does the current reach +100 mA?

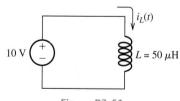

10 V      $i_L(t)$      $L = 50\ \mu H$

**Figure P3.51**

**\*P3.52.** At $t = 0$, the current flowing in a 0.5-H inductance is 4 A. What constant voltage must be applied to reduce the current to 0 at $t = 0.2$ s?

**P3.53.** The current through a 100-mH inductance is given by $i(t) = \exp(-t)\sin(10t)$ in which the angle is in radians. Determine the voltage across the inductance. Then, use the computer program of your choice to plot both the current and the voltage for $0 \le t \le 3$ s.

**P3.54.** Prior to $t = 0$, the current in a 2-H inductance is zero. Starting at $t = 0$, the current is increased linearly with time to 10 A in 5 s. Then, the current remains constant at 10 A. Sketch the voltage, current, power, and stored energy to scale versus time.

**P3.55.** At $t = 0$, a constant 5-V voltage source is applied to a 3-H inductor. Assume an initial current of zero for the inductor. Determine the current, power, and stored energy at $t = 2$ s.

**P3.56.** At $t = t_0$ the current through a certain inductance is zero. A voltage pulse is applied to the inductance between $t_0$ and $t_0 + \Delta t$, and the current through the inductance increases to $I_f$. What can you say about the peak amplitude $V_m$ and area under the pulse waveform (i.e., voltage versus time)? What are the units of the area under the pulse? What must happen to the peak amplitude and area under the pulse as $\Delta t$ approaches zero, assuming that $I_f$ remains the same?

**P3.57.** At $t = 5$ s, the energy stored in a 2-H inductor is 200 J and is increasing at 100 J/s. Determine the voltage magnitude and current magnitude at $t = 5$ s. Does the current enter or leave the positive terminal of the inductor?

**P3.58.** What value of inductance (having zero initial current) corresponds to an open circuit? Explain your answer. Repeat for a short circuit.

**P3.59.** To what circuit element does a very large (ideally, infinite) inductance having an initial current of 10 A correspond? Explain your answer.

**P3.60.** The voltage across an inductance $L$ is given by $v(t) = V_m\cos(\omega t)$. The current is zero at $t = 0$. Suppose that $\omega$ is very large ideally, approaching infinity. For this voltage, does the inductance approximate either an open or a short circuit? Explain.

**Section 3.5: Inductances in Series and Parallel**

**P3.61.** Discuss how inductances are combined in series and in parallel. Compare with how resistances are combined.

**\*P3.62.** Determine the equivalent inductance for each of the series and parallel combinations shown in Figure P3.62.

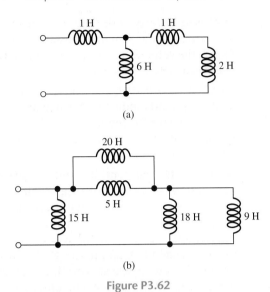

(a)

(b)

**Figure P3.62**

**P3.63.** Find the equivalent inductance for each of the series and parallel combinations shown in Figure P3.63.

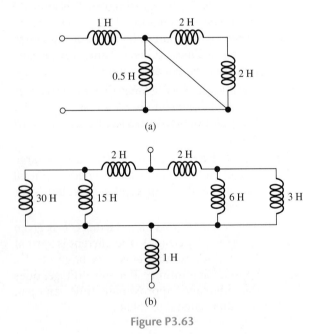

(a)

(b)

**Figure P3.63**

**P3.64.** What is the maximum inductance that can be obtained by connecting four 2-H inductors in series and/or parallel? What is the minimum inductance?

**P3.65.** Suppose we want to combine (in series or in parallel) an inductance $L$ with a 6-H inductance to attain an equivalent inductance of 2 H. Should $L$ be placed in series or in

parallel with the original inductance? What value is required for $L$?

**P3.66.** Repeat Problem P3.65 for an equivalent inductance of 8 H.

*P3.67.** Two inductances $L_1 = 1$ H and $L_2 = 2$ H are connected in parallel, as shown in Figure P3.67. The initial currents are $i_1(0) = 0$ and $i_2(0) = 0$. Find an expression for $i_1(t)$ in terms of $i(t)$, $L_1$, and $L_2$. Repeat for $i_2(t)$. Comment.

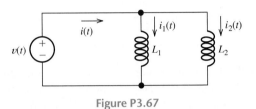

**Figure P3.67**

## Section 3.6: Practical Inductors

**P3.68.** A 10-mH inductor has a parasitic series resistance of $R_s = 1\,\Omega$, as shown in Figure P3.68. **a.** The current is given by $i(t) = 0.1\cos(10^5 t)$. Find $v_R(t)$, $v_L(t)$, and $v(t)$. In this case, for 1-percent accuracy in computing $v(t)$, could the resistance be neglected? **b.** Repeat if $i(t) = 0.1\cos(10t)$.

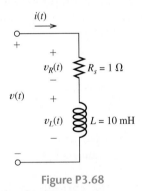

**Figure P3.68**

**P3.69.** Draw the equivalent circuit for a real inductor, including three parasitic effects.

**P3.70.** Suppose that the equivalent circuit shown in Figure 3.24 accurately represents a real inductor. A constant current of 100 mA flows through the inductor, and the voltage across its external terminals is 500 mV. Which of the circuit parameters can be deduced from this information and what is its value?

**P3.71.** Consider the circuit shown in Figure P3.71, in which $v_C(t) = 10\sin(1000t)$ V, with the argument of the sine function in radians. Find $i(t)$, $v_L(t)$, $v(t)$, the energy stored in the capacitance, the energy stored in the inductance, and the total stored energy. Show that the total stored energy is constant with time. Comment on the results.

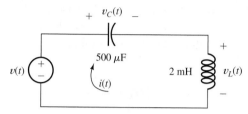

Figure P3.71

**P3.72.** The circuit shown in Figure P3.72 has $i_L(t) = 0.1 \cos(5000t)$ A in which the argument of the cos function is in radians. Find $v(t)$, $i_C(t)$, $i(t)$, the energy stored in the capacitance, the energy stored in the inductance, and the total stored energy. Show that the total stored energy is constant with time. Comment on the results.

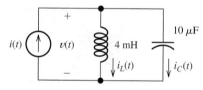

Figure P3.72

## Section 3.7: Mutual Inductance

**P3.73.** Describe briefly the physical basis for mutual inductance.

**P3.74.** The mutually coupled inductances in Figure P3.74 have $L_1 = 1$ H, $L_2 = 2$ H, and $M = 1$ H. Furthermore, $i_1(t) = \sin(10t)$ and $i_2(t) = 0.5 \sin(10t)$. Find expressions for $v_1(t)$ and $v_2(t)$. The arguments of the sine functions are in radians.

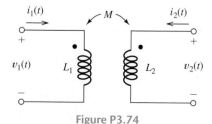

Figure P3.74

*****P3.75.** Repeat Problem P3.74 with the dot placed at the bottom of $L_2$.

*****P3.76.** **a.** Derive an expression for the equivalent inductance for the circuit shown in Figure P3.76. **b.** Repeat if the dot for $L_2$ is moved to the bottom end.

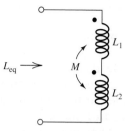

Figure P3.76

**P3.77.** Consider the parallel inductors shown in Figure P3.67, with mutual coupling and the dots at the top ends of $L_1$ and $L_2$. Derive an expression for the equivalent inductance seen by the source in terms of $L_1$, $L_2$, and $M$. [*Hint:* Write the circuit equations and manipulate them to obtain an expression of the form $v(t) = L_{eq}di(t)/dt$ in which $L_{eq}$ is a function of $L_1$, $L_2$, and $M$.]

**P3.78.** Consider the mutually coupled inductors shown in Figure 3.25(a), with a short connected across the terminals of $L_2$. Derive an expression for the equivalent inductance seen looking into the terminals of $L_1$.

**P3.79.** Mutually coupled inductances have

$$L_1 = 2 \text{ H}$$
$$L_2 = 1 \text{ H}$$
$$i_1 = 10 \cos(1000t)$$
$$i_2 = 0$$
$$v_2 = 10^4 \sin(1000t)$$

Find $v_1(t)$ and the magnitude of the mutual inductance. The angles are in radians.

## Section 3.8: Symbolic Integration and Differentiation Using MATLAB

**P3.80.** The current through a 200-mH inductance is given by $i_L(t) = \exp(-2t) \sin(4\pi t)$ A in which the angle is in radians. Using your knowledge of calculus, find an expression for the voltage across the inductance. Then, use MATLAB to verify your answer for the voltage and to plot both the current and the voltage for $0 \le t \le 2$ s.

**P3.81.** A 1-H inductance has $i_L(0) = 0$ and $v_L(t) = t\exp(-t)$ for $0 \le t$. Using your calculus skills, find and an expression for $i_L(t)$. Then, use MATLAB to verify your answer for $i_L(t)$ and to plot $v_L(t)$ and $i_L(t)$ for $0 \le t \le 10$ s.

## Practice Test

Here is a practice test you can use to check your comprehension of the most important concepts in this chapter. Answers can be found in Appendix D and complete solutions are included in the Student Solutions files. See Appendix E for more information about the Student Solutions.

**T3.1.** The current flowing through a 10-$\mu$F capacitor having terminals labeled $a$ and $b$ is $i_{ab} = 0.3\exp(-2000t)$ A for $t \ge 0$. Given that $v_{ab}(0) = 0$, find an expression for $v_{ab}(t)$ for $t \ge 0$. Then, find the energy stored in the capacitor for $t = \infty$.

**T3.2.** Determine the equivalent capacitance $C_{eq}$ for Figure T3.2.

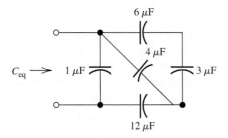

Figure T3.2

**T3.3.** A certain parallel-plate capacitor has plate length of 2 cm and width of 3 cm. The dielectric has a thickness of 0.1 mm and a relative dielectric constant of 80. Determine the capacitance.

**T3.4.** A 2-mH inductance has $i_{ab} = 0.3\sin(2000t)$ A. Find an expression for $v_{ab}(t)$. Then, find the peak energy stored in the inductance.

**T3.5.** Determine the equivalent inductance $L_{eq}$ between terminals $a$ and $b$ in Figure T3.5.

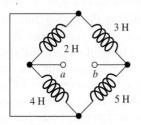

Figure T3.5

**T3.6.** Given that $v_c(t) = 10\sin(1000t)$ V, find $v_s(t)$ in the circuit of Figure T3.6. The argument of the sine function is in radians.

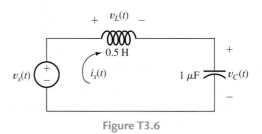

Figure T3.6

**T3.7.** Figure T3.7 has $L_1 = 40$ mH, $M = 20$ mH, and $L_2 = 30$ mH. Find expressions for $v_1(t)$ and $v_2(t)$.

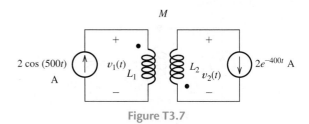

Figure T3.7

**T3.8.** The current flowing through a 20-$\mu$F capacitor having terminals labeled $a$ and $b$ is $i_{ab} = 3 \times 10^5 t^2 \exp(-2000t)$ A for $t \ge 0$. Given that $v_{ab}(0) = 5$ V, write a sequence of MATLAB commands to find the expression for $v_{ab}(t)$ for $t \ge 0$ and to produce plots of the current and voltage for $0 \le t \le 5$ ms.

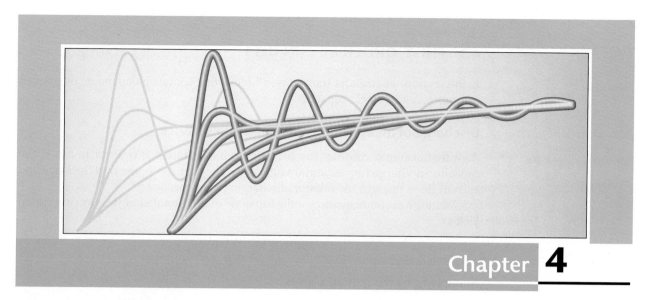

# Chapter 4

# Transients

## Study of this chapter will enable you to:

- Solve first-order $RC$ or $RL$ circuits.

- Understand the concepts of transient response and steady-state response.

- Relate the transient response of first-order circuits to the time constant.

- Solve $RLC$ circuits in dc steady-state conditions.

- Solve second-order circuits.

- Relate the step response of a second-order system to its natural frequency and damping ratio.

- Use the MATLAB Symbolic Toolbox to solve differential equations.

## Introduction to this chapter:

In this chapter, we study circuits that contain sources, switches, resistances, inductances, and capacitances. The time-varying currents and voltages resulting from the sudden application of sources, usually due to switching, are called **transients**.

In transient analysis, we start by writing circuit equations using concepts developed in Chapter 2, such as KCL, KVL, node-voltage analysis, and mesh-current analysis. Because the current–voltage relationships for inductances and capacitances involve integrals and derivatives, we obtain integrodifferential equations. These equations can be converted to pure differential equations by differentiating with respect to time. Thus, the study of transients requires us to solve differential equations.

## 4.1 FIRST-ORDER *RC* CIRCUITS

In this section, we consider transients in circuits that contain independent dc sources, resistances, and a single capacitance.

### Discharge of a Capacitance through a Resistance

As a first example, consider the circuit shown in Figure 4.1(a). Prior to $t = 0$, the capacitor is charged to an initial voltage $V_i$. Then, at $t = 0$, the switch closes and current flows through the resistor, discharging the capacitor.

Writing a current equation at the top node of the circuit after the switch is closed yields

$$C\frac{dv_C(t)}{dt} + \frac{v_C(t)}{R} = 0$$

Multiplying by the resistance gives

$$RC\frac{dv_C(t)}{dt} + v_C(t) = 0 \tag{4.1}$$

As expected, we have obtained a differential equation.

Equation 4.1 indicates that the solution for $v_C(t)$ must be a function that has the same form as its first derivative. Of course, a function with this property is an exponential. Thus, we anticipate that the solution is of the form

$$v_C(t) = Ke^{st} \tag{4.2}$$

in which $K$ and $s$ are constants to be determined.

Using Equation 4.2 to substitute for $v_C(t)$ in Equation 4.1, we have

$$RCKse^{st} + Ke^{st} = 0 \tag{4.3}$$

Solving for $s$, we obtain

$$s = \frac{-1}{RC} \tag{4.4}$$

Equation 4.1 indicates that the solution for $v_C(t)$ must be a function that has the same form as its first derivative. The function with this property is an exponential.

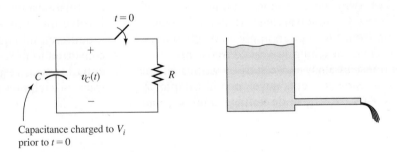

Capacitance charged to $V_i$
prior to $t = 0$

(a) Electrical circuit

(b) Fluid-flow analogy: a filled water tank discharging through a small pipe

**Figure 4.1** A capacitance discharging through a resistance and its fluid-flow analogy. The capacitor is charged to $V_i$ prior to $t = 0$ (by circuitry that is not shown). At $t = 0$, the switch closes and the capacitor discharges through the resistor.

Substituting this into Equation 4.2, we see that the solution is

$$v_C(t) = Ke^{-t/RC} \tag{4.5}$$

Referring to Figure 4.1(a), we reason that the voltage across the capacitor cannot change instantaneously when the switch closes. This is because the current through the capacitance is $i_C(t) = C\,dv_C/dt$. In order for the voltage to change instantaneously, the current would have to be infinite. Since the voltage is finite, the current in the resistance must be finite, and we conclude that the voltage across the capacitor must be continuous. Thus, we write

Because the current is finite, the voltage across the capacitor cannot change instantaneously when the switch closes.

$$v_C(0+) = V_i \tag{4.6}$$

in which $v_C(0+)$ represents the voltage immediately after the switch closes. Substituting into Equation 4.5, we have

$$v_C(0+) = V_i = Ke^0 = K \tag{4.7}$$

Hence, we conclude that the constant $K$ equals the initial voltage across the capacitor. Finally, the solution for the voltage is

$$v_C(t) = V_i e^{-t/RC} \tag{4.8}$$

A plot of the voltage is shown in Figure 4.2. Notice that the capacitor voltage decays exponentially to zero.

The time interval

$$\tau = RC \tag{4.9}$$

The time interval $\tau = RC$ is called the time constant of the circuit.

is called the **time constant** of the circuit. In one time constant, the voltage decays by the factor $e^{-1} \cong 0.368$. After about five time constants, the voltage remaining on the capacitor is negligible compared with the initial value.

An analogous fluid-flow system is shown in Figure 4.1(b). The tank initially filled with water is analogous to the charged capacitor. Furthermore, the small pipe is analogous to the resistor. At first, when the tank is full, the flow is large and the water level drops fast. As the tank empties, the flow decreases.

At one time constant, the voltage across a capacitance discharging through a resistance is $e^{-1} \cong 0.368$ times its initial value. After about three to five time constants, the capacitance is almost totally discharged.

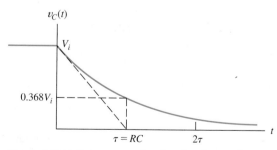

**Figure 4.2** Voltage versus time for the circuit of Figure 4.1(a). When the switch is closed, the voltage across the capacitor decays exponentially to zero. At one time constant, the voltage is equal to 36.8 percent of its initial value.

In the past, engineers have frequently applied *RC* circuits in timing applications. For example, suppose that when a garage door opens or closes, a light is to be turned on and is to remain on for 30 s. To achieve this objective, we could design a circuit consisting of (1) a capacitor that is charged to an initial voltage $V_i$ while the door opener is energized, (2) a resistor through which the capacitor discharges, and (3) a sensing circuit that keeps the light on as long as the capacitor voltage is larger than $0.368 V_i$. If we choose the time constant $\tau = RC$ to be 30 s, the desired operation is achieved.

(In modern designs, a typical garage door opener contains a small computer, known as a microcontroller, and software that counts seconds for timing purposes. We discuss microcontrollers in Chapter 8.)

---

**Example 4.1**    **Capacitance Discharging Through a Resistance**

The circuit of Figure 4.1(a) has $R = 2\ \text{M}\Omega$, $C = 3\ \mu\text{F}$, and $V_i = 100$ V. Determine the value of time $t_x$ for which $v_C(t) = 25$ V.

**Solution**    The voltage is given by Equation 4.8:

$$v_C(t) = V_i e^{-t/RC} \quad \text{for } t > 0$$

in which the time constant is $\tau = RC = (2\ \text{M}\Omega) \times (3\ \mu\text{F}) = 6$ s. Substituting values, we have

$$v_C(t_x) = 25 = 100 e^{-t_x/6}$$

Dividing both sides by 100, we have

$$0.25 = e^{-t_x/6}$$

Then, taking the natural logarithm of both sides, we obtain:

$$\ln(0.25) = -t_x/6$$

$$t_x = -6\ln(0.25)$$

$$t_x = 8.3178 \text{ s}$$

■

---

## Charging a Capacitance from a DC Source through a Resistance

Next, consider the circuit shown in Figure 4.3. The source voltage $V_s$ is constant–in other words, we have a dc source. The source is connected to the *RC* circuit by a switch that closes at $t = 0$. We assume that the initial voltage across the capacitor

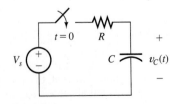

**Figure 4.3** Capacitance charging through a resistance. The switch closes at $t = 0$, connecting the dc source $V_s$ to the circuit.

just before the switch closes is $v_C(0-) = 0$. Let us solve for the voltage across the capacitor as a function of time.

We start by writing a current equation at the node that joins the resistor and the capacitor. This yields

$$C\frac{dv_C(t)}{dt} + \frac{v_C(t) - V_s}{R} = 0 \qquad (4.10)$$

The first term on the left-hand side is the current referenced downward through the capacitor. The second term is the current referenced toward the left through the resistor. KCL requires that the currents leaving the node sum to zero.

Rearranging Equation 4.10, we obtain

$$RC\frac{dv_C(t)}{dt} + v_C(t) = V_s \qquad (4.11)$$

As expected, we have obtained a linear first-order differential equation with constant coefficients. As in the previous circuit, the voltage across the capacitance cannot change instantaneously because the voltages are finite, and thus, the current through the resistance (and therefore through the capacitance) is finite. Infinite current is required to change the voltage across a capacitance in an instant. Thus, we have

$$v_C(0+) = v_C(0-) = 0 \qquad (4.12)$$

Now, we need to find a solution for $v_C(t)$ that (1) satisfies Equation 4.11 and (2) matches the initial conditions of the circuit stated in Equation 4.12. Notice that Equation 4.11 is the same as Equation 4.1, except for the constant on the right-hand side. Thus, we expect the solution to be the same as for Equation 4.1, except for an added constant term. Thus, we are led to try the solution

$$v_C(t) = K_1 + K_2 e^{st} \qquad (4.13)$$

in which $K_1$, $K_2$, and $s$ are constants to be determined.

If we use Equation 4.13 to substitute for $v_C(t)$ in Equation 4.11, we obtain

$$(1 + RCs)K_2 e^{st} + K_1 = V_s \qquad (4.14)$$

For equality, the coefficient of $e^{st}$ must be zero. This leads to

$$s = \frac{-1}{RC} \qquad (4.15)$$

From Equation 4.14, we also have

$$K_1 = V_s \qquad (4.16)$$

Using Equations 4.15 and 4.16 to substitute into Equation 4.13, we obtain

$$v_C(t) = V_s + K_2 e^{-t/RC} \qquad (4.17)$$

in which $K_2$ remains to be determined.

$v_C(0-)$ is the voltage across the capacitor the instant before the switch closes (at $t = 0$). Similarly, $v_C(0+)$ is the voltage across the capacitor the instant after the switch closes.

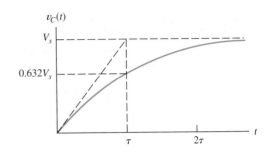

**Figure 4.4** The charging transient for the *RC* circuit of Figure 4.3.

Now, we use the initial condition (Equation 4.12) to find $K_2$. We have

$$v_C(0+) = 0 = V_s + K_2 e^0 = V_s + K_2 \qquad (4.18)$$

from which we find $K_2 = -V_s$. Finally, substituting into Equation 4.17, we obtain the solution

$$v_C(t) = V_s - V_s e^{-t/RC} \qquad (4.19)$$

When a dc source is contained in the circuit, the total response contains two parts: forced (or steady-state) and transient.

The second term on the right-hand side is called the **transient response**, which eventually decays to negligible values. The first term on the right-hand side is the **steady-state response**, also called the **forced response**, which persists after the transient has decayed.

Here again, the product of the resistance and capacitance has units of seconds and is called the time constant $\tau = RC$. Thus, the solution can be written as

$$v_C(t) = V_s - V_s e^{-t/\tau} \qquad (4.20)$$

In the case of a capacitance charging from a dc source through a resistance, a straight line tangent to the start of the transient reaches the final value at one time constant.

A plot of $v_C(t)$ is shown in Figure 4.4. Notice that $v_C(t)$ starts at 0 and approaches the final value $V_s$ asymptotically as $t$ becomes large. After one time constant, $v_C(t)$ has reached 63.2 percent of its final value. For practical purposes, $v_C(t)$ is equal to its final value $V_s$ after about five time constants. Then, we say that the circuit has reached steady state.

It can be shown that if the initial slope of $v_C$ is extended, it intersects the final value at one time constant as shown in Figure 4.4.

We have seen in this section that several time constants are needed to charge or discharge a capacitance. This is the main limitation on the speed at which digital computers can process data. In a typical computer, information is represented by voltages that nominally assume values of either +1.8 or 0 V, depending on the data represented. When the data change, the voltages must change. It is impossible to build circuits that do not have some capacitance that is charged or discharged when voltages change in value. Furthermore, the circuits always have nonzero resistances that limit the currents available for charging or discharging the capacitances. Therefore, a nonzero time constant is associated with each circuit in the computer, limiting its speed. We will learn more about digital computer circuits in later chapters.

*RC* transients are the main limitation on the speed at which computer chips can operate.

**Example 4.2    First-Order *RC* Circuit**

The switch in the circuit of Figure 4.5(a) has been open for a very long time prior to $t = 0$ and closes at $t = 0$. Find an expression for $v_C(t)$ for $t > 0$.

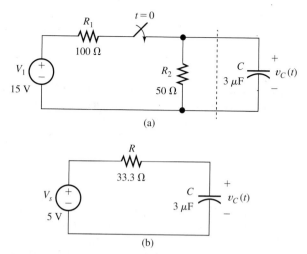

Figure 4.5  Circuit of Example 4.2.

**Solution**   While the switch is open, the capacitance discharges through $R_2$. Because the switch has been open for a very long time, we conclude that $v_C(0-) = 0$. Furthermore, infinite current is not possible in this circuit, so the $v_C(t)$ cannot change instantly. Thus, we conclude that $v_C(0+) = 0$.

We can find the Thévenin equivalent circuit for the portion of the circuit on the left hand side of the dotted line shown in Figure 4.5(a). This is the circuit of Example 2.18 on page 92. The resulting Thévenin equivalent, with some changes in notation, is shown in Figure 4.5(b).

The circuit in Figure 4.5(b) is the same as the circuit of Figure 4.3, and the voltage is given by Equation 4.20:

$$v_C(t) = V_s - V_s e^{(-t/RC)} \quad \text{for} \quad t > 0$$

in which the time constant is $\tau = RC = (33.3\ \Omega) \times (3\ \mu F) = 100\ \mu s$.
Substituting these values, we have

$$v_C(t) = 5 - 5e^{(-10000t)} \quad \text{V} \quad \text{for} \quad t > 0 \qquad \blacksquare$$

**Exercise 4.1**   Suppose that $R = 5000\ \Omega$ and $C = 1\ \mu F$ in the circuit of Figure 4.1(a). Find the time at which the voltage across the capacitor reaches 1 percent of its initial value.
**Answer**   $t = -5\ln(0.01)$ ms $\cong 23$ ms.  □

**Exercise 4.2**   Show that if the initial slope of $v_C(t)$ is extended, it intersects the final value at one time constant, as shown in Figure 4.4. [The expression for $v_C(t)$ is given in Equation 4.20.]  □

## 4.2  DC STEADY STATE

The transient terms in the expressions for currents and voltages in $RLC$ circuits decay to zero with time. (An exception is $LC$ circuits having no resistance.) For dc sources, the steady-state currents and voltages are also constant.

The transient terms in the expressions for currents and voltages in $RLC$ circuits decay to zero with time.

Consider the equation for current through a capacitance:

$$i_C(t) = C\frac{dv_C(t)}{dt}$$

If the voltage $v_C(t)$ is constant, the current is zero. In other words, the capacitance behaves as an open circuit. Thus, we conclude that *for steady-state conditions with dc sources, capacitances behave as open circuits.*

Similarly, for an inductance, we have

$$v_L(t) = L\frac{di_L(t)}{dt}$$

The steps in determining the forced response for *RLC* circuits with dc sources are
1. Replace capacitances with open circuits.
2. Replace inductances with short circuits.
3. Solve the remaining circuit.

When the current is constant, the voltage is zero. Thus, we conclude that *for steady-state conditions with dc sources, inductances behave as short circuits.*

These observations give us another approach to finding the steady-state solutions to circuit equations for *RLC* circuits with constant sources. First, we replace the capacitors by open circuits and the inductors by short circuits. The circuit then consists of dc sources and resistances. Finally, we solve the equivalent circuit for the steady-state currents and voltages.

---

**Example 4.3    Steady-State DC Analysis**

Find $v_x$ and $i_x$ for the circuit shown in Figure 4.6(a) for $t \gg 0$.

**Solution**    After the switch has been closed a long time, we expect the transient response to have decayed to zero. Then the circuit is operating in dc steady-state conditions. We start our analysis by replacing the inductor by a short circuit and the capacitor by an open circuit. The equivalent circuit is shown in Figure 4.6(b).

Steps 1 and 2.

Step 3.

This resistive circuit is readily solved. The resistances $R_1$ and $R_2$ are in series. Thus, we have

$$i_x = \frac{10}{R_1 + R_2} = 1\text{ A}$$

and

$$v_x = R_2 i_x = 5\text{ V}$$ ∎

Sometimes, we are only interested in the steady-state operation of circuits with dc sources. For example, in analyzing the headlight circuits in an automobile, we

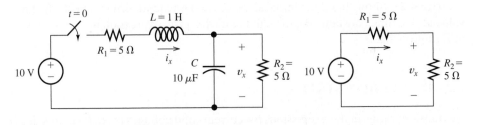

(a) Original circuit                                   (b) Equivalent circuit for steady state

Figure 4.6 The circuit and its dc steady-state equivalent for Example 4.3.

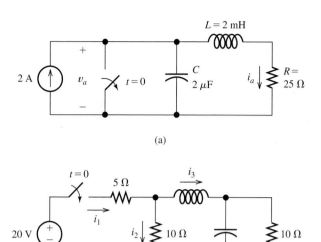

Figure 4.7  Circuits for Exercise 4.3.

are concerned primarily with steady state. On the other hand, we must consider transients in analyzing the operation of the ignition system.

In other applications, we are interested in steady-state conditions with sinusoidal ac sources. For sinusoidal sources, the steady-state currents and voltages are also sinusoidal. In Chapter 5, we study a method for solving sinusoidal steady-state circuits that is similar to the method we have presented here for dc steady state. Instead of short and open circuits, we will replace inductances and capacitances by impedances, which are like resistances, except that impedances can have imaginary values.

**Exercise 4.3**   Solve for the steady-state values of the labeled currents and voltages for the circuits shown in Figure 4.7.
**Answer**   **a.** $v_a = 50$ V, $i_a = 2$ A; **b.** $i_1 = 2$ A, $i_2 = 1$ A, $i_3 = 1$ A.   □

## 4.3   *RL* CIRCUITS

In this section, we consider circuits consisting of dc sources, resistances, and a single inductance. The methods and solutions are very similar to those we studied for *RC* circuits in Section 4.1.

The steps involved in solving simple circuits containing dc sources, resistances, and one energy-storage element (inductance or capacitance) are as follows:

1. Apply Kirchhoff's current and voltage laws to write the circuit equation.
2. If the equation contains integrals, differentiate each term in the equation to produce a pure differential equation.
3. Assume a solution of the form $K_1 + K_2 e^{st}$.
4. Substitute the solution into the differential equation to determine the values of $K_1$ and $s$. (Alternatively, we can determine $K_1$ by solving the circuit in steady state as discussed in Section 4.2.)

**5.** Use the initial conditions to determine the value of $K_2$.

**6.** Write the final solution.

---

**Example 4.4**    *RL Transient Analysis*

Consider the circuit shown in Figure 4.8. Find the current $i(t)$ and the voltage $v(t)$.

**Solution**    First, we find the current $i(t)$. Of course, prior to $t = 0$, the switch is open and the current is zero:

$$i(t) = 0 \quad \text{for } t < 0 \tag{4.21}$$

After the switch is closed, the current increases in value eventually reaching a steady-state value.

Step 1.                Writing a KVL equation around the loop, we have

Step 2 is not needed in this case.

$$Ri(t) + L\frac{di}{dt} = V_s \tag{4.22}$$

Step 3.          This is very similar to Equation 4.11, and we are, therefore, led to try a solution of the same form as that given by Equation 4.13. Thus, our trial solution is

$$i(t) = K_1 + K_2 e^{st} \tag{4.23}$$

Step 4.          in which $K_1$, $K_2$, and $s$ are constants that need to be determined. Following the procedure used in Section 4.1, we substitute the trial solution into the differential equation, resulting in

$$RK_1 + (RK_2 + sLK_2)e^{st} = V_s \tag{4.24}$$

from which we obtain

$$K_1 = \frac{V_s}{R} = 2 \tag{4.25}$$

and

$$s = \frac{-R}{L} \tag{4.26}$$

Substituting these values into Equation 4.23 results in

$$i(t) = 2 + K_2 e^{-tR/L} \tag{4.27}$$

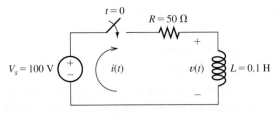

**Figure 4.8** The circuit analyzed in Example 4.4.

Step 5.

Next, we use the initial conditions to determine the value of $K_2$. The current in the inductor is zero prior to $t = 0$ because the switch is open. The applied voltage is finite, and the inductor current must be continuous (because $v_L = L \, di/dt$). Thus, immediately after the switch is closed, the current must be zero. Hence, we have

$$i(0+) = 0 = 2 + K_2e^0 = 2 + K_2 \qquad (4.28)$$

Solving, we find that $K_2 = -2$.

Substituting into Equation 4.27, we find that the solution for the current is

$$i(t) = 2 - 2e^{-t/\tau} \qquad \text{for } t > 0 \qquad (4.29)$$

Step 6.

in which the time constant is given by

$$\tau = \frac{L}{R} \qquad (4.30)$$

A plot of the current versus time is shown in Figure 4.9(a). Notice that the current increases from zero to the steady-state value of 2 A. After five time constants, the current is within 99 percent of the final value. As a check, we verify that the steady-state current is 2 A. (As we saw in Section 4.2, this value can be obtained directly by treating the inductor as a short circuit.)

Now, we consider the voltage $v(t)$. Prior to $t = 0$, with the switch open, the voltage is zero.

$$v(t) = 0 \quad \text{for } t < 0 \qquad (4.31)$$

After $t = 0$, $v(t)$ is equal to the source voltage minus the drop across $R$. Thus, we have

$$v(t) = 100 - 50i(t) \qquad \text{for } t > 0 \qquad (4.32)$$

Substituting the expression found earlier for $i(t)$, we obtain

$$v(t) = 100e^{-t/\tau} \qquad (4.33)$$

A plot of $v(t)$ is shown in Figure 4.9(b).

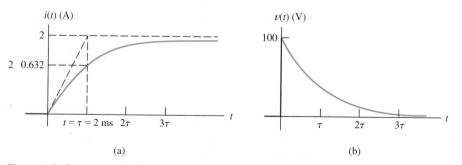

(a)                                                  (b)

**Figure 4.9** Current and voltage versus time for the circuit of Figure 4.8.

At $t = 0$, the voltage across the inductor jumps from 0 to 100 V. As the current gradually increases, the drop across the resistor increases, and the voltage across the inductor falls. In steady state, we have $v(t) = 0$ because the inductor behaves as a short circuit.    ■

After solving several circuits with a single energy-storage element, we can use our experience to skip some of the steps listed earlier in the section. We illustrate this in the next example.

### Example 4.5    *RL* Transient Analysis

Consider the circuit shown in Figure 4.10 in which $V_s$ is a dc source. Assume that the circuit is in steady state with the switch closed prior to $t = 0$. Find expressions for the current $i(t)$ and the voltage $v(t)$.

**First, we use dc steady-state analysis to determine the current before the switch opens.**

**Solution**    Prior to $t = 0$, the inductor behaves as a short circuit. Thus, we have

$$v(t) = 0 \qquad \text{for } t < 0$$

and

$$i(t) = \frac{V_s}{R_1} \qquad \text{for } t < 0$$

Before the switch opens, current circulates clockwise through $V_s$, $R_1$, and the inductance. When the switch opens, current continues to flow through the inductance, but the return path is through $R_2$. Then, a voltage appears across $R_2$ and the inductance, causing the current to decay.

**After the switch opens, the source is disconnected from the circuit, so the steady-state solution for $t > 0$ is zero.**

Since there are no sources driving the circuit after the switch opens, the steady-state solution is zero for $t > 0$. Hence, the solution for $i(t)$ is given by

$$i(t) = Ke^{-t/\tau} \qquad \text{for } t > 0 \tag{4.34}$$

in which the time constant is

$$\tau = \frac{L}{R_2} \tag{4.35}$$

Unless an infinite voltage appears across the inductance, the current must be continuous. Recall that prior to $t = 0$, $i(t) = V_s/R_1$. Consequently, just after the switch opens, we have

$$i(0+) = \frac{V_s}{R_1} = Ke^{-0} = K$$

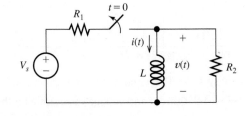

Figure 4.10  The circuit analyzed in Example 4.5.

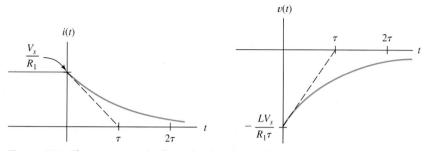

Figure 4.11  The current and voltage for the circuit of Figure 4.10.

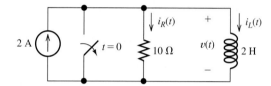

Figure 4.12  The circuit for
Exercise 4.5.

Substituting the value of $K$ into Equation 4.34, we find that the current is

$$i(t) = \frac{V_s}{R_1}e^{-t/\tau} \qquad \text{for } t > 0 \qquad (4.36)$$

The voltage is given by

$$v(t) = L\frac{di(t)}{dt}$$

$$= 0 \qquad \text{for } t < 0$$

$$= -\frac{LV_s}{R_1\tau}e^{-t/\tau} \qquad \text{for } t > 0$$

Plots of the voltage and current are shown in Figure 4.11.    ■

**Exercise 4.4**  For the circuit of Example 4.5 (Figure 4.10), assume that $V_s = 15$ V, $R_1 = 10\ \Omega$, $R_2 = 100\ \Omega$, and $L = 0.1$ H. **a.** What is the value of the time constant (after the switch opens)? **b.** What is the maximum magnitude of $v(t)$? **c.** How does the maximum magnitude of $v(t)$ compare to the source voltage? **d.** Find the time $t$ at which $v(t)$ is one-half of its value immediately after the switch opens.
**Answer**  **a.** $\tau = 1$ ms; **b.** $|v(t)|_{\max} = 150$ V; **c.** the maximum magnitude of $v(t)$ is 10 times the value of $V_s$; **d.** $t = \tau \ln(2) = 0.693$ ms.    □

**Exercise 4.5**  Consider the circuit shown in Figure 4.12, in which the switch opens at $t = 0$. Find expressions for $v(t)$, $i_R(t)$, and $i_L(t)$ for $t > 0$. Assume that $i_L(t)$ is zero before the switch opens.
**Answer**  $v(t) = 20e^{-t/0.2}$, $i_R(t) = 2e^{-t/0.2}$, $i_L(t) = 2 - 2e^{-t/0.2}$.    □

**Exercise 4.6**  Consider the circuit shown in Figure 4.13. Assume that the switch has been closed for a very long time prior to $t = 0$. Find expressions for $i(t)$ and $v(t)$.

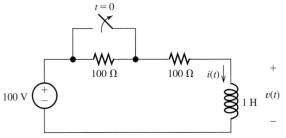

**Figure 4.13** The circuit for Exercise 4.6.

**Answer**

$$i(t) = 1.0 \quad \text{for } t < 0$$

$$= 0.5 + 0.5e^{-t/\tau} \quad \text{for } t > 0$$

$$v(t) = 0 \quad \text{for } t < 0$$

$$= -100e^{-t/\tau} \quad \text{for } t > 0$$

where the time constant is $\tau = 5$ ms.                          □

## 4.4   *RC* AND *RL* CIRCUITS WITH GENERAL SOURCES

Now that we have gained some familiarity with *RL* and *RC* circuits, we discuss their solution in general. In this section, we treat circuits that contain one energy-storage element, either an inductance or a capacitance.

Consider the circuit shown in Figure 4.14(a). The circuit inside the box can be any combination of resistances and sources. The single inductance *L* is shown explicitly. Recall that we can find a Thévenin equivalent for circuits consisting of sources and resistances. The Thévenin equivalent is an independent voltage source $v_t(t)$ in series with the Thévenin resistance *R*. Thus, any circuit composed of sources, resistances, and one inductance has the equivalent circuit shown in Figure 4.14(b). (Of course, we could reduce any circuit containing sources, resistances, and a single capacitance in a similar fashion.)

Writing a KVL equation for Figure 4.14(b), we obtain

$$L\frac{di(t)}{dt} + Ri(t) = v_t(t) \tag{4.37}$$

If we divide through by the resistance *R*, we have

$$\frac{L}{R}\frac{di(t)}{dt} + i(t) = \frac{v_t(t)}{R} \tag{4.38}$$

In general, the equation for any circuit containing one inductance or one capacitance can be put into the form

$$\tau\frac{dx(t)}{dt} + x(t) = f(t), \tag{4.39}$$

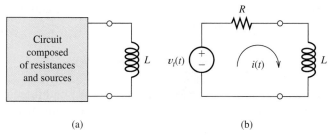

(a)                    (b)

**Figure 4.14** A circuit consisting of sources, resistances, and one inductance has an equivalent circuit consisting of a voltage source and a resistance in series with the inductance.

in which $x(t)$ represents the current or voltage for which we are solving. Then, we need to find solutions to Equation 4.39 that are consistent with the initial conditions (such as the initial current in the inductance).

The constant $\tau$ (which turns out to be the time constant) is a function of only the resistances and the inductance (or capacitance). The sources result in the term $f(t)$, which is called the **forcing function**. If we have a circuit without sources (such as Figure 4.1), the forcing function is zero. For dc sources, the forcing function is constant.

Equation 4.39 is called a first-order differential equation because the highest-order derivative is first order. It is a linear equation because it does not involve powers or other nonlinear functions of $x(t)$ or its derivatives. Thus, to solve an *RL* (or *RC*) circuit, we must find the general solution of a linear first-order differential equation with constant coefficients.

## Solution of the Differential Equation

An important result in differential equations states that the general solution to Equation 4.39 consists of two parts. The first part is called the **particular solution** $x_p(t)$ and is any expression that satisfies Equation 4.39. Thus,

$$\tau \frac{dx_p(t)}{dt} + x_p(t) = f(t) \qquad (4.40)$$

The particular solution is also called the **forced response** because it depends on the forcing function (which in turn is due to the independent sources).

Even though the particular solution satisfies the differential equation, it may not be consistent with the initial conditions, such as the initial voltage on a capacitance or current through an inductance. By adding another term, known as the complementary solution, we obtain a general solution that satisfies both the differential equation and meets the initial conditions.

For the forcing functions that we will encounter, we can often select the form of the particular solution by inspection. Usually, the particular solution includes terms with the same functional forms as the terms found in the forcing function and its derivatives.

Sinusoidal functions of time are one of the most important types of forcing functions in electrical engineering. For example, consider the forcing function

$$f(t) = 10\cos(200t)$$

*The general solution to Equation 4.39 consists of two parts.*

*The particular solution (also called the forced response) is any expression that satisfies the equation.*

*In order to have a solution that satisfies the initial conditions, we must add the complementary solution to the particular solution.*

Because the derivatives of sine and cosine functions are also sine and cosine functions, we would try a particular solution of the form

$$x_p(t) = A\cos(200t) + B\sin(200t)$$

where $A$ and $B$ are constants that must be determined. We find these constants by substituting the proposed solution into the differential equation and requiring the two sides of the equation to be identical. This leads to equations that can be solved for $A$ and $B$. (In Chapter 5, we study shortcut methods for solving for the forced response of circuits with sinusoidal sources.)

The second part of the general solution is called the **complementary solution** $x_c(t)$ and is the solution of the **homogeneous equation**

> The homogeneous equation is obtained by setting the forcing function to zero.

$$\tau\frac{dx_c(t)}{dt} + x_c(t) = 0 \qquad (4.41)$$

> The complementary solution (also called the natural response) is obtained by solving the homogeneous equation.

We obtain the homogeneous equation by setting the forcing function to zero. Thus, the form of the complementary solution does not depend on the sources. It is also called the **natural response** because it depends on the passive circuit elements. The complementary solution must be added to the particular solution in order to obtain a general solution that matches the initial values of the currents and voltages.

We can rearrange the homogeneous equation into this form:

$$\frac{dx_c(t)/dt}{x_c(t)} = \frac{-1}{\tau} \qquad (4.42)$$

Integrating both sides of Equation 4.42, we have

$$\ln[x_c(t)] = \frac{-t}{\tau} + c \qquad (4.43)$$

in which $c$ is the constant of integration. Equation 4.43 is equivalent to

$$x_c(t) = e^{(-t/\tau + c)} = e^c e^{-t/\tau}$$

Then, if we define $K = e^c$, we have the complementary solution

$$x_c(t) = Ke^{-t/\tau} \qquad (4.44)$$

## Step-by-Step Solution

Next, we summarize an approach to solving circuits containing a resistance, a source, and an inductance (or a capacitance):

1. Write the circuit equation and reduce it to a first-order differential equation.
2. Find a particular solution. The details of this step depend on the form of the forcing function. We illustrate several types of forcing functions in examples, exercises, and problems.

**3.** Obtain the complete solution by adding the particular solution to the complementary solution given by Equation 4.44, which contains the arbitrary constant $K$.

**4.** Use initial conditions to find the value of $K$.

We illustrate this procedure with an example.

| Example 4.6 | Transient Analysis of an *RC* Circuit with a Sinusoidal Source |

Solve for the current in the circuit shown in Figure 4.15. The capacitor is initially charged so that $v_C(0+) = 1$ V.

**Solution**    First, we write a voltage equation for $t > 0$. Traveling clockwise and summing voltages, we obtain

*Step 1: Write the circuit equation and reduce it to a first-order differential equation.*

$$Ri(t) + \frac{1}{C}\int_0^t i(t)\, dt + v_C(0) - 2\sin(200t) = 0$$

We convert this to a differential equation by taking the derivative of each term. Of course, the derivative of the integral is simply the integrand. Because $v_C(0)$ is a constant, its derivative is zero. Thus, we have

$$R\frac{di(t)}{dt} + \frac{1}{C}i(t) = 400\cos(200t) \tag{4.45}$$

Multiplying by $C$, we get

$$RC\frac{di(t)}{dt} + i(t) = 400\,C\cos(200t) \tag{4.46}$$

Substituting values for $R$ and $C$, we obtain

$$5\times10^{-3}\frac{di(t)}{dt} + i(t) = 400\times10^{-6}\cos(200t) \tag{4.47}$$

The second step is to find a particular solution $i_p(t)$. Often, we start by guessing at the form of $i_p(t)$, possibly including some unknown constants. Then, we substitute our guess into the differential equation and solve for the constants. In the present case, since the derivatives of $\sin(200t)$ and $\cos(200t)$ are $200\cos(200t)$ and $-200\sin(200t)$, respectively, we try a particular solution of the form

*Step 2: Find a particular solution.*

*The particular solution for a sinusoidal forcing function always has the form given by Equation 4.48.*

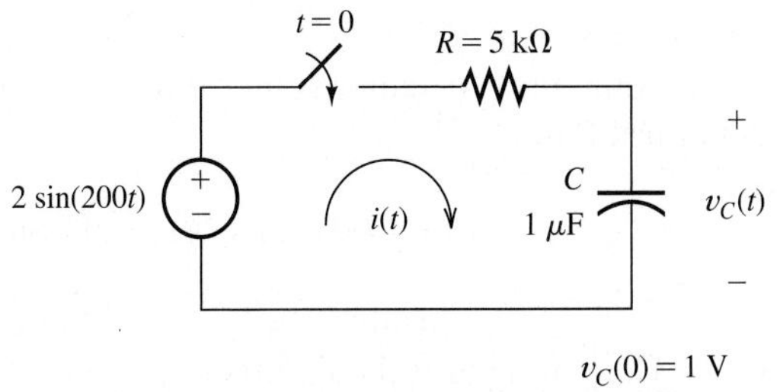

$$i_p(t) = A\cos(200t) + B\sin(200t) \tag{4.48}$$

**Figure 4.15** A first-order *RC* circuit with a sinusoidal source. See Example 4.6.

We substitute Equation 4.48 into the differential equation, and solve for $A$ and $B$.

where $A$ and $B$ are constants to be determined so that $i_p$ is indeed a solution to Equation 4.47.

Substituting the proposed solution into Equation 4.47, we obtain

$$- A \sin(200t) + B \cos(200t) + A \cos(200t) + B \sin(200t)$$
$$= 400 \times 10^{-6} \cos(200t)$$

However, the left-hand side of this equation is required to be identical to the right-hand side. Equating the coefficients of the sine functions, we have

$$-A + B = 0 \tag{4.49}$$

Equating the coefficients of the cosine functions, we get

$$B + A = 400 \times 10^{-6} \tag{4.50}$$

These equations can be readily solved, yielding

$$A = 200 \times 10^{-6} = 200 \ \mu A$$

and

$$B = 200 \times 10^{-6} = 200 \ \mu A$$

Substituting these values into Equation 4.48, we obtain the particular solution

$$i_p(t) = 200 \cos(200t) + 200 \sin(200t) \ \mu A \tag{4.51}$$

which can also be written as

$$i_p(t) = 200\sqrt{2} \cos(200t - 45°)$$

(In Chapter 5, we will learn shortcut methods for combining sine and cosine functions.)

We obtain the homogeneous equation by substituting 0 for the forcing function in Equation 4.46. Thus, we have

$$RC\frac{di(t)}{dt} + i(t) = 0 \tag{4.52}$$

The complementary solution is

$$i_c(t) = Ke^{-t/RC} = Ke^{-t/\tau} \tag{4.53}$$

Step 3: Obtain the complete solution by adding the particular solution to the complementary solution.

Adding the particular solution and the complementary solution, we obtain the general solution

$$i(t) = 200 \cos(200t) + 200 \sin(200t) + Ke^{-t/RC} \ \mu A \tag{4.54}$$

Step 4: Use initial conditions to find the value of $K$.

Finally, we determine the value of the constant $K$ by using the initial conditions. The voltages and currents immediately after the switch closes are shown in Figure 4.16. The source voltage is 0 V and the voltage across the capacitor is

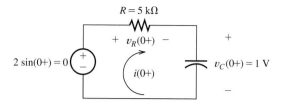

**Figure 4.16** The voltages and currents for the circuit of Figure 4.15 immediately after the switch closes.

$v_C(0+) = 1$. Consequently, the voltage across the resistor must be $v_R(0+) = -1$ V. Thus, we get

$$i(0+) = \frac{v_R(0+)}{R} = \frac{-1}{5000} = -200 \ \mu A$$

Substituting $t = 0$ into Equation 4.54, we obtain

$$i(0+) = -200 = 200 + K \ \mu A \qquad (4.55)$$

Solving, we find that $K = -400 \ \mu A$. Substituting this into Equation 4.54, we have the solution

$$i(t) = 200 \cos(200t) + 200 \sin(200t) - 400e^{-t/RC} \ \mu A \qquad (4.56)$$

Plots of the particular solution and of the complementary solution are shown in Figure 4.17. The time constant for this circuit is $\tau = RC = 5$ ms. Notice that the natural response decays to negligible values in about 25 ms. As expected, the natural response has decayed in about five time constants. Furthermore, notice that for a sinusoidal forcing function, the forced response is also sinusoidal and persists after the natural response has decayed.

A plot of the complete solution is shown in Figure 4.18.  ■

Notice that the forced response is sinusoidal for a sinusoidal forcing function.

**Exercise 4.7**   Repeat Example 4.6 if the source voltage is changed to $2 \cos(200t)$ and the initial voltage on the capacitor is $v_C(0) = 0$. The circuit with these changes is shown in Figure 4.19.

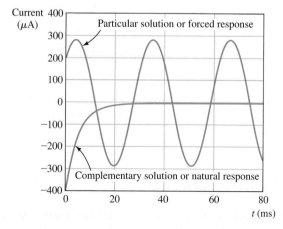

**Figure 4.17** The complementary solution and the particular solution for Example 4.6.

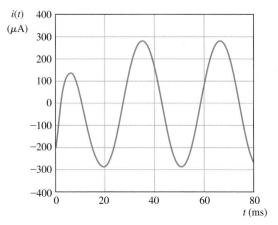

Figure 4.18  The complete solution for Example 4.6.

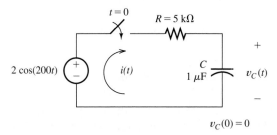

Figure 4.19  The circuit for Exercise 4.7.

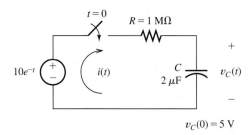

Figure 4.20  The circuit for
Exercise 4.8.

$v_C(0) = 5 \text{ V}$

**Answer**  $i(t) = -200 \sin(200t) + 200 \cos(200t) + 200e^{-t/RC} \; \mu\text{A}$, in which $\tau = RC = 5 \text{ ms}$.

**Exercise 4.8**   Solve for the current in the circuit shown in Figure 4.20 after the switch closes. [*Hint:* Try a particular solution of the form $i_p(t) = Ae^{-t}$.]
**Answer**  $i(t) = 20e^{-t} - 15e^{-t/2} \; \mu\text{A}$.

## 4.5  SECOND-ORDER CIRCUITS

In this section, we consider circuits that contain two energy-storage elements. In particular, we look at circuits that have one inductance and one capacitance, either in series or in parallel.

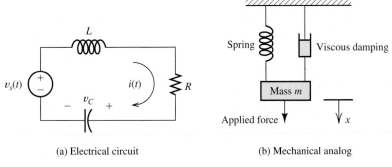

(a) Electrical circuit                    (b) Mechanical analog

**Figure 4.21** The series *RLC* circuit and its mechanical analog.

## Differential Equation

To derive the general form of the equations that we encounter in circuits with two energy-storage elements, consider the series circuit shown in Figure 4.21(a). Writing a KVL equation, we have

$$L\frac{di(t)}{dt} + Ri(t) + \frac{1}{C}\int_0^t i(t)dt + v_C(0) = v_s(t) \tag{4.57}$$

Taking the derivative with respect to time, we get

$$L\frac{d^2i(t)}{dt^2} + R\frac{di(t)}{dt} + \frac{1}{C}i(t) = \frac{dv_s(t)}{dt} \tag{4.58}$$

We convert the integrodifferential equation to a pure differential equation by differentiating with respect to time.

Dividing through by $L$, we obtain

$$\frac{d^2i(t)}{dt^2} + \frac{R}{L}\frac{di(t)}{dt} + \frac{1}{LC}i(t) = \frac{1}{L}\frac{dv_s(t)}{dt} \tag{4.59}$$

Now, we define the **damping coefficient** as

$$\alpha = \frac{R}{2L} \tag{4.60}$$

and the **undamped resonant frequency** as

$$\omega_0 = \frac{1}{\sqrt{LC}} \tag{4.61}$$

The **forcing function** is

$$f(t) = \frac{1}{L}\frac{dv_s(t)}{dt} \tag{4.62}$$

Using these definitions, we find that Equation 4.59 can be written as

$$\frac{d^2i(t)}{dt^2} + 2\alpha\frac{di(t)}{dt} + \omega_0^2 i(t) = f(t) \tag{4.63}$$

This is a linear second-order differential equation with constant coefficients. Thus, we refer to circuits having two energy-storage elements as second-order

If a circuit contains two energy-storage elements (after substituting all possible series or parallel equivalents), the circuit equations can always be reduced to the form given by Equation 4.63.

circuits. (An exception occurs if we can combine the energy-storage elements in series or parallel. For example, if we have two capacitors in parallel, we can combine them into a single equivalent capacitance, and then we would have a first-order circuit.)

## Mechanical Analog

The mechanical analog of the series $RLC$ circuit is shown in Figure 4.21(b). The displacement $x$ of the mass is analogous to electrical charge, the velocity $dx/dt$ is analogous to current, and force is analogous to voltage. The mass plays the role of the inductance, the spring plays the role of the capacitance, and the damper plays the role of the resistance. The equation of motion for the mechanical system can be put into the form of Equation 4.63.

Based on an intuitive consideration of Figure 4.21, we can anticipate that the sudden application of a constant force (dc voltage) can result in a displacement (current) that either approaches steady-state conditions asymptotically or oscillates before settling to the steady-state value. The type of behavior depends on the relative values of the mass, spring constant, and damping coefficient.

## Solution of the Second-Order Equation

We will see that the circuit equations for currents and voltages in circuits having two energy-storage elements can always be put into the form of Equation 4.63. Thus, let us consider the solution of

$$\frac{d^2x(t)}{dt^2} + 2\alpha\frac{dx(t)}{dt} + \omega_0^2 x(t) = f(t) \tag{4.64}$$

where we have used $x(t)$ for the variable, which could represent either a current or a voltage.

Here again, the general solution $x(t)$ to this equation consists of two parts: a particular solution $x_p(t)$ plus the complementary solution $x_c(t)$ and is expressed as

$$x(t) = x_p(t) + x_c(t) \tag{4.65}$$

**Particular Solution.**    The particular solution is any expression $x_p(t)$ that satisfies the differential equation

$$\frac{d^2x_p(t)}{dt^2} + 2\alpha\frac{dx_p(t)}{dt} + \omega_0^2 x_p(t) = f(t) \tag{4.66}$$

The particular solution is also called the **forced response**. (Usually, we eliminate any terms from $x_p(t)$ that produce a zero net result when substituted into the left-hand side of Equation 4.66. In other words, we eliminate any terms that have the same form as the homogeneous solution.)

For dc sources, we can find the particular solution by performing a dc steady-state analysis as discussed in Section 4.2.

We will be concerned primarily with either constant (dc) or sinusoidal (ac) forcing functions. For dc sources, we can find the particular solution directly from the circuit by replacing the inductances by short circuits, replacing the capacitances by open circuits, and solving. This technique was discussed in Section 4.2. In Chapter 5, we will learn efficient methods for finding the forced response due to sinusoidal sources.

**Complementary Solution.**   The complementary solution $x_c(t)$ is found by solving the homogeneous equation, which is obtained by substituting 0 for the forcing function $f(t)$. Thus, the homogeneous equation is

$$\frac{d^2x_c(t)}{dt^2} + 2\alpha\frac{dx_c(t)}{dt} + \omega_0^2 x_c(t) = 0 \qquad (4.67)$$

In finding the solution to the homogeneous equation, we start by substituting the trial solution $x_c(t) = Ke^{st}$. This yields

$$s^2 Ke^{st} + 2\alpha s Ke^{st} + \omega_0^2 Ke^{st} = 0 \qquad (4.68)$$

Factoring, we obtain

$$(s^2 + 2\alpha s + \omega_0^2)Ke^{st} = 0 \qquad (4.69)$$

Since we want to find a solution $Ke^{st}$ that is nonzero, we must have

$$s^2 + 2\alpha s + \omega_0^2 = 0 \qquad (4.70)$$

This is called the **characteristic equation**.
   The **damping ratio** is defined as

$$\zeta = \frac{\alpha}{\omega_0} \qquad (4.71)$$

The form of the complementary solution depends on the value of the damping ratio. The roots of the characteristic equation are given by

$$s_1 = -\alpha + \sqrt{\alpha^2 - \omega_0^2} \qquad (4.72)$$

and

$$s_2 = -\alpha - \sqrt{\alpha^2 - \omega_0^2} \qquad (4.73)$$

We have three cases depending on the value of the damping ratio $\zeta$ compared with unity.

1.  *Overdamped case* ($\zeta > 1$). If $\zeta > 1$ (or equivalently, if $\alpha > \omega_0$), the roots of the characteristic equation are real and distinct. Then the complementary solution is

$$x_c(t) = K_1 e^{s_1 t} + K_2 e^{s_2 t} \qquad (4.74)$$

In this case, we say that the circuit is **overdamped**.
2.  *Critically damped case* ($\zeta = 1$). If $\zeta = 1$ (or equivalently, if $\alpha = \omega_0$), the roots are real and equal. Then, the complementary solution is

$$x_c(t) = K_1 e^{s_1 t} + K_2 t e^{s_1 t} \qquad (4.75)$$

In this case, we say that the circuit is **critically damped**.

*The form of the complementary solution depends on the value of the damping ratio.*

*If the damping ratio is greater than unity, we say that the circuit is overdamped, the roots of the characteristic equation are real, and the complementary solution has the form given in Equation 4.74.*

*If the damping ratio equals unity, the circuit is critically damped, the roots of the characteristic equation are real and equal, and the complementary solution has the form given in Equation 4.75.*

If the damping ratio is less than unity, the roots of the characteristic equation are complex conjugates, and the complementary solution has the form given in Equation 4.77.

**3.** *Underdamped case* ($\zeta < 1$). Finally, if $\zeta < 1$ (or equivalently, if $\alpha < \omega_0$), the roots are complex. (By the term *complex*, we mean that the roots involve the imaginary number $\sqrt{-1}$.) In other words, the roots are of the form

$$s_1 = -\alpha + j\omega_n \quad \text{and} \quad s_2 = -\alpha - j\omega_n$$

in which $j = \sqrt{-1}$ and the **natural frequency** is given by

$$\omega_n = \sqrt{\omega_0^2 - \alpha^2} \tag{4.76}$$

(In electrical engineering, we use $j$ rather than $i$ to stand for the imaginary number $\sqrt{-1}$ because we use $i$ for current.)

For complex roots, the complementary solution is of the form

$$x_c(t) = K_1 e^{-\alpha t} \cos(\omega_n t) + K_2 e^{-\alpha t} \sin(\omega_n t) \tag{4.77}$$

In this case, we say that the circuit is **underdamped**.

---

### Example 4.7    Analysis of a Second-Order Circuit with a DC Source

A dc source is connected to a series *RLC* circuit by a switch that closes at $t = 0$ as shown in Figure 4.22. The initial conditions are $i(0) = 0$ and $v_C(0) = 0$. Write the differential equation for $v_C(t)$. Solve for $v_C(t)$ if $R = 300, 200,$ and $100 \ \Omega$.

First, we write the circuit equations and reduce them to the form given in Equation 4.63.

**Solution**    First, we can write an expression for the current in terms of the voltage across the capacitance:

$$i(t) = C\frac{dv_C(t)}{dt} \tag{4.78}$$

Then, we write a KVL equation for the circuit:

$$L\frac{di(t)}{dt} + Ri(t) + v_C(t) = V_s \tag{4.79}$$

Using Equation 4.78 to substitute for $i(t)$, we get

$$LC\frac{d^2v_C(t)}{dt^2} + RC\frac{dv_C(t)}{dt} + v_C(t) = V_s \tag{4.80}$$

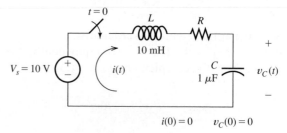

Figure 4.22  The circuit for Example 4.7.

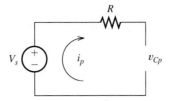

**Figure 4.23** The equivalent circuit for Figure 4.22 under steady-state conditions. The inductor has been replaced by a short circuit and the capacitor by an open circuit.

Dividing through by $LC$, we have

$$\frac{d^2v_C(t)}{dt^2} + \frac{R}{L}\frac{dv_C(t)}{dt} + \frac{1}{LC}v_C(t) = \frac{V_s}{LC} \quad (4.81)$$

As expected, the differential equation for $v_C(t)$ has the same form as Equation 4.63.

Next, we find the particular solution. Since we have a dc source, we can find this part of the solution by replacing the inductance by a short circuit and the capacitance by an open circuit. This is shown in Figure 4.23. Then the current is zero, the drop across the resistance is zero, and the voltage across the capacitance (open circuit) is equal to the dc source voltage. Therefore, the particular solution is

> Next, we find the particular solution by solving the circuit for dc steady-state conditions.

$$v_{Cp}(t) = V_s = 10 \text{ V} \quad (4.82)$$

(It can be verified that this is a particular solution by substituting it into Equation 4.81.) Notice that in this circuit the particular solution for $v_C(t)$ is the same for all three values of resistance.

Next, we find the homogeneous solution and general solution for each value of $R$. For all three cases, we have

> Next, we find the complementary solution for each value of $R$. For each resistance value, we
> 1. Determine the damping ratio and roots of the characteristic equation.

$$\omega_0 = \frac{1}{\sqrt{LC}} = 10^4 \quad (4.83)$$

> 2. Select the appropriate form for the homogeneous solution, depending on the value of the damping ratio.

**Case I** ($R = 300 \ \Omega$)

In this case, we get

$$\alpha = \frac{R}{2L} = 1.5 \times 10^4 \quad (4.84)$$

> 3. Add the homogeneous solution to the particular solution and determine the values of the coefficients ($K_1$ and $K_2$), based on the initial conditions.

The damping ratio is $\zeta = \alpha/\omega_0 = 1.5$. Because we have $\zeta > 1$, this is the overdamped case. The roots of the characteristic equation are given by Equations 4.72 and 4.73. Substituting values, we find that

$$s_1 = -\alpha + \sqrt{\alpha^2 - \omega_0^2}$$

$$= -1.5 \times 10^4 + \sqrt{(1.5 \times 10^4)^2 - (10^4)^2}$$

$$= -0.3820 \times 10^4$$

and

$$s_2 = -\alpha - \sqrt{\alpha^2 - \omega_0^2}$$

$$= -2.618 \times 10^4$$

The homogeneous solution has the form of Equation 4.74. Adding the particular solution given by Equation 4.82 to the homogeneous solution, we obtain the general solution

$$v_C(t) = 10 + K_1 e^{s_1 t} + K_2 e^{s_2 t} \tag{4.85}$$

Now, we must find values of $K_1$ and $K_2$ so the solution matches the known initial conditions in the circuit. It was given that the initial voltage on the capacitance is zero. Hence,

$$v_C(0) = 0$$

Evaluating Equation 4.85 at $t = 0$, we obtain

$$10 + K_1 + K_2 = 0 \tag{4.86}$$

Furthermore, the initial current was given as $i(0) = 0$. Since the current through the capacitance is given by

$$i(t) = C \frac{dv_C(t)}{dt}$$

we conclude that

$$\frac{dv_C(0)}{dt} = 0$$

Taking the derivative of Equation 4.85 and evaluating at $t = 0$, we have

$$s_1 K_1 + s_2 K_2 = 0 \tag{4.87}$$

Now, we can solve Equations 4.86 and 4.87 for the values of $K_1$ and $K_2$. The results are $K_1 = -11.708$ and $K_2 = 1.708$. Substituting these values into Equation 4.85, we have the solution

$$v_C(t) = 10 - 11.708 e^{s_1 t} + 1.708 e^{s_2 t}$$

Plots of each of the terms of this equation and the complete solution are shown in Figure 4.24.

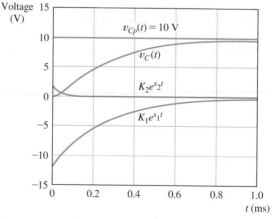

Figure 4.24 Solution for $R = 300 \ \Omega$.

**Case II** ($R = 200\ \Omega$)

In this case, we get

Now, we repeat the steps for $R = 200\ \Omega$.

$$\alpha = \frac{R}{2L} = 10^4 \qquad (4.88)$$

Because $\zeta = \alpha/\omega_0 = 1$, this is the critically damped case. The roots of the characteristic equation are given by Equations 4.72 and 4.73. Substituting values, we have

$$s_1 = s_2 = -\alpha + \sqrt{\alpha^2 - \omega_0^2} = -\alpha = -10^4$$

The homogeneous solution has the form of Equation 4.75. Adding the particular solution (Equation 4.82) to the homogeneous solution, we find that

$$v_C(t) = 10 + K_1 e^{s_1 t} + K_2 t e^{s_1 t} \qquad (4.89)$$

As in case I, the initial conditions require $v_C(0) = 0$ and $dv_C(0)/dt = 0$. Thus, substituting $t = 0$ into Equation 4.89, we get

$$10 + K_1 = 0 \qquad (4.90)$$

Differentiating Equation 4.89 and substituting $t = 0$ yields

$$s_1 K_1 + K_2 = 0 \qquad (4.91)$$

Solving Equations 4.90 and 4.91 yields $K_1 = -10$ and $K_2 = -10^5$. Thus, the solution is

$$v_C(t) = 10 - 10 e^{s_1 t} - 10^5 t e^{s_1 t} \qquad (4.92)$$

Plots of each of the terms of this equation and the complete solution are shown in Figure 4.25.

**Case III** ($R = 100\ \Omega$)

For this value of resistance, we have

$$\alpha = \frac{R}{2L} = 5000 \qquad (4.93)$$

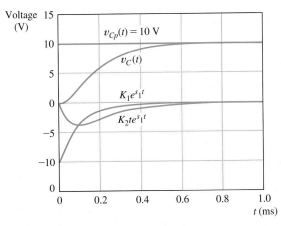

**Figure 4.25** Solution for $R = 200\ \Omega$.

Finally, we repeat the solution for $R = 100\ \Omega$.

Because $\zeta = \alpha/\omega_0 = 0.5$, this is the underdamped case. Using Equation 4.76, we compute the natural frequency:

$$\omega_n = \sqrt{\omega_0^2 - \alpha^2} = 8660 \tag{4.94}$$

The homogeneous solution has the form of Equation 4.77. Adding the particular solution found earlier to the homogeneous solution, we obtain the general solution:

$$v_C(t) = 10 + K_1 e^{-\alpha t} \cos(\omega_n t) + K_2 e^{-\alpha t} \sin(\omega_n t) \tag{4.95}$$

As in the previous cases, the initial conditions are $v_C(0) = 0$ and $dv_C(0)/dt = 0$. Evaluating Equation 4.95 at $t = 0$, we obtain

$$10 + K_1 = 0 \tag{4.96}$$

Differentiating Equation 4.95 and evaluating at $t = 0$, we have

$$-\alpha K_1 + \omega_n K_2 = 0 \tag{4.97}$$

Solving Equations 4.96 and 4.97, we obtain $K_1 = -10$ and $K_2 = -5.774$. Thus, the complete solution is

$$v_C(t) = 10 - 10e^{-\alpha t} \cos(\omega_n t) - 5.774 e^{-\alpha t} \sin(\omega_n t) \tag{4.98}$$

Plots of each of the terms of this equation and the complete solution are shown in Figure 4.26.

Figure 4.27 shows the complete response for all three values of resistance. ∎

## Normalized Step Response of Second-Order Systems

When we suddenly apply a constant source to a circuit, we say that the forcing function is a **step function**. A unit step function, denoted by $u(t)$, is shown in Figure 4.28. By definition, we have

$$u(t) = 0 \quad t < 0$$
$$\qquad = 1 \quad t \geq 0$$

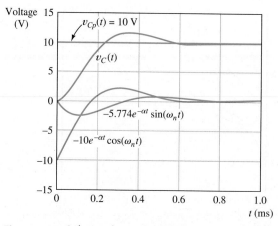

Figure 4.26 Solution for $R = 100\ \Omega$.

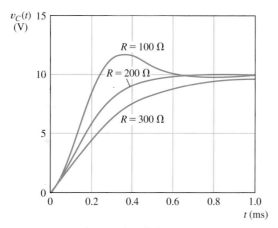

Figure 4.27 Solutions for all three resistances.

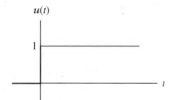

Figure 4.28 A unit step function
$u(t)$. For $t < 0$, $u(t) = 0$. For
$t \geq 0$, $u(t) = 1$.

For example, if we apply a dc voltage of $A$ volts to a circuit by closing a switch, the applied voltage is a step function, given by

$$v(t) = Au(t)$$

This is illustrated in Figure 4.29.

We often encounter situations, such as Example 4.7, in which step forcing functions are applied to second-order systems described by a differential equation of the form

$$\frac{d^2x(t)}{dt^2} + 2\alpha\frac{dx(t)}{dt} + \omega_0^2 x(t) = Au(t) \tag{4.99}$$

The differential equation is characterized by its undamped resonant frequency $\omega_0$ and damping ratio $\zeta = \alpha/\omega_0$. [Of course, the solution for $x(t)$ also depends on the initial conditions.] Normalized solutions are shown in Figure 4.30 for the initial conditions $x(0) = 0$ and $x'(0) = 0$.

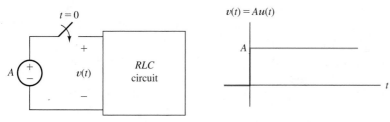

Figure 4.29 Applying a dc voltage by closing a switch results in a forcing function that is a step function.

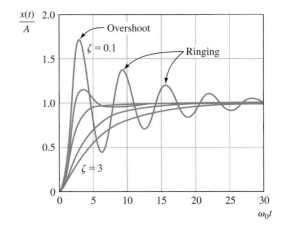

**Figure 4.30** Normalized step responses for second-order systems described by Equation 4.99 with damping ratios of $\zeta$ = 0.1, 0.5, 1, 2, and 3. The initial conditions are assumed to be $x(0)$ = 0 and $x'(0)$ = 0.

The system response for small values of the damping ratio $\zeta$ displays **overshoot** and **ringing** before settling to the steady-state value. On the other hand, if the damping ratio is large (compared to unity), the response takes a relatively long time to closely approach the final value.

Sometimes, we want to design a second-order system that quickly settles to steady state. Then we try to design for a damping ratio close to unity. For example, the control system for a robot arm could be a second-order system. When a step signal calls for the arm to move, we probably want it to achieve the final position in the minimum time without excessive overshoot and ringing.

Frequently, electrical control systems and mechanical systems are best designed with a damping ratio close to unity. For example, when the suspension system on your automobile becomes severely underdamped, it is time for new shock absorbers.

### Circuits with Parallel $L$ and $C$

The solution of circuits having an inductance and capacitance in parallel is very similar to the series case. Consider the circuit shown in Figure 4.31(a). The circuit inside the box is assumed to consist of sources and resistances. As we saw in Section 2.6, we can find a Norton equivalent circuit for any two-terminal circuit composed of resistances and sources. The equivalent circuit is shown in Figure 4.31(b).

We can analyze this circuit by writing a KCL equation at the top node of Figure 4.31(b) which results in

$$C\frac{dv(t)}{dt} + \frac{1}{R}v(t) + \frac{1}{L}\int_0^t v(t)\,dt + i_L(0) = i_n(t) \tag{4.100}$$

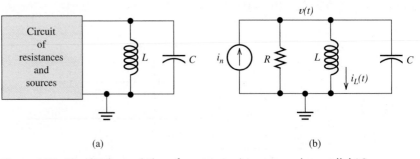

(a)                    (b)

**Figure 4.31** Any circuit consisting of sources, resistances, and a parallel $LC$ combination can be reduced to the equivalent circuit shown in (b).

This can be converted into a pure differential equation by taking the derivative with respect to time:

$$C\frac{d^2v(t)}{dt^2} + \frac{1}{R}\frac{dv(t)}{dt} + \frac{1}{L}v(t) = \frac{di_n(t)}{dt} \tag{4.101}$$

Dividing through by the capacitance, we have

$$\frac{d^2v(t)}{dt^2} + \frac{1}{RC}\frac{dv(t)}{dt} + \frac{1}{LC}v(t) = \frac{1}{C}\frac{di_n(t)}{dt} \tag{4.102}$$

Now, if we define the damping coefficient

$$\alpha = \frac{1}{2RC} \tag{4.103}$$

the undamped resonant frequency

$$\omega_0 = \frac{1}{\sqrt{LC}} \tag{4.104}$$

and the forcing function

$$f(t) = \frac{1}{C}\frac{di_n(t)}{dt} \tag{4.105}$$

the differential equation can be written as

$$\frac{d^2v(t)}{dt^2} + 2\alpha\frac{dv(t)}{dt} + \omega_0^2v(t) = f(t) \tag{4.106}$$

This equation has exactly the same form as Equation 4.64. Therefore, transient analysis of circuits with parallel $LC$ elements is very similar to that of series $LC$ circuits. However, notice that the equation for the damping coefficient $\alpha$ is different for the parallel circuit (in which $\alpha = 1/2RC$) than for the series circuit (in which $\alpha = R/2L$).

Notice that the equation for the damping coefficient of the parallel *RLC circuit is different* from that for the series circuit.

**Exercise 4.9**  Consider the circuit shown in Figure 4.32 with $R = 25\ \Omega$. **a.** Compute the undamped resonant frequency, the damping coefficient, and the damping ratio. **b.** The initial conditions are $v(0-) = 0$ and $i_L(0-) = 0$. Show that this requires that $v'(0+) = 10^6$ V/s. **c.** Find the particular solution for $v(t)$. **d.** Find the general solution for $v(t)$, including the numerical values of all parameters.
**Answer**  **a.** $\omega_0 = 10^5$, $\alpha = 2 \times 10^5$, and $\zeta = 2$; **b.** KCL requires that $i_C(0) = 0.1$ A $= Cv'(0)$, thus $v'(0) = 10^6$; **c.** $v_p(t) = 0$; **d.** $v(t) = 2.89(e^{-0.268\times10^5 t} - e^{-3.73\times10^5 t})$. ☐

v(0−) and $i_L$(0−) are the voltage and current values immediately before the switch opens.

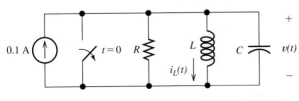

$$L = 1\ \text{mH} \quad C = 0.1\ \mu\text{F}$$

**Figure 4.32**  Circuit for Exercises 4.9, 4.10, and 4.11.

## PRACTICAL APPLICATION   4.1

### Electronics and the Art of Automotive Maintenance

Throughout the early history of the automobile, ignition systems were designed as a straightforward application of electrical transients. The basic ignition system used for many years is shown in Figure PA4.1. The coil is a pair of mutually coupled inductors known as the primary and the secondary. The points form a switch that opens and closes as the engine rotates, opening at the instant that an ignition spark is needed by one of the cylinders. While the points are closed, current builds up relatively slowly in the primary winding of the coil. Then, when the points open, the current is rapidly interrupted. The resulting high rate of change of current induces a large voltage across the secondary winding, which is connected to the appropriate spark plug by the distributor. The resistance is needed to limit the current in case the engine stops with the points closed.

The capacitor prevents the voltage across the points from rising too rapidly when they open. (Recall that the voltage across a capacitance cannot change instantaneously.) Otherwise, arcing would occur across the points, causing them to become burned and pitted. By slowing the rise of voltage, the capacitor gives the gap between the points time to become wide enough to withstand the voltage across them. (Even so, the peak voltage across the points is many times the battery voltage.)

The primary inductance, current-limiting resistance, and capacitance form an underdamped series RLC circuit. Thus, an oscillatory current flows through the primary when the points open, inducing the requisite voltage in the secondary.

In its early forms, the ignition system had mechanical or vacuum systems to make adjustments to the timing, depending on engine speed and throttle setting. In more recent years, the availability of complex electronics at reasonable costs plus the desire to adjust the ignition to obtain good performance and low pollution levels with varying air temperature, fuel quality, air pressure, engine temperature, and other factors have greatly affected the design of ignition systems. The basic principles remain the same as in the days of the classic automobile, but a complex network of electrical sensors, a digital computer, and an electronic switch have replaced the points and simple vacuum advance.

The complexity of modern engineering designs has become somewhat intimidating, even to practicing engineers. In the 1960s, as a new engineering graduate, one could study the design of an ignition system, a radio, or a home appliance, readily spotting and repairing malfunctions with the aid of a few tools and standard parts. Nowadays, if my car should fail to start due to ignition malfunction, at the end of a fishing trip into the backwoods of northern Michigan, I might very well have to walk back to civilization. Nevertheless, the improvements in performance provided by modern electronics make up for its difficulty of repair.

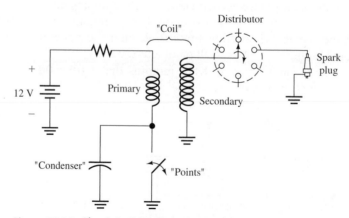

**Figure PA4.1**  Classic ignition for an internal-combustion engine.

**Exercise 4.10**   Repeat Exercise 4.9 for $R = 50 \, \Omega$.
**Answer**   **a.** $\omega_0 = 10^5$, $\alpha = 10^5$, and $\zeta = 1$; **b.** KCL requires that $i_C(0) = 0.1$
$A = Cv'(0)$, thus $v'(0) = 10^6$; **c.** $v_p(t) = 0$; d. $v(t) = 10^6 t e^{-10^5 t}$.

**Exercise 4.11**   Repeat Exercise 4.9 for $R = 250 \, \Omega$.
**Answer**   **a.** $\omega_0 = 10^5$, $\alpha = 0.2 \times 10^5$, and $\zeta = 0.2$; **b.** KCL requires that
$i_C(0) = 0.1 \, A = Cv'(0)$, thus $v'(0) = 10^6$; **c.** $v_p(t) = 0$; **d.** $v(t) = 10.21 e^{-2 \times 10^4 t} \sin(97.98 \times 10^3 t)$.

## 4.6 TRANSIENT ANALYSIS USING THE MATLAB SYMBOLIC TOOLBOX

The MATLAB Symbolic Toolbox greatly facilitates the solution of transients in electrical circuits. It makes the solution of systems of differential equations almost as easy as arithmetic using a calculator. A step-by-step process for solving a circuit in this manner is

1. Write the differential-integral equations for the mesh currents, node voltages, or other circuit variables of interest.

2. If necessary, differentiate the equations to eliminate integrals.

3. Analyze the circuit at $t = 0+$ (i.e., immediately after switches operate) to determine initial conditions for the circuit variables and their derivatives. For a first-order equation, we need the initial value of the circuit variable. For a second-order equation we need the initial values of the circuit variable and its first derivative.

4. Enter the equations and initial values into the dsolve command in MATLAB.

We illustrate with a few examples.

---

**Example 4.8**   **Computer-Aided Solution of a First-Order Circuit**

Solve for $v_L(t)$ in the circuit of Figure 4.33(a). (Note: The argument of the cosine function is in radians.)

**Solution**   First, we write a KCL equation at the node joining the resistance and inductance.

$$\frac{v_L(t) - 20 \cos(100t)}{R} + \frac{1}{L} \int_0^t v_L(t) \, dt + i_L(0) = 0$$

Taking the derivative of the equation to eliminate the integral, multiplying each term by $R$, and substituting values, we eventually obtain

$$\frac{dv_L(t)}{dt} + 100 v_L(t) = -2000 \sin(100t)$$

Next, we need to determine the initial value of $v_L$. Because the switch is open prior to $t = 0$, the initial current in the inductance is zero prior to $t = 0$. Furthermore, the current cannot change instantaneously in this circuit. Thus, we have $i_L(0+) = 0$.

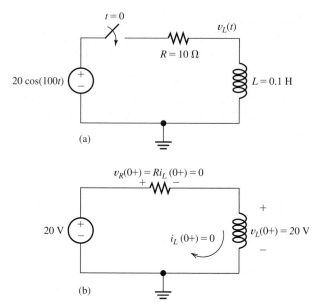

**Figure 4.33** (a) Circuit of Example 4.8. (b) Circuit conditions at $t = 0+$.

Immediately after the switch closes, the voltage source has a value of 20 V, and the current flowing in the circuit is zero, resulting in zero volts across the resistor. Then KVL yields $v_L(0+) = 20$ V. This is illustrated in Figure 4.33(b).

Now, we can write the MATLAB commands. As usual, we show the commands in **boldface**, comments in regular font, and MATLAB responses in color.

```
>> clear all
>> syms VL t
>> % Enter the equation and initial value in the dsolve command.
>> % DVL represents the derivative of VL with respect to time.
>> VL = dsolve('DVL + 100*VL = -2000*sin(100*t)', 'VL(0) = 20');
>> % Print answer with 4 decimal place accuracy for the constants:
>> vpa(VL,4)
     ans =
       10.0*cos(100.0*t)-10.0*sin(100.0*t)+10.0*exp(-100.0*t)
```

In standard mathematical notation, the result becomes

$$v_L(t) = 10 \cos(100t) - 10 \sin(100t) + 10 \exp(-100t)$$

This can be shown to be equivalent to

$$v_L(t) = 14.14 \cos(100t + 0.7854) + 10 \exp(-100t)$$

in which the argument of the cosine function is in radians. Some versions of MATLAB may give this result. **Keep in mind that different versions of the software may give results with different appearances that are mathematically equivalent.**

An m-file named Example_4_8 containing the commands for this example can be found in the MATLAB folder. (See Appendix E for information about access to this folder.) ∎

**Example 4.9**   **Computer-Aided Solution of a Second-Order Circuit**

The switch in the circuit of Figure 4.34(a) is closed for a long time prior to $t = 0$. Assume that $i_L(0+) = 0$. Use MATLAB to solve for $i_L(t)$ and plot the result for $0 \le t \le 2$ ms.

**Solution**   Because this circuit contains two nodes and three meshes, node-voltage analysis is simpler than mesh analysis. We will solve for $v(t)$ and then take $1/L$ times the integral of the voltage to obtain the current through the inductance.

We start the node-voltage analysis by writing the KCL equation at the top node of the circuit (with the switch open).

$$C\frac{dv(t)}{dt} + \frac{v(t)}{R} + \frac{1}{L}\int_0^t v(t)\,dt + i_L(0+) = 0.2\,\exp(-1000t)$$

Taking the derivative of the equation to eliminate the integral and substituting values, we eventually obtain

$$10^{-6}\frac{d^2v(t)}{dt^2} + 4 \times 10^{-3}\frac{dv(t)}{dt} + 250v(t) = -200\,\exp(-1000t)$$

Because this is a second-order equation, we need the initial value for both $v(t)$ and its first derivative. The circuit conditions at $t = 0+$ are shown in Figure 4.34(b). The problem states that the initial current in the inductance is zero. The initial voltage $v(0+)$ is zero, because, with the switch closed, the capacitor is shorted. When the switch opens, the voltage remains zero, because an infinite current would be required to change the capacitor voltage instantaneously. Furthermore, the current flowing through the resistor is zero because the voltage across it is zero. Thus, the 0.2 A from the source must flow through the capacitor, and we have

$$C\frac{dv(0+)}{dt} = 0.2$$

We have established that $v(0+) = 0$ and $v'(0+) = dv(0+)/dt = 0.2 \times 10^6$ V/s.

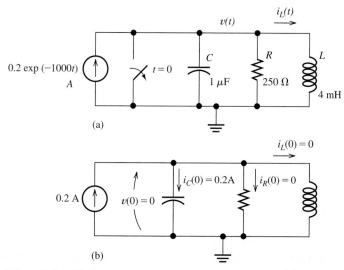

(a)

(b)

**Figure 4.34** (a) Circuit of Example 4.9. (b) Circuit conditions at $t = 0+$.

After the voltage is found, the current is given by

$$i_L(t) = \frac{1}{L}\int_0^t v(t)\,dt = 250\int_0^t v(t)\,dt$$

We use the following MATLAB commands to obtain the solution:

```
>> clear all
>> syms IL V t
>> % Enter the equation and initial values in the dsolve command.
>> % D2V represents the second derivative of V.
>> V = dsolve('(1e-6)*D2V + (4e-3)*DV + 250*V = -200*exp(-1000*t)', ...
             'DV(0)=0.2e6', 'V(0)=0');
>> % Calculate the inductor current by integrating V with respect to t
>> % from 0 to t and multiplying by 1/L:
>> IL = (250)*int(V,t,0,t);
>> % Display the expression for current to 4 decimal place accuracy:
>> vpa(IL,4)
     ans =
     -(0.0008229*(246.0*cos(15688.0*t) - 246.0*exp(1000.0*t) +
         15.68*sin(15688.0*t)))/exp(2000.0*t)
>> ezplot(IL,[0 2e-3])
```

In standard mathematical notation, the result is

$$\begin{aligned} i_L(t) = {} & -0.2024\exp(-2000t)\cos(15680t) - \\ & 0.01290\exp(-2000t)\sin(15680t) + 0.2024\exp(-1000t) \end{aligned}$$

The plot (after some editing to dress it up) is shown in Figure 4.35. An m-file named Example_4_9 containing the commands for this example can be found in the MATLAB folder. (See Appendix E for information about accessing this folder.) ∎

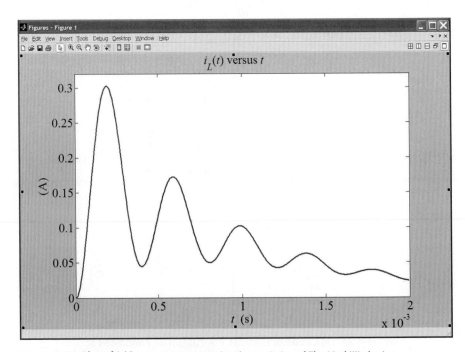

Figure 4.35 Plot of $i_L(t)$ versus $t$. Reprinted with permission of The MathWorks, Inc.

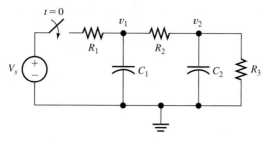

**Figure 4.36**  Circuit of Example 4.10.

$$V_s = 10 \text{ V} \quad R_1 = R_2 = R_3 = 1 \text{ M}\Omega \quad C_1 = C_2 = 1 \text{ }\mu\text{F}$$

## Solving Systems of Linear Differential Equations

So far in this chapter, each of our examples has involved a single differential equation. Circuits that require two or more circuit variables (such as node voltages or mesh currents) result in systems of differential equations. While these systems can be rather formidable to solve by traditional methods, the MATLAB Symbolic Toolbox can solve them with relative ease.

---

**Example 4.10**   Computer-Aided Solution of a System of Differential Equations

Use MATLAB to solve for the node voltages in the circuit of Figure 4.36. The circuit has been connected for a long time prior to $t = 0$ with the switch open, so the initial values of the node voltages are zero.

**Solution**   First, we write the KCL equations at nodes 1 and 2.

$$C_1 \frac{dv_1(t)}{dt} + \frac{v_1(t) - V_s}{R_1} + \frac{v_1(t) - v_2(t)}{R_2} = 0$$

$$C_2 \frac{dv_2(t)}{dt} + \frac{v_2(t) - v_1(t)}{R_2} + \frac{v_2(t)}{R_3} = 0$$

Now substituting values, multiplying each term by $10^6$, and rearranging terms, we have

$$\frac{dv_1(t)}{dt} + 2v_1(t) - v_2(t) = 10$$

$$\frac{dv_2(t)}{dt} + 2v_2(t) - v_1(t) = 0$$

The MATLAB commands and results are:

```
>> clear all
>> syms v1 v2 t
>> [v1 v2] = dsolve('Dv1 + 2*v1 - v2 = 10','Dv2 + 2*v2 -v1 = 0', . . .
                    'v1(0) = 0','v2(0)= 0');
>> v1
        v1 =
        exp(-t)*(5*exp(t) - 5) + exp(-3*t)*((5*exp(3*t))/3 - 5/3)
>> v2
        v2 =
        exp(-t)*(5*exp(t) - 5) - exp(-3*t)*((5*exp(3*t))/3 - 5/3)
```

As usual, keep in mind that different versions of the software can give results different in appearance but mathematically equivalent to that shown here. In standard mathematical notation, the results can be put into the form:

$$v_1(t) = 20/3 - 5 \exp(-t) - (5/3) \exp(-3t)$$
$$v_2(t) = 10/3 - 5 \exp(-t) + (5/3) \exp(-3t)$$

It is always a good idea to perform a few checks on our answers. First, we can verify that the MATLAB results are both zero at $t = 0$ as required by the initial conditions. Furthermore, at $t = \infty$, the capacitors act as open circuits, and the voltage division principle yields $v_1(\infty) = 20/3$ V and $v_2(\infty) = 10/3$. The expressions delivered by MATLAB also yield these values. ∎

**Exercise 4.12**   Use the MATLAB Symbolic Toolbox to solve Example 4.6, obtaining the result given in Equation 4.56 and a plot similar to Figure 4.18 on page 186.

**Answer**   A sequence of commands that produces the solution and the plot is:

```
clear all
syms ix t R C vCinitial w
ix = dsolve('(R*C)*Dix + ix = (w*C)*2*cos(w*t)', 'ix(0)=-vCinitial/R');
ians = subs(ix,[R C vCinitial w],[5000 1e-6 1 200]);
vpa(ians, 4)
ezplot(ians,[0 80e-3])
```

An m-file named Exercise_4_12 containing these commands can be found in the MATLAB folder. (See Appendix E for information about accessing this folder.) □

**Exercise 4.13**   Use the MATLAB Symbolic Toolbox to solve Example 4.7 obtaining the results given in the example for $v_C(t)$ and a plot similar to Figure 4.27 on page 195.

**Answer**   A list of commands that produces the solution and the plot is:

```
clear all
syms vc t
% Case I, R = 300:
vc = dsolve('(1e-8)*D2vc + (1e-6)*300*Dvc+ vc =10', 'vc(0) = 0','Dvc(0)=0');
vpa(vc,4)
ezplot(vc, [0 1e-3])
hold on % Turn hold on so all plots are on the same axes
% Case II, R = 200:
vc = dsolve('(1e-8)*D2vc + (1e-6)*200*Dvc+ vc =10', 'vc(0) = 0','Dvc(0)=0');
vpa(vc,4)
ezplot(vc, [0 1e-3])
% Case III, R = 100:
vc = dsolve('(1e-8)*D2vc + (1e-6)*100*Dvc+ vc =10', 'vc(0) = 0','Dvc(0)=0');
vpa(vc,4)
ezplot(vc, [0 1e-3])
```

An m-file named Exercise_4_13 containing these commands resides in the MATLAB folder. (See Appendix E for information about accessing this folder.) □

## Summary

1. The transient part of the response for a circuit containing sources, resistances, and a single energy-storage element ($L$ or $C$) is of the form $Ke^{-t/\tau}$. The time constant is given by $\tau = RC$ or by $\tau = L/R$, where $R$ is the Thévenin resistance seen looking back into the circuit from the terminals of the energy-storage element.

2. In dc steady-state conditions, inductors behave as short circuits and capacitors behave as open circuits. We can find the steady-state (forced) response for dc sources by analyzing the dc equivalent circuit.

3. To find the transient currents and voltages, we must solve linear differential equations with constant coefficients. The solutions are the sum of two parts. The particular solution, also called the forced response, depends on the sources, as well as the other circuit elements. The homogeneous solution, also called the natural response, depends on the passive elements ($R$, $L$, and $C$), but not on the sources. In circuits that contain resistances, the natural response eventually decays to zero.

4. The natural response of a second-order circuit containing a series or parallel combination of inductance and capacitance depends on the damping ratio and undamped resonant frequency.

   If the damping ratio is greater than unity, the circuit is overdamped, and the natural response is of the form

   $$x_c(t) = K_1 e^{s_1 t} + K_2 e^{s_2 t}$$

   If the damping ratio equals unity, the circuit is critically damped, and the natural response is of the form

   $$x_c(t) = K_1 e^{s_1 t} + K_2 t e^{s_1 t}$$

   If the damping ratio is less than unity, the circuit is underdamped, and the natural response is of the form

   $$x_c(t) = K_1 e^{-\alpha t} \cos(\omega_n t) + K_2 e^{-\alpha t} \sin(\omega_n t)$$

   The normalized step response for second-order systems is shown in Figure 4.30 on page 196 for several values of the damping ratio.

5. The MATLAB Symbolic Toolbox is a powerful tool for solving the equations for transient circuits. A step-by-step procedure is given on page 199.

## Problems

### Section 4.1: First-Order RC Circuits

**P4.1.** Suppose we have a capacitance $C$ discharging through a resistance $R$. Define and give an expression for the time constant. To attain a long time constant, do we need large or small values for $R$? For $C$?

**\*P4.2.** The dielectric materials used in real capacitors are not perfect insulators. A resistance called a leakage resistance in parallel with the capacitance can model this imperfection. A $100\text{-}\mu\text{F}$ capacitor is initially charged to $100$ V. We want $90$ percent of the initial energy to remain after one minute. What is the limit on the leakage resistance for this capacitor?

**\*P4.3.** The initial voltage across the capacitor shown in Figure P4.3 is $v_C(0+) = -10$ V. Find an expression for the voltage across the capacitor as a function of time. Also, determine the time $t_0$ at which the voltage crosses zero.

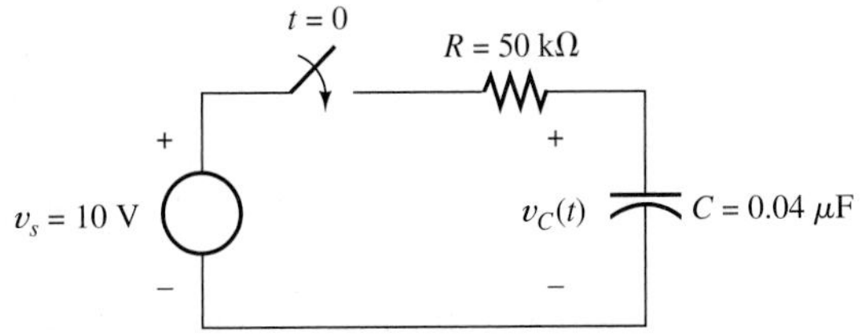

$t = 0$     $R = 50 \text{ k}\Omega$

$v_s = 10$ V     $v_C(t)$     $C = 0.04 \ \mu\text{F}$

Figure P4.3

---

\*Denotes that answers are contained in the Student Solutions files. See Appendix E for more information about accessing the Student Solutions.

**\*P4.4.** A 100-$\mu$F capacitance is initially charged to 1000 V. At $t = 0$, it is connected to a 1-k$\Omega$ resistance. At what time $t_2$ has 50 percent of the initial energy stored in the capacitance been dissipated in the resistance?

**\*P4.5.** At $t = 0$, a charged 10-$\mu$F capacitance is connected to a voltmeter, as shown in Figure P4.5. The meter can be modeled as a resistance. At $t = 0$, the meter reads 50 V. At $t = 30$ s, the reading is 25 V. Find the resistance of the voltmeter.

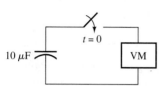

Figure P4.5

**\*P4.6.** At time $t_1$, a capacitance $C$ is charged to a voltage of $V_1$. Then, the capacitance discharges through a resistance $R$. Write an expression for the voltage across the capacitance as a function of time for $t > t_1$ in terms of $R, C, V_1,$ and $t_1$.

**P4.7.** Given an initially charged capacitance that begins to discharge through a resistance at $t = 0$, what percentage of the initial voltage remains at two time constants? What percentage of the initial stored energy remains?

**P4.8.** The initial voltage across the capacitor shown in Figure P4.3 is $v_C(0+) = 0$. Find an expression for the voltage across the capacitor as a function of time, and sketch to scale versus time.

**P4.9.** In physics, the half-life is often used to characterize exponential decay of physical quantities such as radioactive substances. The half-life is the time required for the quantity to decay to half of its initial value. The time constant for the voltage on a capacitance discharging through a resistance is $\tau = RC$. Find an expression for the half-life of the voltage in terms of $R$ and $C$.

**P4.10.** We know that a 50-$\mu$F capacitance is charged to an unknown voltage $V_i$ at $t = 0$. The capacitance is in parallel with a 3-k$\Omega$ resistance. At $t = 100$ ms, the voltage across the capacitance is 5 V. Determine the value of $V_i$.

**P4.11.** We know that the capacitor shown in Figure P4.11 is charged to a voltage of 10 V prior to

$t = 0$. **a.** Find expressions for the voltage across the capacitor $v_C(t)$ and the voltage across the resistor $v_R(t)$ for all time. **b.** Find an expression for the power delivered to the resistor. **c.** Integrate the power from $t = 0$ to $t = \infty$ to find the energy delivered. **d.** Show that the energy delivered to the resistor is equal to the energy stored in the capacitor prior to $t = 0$.

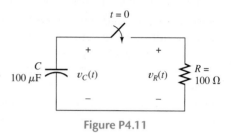

Figure P4.11

**P4.12.** The purchasing power $P$ of a certain unit of currency declines by 3 percent per year. Determine the time constant associated with the purchasing power of this currency.

**P4.13.** Derive an expression for $v_C(t)$ in the circuit of Figure P4.13 and sketch $v_C(t)$ to scale versus time.

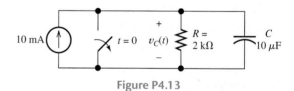

Figure P4.13

**P4.14.** Suppose that at $t = 0$, we connect an uncharged 10-$\mu$F capacitor to a charging circuit consisting of a 2500-V voltage source in series with a 2-M$\Omega$ resistance. At $t = 40$ s, the capacitor is disconnected from the charging circuit and connected in parallel with a 5-M$\Omega$ resistor. Determine the voltage across the capacitor at $t = 40$ s and at $t = 100$ s. (*Hint:* You may find it convenient to redefine the time variable to be $t' = t - 40$ for the discharge interval so that the discharge starts at $t' = 0$.)

**P4.15.** Suppose we have a capacitance $C$ that is charged to an initial voltage $V_i$. Then at $t = 0$, a resistance $R$ is connected across the capacitance. Write an expression for the current. Then, integrate the current from $t = 0$ to $t = \infty$, and show that the result is equal to the initial charge stored on the capacitance.

**P4.16.** A person shuffling across a dry carpet can be approximately modeled as a charged 100-pF capacitance with one end grounded. If the person touches a grounded metallic object such as a water faucet, the capacitance is discharged and the person experiences a brief shock. Typically, the capacitance may be charged to 20,000 V and the resistance (mainly of one's finger) is 100 Ω. Determine the peak current during discharge and the time constant of the shock.

**P4.17.** Consider the circuit of Figure P4.17, in which the switch instantaneously moves back and forth between contacts $A$ and $B$, spending 2 seconds in each position. Thus, the capacitor repeatedly charges for 2 seconds and then discharges for 2 seconds. Assume that $v_C(0) = 0$ and that the switch moves to position $A$ at $t = 0$. Determine $v_C(2)$, $v_C(4)$, $v_C(6)$, and $v_C(8)$.

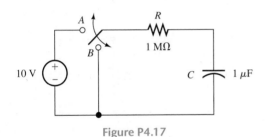

Figure P4.17

**P4.18.** Consider the circuit shown in Figure P4.18. Prior to $t = 0$, $v_1 = 100$ V, and $v_2 = 0$. **a.** Immediately after the switch is closed, what is the value of the current [i.e., what is the value of $i(0+)$]? **b.** Write the KVL equation for the circuit in terms of the current and initial voltages. Take the derivative to obtain a differential equation. **c.** What is the value of the time constant in this circuit? **d.** Find an expression for the current as a function of time. **e.** Find the value that $v_2$ approaches as $t$ becomes very large.

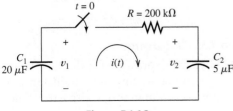

Figure P4.18

## Section 4.2: DC Steady State

**P4.19.** List the steps for dc steady-state analysis of $RLC$ circuits.

**P4.20.** Explain why we replace capacitances with open circuits and inductances with short circuits in dc steady-state analysis.

*__P4.21.__ Solve for the steady-state values of $i_1$, $i_2$, and $i_3$ for the circuit shown in Figure P4.21.

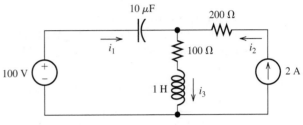

Figure P4.21

*__P4.22.__ Consider the circuit shown in Figure P4.22. What is the steady-state value of $v_C$ after the switch opens? Determine how long it takes after the switch opens before $v_C$ is within 1 percent of its steady-state value.

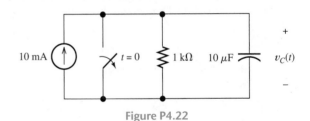

Figure P4.22

*__P4.23.__ In the circuit of Figure P4.23, the switch is in position $A$ for a long time prior to $t = 0$. Find expressions for $v_R(t)$ and sketch it to scale for $-2 \leq t \leq 10$ s.

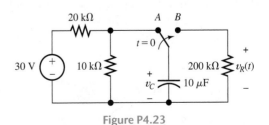

Figure P4.23

**P4.24.** The circuit shown in Figure P4.24 has been set up for a long time prior to $t = 0$ with the switch closed. Find the value of $v_C$ prior to $t = 0$. Find the steady-state value of $v_C$ after the switch has been opened for a long time.

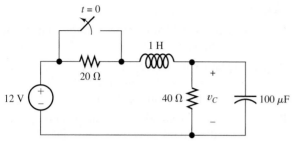

**Figure P4.24**

**P4.25.** Solve for the steady-state values of $i_1$, $i_2$, $i_3$, $i_4$, and $v_C$ for the circuit shown in Figure P4.25, assuming that the switch has been closed for a long time.

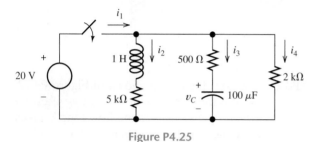

**Figure P4.25**

**P4.26.** The circuit shown in Figure P4.26 is operating in steady state. Determine the values of $i_L$, $v_x$, and $v_C$.

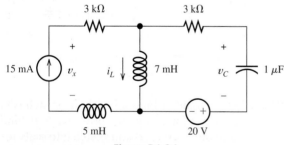

**Figure P4.26**

**P4.27.** The circuit of Figure P4.27 has been connected for a very long time. Determine the values of $v_C$ and $i_R$.

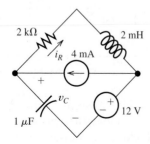

**Figure P4.27**

**P4.28.** Consider the circuit of Figure P4.28 in which the switch has been closed for a long time prior to $t = 0$. Determine the values of $v_C(t)$ before $t = 0$ and a long time after $t = 0$. Also, determine the time constant after the switch opens and expressions for $v_C(t)$. Sketch $v_C(t)$ to scale versus time for $-0.2 \le t \le 0.5$ s.

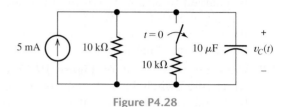

**Figure P4.28**

**P4.29.** For the circuit shown in Figure P4.29, the switch is closed for a long time prior to $t = 0$. Find expressions for $v_C(t)$ and sketch it to scale for $-80 \le t \le 160$ ms.

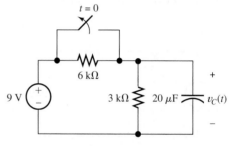

**Figure P4.29**

**P4.30.** Consider the circuit of Figure P4.30 in which the switch has been closed for a long time prior to $t = 0$. Determine the values of $v_C(t)$ before $t = 0$ and a long time after $t = 0$. Also, determine the time constant after the switch opens and expressions for $v_C(t)$. Sketch $v_C(t)$ to scale versus time for $-4 \le t \le 16$ s.

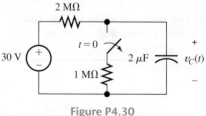

**Figure P4.30**

**Section 4.3:** *RL* **Circuits**

**P4.31.** Give the expression for the time constant of a circuit consisting of an inductance with an initial current in series with a resistance $R$. To attain a long time constant, do we need large or small values for $R$? For $L$?

**P4.32.** A circuit consists of switches that open or close at $t = 0$, resistances, dc sources, and a single energy storage element, either an inductance or a capacitance. We wish to solve for a current or a voltage $x(t)$ as a function of time for $t \geq 0$. Write the general form for the solution. How is each unknown in the solution determined?

**\*P4.33.** The circuit shown in Figure P4.33 is operating in steady state with the switch closed prior to $t = 0$. Find $i(t)$ for $t < 0$ and for $t \geq 0$.

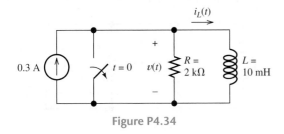

**Figure P4.33**

**\*P4.34.** Consider the circuit shown in Figure P4.34. The initial current in the inductor is $i_L(0-) = -0.2$ A. Find expressions for $i_L(t)$ and $v(t)$ for $t \geq 0$ and sketch to scale versus time.

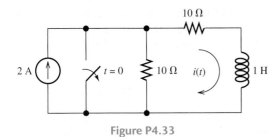

**Figure P4.34**

**P4.35.** Repeat Problem P4.34 given $i_L(0-) = 0$ A.

**\*P4.36.** Real inductors have series resistance associated with the wire used to wind the coil. Suppose that we want to store energy in a 10-H inductor. Determine the limit on the series resistance so the energy remaining after one hour is at least 75 percent of the initial energy.

**P4.37.** Determine expressions for and sketch $i_s(t)$ to scale versus time for $-0.2 \leq t \leq 1.0$ s for the circuit of Figure P4.37.

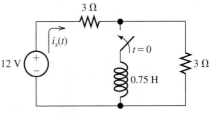

**Figure P4.37**

**P4.38.** For the circuit shown in Figure P4.38, find an expression for the current $i_L(t)$ and sketch it to scale versus time. Also, find an expression for $v_L(t)$ and sketch it to scale versus time.

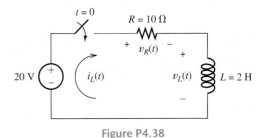

**Figure P4.38**

**P4.39.** The circuit shown in Figure P4.39 is operating in steady state with the switch closed prior to $t = 0$. Find expressions for $i_L(t)$ for $t < 0$ and for $t \geq 0$. Sketch $i_L(t)$ to scale versus time.

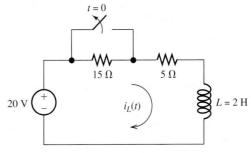

**Figure P4.39**

**P4.40.** Consider the circuit shown in Figure P4.40. A voltmeter (VM) is connected across the inductance. The switch has been closed for a long time. When the switch is opened, an arc appears across the switch contacts. Explain why. Assuming an ideal switch and inductor, what voltage appears across the inductor when the switch is opened? What could happen to the voltmeter when the switch opens?

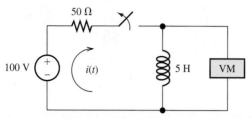

Figure P4.40

**P4.41.** Due to components not shown in the figure, the circuit of Figure P4.41 has $i_L(0) = I_i$.
**a.** Write an expression for $i_L(t)$ for $t \geq 0$.
**b.** Find an expression for the power delivered to the resistance as a function of time. **c.** Integrate the power delivered to the resistance from $t = 0$ to $t = \infty$, and show that the result is equal to the initial energy stored in the inductance.

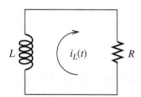

Figure P4.41

**P4.42.** The switch shown in Figure P4.42 has been closed for a long time prior to $t = 0$, then it opens at $t = 0$ and closes again at $t = 1$ s. Find $i_L(t)$ for all $t$.

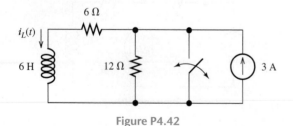

Figure P4.42

**P4.43.** Determine expressions for and sketch $v_R(t)$ to scale versus time for the circuit of Figure P4.43. The circuit is operating in steady state with the switch closed prior to $t = 0$. Consider the time interval $-1 \leq t \leq 5$ ms.

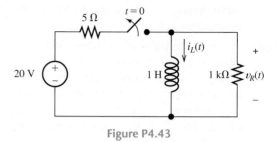

Figure P4.43

### Section 4.4: *RC* and *RL* Circuits with General Sources

**P4.44.** What are the steps in solving a circuit having a resistance, a source, and an inductance (or capacitance)?

*\*P4.45.** Write the differential equation for $i_L(t)$ and find the complete solution for the circuit of Figure P4.45. [*Hint:* Try a particular solution of the form $i_{Lp}(t) = Ae^{-t}$.]

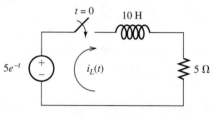

Figure P4.45

*\*P4.46.** Solve for $v_C(t)$ for $t > 0$ in the circuit of Figure P4.46. [*Hint:* Try a particular solution of the form $v_{Cp}(t) = Ae^{-3t}$.]

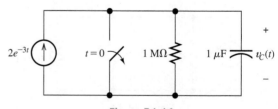

Figure P4.46

*\*P4.47.** Solve for $v(t)$ for $t > 0$ in the circuit of Figure P4.47, given that the inductor current is zero prior to $t = 0$. [*Hint:* Try a particular solution of the form $v_p = A \cos(10t) + B \sin(10t)$.]

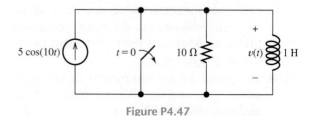

Figure P4.47

**P4.48.** Solve for $i_L(t)$ for $t > 0$ in the circuit of Figure P4.48. You will need to make an educated guess as to the form of the particular solution. [*Hint:* The particular solution includes terms with the same functional forms as the terms found in the forcing function and its derivatives.]

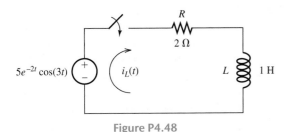

Figure P4.48

**P4.49.** Consider the circuit shown in Figure P4.49. The voltage source is known as a **ramp function**, which is defined by

$$v(t) = \begin{cases} 0 & \text{for } t < 0 \\ t & \text{for } t \geq 0 \end{cases}$$

Assume that $v_C(0) = 0$. Derive an expression for $v_C(t)$ for $t \geq 0$. Sketch $v_C(t)$ to scale versus time. [*Hint:* Write the differential equation for $v_C(t)$ and assume a particular solution of the form $v_{Cp}(t) = A + Bt$.]

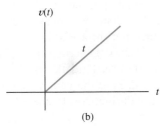

(a)

(b)

Figure P4.49

**P4.50.** Consider the circuit shown in Figure P4.50. The initial current in the inductor is $i_s(0+) = 0$. Write the differential equation for $i_s(t)$ and solve. [*Hint:* Try a particular solution of the form $i_{sp}(t) = A \cos(300t) + B \sin(300t)$.]

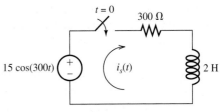

Figure P4.50

**P4.51.** The voltage source shown in Figure P4.51 is called a ramp function. Assume that $i_L(0) = 0$. Write the differential equation for $i_L(t)$, and find the complete solution. [*Hint:* Try a particular solution of the form $i_p(t) = A + Bt$.]

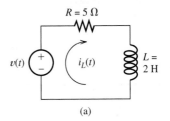

(a)

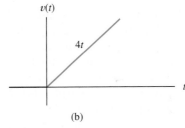

(b)

Figure P4.51

**P4.52.** Determine the form of the particular solution for the differential equation

$$2\frac{dv(t)}{dt} + v(t) = 5t \sin(t)$$

Then, find the particular solution. [*Hint:* The particular solution includes terms with the same functional forms as the terms found in the forcing function and its derivatives.]

**P4.53.** Determine the form of the particular solution for the differential equation

$$\frac{dv(t)}{dt} + 3v(t) = t^2 \exp(-t)$$

Then, find the particular solution. [*Hint:* The particular solution includes terms with the same functional forms as the terms found in the forcing function and its derivatives.]

**P4.54.** Consider the circuit shown in Figure P4.54.
  **a.** Write the differential equation for $i(t)$.
  **b.** Find the time constant and the form of the complementary solution.
  **c.** Usually, for an exponential forcing function like this, we would try a particular solution of the form $i_p(t) = K\exp(-3t)$. Why doesn't that work in this case?
  **d.** Find the particular solution. [*Hint:* Try a particular solution of the form $i_p(t) = Kt\exp(-3t)$.]
  **e.** Find the complete solution for $i(t)$.

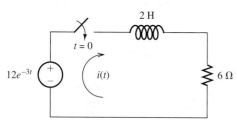

**Figure P4.54**

**P4.55.** Consider the circuit shown in Figure P4.55.
  **a.** Write the differential equation for $v(t)$.
  **b.** Find the time constant and the form of the complementary solution.
  **c.** Usually, for an exponential forcing function like this, we would try a particular solution of the form $v_p(t) = K\exp(-10t)$. Why doesn't that work in this case?
  **d.** Find the particular solution. [*Hint:* Try a particular solution of the form $v_p(t) = Kt\exp(-10t)$.]
  **e.** Find the complete solution for $v(t)$.

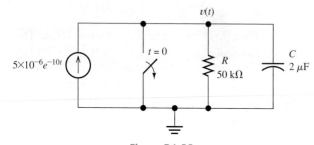

**Figure P4.55**

## Section 4.5: Second-Order Circuits

**P4.56.** How can first- or second-order circuits be identified by inspecting the circuit diagrams?

**P4.57.** How can an underdamped second-order system be identified? What form does its

complementary solution take? Repeat for a critically damped system and for an over-damped system.

**P4.58.** What is a unit step function?

**P4.59.** Discuss two methods that can be used to determine the particular solution of a circuit with constant dc sources.

**P4.60.** Sketch a step response for a second-order system that displays considerable overshoot and ringing. In what types of circuits do we find pronounced overshoot and ringing?

*\***P4.61.** A dc source is connected to a series $RLC$ circuit by a switch that closes at $t = 0$, as shown in Figure P4.61. The initial conditions are $i(0+) = 0$ and $v_C(0+) = 0$. Write the differential equation for $v_C(t)$. Solve for $v_C(t)$, if $R = 80\ \Omega$.

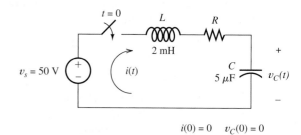

$i(0) = 0 \quad v_C(0) = 0$

**Figure P4.61**

*\***P4.62.** Repeat Problem P4.61 for $R = 40\ \Omega$.

*\***P4.63.** Repeat Problem P4.61 for $R = 20\ \Omega$.

**P4.64.** Consider the circuit shown in Figure P4.64 in which the switch has been open for a long time prior to $t = 0$ and we are given $R = 25\ \Omega$.
  **a.** Compute the undamped resonant frequency, the damping coefficient, and the damping ratio of the circuit after the switch closes. **b.** Assume that the capacitor is initially charged by a 25-V dc source not shown in the figure, so we have $v(0+) = 25$ V. Determine the values of $i_L(0+)$ and $v'(0+)$. **c.** Find the particular solution for $v(t)$. **d.** Find the general solution for $v(t)$, including the numerical values of all parameters.

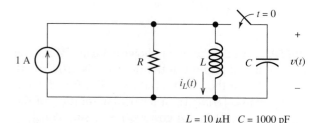

$L = 10\ \mu H \quad C = 1000\ pF$

**Figure P4.64**

**P4.65.** Repeat Problem P4.64 for $R = 50 \ \Omega$.

**P4.66.** Repeat Problem P4.64 for $R = 500 \ \Omega$.

**P4.67.** Solve for $i(t)$ for $t > 0$ in the circuit of Figure P4.67, with $R = 50 \ \Omega$, given that $i(0+) = 0$ and $v_C(0+) = 20$ V. [*Hint:* Try a particular solution of the form $i_p(t) = A \cos(100t) + B \sin(100t)$.]

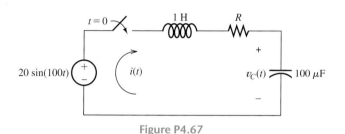

**Figure P4.67**

**P4.68.** Repeat Problem P4.67 with $R = 200 \ \Omega$.

**P4.69.** Repeat Problem P4.67 with $R = 400 \ \Omega$.

**P4.70.** Consider the circuit shown in Figure P4.70.
  **a.** Write the differential equation for $v(t)$.
  **b.** Find the damping coefficient, the natural frequency, and the form of the complementary solution.
  **c.** Usually, for a sinusoidal forcing function, we try a particular solution of the form $v_p(t) = A \cos(10^4 t) + B \sin(10^4 t)$. Why doesn't that work in this case?
  **d.** Find the particular solution. [*Hint:* Try a particular solution of the form $v_p(t) = At \cos(10^4 t) + B t \sin(10^4 t)$.]
  **e.** Find the complete solution for $v(t)$.

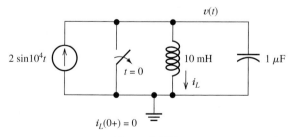

**Figure P4.70**

**Section 4.6:** Transient Analysis Using the MATLAB Symbolic Toolbox

**P4.71.** Use MATLAB to derive an expression for $v_C(t)$ in the circuit of Figure P4.13 and plot $v_C(t)$ versus time for $0 < t < 100$ ms.

**P4.72.** Consider the circuit shown in Figure P4.49. The voltage source is known as a **ramp function**, which is defined by

$$v(t) = \begin{cases} 0 & \text{for } t < 0 \\ t & \text{for } t \geq 0 \end{cases}$$

Use MATLAB to derive an expression for $v_C(t)$ in terms of $R$, $C$, and $t$. Next, substitute $R = 1 \ \text{M}\Omega$ and $C = 1 \ \mu\text{F}$. Then, plot $v_C(t)$ and $v(t)$ on the same axes for $0 < t < 5$ s.

**P4.73.** Consider the circuit shown in Figure P4.50 in which the switch is open for a long time prior to $t = 0$. The initial current is $i_s(0+) = 0$. Write the differential equation for $i_s(t)$ and use MATLAB to plot $i_s(t)$ for $t$ ranging from 0 to 80 ms. [*Hint:* Avoid using lowercase "i" as the first letter of the dependent variable, instead use "Is" for the current in MATLAB.]

**P4.74.** Consider the circuit shown in Figure P4.64 in which the switch has been open for a long time prior to $t = 0$ and we are given $R = 25 \ \Omega$.
  **a.** Write the differential equation for $v(t)$.
  **b.** Assume that the capacitor is initially charged by a 50-V dc source not shown in the figure, so we have $v(0+) = 50$ V. Determine the values of $i_L(0+)$ and $v'(0+)$. **c.** Use MATLAB to find the general solution for $v(t)$.

**P4.75.** Consider the circuit shown in Figure P4.70. **a.** Write the differential equation for $v(t)$. **b.** Determine the values for $v(0+)$ and $v'(0+)$. **c.** Use MATLAB to find the complete solution for $v(t)$. Then plot $v(t)$ for $0 \leq t \leq 10$ ms.

**P4.76.** Use MATLAB to solve for the mesh currents in the circuit of Figure P4.76. The circuit has been connected for a long time prior to $t = 0$ with the switch open, so the initial values of the inductor currents are zero.

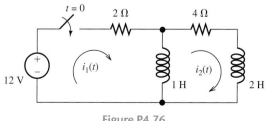

**Figure P4.76**

## Practice Test

Here is a practice test you can use to check your comprehension of the most important concepts in this chapter. Answers can be found in Appendix D and complete solutions are included in the Student Solutions files. See Appendix E for more information about the Student Solutions.

**T4.1.** The switch in the circuit shown in Figure T4.1 is closed prior to $t = 0$. The switch opens at $t = 0$. Determine the time $t_x$ at which $v_C(t)$ reaches 15 V.

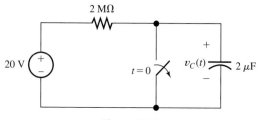

**Figure T4.1**

**T4.2.** Consider the circuit shown in Figure T4.2. The circuit has been operating for a long time with the switch closed prior to $t = 0$. **a.** Determine the values of $i_L$, $i_1$, $i_2$, $i_3$, and $v_C$ just before the switch opens. **b.** Determine the values of $i_L$, $i_1$, $i_2$, $i_3$, and $v_C$ immediately after the switch opens. **c.** Find $i_L(t)$ for $t > 0$. **d.** Find $v_C(t)$ for $t > 0$.

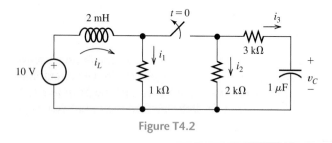

**Figure T4.2**

**T4.3.** Consider the circuit shown in Figure T4.3.
   **a.** Write the differential equation for $i(t)$.
   **b.** Find the time constant and the form of the complementary solution.
   **c.** Find the particular solution.
   **d.** Find the complete solution for $i(t)$.

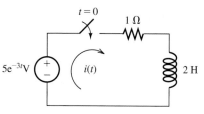

**Figure T4.3**

**T4.4.** Consider the circuit shown in Figure T4.4 in which the initial inductor current and capacitor voltage are both zero.
   **a.** Write the differential equation for $v_C(t)$.
   **b.** Find the particular solution.
   **c.** Is this circuit overdamped, critically damped, or underdamped? Find the form of the complementary solution.
   **d.** Find the complete solution for $v_C(t)$.

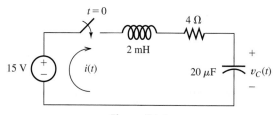

**Figure T4.4**

**T4.5.** Write the MATLAB commands to obtain the solution for the differential equation of question T4.4 with four decimal place accuracy for the constants.

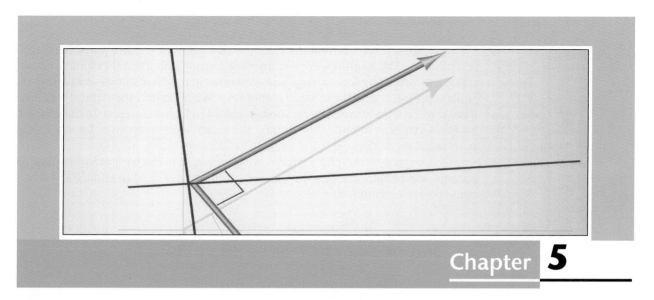

# Chapter 5

# Steady-State Sinusoidal Analysis

## Study of this chapter will enable you to:

- Identify the frequency, angular frequency, peak value, rms value, and phase of a sinusoidal signal.

- Determine the root-mean-square (rms) value of any periodic current or voltage.

- Solve steady-state ac circuits, using phasors and complex impedances.

- Compute power for steady-state ac circuits.

- Find Thévenin and Norton equivalent circuits.

- Determine load impedances for maximum power transfer.

- Discuss the advantages of three-phase power distribution.

- Solve balanced three-phase circuits.

- Use MATLAB to facilitate ac circuit calculations.

## Introduction to this chapter:

Circuits with sinusoidal sources have many important applications. For example, electric power is distributed to residences and businesses by sinusoidal currents and voltages. Furthermore, sinusoidal signals have many uses in radio communication. Finally, a branch of mathematics known as Fourier analysis shows that all signals of practical interest are composed of sinusoidal components. Thus, the study of circuits with sinusoidal sources is a central theme in electrical engineering.

In Chapter 4, we saw that the response of a network has two parts: the forced response and the natural response. In most circuits, the natural response decays rapidly to zero. The forced response for sinusoidal sources persists indefinitely and, therefore, is called the steady-state response. Because the natural response quickly decays, the steady-state response is often of highest interest. In this chapter, we learn efficient methods for finding the steady-state responses for sinusoidal sources.

We also study three-phase circuits, which are used in electric power-distribution systems. Most engineers who work in industrial settings need to understand three-phase power distribution.

## 5.1  SINUSOIDAL CURRENTS AND VOLTAGES

A sinusoidal voltage is shown in Figure 5.1 and is given by

$$v(t) = V_m \cos(\omega t + \theta) \tag{5.1}$$

where $V_m$ is the **peak value** of the voltage, $\omega$ is the **angular frequency** in radians per second, and $\theta$ is the **phase angle**.

Sinusoidal signals are periodic, repeating the same pattern of values in each **period** $T$. Because the cosine (or sine) function completes one cycle when the angle increases by $2\pi$ radians, we get

$$\omega T = 2\pi \tag{5.2}$$

The **frequency** of a periodic signal is the number of cycles completed in one second. Thus, we obtain

$$f = \frac{1}{T} \tag{5.3}$$

We refer to $\omega$ as angular frequency with units of radians per second and $f$ simply as frequency with units of hertz (Hz).

The units of frequency are hertz (Hz). (Actually, the physical units of hertz are equivalent to inverse seconds.) Solving Equation 5.2 for the angular frequency, we have

$$\omega = \frac{2\pi}{T} \tag{5.4}$$

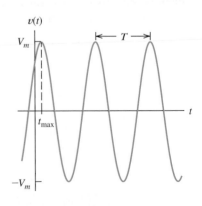

**Figure 5.1**  A sinusoidal voltage waveform given by $v(t) = V_m \cos(\omega t + \theta)$. *Note:* Assuming that $\theta$ is in degrees, we have $t_{max} = \frac{-\theta}{360} \times T$. For the waveform shown, $\theta$ is $-45°$.

Using Equation 5.3 to substitute for $T$, we find that

$$\omega = 2\pi f \tag{5.5}$$

Throughout our discussion, the argument of the cosine (or sine) function is of the form

$$\omega t + \theta$$

We assume that the angular frequency $\omega$ has units of radians per second (rad/s). However, we sometimes give the phase angle $\theta$ in degrees. Then, the argument has mixed units. If we wanted to evaluate $\cos(\omega t + \theta)$ for a particular value of time, we would have to convert $\theta$ to radians before adding the terms in the argument. Usually, we find it easier to visualize an angle expressed in degrees, and mixed units are not a problem.

> Electrical engineers often write the argument of a sinusoid in mixed units: $\omega t$ is in radians and the phase angle $\theta$ is in degrees.

For uniformity, we express sinusoidal functions by using the cosine function rather than the sine function. The functions are related by the identity

$$\sin(z) = \cos(z - 90°) \tag{5.6}$$

For example, when we want to find the phase angle of

$$v_x(t) = 10 \sin(200t + 30°)$$

we first write it as

$$v_x(t) = 10 \cos(200t + 30° - 90°)$$
$$= 10 \cos(200t - 60°)$$

Thus, we state that the phase angle of $v_x(t)$ is $-60°$.

## Root-Mean-Square Values

Consider applying a periodic voltage $v(t)$ with period $T$ to a resistance $R$. The power delivered to the resistance is given by

$$p(t) = \frac{v^2(t)}{R} \tag{5.7}$$

Furthermore, the energy delivered in one period is given by

$$E_T = \int_0^T p(t)\, dt \tag{5.8}$$

The average power $P_{avg}$ delivered to the resistance is the energy delivered in one cycle divided by the period. Thus,

$$P_{avg} = \frac{E_T}{T} = \frac{1}{T} \int_0^T p(t)\, dt \tag{5.9}$$

Using Equation 5.7 to substitute into Equation 5.9, we obtain

$$P_{avg} = \frac{1}{T} \int_0^T \frac{v^2(t)}{R}\, dt \tag{5.10}$$

This can be rearranged as

$$P_{\text{avg}} = \frac{\left[\sqrt{\frac{1}{T} \int_0^T v^2(t)\, dt}\right]^2}{R} \tag{5.11}$$

Now, we define the **root-mean-square** (rms) value of the periodic voltage $v(t)$ as

$$V_{\text{rms}} = \sqrt{\frac{1}{T} \int_0^T v^2(t)\, dt} \tag{5.12}$$

Using this equation to substitute into Equation 5.11, we get

$$P_{\text{avg}} = \frac{V_{\text{rms}}^2}{R} \tag{5.13}$$

Thus, if the rms value of a periodic voltage is known, it is relatively easy to compute the average power that the voltage can deliver to a resistance. The rms value is also called the **effective value**.

Power calculations are facilitated by using rms values for voltage or current.

Similarly for a periodic current $i(t)$, we define the rms value as

$$I_{\text{rms}} = \sqrt{\frac{1}{T} \int_0^T i^2(t)\, dt} \tag{5.14}$$

and the average power delivered if $i(t)$ flows through a resistance is given by

$$P_{\text{avg}} = I_{\text{rms}}^2 R \tag{5.15}$$

## RMS Value of a Sinusoid

Consider a sinusoidal voltage given by

$$v(t) = V_m \cos(\omega t + \theta) \tag{5.16}$$

To find the rms value, we substitute into Equation 5.12, which yields

$$V_{\text{rms}} = \sqrt{\frac{1}{T} \int_0^T V_m^2 \cos^2(\omega t + \theta)\, dt} \tag{5.17}$$

Next, we use the trigonometric identity

$$\cos^2(z) = \frac{1}{2} + \frac{1}{2} \cos(2z) \tag{5.18}$$

to write Equation 5.17 as

$$V_{\text{rms}} = \sqrt{\frac{V_m^2}{2T} \int_0^T [1 + \cos(2\omega t + 2\theta)]\, dt} \tag{5.19}$$

Integrating, we get

$$V_{\text{rms}} = \sqrt{\frac{V_m^2}{2T} \left[ t + \frac{1}{2\omega} \sin(2\omega t + 2\theta) \right]_0^T} \tag{5.20}$$

Evaluating, we have

$$V_{\text{rms}} = \sqrt{\frac{V_m^2}{2T} \left[ T + \frac{1}{2\omega} \sin(2\omega T + 2\theta) - \frac{1}{2\omega} \sin(2\theta) \right]} \tag{5.21}$$

Referring to Equation 5.2, we see that $\omega T = 2\pi$. Thus, we obtain

$$\frac{1}{2\omega} \sin(2\omega T + 2\theta) - \frac{1}{2\omega} \sin(2\theta) = \frac{1}{2\omega} \sin(4\pi + 2\theta) - \frac{1}{2\omega} \sin(2\theta)$$

$$= \frac{1}{2\omega} \sin(2\theta) - \frac{1}{2\omega} \sin(2\theta)$$

$$= 0$$

Therefore, Equation 5.21 reduces to

$$V_{\text{rms}} = \frac{V_m}{\sqrt{2}} \tag{5.22}$$

This is a useful result that we will use many times in dealing with sinusoids.

Usually in discussing sinusoids, the rms or effective value is given rather than the peak value. For example, ac power in residential wiring is distributed as a 60-Hz 115-V rms sinusoid (in the United States). Most people are aware of this, but probably few know that 115 V is the rms value and that the peak value is $V_m = V_{\text{rms}} \times \sqrt{2} = 115 \times \sqrt{2} \cong 163$ V. (Actually, 115 V is the nominal residential distribution voltage. It can vary from approximately 105 to 130 V.)

Keep in mind that $V_{\text{rms}} = V_m/\sqrt{2}$ applies to sinusoids. To find the rms value of other periodic waveforms, we would need to employ the definition given by Equation 5.12.

The rms value for a sinusoid is the peak value divided by the square root of two. This is not true for other periodic waveforms such as square waves or triangular waves.

---

### Example 5.1   Power Delivered to a Resistance by a Sinusoidal Source

Suppose that a voltage given by $v(t) = 100 \cos(100\pi t)$ V is applied to a 50-$\Omega$ resistance. Sketch $v(t)$ to scale versus time. Find the rms value of the voltage and the average power delivered to the resistance. Find the power as a function of time and sketch to scale.

**Solution**   By comparison of the expression given for $v(t)$ with Equation 5.1, we see that $\omega = 100\pi$. Using Equation 5.5, we find that the frequency is $f = \omega/2\pi = 50$ Hz. Then, the period is $T = 1/f = 20$ ms. A plot of $v(t)$ versus time is shown in Figure 5.2(a).

The peak value of the voltage is $V_m = 100$ V. Thus, the rms value is $V_{\text{rms}} = V_m/\sqrt{2} = 70.71$ V. Then, the average power is

$$P_{\text{avg}} = \frac{V_{\text{rms}}^2}{R} = \frac{(70.71)^2}{50} = 100 \text{ W}$$

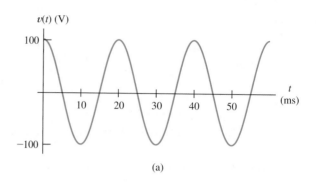

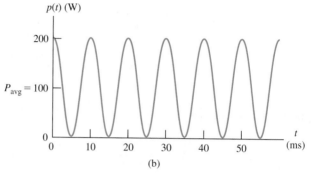

Figure 5.2 Voltage and power versus time for Example 5.1.

The power as a function of time is given by

$$p(t) = \frac{v^2(t)}{R} = \frac{100^2 \cos^2(100\pi t)}{50} = 200 \cos^2(100\pi t) \text{ W}$$

A plot of $p(t)$ versus time is shown in Figure 5.2(b). Notice that the power fluctuates from 0 to 200 W. However, the average power is 100 W, as we found by using the rms value. ■

For a sinusoidal current flowing in a resistance, power fluctuates periodically from zero to twice the average value.

## RMS Values of Nonsinusoidal Voltages or Currents

Sometimes we need to determine the rms values of periodic currents or voltages that are not sinusoidal. We can accomplish this by applying the definition given by Equation 5.12 or 5.14 directly.

### Example 5.2    RMS Value of a Triangular Voltage

The voltage shown in Figure 5.3(a) is known as a triangular waveform. Determine its rms value.

**Solution**  First, we need to determine the equations describing the waveform between $t = 0$ and $t = T = 2$ s. As illustrated in Figure 5.3(b), the equations for the first period of the triangular wave are

$$v(t) = \begin{cases} 3t & \text{for} \quad 0 \le t \le 1 \\ 6 - 3t & \text{for} \quad 1 \le t \le 2 \end{cases}$$

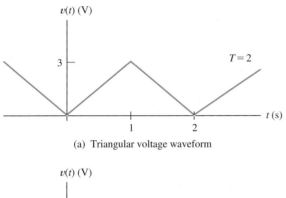

(a)  Triangular voltage waveform

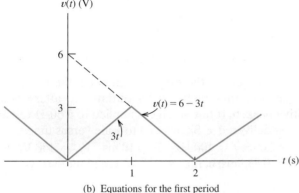

(b)  Equations for the first period

**Figure 5.3** Triangular voltage waveform of Example 5.2.

Equation 5.12 gives the rms value of the voltage.

$$V_{\text{rms}} = \sqrt{\frac{1}{T} \int_0^T v^2(t)\,dt}$$

Dividing the interval into two parts and substituting for $v(t)$, we have

$$V_{\text{rms}} = \sqrt{\frac{1}{2}\left[ \int_0^1 9t^2\,dt + \int_1^2 (6-3t)^2\,dt \right]}$$

$$V_{\text{rms}} = \sqrt{\frac{1}{2}\,[3t^3|_{t=0}^{t=1} + (36t - 18t^2 + 3t^3)|_{t=1}^{t=2}]}$$

Evaluating, we find

$$V_{\text{rms}} = \sqrt{\frac{1}{2}\,[3 + (72 - 36 - 72 + 18 + 24 - 3)]} = \sqrt{3}\ \text{V} \qquad \blacksquare$$

The integrals in this example are easy to carry out manually. However, when the integrals are more difficult, we can sometimes obtain answers using the MATLAB Symbolic Toolbox. Here are the MATLAB commands needed to perform the integrals in this example:

```
>> syms Vrms t
>> Vrms = sqrt((1/2)*(int(9*t^2,t,0,1) + int((6-3*t)^2,t,1,2)))
   Vrms =
   3^(1/2)
```

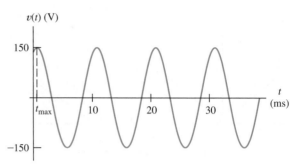

**Figure 5.4** Answer for Exercise 5.1(c).

**Exercise 5.1**  Suppose that a sinusoidal voltage is given by

$$v(t) = 150 \cos(200\pi t - 30°) \text{ V}$$

**a.** Find the angular frequency, the frequency in hertz, the period, the peak value, and the rms value. Also, find the first value of time $t_{max}$ after $t = 0$ such that $v(t)$ attains its positive peak. **b.** If this voltage is applied to a 50-$\Omega$ resistance, compute the average power delivered. **c.** Sketch $v(t)$ to scale versus time.

**Answer  a.** $\omega = 200\pi, f = 100 \text{ Hz}, \ T = 10 \text{ ms}, \ V_m = 150 \text{ V}, \ V_{rms} = 106.1 \text{ V},$ $t_{max} = \frac{30°}{360°} \times T = 0.833 \text{ ms}$; **b.** $P_{avg} = 225 \text{ W}$; **c.** a plot of $v(t)$ versus time is shown in Figure 5.4.    □

**Exercise 5.2**  Express $v(t) = 100 \sin(300\pi t + 60°)$ V as a cosine function.
**Answer**  $v(t) = 100 \cos(300\pi t - 30°)$ V.    □

**Exercise 5.3**  Suppose that the ac line voltage powering a computer has an rms value of 110 V and a frequency of 60 Hz, and the peak voltage is attained at $t = 5$ ms. Write an expression for this ac voltage as a function of time.
**Answer**  $v(t) = 155.6 \cos(377t - 108°)$ V.    □

## 5.2  PHASORS

In the next several sections, we will see that sinusoidal steady-state analysis is greatly facilitated if the currents and voltages are represented as vectors (called **phasors**) in the complex-number plane. In preparation for this material, you may wish to study the review of complex-number arithmetic in Appendix A.

We start with a study of convenient methods for adding (or subtracting) sinusoidal waveforms. We often need to do this in applying Kirchhoff's voltage law (KVL) or Kirchhoff's current law (KCL) to ac circuits. For example, in applying KVL to a network with sinusoidal voltages, we might obtain the expression

$$v(t) = 10 \cos(\omega t) + 5 \sin(\omega t + 60°) + 5 \cos(\omega t + 90°) \qquad (5.23)$$

To obtain the peak value of $v(t)$ and its phase angle, we need to put Equation 5.23 into the form

$$v(t) = V_m \cos(\omega t + \theta) \qquad (5.24)$$

This could be accomplished by repeated substitution, using standard trigonometric identities. However, that method is too tedious for routine work. Instead, we will see that we can represent each term on the right-hand side of Equation 5.23 by a vector in the complex-number plane known as a **phasor**. Then, we can add the phasors with relative ease and convert the sum into the desired form.

## Phasor Definition

For a sinusoidal voltage of the form

$$v_1(t) = V_1 \cos(\omega t + \theta_1)$$

we define the phasor as

$$\mathbf{V}_1 = V_1 \underline{/\theta_1}$$

Thus, the phasor for a sinusoid is a complex number having a magnitude equal to the peak value and having the same phase angle as the sinusoid. We use boldface letters for phasors. (Actually, engineers are not consistent in choosing the magnitudes of phasors. In this chapter and in Chapter 6, we take the peak values for the magnitudes of phasors, which is the prevailing custom in circuit-analysis courses for electrical engineers. However, later in Chapters 14 and 15, we will take the rms values for the phasor magnitudes as power-system engineers customarily do. We will take care to label rms phasors as such when we encounter them. In this book, if phasors are not labeled as rms, you can assume that they are peak values.)

Phasors are complex numbers that represent sinusoidal voltages or currents. The magnitude of a phasor equals the peak value and the angle equals the phase of the sinusoid (written as a cosine).

If the sinusoid is of the form

$$v_2(t) = V_2 \sin(\omega t + \theta_2)$$

we first convert to a cosine function by using the trigonometric identity

$$\sin(z) = \cos(z - 90°) \tag{5.25}$$

Thus, we have

$$v_2(t) = V_2 \cos(\omega t + \theta_2 - 90°)$$

and the phasor is

$$\mathbf{V}_2 = V_2 \underline{/\theta_2 - 90°}$$

Phasors are obtained for sinusoidal currents in a similar fashion. Thus, for the currents

$$i_1(t) = I_1 \cos(\omega t + \theta_1)$$

and

$$i_2(t) = I_2 \sin(\omega t + \theta_2)$$

the phasors are

$$\mathbf{I}_1 = I_1 \underline{/\theta_1}$$

and

$$\mathbf{I}_2 = I_2 \underline{/\theta_2 - 90°}$$

respectively.

### Adding Sinusoids Using Phasors

Now, we illustrate how we can use phasors to combine the terms of the right-hand side of Equation 5.23. In this discussion, we proceed in small logical steps to illustrate clearly why sinusoids can be added by adding their phasors. Later, we streamline the procedure for routine work.

Our first step in combining the terms in Equation 5.23 is to write all the sinusoids as cosine functions by using Equation 5.25. Thus, Equation 5.23 can be written as

$$v(t) = 10\cos(\omega t) + 5\cos(\omega t + 60° - 90°) + 5\cos(\omega t + 90°) \tag{5.26}$$

$$v(t) = 10\cos(\omega t) + 5\cos(\omega t - 30°) + 5\cos(\omega t + 90°) \tag{5.27}$$

Referring to Euler's formula (Equation A.8) in Appendix A, we see that we can write

$$\cos(\theta) = \text{Re}(e^{j\theta}) = \text{Re}[\cos(\theta) + j\sin(\theta)] \tag{5.28}$$

where the notation Re() means that we retain only the real part of the quantity inside the parentheses. Thus, we can rewrite Equation 5.27 as

$$v(t) = 10\,\text{Re}\left[\,e^{j\omega t}\right] + 5\,\text{Re}\left[e^{j(\omega t - 30°)}\right] + 5\,\text{Re}\left[e^{j(\omega t + 90°)}\right] \tag{5.29}$$

When we multiply a complex number $Z$ by a real number $A$, both the real and imaginary parts of $Z$ are multiplied by $A$. Thus, Equation 5.29 becomes

$$v(t) = \text{Re}\left[10e^{j\omega t}\right] + \text{Re}\left[5e^{j(\omega t - 30°)}\right] + \text{Re}\left[5e^{j(\omega t + 90°)}\right] \tag{5.30}$$

Next, we can write

$$v(t) = \text{Re}\left[10e^{j\omega t} + 5e^{j(\omega t - 30°)} + 5e^{j(\omega t + 90°)}\right] \tag{5.31}$$

because the real part of the sum of several complex quantities is equal to the sum of the real parts. If we factor out the common term $e^{j\omega t}$, Equation 5.31 becomes

$$v(t) = \text{Re}\left[(10 + 5e^{-j30°} + 5^{j90°})\,e^{j\omega t}\right] \tag{5.32}$$

Putting the complex numbers into polar form, we have

$$v(t) = \text{Re}\left[(10\underline{/0°} + 5\underline{/-30°} + 5\underline{/90°})e^{j\omega t}\right] \tag{5.33}$$

Now, we can combine the complex numbers as

$$10\underline{/0°} + 5\underline{/-30°} + 5\underline{/90°} = 10 + 4.33 - j2.50 + j5$$
$$= 14.33 + j2.5$$
$$= 14.54\underline{/9.90°}$$
$$= 14.54e^{j9.90°} \tag{5.34}$$

Using this result in Equation 5.33, we have

$$v(t) = \text{Re}\left[(14.54e^{j9.90°})\,e^{j\omega t}\right]$$

which can be written as

$$v(t) = \text{Re}\left[14.54e^{j(\omega t+9.90°)}\right] \tag{5.35}$$

Now, using Equation 5.28, we can write this as

$$v(t) = 14.54\cos(\omega t + 9.90°) \tag{5.36}$$

Thus, we have put the original expression for $v(t)$ into the desired form. The terms on the left-hand side of Equation 5.34 are the phasors for the terms on the right-hand side of the original expression for $v(t)$. Notice that the essential part of the work needed to combine the sinusoids is to add the phasors.

## Streamlined Procedure for Adding Sinusoids

From now on, to add sinusoids, we will first write the phasor for each term in the sum, add the phasors by using complex-number arithmetic, and then write the simplified expression for the sum.

To add sinusoids, we find the phasor for each term, add the phasors by using complex-number arithmetic, express the sum in polar form, and then write the corresponding sinusoidal time function.

---

**Example 5.3    Using Phasors to Add Sinusoids**

Suppose that

$$v_1(t) = 20\cos(\omega t - 45°)$$
$$v_2(t) = 10\sin(\omega t + 60°)$$

Reduce the sum $v_s(t) = v_1(t) + v_2(t)$ to a single term.

**Solution**    The phasors are

$$\mathbf{V}_1 = 20\underline{/-45°}$$
$$\mathbf{V}_2 = 10\underline{/-30°}$$

In using phasors to add sinusoids, all of the terms must have the same frequency.

Step 1: Determine the phasor for each term.

Notice that we have subtracted 90° to find the phase angle for $\mathbf{V}_2$ because $v_2(t)$ is a sine function rather than a cosine function.

Next, we use complex-number arithmetic to add the phasors and convert the sum to polar form:

$$\begin{aligned}
\mathbf{V}_s &= \mathbf{V}_1 + \mathbf{V}_2 \\
&= 20\underline{/-45°} + 10\underline{/-30°} \\
&= 14.14 - j14.14 + 8.660 - j5 \\
&= 22.80 - j19.14 \\
&= 29.77\underline{/-40.01°}
\end{aligned}$$

Step 2: Use complex arithmetic to add the phasors.

Step 3: Convert the sum to polar form.

Now, we write the time function corresponding to the phasor $\mathbf{V}_s$.

$$v_s(t) = 29.77\cos(\omega t - 40.01°)$$

Step 4: Write the result as a time function.

■

**Exercise 5.4**    Reduce the following expressions by using phasors:

$$v_1(t) = 10 \cos(\omega t) + 10 \sin(\omega t)$$

$$i_1(t) = 10 \cos(\omega t + 30°) + 5 \sin(\omega t + 30°)$$

$$i_2(t) = 20 \sin(\omega t + 90°) + 15 \cos(\omega t - 60°)$$

**Answer**

$$v_1(t) = 14.14 \cos(\omega t - 45°)$$

$$i_1(t) = 11.18 \cos(\omega t + 3.44°)$$

$$i_2(t) = 30.4 \cos(\omega t - 25.3°)$$

□

## Phasors as Rotating Vectors

Consider a sinusoidal voltage given by

$$v(t) = V_m \cos(\omega t + \theta)$$

In developing the phasor concept, we write

$$v(t) = \text{Re} \left[ V_m e^{j(\omega t + \theta)} \right]$$

The complex quantity inside the brackets is

$$V_m e^{j(\omega t + \theta)} = V_m \underline{/\omega t + \theta}$$

Sinusoids can be visualized as the real-axis projection of vectors rotating in the complex plane. The phasor for a sinusoid is a snapshot of the corresponding rotating vector at $t = 0$.

This can be visualized as a vector of length $V_m$ that rotates counterclockwise in the complex plane with an angular velocity of $\omega$ rad/s. Furthermore, the voltage $v(t)$ is the real part of the vector, which is illustrated in Figure 5.5. As the vector rotates, its projection on the real axis traces out the voltage as a function of time. The phasor is simply a "snapshot" of this rotating vector at $t = 0$.

## Phase Relationships

We will see that the phase relationships between currents and voltages are often important. Consider the voltages

$$v_1(t) = 3 \cos(\omega t + 40°)$$

and

$$v_2(t) = 4 \cos(\omega t - 20°)$$

To determine phase relationships from a phasor diagram, consider that the phasors rotate counterclockwise. Then, when standing at a fixed point, if $V_1$ arrives first followed by $V_2$ after a rotation of $\theta$, we say that $V_1$ leads $V_2$ by $\theta$. Alternatively, we could say that $V_2$ lags $V_1$ by $\theta$. (Usually, we take $\theta$ as the smaller angle between the two phasors.)

The corresponding phasors are

$$\mathbf{V}_1 = 3 \underline{/40°}$$

and

$$\mathbf{V}_2 = 4 \underline{/-20°}$$

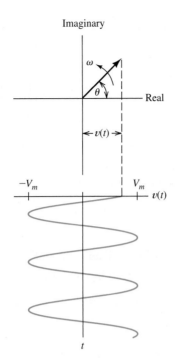

**Figure 5.5** A sinusoid can be represented as the real part of a vector rotating counterclockwise in the complex plane.

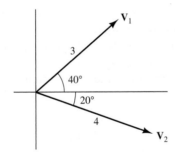

**Figure 5.6** Because the vectors rotate counterclockwise, $\mathbf{V}_1$ leads $\mathbf{V}_2$ by 60° (or, equivalently, $\mathbf{V}_2$ lags $\mathbf{V}_1$ by 60°).

The phasor diagram is shown in Figure 5.6. Notice that the angle between $\mathbf{V}_1$ and $\mathbf{V}_2$ is 60°. Because the complex vectors rotate counterclockwise, we say that $\mathbf{V}_1$ *leads* $\mathbf{V}_2$ by 60°. (An alternative way to state the phase relationship is to state that $\mathbf{V}_2$ *lags* $\mathbf{V}_1$ by 60°.)

We have seen that the voltages versus time can be obtained by tracing the real part of the rotating vectors. The plots of $v_1(t)$ and $v_2(t)$ versus $\omega t$ are shown in Figure 5.7. Notice that $v_1(t)$ reaches its peak 60° earlier than $v_2(t)$. This is the meaning of the statement that $v_1(t)$ leads $v_2(t)$ by 60°.

To determine phase relationships between sinusoids from their plots versus time, find the shortest time interval $t_p$ between positive peaks of the two waveforms. Then, the phase angle is $\theta = (t_p/T) \times 360°$. If the peak of $v_1(t)$ occurs first, we say that $v_1(t)$ leads $v_2(t)$ or that $v_2(t)$ lags $v_1(t)$.

**Exercise 5.5**   Consider the voltages given by

$$v_1(t) = 10 \cos(\omega t - 30°)$$
$$v_2(t) = 10 \cos(\omega t + 30°)$$
$$v_3(t) = 10 \sin(\omega t + 45°)$$

State the phase relationship between each pair of the voltages. (*Hint:* Find the phasor for each voltage and draw the phasor diagram.)

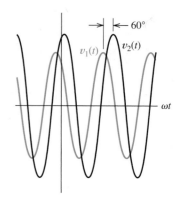

**Figure 5.7** The peaks of $v_1(t)$ occur 60° before the peaks of $v_2(t)$. In other words, $v_1(t)$ leads $v_2(t)$. by 60°.

**Answer**

$v_1$ lags $v_2$ by 60° (or $v_2$ leads $v_1$ by 60°)

$v_1$ leads $v_3$ by 15° (or $v_3$ lags $v_1$ by 15°)

$v_2$ leads $v_3$ by 75° (or $v_3$ lags $v_2$ by 75°)

## 5.3 COMPLEX IMPEDANCES

In this section, we learn that by using phasors to represent sinusoidal voltages and currents, we can solve sinusoidal steady-state circuit problems with relative ease compared with the methods of Chapter 4. Except for the fact that we use complex arithmetic, sinusoidal steady-state analysis is virtually the same as the analysis of resistive circuits, which we studied in Chapter 2.

### Inductance

Consider an inductance in which the current is a sinusoid given by

$$i_L(t) = I_m \sin(\omega t + \theta) \tag{5.37}$$

Recall that the voltage across an inductance is

$$v_L(t) = L\frac{di_L(t)}{dt} \tag{5.38}$$

Substituting Equation 5.37 into Equation 5.38 and reducing, we obtain

$$v_L(t) = \omega L I_m \cos(\omega t + \theta) \tag{5.39}$$

Now, the phasors for the current and voltage are

$$\mathbf{I}_L = I_m \underline{/\theta - 90°} \tag{5.40}$$

and

$$\mathbf{V}_L = \omega L I_m \underline{/\theta} = V_m \underline{/\theta} \tag{5.41}$$

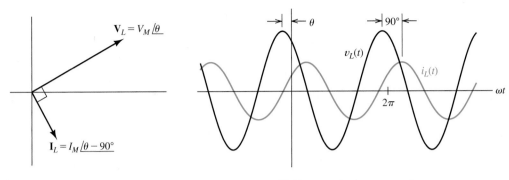

(a) Phasor diagram                    (b) Current and voltage versus time

**Figure 5.8** Current lags voltage by 90° in a pure inductance.

The phasor diagram of the current and voltage is shown in Figure 5.8(a). The corresponding waveforms of current and voltage are shown in Figure 5.8(b). *Notice that the current lags the voltage by 90° for a pure inductance.*

Equation 5.41 can be written in the form

$$\mathbf{V}_L = (\omega L \,\underline{/90°}) \times I_m \,\underline{/\theta - 90°} \tag{5.42}$$

Using Equation 5.40 to substitute into Equation 5.42, we find that

$$\mathbf{V}_L = (\omega L \,\underline{/90°}) \times \mathbf{I}_L \tag{5.43}$$

which can also be written as

$$\mathbf{V}_L = j\omega L \times \mathbf{I}_L \tag{5.44}$$

We refer to the term $j\omega L = \omega L \,\underline{/90°}$ as the **impedance** of the inductance and denote it as $Z_L$. Thus, we have

$$Z_L = j\omega L = \omega L \,\underline{/90°} \tag{5.45}$$

and

$$\mathbf{V}_L = Z_L \mathbf{I}_L \tag{5.46}$$

Thus, the phasor voltage is equal to the impedance times the phasor current. This is Ohm's law in phasor form. However, for an inductance, the impedance is an imaginary number, whereas resistance is a real number. (Impedances that are pure imaginary are also called **reactances**.)

## Capacitance

In a similar fashion for a capacitance, we can show that if the current and voltage are sinusoidal, the phasors are related by

$$\mathbf{V}_C = Z_C \mathbf{I}_C \tag{5.47}$$

Current lags voltage by 90° for a pure inductance.

Equation 5.46 shows that phasor voltage and phasor current for an inductance are related in a manner analogous to Ohm's law.

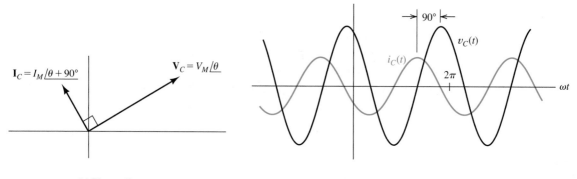

(a) Phasor diagram                              (b) Current and voltage versus time

**Figure 5.9** Current leads voltage by 90° in a pure capacitance.

in which the impedance of the capacitance is

$$Z_C = -j\frac{1}{\omega C} = \frac{1}{j\omega C} = \frac{1}{\omega C}\underline{/-90°} \tag{5.48}$$

Notice that the impedance of a capacitance is also a pure imaginary number. Suppose that the phasor voltage is

$$\mathbf{V}_C = V_m\underline{/\theta}$$

Then, the phasor current is

$$\mathbf{I}_C = \frac{\mathbf{V}_C}{Z_C} = \frac{V_m\underline{/\theta}}{(1/\omega C) - \underline{/90°}} = \omega C V_m\underline{/\theta + 90°}$$

$$\mathbf{I}_C = I_m\underline{/\theta + 90°}$$

Current leads voltage by 90° for a pure capacitance.

where $I_m = \omega C V_m$. The phasor diagram for current and voltage in a pure capacitance is shown in Figure 5.9(a). The corresponding plots of current and voltage versus time are shown in Figure 5.9(b). Notice that the current leads the voltage by 90°. (On the other hand, current lags voltage for an inductance. This is easy to remember if you know *ELI* the *ICE* man. The letter *E* is sometimes used to stand for *electromotive force*, which is another term for voltage, *L* and *C* are used for inductance and capacitance, respectively, and *I* is used for current.)

## Resistance

For a resistance, the phasors are related by

Current and voltage are in phase for a resistance.

$$\mathbf{V}_R = R\mathbf{I}_R \tag{5.49}$$

Because resistance is a real number, the current and voltage are in phase, as illustrated in Figure 5.10.

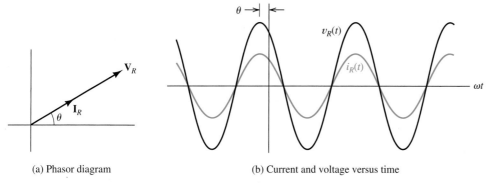

(a) Phasor diagram                    (b) Current and voltage versus time

**Figure 5.10** For a pure resistance, current and voltage are in phase.

## Complex Impedances in Series and Parallel

Impedances of inductances, capacitances, and resistances are combined in series and parallel in the same manner as resistances. (Recall that we combine capacitances in series as we do resistances in parallel. However, the **impedances** of capacitances are combined in the same manner as resistances.)

---

**Example 5.4**    Combining Impedances in Series and Parallel

Determine the complex impedance between terminals shown in Figure 5.11(a) for $\omega = 1000$ rad/s.

**Solution**    First, the impedance of the inductance is $j\omega L = j100 \ \Omega$, and the impedance of the capacitance is $-j/(\omega C) = -j80 \ \Omega$. These values are shown in Figure 5.11(b).

Next, we observe that the 200-$\Omega$ resistance is in parallel with the series impedance $100 + j100 \ \Omega$. The impedance of this parallel combination is

$$\frac{1}{1/100 + 1/(100 + j100)} = 80 + j40 \ \Omega$$

The resulting equivalent is shown in Figure 5.11(c). (We use rectangular boxes to represent the combined impedances of dissimilar types of components.)

Then, notice that the impedances in Figure 5.1(c) are in series, and they are combined by adding them resulting in:

$$-j80 + 80 + j40 = 80 - j40 = 89.44 - 26.57 \ \Omega$$

This is shown in Figure 5.11(d).    ■

---

**Exercise 5.6**    A voltage $v_L(t) = 100 \cos(200t)$ is applied to a 0.25-H inductance. (Notice that $\omega = 200$.) **a.** Find the impedance of the inductance, the phasor current, and the phasor voltage. **b.** Draw the phasor diagram.
**Answer    a.** $Z_L = j50 = 50 \ \underline{/90°}$, $I_L = 2 \ \underline{/-90°}$, $V_L = 100 \ \underline{/0°}$; **b.** the phasor diagram is shown in Figure 5.12(a).    □

---

**Exercise 5.7**    A voltage $v_C(t) = 100 \cos(200t)$ is applied to a 100-$\mu$F capacitance. **a.** Find the impedance of the capacitance, the phasor current, and the phasor voltage. **b.** Draw the phasor diagram.

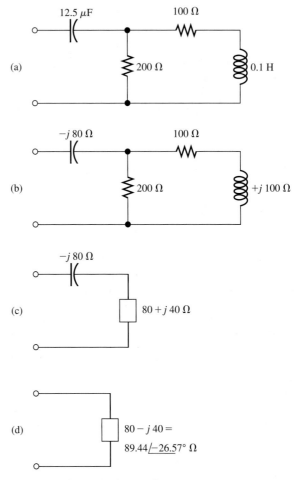

**Figure 5.11** Circuit of Example 5.4.

**Answer    a.** $Z_C = -j50 = 50\,\underline{/-90°}$, $\mathbf{I}_C = 2\,\underline{/90°}$, $\mathbf{V}_C = 100\,\underline{/0°}$; **b.** the phasor diagram is shown in Figure 5.12(b).    □

**Exercise 5.8**    A voltage $v_R(t) = 100\cos(200t)$ is applied to a 50-Ω resistance. **a.** Find the phasor for the current and the phasor voltage. **b.** Draw the phasor diagram. **Answer    a.** $\mathbf{I}_R = 2\,\underline{/0°}$, $\mathbf{V}_R = 100\,\underline{/0°}$; **b.** the phasor diagram is shown in Figure 5.12(c).    □

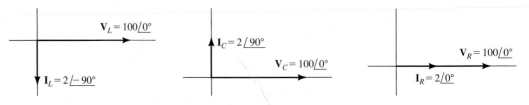

(a) Exercise 5.6 (0.25 H inductance)    (b) Exercise 5.7 (100 μF capacitance)    (c) Exercise 5.8 (50 Ω resistance)

**Figure 5.12** Answers for Exercises 5.6, 5.7, and 5.8. The scale has been expanded for the currents compared with the voltages so the current phasors can be easily seen.

# 5.4 CIRCUIT ANALYSIS WITH PHASORS AND COMPLEX IMPEDANCES

## Kirchhoff's Laws in Phasor Form

Recall that KVL requires that the voltages sum to zero for any closed path in an electrical network. A typical KVL equation is

$$v_1(t) + v_2(t) - v_3(t) = 0 \qquad (5.50)$$

If the voltages are sinusoidal, they can be represented by phasors. Then, Equation 5.50 becomes

$$\mathbf{V}_1 + \mathbf{V}_2 - \mathbf{V}_3 = 0 \qquad (5.51)$$

Thus, we can apply KVL directly to the phasors. The sum of the phasor voltages equals zero for any closed path.

Similarly, KCL can be applied to currents in phasor form. The sum of the phasor currents entering a node must equal the sum of the phasor currents leaving.

## Circuit Analysis Using Phasors and Impedances

We have seen that phasor currents and voltages are related by complex impedances, and Kirchhoff's laws apply in phasor form. Except for the fact that the voltages, currents, and impedances can be complex, the equations are exactly like those of resistive circuits.

A step-by-step procedure for steady-state analysis of circuits with sinusoidal sources is

1. Replace the time descriptions of the voltage and current sources with the corresponding phasors. (All of the sources must have the same frequency.)
2. Replace inductances by their complex impedances $Z_L = j\omega L = \omega L \,\underline{/90°}$. Replace capacitances by their complex impedances $Z_C = 1/(j\omega C) = (1/\omega C) \,\underline{/-90°}$. Resistances have impedances equal to their resistances.
3. Analyze the circuit by using any of the techniques studied in Chapter 2, and perform the calculations with complex arithmetic.

---

**Example 5.5**   **Steady-State AC Analysis of a Series Circuit**

Find the steady-state current for the circuit shown in Figure 5.13(a). Also, find the phasor voltage across each element and construct a phasor diagram.

**Solution**   From the expression given for the source voltage $v_s(t)$, we see that the peak voltage is 100 V, the angular frequency is $\omega = 500$, and the phase angle is 30°. The phasor for the voltage source is

> Step 1: Replace the time description of the voltage source with the corresponding phasor.

$$\mathbf{V}_s = 100 \,\underline{/30°}$$

The complex impedances of the inductance and capacitance are

> Step 2: Replace inductances and capacitances with their complex impedances.

$$Z_L = j\omega L = j500 \times 0.3 = j150 \ \Omega$$

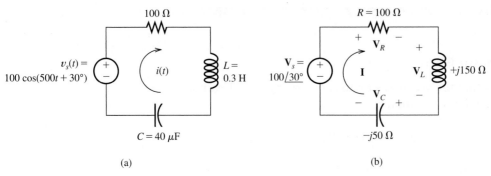

**Figure 5.13**  Circuit for Example 5.5.

and

$$Z_C = -j\frac{1}{\omega C} = -j\frac{1}{500 \times 40 \times 10^{-6}} = -j50 \ \Omega$$

**Step 3: Use complex arithmetic to analyze the circuit.**

The transformed circuit is shown in Figure 5.13(b). All three elements are in series. Thus, we find the equivalent impedance of the circuit by adding the impedances of all three elements:

$$Z_{eq} = R + Z_L + Z_C$$

Substituting values, we have

$$Z_{eq} = 100 + j150 - j50 = 100 + j100$$

Converting to polar form, we obtain

$$Z_{eq} = 141.4 \ \underline{/45°}$$

Now, we can find the phasor current by dividing the phasor voltage by the equivalent impedance, resulting in

$$\mathbf{I} = \frac{\mathbf{V}_s}{Z} = \frac{100 \ \underline{/30°}}{141.4 \ \underline{/45°}} = 0.707 \ \underline{/-15°}$$

As a function of time, the current is

$$i(t) = 0.707 \cos(500t - 15°)$$

Next, we can find the phasor voltage across each element by multiplying the phasor current by the respective impedance:

$$\mathbf{V}_R = R \times \mathbf{I} = 100 \times 0.707 \ \underline{/-15°} = 70.7 \ \underline{/-15°}$$

$$\mathbf{V}_L = j\omega L \times \mathbf{I} = \omega L \ \underline{/90°} \times \mathbf{I} = 150 \ \underline{/90°} \times 0.707 \ \underline{/-15°}$$

$$= 106.1 \ \underline{/75°}$$

$$\mathbf{V}_C = -j\frac{1}{\omega C} \times \mathbf{I} = \frac{1}{\omega C} \ \underline{/-90°} \times \mathbf{I} = 50 \ \underline{/-90°} \times 0.707 \ \underline{/-15°}$$

$$= 35.4 \ \underline{/-105°}$$

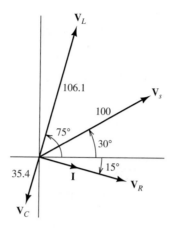

**Figure 5.14** Phasor diagram for Example 5.5.

The phasor diagram for the current and voltages is shown in Figure 5.14. Notice that the current **I** lags the source voltage $\mathbf{V}_s$ by 45°. As expected, the voltage $\mathbf{V}_R$ and current **I** are in phase for the resistance. For the inductance, the voltage $\mathbf{V}_L$ leads the current **I** by 90°. For the capacitance, the voltage $\mathbf{V}_C$ lags the current by 90°.    ■

---

### Example 5.6    Series and Parallel Combinations of Complex Impedances

Consider the circuit shown in Figure 5.15(a). Find the voltage $v_C(t)$ in steady state. Find the phasor current through each element, and construct a phasor diagram showing the currents and the source voltage.

**Solution**    The phasor for the voltage source is $\mathbf{V}_s = 10 \angle{-90°}$. [Notice that $v_s(t)$ is a sine function rather than a cosine function, and it is necessary to subtract 90° from the phase.] The angular frequency of the source is $\omega = 1000$. The impedances of the inductance and capacitance are

*Step 1: Replace the time description of the voltage source with the corresponding phasor.*

$$Z_L = j\omega L = j1000 \times 0.1 = j100 \ \Omega$$

and

$$Z_C = -j\frac{1}{\omega C} = -j\frac{1}{1000 \times 10 \times 10^{-6}} = -j100 \ \Omega$$

*Step 2: Replace inductances and capacitances with their complex impedances.*

The transformed network is shown in Figure 5.15(b).

To find $\mathbf{V}_C$, we will first combine the resistance and the impedance of the capacitor in parallel. Then, we will use the voltage-division principle to compute the voltage across the $RC$ combination. The impedance of the parallel $RC$ circuit is

*Step 3: Use complex arithmetic to analyze the circuit.*

$$Z_{RC} = \frac{1}{1/R + 1/Z_C} = \frac{1}{1/100 + 1/(-j100)}$$

$$= \frac{1}{0.01 + j0.01} = \frac{1 \angle{0°}}{0.01414 \angle{45°}} = 70.71 \angle{-45°}$$

Converting to rectangular form, we have

$$Z_{RC} = 50 - j50$$

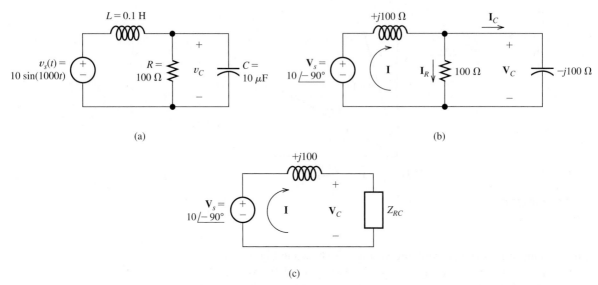

(a)

(b)

(c)

**Figure 5.15** Circuit for Example 5.6.

The equivalent network is shown in Figure 5.15(c).
Now, we use the voltage-division principle to obtain

$$\mathbf{V}_C = \mathbf{V}_s \frac{Z_{RC}}{Z_L + Z_{RC}} = 10\,\underline{/-90°}\,\frac{70.71\,\underline{/-45°}}{-j100 + 50 - j50}$$

$$= 10\,\underline{/-90°}\,\frac{70.71\,\underline{/-45°}}{50 + j50} = 10\,\underline{/-90°}\,\frac{70.71\,\underline{/-45°}}{70.71\,\underline{/45°}}$$

$$= 10\,\underline{/-180°}$$

Converting the phasor to a time function, we have

$$v_C(t) = 10\cos(1000t - 180°) = -10\cos(1000t)$$

Next, we compute the current in each element yielding

$$\mathbf{I} = \frac{\mathbf{V}_s}{Z_L + Z_{RC}} = \frac{10\,\underline{/-90°}}{j100 + 50 - j50} = \frac{10\,\underline{/-90°}}{50 + j50}$$

$$= \frac{10\,\underline{/-90°}}{70.71\,\underline{/45°}} = 0.1414\,\underline{/-135°}$$

$$\mathbf{I}_R = \frac{\mathbf{V}_C}{R} = \frac{10\,\underline{/-180°}}{100} = 0.1\,\underline{/-180°}$$

$$\mathbf{I}_C = \frac{\mathbf{V}_C}{Z_C} = \frac{10\,\underline{/-180°}}{-j100} = \frac{10\,\underline{/-180°}}{100\,\underline{/-90°}} = 0.1\,\underline{/-90°}$$

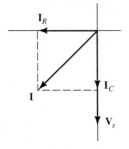

**Figure 5.16** Phasor diagram for Example 5.6.

The phasor diagram is shown in Figure 5.16.

## Node-Voltage Analysis

We can perform node-voltage analysis by using phasors in a manner parallel to that of Chapter 2. We illustrate with an example.

---

**Example 5.7**   **Steady-State AC Node-Voltage Analysis**

Use the node-voltage technique to find $v_1(t)$ in steady state for the circuit shown in Figure 5.17(a).

**Solution**   The transformed network is shown in Figure 5.17(b). We obtain two equations by applying KCL at node 1 and at node 2. This yields

$$\frac{\mathbf{V}_1}{10} + \frac{\mathbf{V}_1 - \mathbf{V}_2}{-j5} = 2 \underline{/-90°}$$

$$\frac{\mathbf{V}_2}{j10} + \frac{\mathbf{V}_2 - \mathbf{V}_1}{-j5} = 1.5 \underline{/0°}$$

These equations can be put into the standard form

$$(0.1 + j0.2)\mathbf{V}_1 - j0.2\mathbf{V}_2 = -j2$$

$$-j0.2\mathbf{V}_1 + j0.1\mathbf{V}_2 = 1.5$$

Now, we solve for $\mathbf{V}_1$ yielding

$$\mathbf{V}_1 = 16.1 \underline{/29.7°}$$

Then, we convert the phasor to a time function and obtain

$$v_1(t) = 16.1 \cos(100t + 29.7°)$$

∎

## Mesh-Current Analysis

In a similar fashion, you can use phasors to carry out mesh-current analysis in ac circuits.

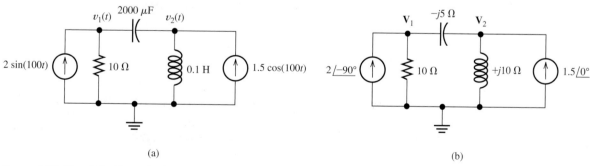

(a)                                                                (b)

**Figure 5.17**  Circuit for Example 5.7.

| Example 5.8 | Steady-State AC Mesh-Current Analysis |
|---|---|

Use the mesh-current technique to find $i_1(t)$ in steady state for the circuit shown in Figure 5.18(a).

**Solution**   First, we note that $\omega = 1000$ rad/s for both of the sources in this circuit. The impedance of the inductance is $j\omega L = j50\ \Omega$, and the impedance of the capacitance is $-j/(\omega C) = -j100\ \Omega$. The transformed network is shown in Figure 5.18(b).

Next, we write KVL equations. We cannot write equations around either mesh 1 or mesh 2 because we do not know the voltage across the current source. The only option is to write a KVL equation around the outside of the network, which yields:

$$j100 + 50\mathbf{I}_1 + 100\mathbf{I}_2 - j100\ \mathbf{I}_2 + j50\ \mathbf{I}_2 = 0$$

The current flowing upward through the current source is

$$\mathbf{I}_2 - \mathbf{I}_1 = 1$$

In standard form, these equations become:

$$50\mathbf{I}_1 + (100 - j50)\ \mathbf{I}_2 = -j100$$
$$-\mathbf{I}_1 + \mathbf{I}_2 = 1$$

Solving these equations results in:

$$\mathbf{I}_1 = 0.7071\ \underline{/-135°} \quad \text{or} \quad i_1(t) = 0.7071\cos(1000t - 135°)\ \text{V} \qquad \blacksquare$$

**Exercise 5.9**   Consider the circuit shown in Figure 5.19(a). **a.** Find $i(t)$. **b.** Construct a phasor diagram showing all three voltages and the current. **c.** What is the phase relationship between $v_s(t)$ and $i(t)$?
**Answer**   **a.** $i(t) = 0.0283\cos(500t - 135°)$; **b.** the phasor diagram is shown in Figure 5.19(b); **c.** $i(t)$ lags $v_s(t)$ by $45°$. □

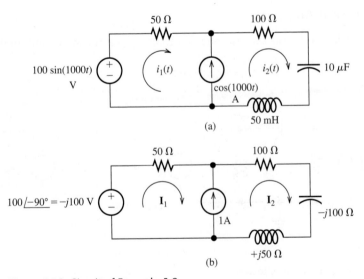

(a)

(b)

Figure 5.18 Circuit of Example 5.8.

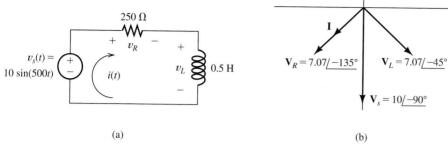

(a)

(b)

**Figure 5.19**  Circuit and phasor diagram for Exercise 5.9.

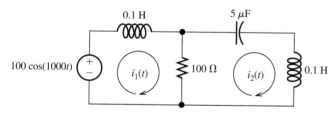

**Figure 5.20**  Circuit for Exercise 5.10.

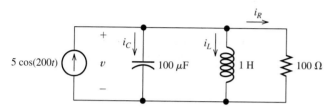

**Figure 5.21**  Circuit for Exercise 5.11.

**Exercise 5.10**   Find the phasor voltage and the phasor current through each element in the circuit of Figure 5.20.
**Answer**   $\mathbf{V} = 277 \,\underline{/-56.3°}$, $\mathbf{I}_C = 5.55 \,\underline{/33.7°}$, $\mathbf{I}_L = 1.39 \,\underline{/-146.3°}$, $\mathbf{I}_R = 2.77 \,\underline{/-56.3°}$. ☐

**Exercise 5.11**   Solve for the mesh currents shown in Figure 5.21.
**Answer**   $i_1(t) = 1.414 \cos(1000t - 45°)$, $i_2(t) = \cos(1000t)$. ☐

## 5.5  POWER IN AC CIRCUITS

Consider the situation shown in Figure 5.22. A voltage $v(t) = V_m \cos(\omega t)$ is applied to a network composed of resistances, inductances, and capacitances (i.e., an *RLC* network). The phasor for the voltage source is $\mathbf{V} = V_m \,\underline{/0°}$, and the equivalent impedance of the network is $Z = |Z| \,\underline{/\theta} = R + jX$. The phasor current is

$$\mathbf{I} = \frac{\mathbf{V}}{Z} = \frac{V_m \,\underline{/0°}}{|Z| \,\underline{/\theta}} = I_m \,\underline{/-\theta} \tag{5.52}$$

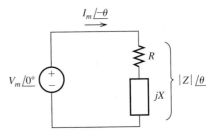

**Figure 5.22** A voltage source delivering power to a load impedance $Z = R + jX$.

where we have defined

$$I_m = \frac{V_m}{|Z|} \tag{5.53}$$

Before we consider the power delivered by the source to a general load, it is instructive to consider a pure resistive load, a pure inductive load, and a pure capacitive load.

### Current, Voltage, and Power for a Resistive Load

First, consider the case in which the network is a pure resistance. Then, $\theta = 0$, and we have

$$v(t) = V_m \cos(\omega t)$$

$$i(t) = I_m \cos(\omega t)$$

$$p(t) = v(t)i(t) = V_m I_m \cos^2(\omega t)$$

Plots of these quantities are shown in Figure 5.23. Notice that the current is in phase with the voltage (i.e., they both reach their peak values at the same time). Because $p(t)$ is positive at all times, we conclude that energy flows continually in the direction

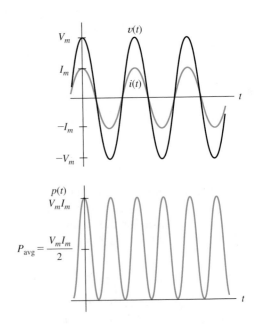

**Figure 5.23** Current, voltage, and power versus time for a purely resistive load.

from the source to the load (where it is converted to heat). Of course, the value of the power rises and falls with the voltage (and current) magnitude.

## Current, Voltage, and Power for an Inductive Load

Next, consider the case in which the load is a pure inductance for which $Z = \omega L \underline{/90°}$. Thus, $\theta = 90°$, and we get

$$v(t) = V_m \cos(\omega t)$$

$$i(t) = I_m \cos(\omega t - 90°) = I_m \sin(\omega t)$$

$$p(t) = v(t)i(t) = V_m I_m \cos(\omega t) \sin(\omega t)$$

Using the trigonometric identity $\cos(x) \sin(x) = (1/2) \sin(2x)$, we find that the expression for the power becomes

$$p(t) = \frac{V_m I_m}{2} \sin(2\omega t)$$

Average power is absorbed by resistances in ac circuits.

Power surges into and out of inductances in ac circuits. The average power absorbed by inductances is zero.

Plots of the current, voltage, and power are shown in Figure 5.24(a). Notice that the current lags the voltage by 90°. Half of the time the power is positive, showing that energy is delivered to the inductance, where it is stored in the magnetic field. For the other half of the time, power is negative, showing that the inductance returns energy to the source. Notice that the average power is zero. In this case, we say that **reactive power** flows from the source to the load.

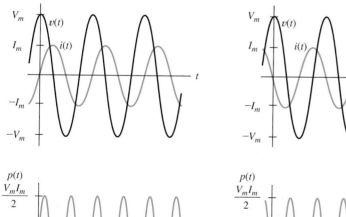

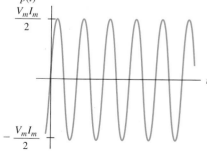

(a) Pure inductive load

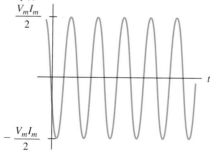

(b) Pure capacitive load

**Figure 5.24** Current, voltage, and power versus time for pure energy-storage elements.

## Current, Voltage, and Power for a Capacitive Load

Next, consider the case in which the load is a pure capacitance for which $Z = (1/\omega C)\,\underline{/-90°}$. Then, $\theta = -90°$, and we have

$$v(t) = V_m \cos(\omega t)$$

$$i(t) = I_m \cos(\omega t + 90°) = -I_m \sin(\omega t)$$

$$p(t) = v(t)i(t) = -V_m I_m \cos(\omega t) \sin(\omega t)$$

$$= -\frac{V_m I_m}{2} \sin(2\omega t)$$

Power surges into and out of capacitances in ac circuits. The average power absorbed by capacitances is zero.

    Plots of the current, voltage, and power are shown in Figure 5.24(b). Here again, the average power is zero, and we say that reactive power flows. Notice, however, that the power for the capacitance carries the opposite sign as that for the inductance. Thus, we say that reactive power is positive for an inductance and is negative for a capacitance. If a load contains both inductance and capacitance with reactive powers of equal magnitude, the reactive powers cancel.

## Importance of Reactive Power

The power flow back and forth to inductances and capacitances is called reactive power. Reactive power flow is important because it causes power dissipation in the lines and transformers of a power distribution system.

Even though no average power is consumed by a pure energy-storage element (inductance or capacitance), reactive power is still of concern to power-system engineers because transmission lines, transformers, fuses, and other elements must be capable of withstanding the current associated with reactive power. It is possible to have loads composed of energy-storage elements that draw large currents requiring heavy-duty wiring, even though little average power is consumed. Therefore, electric-power companies charge their industrial customers for reactive power (but at a lower rate) as well as for total energy delivered.

## Power Calculations for a General Load

Now, let us consider the voltage, current, and power for a general $RLC$ load for which the phase $\theta$ can be any value from $-90°$ to $+90°$. We have

$$v(t) = V_m \cos(\omega t) \tag{5.54}$$

$$i(t) = I_m \cos(\omega t - \theta) \tag{5.55}$$

$$p(t) = V_m I_m \cos(\omega t) \cos(\omega t - \theta) \tag{5.56}$$

Using the trigonometric identity

$$\cos(\omega t - \theta) = \cos(\theta)\cos(\omega t) + \sin(\theta)\sin(\omega t)$$

we can put Equation 5.56 into the form

$$p(t) = V_m I_m \cos(\theta)\cos^2(\omega t) + V_m I_m \sin(\theta)\cos(\omega t)\sin(\omega t) \tag{5.57}$$

Using the identities

$$\cos^2(\omega t) = \frac{1}{2} + \frac{1}{2}\cos(2\omega t)$$

and

$$\cos(\omega t) \sin(\omega t) = \frac{1}{2} \sin(2\omega t)$$

we find that Equation 5.57 can be written as

$$p(t) = \frac{V_m I_m}{2} \cos(\theta)[1 + \cos(2\omega t)] + \frac{V_m I_m}{2} \sin(\theta) \sin(2\omega t) \qquad (5.58)$$

Notice that the terms involving $\cos(2\omega t)$ and $\sin(2\omega t)$ have average values of zero. Thus, the average power $P$ is given by

$$P = \frac{V_m I_m}{2} \cos(\theta) \qquad (5.59)$$

Using the fact that $V_{rms} = V_m/\sqrt{2}$ and $I_{rms} = I_m/\sqrt{2}$, we can write the expression for average power as

$$P = V_{rms} I_{rms} \cos(\theta) \qquad (5.60)$$

As usual, the units of power are watts (W).

## Power Factor

The term $\cos(\theta)$ is called the **power factor**:

$$\mathbf{PF} = \cos(\theta) \qquad (5.61)$$

To simplify our discussion, we assumed a voltage having zero phase. In general, the phase of the voltage may have a value other than zero. Then, $\theta$ should be taken as the phase of the voltage $\theta_v$ minus the phase of the current $\theta_i$, or

$$\theta = \theta_v - \theta_i \qquad (5.62)$$

Power factor is the cosine of the angle $\theta$ by which the current lags the voltage. (If the current leads the voltage, the angle is negative.)

Sometimes, $\theta$ is called the **power angle**.

Often, power factor is stated as a percentage. Also, it is common to state whether the current leads (capacitive load) or lags (inductive load) the voltage. A typical power factor would be stated to be 90 percent lagging, which means that $\cos(\theta) = 0.9$ and that the current lags the voltage.

Often, power factor is expressed as a percentage.

If the current lags the voltage, the power factor is said to be inductive or lagging. If the current leads the voltage, the power factor is said to be capacitive or leading.

## Reactive Power

In ac circuits, energy flows into and out of energy storage elements (inductances and capacitances). For example, when the voltage magnitude across a capacitance is increasing, energy flows into it, and when the voltage magnitude decreases, energy flows out. Similarly, energy flows into an inductance when the current flowing through it increases in magnitude. Although instantaneous power can be very large, the net energy transferred per cycle is zero for either an ideal capacitance or inductance.

When a capacitance and an inductance are in parallel (or series) energy flows into one, while it flows out of the other. Thus, the power flow of a capacitance tends to cancel that of an inductance at each instant in time.

The peak instantaneous power associated with the energy storage elements contained in a general load is called **reactive power** and is given by

$$Q = V_{rms}I_{rms} \sin(\theta) \tag{5.63}$$

where $\theta$ is the power angle given by Equation 5.62, $V_{rms}$ is the effective (or rms) voltage across the load, and $I_{rms}$ is the effective current through the load. (Notice that if we had a purely resistive load, we would have $\theta = 0$ and $Q = 0$.)

The physical units of reactive power are watts. However, to emphasize the fact that $Q$ does not represent the flow of net energy, its units are usually given as *Volt Amperes Reactive* (VARs).

> The units of reactive power $Q$ are VARs.

## Apparent Power

Another quantity of interest is the **apparent power**, which is defined as the product of the effective voltage and the effective current, or

$$\text{apparent power} = V_{rms}I_{rms}$$

> Apparent power equals the product of rms current and rms voltage. The units for apparent power are stated as volt-amperes (VA).

Its units are volt-amperes (VA).

Using Equations 5.60 and 5.63, we can write

$$P^2 + Q^2 = (V_{rms}I_{rms})^2 \cos^2(\theta) + (V_{rms}I_{rms})^2 \sin^2(\theta)$$

However, $\cos^2(\theta) + \sin^2(\theta) = 1$, so we have

$$P^2 + Q^2 = (V_{rms}I_{rms})^2 \tag{5.64}$$

## Units

Often, the units given for a quantity indicate whether the quantity is power (W), reactive power (VAR), or apparent power (VA). For example, if we say that we have a 5-kW load, this means that $P = 5$ kW. On the other hand, if we have a 5-kVA load, $V_{rms}I_{rms} = 5$ kVA. If we say that a load absorbs 5 kVAR, then $Q = 5$ kVAR.

## Power Triangle

> The power triangle is a compact way to represent ac power relationships.

The relationships between real power $P$, reactive power $Q$, apparent power $V_{rms}I_{rms}$, and the power angle $\theta$ can be represented by the **power triangle**. The power triangle is shown in Figure 5.25(a) for an inductive load, in which case $\theta$ and $Q$ are positive. The power triangle for a capacitive load is shown in Figure 5.25(b), in which case $\theta$ and $Q$ are negative.

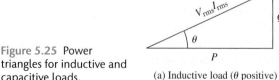

**Figure 5.25** Power triangles for inductive and capacitive loads.

(a) Inductive load ($\theta$ positive)

(b) Capacitive load ($\theta$ negative)

## Additional Power Relationships

The impedance $Z$ is

$$Z = |Z| \underline{/\theta} = R + jX$$

in which $R$ is the resistance of the load and $X$ is the reactance. This is illustrated in Figure 5.26. We can write

$$\cos(\theta) = \frac{R}{|Z|} \tag{5.65}$$

and

$$\sin(\theta) = \frac{X}{|Z|} \tag{5.66}$$

Substituting Equation 5.65 into Equation 5.59, we find that

$$P = \frac{V_m I_m}{2} \times \frac{R}{|Z|} \tag{5.67}$$

However, Equation 5.53 states that $I_m = V_m/|Z|$, so we have

$$P = \frac{I_m^2}{2}R \tag{5.68}$$

Using the fact that $I_{rms} = I_m/\sqrt{2}$, we get

$$P = I_{rms}^2 R \tag{5.69}$$

In a similar fashion, we can show that

$$Q = I_{rms}^2 X \tag{5.70}$$

In applying Equation 5.70, we retain the algebraic sign of $X$. For an inductive load, $X$ is positive, whereas for a capacitive load, $X$ is negative. This is not hard to remember if we keep in mind that $Q$ is positive for inductive loads and negative for capacitive loads.

Furthermore, in Section 5.1, we showed that the average power delivered to a resistance is

$$P = \frac{V_{Rrms}^2}{R} \tag{5.71}$$

In Equation 5.69, $R$ is the real part of the impedance through which the current flows.

In Equation 5.70, $X$ is the imaginary part (including the algebraic sign) of the impedance through which the current flows.

Reactive power $Q$ is positive for inductive loads and negative for capacitive loads.

In Equation 5.71, $V_{Rrms}$ is the rms voltage across the resistance.

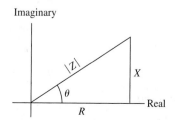

**Figure 5.26** The load impedance in the complex plane.

where $V_{R\text{rms}}$ is the rms value of the voltage *across the resistance.* (Notice in Figure 5.22 that the source voltage does not appear across the resistance, because the reactance is in series with the resistance.)

Similarly, we have

In Equation 5.72, $V_{X\text{rms}}$ is the rms voltage across the reactance.

$$Q = \frac{V_{X\text{rms}}^2}{X} \tag{5.72}$$

where $V_{X\text{rms}}$ is the rms value of the voltage *across the reactance.* Here again, $X$ is positive for an inductance and negative for a capacitance.

## Complex Power

Consider the portion of a circuit shown in Figure 5.27. The **complex power**, denoted as **S**, delivered to this circuit is defined as one half the product of the phasor voltage **V** and the complex conjugate of the phasor current **I**\*.

$$\mathbf{S} = \frac{1}{2}\mathbf{VI}^* \tag{5.73}$$

The phasor voltage is $\mathbf{V} = V_m \underline{/\theta_v}$ in which $V_m$ is the peak value of the voltage and $\theta_v$ is the phase angle of the voltage. Furthermore, the phasor current is $\mathbf{I} = I_m \underline{/\theta_i}$ where $I_m$ is the peak value and $\theta_i$ is the phase angle of the current. Substituting into Equation 5.73, we have

$$\mathbf{S} = \frac{1}{2}\mathbf{VI}^* = \frac{1}{2}(V_m \underline{/\theta_v}) \times (I_m \underline{/-\theta_i}) = \frac{V_m I_m}{2}\underline{/\theta_v - \theta_i} = \frac{V_m I_m}{2}\underline{/\theta} \tag{5.74}$$

where, as before, $\theta = \theta_v - \theta_i$ is the power angle. Expanding the right-hand term of Equation 5.74 into real and imaginary parts, we have

$$\mathbf{S} = \frac{V_m I_m}{2}\cos(\theta) + j\frac{V_m I_m}{2}\sin(\theta)$$

However, the first term on the right-hand side is the average power $P$ delivered to the circuit and the second term is $j$ times the reactive power. Thus, we can write:

$$\mathbf{S} = \frac{1}{2}\mathbf{VI}^* = P + jQ \tag{5.75}$$

If we know the complex power **S**, then we can find the power, reactive power, and apparent power:

$$P = \text{Re}(\mathbf{S}) = \text{Re}\left(\frac{1}{2}\mathbf{VI}^*\right) \tag{5.76}$$

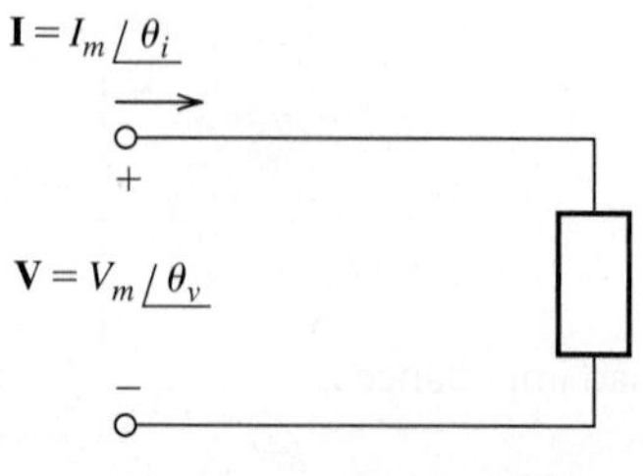

$\mathbf{I} = I_m \underline{/\theta_i}$

$\mathbf{V} = V_m \underline{/\theta_v}$

**Figure 5.27** The complex power delivered to this circuit element is $\mathbf{S} = \frac{1}{2}\mathbf{VI}^*$.

$$Q = \text{Im}(\mathbf{S}) = \text{Im}\left(\frac{1}{2}\mathbf{VI}^*\right) \tag{5.77}$$

$$\text{apparent power} = |\mathbf{S}| = \left|\frac{1}{2}\mathbf{VI}^*\right| \tag{5.78}$$

where $\text{Re}(\mathbf{S})$ denotes the real part of $\mathbf{S}$ and $\text{Im}(\mathbf{S})$ denotes the imaginary part of $\mathbf{S}$.

---

**Example 5.9**   **AC Power Calculations**

Compute the power and reactive power taken from the source for the circuit of Example 5.6. Also, compute the power and reactive power delivered to each element in the circuit. For convenience, the circuit and the currents that were computed in Example 5.6 are shown in Figure 5.28.

**Solution**   To find the power and reactive power for the source, we must first find the power angle which is given by Equation 5.62:

$$\theta = \theta_v - \theta_i$$

The angle of the source voltage is $\theta_v = -90°$, and the angle of the current delivered by the source is $\theta_i = -135°$. Therefore, we have

$$\theta = -90° - (-135°) = 45°$$

The rms source voltage and current are

$$V_{srms} = \frac{|\mathbf{V}_s|}{\sqrt{2}} = \frac{10}{\sqrt{2}} = 7.071 \text{ V}$$

$$I_{rms} = \frac{|\mathbf{I}|}{\sqrt{2}} = \frac{0.1414}{\sqrt{2}} = 0.1 \text{ A}$$

Now, we use Equations 5.60 and 5.63 to compute the power and reactive power delivered by the source:

$$P = V_{srms}I_{rms}\cos(\theta)$$
$$= 7.071 \times 0.1 \cos(45°) = 0.5 \text{ W}$$
$$Q = V_{srms}I_{rms}\sin(\theta)$$
$$= 7.071 \times 0.1 \sin(45°) = 0.5 \text{ VAR}$$

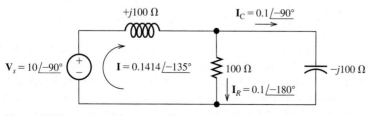

**Figure 5.28** Circuit and currents for Example 5.9.

An alternative and more compact method for computing $P$ and $Q$ is to first find the complex power and then take the real and imaginary parts:

$$\mathbf{S} = \frac{1}{2}\mathbf{V_sI^*} = \frac{1}{2}(10\ \underline{/-90°})(0.1414\ \underline{/135°}) = 0.707\ \underline{/45°} = 0.5 + j0.5$$

$$P = \text{Re}(\mathbf{S}) = 0.5\ \text{W}$$

$$Q = \text{Im}(\mathbf{S}) = 0.5\ \text{VAR}$$

We can use Equation 5.70 to compute the reactive power delivered to the inductor, yielding

$$Q_L = I^2_{\text{rms}}X_L = (0.1)^2(100) = 1.0\ \text{VAR}$$

For the capacitor, we have

$$Q_C = I^2_{\text{Crms}}X_C = \left(\frac{0.1}{\sqrt{2}}\right)^2(-100) = -0.5\ \text{VAR}$$

Notice that we have used the rms value of the current through the capacitor in this calculation. Furthermore, notice that the reactance $X_C$ of the capacitance is negative. As expected, the reactive power is negative for a capacitance. The reactive power for the resistance is zero. As a check, we can verify that the reactive power delivered by the source is equal to the sum of the reactive powers absorbed by the inductance and capacitance. This is demonstrated by

$$Q = Q_L + Q_C$$

The power delivered to the resistance is

$$P_R = I^2_{R\ \text{rms}}R = \left(\frac{|\mathbf{I_R}|}{\sqrt{2}}\right)^2 R = \left(\frac{0.1}{\sqrt{2}}\right)^2 100$$

$$= 0.5\ \text{W}$$

The power absorbed by the capacitance and inductance is given by

$$P_L = 0$$

$$P_C = 0$$

Thus, all of the power delivered by the source is absorbed by the resistance.    ■

In power distribution systems, we typically encounter much larger values of power, reactive power, and apparent power than the small values of the preceding example. For example, a large power plant may generate 1000 MW. A 100-hp motor used in an industrial application absorbs approximately 85 kW of electrical power under full load.

A typical residence absorbs a *peak* power in the range of 10 to 40 kW. The *average* power for my home (which is of average size, has two residents, and does not use electrical heating) is approximately 600 W. It is interesting to keep your average power consumption and the power used by various appliances in mind because it gives you a clear picture of the economic and environmental impact of turning off lights, computers, and so on, that are not being used.

**Example 5.10**   **Using Power Triangles**

Consider the situation shown in Figure 5.29. Here, a voltage source delivers power to two loads connected in parallel. Find the power, reactive power, and power factor for the source. Also, find the phasor current **I**.

**Solution**   By the units given in the figure, we see that load $A$ has an *apparent power* of 10 kVA. On the other hand, the *power* for load $B$ is specified as 5 kW.

Furthermore, load $A$ has a power factor of 0.5 leading, which means that the current leads the voltage in load $A$. Another way to say this is that load $A$ is capacitive. Similarly, load $B$ has a power factor of 0.7 lagging (or inductive).

Our approach is to find the power and reactive power for each load. Then, we add these values to find the power and reactive power for the source. Finally, we compute the power factor for the source and then find the current.

Because load $A$ has a leading (capacitive) power factor, we know that the reactive power $Q_A$ and power angle $\theta_A$ are negative. The power triangle for load $A$ is shown in Figure 5.30(a). The power factor is

*Calculations for load A*

$$\cos(\theta_A) = 0.5$$

The power is

$$P_A = V_{\text{rms}}I_{\text{Arms}}\cos(\theta_A) = 10^4(0.5) = 5\text{ kW}$$

Solving Equation 5.64 for reactive power, we have

$$Q_A = \sqrt{(V_{\text{rms}}I_{\text{Arms}})^2 - P_A^2}$$
$$= \sqrt{(10^4)^2 - (5000)^2}$$
$$= -8.660\text{ kVAR}$$

Notice that we have selected the negative value for $Q_A$, because we know that reactive power is negative for a capacitive (leading) load.

The power triangle for load $B$ is shown in Figure 5.30(b). Since load $B$ has a lagging (inductive) power factor, we know that the reactive power $Q_B$ and power angle $\theta_B$ are positive. Thus,

*Calculations for load B*

$$\theta_B = \arccos(0.7) = 45.57°$$

Applying trigonometry, we can write

$$Q_B = P_B\tan(\theta_B) = 5000\tan(45.57°)$$
$$Q_B = 5.101\text{ kVAR}$$

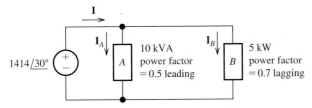

**Figure 5.29** Circuit for Example 5.10.

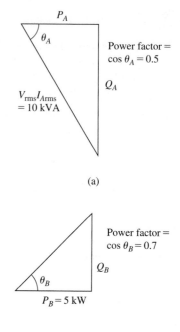

(a)

Power factor =
cos $\theta_B$ = 0.7

(b)

**Figure 5.30** Power triangles for loads
A and B of Example 5.10.

Total power is obtained by adding the powers for the various loads. Similarly, the reactive powers are added.

At this point, as shown here we can find the power and reactive power delivered by the source:

$$P = P_A + P_B = 5 + 5 = 10 \text{ kW}$$

$$Q = Q_A + Q_B = -8.660 + 5.101 = -3.559 \text{ kVAR}$$

Power calculations for the source.

Because $Q$ is negative, we know that the power angle is negative. Thus, we have

$$\theta = \arctan\left(\frac{Q}{P}\right) = \arctan\left(\frac{-3.559}{10}\right) = -19.59°$$

The power factor is

$$\cos(\theta) = 0.9421$$

Power-system engineers frequently express power factors as percentages and would state this power factor as 94.21 percent leading.

The complex power delivered by the source is

$$\mathbf{S} = P + jQ = 10 - j3.559 = 10.61 \underline{/-19.59°} \text{ kVA}$$

Thus, we have

$$\mathbf{S} = \frac{1}{2}\mathbf{V}_s\mathbf{I}^* = \frac{1}{2}(1414 \underline{/30°})\mathbf{I}^* = 10.61 \times 10^3 \underline{/-19.59°} \text{ kVA}$$

Solving for the phasor current, we obtain:

$$\mathbf{I} = 15.0 \underline{/49.59°} \text{ A}$$

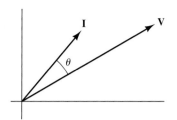

Figure 5.31 Phasor diagram for Example 5.10.

The phasor diagram for the current and voltage is shown in Figure 5.31. Notice that the current is leading the voltage. ■

## Power-Factor Correction

We have seen that large currents can flow in energy-storage devices (inductance and capacitance) without average power being delivered. In heavy industry, many loads are partly inductive, and large amounts of reactive power flow. This reactive power causes higher currents in the power distribution system. Consequently, the lines and transformers must have higher ratings than would be necessary to deliver the same average power to a resistive (100 percent power factor) load.

Energy rates charged to industry depend on the power factor, with higher charges for energy delivered at lower power factors. (Power factor is not taken into account for residential customers.) Therefore, it is advantageous to choose loads that operate at near unity power factor. A common approach is to place capacitors in parallel with an inductive load to increase the power factor.

Power-factor correction can provide a significant economic advantage for consumers of large amounts of electrical energy.

### Example 5.11   Power-Factor Correction

A 50-kW load operates from a 60-Hz 10-kV-rms line with a power factor of 60 percent lagging. Compute the capacitance that must be placed in parallel with the load to achieve a 90 percent lagging power factor.

Solution   First, we find the load power angle:

$$\theta_L = \arccos(0.6) = 53.13°$$

Then, we use the power-triangle concept to find the reactive power of the load. Hence,

$$Q_L = P_L \tan(\theta_L) = 66.67 \text{ kVAR}$$

After adding the capacitor, the power will still be 50 kW and the power angle will become

$$\theta_{new} = \arccos(0.9) = 25.84°$$

The new value of the reactive power will be

$$Q_{new} = P_L \tan(\theta_{new}) = 24.22 \text{ kVAR}$$

Thus, the reactive power of the capacitance must be

$$Q_C = Q_{new} - Q_L = -42.45 \text{ kVAR}$$

Now, we find that the reactance of the capacitor is

$$X_C = -\frac{V_{\text{rms}}^2}{Q_C} = \frac{(10^4)^2}{42,450} = -2356 \ \Omega$$

Finally, the angular frequency is

$$\omega = 2\pi 60 = 377.0$$

and the required capacitance is

$$C = \frac{1}{\omega|X_C|} = \frac{1}{377 \times 2356} = 1.126 \ \mu\text{F}$$  ■

**Exercise 5.12**   **a.** A voltage source $\mathbf{V} = 707.1 \ \underline{/40°}$ delivers 5 kW to a load with a power factor of 100 percent. Find the reactive power and the phasor current. **b.** Repeat if the power factor is 20 percent lagging. **c.** For which power factor would the current ratings of the conductors connecting the source to the load be higher? In which case could the wiring be a lower cost?
**Answer**   **a.** $Q = 0$, $\mathbf{I} = 14.14 \ \underline{/40°}$; **b.** $Q = 24.49$ kVAR, $\mathbf{I} = 70.7 \ \underline{/-38.46°}$; **c.** The current ratings for the conductors would need to be five times higher for part (b) than for part (a). Clearly, the wiring could be a lower cost for 100 percent power factor. ◻

**Exercise 5.13**   A 1-kV-rms 60-Hz voltage source delivers power to two loads in parallel. The first load is a 10-$\mu$F capacitor, and the second load absorbs an apparent power of 10 kVA with an 80 percent lagging power factor. Find the total power, the total reactive power, the power factor for the source, and the rms source current.
**Answer**   $P = 8$ kW, $Q = 2.23$ kVAR, PF = 96.33 percent lagging, $I_{\text{rms}} = 8.305$ A. ◻

## 5.6 THÉVENIN AND NORTON EQUIVALENT CIRCUITS

### Thévenin Equivalent Circuits

In Chapter 2, we saw that a two-terminal network composed of sources and resistances has a Thévenin equivalent circuit consisting of a voltage source in series with a resistance. We can apply this concept to circuits composed of sinusoidal sources (all having a common frequency), resistances, inductances, and capacitances. Here, the Thévenin equivalent consists of a phasor voltage source in series with a complex impedance as shown in Figure 5.32. Recall that phasors and complex impedances apply only for steady-state operation; therefore, these Thévenin equivalents are valid for only steady-state operation of the circuit.

As in resistive circuits, the Thévenin voltage is equal to the open-circuit voltage of the two-terminal circuit. In ac circuits, we use phasors, so we can write

*The Thévenin voltage is equal to the open-circuit phasor voltage of the original circuit.*

$$\mathbf{V}_t = \mathbf{V}_{\text{oc}} \tag{5.79}$$

The Thévenin impedance $Z_t$ can be found by zeroing the *independent* sources and looking back into the terminals to find the equivalent impedance. (Recall that in

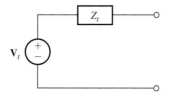

Figure 5.32 The Thévenin equivalent for an ac circuit consists of a phasor voltage source $\mathbf{V}_t$ in series with a complex impedance $Z_t$.

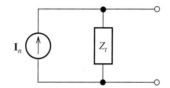

Figure 5.33 The Norton equivalent circuit consists of a phasor current source $\mathbf{I}_n$ in parallel with the complex impedance $Z_t$.

zeroing a voltage source, we reduce its voltage to zero, and it becomes a short circuit. On the other hand, in zeroing a current source, we reduce its current to zero, and it becomes an open circuit.) Also, keep in mind that we must not zero the *dependent* sources.

We can find the Thévenin impedance by zeroing the independent sources and determining the impedance looking into the circuit terminals.

Another approach to determining the Thévenin impedance is first to find the short-circuit phasor current $\mathbf{I}_{sc}$ and the open-circuit voltage $\mathbf{V}_{oc}$. Then, the Thévenin impedance is given by

$$Z_t = \frac{\mathbf{V}_{oc}}{\mathbf{I}_{sc}} = \frac{\mathbf{V}_t}{\mathbf{I}_{sc}} \tag{5.80}$$

The Thévenin impedance equals the open-circuit voltage divided by the short-circuit current.

Thus, except for the use of phasors and complex impedances, the concepts and procedures for Thévenin equivalents of steady-state ac circuits are the same as for resistive circuits.

## Norton Equivalent Circuits

Another equivalent for a two-terminal steady-state ac circuit is the Norton equivalent, which consists of a phasor current source $\mathbf{I}_n$ in parallel with the Thévenin impedance. This is shown in Figure 5.33. The Norton current is equal to the short-circuit current of the original circuit:

$$\mathbf{I}_n = \mathbf{I}_{sc} \tag{5.81}$$

### Example 5.12   Thévenin and Norton Equivalents

Find the Thévenin and Norton equivalent circuits for the circuit shown in Figure 5.34(a).

**Solution**   We must find two of the three quantities: $\mathbf{V}_{oc}$, $\mathbf{I}_{sc}$, or $Z_t$. Often, it pays to look for the two that can be found with the least amount of work. In this case, we elect to start by zeroing the sources to find $Z_t$. After that part of the problem is finished, we will find the short-circuit current.

First, look to see which two of the three quantities $\mathbf{V}_{oc}$, $\mathbf{I}_{sc}$, or $Z_t$ are easiest to determine.

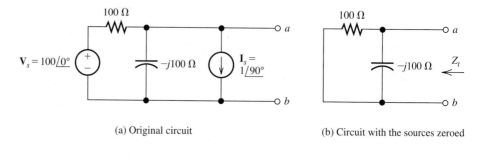

(a) Original circuit                              (b) Circuit with the sources zeroed

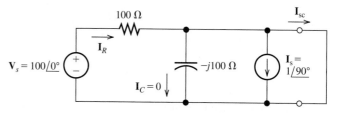

(c) Circuit with a short circuit

**Figure 5.34** Circuit of Example 5.12.

If we zero the sources, we obtain the circuit shown in Figure 5.34(b). The Thévenin impedance is the impedance seen looking back into terminals $a-b$. This is the parallel combination of the resistance and the impedance of the capacitance. Thus, we have

$$Z_t = \frac{1}{1/100 + 1/(-j100)}$$

$$= \frac{1}{0.01 + j0.01}$$

$$= \frac{1}{0.01414 \,\underline{/45°}}$$

$$= 70.71 \,\underline{/-45°}$$

$$= 50 - j50 \ \Omega$$

Now, we apply a short circuit to terminals $a-b$ and find the current, which is shown in Figure 5.34(c). With a short circuit, the voltage across the capacitance is zero. Therefore, $\mathbf{I}_C = 0$. Furthermore, the source voltage $\mathbf{V}_s$ appears across the resistance, so we have

$$I_R = \frac{\mathbf{V}_s}{100} = \frac{100}{100} = 1 \,\underline{/0°} \ \text{A}$$

Then applying KCL, we can write

$$\mathbf{I}_{sc} = \mathbf{I}_R - \mathbf{I}_s = 1 - 1 \,\underline{/90°} = 1 - j = 1.414 \,\underline{/-45°} \ \text{A}$$

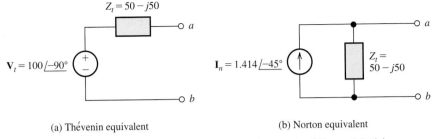

(a) Thévenin equivalent                    (b) Norton equivalent

**Figure 5.35** Thévenin and Norton equivalents for the circuit of Figure 5.34(a).

Next, we can solve Equation 5.80 for the Thévenin voltage:

$$\mathbf{V}_t = \mathbf{I}_{sc}Z_t = 1.414\ \underline{/-45°} \times 70.71\ \underline{/-45°} = 100\ \underline{/-90°}\ \text{V}$$

Finally, we can draw the Thévenin and Norton equivalent circuits, which are shown in Figure 5.35. ∎

## Maximum Average Power Transfer

Sometimes we are faced with the problem of adjusting a load impedance to extract the maximum average power from a two-terminal circuit. This situation is shown in Figure 5.36, in which we have represented the two-terminal circuit by its Thévenin equivalent. Of course, the power delivered to the load depends on the load impedance. A short-circuit load receives no power because the voltage across it is zero. Similarly, an open-circuit load receives no power because the current through it is zero. Furthermore, a pure reactive load (inductance or capacitance) receives no power because the load power factor is zero.

Two situations are of interest. First, suppose that the load impedance can take any complex value. Then, it turns out that the load impedance for maximum-power transfer is the complex conjugate of the Thévenin impedance:

$$Z_{\text{load}} = Z_t^*$$

Let us consider why this is true. Suppose that the Thévenin impedance is

$$Z_t = R_t + jX_t$$

Then, the load impedance for maximum-power transfer is

$$Z_{\text{load}} = Z_t^* = R_t - jX_t$$

Of course, the total impedance seen by the Thévenin source is the sum of the Thévenin impedance and the load impedance:

$$\begin{aligned}
Z_{\text{total}} &= Z_t + Z_{\text{load}} \\
&= R_t + jX_t + R_t - jX_t \\
&= 2R_t
\end{aligned}$$

Thus, the reactance of the load cancels the internal reactance of the two-terminal circuit. Maximum power is transferred to a given load resistance by maximizing the current. For given resistances, maximum current is achieved by choosing the

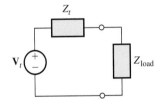

**Figure 5.36** The Thévenin equivalent of a two-terminal circuit delivering power to a load impedance.

If the load can take on any complex value, maximum-power transfer is attained for a load impedance equal to the complex conjugate of the Thévenin impedance.

reactance to minimize the total impedance magnitude. Of course, for fixed resistances, the minimum impedance magnitude occurs for zero total reactance.

> If the load is required to be a pure resistance, maximum-power transfer is attained for a load resistance equal to the magnitude of the Thévenin impedance.

Having established the fact that the total reactance should be zero, we have a resistive circuit. We considered this resistive circuit in Chapter 2, where we showed that maximum power is transferred for. $R_{\text{load}} = R_t$.

The second case of interest is a load that is constrained to be a pure resistance. In this case, it can be shown that the load resistance for maximum-power transfer is equal to the magnitude of the Thévenin impedance:

$$Z_{\text{load}} = R_{\text{load}} = |Z_t|$$

### Example 5.13    Maximum Power Transfer

Determine the maximum power that can be delivered to a load by the two-terminal circuit of Figure 5.34(a) if **a.** the load can have any complex value and **b.** the load must be a pure resistance.

**Solution**    In Example 5.12, we found that the circuit has the Thévenin equivalent shown in Figure 5.35(a). The Thévenin impedance is

$$Z_t = 50 - j50 \ \Omega$$

**a.** The complex load impedance that maximizes power transfer is

$$Z_{\text{load}} = Z_t^* = 50 + j50$$

The Thévenin equivalent with this load attached is shown in Figure 5.37(a). The current is

$$\mathbf{I}_a = \frac{\mathbf{V}_t}{Z_t + Z_{\text{load}}}$$

$$= \frac{100 \ /{-90°}}{50 - j50 + 50 + j50}$$

$$= 1 \ /{-90°} \ \text{A}$$

The rms load current is $I_{a\text{rms}} = 1/\sqrt{2}$. Finally, the power delivered to the load is

$$P = I_{a\text{rms}}^2 R_{\text{load}} = \left(\frac{1}{\sqrt{2}}\right)^2 (50) = 25 \ \text{W}$$

**b.** The purely resistive load for maximum power transfer is

$$R_{\text{load}} = |Z_t|$$
$$= |50 - j50|$$
$$= \sqrt{50^2 + (-50)^2}$$
$$= 70.71 \ \Omega$$

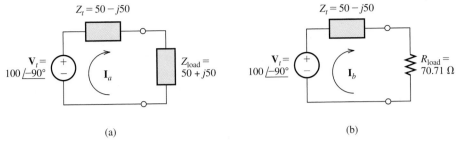

$Z_t = 50 - j50$

$V_t = 100\ \underline{/-90°}$

$Z_{\text{load}} = 50 + j50$

$I_a$

$Z_t = 50 - j50$

$V_t = 100\ \underline{/-90°}$

$R_{\text{load}} = 70.71\ \Omega$

$I_b$

(a)                                        (b)

Figure 5.37 Thévenin equivalent circuit and loads of Example 5.13.

The Thévenin equivalent with this load attached is shown in Figure 5.37(b). The current is

$$\mathbf{I}_b = \frac{\mathbf{V}_t}{Z_t + Z_{\text{load}}}$$

$$= \frac{100\ \underline{/-90°}}{50 - j50 + 70.71}$$

$$= \frac{100\ \underline{/-90°}}{130.66\ \underline{/-22.50°}}$$

$$= 0.7654\ \underline{/-67.50°}\ \text{A}$$

The power delivered to this load is

$$P = I_{b\text{rms}}^2 R_{\text{load}}$$

$$= \left(\frac{0.7653}{\sqrt{2}}\right)^2 70.71$$

$$= 20.71\ \text{W}$$

Notice that the power available to a purely resistive load is less than that for a complex load. ■

**Exercise 5.14**   Find the Thévenin impedance, the Thévenin voltage, and the Norton current for the circuit shown in Figure 5.38.
**Answer**   $Z_t = 100 + j25\ \Omega$, $\mathbf{V}_t = 70.71\ \underline{/-45°}$, $\mathbf{I}_n = 0.686\ \underline{/-59.0°}$.

**Exercise 5.15**   Determine the maximum power that can be delivered to a load by the two-terminal circuit of Figure 5.38 if **a.** the load can have any complex value and **b.** the load must be a pure resistance.
**Answer**   **a.** 6.25 W; **b.** 6.16 W.

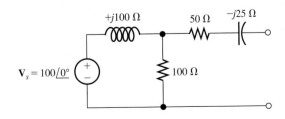

$+j100\ \Omega$          $50\ \Omega$      $-j25\ \Omega$

$V_s = 100\underline{/0°}$          $100\ \Omega$

Figure 5.38 Circuit of Exercises 5.14 and 5.15.

## 5.7  BALANCED THREE-PHASE CIRCUITS

Much of the power used by business and industry is supplied by three-phase distribution systems. Plant engineers need to be familiar with three-phase power.

We will see that there are important advantages in generating and distributing power with multiple ac voltages having different phases. We consider the most common case: three equal-amplitude ac voltages having phases that are 120° apart. This is known as a **balanced three-phase source**, an example of which is illustrated in Figure 5.39. [Recall that in double-subscript notation for voltages the first subscript is the positive reference. Thus, $v_{an}(t)$ is the voltage between nodes $a$ and $n$ with the positive reference at node $a$.] In Chapter 16, we will learn how three-phase voltages are generated.

The source shown in Figure 5.39(a) is said to be **wye connected (Y connected)**. Later in this chapter, we consider another configuration, known as the delta (Δ) connection.

The three voltages shown in Figure 5.39(b) are given by

$$v_{an}(t) = V_Y \cos(\omega t) \tag{5.82}$$

$$v_{bn}(t) = V_Y \cos(\omega t - 120°) \tag{5.83}$$

$$v_{cn}(t) = V_Y \cos(\omega t + 120°) \tag{5.84}$$

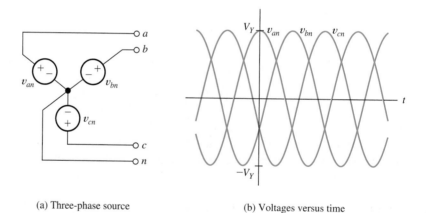

(a) Three-phase source                    (b) Voltages versus time

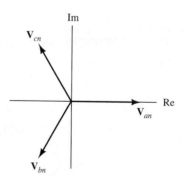

(c) Phasor diagram

**Figure 5.39** A balanced three-phase voltage source.

where $V_Y$ is the magnitude of each source in the wye-connected configuration. The corresponding phasors are

$$\mathbf{V}_{an} = V_Y \underline{/0°} \qquad (5.85)$$

$$\mathbf{V}_{bn} = V_Y \underline{/-120°} \qquad (5.86)$$

$$\mathbf{V}_{cn} = V_Y \underline{/120°} \qquad (5.87)$$

The phasor diagram is shown in Figure 5.39(c).

## Phase Sequence

This set of voltages is said to have a **positive phase sequence** because the voltages reach their peak values in the order *abc*. Refer to Figure 5.39(c) and notice that $v_{an}$ leads $v_{bn}$, which in turn leads $v_{cn}$. (Recall that we think of the phasors as rotating counterclockwise in determining phase relationships.) If we interchanged *b* and *c*, we would have a **negative phase sequence**, in which the order is *acb*.

Phase sequence can be important. For example, if we have a three-phase induction motor, the direction of rotation is opposite for the two phase sequences. To reverse the direction of rotation of such a motor, we would interchange the *b* and *c* connections. (You may find this piece of information useful if you ever work with three-phase motors, which are very common in industry.) Because circuit analysis is very similar for both phase sequences, we consider only the positive phase sequence in most of the discussion that follows.

*Three-phase sources can have either a positive or negative phase sequence.*

*We will see later in the book that the direction of rotation of certain three-phase motors can be reversed by changing the phase sequence.*

## Wye–Wye Connection

Consider the three-phase source connected to a balanced three-phase load shown in Figure 5.40. The wires $a-A$, $b-B$, and $c-C$ are called **lines**, and the wire $n-N$ is called the **neutral**. This configuration is called a wye–wye (Y–Y) connection with neutral. By the term *balanced load*, we mean that the three load impedances are equal. (In this book, we consider only balanced loads.)

*Three-phase sources and loads can be connected either in a wye configuration or in a delta configuration.*

Later, we will see that other configurations are useful. For example, the neutral wire $n-N$ can be omitted. Furthermore, the source and load can be connected in the form of a delta. We will see that currents, voltages, and power can be computed for these other configurations by finding an equivalent wye–wye circuit. Thus, the key to understanding three-phase circuits is a careful examination of the wye–wye circuit.

*The key to understanding the various three-phase configurations is a careful examination of the wye–wye circuit.*

Often, we use the term *phase* to refer to part of the source or the load. Thus, phase A of the source is $v_{an}(t)$, and phase A of the load is the impedance connected between A and N. We refer to $V_Y$ as the **phase voltage** or as the **line-to-neutral voltage** of the wye-connected source. (Power-systems engineers usually specify rms values rather than peak magnitudes. Unless stated otherwise, we use phasors having magnitudes equal to the peak values rather than the rms values.) Furthermore, $\mathbf{I}_{aA}$, $\mathbf{I}_{bB}$, and $\mathbf{I}_{cC}$ are called **line currents**. (Recall that in the double-subscript notation for currents, the reference direction is from the first subscript to the second. Thus, $\mathbf{I}_{aA}$ is the current referenced from node *a* to node *A*, as illustrated in Figure 5.38.)

The current in phase A of the load is given by

*In Chapters 5 and 6, we take the magnitude of a phasor to be the peak value. Power-systems engineers often use the rms value as the magnitude for phasors, which we do in Chapters 14 and 15. We will label rms phasors as rms.*

$$I_{aA} = \frac{\mathbf{V}_{an}}{Z \underline{/\theta}} = \frac{V_Y \underline{/0°}}{Z \underline{/\theta}} = I_L \underline{/-\theta}$$

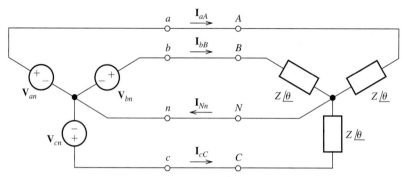

**Figure 5.40** A three-phase wye–wye connection with neutral.

where $I_L = V_Y/Z$ is the magnitude of the line current. Because the load impedances are equal, all of the line currents are the same, except for phase. Thus, the currents are given by

$$i_{aA}(t) = I_L \cos(\omega t - \theta) \tag{5.88}$$

$$i_{bB}(t) = I_L \cos(\omega t - 120° - \theta) \tag{5.89}$$

$$i_{cC}(t) = I_L \cos(\omega t + 120° - \theta) \tag{5.90}$$

The neutral current in Figure 5.40 is given by

$$i_{Nn}(t) = i_{aA}(t) + i_{bB}(t) + i_{cC}(t)$$

In terms of phasors, this is

$$\begin{aligned}
\mathbf{I}_{Nn} &= \mathbf{I}_{aA} + \mathbf{I}_{bB} + \mathbf{I}_{cC} \\
&= I_L\,\underline{/-\theta} + I_L\,\underline{/-120° - \theta} + I_L\,\underline{/120° - \theta} \\
&= I_L\,\underline{/-\theta} \times (1 + 1\,\underline{/-120°} + 1\,\underline{/120°}) \\
&= I_L\,\underline{/-\theta} \times (1 - 0.5 - j0.866 - 0.5 + j0.866) \\
&= 0
\end{aligned}$$

The sum of three equal magnitude phasors 120° apart in phase is zero.

Thus, the sum of three phasors with equal magnitudes and 120° (We make use of this fact again later in this section.)

We have shown that the neutral current is zero in a balanced three-phase system. Consequently, the neutral wire can be eliminated without changing any of the voltages or currents. Then, the three source voltages are delivered to the three load impedances with three wires.

The neutral current is zero in a balanced wye–wye system. Thus in theory, the neutral wire can be inserted or removed without affecting load currents or voltages. This is *not* true if the load is unbalanced, which is often the case in real power distribution systems.

An important advantage of three-phase systems compared with single phase is that the wiring for connecting the sources to the loads is less expensive. As shown in Figure 5.41, it would take six wires to connect three single-phase sources to three loads separately, whereas only three wires (four if the neutral wire is used) are needed for the three-phase connection to achieve the same power transfer.

## Power

Another advantage of balanced three-phase systems, compared with single-phase systems, is that the total power is constant (as a function of time) rather than pulsating.

(Refer to Figure 5.2 on page 220 to see that power pulsates in the single-phase case.) To show that the power is constant for the balanced wye–wye connection shown in Figure 5.40, we write an expression for the total power. The power delivered to phase A of the load is $v_{an}(t)i_{aA}(t)$. Similarly, the power for each of the other phases of the load is the product of the voltage and the current. Thus, the total power is

$$p(t) = v_{an}(t)i_{aA}(t) + v_{bn}(t)i_{bB}(t) + v_{cn}(t)i_{cC}(t) \tag{5.91}$$

Using Equations 5.82, 5.83, and 5.84 to substitute for the voltages and Equations 5.88, 5.89, and 5.90 to substitute for the currents, we obtain

$$\begin{aligned} p(t) = \ &V_Y \cos(\omega t)I_L \cos(\omega t - \theta) \\ &+ V_Y \cos(\omega t - 120°)I_L \cos(\omega t - \theta - 120°) \\ &+ V_Y \cos(\omega t + 120°)I_L \cos(\omega t - \theta + 120°) \end{aligned} \tag{5.92}$$

Using the trigonometric identity

$$\cos(x)\cos(y) = \frac{1}{2}\cos(x - y) + \frac{1}{2}\cos(x + y)$$

we find that Equation 5.92 can be written as

$$\begin{aligned} p(t) = \ &3\frac{V_Y I_L}{2}\cos(\theta) + \frac{V_Y I_L}{2}[\cos(2\omega t - \theta) \\ &+ \cos(2\omega t - \theta - 240°) + \cos(2\omega t - \theta + 480°)] \end{aligned} \tag{5.93}$$

However, the term in brackets is

$$\begin{aligned} &\cos(2\omega t - \theta) + \cos(2\omega t - \theta - 240°) + \cos(2\omega t - \theta + 480°) \\ &= \cos(2\omega t - \theta) + \cos(2\omega t - \theta + 120°) + \cos(2\omega t - \theta - 120°) \\ &= 0 \end{aligned}$$

(Here, we have used the fact, established earlier, that the sum is zero for three sine waves of equal amplitude and 120° apart in phase.) Thus, the expression for power becomes

$$p(t) = 3\frac{V_Y I_L}{2}\cos(\theta) \tag{5.94}$$

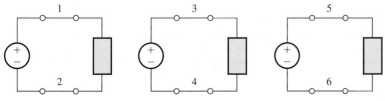

**Figure 5.41**  Six wires are needed to connect three single-phase sources to three loads. In a three-phase system, the same power transfer can be accomplished with three wires.

In balanced three-phase systems, total power flow is constant with respect to time.

Notice that the total power is constant with respect to time. A consequence of this fact is that the torque required to drive a three-phase generator connected to a balanced load is constant, and vibration is lessened. Similarly, the torque produced by a three-phase motor is constant rather than pulsating as it is for a single-phase motor.

The rms voltage from each line to neutral is

$$V_{Yrms} = \frac{V_Y}{\sqrt{2}} \qquad (5.95)$$

Similarly, the rms value of the line current is

$$I_{Lrms} = \frac{I_L}{\sqrt{2}} \qquad (5.96)$$

In Equations 5.97 and 5.98, $V_{Yrms}$ is the rms line-to-neutral voltage, $I_{Lrms}$ is the rms line current, and $\theta$ is the angle of the load impedances.

Using Equations 5.95 and 5.96 to substitute into Equation 5.94, we find that

$$P_{avg} = p(t) = 3V_{Yrms}I_{Lrms}\cos(\theta) \qquad (5.97)$$

### Reactive Power

As in single-phase circuits, power flows back and forth between the sources and energy-storage elements contained in a three-phase load. This power is called *reactive power*. The higher currents that result because of the presence of reactive power require wiring and other power-distribution components having higher ratings. The reactive power delivered to a balanced three-phase load is given by

$$Q = 3\frac{V_Y I_Y}{2}\sin(\theta) = 3V_{Y\,rms}I_{Lrms}\sin(\theta) \qquad (5.98)$$

### Line-to-Line Voltages

As we have mentioned earlier, the voltages between terminals $a$, $b$, or $c$ and the neutral point $n$ are called **line-to-neutral voltages**. On the other hand, voltages between $a$ and $b$, $b$ and $c$, or $a$ and $c$ are called **line-to-line voltages** or, more simply, **line voltages**. Thus $\mathbf{V}_{an}$, $\mathbf{V}_{bn}$, and $\mathbf{V}_{cn}$ are line-to-neutral voltages, whereas $\mathbf{V}_{ab}$, $\mathbf{V}_{bc}$, and $\mathbf{V}_{ca}$ are line-to-line voltages. (For consistency, we choose the subscripts cyclically in the order *abcabc*.) Let us consider the relationships between line-to-line voltages and line-to-neutral voltages.

We can obtain the following relationship by applying KVL to Figure 5.40:

$$\mathbf{V}_{ab} = \mathbf{V}_{an} - \mathbf{V}_{bn}$$

Using Equations 5.85 and 5.86 to substitute for $\mathbf{V}_{an}$ and $\mathbf{V}_{bn}$, we obtain

$$\mathbf{V}_{ab} = V_Y\underline{/0°} - V_Y\underline{/-120°} \qquad (5.99)$$

which is equivalent to

$$\mathbf{V}_{ab} = V_Y\underline{/0°} + V_Y\underline{/60°} \qquad (5.100)$$

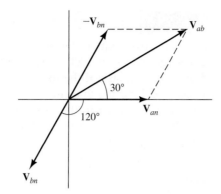

**Figure 5.42** Phasor diagram showing the relationship between the line-to-line voltage $\mathbf{V}_{ab}$ and the line-to-neutral voltages $\mathbf{V}_{an}$ and $\mathbf{V}_{bn}$.

This relationship is illustrated in Figure 5.42. It can be shown that Equation 5.100 reduces to

$$\mathbf{V}_{ab} = \sqrt{3}V_Y \underline{/30°} \tag{5.101}$$

We denote the magnitude of the line-to-line voltage as $V_L$. The magnitude of the line-to-line voltage is $\sqrt{3}$ times the magnitude of the line-to-neutral voltage:

$$V_L = \sqrt{3}V_Y \tag{5.102}$$

Thus, the relationship between the line-to-line voltage $\mathbf{V}_{ab}$ and the line-to-neutral voltage $\mathbf{V}_{an}$ is

$$\mathbf{V}_{ab} = \mathbf{V}_{an} \times \sqrt{3} \underline{/30°} \tag{5.103}$$

Similarly, it can be shown that

$$\mathbf{V}_{bc} = \mathbf{V}_{bn} \times \sqrt{3} \underline{/30°} \tag{5.104}$$

and

$$\mathbf{V}_{ca} = \mathbf{V}_{cn} \times \sqrt{3} \underline{/30°} \tag{5.105}$$

These voltages are shown in Figure 5.43.

Figure 5.43(b) provides a convenient way to remember the phase relationships between line-to-line and line-to-neutral voltages.

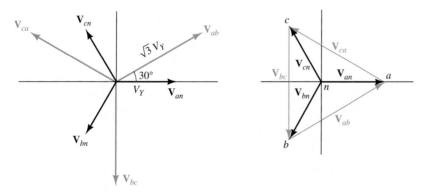

(a) All phasors starting from the origin

(b) A more intuitive way to draw the phasor diagram

**Figure 5.43** Phasor diagram showing line-to-line voltages and line-to-neutral voltages

### Example 5.14    Analysis of a Wye–Wye System

A balanced positive-sequence wye-connected 60-Hz three-phase source has line-to-neutral voltages of $V_Y = 1000$ V. This source is connected to a balanced wye-connected load. Each phase of the load consists of a 0.1-H inductance in series with a 50-$\Omega$ resistance. Find the line currents, the line-to-line voltages, the power, and the reactive power delivered to the load. Draw a phasor diagram showing the line-to-neutral voltages, the line-to-line voltages, and the line currents. Assume that the phase angle of $\mathbf{V}_{an}$ is zero.

**Solution**    First, by computing the complex impedance of each phase of the load, we find that

$$Z = R + j\omega L = 50 + j2\pi(60)(0.1) = 50 + j37.70$$

$$= 62.62 \underline{/37.02°}$$

Next, we draw the circuit as shown in Figure 5.44(a). In balanced wye–wye calculations, we can assume that $n$ and $N$ are connected. (The currents and voltages are the same whether or not the neutral connection actually exists.) Thus, $\mathbf{V}_{an}$ appears across phase $A$ of the load, and we can write

$$\mathbf{I}_{aA} = \frac{\mathbf{V}_{an}}{Z} = \frac{1000 \underline{/0°}}{62.62 \underline{/37.02°}} = 15.97 \underline{/-37.02°}$$

Similarly,

$$\mathbf{I}_{bB} = \frac{\mathbf{V}_{bn}}{Z} = \frac{1000 \underline{/-120°}}{62.62 \underline{/37.02°}} = 15.97 \underline{/-157.02°}$$

$$\mathbf{I}_{cC} = \frac{\mathbf{V}_{cn}}{Z} = \frac{1000 \underline{/120°}}{62.62 \underline{/37.02°}} = 15.97 \underline{/82.98°}$$

We use Equations 5.103, 5.104, and 5.105 to find the line-to-line phasors:

$$\mathbf{V}_{ab} = \mathbf{V}_{an} \times \sqrt{3} \underline{/30°} = 1732 \underline{/30°}$$

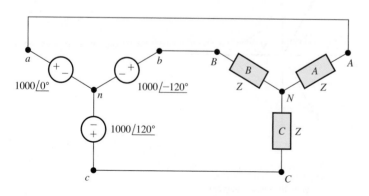

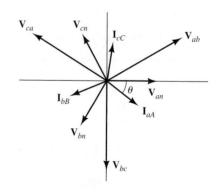

(a) Circuit diagram                              (b) Phasor diagram

**Figure 5.44**  Circuit and phasor diagram for Example 5.14.

$$\mathbf{V}_{bc} = \mathbf{V}_{bn} \times \sqrt{3} \,\underline{/30°} = 1732 \,\underline{/-90°}$$

$$\mathbf{V}_{ca} = \mathbf{V}_{cn} \times \sqrt{3} \,\underline{/30°} = 1732 \,\underline{/150°}$$

The power delivered to the load is given by Equation 5.94:

$$P = 3\frac{V_Y I_L}{2}\cos(\theta) = 3\left(\frac{1000 \times 15.97}{2}\right)\cos(37.02°) = 19.13 \text{ kW}$$

The reactive power is given by Equation 5.98:

$$Q = 3\frac{V_Y I_L}{2}\sin(\theta) = 3\left(\frac{1000 \times 15.97}{2}\right)\sin(37.02°) = 14.42 \text{ kVAR}$$

The phasor diagram is shown in Figure 5.44(b). As usual, we have chosen a different scale for the currents than for the voltages. ∎

**Exercise 5.16**　A balanced positive-sequence wye-connected 60-Hz three-phase source has line-to-line voltages of $V_L = 1000$ V. This source is connected to a balanced wye-connected load. Each phase of the load consists of a 0.2-H inductance in series with a 100-$\Omega$ resistance. Find the line-to-neutral voltages, the line currents, the power, and the reactive power delivered to the load. Assume that the phase of $\mathbf{V}_{an}$ is zero.

**Answer**　$\mathbf{V}_{an} = 577.4 \,\underline{/0°}$, $\mathbf{V}_{bn} = 577.4 \,\underline{/-120°}$, $\mathbf{V}_{cn} = 577.4 \,\underline{/120°}$;
$\mathbf{I}_{aA} = 4.61 \,\underline{/-37°}$, $\mathbf{I}_{bB} = 4.61 \,\underline{/-157°}$, $\mathbf{I}_{cC} = 4.61 \,\underline{/83°}$; $P = 3.19$ kW;
$Q = 2.40$ kVAR.　□

## Delta-Connected Sources

A set of balanced three-phase voltage sources can be connected in the form of a delta, as shown in Figure 5.45. Ordinarily, we avoid connecting voltage sources in closed loops. However, in this case, it turns out that the sum of the voltages is zero:

$$\mathbf{V}_{ab} + \mathbf{V}_{bc} + \mathbf{V}_{ca} = 0$$

Thus, the current circulating in the delta is zero. (Actually, this is a first approximation. There are many subtleties of power distribution systems that are beyond the scope of our discussion. For example, the voltages in actual power distribution systems are not exactly sinusoidal; instead, they are the sum of several harmonic components. The behavior of harmonic components is an important factor in making a choice between wye- and delta-connected sources or loads.)

For a given delta-connected source, we can find an equivalent wye-connected source (or vice versa) by using Equations 5.103 through 5.105. Clearly, a delta-connected source has no neutral point, so a four-wire connection is possible for only a wye-connected source.

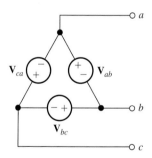

Figure 5.45　Delta-connected three-phase source.

## Wye- and Delta-Connected Loads

Load impedances can be either wye connected or delta connected, as shown in Figure 5.46. It can be shown that the two loads are equivalent if

$$Z_\Delta = 3Z_Y \tag{5.106}$$

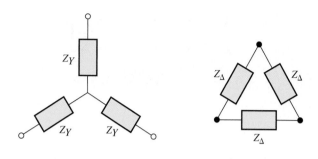

(a) Wye-connected load          (b) Delta-connected load

**Figure 5.46** Loads can be either wye connected or delta connected.

Thus, we can convert a delta-connected load to an equivalent wye-connected load, or vice versa.

### Delta–Delta Connection

Figure 5.47 shows a delta-connected source delivering power to a delta-connected load. We assume that the source voltages are given by

$$\mathbf{V}_{ab} = V_L \,\underline{/30°} \tag{5.107}$$

$$\mathbf{V}_{bc} = V_L \,\underline{/-90°} \tag{5.108}$$

$$\mathbf{V}_{ca} = V_L \,\underline{/150°} \tag{5.109}$$

These phasors are shown in Figure 5.43. (We have chosen the phase angles of the delta-connected source to be consistent with our earlier discussion.)

If the impedances of the connecting wires are zero, the line-to-line voltages at the load are equal to those at the source. Thus $\mathbf{V}_{AB} = \mathbf{V}_{ab}$, $\mathbf{V}_{BC} = \mathbf{V}_{bc}$, and $\mathbf{V}_{CA} = \mathbf{V}_{ca}$.

We assume that the impedance of each phase of the load is $Z_\Delta \,\underline{/\theta}$. Then, the load current for phase $AB$ is

$$\mathbf{I}_{AB} = \frac{\mathbf{V}_{AB}}{Z_\Delta \,\underline{/\theta}} = \frac{\mathbf{V}_{ab}}{Z_\Delta \,\underline{/\theta}} = \frac{V_L \,\underline{/30°}}{Z_\Delta \,\underline{/\theta}} = \frac{V_L}{Z_\Delta} \,\underline{/30° - \theta}$$

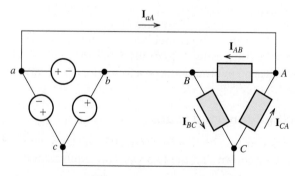

**Figure 5.47** A delta-connected source delivering power to a delta-connected load.

We define the magnitude of the current as

$$I_\Delta = \frac{V_L}{Z_\Delta} \tag{5.110}$$

Hence,

$$\mathbf{I}_{AB} = I_\Delta \, \underline{/30° - \theta} \tag{5.111}$$

Similarly,

$$\mathbf{I}_{BC} = I_\Delta \, \underline{/-90° - \theta} \tag{5.112}$$

$$\mathbf{I}_{CA} = I_\Delta \, \underline{/150° - \theta} \tag{5.113}$$

The current in line $a-A$ is

$$
\begin{aligned}
\mathbf{I}_{aA} &= \mathbf{I}_{AB} - \mathbf{I}_{CA} \\
&= I_\Delta \, \underline{/30° - \theta} - I_\Delta \, \underline{/150° - \theta} \\
&= (I_\Delta \, \underline{/30° - \theta}) \times (1 - 1 \, \underline{/120°}) \\
&= (I_\Delta \, \underline{/30° - \theta}) \times (1.5 - j0.8660) \\
&= (I_\Delta \, \underline{/30° - \theta}) \times (\sqrt{3} \, \underline{/-30°}) \\
&= I_{AB} \times \sqrt{3} \, \underline{/-30°}
\end{aligned}
$$

The magnitude of the line current is

$$I_L = \sqrt{3} I_\Delta \tag{5.114}$$

*For a balanced delta-connected load, the line-current magnitude is equal to the square root of three times the current magnitude in any arm of the delta.*

## Example 5.15   Analysis of a Balanced Delta–Delta System

Consider the circuit shown in Figure 5.48(a). A delta-connected source supplies power to a delta-connected load through wires having impedances of $Z_{\text{line}} = 0.3 + j0.4 \, \Omega$. The load impedances are $Z_\Delta = 30 + j6$. The source voltages are

$$\mathbf{V}_{ab} = 1000 \, \underline{/30°}$$

$$\mathbf{V}_{bc} = 1000 \, \underline{/-90°}$$

$$\mathbf{V}_{ca} = 1000 \, \underline{/150°}$$

Find the line current, the line-to-line voltage at the load, the current in each phase of the load, the power delivered to the load, and the power dissipated in the line.

**Solution**   First, we find the wye-connected equivalents for the source and the load. (Actually, we only need to work with one third of the circuit because the other two thirds are the same except for phase angles.) We choose to work with the $A$ phase of the wye-equivalent circuit. Solving Equation 5.103 for $\mathbf{V}_{an}$, we find that

*Often, it is convenient to start an analysis by finding the wye–wye equivalent of a system.*

$$\mathbf{V}_{an} = \frac{\mathbf{V}_{ab}}{\sqrt{3} \, \underline{/30°}} = \frac{1000 \, \underline{/30°}}{\sqrt{3} \, \underline{/30°}} = 577.4 \, \underline{/0°}$$

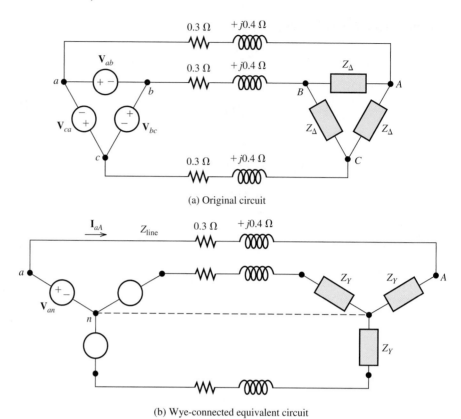

(a) Original circuit

(b) Wye-connected equivalent circuit

**Figure 5.48**  Circuit of Example 5.15.

Using Equation 5.106, we have

$$Z_Y = \frac{Z_\Delta}{3} = \frac{30 + j6}{3} = 10 + j2$$

Now, we can draw the wye-equivalent circuit, which is shown in Figure 5.48(b).

In a balanced wye–wye system, we can consider the neutral points to be connected together as shown by the dashed line in Figure 5.48(b). This reduces the three-phase circuit to three single-phase circuits. For phase $A$ of Figure 5.48(b), we can write

$$\mathbf{V}_{an} = (Z_{\text{line}} + Z_Y)\mathbf{I}_{aA}$$

Therefore,

$$\mathbf{I}_{aA} = \frac{\mathbf{V}_{an}}{Z_{\text{line}} + Z_Y} = \frac{577.4 \underline{/0°}}{0.3 + j0.4 + 10 + j2}$$

$$= \frac{577.4 \underline{/0°}}{10.3 + j2.4} = \frac{577.4 \underline{/0°}}{10.58 \underline{/13.12°}}$$

$$= 54.60 \underline{/-13.12°}$$

To find the line-to-neutral voltage at the load, we write

$$\mathbf{V}_{An} = \mathbf{I}_{Aa}Z_Y = 54.60\,\underline{/-13.12°} \times (10 + j2)$$
$$= 54.60\,\underline{/-13.12°} \times 10.20\,\underline{/11.31°}$$
$$= 556.9\,\underline{/-1.81°}$$

Now, we compute the line-to-line voltage at the load:

$$\mathbf{V}_{AB} = \mathbf{V}_{An} \times \sqrt{3}\,\underline{/30°} = 556.9\,\underline{/-1.81°} \times \sqrt{3}\,\underline{/30°}$$
$$= 964.6\,\underline{/28.19°}$$

The current through phase $AB$ of the load is

$$\mathbf{I}_{AB} = \frac{\mathbf{V}_{AB}}{Z_\Delta} = \frac{964.6\,\underline{/28.19°}}{30 + j6} = \frac{964.6\,\underline{/28.19°}}{30.59\,\underline{/11.31°}}$$
$$= 31.53\,\underline{/16.88°}$$

The power delivered to phase $AB$ of the load is the rms current squared times the resistance:

$$P_{AB} = I_{ABrms}^2 R = \left(\frac{31.53}{\sqrt{2}}\right)^2 (30) = 14.91\text{ kW}$$

The powers delivered to the other two phases of the load are the same, so the total power is

$$P = 3P_{AB} = 44.73\text{ kW}$$

The power lost in line $A$ is

$$P_{lineA} = I_{aArms}^2 R_{line} = \left(\frac{54.60}{\sqrt{2}}\right)^2 (0.3) = 0.447\text{ kW}$$

The power lost in the other two lines is the same, so the total line loss is

$$P_{line} = 3 \times P_{lineA} = 1.341\text{ kW}$$ ∎

**Exercise 5.17**   A delta-connected source has voltages given by

$$\mathbf{V}_{ab} = 1000\,\underline{/30°}$$
$$\mathbf{V}_{bc} = 1000\,\underline{/-90°}$$
$$\mathbf{V}_{ca} = 1000\,\underline{/150°}$$

This source is connected to a delta-connected load consisting of 50-$\Omega$ resistances. Find the line currents and the power delivered to the load.

**Answer**   $\mathbf{I}_{aA} = 34.6\,\underline{/0°}$, $\mathbf{I}_{bB} = 34.6\,\underline{/-120°}$, $\mathbf{I}_{cC} = 34.6\,\underline{/120°}$; $P = 30\text{ kW}$.   □

## 5.8  AC ANALYSIS USING MATLAB

In this section, we will illustrate how MATLAB can greatly facilitate the analysis of complicated ac circuits. In fact, a practicing engineer working at a computer might have little use for a calculator, as it is easy to keep a MATLAB window open for all sorts of engineering calculations. Of course, you will probably need to use calculators for course exams and when you take the Professional Engineer (PE) exams. The PE exams allow only fairly simple scientific calculators, and you should practice with one of those allowed before attempting the exams.

### Complex Data in MATLAB

By default, MATLAB assumes that $i = j = \sqrt{-1}$. However, I have encountered at least one bug in the software attributable to using $j$ instead of $i$, and therefore I recommend using $i$ in MATLAB and the Symbolic Toolbox. We need to be careful to avoid using $i$ for other purposes when using MATLAB to analyze ac circuits. For example, if we were to use $i$ as the name of a current or other variable, we would later experience errors if we also used $i$ for the imaginary unit without reassigning its value.

Complex numbers are represented in rectangular form (such as 3 + 4i or alternatively 3 + i*4) in MATLAB.

We can use the fact that $M \underline{/\theta} = M \exp(j\theta)$ to enter polar data. In MATLAB, angles are assumed to be in radians, so we need to multiply angles that are expressed in degrees by $\pi/180$ to convert to radians before entering them. For example, we use the following command to enter the voltage $V_s = 5\sqrt{2} \underline{/45°}$:

```
>> Vs = 5*sqrt(2)*exp(i*45*pi/180)
Vs =
    5.0000 + 5.0000i
```

We can readily verify that MATLAB has correctly computed the rectangular form of $5\sqrt{2} \underline{/45°}$.

Alternatively, we could use Euler's formula

$$M \underline{/\theta} = M \exp(j\theta) = M \cos(\theta) + jM \sin(\theta)$$

to enter polar data, again with angles in radians. For example, $V_s = 5\sqrt{2} \underline{/45°}$ can be entered as:

```
>> Vs = 5*sqrt(2)*cos(45*pi/180) + i*5*sqrt(2)*sin(45*pi/180)
Vs =
    5.0000 + 5.0000i
```

Values that are already in rectangular form can be entered directly. For example, to enter $Z = 3 + j4$, we use the command:

```
>> Z = 3 + i*4
Z =
    3.0000 + 4.0000i
```

Then, if we enter

```
>> Ix = Vs/Z
Ix =
    1.4000 - 0.2000i
```

MATLAB performs the complex arithmetic and gives the answer in rectangular form.

## Finding the Polar Form of MATLAB Results

Frequently, we need the polar form of a complex value calculated by MATLAB. We can find the magnitude using the abs command and the angle in radians using the angle command. To obtain the angle in degrees, we must convert the angle from radians by multiplying by $180/\pi$. Thus, to obtain the magnitude and angle in degrees for Vs, we would enter the following commands:

```
>> abs(Vs) % Find the magnitude of Vs.
ans =
    7.0711
>> (180/pi)*angle(Vs) % Find the angle of Vs in degrees.
ans =
    45.0000
```

## Adding New Functions to MATLAB

Because we often want to enter values or see results in polar form with the angles in degrees, it is convenient to add two new functions to MATLAB. Thus, we write an m-file, named pin.m, containing the commands to convert from polar to rectangular form, and store it in our working MATLAB folder. The commands in the m-file are:

```
function z = pin(magnitude, angleindegrees)
z = magnitude*exp(i*angleindegrees*pi/180)
```

Then, we can enter $\text{Vs} = 5\sqrt{2} \underline{/45°}$ simply by typing the command:

```
>> Vs = pin(5*sqrt(2),45)
Vs =
    5.0000 + 5.0000i
```

We have chosen pin as the name of this new function to suggest "polar input." This file is included in the MATLAB folder. (See Appendix E for information about accessing this folder.)

Similarly, to obtain the polar form of an answer, we create a new function, named pout (to suggest "polar out"), with the commands:

```
function [y] = pout(x);
magnitude = abs(x);
angleindegrees = (180/pi)*angle(x);
y = [magnitude angleindegrees];
```

which are stored in the m-file named pout.m. Then, to find the polar form of a result, we can use the new function. For example,

```
>> pout(Vs)
ans =
    7.0711    45.0000
```

Here is another simple example:

```
>> pout(i*200)
ans =
    200    90
```

### Solving Network Equations with MATLAB

We can readily solve node voltage or mesh equations and perform other calculations for ac circuits in MATLAB. The steps are:

1. Write the mesh current or node voltage equations.

2. Put the equations into matrix form, which is $\mathbf{ZI} = \mathbf{V}$ for mesh currents, in which $\mathbf{Z}$ is the coefficient matrix, $\mathbf{I}$ is the column vector of mesh current variables to be found, and $\mathbf{V}$ is the column vector of constant terms. For node voltages, the matrix equations take the form $\mathbf{YV} = \mathbf{I}$ in which $\mathbf{Y}$ is the coefficient matrix, $\mathbf{V}$ is the column vector of node voltage variables to be determined, and $\mathbf{I}$ is the column vector of constants.

3. Enter the matrices into MATLAB and compute the mesh currents or node voltages using the inverse matrix approach. $\mathbf{I} = \text{inv}(\mathbf{Z}) \times \mathbf{V}$ for mesh currents or $\mathbf{V} = \text{inv}(\mathbf{Y}) \times \mathbf{I}$ for node voltages, where inv denotes the matrix inverse.

4. Use the results to compute any other quantities of interest.

---

### Example 5.16    Phasor Mesh-Current Analysis with MATLAB

Determine the values for the mesh currents, the real power supplied by $\mathbf{V}_1$, and the reactive power supplied by $\mathbf{V}_1$ in the circuit of Figure 5.49.

**Solution**    First, we apply KVL to each loop obtaining the mesh-current equations:

$$(5 + j3)\mathbf{I}_1 + (50 \underline{/-10°})(\mathbf{I}_1 - \mathbf{I}_2) = 2200\sqrt{2}$$

$$(50 \underline{/-10°})(\mathbf{I}_2 - \mathbf{I}_1) + (4 + j)\mathbf{I}_2 + 2000\sqrt{2} \underline{/30} = 0$$

In matrix form, these equations become

$$\begin{bmatrix} (5 + j3 + 50 \underline{/-10°}) & -50 \underline{/-10°} \\ -50 \underline{/-10°} & (4 + j + 50 \underline{/-10°}) \end{bmatrix} \begin{bmatrix} \mathbf{I}_1 \\ \mathbf{I}_2 \end{bmatrix} = \begin{bmatrix} 2200\sqrt{2} \\ -2000\sqrt{2} \underline{/-10°} \end{bmatrix}$$

We will solve these equations for $\mathbf{I}_1$ and $\mathbf{I}_2$. Then, we will compute the complex power delivered by $\mathbf{V}_1$

$$\mathbf{S}_1 = \frac{1}{2}\mathbf{V}_1\mathbf{I}_1^*$$

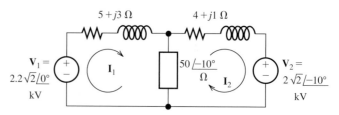

Figure 5.49  Circuit for Example 5.16.

Finally, the power is the real part of $\mathbf{S}_1$ and the reactive power is the imaginary part.

We enter the coefficient matrix $\mathbf{Z}$ and the voltage matrix $\mathbf{V}$ into MATLAB, making use of our new pin function to enter polar values. Then, we calculate the current matrix.

```
>> Z = [(5 + i*3 + pin(50,-10)) (-pin(50,-10));...
(-pin(50,-10)) (4 + i + pin(50,-10))];
>> V = [2200*sqrt(2); -pin(2000*sqrt(2),-10)];
>> I = inv(Z)*V
I =
   74.1634 + 29.0852i
   17.1906 + 26.5112i
```

This has given us the values of the mesh currents in rectangular form. Next, we obtain the polar form for the mesh currents, making use of our new pout function:

```
>> pout(I(1))
ans =
      79.6628    21.4140
>> pout(I(2))
ans =
      31.5968    57.0394
```

Thus, the currents are $\mathbf{I}_1 = 79.66 \,\underline{/21.41°}$ A and $\mathbf{I}_2 = 31.60 \,\underline{/57.04°}$ A, rounded to two decimal places. Next, we compute the complex power, real power, and reactive power for the first source.

$$\mathbf{S}_1 = \frac{1}{2}\mathbf{V}_1\mathbf{I}_1^*$$

```
>> S1 = (1/2)*(2200*sqrt(2))*conj(I(1));
>> P1 = real(S1)
P1 =
    1.1537e + 005
>> Q1 = imag(S1)
Q1 =
   -4.5246e + 004
```

Thus, the power supplied by $\mathbf{V}_1$ is 115.37 kW and the reactive power is $-45.25$ kVAR. The commands for this example appear in the m-file named Example_5_16.    ∎

**Exercise 5.18**   Use MATLAB to solve for the phasor node voltages in polar form for the circuit of Figure 5.50.

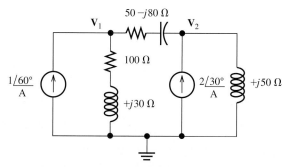

Figure 5.50  Circuit for Exercise 5.18.

**Answer**    The MATLAB commands are:

```
clear all
Y = [(1/(100+i*30)+1/(50-i*80))  (-1/(50-i*80));...
(-1/(50-i*80))  (1/(i*50)+1/(50-i*80))];
I = [pin(1,60); pin(2,30)];
V = inv(Y)*I;
pout(V(1))
pout(V(2))
```

and the results are $\mathbf{V}_1 = 79.98 \,\underline{/106.21°}$ and $\mathbf{V}_2 = 124.13 \,\underline{/116.30°}$.    □

## Summary

1. A sinusoidal voltage is given by $v(t) = V_m \cos(\omega t + \theta)$, where $V_m$ is the peak value of the voltage, $\omega$ is the angular frequency in radians per second, and $\theta$ is the phase angle. The frequency in hertz is $f = 1/T$, where $T$ is the period. Furthermore, $\omega = 2\pi f$.

2. For uniformity, we express sinusoidal voltages in terms of the cosine function. A sine function can be converted to a cosine function by use of the identity $\sin(z) = \cos(z - 90°)$.

3. The root-mean-square (rms) value (or effective value) of a periodic voltage $v(t)$ is

$$V_{rms} = \sqrt{\frac{1}{T} \int_0^T v^2(t)\, dt}$$

The average power delivered to a resistance by $v(t)$ is

$$P_{avg} = \frac{V_{rms}^2}{R}$$

Similarly, for a current $i(t)$, we have

$$I_{rms} = \sqrt{\frac{1}{T} \int_0^T i^2(t)\, dt}$$

and the average power delivered if $i(t)$ flows through a resistance is

$$P_{avg} = I_{rms}^2 R$$

For a sinusoid, the rms value is the peak value divided by $\sqrt{2}$.

4. We can represent sinusoids with phasors. The magnitude of the phasor is the peak value of the sinusoid. The phase angle of the phasor is the phase angle of the sinusoid (assuming that we have written the sinusoid in terms of a cosine function).

5. We can add (or subtract) sinusoids by adding (or subtracting) their phasors.

6. The phasor voltage for a passive circuit is the phasor current times the complex impedance of the circuit. For a resistance, $\mathbf{V}_R = R\mathbf{I}_R$, and the voltage is in phase with the current. For an inductance, $\mathbf{V}_L = j\omega L \mathbf{I}_L$, and the voltage leads the current by 90°. For a capacitance, $\mathbf{V}_C = -j(1/\omega C)\mathbf{I}_C$, and the voltage lags the current by 90°.

7. Many techniques learned in Chapter 2 for resistive circuits can be applied directly to sinusoidal circuits if the currents and voltages are replaced by phasors and the passive circuit elements are replaced by their complex impedances. For example, complex impedances can be combined in series or parallel in the same way as resistances (except that complex arithmetic must be used). Node voltages, the current-division principle, and the voltage-division principle also apply to ac circuits.

8. When a sinusoidal current flows through a sinusoidal voltage, the average power delivered is $P = V_{rms}I_{rms} \cos(\theta)$, where $\theta$ is the power angle, which is found by subtracting the phase angle of the current from the phase angle of the voltage (i.e., $\theta = \theta_v - \theta_i$). The power factor is $\cos(\theta)$.

9. Reactive power is the flow of energy back and forth between the source and energy-storage elements ($L$ and $C$). We define reactive power to be positive for an inductance and negative for a capacitance. The net energy transferred per cycle

by reactive power flow is zero. Reactive power is important because a power distribution system must have higher current ratings if reactive power flows than would be required for zero reactive power.

10. Apparent power is the product of rms voltage and rms current. Many useful relationships between power, reactive power, apparent power, and the power angle can be obtained from the power triangles shown in Figure 5.25 on page 244.

11. In steady state, a network composed of resistances, inductances, capacitances, and sinusoidal sources (all of the same frequency) has a Thévenin equivalent consisting of a phasor voltage source in series with a complex impedance. The Norton equivalent consists of a phasor current source in parallel with the Thévenin impedance.

12. For maximum-power transfer from a two-terminal ac circuit to a load, the load impedance is selected to be the complex conjugate of the Thévenin impedance. If the load is constrained to be a pure resistance, the value for maximum power transfer is equal to the magnitude of the Thévenin impedance.

13. Because of savings in wiring, three-phase power distribution is more economical than single phase. The power flow in balanced three-phase systems is smooth, whereas power pulsates in single-phase systems. Thus, three-phase motors generally have the advantage of producing less vibration than single-phase motors.

## Problems

### Section 5.1: Sinusoidal Currents and Voltages

**P5.1.** Consider the plot of the sinusoidal voltage $v(t) = V_m \cos(\omega t + \theta)$ shown in Figure 5.1 on page 216 and the following statements:

1. Stretches the sinusoidal curve vertically.
2. Compresses the sinusoidal curve vertically.
3. Stretches the sinusoidal curve horizontally.
4. Compresses the sinusoidal curve horizontally.
5. Translates the sinusoidal curve to the right.
6. Translates the sinusoidal curve to the left.

Which statement best describes

   a. Increasing the peak amplitude $V_m$?
   b. Increasing the frequency $f$?
   c. Decreasing $\theta$?
   d. Decreasing the angular frequency $\omega$?
   e. Increasing the period?

**P5.2.** What are the units for angular frequency $\omega$? For frequency $f$? What is the relationship between them?

**\*P5.3.** A voltage is given by $v(t) = 10 \sin(1000\pi t + 30°)$. First, use a cosine function to express $v(t)$. Then, find the angular frequency, the frequency in hertz, the phase angle, the period, and the rms value. Find the power that this voltage delivers to a 50-$\Omega$ resistance. Find the first value of time after $t = 0$ that $v(t)$ reaches its peak value. Sketch $v(t)$ to scale versus time.

**P5.4.** Repeat Problem P5.3 for $v(t) = 50 \sin(500\pi t + 120°)$.

**\*P5.5.** A sinusoidal voltage $v(t)$ has an rms value of 20 V, has a period of 100 $\mu$s, and reaches a positive peak at $t = 20$ $\mu$s. Write an expression for $v(t)$.

**P5.6.** A sinusoidal voltage has a peak value of 15 V, has a frequency of 125 Hz, and crosses zero with positive slope at $t = 1$ ms. Write an expression for the voltage.

**P5.7.** A current $i(t) = 10 \cos(2000\pi t)$ flows through a 100-$\Omega$ resistance. Sketch $i(t)$ and $p(t)$ to scale versus time. Find the average power delivered to the resistance.

**P5.8.** We have a voltage $v(t) = 1000 \sin(500\pi t)$ across a 500-$\Omega$ resistance. Sketch $v(t)$ and $p(t)$

---

\* Denotes that answers are contained in the Student Solutions files. See See Appendix E for more information about accessing the Student Solutions.

to scale versus time. Find the average power delivered to the resistance.

**P5.9.** Suppose we have a sinusoidal current $i(t)$ that has an rms value of 5 A, has a period of 10 ms, and reaches a positive peak at $t = 3$ ms. Write an expression for $i(t)$.

**P5.10.** A **Lissajous figure** results if one sinusoid is plotted versus another. Consider $x(t) = \cos(\omega_x t)$ and $y(t) = \cos(\omega_y t + \theta)$. Use a computer program of your choice to generate values of $x$ and $y$ for 20 seconds at 100 points per second and obtain a plot of $y$ versus $x$ for **a.** $\omega_x = \omega_y = 2\pi$ and $\theta = 90°$; **b.** $\omega_x = \omega_y = 2\pi$ and $\theta = 45°$; **c.** $\omega_x = \omega_y = 2\pi$ and $\theta = 0°$; **d.** $\omega_x = 2\pi$, $\omega_y = 4\pi$, and $\theta = 0°$.

**\*P5.11.** Find the rms value of the voltage waveform shown in Figure P5.11.

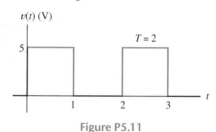

Figure P5.11

**\*P5.12.** Calculate the rms value of the half-wave rectified sine wave shown in Figure P5.12.

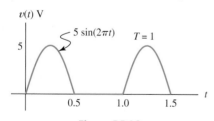

Figure P5.12

**\*P5.13.** Find the rms value of the current waveform shown in Figure P5.13.

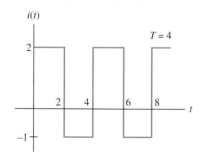

Figure P5.13

**P5.14.** Determine the rms value of $v(t) = A\cos(2\pi t) + B\sin(2\pi t)$.

**P5.15.** Determine the rms value of $v(t) = 5 + 10\cos(20\pi t)$.

**P5.16.** Compute the rms value of the periodic waveform shown in Figure P5.16.

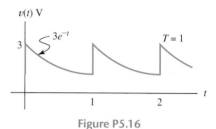

Figure P5.16

**P5.17.** Find the rms value of the voltage waveform shown in Figure P5.17.

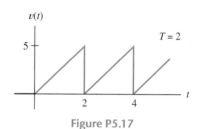

Figure P5.17

**P5.18.** Is the rms value of a periodic waveform always equal to the peak value divided by the square root of two? When is it?

## Section 5.2: Phasors

**P5.19.** What steps do we follow in adding sinusoidal currents or voltages? What must be true of the sinusoids?

**P5.20.** Describe two methods to determine the phase relationship between two sinusoids of the same frequency.

**\*P5.21.** Suppose that $v_1(t) = 100\cos(\omega t)$ and $v_2(t) = 100\sin(\omega t)$. Use phasors to reduce the sum $v_s(t) = v_1(t) + v_2(t)$ to a single term of the form $V_m\cos(\omega t + \theta)$. Draw a phasor diagram, showing $\mathbf{V}_1$, $\mathbf{V}_2$, and $\mathbf{V}_s$. State the phase relationships between each pair of these phasors.

**\*P5.22.** Consider the phasors shown in Figure P5.22. The frequency of each signal is $f = 200$ Hz. Write a time-domain expression for each

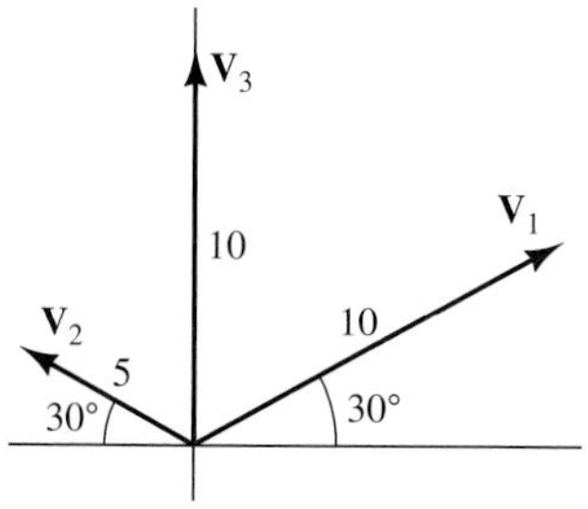

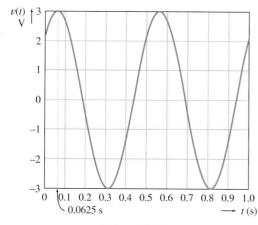

Figure P5.22

voltage in the form $V_m \cos(\omega t + \theta)$. State the phase relationships between pairs of these phasors.

**\*P5.23.** Reduce $5\cos(\omega t + 75°) - 3\cos(\omega t - 75°) + 4\sin(\omega t)$ to the form $V_m \cos(\omega t + \theta)$.

**P5.24.** Two sinusoidal voltages of the same frequency have rms values of 8 V and 3 V. What is the smallest rms value that the sum of these voltages could have? The largest? Justify your answers.

**P5.25.** Suppose that $v_1(t) = 100 \cos(\omega t + 45°)$ and $v_2(t) = 150 \sin(\omega t + 60°)$. Use phasors to reduce the sum $v_s(t) = v_1(t) + v_2(t)$ to a single term of the form $V_m \cos(\omega t + \theta)$. Draw a phasor diagram showing $\mathbf{V}_1$, $\mathbf{V}_2$, and $\mathbf{V}_s$. State the phase relationships between each pair of these phasors.

**P5.26.** Write an expression for the sinusoid shown in Figure P5.26 of the form $v(t) = V_m \cos(\omega t + \theta)$, giving the numerical values of $V_m$, $\omega$, and $\theta$. Also, determine the phasor and the rms value of $v(t)$.

Figure P5.26

**P5.27.** We have $v_1(t) = 10 \cos(\omega t + 30°)$. The current $i_1(t)$ has an rms value of 5 A and leads $v_1(t)$ by 20°. (The current and the voltage have the same frequency.) Draw a phasor diagram showing the phasors. Write an expression for $i_1(t)$ of the form $I_m \cos(\omega t + \theta)$.

**P5.28.** Reduce $5 \sin(\omega t) + 5 \cos(\omega t + 30°) + 5 \cos(\omega t + 150°)$ to the form $V_m \cos(\omega t + \theta)$.

**P5.29.** Using a computer program of your choice, obtain a plot of $v(t) = \cos(19\pi t) + \cos(21\pi t)$ for $t$ ranging from 0 to 2 s in 0.01-s increments. (Notice that because the terms have different frequencies, they cannot be combined by using phasors.) Then, considering that the two terms can be represented as the real projection of the sum of two vectors rotating (at different speeds) in the complex plane, comment on the plot.

**P5.30.** A sinusoidal current $i_1(t)$ has a phase angle of 30°. Furthermore, $i_1(t)$ attains its positive peak 2 ms earlier than current $i_2(t)$ does. Both currents have a frequency of 200 Hz. Determine the phase angle of $i_2(t)$.

### Section 5.3: Complex Impedances

**P5.31.** Write the relationship between the phasor voltage and phasor current for an inductance. Repeat for capacitance.

**P5.32.** State the phase relationship between current and voltage for a resistance, for an inductance, and for a capacitance.

**\*P5.33.** A voltage $v_L(t) = 10 \cos(2000\pi t)$ is applied to a 100-mH inductance. Find the complex impedance of the inductance. Find the phasor voltage and current, and construct a phasor diagram. Write the current as a function of time. Sketch the voltage and current to scale versus time. State the phase relationship between the current and voltage.

**\*P5.34.** A voltage $v_C(t) = 10 \cos(2000\pi t)$ is applied to a 10-$\mu$F capacitance. Find the complex impedance of the capacitance. Find the phasor voltage and current, and construct a phasor diagram. Write the current as a function of time. Sketch the voltage and current to scale versus time. State the phase relationship between the current and voltage.

**P5.35.** A certain circuit element is known to be a resistance, an inductance, or a capacitance. Determine the type and value (in ohms, henrys, or farads) of the element if the voltage and current for the element are given by **a.** $v(t) = 100 \sin(200t + 30°)$ V, $i(t) = \cos(200t + 30°)$ A; **b.** $v(t) = 500 \cos(100t + 50°)$ V, $i(t) = 2 \cos(100t + 50°)$ A; **c.** $v(t) = 100 \cos(400t + 30°)$ V, $i(t) = \sin(400t + 30°)$ A.

**P5.36.** Sketch plots of the magnitudes of the impedances of a 10-mH inductance, a 10-$\mu$F capacitance, and a 50-$\Omega$ resistance to scale versus frequency for the range from zero to 1000 Hz.

**P5.37. a.** A certain element has a phasor voltage of $\mathbf{V} = 100\angle 30°$ V and current of $\mathbf{I} = 5\angle 120°$ A. The angular frequency is 500 rad/s. Determine the nature and value of the element. **b.** Repeat for $\mathbf{V} = 20\angle -45°$ V and current of $\mathbf{I} = 5\angle -135°$ A. **c.** Repeat for $\mathbf{V} = 5\angle 45°$ V and current of $\mathbf{I} = 5\angle 45°$ A.

**P5.38. a.** The current and voltage for a certain circuit element are shown in Figure P5.38(a). Determine the nature and value of the element. **b.** Repeat for Figure P5.38(b).

**Section 5.4: Circuit Analysis with Phasors and Complex Impedances**

**P5.39.** Give a step-by-step procedure for steady-state analysis of circuits with sinusoidal sources. What condition must be true of the sources?

**\*P5.40.** Find the complex impedance in polar form of the network shown in Figure P5.40 for $\omega = 500$. Repeat for $\omega = 1000$ and $\omega = 2000$.

**\*P5.41.** Find the phasors for the current and for the voltages of the circuit shown in Figure P5.41. Construct a phasor diagram showing $\mathbf{V}_s$, $\mathbf{I}$, $\mathbf{V}_R$, and $\mathbf{V}_L$. What is the phase relationship between $\mathbf{V}_s$ and $\mathbf{I}$?

**P5.42.** Change the inductance to 0.1 H, and repeat Problem P5.41.

**P5.43.** Find the complex impedance of the network shown in Figure P5.43 for $\omega = 500$. Repeat for $\omega = 1000$ and $\omega = 2000$.

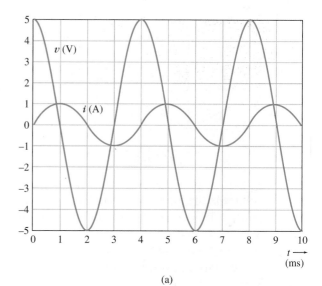

(a)

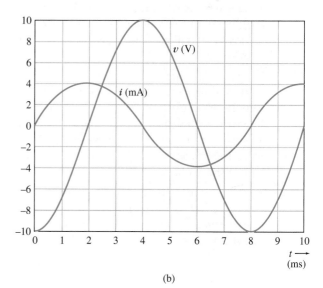

(b)

**Figure P5.38**

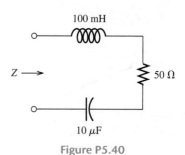

**Figure P5.40**

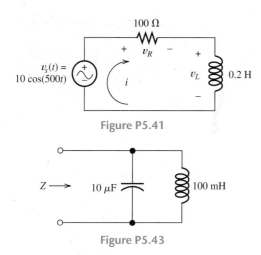

Figure P5.41

Figure P5.43

**P5.44.** A 10-mH inductance, a 100-$\Omega$ resistance, and a 100-$\mu$F capacitance are connected in parallel. Calculate the impedance of the combination for angular frequencies of 500, 1000, and 2000 radians per second. For each frequency, state whether the impedance is inductive, purely resistive, or capacitive.

**P5.45.** Find the phasors for the current and the voltages for the circuit shown in Figure P5.45. Construct a phasor diagram showing $\mathbf{V}_s$, $\mathbf{I}$, $\mathbf{V}_R$, and $\mathbf{V}_C$. What is the phase relationship between $\mathbf{V}_s$ and $\mathbf{I}$?

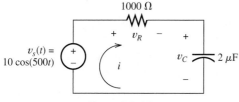

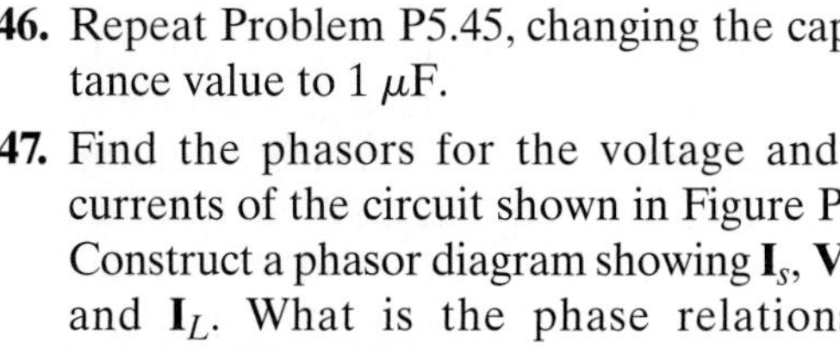

Figure P5.45

**\*P5.46.** Repeat Problem P5.45, changing the capacitance value to 1 $\mu$F.

**P5.47.** Find the phasors for the voltage and the currents of the circuit shown in Figure P5.47. Construct a phasor diagram showing $\mathbf{I}_s$, $\mathbf{V}$, $\mathbf{I}_R$, and $\mathbf{I}_L$. What is the phase relationship between $\mathbf{V}$ and $\mathbf{I}_s$?

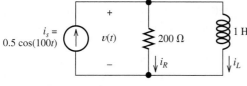

Figure P5.47

**\*P5.48.** Find the phasors for the voltage and the currents for the circuit shown in Figure P5.48. Construct a phasor diagram showing $\mathbf{I}_s$, $\mathbf{V}$, $\mathbf{I}_R$, and $\mathbf{I}_C$. What is the phase relationship between $\mathbf{V}$ and $\mathbf{I}_s$?

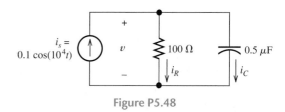

Figure P5.48

**\*P5.49.** Consider the circuit shown in Figure P5.49. Find the phasors $\mathbf{I}_s$, $\mathbf{V}$, $\mathbf{I}_R$, $\mathbf{I}_L$, and $\mathbf{I}_C$. Compare the peak value of $i_L(t)$ with the peak value of $i_s(t)$. Do you find the answer surprising? Explain.

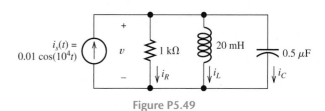

Figure P5.49

**P5.50.** Consider the circuit shown in Figure P5.50. Find the phasors $\mathbf{V}_s$, $\mathbf{I}$, $\mathbf{V}_L$, $\mathbf{V}_R$, and $\mathbf{V}_C$. Compare the peak value of $v_L(t)$ with the peak value of $v_s(t)$. Do you find the answer surprising? Explain.

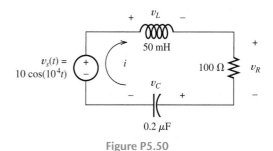

Figure P5.50

**P5.51.** Consider the circuit shown in Figure P5.51. Find the phasors $\mathbf{V}_1$, $\mathbf{V}_2$, $\mathbf{V}_R$, $\mathbf{V}_L$, and $\mathbf{I}$. Draw the phasor diagram to scale. What is the phase relationship between $\mathbf{I}$ and $\mathbf{V}_1$? Between $\mathbf{I}$ and $\mathbf{V}_L$?

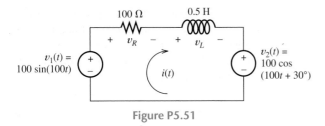

**Figure P5.51**

**P5.52.** Consider the circuit shown in Figure P5.52. Find the phasors I, $\mathbf{I}_R$, and $\mathbf{I}_C$. Construct the phasor diagram.

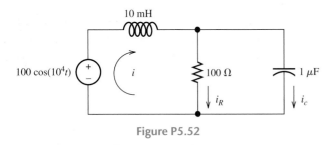

**Figure P5.52**

*__P5.53.__ Solve for the node voltages shown in Figure P5.53.

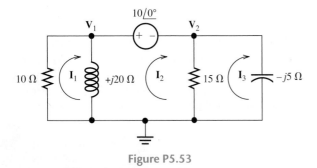

**Figure P5.53**

**P5.54.** Solve for the node voltage shown in Figure P5.54.

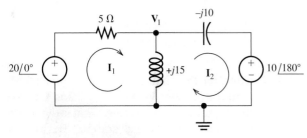

**Figure P5.54**

**P5.55.** Solve for the node voltage shown in Figure P5.55.

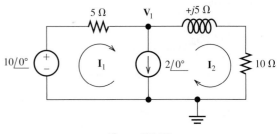

**Figure P5.55**

**P5.56.** Solve for the node voltages shown in Figure P5.56.

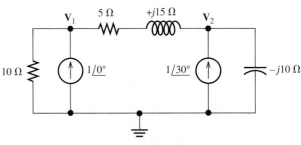

**Figure P5.56**

*__P5.57.__ Solve for the mesh currents shown in Figure P5.54.

**P5.58.** Solve for the mesh currents shown in Figure P5.55.

**P5.59.** Solve for the mesh currents shown in Figure P5.53.

**P5.60. a.** A 20-mH inductance is in series with a 50-$\mu$F capacitance. Sketch or use the computer program of your choice to produce a plot of the impedance magnitude versus angular frequency. Allow $\omega$ to range from zero to 2000 rad/s and the vertical axis to range from 0 to 100 $\Omega$. **b.** Repeat with the inductance and capacitance in parallel.

**P5.61. a.** A 20-mH inductance is in series with a 50-$\Omega$ resistance. Sketch or use the computer program of your choice to produce a plot of the impedance magnitude versus angular frequency. Allow $\omega$ to range from zero to 5000 rad/s. **b.** Repeat with the inductance and resistance in parallel.

## Section 5.5: Power in AC Circuits

**P5.62.** What are the customary units for real power? For reactive power? For apparent power?

**P5.63.** How are power factor and power angle related?

**P5.64.** Assuming that a nonzero ac source is applied, state whether the power and reactive power are positive, negative, or zero for **a.** a pure resistance; **b.** a pure inductance; **c.** a pure capacitance.

**P5.65.** A load is said to have a leading power factor. Is it capacitive or inductive? Is the reactive power positive or negative? Repeat for a load with lagging power factor.

**P5.66. a.** Sketch a power triangle for an inductive load, label the sides, and show the power angle. **b.** Repeat for a capacitive load.

**P5.67.** Discuss why power plant and distribution system engineers are concerned **a.** with the real power absorbed by a load; **b.** with the reactive power.

**P5.68.** Define what we mean by "power-factor correction." For power-factor correction of an inductive load, what type of element should we place in parallel with the load?

**\*P5.69.** Consider a load that has an impedance given by $Z = 100 - j50 \ \Omega$. The current flowing through this load is $\mathbf{I} = 15\sqrt{2} \ \angle 30° $ A. Is the load inductive or capacitive? Determine the power factor, power, reactive power, and apparent power delivered to the load.

**P5.70.** We have a load with an impedance given by $Z = 30 + j40 \ \Omega$. The voltage across this load is $\mathbf{V} = 1500\sqrt{2} \ \angle 30°$ V. Is the load inductive or capacitive? Determine the power factor, power, reactive power, and apparent power delivered to the load.

**P5.71.** The phasor voltage across a certain load is $\mathbf{V} = 1000\sqrt{2} \ \angle 30°$ V, and the phasor current through it is $\mathbf{I} = 15\sqrt{2} \ \angle 60°$ A. Determine the power factor, power, reactive power, apparent power, and impedance. Is the power factor leading or lagging?

**P5.72.** The voltage across a load is $v(t) = 10^4\sqrt{2} \cos(\omega t + 10°)$ V, and the current through the load is $i(t) = 20\sqrt{2} \cos(\omega t - 20°)$ A. The reference direction for the current points into the positive reference for the voltage. Determine the power factor, the power, the

reactive power, and the apparent power for the load. Is this load inductive or capacitive?

**P5.73.** Assuming that a nonzero ac voltage source is applied, state whether the power and reactive power are positive, negative, or zero for **a.** a resistance in series with an inductance; **b.** a resistance in series with a capacitance. (Assume that the resistances, inductance, and capacitance are nonzero and finite in value.)

**P5.74.** Assuming that a nonzero ac voltage source is applied, what can you say about whether the power and reactive power are positive, negative, or zero for a pure capacitance in series with a pure inductance? Consider cases in which the impedance magnitude of the capacitance is greater than, equal to, or less than the impedance magnitude of the inductance.

**P5.75.** Repeat Problem P5.74 for the inductance and capacitance in parallel.

**P5.76.** Determine the power for each source shown in Figure P5.76. Also, state whether each source is delivering or absorbing energy.

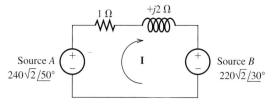

**Figure P5.76**

**P5.77.** Determine the power for each source shown in Figure P5.77. Also, state whether each source is delivering or absorbing energy.

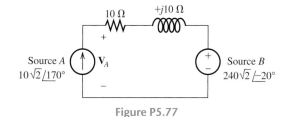

**Figure P5.77**

**P5.78.** A 60-Hz 220-V-rms source supplies power to a load consisting of a resistance in series with an inductance. The real power is 1500 W, and the apparent power is 2500 VAR. Determine the value of the resistance and the value of the inductance.

**P5.79.** Consider the circuit shown in Figure P5.79. Find the phasor current I. Find the power, reactive power, and apparent power delivered by the source. Find the power factor and state whether it is lagging or leading.

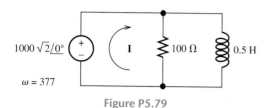

Figure P5.79

**\*P5.80.** Repeat Problem P5.79, replacing the inductance by a 10-$\mu$F capacitance.

**\*P5.81.** Two loads, $A$ and $B$, are connected in parallel across a 1-kV-rms 60-Hz line, as shown in Figure P5.81. Load $A$ consumes 10 kW with a 90 percent lagging power factor. Load $B$ has an apparent power of 15 kVA with an 80 percent lagging power factor. Find the power, reactive power, and apparent power delivered by the source. What is the power factor seen by the source?

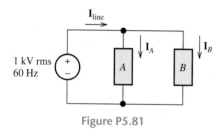

Figure P5.81

**P5.82.** Repeat Problem P5.81 if load $A$ consumes 5 kW with a 90 percent lagging power factor and load $B$ consumes 10 kW with an 80 percent leading power factor.

**P5.83.** Find the power, reactive power, and apparent power delivered by the source in Figure P5.83. Find the power factor and state whether it is leading or lagging.

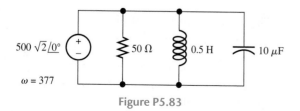

Figure P5.83

**P5.84.** Repeat Problem P5.83 with the resistance, inductance, and capacitance connected in series rather than in parallel.

**\*P5.85.** Consider the situation shown in Figure P5.85. A 1000-V-rms source delivers power to a load. The load consumes 100 kW with a power factor of 25 percent lagging. **a.** Find the phasor I, assuming that the capacitor is not connected to the circuit. **b.** Find the value of the capacitance that must be connected in parallel with the load to achieve a power factor of 100 percent. Usually, power-systems engineers rate capacitances used for power-factor correction in terms of their reactive power rating. What is the rating of this capacitance in kVAR? Assuming that this capacitance is connected, find the new value for the phasor I. **c.** Suppose that the source is connected to the load by a long distance. What are the potential advantages and disadvantages of connecting the capacitance across the load?

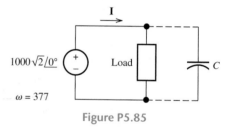

Figure P5.85

### Section 5.6: Thévenin and Norton Equivalent Circuits

**P5.86.** Of what does an ac steady-state Thévenin equivalent circuit consist? A Norton equivalent circuit? How are the values of the parameters of these circuits determined?

**P5.87.** To attain maximum power delivered to a load, what value of load impedance is required if **a.** the load can have any complex value; **b.** the load must be pure resistance?

**P5.88.** For an ac circuit consisting of a load connected to a Thévenin circuit, is it possible for the load voltage to exceed the Thévenin voltage in magnitude? If not, why not? If so, under what conditions is it possible? Explain.

**\*P5.89.** **a.** Find the Thévenin and Norton equivalent circuits for the circuit shown in Figure P5.89. **b.** Find the maximum power that this circuit can deliver to a load if the load can have any complex impedance. **c.** Repeat if the load is purely resistive.

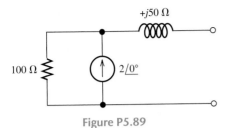

Figure P5.89

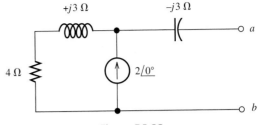

Figure P5.92

**P5.90.** **a.** Find the Thévenin and Norton equivalent circuits for the circuit shown in Figure P5.90. **b.** Find the maximum power that this circuit can deliver to a load if the load can have any complex impedance. **c.** Repeat if the load must be purely resistive.

**P5.93.** The Thévenin equivalent of a two-terminal network is shown in Figure P5.93. The frequency is $f = 60$ Hz. We wish to connect a load across terminals $a-b$ that consists of a resistance and a capacitance in series such that the power delivered to the resistance is maximized. Find the value of the resistance and the value of the capacitance.

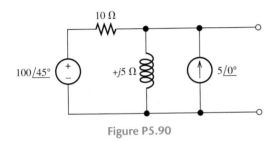

Figure P5.90

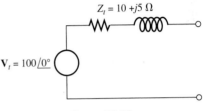

Figure P5.93

**P5.91.** Draw the Thévenin and Norton equivalent circuits for Figure P5.91, labeling the elements and terminals.

**\*P5.94.** Repeat Problem P5.93 with the load required to consist of a resistance and a capacitance in parallel.

**Section 5.7: Balanced Three-Phase Circuits**

**P5.95.** A balanced positive-sequence three-phase source has

$$v_{an}(t) = 100 \cos(377t + 90°) \text{ V}$$

**a.** Find the frequency of this source in Hz.
**b.** Give expressions for $v_{bn}(t)$ and $v_{cn}(t)$.
**c.** Repeat part (b) for a negative-sequence source.

**P5.96.** A three-phase source has

$$v_{an}(t) = 100 \cos(\omega t - 60°)$$
$$v_{bn}(t) = 100 \cos(\omega t + 60°)$$
$$v_{cn}(t) = -100 \cos(\omega t)$$

Is this a positive-sequence or a negative-sequence source? Find time-domain expressions for $v_{ab}(t)$, $v_{bc}(t)$, and $v_{ca}(t)$.

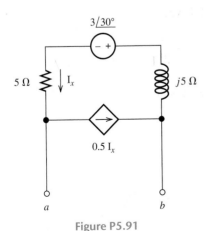

Figure P5.91

**P5.92.** Draw the Thévenin and Norton equivalent circuits for Figure P5.92, labeling the elements and terminals.

**\*P5.97.** A balanced wye-connected three-phase source has line-to-neutral voltages of 440 V rms. Find the rms line-to-line voltage magnitude. If this source is applied to a wye-connected load composed of three 30-$\Omega$ resistances, find the rms line-current magnitude and the total power delivered.

**\*P5.98.** Each phase of a wye-connected load consists of a 50-$\Omega$ resistance in parallel with a 100-$\mu$F capacitance. Find the impedance of each phase of an equivalent delta-connected load. The frequency of operation is 60 Hz.

**P5.99.** What can you say about the flow of power as a function of time between a balanced three-phase source and a balanced load? Is this true of a single-phase source and a load? How is this a potential advantage for the three-phase system? What is another advantage of three-phase power distribution compared with single-phase?

**P5.100.** A delta-connected source delivers power to a delta-connected load, as shown in Figure P5.100. The rms line-to-line voltage at the source is $V_{abrms} = 440$ V. The load impedance is $Z_\Delta = 10 - j2$. Find $\mathbf{I}_{aA}$, $\mathbf{V}_{AB}$, $\mathbf{I}_{AB}$, the total power delivered to the load, and the power lost in the line.

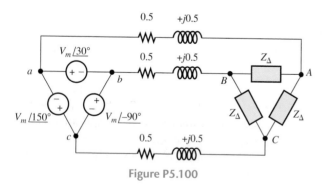

Figure P5.100

**\*P5.101.** Repeat Problem P5.100, with $Z_\Delta = 5 - j2$.

**P5.102.** A negative-sequence wye-connected source has line-to-neutral voltages $\mathbf{V}_{an} = V_Y \angle 0°$, $\mathbf{V}_{bn} = V_Y \angle 120°$, and $\mathbf{V}_{cn} = V_Y \angle -120°$. Find the line-to-line voltages $\mathbf{V}_{ab}$, $\mathbf{V}_{bc}$, and $\mathbf{V}_{ca}$. Construct a phasor diagram showing both sets of voltages and compare with Figure 5.41 on page 261.

**P5.103.** A balanced positive-sequence wye-connected 60-Hz three-phase source has line-to-line voltages of $V_L = 440$ V rms. This source is connected to a balanced wye-connected load. Each phase of the load consists of a 0.3-H inductance in series with a 50-$\Omega$ resistance. Find the line-to-neutral voltage phasors, the line-to-line voltage phasors, the line-current phasors, the power, and the reactive power delivered to the load. Assume that the phase of $\mathbf{V}_{an}$ is zero.

**P5.104.** A balanced wye-connected three-phase source has line-to-neutral voltages of 240 V rms. Find the rms line-to-line voltage. This source is applied to a delta-connected load, each arm of which consists of a 10-$\Omega$ resistance in parallel with a $+j5$-$\Omega$ reactance. Determine the rms line current magnitude, the power factor, and the total power delivered.

**P5.105.** In this chapter, we have considered balanced loads only. However, it is possible to determine an equivalent wye for an unbalanced delta, and vice versa. Consider the equivalent circuits shown in Figure P5.105. Derive formulas for the impedances of the wye in terms of the impedances of the delta. [*Hint*: Equate the impedances between corresponding pairs of terminals of the two circuits with the third terminal open. Then, solve the equations for $Z_a$, $Z_b$, and $Z_c$ in terms of $Z_A$, $Z_B$, and $Z_C$. Take care in distinguishing between upper- and lowercase subscripts.]

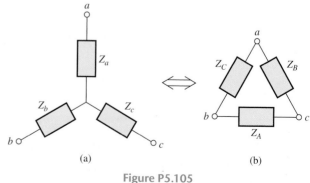

Figure P5.105

**P5.106.** Repeat Problem P5.105, but solve for the impedances of the delta in terms of those of the wye. [*Hint:* Start by working in terms of the admittances of the delta ($Y_A$, $Y_B$, and $Y_C$) and the impedances of the wye ($Z_a$, $Z_b$, and $Z_c$). Short terminals $b$ and $c$ for each circuit. Then equate the admittances between terminal $a$ and the shorted terminals for the two circuits. Repeat this twice more with shorts between the remaining two pairs of terminals. Solve the equations to determine $Y_A$, $Y_B$, and $Y_C$ in terms of $Z_a$, $Z_b$, and $Z_c$. Finally, invert the equations for $Y_A$, $Y_B$, and $Y_C$ to obtain equations relating the impedances. Take care in distinguishing between upper- and lowercase subscripts.]

### Section 5.8: AC Analysis Using MATLAB

*P5.107 Use MATLAB to solve for the node voltages shown in Figure P5.107.

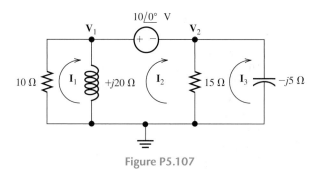

Figure P5.107

**P5.108** Use MATLAB to solve for the mesh currents shown in Figure P5.107.

*P5.109 Use MATLAB to solve for the mesh currents shown in Figure P5.109.

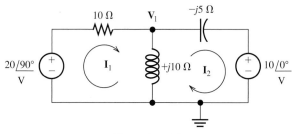

Figure P5.109

**P5.110** Use MATLAB to solve for the mesh currents shown in Figure P5.110.

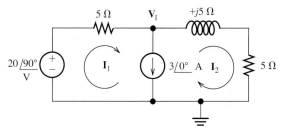

Figure P5.110

**P5.111** Use MATLAB to solve for the node voltages shown in Figure P5.111.

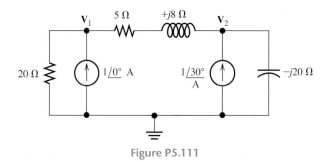

Figure P5.111

**P5.112** Use the MATLAB Symbolic Toolbox to determine the rms value of $v(t)$ which has a period of 1 s and is given by $v(t) = 10 \exp(-5t) \sin(20\pi t)$ V for $0 \leq t \leq 1$ s.

## Practice Test

Here is a practice test you can use to check your comprehension of the most important concepts in this chapter. Answers can be found in Appendix D and complete solutions are included in the Student Solutions files.

See Appendix E for more information about the Student Solutions.

**T5.1.** Determine the rms value of the current shown in Figure T5.1 and the average power delivered to the 50-$\Omega$ resistance.

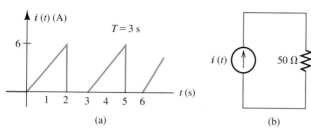

(a)                          (b)

**Figure T5.1**

**T5.2.** Reduce the expression

$$v(t) = 5 \sin(\omega t + 45°) + 5 \cos(\omega t - 30°)$$

to the form $V_m \cos(\omega t + \theta)$.

**T5.3.** We have two voltages $v_1(t) = 15 \sin(400\pi t + 45°)$ V and $v_2(t) = 5 \cos(400\pi t - 30°)$ V. Determine (including units): **a.** the rms value of $v_1(t)$; **b.** the frequency of the voltages; **c.** the angular frequency of the voltages; **d.** the period of the voltages; **e.** the phase relationship between $v_1(t)$ and $v_2(t)$.

**T5.4.** Find the phasor values of $\mathbf{V}_R$, $\mathbf{V}_L$, and $\mathbf{V}_C$ in polar form for the circuit of Figure T5.4.

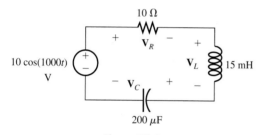

**Figure T5.4**

**T5.5.** Use the node-voltage approach to solve for $v_1(t)$ under steady-state conditions in the circuit of Figure T5.5.

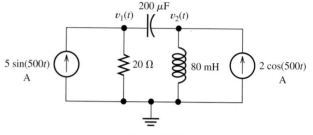

**Figure T5.5**

**T5.6.** Determine the complex power, power, reactive power, and apparent power absorbed by the load in Figure T5.6. Also, determine the power factor for the load.

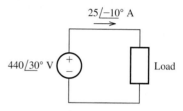

**Figure T5.6**

**T5.7.** Determine the line current $\mathbf{I}_{aA}$ in polar form for the circuit of Figure T5.7. This is a positive-sequence, balanced, three-phase system with $\mathbf{V}_{an} = 208\ \underline{/30°}$V and $Z_\Delta = 6 + j8\ \Omega$.

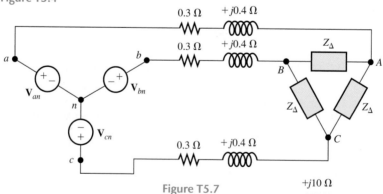

**Figure T5.7**

**T5.8.** Write the MATLAB commands to obtain the values of the mesh currents of Figure T5.8 in polar form. You may use the pin and pout functions defined in this chapter if you wish.

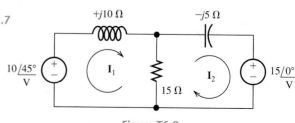

**Figure T5.8**

# Chapter 6

# Frequency Response, Bode Plots, and Resonance

## Study of this chapter will enable you to:

- State the fundamental concepts of Fourier analysis.

- Use a filter's transfer function to determine its output for a given input consisting of sinusoidal components.

- Use circuit analysis to determine the transfer functions of simple circuits.

- Draw first-order lowpass or highpass filter circuits and sketch their transfer functions.

- Understand decibels, logarithmic frequency scales, and Bode plots.

- Draw the Bode plots for transfer functions of first-order filters.

- Calculate parameters for series- and parallel-resonant circuits.

- Select and design simple filter circuits.

- Use MATLAB to derive and plot network functions.

- Design simple digital signal-processing systems.

## Introduction to this chapter:

Much of electrical engineering is concerned with information-bearing currents and voltages that we call **signals**. For example, transducers on an internal combustion engine provide electrical signals that represent temperature, speed, throttle position, and the rotational position of the crankshaft. These signals are **processed** (by electrical circuits) to determine the optimum firing instant for each cylinder. Finally, electrical pulses are generated for each spark plug.

Surveyors can measure distances by using an instrument that emits a pulse of light that is reflected by a mirror at the point of interest. The return light pulse is converted to an electrical signal that is processed by circuits to determine the round-trip time delay between the instrument and the mirror. Finally, the delay is converted to distance and displayed.

Another example of signal processing is the electrocardiogram, which is a plot of the electrical signal generated by the human heart. In a cardiac-care unit, circuits and computers are employed to extract information concerning the behavior of a patient's heart. A physician or nurse is alerted when the patient needs attention.

In general, **signal processing** is concerned with manipulating signals to extract information and using that information to generate other useful electrical signals. It is an important and far-reaching subject. In this chapter, we consider several simple but, nevertheless, useful circuits from a signal-processing point of view.

Recall that in Chapter 5 we learned how to analyze circuits containing sinusoidal sources, all of which have a common frequency. An important application is electrical power systems. However, most real-world information-bearing electrical signals are not sinusoidal. Nevertheless, we will see that phasor concepts can be very useful in understanding how circuits respond to nonsinusoidal signals. This is true because nonsinusoidal signals can be considered to be the sum of sinusoidal components having various frequencies, amplitudes, and phases.

## 6.1 FOURIER ANALYSIS, FILTERS, AND TRANSFER FUNCTIONS

### Fourier Analysis

As mentioned in the introduction to this chapter, most information-bearing signals are not sinusoidal. For example, the waveform produced by a microphone for speech or music is a complex nonsinusoidal waveform that is not predictable in advance. Figure 6.1(a) shows a (very) short segment of a music signal.

Even though many interesting signals are not sinusoidal, it turns out that we can construct any waveform by adding sinusoids that have the proper amplitudes, frequencies, and phases. For illustration, the waveform shown in Figure 6.1(a) is the sum of the sinusoids shown in Figure 6.1(b). The waveform shown in Figure 6.1 is relatively simple because it is composed of only three components. Most natural

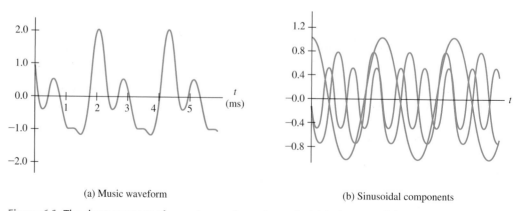

(a) Music waveform                                   (b) Sinusoidal components

**Figure 6.1** The short segment of a music waveform shown in (a) is the sum of the sinusoidal components shown in (b).

signals contain thousands of components. (In theory, the number is infinite in many cases.)

When we listen to music, our ears respond differently to the various frequency components. Some combinations of amplitudes and frequencies are pleasing, whereas other combinations are not. Thus, in the design of signal-processing circuits (such as amplifiers) for audio signals, we must consider how the circuits respond to components having different frequencies.

**Fourier analysis** is a mathematical technique for finding the amplitudes, frequencies, and phases of the components of a given waveform. Aside from mentioning some of the results of Fourier analysis, we will not develop the theory in detail. *The important point is that all real-world signals are sums of sinusoidal components.*

The range of the frequencies of the components depends on the type of signal under consideration. The frequency ranges for several types of signals are given in Table 6.1. Thus, electrocardiograms are composed of numerous sinusoidal components with frequencies ranging from 0.05 Hz to 100 Hz.

> All real-world signals are sums of sinusoidal components having various frequencies, amplitudes, and phases.

**Fourier Series of a Square Wave.**   As another example, consider the signal shown in Figure 6.2(a), which is called a **square wave**. Fourier analysis shows that the square wave can be written as an infinite series of sinusoidal components,

$$v_{sq}(t) = \frac{4A}{\pi} \sin(\omega_0 t) + \frac{4A}{3\pi} \sin(3\omega_0 t) + \frac{4A}{5\pi} \sin(5\omega_0 t) + \cdots \qquad (6.1)$$

in which $\omega_0 = 2\pi/T$ is the called the **fundamental angular frequency** of the square wave.

Figure 6.2(b) shows several of the terms in this series and the result of summing the first five terms. Clearly, even the sum of the first five terms is a fairly good approximation to the square wave, and the approximation becomes better as more components are added. Thus, the square wave is composed of an infinite number of sinusoidal components. The frequencies of the components are odd integer multiples of the fundamental frequency, the amplitudes decline with increasing frequency, and the phases of all components are $-90°$. Unlike the square wave, the components of real-world signals are confined to finite ranges of frequency, and their amplitudes are not given by simple mathematical expressions.

> The components of real-world signals are confined to finite ranges of frequency.

Sometimes a signal contains a component that has a frequency of zero. For zero frequency, a general sinusoid of the form $A \cos(\omega t + \theta)$ becomes simply $A \cos(\theta)$, which is constant for all time. Recall that we refer to constant voltages as dc, so zero frequency corresponds to dc.

> Zero frequency corresponds to dc.

**Table 6.1**  **Frequency Ranges of Selected Signals**

| | |
|---|---|
| Electrocardiogram | 0.05 to 100 Hz |
| Audible sounds | 20 Hz to 15 kHz |
| AM radio broadcasting | 540 to 1600 kHz |
| HD component video signals | Dc to 25 MHz |
| FM radio broadcasting | 88 to 108 MHz |
| Cellular phone | 824 to 894 MHz and 1850 to 1990 MHz |
| Satellite television downlinks (C-band) | 3.7 to 4.2 GHz |
| Digital satellite television | 12.2 to 12.7 GHz |

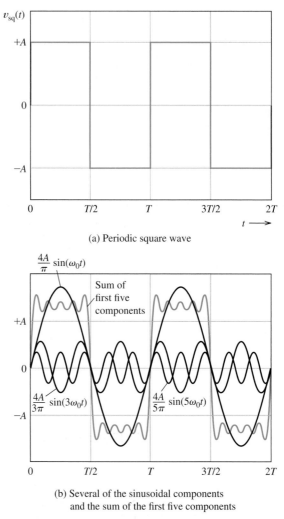

(a) Periodic square wave

(b) Several of the sinusoidal components
and the sum of the first five components

**Figure 6.2** A square wave and some of its components.

In sum, the fact that all signals are composed of sinusoidal components is a fundamental idea in electrical engineering. The frequencies of the components, as well as their amplitudes and phases, for a given signal can be determined by theoretical analysis or by laboratory measurements (using an instrument called a *spectrum analyzer*). Very often, the design of a system for processing information-bearing signals is based on considerations of how the system should respond to components of various frequencies.

> ... the fact that all signals are composed of sinusoidal components is a fundamental idea in electrical engineering.

### Filters

There are many applications in which we want to retain components in a given range of frequencies and discard the components in another range. This can be accomplished by the use of electrical circuits called **filters**. (Actually, filters can take many forms, but we limit our discussion to a few relatively simple *RLC* circuits.)

Usually, filter circuits are **two-port networks**, an example of which is illustrated in Figure 6.3. The signal to be filtered is applied to the input port and (ideally) only the components in the frequency range of interest appear at the output port.

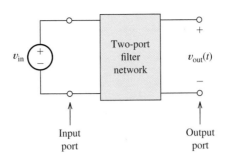

**Figure 6.3** When an input signal $v_{in}(t)$ is applied to the input port of a filter, some components are passed to the output port while others are not, depending on their frequencies. Thus, $v_{out}(t)$ contains some of the components of $v_{in}(t)$, but not others. Usually, the amplitudes and phases of the components are altered in passing through the filter.

For example, an FM radio antenna produces a voltage composed of signals from many transmitters. By using a filter that retains the components in the frequency range from 88 to 108 MHz and discards everything else, we can select the FM radio signals and reject other signals that could interfere with the process of extracting audio information.

*Filters process the sinusoid components of an input signal differently depending on the frequency of each component. Often, the goal of the filter is to retain the components in certain frequency ranges and to reject components in other ranges.*

As we learned in Chapter 5, the impedances of inductances and capacitances change with frequency. For example, the impedance of an inductance is $Z_L = \omega L \underline{/90°} = 2\pi f L \underline{/90°}$. Thus, the high-frequency components of a voltage signal applied to an inductance experience a higher impedance magnitude than do the low-frequency components. Consequently, electrical circuits can respond selectively to signal components, depending on their frequencies. Thus, $RLC$ circuits provide one way to realize electrical filters. We consider several specific examples later in this chapter.

*$RLC$ circuits provide one way to realize filters.*

## Transfer Functions

Consider the two-port network shown in Figure 6.3. Suppose that we apply a sinusoidal input signal having a frequency denoted as $f$ and having a phasor $\mathbf{V}_{in}$. In steady state, the output signal is sinusoidal and has the same frequency as the input. The output phasor is denoted as $\mathbf{V}_{out}$.

The **transfer function** $H(f)$ of the two-port filter is defined to be the ratio of the phasor output voltage to the phasor input voltage as a function of frequency:

$$H(f) = \frac{\mathbf{V}_{out}}{\mathbf{V}_{in}} \tag{6.2}$$

Because phasors are complex, the transfer function is a complex quantity having both magnitude and phase. Furthermore, both the magnitude and the phase can be functions of frequency.

*The transfer function $H(f)$ of the two-port filter is defined to be the ratio of the phasor output voltage to the phasor input voltage as a function of frequency.*

The transfer-function magnitude is the ratio of the output amplitude to the input amplitude. The phase of the transfer function is the output phase minus the input phase. Thus, the magnitude of the transfer function shows how the amplitude of each frequency component is affected by the filter. Similarly, the phase of the transfer function shows how the phase of each frequency component is affected by the filter.

*The magnitude of the transfer function shows how the amplitude of each frequency component is affected by the filter. Similarly, the phase of the transfer function shows how the phase of each frequency component is affected by the filter.*

---

**Example 6.1   Using the Transfer Function to Determine the Output**

The transfer function $H(f)$ of a filter is shown in Figure 6.4. [Notice that the magnitude $|H(f)|$ and phase $\underline{/H(f)}$ are shown separately in the figure.] If the input signal is given by

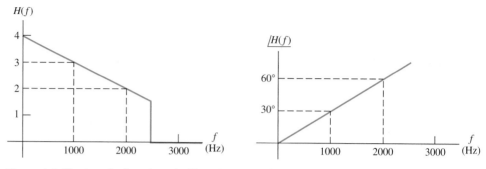

Figure 6.4  The transfer function of a filter. See Examples 6.1 and 6.2

$$v_{in}(t) = 2 \cos(2000\pi t + 40°)$$

find an expression (as a function of time) for the output of the filter.

**Solution**    By inspection, the frequency of the input signal is $f = 1000$ Hz. Referring to Figure 6.4, we see that the magnitude and phase of the transfer function are $|H(1000)| = 3$ and $\underline{/H(1000)} = 30°$, respectively. Thus, we have

$$H(1000) = 3\underline{/30°} = \frac{\mathbf{V}_{out}}{\mathbf{V}_{in}}$$

The phasor for the input signal is $\mathbf{V}_{in} = 2\underline{/40°}$, and we get

$$\mathbf{V}_{out} = H(1000) \times \mathbf{V}_{in} = 3\underline{/30°} \times 2\underline{/40°} = 6\underline{/70°}$$

Thus, the output signal is

$$v_{out}(t) = 6 \cos(2000\pi t + 70°)$$

In this case, the amplitude of the input is tripled by the filter. Furthermore, the signal is phase shifted by 30°. Of course, this is evident from the values shown in the plots of the transfer function at $f = 1000$.    ■

**Exercise 6.1**    Repeat Example 6.1 if the input signal is given by: **a.** $v_{in}(t) = 2 \cos(4000\pi t)$; and **b.** $v_{in}(t) = 1 \cos(6000\pi t - 20°)$.
**Answer**    **a.** $v_{out}(t) = 4 \cos(4000\pi t + 60°)$; **b.** $v_{out}(t) = 0$.    ☐

Notice that the effect of the filter on the magnitude and phase of the signal depends on signal frequency.

**Example: Graphic Equalizer.**    You may own a stereo audio system that has a *graphic equalizer*, which is a filter that has an adjustable transfer function. Usually, the controls of the equalizer are arranged so their positions give an approximate representation of the transfer-function magnitude versus frequency. (Actually, the equalizer in a stereo system contains two filters one for the left channel and one for the right channel-and the controls are ganged together.) Users can adjust the transfer function to achieve the mix of amplitudes versus frequency that is most pleasing to them.

**Input Signals with Multiple Components.**    If the input signal to a filter contains several frequency components, we can find the output for each input component separately and then add the output components. This is an application of the superposition principle first introduced in Section 2.7.

A step-by-step procedure for determining the output of a filter for an input with multiple components is as follows:

1. Determine the frequency and phasor representation for each input component.

2. Determine the (complex) value of the transfer function for each component.

3. Obtain the phasor for each output component by multiplying the phasor for each input component by the corresponding transfer-function value.

4. Convert the phasors for the output components into time functions of various frequencies. Add these time functions to produce the output.

---

**Example 6.2    Using the Transfer Function with Several Input Components**

Suppose that the input signal for the filter of Figure 6.4 is given by

$$v_{in}(t) = 3 + 2\cos(2000\pi t) + \cos(4000\pi t - 70°)$$

Find an expression for the output signal.

**Solution**    We start by breaking the input signal into its components. The first    Step 1.
component is

$$v_{in1}(t) = 3$$

the second component is

$$v_{in2}(t) = 2\cos(2000\pi t)$$

and the third component is

$$v_{in3}(t) = \cos(4000\pi t - 70°)$$

By inspection, we see that the frequencies of the components are 0, 1000, and 2000    Step 2.
Hz, respectively. Referring to the transfer function shown in Figure 6.4, we find that

$$H(0) = 4$$

$$H(1000) = 3\underline{/30°}$$

and

$$H(2000) = 2\underline{/60°}$$

The constant (dc) output term is simply $H(0)$ times the dc input:

$$v_{out1} = H(0)v_{in1} = 4 \times 3 = 12$$

The phasor outputs for the two input sinusoids are    Step 3.

$$\mathbf{V}_{out2} = H(1000) \times \mathbf{V}_{in2} = 3\underline{/30°} \times 2\underline{/0°} = 6\underline{/30°}$$

$$\mathbf{V}_{out3} = H(2000) \times \mathbf{V}_{in3} = 2\underline{/60°} \times 1\underline{/-70°} = 2\underline{/-10°}$$

Next, we can write the output components as functions of time:

$$v_{out1}(t) = 12$$

$$v_{out2}(t) = 6\cos(2000\pi t + 30°)$$

and

$$v_{out3}(t) = 2\cos(4000\pi t - 10°)$$

Finally, we add the output components to find the output voltage:

$$v_{out}(t) = v_{out1}(t) + v_{out2}(t) + v_{out3}(t)$$

and

$$v_{out}(t) = 12 + 6\cos(2000\pi t + 30°) + 2\cos(4000\pi t - 10°) \qquad ■$$

Notice that we did not add the phasors $\mathbf{V}_{out2}$ and $\mathbf{V}_{out3}$ in Example 6.2. The phasor concept was developed for sinusoids, all of which have the same frequency. *Hence, convert the phasors back into time-dependent signals before adding the components.*

Real-world information-bearing signals contain thousands of components. In principle, the output of a given filter for any input signal could be found by using the procedure of Example 6.2. However, it would usually be much too tedious to carry out. Fortunately, we will not need to do this. *It is the principle that is most important.* In summary, we can say that linear circuits (or any other systems for which the relationship between input and output can be described by linear time-invariant differential equations) behave as if they

**We must convert the phasors back into time-dependent signals before adding the components.**

1. Separate the input signal into components having various frequencies.
2. Alter the amplitude and phase of each component depending on its frequency.
3. Add the altered components to produce the output signal.

This process is illustrated in Figure 6.5.

The transfer function of a filter is important because it shows how the components are altered in amplitude and phase.

**Experimental Determination of the Transfer Function.**   To determine the transfer function of a filter experimentally, we connect a sinusoidal source to the input port, measure the amplitudes and phases of both the input signal and the resulting output signal, and divide the output phasor by the input phasor. This is repeated for each frequency of interest. The experimental setup is illustrated in Figure 6.6. Various instruments, such as voltmeters and oscilloscopes, can be employed to measure the amplitudes and phases.

In the next few sections of this chapter, we use mathematical analysis to investigate the transfer functions of several relatively simple electrical circuits.

**Exercise 6.2**   Consider the transfer function shown in Figure 6.4. The input signal is given by

$$v_{in}(t) = 2\cos(1000\pi t + 20°) + 3\cos(3000\pi t)$$

Find an expression for the output signal.

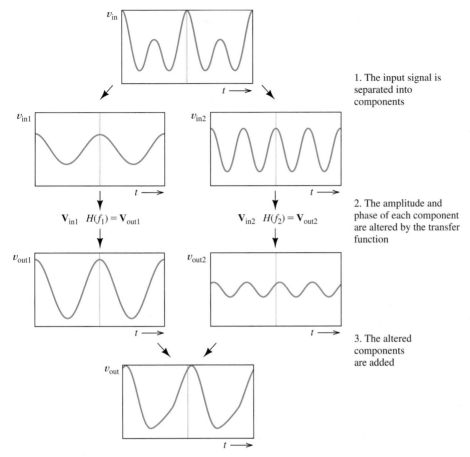

1. The input signal is separated into components

2. The amplitude and phase of each component are altered by the transfer function

3. The altered components are added

**Figure 6.5** Filters behave as if they separate the input into components, modify the amplitudes and phases of the components, and add the altered components to produce the output.

**Figure 6.6** To measure the transfer function we apply a sinusoidal input signal, measure the amplitudes and phases of input and output in steady state, and then divide the phasor output by the phasor input. The procedure is repeated for each frequency of interest.

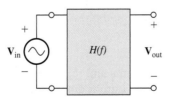

**Answer**    $v_{out}(t) = 7 \cos(1000\pi t + 35°) + 7.5 \cos(3000\pi t + 45°)$.    □

**Exercise 6.3**    Consider the transfer function shown in Figure 6.4. The input signal is given by

$$v_{in}(t) = 1 + 2 \cos(2000\pi t) + 3 \cos(6000\pi t)$$

Find an expression for the output signal.

**Answer**    $v_{out}(t) = 4 + 6 \cos(2000\pi t + 30°)$. Notice that the 3 kHz component is totally eliminated (rejected) by the filter.    □

**PRACTICAL APPLICATION   6.1**

**Active Noise Cancellation**

Noise and vibration are annoying to passengers in helicopters and other aircraft. Traditional sound-absorbing materials can be very effective in reducing noise levels, but are too bulky and massive for application in aircraft. An alternative approach is an electronic system that cancels noise. The diagram of such a system is shown in Figure PA6.1. A microphone near the sources of the noise, such as the engines, samples the noise before it enters the passenger area. The resulting electrical signal passes through a filter whose transfer function is continuously adjusted by a special-purpose computer to match the transfer function of the sound path. Finally, an inverted version of the signal is applied to loudspeakers. The sound waves from the speaker are out of phase with those from the noise source, resulting in partial cancellation. Another set of microphones on the headrest monitors the sound experienced by the passenger so that the computer can determine the filter adjustments needed to best cancel the sound.

Recently, noise-canceling systems based on these principles have appeared that contain all of the system elements in a lightweight headset. Many passengers on commercial aircraft wear these headsets to provide themselves with a quieter, more restful trip.

Active noise cancellation systems can effectively replace sound absorbing materials weighing a great deal more. As a result, active noise cancellation is very attractive for use in aircraft and automobiles. You can find many research reports and popular articles on this topic with an Internet search.

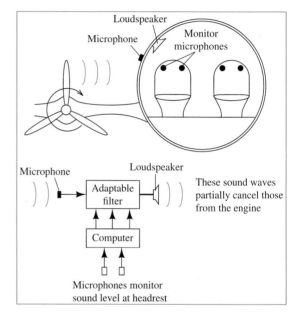

**Figure PA6.1**

## 6.2  FIRST-ORDER LOWPASS FILTERS

Consider the circuit shown in Figure 6.7. We will see that this circuit tends to pass low-frequency components and reject high-frequency components. (In other words, for low frequencies, the output amplitude is nearly the same as the input. For high frequencies, the output amplitude is much less than the input.) In Chapter 4, we saw that a first-order differential equation describes this circuit. Because of these facts, the circuit is called a **first-order lowpass filter**.

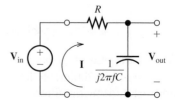

**Figure 6.7**  A first-order lowpass filter.

To determine the transfer function, we apply a sinusoidal input signal having a phasor $\mathbf{V}_{in}$, and then we analyze the behavior of the circuit as a function of the source frequency $f$.

The phasor current is the input voltage divided by the complex impedance of the circuit. This is given by

We can determine the transfer functions of *RLC* circuits by using steady-state analysis with complex impedances as a function of frequency.

$$\mathbf{I} = \frac{\mathbf{V}_{in}}{R + 1/j2\pi fC} \tag{6.3}$$

The phasor for the output voltage is the product of the phasor current and the impedance of the capacitance, illustrated by

$$\mathbf{V}_{out} = \frac{1}{j2\pi fC}\mathbf{I} \tag{6.4}$$

Using Equation 6.3 to substitute for $\mathbf{I}$, we have

$$\mathbf{V}_{out} = \frac{1}{j2\pi fC} \times \frac{\mathbf{V}_{in}}{R + 1/j2\pi fC} \tag{6.5}$$

Recall that the transfer function $H(f)$ is defined to be the ratio of the output phasor to the input phasor:

$$H(f) = \frac{\mathbf{V}_{out}}{\mathbf{V}_{in}} \tag{6.6}$$

Rearranging Equation 6.5, we have

$$H(f) = \frac{\mathbf{V}_{out}}{\mathbf{V}_{in}} = \frac{1}{1 + j2\pi fRC} \tag{6.7}$$

Next, we define the parameter:

$$f_B = \frac{1}{2\pi RC} \tag{6.8}$$

Then, the transfer function can be written as

$$H(f) = \frac{1}{1 + j(f/f_B)} \tag{6.9}$$

## Magnitude and Phase Plots of the Transfer Function

As expected, the transfer function $H(f)$ is a complex quantity having a magnitude and phase angle. Referring to the expression on the right-hand side of Equation 6.9, the magnitude of $H(f)$ is the magnitude of the numerator (which is unity) over the magnitude of the denominator. Recall that the magnitude of a complex quantity is the square root of the sum of the real part squared and the imaginary part squared.

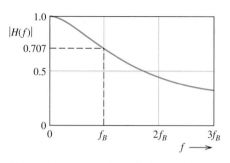
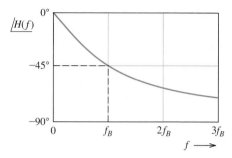

**Figure 6.8** Magnitude and phase of the first-order lowpass transfer function versus frequency.

Thus, the magnitude is given by

$$|H(f)| = \frac{1}{\sqrt{1 + (f/f_B)^2}} \tag{6.10}$$

Referring to the expression on the right-hand side of Equation 6.9, the phase angle of the transfer function is the phase of the numerator (which is zero) minus the phase of the denominator. This is given by

$$\angle H(f) = -\arctan\left(\frac{f}{f_B}\right) \tag{6.11}$$

Plots of the magnitude and phase of the transfer function are shown in Figure 6.8. For low frequencies ($f$ approaching zero), the magnitude is approximately unity and the phase is nearly zero, which means that the amplitudes and phases of low-frequency components are affected very little by this filter. The low-frequency components are passed to the output almost unchanged in amplitude or phase.

On the other hand, for high frequencies ($f \gg f_B$), the magnitude of the transfer function approaches zero. Thus, the amplitude of the output is much smaller than the amplitude of the input for the high-frequency components. We say that the high-frequency components are rejected by the filter. Furthermore, at high frequencies, the phase of the transfer function approaches $-90°$. Thus, as well as being reduced in amplitude, the high-frequency components are phase shifted.

Notice that for $f = f_B$, the magnitude of the output is $1/\sqrt{2} \cong 0.707$ times the magnitude of the input signal. When the amplitude of a voltage is multiplied by a factor of $1/\sqrt{2}$, the power that the voltage can deliver to a given resistance is multiplied by a factor of one-half (because power is proportional to voltage squared). Thus, $f_B$ is called the **half-power frequency**.

At the half-power frequency, the transfer-function magnitude is $1/\sqrt{2} \cong 0.707$ times its maximum value.

### Applying the Transfer Function

As we saw in Section 6.1, if an input signal to a filter consists of several components of different frequencies, we can use the transfer function to compute the output for each component separately. Then, we can find the complete output by adding the separate components.

**Example 6.3**   **Calculation of *RC* Lowpass Output**

Suppose that an input signal given by

$$v_{in}(t) = 5\cos(20\pi t) + 5\cos(200\pi t) + 5\cos(2000\pi t)$$

is applied to the lowpass *RC* filter shown in Figure 6.9. Find an expression for the output signal.

**Solution**   The filter has the form of the lowpass filter analyzed in this section. The half-power frequency is given by

$$f_B = \frac{1}{2\pi RC} = \frac{1}{2\pi \times (1000/2\pi) \times 10 \times 10^{-6}} = 100 \text{ Hz}$$

The first component of the input signal is

$$v_{in1}(t) = 5\cos(20\pi t)$$

For this component, the phasor is $\mathbf{V}_{in1} = 5\underline{/0°}$, and the angular frequency is $\omega = 20\pi$. Therefore, $f = \omega/2\pi = 10$. The transfer function of the circuit is given by Equation 6.9, which is repeated here for convenience:

$$H(f) = \frac{1}{1 + j(f/f_B)}$$

Evaluating the transfer function for the frequency of the first component ($f = 10$), we have

$$H(10) = \frac{1}{1 + j(10/100)} = 0.9950\underline{/-5.71°}$$

The output phasor for the $f = 10$ component is simply the input phasor times the transfer function. Thus, we obtain

$$\mathbf{V}_{out1} = H(10) \times \mathbf{V}_{in1}$$

$$= (0.9950\underline{/-5.71°}) \times (5\underline{/0°}) = 4.975\underline{/-5.71°}$$

Hence, the output for the first component of the input signal is

$$v_{out1}(t) = 4.975\cos(20\pi t - 5.71°)$$

Similarly, the second component of the input signal is

$$v_{in2}(t) = 5\cos(200\pi t)$$

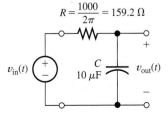

$$R = \frac{1000}{2\pi} = 159.2 \ \Omega$$

**Figure 6.9** Circuit of Example 6.3. The resistance has been picked so the break frequency turns out to be a convenient value.

$v_{in}(t)$

$C$
$10 \ \mu F$

$v_{out}(t)$

and we have

$$\mathbf{V}_{in2} = 5\underline{/0°}$$

The frequency of the second component is $f = 100$:

$$H(100) = \frac{1}{1 + j(100/100)} = 0.7071\underline{/-45°}$$

$$\mathbf{V}_{out2} = H(100) \times \mathbf{V}_{in2}$$

$$= (0.7071\underline{/-45°}) \times (5\underline{/0°}) = 3.535\underline{/-45°}$$

Therefore, the output for the second component of the input signal is

$$v_{out2}(t) = 3.535 \cos(200\pi t - 45°)$$

Finally, for the third and last component, we have

$$v_{in3}(t) = 5 \cos(2000\pi t)$$

$$\mathbf{V}_{in3} = 5\underline{/0°}$$

$$H(1000) = \frac{1}{1 + j(1000/100)} = 0.0995\underline{/-84.29°}$$

$$\mathbf{V}_{out3} = H(1000) \times \mathbf{V}_{in3}$$

$$= (0.0995\underline{/-84.29°}) \times (5\underline{/0°}) = 0.4975\underline{/-84.29°}$$

Consequently, the output for the third component of the input signal is

$$v_{out3}(t) = 0.4975 \cos(2000\pi t - 84.29°)$$

Now, we can write an expression for the output signal by adding the output components:

$$v_{out}(t) = 4.975 \cos(20\pi t - 5.71°) + 3.535 \cos(200\pi t - 45°)$$

$$+ 0.4975 \cos(2000\pi t - 84.29°)$$

Notice that each component of the input signal $v_{in}(t)$ is treated differently by this filter. The $f = 10$ component is nearly unaffected in amplitude and phase. The $f = 100$ component is reduced in amplitude by a factor of 0.7071 and phase shifted by $-45°$. The amplitude of the $f = 1000$ component is reduced by approximately an order of magnitude. Thus, the filter discriminates against the high-frequency components. ∎

### Application of the First-Order Lowpass Filter

A simple application of the first-order lowpass filter is the tone control on an old-fashion AM radio. The tone control adjusts the resistance and, therefore, the break frequency of the filter. Suppose that we are listening to an interesting news item from a distant radio station with an AM radio and lightning storms are causing electrical noise. It turns out that the components of voice signals are concentrated in the low end of the audible-frequency range. On the other hand, the noise caused by lightning has roughly equal-amplitude components at all frequencies. In this situation, we could adjust the tone control to lower the break frequency. Then, the high-frequency

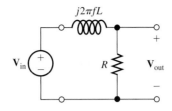

Figure 6.10 Another first-order lowpass filter; see Exercise 6.4.

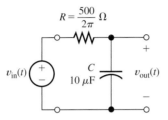

Figure 6.11 Circuit for Exercise 6.5.

noise components would be rejected, while most of the voice components would be passed. In this way, we can improve the ratio of desired signal power to noise power produced by the loudspeaker and make the news more intelligible.

## Using Phasors with Components of Different Frequencies

Recall that phasors can be combined only for sinusoids with the same frequency. It is important to understand that *we should not add the phasors for components with different frequencies*. Thus, in the preceding example, we used phasors to find the output components as functions of time, which we then added.

We should not add the phasors for components with different frequencies.

**Exercise 6.4**   Derive an expression for the transfer function $H(f) = \mathbf{V}_{\text{out}}/\mathbf{V}_{\text{in}}$ of the filter shown in Figure 6.10. Show that $H(f)$ takes the same form as Equation 6.9 if we define $f_B = R/2\pi L$. ☐

**Exercise 6.5**   Suppose that the input signal for the circuit shown in Figure 6.11 is given by

$$v_{\text{in}}(t) = 10\cos(40\pi t) + 5\cos(1000\pi t) + 5\cos(2\pi 10^4 t)$$

Find an expression for the output signal $v_{\text{out}}(t)$.
**Answer**

$$v_{\text{out}}(t) = 9.95\cos(40\pi t - 5.71°) + 1.86\cos(1000\pi t - 68.2°)$$
$$+ 0.100\cos(2\pi 10^4 t - 88.9°)$$
☐

## 6.3  DECIBELS, THE CASCADE CONNECTION, AND LOGARITHMIC FREQUENCY SCALES

In comparing the performance of various filters, it is helpful to express the magnitudes of the transfer functions in **decibels**. To convert a transfer-function magnitude to decibels, we multiply the common logarithm (base 10) of the transfer-function magnitude by 20:

$$|H(f)|_{\text{dB}} = 20\log|H(f)| \tag{6.12}$$

**Table 6.2** Transfer-Function Magnitudes and Their Decibel Equivalents

| $\lvert H(f)\rvert$ | $\lvert H(f)\rvert_{dB}$ |
| --- | --- |
| 100 | 40 |
| 10 | 20 |
| 2 | 6 |
| $\sqrt{2}$ | 3 |
| 1 | 0 |
| $1/\sqrt{2}$ | −3 |
| 1/2 | −6 |
| 0.1 | −20 |
| 0.01 | −40 |

(A transfer function is a ratio of voltages and is converted to decibels as 20 times the logarithm of the ratio. On the other hand, ratios of powers are converted to decibels by taking 10 times the logarithm of the ratio.)

Table 6.2 shows the decibel equivalents for selected values of transfer-function magnitude. Notice that the decibel equivalents are positive for magnitudes greater than unity, whereas the decibel equivalents are negative for magnitudes less than unity.

In many applications, the ability of a filter to strongly reject signals in a given frequency band is of primary importance. For example, a common problem associated with audio signals is that a small amount of the ac power line voltage can inadvertently be added to the signal. When applied to a loudspeaker, this 60-Hz component produces a disagreeable hum. (Actually, this problem is rapidly becoming a thing of the past as digital technologies replace analog.)

Usually, we approach this problem by trying to eliminate the electrical path by which the power line voltage is added to the desired audio signal. However, this is sometimes not possible. Then, we could try to design a filter that rejects the 60-Hz component and passes components at other frequencies. The magnitude of a filter transfer function to accomplish this is shown in Figure 6.12(a). A filter such as this, designed to eliminate components in a narrow range of frequencies, is called a **notch filter**.

It turns out that to reduce a loud hum (as loud as a heated conversation) to be barely audible, the transfer function must be −80 dB or less for the 60-Hz component, which corresponds to $\lvert H(f)\rvert = 10^{-4}$ or smaller. On the other hand, the transfer-function magnitude should be close to unity for the components to be passed by the filter. We refer to the range of frequencies to be passed as the **passband**.

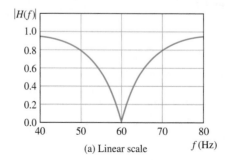

(a) Linear scale

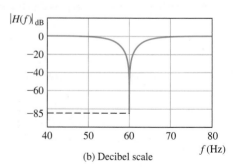

(b) Decibel scale

**Figure 6.12** Transfer-function magnitude of a notch filter used to reduce hum in audio signals.

When we plot $|H(f)|$ without converting to decibels, it is difficult to show both values clearly on the same plot. If we choose a scale that shows the passband magnitude, we cannot see whether the magnitude is sufficiently small at 60 Hz. This is the case for the plot shown in Figure 6.12(a). On the other hand, if we choose a linear scale that clearly shows the magnitude at 60 Hz, the magnitude would be way off scale at other frequencies of interest.

However, when the magnitude is converted to decibels, both parts of the magnitude are readily seen. For example, Figure 6.12(b) shows the decibel equivalent for the magnitude plot shown in Figure 6.12(a). On this plot, we can see that the passband magnitude is approximately unity (0 dB) and that at 60 Hz, the magnitude is sufficiently small (less than $-80$ dB).

Thus, one of the advantages of converting transfer-function magnitudes to decibels before plotting is that very small and very large magnitudes can be displayed clearly on a single plot. We will see that another advantage is that decibel plots for many filter circuits can be approximated by straight lines (provided that a logarithmic scale is used for frequency). Furthermore, to understand some of the jargon used by electrical engineers, we must be familiar with decibels.

> One of the advantages of converting transfer-function magnitudes to decibels before plotting is that very small and very large magnitudes can be displayed clearly on a single plot.

## Cascaded Two-Port Networks

When we connect the output terminals of one two-port circuit to the input terminals of another two-port circuit, we say that we have a **cascade** connection. This is illustrated in Figure 6.13. Notice that the output voltage of the first two-port network is the input voltage of the second two-port. The overall transfer function is

> In the cascade connection, the output of one filter is connected to the input of a second filter.

$$H(f) = \frac{\mathbf{V}_{\text{out}}}{\mathbf{V}_{\text{in}}}$$

However, the output voltage of the cascade is the output of the second two-port (i.e., $\mathbf{V}_{\text{out}} = \mathbf{V}_{\text{out2}}$). Furthermore, the input to the cascade is the input to the first two-port (i.e., $\mathbf{V}_{\text{in}} = \mathbf{V}_{\text{in1}}$). Thus,

$$H(f) = \frac{\mathbf{V}_{\text{out2}}}{\mathbf{V}_{\text{in1}}}$$

Multiplying and dividing by $\mathbf{V}_{\text{out1}}$, we have

$$H(f) = \frac{\mathbf{V}_{\text{out1}}}{\mathbf{V}_{\text{in1}}} \times \frac{\mathbf{V}_{\text{out2}}}{\mathbf{V}_{\text{out1}}}$$

Now, the output voltage of the first two-port is the input to the second two-port (i.e., $\mathbf{V}_{\text{out1}} = \mathbf{V}_{\text{in2}}$). Hence,

$$H(f) = \frac{\mathbf{V}_{\text{out1}}}{\mathbf{V}_{\text{in1}}} \times \frac{\mathbf{V}_{\text{out2}}}{\mathbf{V}_{\text{in2}}}$$

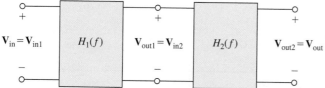

Figure 6.13 Cascade connection of two two-port circuits.

Finally, we can write

$$H(f) = H_1(f) \times H_2(f) \tag{6.13}$$

*Thus, the transfer function of the cascade connection is the product of the transfer functions of the individual two-port networks.* This fact can be extended to three or more two-ports connected in cascade.

A potential source of difficulty in applying Equation 6.13 is that the transfer function of a two-port usually depends on what is attached to its output terminals. *Thus, in applying Equation 6.13, we must find $H_1(f)$ with the second two-port attached.*

> In applying Equation 6.13, we must find $H_1(f)$ with the second two-port attached.

Taking the magnitudes of the terms on both sides of Equation 6.13 and expressing in decibels, we have

$$20 \log|H(f)| = 20 \log[|H_1(f)| \times |H_2(f)|] \tag{6.14}$$

Using the fact that the logarithm of a product is equal to the sum of the logarithms of the terms in the product, we have

$$20 \log|H(f)| = 20 \log|H_1(f)| + 20 \log|H_2(f)| \tag{6.15}$$

which can be written as

$$|H(f)|_{dB} = |H_1(f)|_{dB} + |H_2(f)|_{dB} \tag{6.16}$$

> In decibels, the individual transfer-function magnitudes are added to find the overall transfer-function magnitude for a cascade connection.

*Thus, in decibels, the individual transfer-function magnitudes are added to find the overall transfer-function magnitude for a cascade connection.*

## Logarithmic Frequency Scales

We often use a **logarithmic scale** for frequency when plotting transfer functions. On a logarithmic scale, the variable is *multiplied* by a given factor for equal increments of length along the axis. (On a linear scale, equal lengths on the scale correspond to *adding* a given amount to the variable.) For example, a logarithmic frequency scale is shown in Figure 6.14.

> On a logarithmic scale, the variable is multiplied by a given factor for equal increments of length along the axis.

A **decade** is a range of frequencies for which the ratio of the highest frequency to the lowest is 10. The frequency range from 2 to 20 Hz is one decade. Similarly, the range from 50 to 5000 Hz is two decades. (50 to 500 Hz is one decade, and 500 to 5000 Hz is another decade.)

An **octave** is a two-to-one change in frequency. For example, the range 10 to 20 Hz is one octave. The range 2 to 16 kHz is three octaves.

Suppose that we have two frequencies $f_1$ and $f_2$ for which $f_2 > f_1$. The number of decades between $f_1$ and $f_2$ is given by

$$\text{number of decades} = \log\left(\frac{f_2}{f_1}\right) \tag{6.17}$$

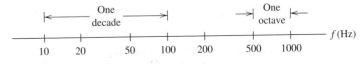

Figure 6.14 Logarithmic frequency scale.

in which we assume that the logarithm is base 10. The number of octaves between the two frequencies is

$$\text{number of octaves} = \log_2\left(\frac{f_2}{f_1}\right) = \frac{\log(f_2/f_1)}{\log(2)} \tag{6.18}$$

The advantage of a logarithmic frequency scale compared with a linear scale is that the variations in the magnitude or phase of a transfer function for a low range of frequency such as 10 to 20 Hz, as well as the variations in a high range such as 10 to 20 MHz, can be clearly shown on a single plot. With a linear scale, either the low range would be severely compressed or the high range would be off scale.

---

**Example 6.4**   **Decibels and Logarithmic Frequency Scales**

The transfer function magnitude of a certain filter is given by

$$|H(f)| = \frac{10}{\sqrt{1 + (f/5000)^6}}$$

a. What is the value of the transfer function magnitude in decibels for very low frequencies?

b. At what frequency $f_{3dB}$ is the transfer function magnitude 3 dB less than the value at very low frequencies?

c. At what frequency $f_{60dB}$ is the transfer function magnitude 60 dB less than the value at very low frequencies?

d. How many decades are between $f_{3dB}$ and $f_{60dB}$? How many octaves?

**Solution**

a. Very low frequencies are those approaching zero. For $f = 0$, we have $|H(0)| = 10$. Then, we have $|H(0)|_{dB} = 20 \log(10) = 20$ dB.

b. Because $-3$ dB corresponds to $1/\sqrt{2}$, we have

$$|H(f_{3dB})| = \frac{10}{\sqrt{2}} = \frac{10}{\sqrt{1 + (f_{3dB}/5000)^6}}$$

from which we find that $f_{3dB} = 5000$ Hz.

c. Also, because $-60$ dB corresponds to $1/1000$, we have

$$|H(f_{60dB})| = \frac{10}{1000} = \frac{10}{\sqrt{1 + (f_{60dB}/5000)^6}}$$

from which we find that $f_{60dB} = 50$ kHz.

d. Clearly, $f_{60dB} = 50$ kHz is one decade higher than $f_{3dB} = 5$ kHz. Using Equation 6.18, we find that the number of octaves between the two frequencies is

$$\frac{\log(50/5)}{\log(2)} = \frac{1}{\log(2)} = 3.32 \qquad \blacksquare$$

**Exercise 6.6**   Suppose that $|H(f)| = 50$. Find the decibel equivalent.
**Answer**   $|H(f)|_{dB} = 34$ dB.      □

**Exercise 6.7**   **a.** Suppose that $|H(f)|_{dB} = 15$ dB. Find $|H(f)|$. **b.** Repeat for $|H(f)|_{dB} = 30$ dB.
**Answer**   **a.** $|H(f)| = 5.62$; **b.** $|H(f)| = 31.6$.      □

**Exercise 6.8**   **a.** What frequency is two octaves higher than 1000 Hz? **b.** Three octaves lower? **c.** Two decades higher? **d.** One decade lower?
**Answer**   **a.** 4000 Hz is two octaves higher than 1000 Hz; **b.** 125 Hz is three octaves lower than 1000 Hz; **c.** 100 kHz is two decades higher than 1000 Hz; **d.** 100 Hz is one decade lower than 1000 Hz.      □

**Exercise 6.9**   **a.** What frequency is halfway between 100 and 1000 Hz on a logarithmic frequency scale? **b.** On a linear frequency scale?
**Answer**   **a.** 316.2 Hz is halfway between 100 and 1000 Hz on a logarithmic scale; **b.** 550 Hz is halfway between 100 and 1000 Hz on a linear frequency scale.      □

**Exercise 6.10**   **a.** How many decades are between $f_1 = 20$ Hz and $f_2 = 15$ kHz? (This is the approximate range of audible frequencies.) **b.** How many octaves?
**Answer**

**a.** Number of decades $= \log\left(\dfrac{15\text{ kHz}}{20\text{ Hz}}\right) = 2.87$

**b.** Number of octaves $= \dfrac{\log(15000/20)}{\log(2)} = 9.55$      □

## 6.4   BODE PLOTS

A Bode plot is a plot of the decibel magnitude of a network function versus frequency using a logarithmic scale for frequency.

A **Bode plot** is a plot of the decibel magnitude of a network function versus frequency using a logarithmic scale for frequency. Because it can clearly illustrate very large and very small magnitudes for a wide range of frequencies on one plot, the Bode plot is particularly useful for displaying transfer functions. Furthermore, it turns out that Bode plots of network functions can often be closely approximated by straight-line segments, so they are relatively easy to draw. (Actually, we now use computers to plot functions, so this advantage is not as important as it once was.) Terminology related to these plots is frequently encountered in signal-processing literature. Finally, an understanding of Bode plots enables us to make estimates quickly when dealing with transfer functions.

To illustrate Bode plot concepts, we consider the first-order lowpass transfer function of Equation 6.9, repeated here for convenience:

$$H(f) = \frac{1}{1 + j(f/f_B)}$$

The magnitude of this transfer function is given by Equation 6.10, which is

$$|H(f)| = \frac{1}{\sqrt{1 + (f/f_B)^2}}$$

To convert the magnitude to decibels, we take 20 times the logarithm of the magnitude:

$$|H(f)|_{dB} = 20 \log|H(f)|$$

Substituting the expression for the transfer-function magnitude, we get

$$|H(f)|_{dB} = 20 \log \frac{1}{\sqrt{1 + (f/f_B)^2}}$$

Using the properties of the logarithm, we obtain

$$|H(f)|_{dB} = 20 \log(1) - 20 \log\sqrt{1 + \left(\frac{f}{f_B}\right)^2}$$

Of course, the logarithm of unity is zero. Therefore,

$$|H(f)|_{dB} = -20 \log\sqrt{1 + \left(\frac{f}{f_B}\right)^2}$$

Finally, since $\log(\sqrt{x}) = \frac{1}{2}\log(x)$, we have

$$|H(f)|_{dB} = -10 \log\left[1 + \left(\frac{f}{f_B}\right)^2\right] \tag{6.19}$$

Notice that the value given by Equation 6.19 is approximately 0 dB for $f \ll f_B$. Thus, for low frequencies, the transfer-function magnitude is approximated by the horizontal straight line shown in Figure 6.15, labeled as the **low-frequency asymptote**. On the other hand, for $f \gg f_B$, Equation 6.19 is approximately

The low-frequency asymptote is constant at 0 dB.

$$|H(f)|_{dB} \cong -20 \log\left(\frac{f}{f_B}\right) \tag{6.20}$$

Evaluating for various values of $f$, we obtain the results shown in Table 6.3. Plotting these values results in the straight line shown sloping downward on the right-hand side of Figure 6.15, labeled as the **high-frequency asymptote**. Notice that the two straight-line asymptotes intersect at the half-power frequency $f_B$. For this reason, $f_B$ is also known as the **corner frequency** or as the **break frequency**.

The high-frequency asymptote slopes downward at 20 dB/decade, starting from 0 dB at $f_B$.

Also, notice that the slope of the high-frequency asymptote is $-20$ dB per decade of frequency. (This slope can also be stated as $-6$ dB per octave.)

If we evaluate Equation 6.19 at $f = f_B$, we find that

Notice that the two straight-line asymptotes intersect at the half-power frequency $f_B$.

$$|H(f_B)|_{dB} = -3 \text{ dB}$$

Thus, the asymptotes are in error by only 3 dB at the corner frequency. The actual curve for $|H(f)|_{dB}$ is also shown in Figure 6.15.

The asymptotes are in error by only 3 dB at the corner frequency $f_B$.

**Table 6.3** Values of the Approximate Expression (Equation 6.20) for Selected Frequencies

| $f$ | $|H(f)|_{dB}$ |
| --- | --- |
| $f_B$ | 0 |
| $2f_B$ | $-6$ |
| $10f_B$ | $-20$ |
| $100f_B$ | $-40$ |
| $1000f_B$ | $-60$ |

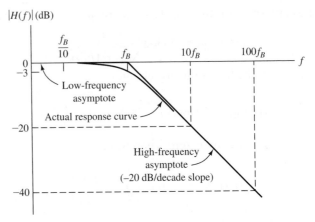

**Figure 6.15** Magnitude Bode plot for the first-order lowpass filter.

## Phase Plot

The phase of the first-order lowpass transfer function is given by Equation 6.11, which is repeated here for convenience:

$$\angle H(f) = -\arctan\left(\frac{f}{f_B}\right)$$

Evaluating, we find that the phase approaches zero at very low frequencies, equals $-45°$ at the break frequency, and approaches $-90°$ at high frequencies.

Figure 6.16 shows a plot of phase versus frequency. Notice that the curve can be approximated by the following straight-line segments:

1. A horizontal line at zero for $f < f_B/10$.
2. A sloping line from zero phase at $f_B/10$ to $-90°$ at $10f_B$.
3. A horizontal line at $-90°$ for $f > 10f_B$.

The actual phase curve departs from these straight-line approximations by less than $6°$. Hence, working by hand, we could easily construct an approximate plot of phase.

Many circuit functions can be plotted by the methods we have demonstrated for the simple lowpass $RC$ circuit; however, we will not try to develop your skill at

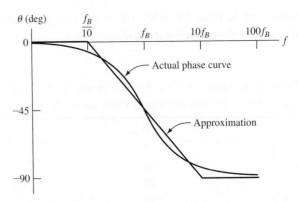

**Figure 6.16** Phase Bode plot for the first-order lowpass filter.

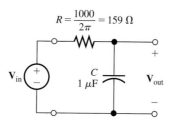

Figure 6.17  Circuit for Exercise 6.11.

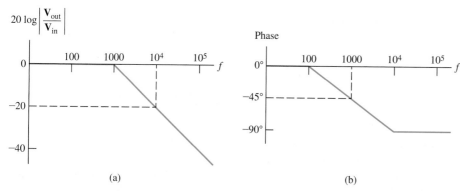

(a)                                                      (b)

Figure 6.18  Answers for Exercise 6.11.

this to a high degree. Bode plots of amplitude and phase for $RLC$ circuits are easily produced by computer programs. We have shown the manual approach to analyzing and drawing the Bode plot for the $RC$ lowpass filter mainly to present the concepts and terminology.

**Exercise 6.11**   Sketch the approximate straight-line Bode magnitude and phase plots to scale for the circuit shown in Figure 6.17.
**Answer**   See Figure 6.18.   □

## 6.5  FIRST-ORDER HIGHPASS FILTERS

The circuit shown in Figure 6.19 is called a **first-order highpass filter**. It can be analyzed in much the same manner as the lowpass circuit considered earlier in this chapter. The resulting transfer function is given by

$$H(f) = \frac{\mathbf{V}_{out}}{\mathbf{V}_{in}} = \frac{j(f/f_B)}{1 + j(f/f_B)} \tag{6.21}$$

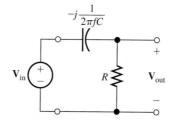

Figure 6.19  First-order highpass filter.

in which

$$f_B = \frac{1}{2\pi RC} \tag{6.22}$$

**Exercise 6.12**    Use circuit analysis to derive the transfer function for the circuit of Figure 6.19, and show that it can be put into the form of Equations 6.21 and 6.22.    ☐

### Magnitude and Phase of the Transfer Function

The magnitude of the transfer function is given by

$$|H(f)| = \frac{f/f_B}{\sqrt{1 + (f/f_B)^2}} \tag{6.23}$$

This is plotted in Figure 6.20(a). Notice that the transfer-function magnitude goes to zero for dc ($f = 0$). For high frequencies ($f \gg f_B$), the transfer-function magnitude approaches unity. Thus, this filter passes high-frequency components and tends to reject low-frequency components. That is why the circuit is called a highpass filter.

Highpass filters are useful whenever we want to retain high-frequency components and reject low-frequency components. For example, suppose that we want to record warbler songs in a noisy environment. It turns out that bird calls fall in the high-frequency portion of the audible range. The audible range of frequencies is from 20 Hz to 15 kHz (approximately), and the calls of warblers fall (mainly) in the range above 2 kHz. On the other hand, the noise may be concentrated at lower frequencies. For example, heavy trucks rumbling down a bumpy road would produce strong noise components lower in frequency than 2 kHz. To record singing warblers in the vicinity of such a noise source, a highpass filter would be helpful. We would select $R$ and $C$ to achieve a half-power frequency $f_B$ of approximately 2 kHz. Then, the filter would pass the songs and reject some of the noise.

Recall that if the amplitude of a component is multiplied by a factor of $1/\sqrt{2}$, the power that the component can deliver to a resistance is multiplied by a factor of 1/2. For $f = f_B$, $|H(f)| = 1/\sqrt{2} \cong 0.707$, so that, as in the case of the lowpass filter, $f_B$

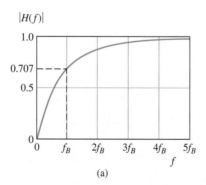

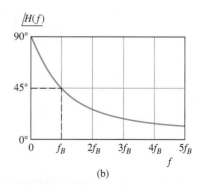

(a)                                        (b)

**Figure 6.20** Magnitude and phase for the first-order highpass transfer function.

is called the *half-power frequency*. (Here again, several alternative names are *corner frequency*, *3-dB frequency*, and *break frequency*.)

The phase of the highpass transfer function (Equation 6.21) is given by

$$\underline{/H(f)} = 90° - \arctan\left(\frac{f}{f_B}\right) \qquad (6.24)$$

A plot of the phase shift of the highpass filter is shown in Figure 6.20(b).

## Bode Plots for the First-Order Highpass Filter

As we have seen, a convenient way to plot transfer functions is to use the Bode plot, in which the magnitude is converted to decibels and a logarithmic frequency scale is used. In decibels, the magnitude of the highpass transfer function is

$$|H(f)|_{dB} = 20\log\frac{f/f_B}{\sqrt{1 + (f/f_B)^2}}$$

This can be written as

$$|H(f)|_{dB} = 20\log\left(\frac{f}{f_B}\right) - 10\log\left[1 + \left(\frac{f}{f_B}\right)^2\right] \qquad (6.25)$$

For $f \ll f_B$, the second term on the right-hand side of Equation 6.25 is approximately zero. Thus, for $f \ll f_B$, we have

$$|H(f)|_{dB} \cong 20\log\left(\frac{f}{f_B}\right) \qquad \text{for } f \ll f_B \qquad (6.26)$$

Evaluating this for selected values of $f$, we find the values given in Table 6.4. Plotting these values, we obtain the low-frequency asymptote shown on the left-hand side of Figure 6.21(a). Notice that the low-frequency asymptote slopes downward to the left at a rate of 20 dB per decade.

For $f \gg f_B$, the magnitude given by Equation 6.25 is approximately 0 dB. Hence,

$$|H(f)|_{dB} \cong 0 \qquad \text{for } f \gg f_B \qquad (6.27)$$

**Table 6.4** Values of the Approximate Expression Given in Equation 6.26 for Selected Frequencies

| $f$ | $|H(f)|_{dB}$ |
|---|---|
| $f_B$ | 0 |
| $f_B/2$ | −6 |
| $f_B/10$ | −20 |
| $f_B/100$ | −40 |

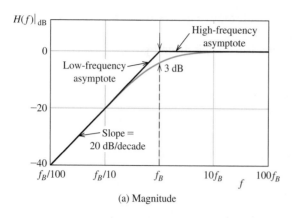

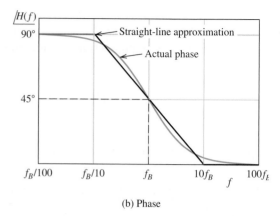

(a) Magnitude

(b) Phase

Figure 6.21 Bode plots for the first-order highpass filter.

This is plotted as the high-frequency asymptote in Figure 6.21(a). Notice that the high-frequency asymptote and the low-frequency asymptote meet at $f = f_B$. (That is why $f_B$ is sometimes called the *break frequency*.)

The actual values of $|H(f)|_{dB}$ are also plotted in Figure 6.21(a). Notice that the actual value at $f = f_B$ is $|H(f_B)|_{dB} = -3$ dB. Thus, the actual curve is only 3 dB from the asymptotes at $f = f_B$. For other frequencies, the actual curve is closer to the asymptotes. The Bode phase plot is shown in Figure 6.21(b) along with straight-line approximations.

### Example 6.5    Determination of the Break Frequency for a Highpass Filter

Suppose that we want a first-order highpass filter that has a transfer-function magnitude of $-30$ dB at $f = 60$ Hz. Find the break frequency for this filter.

**Solution**    Recall that the low-frequency asymptote slopes at a rate of 20 dB/decade. Thus, we must select $f_B$ to be

$$\frac{30 \text{ dB}}{20 \text{ dB/decade}} = 1.5 \text{ decades}$$

higher than 60 Hz. Employing Equation 6.17, we have

$$\log\left(\frac{f_B}{60}\right) = 1.5$$

This is equivalent to

$$\frac{f_B}{60} = 10^{1.5} = 31.6$$

which yields

$$f_B \cong 1900 \text{ Hz} \qquad \blacksquare$$

We often need a filter that greatly reduces the amplitude of a component at a given frequency, but has a negligible effect on components at nearby frequencies. The preceding example shows that to reduce the amplitude of a given component

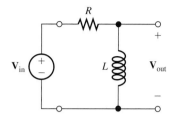

Figure 6.22   Circuit for Exercise 6.13.

by a large factor by using a first-order filter, we must place the break frequency far from the component to be rejected. Then, components at other frequencies are also affected. This is a problem that can only be solved by using more complex (higher order) filter circuits. We consider second-order filters later in the chapter.

**Exercise 6.13**   Consider the circuit shown in Figure 6.22. Show that the transfer function of this filter is given by Equation 6.21 if the half-power frequency is defined to be $f_B = R/2\pi L$.   □

**Exercise 6.14**   Suppose that we need a first-order $RC$ highpass filter that reduces the amplitude of a component at a frequency of 1 kHz by 50 dB. The resistance is to be 1 k$\Omega$. Find the half-power frequency and the capacitance.
**Answer**   $f_B = 316$ kHz, $C = 503$ pF.   □

## 6.6  SERIES RESONANCE

In this section and the next, we consider resonant circuits. These circuits form the basis for filters that have better performance (in passing desired signals and rejecting undesired signals that are relatively close in frequency) than first-order filters. Such filters are useful in radio receivers, for example. Another application is a notch filter to remove 60-Hz interference from audio signals. Resonance is a phenomenon that can be observed in mechanical systems as well as in electrical circuits. For example, a guitar string is a resonant mechanical system.

> Resonance is a phenomenon that can be observed in mechanical systems and electrical circuits.

We will see that when a sinusoidal source of the proper frequency is applied to a resonant circuit, voltages much larger than the source voltage can appear in the circuit. The familiar story of opera singers using their voices to break wine goblets is an example of a mechanically resonant structure (the goblet) driven by an approximately sinusoidal source (the sound), resulting in vibrations in the glass of sufficient magnitude to cause fracture. Another example is the Tacoma Narrows Bridge collapse in 1940. Driven by wind forces, a resonance of the bridge structure resulted in oscillations that tore the bridge apart. Some other examples of mechanical resonant systems are the strings of musical instruments, bells, the air column in an organ pipe, and a mass suspended by a spring.

> You can find a short video clip of the bridge in motion on the internet.

Consider the series circuit shown in Figure 6.23. The impedance seen by the source in this circuit is given by

$$Z_s(f) = j2\pi f L + R - j\frac{1}{2\pi f C} \tag{6.28}$$

> The resonant frequency $f_0$ is defined to be the frequency at which the impedance is purely resistive (i.e., the total reactance is zero).

The **resonant frequency** $f_0$ is defined to be the frequency at which the impedance is purely resistive (i.e., the total reactance is zero). For the reactance to equal zero, the

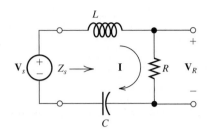

**Figure 6.23** The series resonant circuit.

impedance of the inductance must equal the impedance of the capacitance in magnitude. Thus, we have

$$2\pi f_0 L = \frac{1}{2\pi f_0 C} \tag{6.29}$$

Solving for the resonant frequency, we get

$$f_0 = \frac{1}{2\pi \sqrt{LC}} \tag{6.30}$$

The **quality factor** $Q_s$ is defined to be the ratio of the reactance of the inductance at the resonant frequency to the resistance:

The quality factor $Q_s$ of a series circuit is defined to be the ratio of the reactance of the inductance at the resonant frequency to the resistance.

$$Q_s = \frac{2\pi f_0 L}{R} \tag{6.31}$$

Solving Equation 6.29 for $L$ and substituting into Equation 6.31, we obtain

$$Q_s = \frac{1}{2\pi f_0 C R} \tag{6.32}$$

Using Equations 6.30 and 6.31 to substitute into Equation 6.28, we can eventually reduce the equation for the impedance to

$$Z_s(f) = R\left[1 + jQ_s\left(\frac{f}{f_0} - \frac{f_0}{f}\right)\right] \tag{6.33}$$

Thus, the series resonant circuit is characterized by its quality factor $Q_s$ and resonant frequency $f_0$.

Plots of the normalized magnitude and the phase of the impedance versus normalized frequency $f/f_0$ are shown in Figure 6.24. Notice that the impedance magnitude is minimum at the resonant frequency. As the quality factor becomes larger, the minimum becomes sharper.

### Series Resonant Circuit as a Bandpass Filter

Referring to Figure 6.23, the current is given by

$$I = \frac{V_s}{Z_s(f)}$$

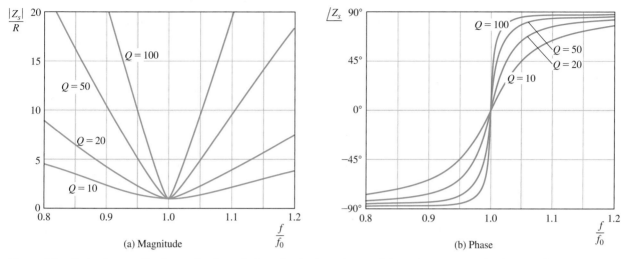

(a) Magnitude

(b) Phase

**Figure 6.24** Plots of normalized magnitude and phase for the impedance of the series resonant circuit versus frequency.

Using Equation 6.33 to substitute for the impedance, we have

$$\mathbf{I} = \frac{\mathbf{V}_s/R}{1 + jQ_s(f/f_0 - f_0/f)}$$

The voltage across the resistance is

$$\mathbf{V}_R = R\mathbf{I} = \frac{\mathbf{V}_s}{1 + jQ_s(f/f_0 - f_0/f)}$$

Dividing by $\mathbf{V}_s$, we obtain the transfer function

$$\frac{\mathbf{V}_R}{\mathbf{V}_s} = \frac{1}{1 + jQ_s(f/f_0 - f_0/f)}$$

Plots of the magnitude of $\mathbf{V}_R/\mathbf{V}_s$ versus $f$ are shown in Figure 6.25 for various values of $Q_s$.

Consider a (sinusoidal) source of constant amplitude and variable frequency. At low frequencies, the impedance magnitude of the capacitance is large, the current $\mathbf{I}$ is small in magnitude, and $\mathbf{V}_R$ is small in magnitude (compared with $\mathbf{V}_s$). At resonance, the total impedance magnitude reaches a minimum (because the reactances of the inductance and the capacitance cancel), the current magnitude is maximum, and $\mathbf{V}_R = \mathbf{V}_s$. At high frequencies, the impedance of the inductance is large, the current magnitude is small, and $\mathbf{V}_R$ is small in magnitude.

Now, suppose that we apply a source signal having components ranging in frequency about the resonant frequency. The components of the source that are close to the resonant frequency appear across the resistance with little change in amplitude. However, components that are higher or lower in frequency are significantly reduced in amplitude. Thus, a band of components centered at the resonant frequency is passed while components farther from the resonant frequency are (partly) rejected. We say that the resonant circuit behaves as a **bandpass filter**.

Recall that the half-power frequencies of a filter are the frequencies for which the transfer-function magnitude has fallen from its maximum by a factor of

The resonant circuit behaves as a bandpass filter.

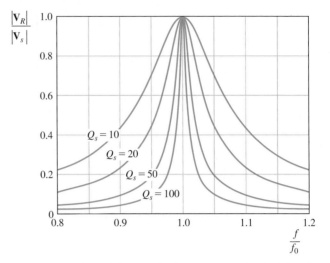

**Figure 6.25** Plots of the transfer-function magnitude $|\mathbf{V}_R/\mathbf{V}_s|$ for the series resonant bandpass-filter circuit.

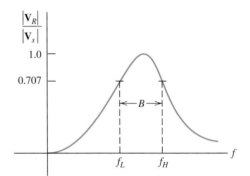

**Figure 6.26** The bandwidth $B$ is equal to the difference between the half-power frequencies.

$1/\sqrt{2} \cong 0.707$. For the series resonant circuit, there are two half-power frequencies $f_L$ and $f_H$. This is illustrated in Figure 6.26.

The **bandwidth** $B$ of this filter is the difference between the half-power frequencies:

$$B = f_H - f_L \tag{6.34}$$

For the series resonant circuit, it can be shown that

$$B = \frac{f_0}{Q_s} \tag{6.35}$$

Furthermore, for $Q_s \gg 1$, the half-power frequencies are given by the approximate expressions

$$f_H \cong f_0 + \frac{B}{2} \tag{6.36}$$

and

$$f_L \cong f_0 - \frac{B}{2} \tag{6.37}$$

Example 6.6   **Series Resonant Circuit**

Consider the series resonant circuit shown in Figure 6.27. Compute the resonant frequency, the bandwidth, and the half-power frequencies. Assuming that the frequency of the source is the same as the resonant frequency, find the phasor voltages across the elements and draw a phasor diagram.

**Solution**   First, we use Equation 6.30 to compute the resonant frequency:

$$f_0 = \frac{1}{2\pi\sqrt{LC}} = \frac{1}{2\pi\sqrt{0.1592 \times 0.1592 \times 10^{-6}}} = 1000 \text{ Hz}$$

The quality factor is given by Equation 6.31

$$Q_s = \frac{2\pi f_0 L}{R} = \frac{2\pi \times 1000 \times 0.1592}{100} = 10$$

The bandwidth is given by Equation 6.35

$$B = \frac{f_0}{Q_s} = \frac{1000}{10} = 100 \text{ Hz}$$

Next, we use Equations 6.36 and 6.37 to find the approximate half-power frequencies:

$$f_H \cong f_0 + \frac{B}{2} = 1000 + \frac{100}{2} = 1050 \text{ Hz}$$

$$f_L \cong f_0 - \frac{B}{2} = 1000 - \frac{100}{2} = 950 \text{ Hz}$$

At resonance, the impedance of the inductance and capacitance are

$$Z_L = j2\pi f_0 L = j2\pi \times 1000 \times 0.1592 = j1000 \ \Omega$$

$$Z_C = -j\frac{1}{2\pi f_0 C} = -j\frac{1}{2\pi \times 1000 \times 0.1592 \times 10^{-6}} = -j1000 \ \Omega$$

As expected, the reactances are equal in magnitude at the resonant frequency. The total impedance of the circuit is

$$Z_s = R + Z_L + Z_C = 100 + j1000 - j1000 = 100 \ \Omega$$

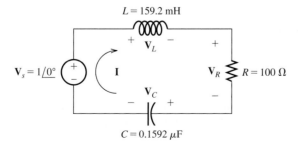

**Figure 6.27** Series resonant circuit of Example 6.6. (The component values have been selected so the resonant frequency and $Q_s$ turn out to be round numbers.)

The phasor current is given by

$$\mathbf{I} = \frac{\mathbf{V}_s}{Z_s} = \frac{1\underline{/0°}}{100} = 0.01\underline{/0°}$$

The voltages across the elements are

$$\mathbf{V}_R = R\mathbf{I} = 100 \times 0.01\underline{/0°} = 1\underline{/0°}$$
$$\mathbf{V}_L = Z_L\mathbf{I} = j1000 \times 0.01\underline{/0°} = 10\underline{/90°}$$
$$\mathbf{V}_C = Z_C\mathbf{I} = -j1000 \times 0.01\underline{/0°} = 10\underline{/-90°}$$

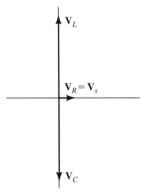

**Figure 6.28** Phasor diagram for Example 6.6.

The phasor diagram is shown in Figure 6.28. Notice that the voltages across the inductance and capacitance are much larger than the source voltage in magnitude. Nevertheless, Kirchhoff's voltage law is satisfied because $\mathbf{V}_L$ and $\mathbf{V}_C$ are out of phase and cancel.    ■

In Example 6.6, we found that the voltage magnitudes across the inductance and capacitance are $Q_s$ times higher than the source voltage. Thus, a higher quality factor leads to higher voltage magnification. This is similar to the large vibrations that can be caused in a wine goblet by an opera singer's voice.

**Exercise 6.15**    Determine the $R$ and $C$ values for a series resonant circuit that has $L = 10\ \mu\text{H}$, $f_0 = 1$ MHz, and $Q_s = 50$. Find the bandwidth and approximate half-power frequencies of the circuit.
**Answer**    $C = 2533$ pF, $R = 1.257\ \Omega$, $B = 20$ kHz, $f_L \cong 990$ kHz, $f_H \cong 1010$ kHz.  □

**Exercise 6.16**    Suppose that a voltage $\mathbf{V}_s = 1\underline{/0°}$ at a frequency of 1 MHz is applied to the circuit of Exercise 6.15. Find the phasor voltages across the resistance, capacitance, and inductance.
**Answer**    $\mathbf{V}_R = 1\underline{/0°}$, $\mathbf{V}_C = 50\underline{/-90°}$, $\mathbf{V}_L = 50\underline{/90°}$.    □

**Exercise 6.17**    Find the $R$ and $L$ values for a series resonant circuit that has $C = 470$ pF, a resonant frequency of 5 MHz, and a bandwidth of 200 kHz.
**Answer**    $R = 2.709\ \Omega$, $L = 2.156\ \mu\text{H}$.    □

## 6.7 PARALLEL RESONANCE

Another type of resonant circuit known as a **parallel resonant circuit** is shown in Figure 6.29. The impedance of this circuit is given by

$$Z_p = \frac{1}{1/R + j2\pi fC - j(1/2\pi fL)} \tag{6.38}$$

As in the series resonant circuit, the **resonant frequency** $f_0$ is the frequency for which the impedance is purely resistive. This occurs when the imaginary parts of the denominator of Equation 6.38 cancel. Thus, we have

$$2\pi f_0 C = \frac{1}{2\pi f_0 L} \tag{6.39}$$

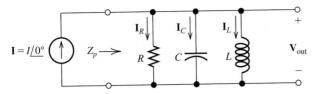

**Figure 6.29** The parallel resonant circuit.

Solving for the resonant frequency, we get

$$f_0 = \frac{1}{2\pi\sqrt{LC}} \tag{6.40}$$

which is exactly the same as the expression for the resonant frequency of the series circuit discussed in Section 6.6.

For the parallel circuit, we define the quality factor $Q_p$ as the ratio of the resistance to the reactance of the inductance at resonance, given by

$$Q_p = \frac{R}{2\pi f_0 L} \tag{6.41}$$

Notice that this is the reciprocal of the expression for the quality factor $Q_s$ of the series resonant circuit. Solving Equation 6.40 for $L$ and substituting into Equation 6.41, we obtain another expression for the quality factor:

Notice that the formula for $Q_p$ of a parallel circuit in terms of the circuit elements is the reciprocal of the formula for $Q_s$ of a series circuit.

$$Q_p = 2\pi f_0 CR \tag{6.42}$$

If we solve Equations 6.41 and 6.42 for $L$ and $C$, respectively, and then substitute into Equation 6.38, we eventually obtain

$$Z_p = \frac{R}{1 + jQ_p(f/f_0 - f_0/f)} \tag{6.43}$$

The voltage across the parallel resonant circuit is the product of the phasor current and the impedance:

$$\mathbf{V}_{out} = \frac{\mathbf{I}R}{1 + jQ_p(f/f_0 - f_0/f)} \tag{6.44}$$

Suppose that we hold the current constant in magnitude and change the frequency. Then, the magnitude of the voltage is a function of frequency. A plot of voltage magnitude for the parallel resonant circuit is shown in Figure 6.30. Notice that the voltage magnitude reaches its maximum $V_{omax} = RI$ at the resonant frequency. These curves have the same shape as the curves shown in Figures 6.25 and 6.26 for the voltage transfer function of the series resonant circuit.

The half-power frequencies $f_L$ and $f_H$ are defined to be the frequencies at which the voltage magnitude reaches the maximum value times $1/\sqrt{2}$. The bandwidth of the circuit is given by

$$B = f_H - f_L \tag{6.45}$$

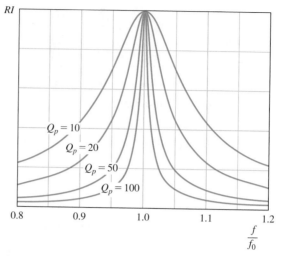

Figure 6.30 Voltage across the parallel resonant circuit for a constant-amplitude variable-frequency current source.

It can be shown that the bandwidth is related to the resonant frequency and quality factor by the expression

$$B = \frac{f_0}{Q_p} \tag{6.46}$$

### Example 6.7    Parallel Resonant Circuit

Find the $L$ and $C$ values for a parallel resonant circuit that has $R = 10 \text{ k}\Omega$, $f_0 = 1 \text{ MHz}$, and $B = 100 \text{ kHz}$. If $\mathbf{I} = 10^{-3}\underline{/0°}$ A, draw the phasor diagram showing the currents through each of the elements in the circuit at resonance.

**Solution**    First, we compute the quality factor of the circuit. Rearranging Equation 6.46 and substituting values, we have

$$Q_p = \frac{f_0}{B} = \frac{10^6}{10^5} = 10$$

Solving Equation 6.41 for the inductance and substituting values, we get

$$L = \frac{R}{2\pi f_0 Q_p} = \frac{10^4}{2\pi \times 10^6 \times 10} = 159.2 \ \mu\text{H}$$

Similarly, using Equation 6.42, we find that

$$C = \frac{Q_p}{2\pi f_0 R} = \frac{10}{2\pi \times 10^6 \times 10^4} = 159.2 \text{ pF}$$

At resonance, the voltage is given by

$$\mathbf{V}_{\text{out}} = \mathbf{I}R = (10^{-3}\underline{/0°}) \times 10^4 = 10\underline{/0°} \text{ V}$$

and the currents are given by

$$\mathbf{I}_R = \frac{\mathbf{V}_{\text{out}}}{R} = \frac{10 \underline{/0°}}{10^4} = 10^{-3} \underline{/0°} \text{ A}$$

$$\mathbf{I}_L = \frac{\mathbf{V}_{\text{out}}}{j2\pi f_0 L} = \frac{10 \underline{/0°}}{j10^3} = 10^{-2} \underline{/-90°} \text{ A}$$

$$\mathbf{I}_C = \frac{\mathbf{V}_{\text{out}}}{-j/2\pi f_0 C} = \frac{10 \underline{/0°}}{-j10^3} = 10^{-2} \underline{/90°} \text{ A}$$

The phasor diagram is shown in Figure 6.31. Notice that the currents through the inductance and capacitance are larger in magnitude than the applied source current. However, since $\mathbf{I}_C$ and $\mathbf{I}_L$ are out of phase, they cancel. ■

**Exercise 6.18**  A parallel resonant circuit has $R = 10 \text{ k}\Omega$, $L = 100 \mu\text{H}$, and $C = 500 \text{ pF}$. Find the resonant frequency, quality factor, and bandwidth.
**Answer**  $f_0 = 711.8 \text{ kHz}$, $Q_p = 22.36$, $B = 31.83 \text{ kHz}$. □

**Exercise 6.19**  A parallel resonant circuit has $f_0 = 10 \text{ MHz}$, $B = 200 \text{ kHz}$, and $R = 1 \text{ k}\Omega$. Find $L$ and $C$.
**Answer**  $L = 0.3183 \mu\text{H}$, $C = 795.8 \text{ pF}$. □

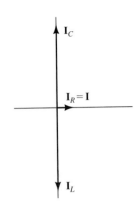

**Figure 6.31** Phasor diagram for Example 6.7.

## 6.8  IDEAL AND SECOND-ORDER FILTERS

### Ideal Filters

In discussing filter performance, it is helpful to consider ideal filters. An ideal filter passes components in the desired frequency range with no change in amplitude or phase and totally rejects the components in the undesired frequency range. Depending on the locations of the frequencies to be passed and rejected, we have different types of filters: lowpass, highpass, bandpass, and band reject. The transfer functions $H(f) = \mathbf{V}_{\text{out}}/\mathbf{V}_{\text{in}}$ of the four types of ideal filters are shown in Figure 6.32.

■ An **ideal lowpass filter** [Figure 6.32(a)] passes components below its cutoff frequency $f_H$ and rejects components higher in frequency than $f_H$.

■ An **ideal highpass filter** [Figure 6.32(b)] passes components above its cutoff frequency $f_L$ and rejects components lower in frequency than $f_L$.

■ An **ideal bandpass filter** [Figure 6.32(c)] passes components that lie between its cutoff frequencies ($f_L$ and $f_H$) and rejects components outside that range.

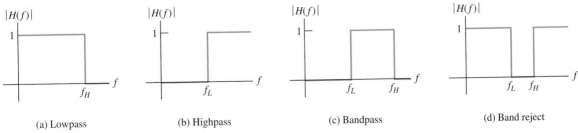

(a) Lowpass          (b) Highpass          (c) Bandpass          (d) Band reject

**Figure 6.32** Transfer functions of ideal filters.

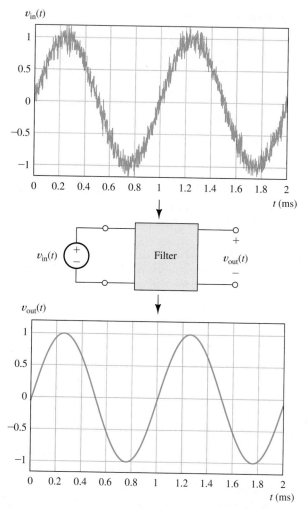

**Figure 6.33** The input signal $v_{in}$ consists of a 1-kHz sine wave plus high-frequency noise. By passing $v_{in}$ through an ideal lowpass filter with the proper cutoff frequency, the sine wave is passed and the noise is rejected, resulting in a clean output signal.

- An **ideal band-reject filter** [Figure 6.32(d)], which is also called a **notch filter**, rejects components that lie between its cutoff frequencies ($f_L$ and $f_H$) and passes components outside that range.

As we have seen earlier in this chapter, filters are useful whenever a signal contains desired components in one range of frequency and undesired components in another range of frequency. For example, Figure 6.33(a) shows a 1-kHz sine wave that has been corrupted by high-frequency noise. By passing this noisy signal through a lowpass filter, the noise is eliminated.

---

**Example 6.8**    Cascaded Ideal Filters

Electrocardiographic (ECG) signals are voltages between electrodes placed on the torso, arms, or legs of a medical patent. ECG signals are used by cardiologists to help diagnose various types of heart disease.

Unfortunately, the voltages between the electrodes can contain undesirable noises (called "artifacts" in medical jargon). The undesirable components are dc and frequency components below 0.5 Hz known as "baseline wander," a large 60-Hz sinewave due to power-line interference, and "muscle noise" with components above about 100 Hz caused by muscle movement, such as when the patient is on a treadmill. The part of the ECG signal of interest to cardiologists lies between about 0.5 Hz and 100 Hz.

We wish to design a cascade connection of ideal filters to eliminate the noise and preserve the ECG signal components of interest.

**Solution**   First, we can use an ideal highpass filter having a transfer function magnitude $|H_1(f)|$ as shown in Figure 6.34(a) to eliminate the dc and baseline wander with components below 0.5 Hz. Notice that we have used logarithmic frequency

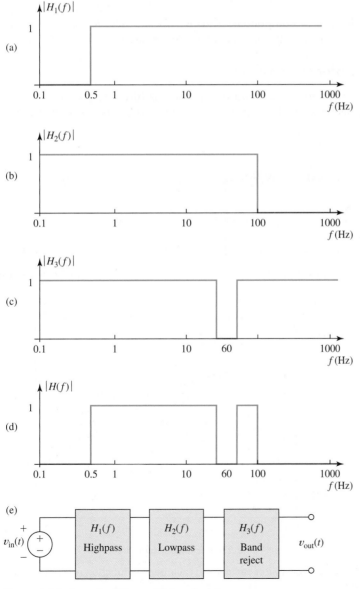

**Figure 6.34** Cascaded filters of Example 6.8.

scales in Figure 6.34 to show low and high frequencies more clearly than could be accomplished with a linear frequency scale.

Next, we can use an ideal lowpass filter having a transfer function magnitude $|H_2(f)|$ as shown in Figure 6.34(b) to eliminate the muscle noise components above 100 Hz.

Finally, we employ a band-reject filter having a transfer function magnitude $|H_3(f)|$ as shown in Figure 6.34(c) with cutoff frequencies slightly above 60 Hz and slightly below 60 Hz to eliminate the power line interference. We should strive to keep the cutoff frequencies of the band-reject filter very close to 60 Hz to avoid removing too many components of the ECG.

The overall transfer function magnitude $|H(f)| = |H_1(f)| \times |H_2(f)| \times |H_3(f)|$ is shown in Figure 6.34(d) and the cascaded filters are shown in Figure 6.34(e). ∎

Unfortunately, it is not possible to construct ideal filters—they can only be approximated by real circuits. As the circuits are allowed to increase in complexity, it is possible to design filters that do a better job of rejecting unwanted components and retaining the desired components. Thus, we will see that second-order circuits perform better (i.e., closer to ideal) than the first-order circuits considered earlier in this chapter.

### Second-Order Lowpass Filter

Figure 6.35(a) shows a second-order lowpass filter based on the series resonant circuit of Section 6.6. The filter is characterized by its resonant frequency $f_0$ and

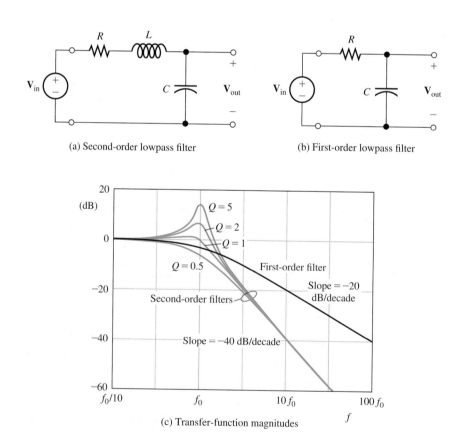

(a) Second-order lowpass filter    (b) First-order lowpass filter

(c) Transfer-function magnitudes

**Figure 6.35** Lowpass filter circuits and their transfer-function magnitudes versus frequency.

quality factor $Q_s$, which are given by Equations 6.30 and 6.31. It can be shown that the transfer function for this circuit is given by

$$H(f) = \frac{\mathbf{V}_{out}}{\mathbf{V}_{in}} = \frac{-jQ_s(f_0/f)}{1 + jQ_s(f/f_0 - f_0/f)} \qquad (6.47)$$

Bode plots of the transfer-function magnitude are shown in Figure 6.35(c). Notice that for $Q_s \gg 1$, the transfer-function magnitude reaches a high peak in the vicinity of the resonant frequency. Usually, in designing a filter, we want the gain to be approximately constant in the passband, and we select $Q_s \cong 1$. (Actually, $Q_s = 0.707$ is the highest value for which the transfer-function magnitude does not display an increase before rolling off. The transfer function for this value of $Q_s$ is said to be *maximally flat*, is also known as a *Butterworth function*, and is often used for lowpass filters.)

## Comparison of First- and Second-Order Filters

For comparison, a first-order lowpass filter is shown in Figure 6.35(b), and the Bode plot of its transfer function is shown in Figure 6.35(c). The first-order circuit is characterized by its half-power frequency $f_B = 1/(2\pi RC)$. (We have selected $f_B = f_0$ in making the comparison.) Notice that above $f_0$ the magnitude of the transfer function falls more rapidly for the second-order filter than for the first-order filter ($-40$ dB/decade versus $-20$ dB/decade).

The transfer-function magnitude of a second-order lowpass filter declines 40 dB per decade well above the break frequency, whereas the transfer-function magnitude for the first-order filter declines at only 20 dB per decade. Thus, the second-order filter is a better approximation to an ideal lowpass filter.

## Second-Order Highpass Filter

A second-order highpass filter is shown in Figure 6.36(a), and its magnitude Bode plot is shown in Figure 6.36(b). Here again, we usually want the magnitude to be as nearly constant as possible in the passband, so we select $Q_s \cong 1$. (In other words, we usually want to design the filter to approximate an ideal filter as closely as possible.)

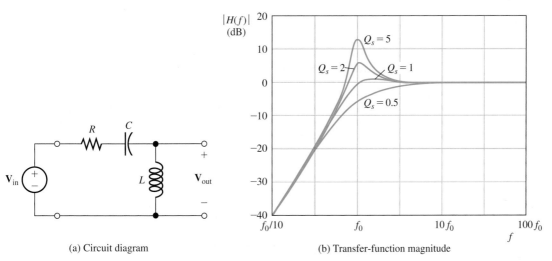

(a) Circuit diagram

(b) Transfer-function magnitude

**Figure 6.36** Second-order highpass filter and its transfer-function magnitude versus frequency for several values of $Q_s$.

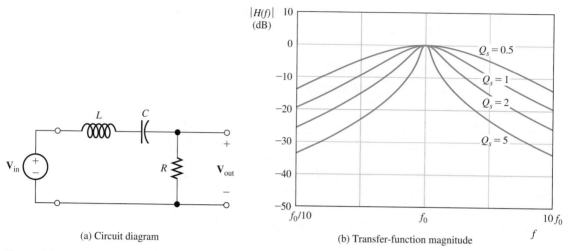

(a) Circuit diagram

(b) Transfer-function magnitude

**Figure 6.37** Second-order bandpass filter and its transfer-function magnitude versus frequency for several values of $Q_s$.

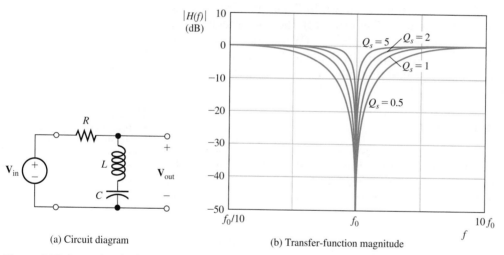

(a) Circuit diagram

(b) Transfer-function magnitude

**Figure 6.38** Second-order band-reject filter and its transfer-function magnitude versus frequency for several values of $Q_s$.

## Second-Order Bandpass Filter

A second-order bandpass filter is shown in Figure 6.37(a), and its magnitude Bode plot is shown in Figure 6.37(b). The half-power bandwidth $B$ is given by Equations 6.34 and 6.35, which state that

$$B = f_H - f_L$$

and

$$B = \frac{f_0}{Q_s}$$

## Second-Order Band-Reject (Notch) Filter

A second-order band-reject filter is shown in Figure 6.38(a) and its magnitude Bode plot is shown in Figure 6.38(b). In theory, the magnitude of the transfer function is zero for $f = f_0$. [In decibels, this corresponds to $|H(f_0)| = -\infty$ dB.] However, real inductors contain series resistance, so rejection of the $f_0$ component is not perfect for actual circuits.

---

### Example 6.9   Filter Design

Suppose that we need a filter that passes components higher in frequency than 1 kHz and rejects components lower than 1 kHz. Select a suitable second-order circuit configuration, choose $L = 50$ mH, and specify the values required for the other components.

**Solution**   We need to pass high-frequency components and reject low-frequency components. Therefore, we need a highpass filter. The circuit diagram for a second-order highpass filter is shown in Figure 6.36(a), and the corresponding transfer-function magnitude plots are shown in Figure 6.36(b). Usually, we want the transfer function to be approximately constant in the passband. Thus, we choose $Q_s \cong 1$. We select $f_0 \cong 1$ kHz, so the components above 1 kHz are passed, while lower-frequency components are (at least partly) rejected. Solving Equation 6.30, for the capacitance and substituting values, we have

$$C = \frac{1}{(2\pi)^2 f_0^2 L} = \frac{1}{(2\pi)^2 \times 10^6 \times 50 \times 10^{-3}}$$
$$= 0.507 \ \mu\text{F}$$

Solving Equation 6.31 for the resistance and substituting values, we get

$$R = \frac{2\pi f_0 L}{Q_s} = \frac{2\pi \times 1000 \times 50 \times 10^{-3}}{1} = 314.1 \ \Omega$$

The circuit and values are shown in Figure 6.39.                                              ∎

There are several reasons why we might not use the exact values that we calculated for the components in the last example. First, fixed-value capacitors and resistors are readily available only in certain standard values. Furthermore, the design called for a filter to reject components lower than 1 kHz and pass components higher than 1 kHz. We arbitrarily selected $f_0 = 1$ kHz. Depending on whether it is more important to reject the low frequencies or to pass the high frequencies without change in amplitude, a slightly higher or lower value for $f_0$ could be better. Finally, our choice of $Q_s$ was somewhat arbitrary. In practice, we could choose variable components by using the calculations as a starting point. Then, we would adjust the filter experimentally for the most satisfactory performance.

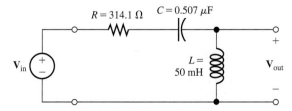

**Figure 6.39** Filter designed in Example 6.9.

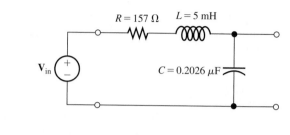

**Figure 6.40** Answer for
Exercise 6.20.

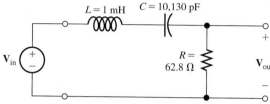

**Figure 6.41** Answer for
Exercise 6.21.

**Exercise 6.20**   Suppose that we need a filter that passes components lower in frequency than 5 kHz and rejects components higher than 5 kHz. Select a suitable second-order circuit configuration, choose $L = 5$ mH, and specify the values required for the other components.
**Answer**   See Figure 6.40.   □

**Exercise 6.21**   Suppose that we want a filter that passes components between $f_L = 45$ kHz and $f_H = 55$ kHz. Higher and lower frequencies are to be rejected. Design a circuit using a 1-mH inductance.
**Answer**   We need a bandpass filter with $f_0 \cong 50$ kHz and $Q_s = 5$. The resulting circuit is shown in Figure 6.41.   □

## 6.9   BODE PLOTS WITH MATLAB

So far in this chapter, we have used manual methods to illustrate Bode-plot concepts for simple filters. While manual methods can be extended to more complex circuits, it is often quicker and more accurate to use computer software to produce Bode plots.

Because subtle programming errors can result in grossly erroneous results, it is good practice to employ independent checks on computer-generated Bode plots. For example, a complex circuit can often be readily analyzed manually at very high and at very low frequencies. At very low frequencies, the inductances behave as short circuits and the capacitances behave as open circuits, as we discussed in Section 4.2. Thus, we can replace the inductances by shorts and the capacitances by opens and analyze the simplified circuit to determine the value of the transfer function at low frequencies, providing an independent check on the plots produced by a computer.

Similarly, at very high frequencies, the inductances become open circuits, and the capacitances become shorts. Next, we illustrate this approach with an example.

Manual analysis at dc and very high frequencies often provides some easy checks on computer-aided Bode plots.

| Example 6.10 | **Computer-Generated Bode Plot** |

The circuit of Figure 6.42 is a notch filter. Use MATLAB to generate a magnitude Bode plot of the transfer function $H(f) = \mathbf{V}_{\text{out}}/\mathbf{V}_{\text{in}}$ with frequency ranging from 10 Hz to 100 kHz. Then, analyze the circuit manually at very high and very low

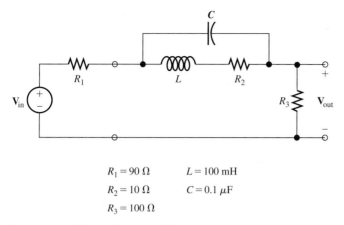

$$R_1 = 90\ \Omega \qquad L = 100\ \text{mH}$$
$$R_2 = 10\ \Omega \qquad C = 0.1\ \mu\text{F}$$
$$R_3 = 100\ \Omega$$

**Figure 6.42** Filter of Example 6.10.

frequencies to provide checks on the plot. Use the plot to determine the frequency of maximum attenuation and the value of the transfer function at that frequency.

**Solution**   Using the voltage-divider principle, we can write the transfer function for the filter as

$$H(f) = \frac{\mathbf{V}_{\text{out}}}{\mathbf{V}_{\text{in}}} = \frac{R_3}{R_1 + R_3 + 1/[j\omega C + 1/(R_2 + j\omega L)]}$$

A MATLAB m-file that produces the Bode plot is:

```
clear
% Enter the component values:
R1 = 90; R2 = 10; R3 = 100;
L = 0.1; C = 1e-7;
% The following command generates 1000 frequency values
% per decade, evenly spaced from 10^1 to 10^5 Hz
% on a logarithmic scale:
f = logspace(1,5,4000);
w = 2*pi*f;
% Evaluate the transfer function for each frequency.
% As usual, we are using i in place of j:
H = R3./(R1+R3+1./(i*w*C + 1./(R2 + i*w*L)));
% Convert the magnitude values to decibels and plot:
semilogx(f,20*log10(abs(H)))
```

The resulting plot is shown in Figure 6.43. This circuit is called a notch filter because it strongly rejects components in the vicinity of 1591 Hz while passing higher and lower frequencies. The maximum attenuation is 60 dB.

The m-file is named Example_6_10 and appears in the MATLAB folder, and if you have access to MATLAB, you can run it to see the result. (See Appendix E for information on how to access the MATLAB folder.) Then, you can use the toolbar on the figure screen to magnify a portion of the plot and obtain the notch frequency and maximum attenuation with excellent accuracy.

The command

```
f = logspace(1,5,4000)
```

generates an array of 4000 frequency values, starting at $10^1$ Hz and ending at $10^5$ Hz, evenly spaced on a logarithmic scale with 1000 points per decade. (Typically, we

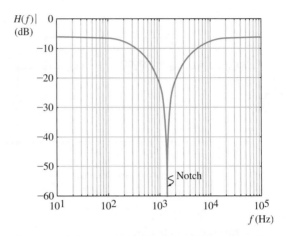

Figure 6.43 Bode plot for Example 6.10 produced using MATLAB.

might start with 100 points per decade, but this transfer function changes very rapidly in the vicinity of 1590 Hz, so we increased the number of points to more accurately determine the location and depth of the notch.)

As a partial check on our analysis and program, we analyze the circuit at $f = 0$ (dc) to determine the transfer function at very low frequencies. To do so, we replace the inductance by a short and the capacitance by an open circuit. Then, the circuit becomes a simple resistive voltage divider consisting of $R_1$, $R_2$, and $R_3$. Therefore, we have

$$H(0) = \frac{\mathbf{V}_{\text{out}}}{\mathbf{V}_{\text{in}}} = \frac{R_3}{R_1 + R_2 + R_3} = 0.5$$

In decibels, this becomes

$$H_{\text{dB}}(0) = 20 \log(0.5) = -6 \text{ dB}$$

which agrees very well with the plotted value at 10 Hz.

For a second check, we replace the capacitance by a short circuit and the inductance by an open circuit to determine the value of the transfer function at very high frequencies. Then, the circuit again becomes a simple resistive voltage divider consisting of $R_1$ and $R_3$. Thus, we have

$$H(\infty) = \frac{R_3}{R_1 + R_3} = 0.5263$$

In decibels, this becomes

$$H_{\text{dB}}(\infty) = 20 \log(0.5263) = -5.575 \text{ dB}$$

which agrees very closely with the value plotted at 100 kHz.    ■

**Exercise 6.22**   If you have access to MATLAB, run the m-file Example_6_10 that is contained in the MATLAB folder.
**Answer**   The resulting plot should be very similar to Figure 6.43.    □

## 6.10   DIGITAL SIGNAL PROCESSING

So far, we have introduced the concepts related to filters in the context of *RLC* circuits. However, many modern systems make use of a more sophisticated technology called **digital signal processing** (DSP). In using DSP to filter a signal, the analog input signal $x(t)$ is converted to digital form (a sequence of numbers) by an **analog-to-digital converter** (ADC). A digital computer then uses the digitized input signal to compute a sequence of values for the output signal. Finally, if desired, the computed values are converted to analog form by a **digital-to-analog converter** (DAC) to produce the output signal $y(t)$. The generic block diagram of a DSP system is shown in Figure 6.44.

Besides filtering, many other operations, such as speech recognition, can be performed by DSP systems. DSP was used in the early days of the Space Telescope to focus blurry images resulting from an error in the telescope's design. High-definition televisions, digital cell phones, and MP3 music players are examples of products that have been made possible by DSP technology.

DSP is a large and rapidly evolving field that will continue to produce novel products. We discuss digital filters very briefly to give you a glimpse of this exciting field.

### Conversion of Signals from Analog to Digital Form

Analog signals are converted to digital form by a DAC in a two-step process. First, the analog signal is sampled (i.e., measured) at periodic points in time. Then, a code word is assigned to represent the approximate value of each sample. Usually, the code words consist of binary symbols. This process is illustrated in Figure 6.45, in which each sample value is represented by a three-bit code word corresponding to the amplitude zone into which the sample falls. Thus, each sample value is converted into a code word, which in turn can be represented by a digital waveform as shown in the figure.

The rate $f_s$ at which a signal must be sampled depends on the frequencies of the signal components. We have seen that all real signals can be considered to consist of sinusoidal components having various frequencies, amplitudes, and phases. If a signal contains no components with frequencies higher than $f_H$, the signal can (in theory) be exactly reconstructed from its samples, provided that the sampling frequency $f_s$ is selected to be more than twice $f_H$:

$$f_s > 2f_H \tag{6.48}$$

For example, high-fidelity audio signals have a highest frequency of about 15 kHz. Therefore, the minimum sampling rate that should be used for audio signals is 30 kHz. Practical considerations dictate a sampling frequency somewhat higher than the theoretical minimum. For instance, audio compact-disc technology converts audio signals to digital form with a sampling rate of 44.1 kHz. Naturally, it is desirable to use the lowest practical sampling rate to minimize the amount of data (in the form of code words) that must be stored or manipulated by the DSP system.

If a signal contains no components with frequencies higher than $f_H$, the signal can be exactly reconstructed from its samples, provided that the sampling rate $f_s$ is selected to be more than twice $f_H$.

Figure 6.44 Generic block diagram of a digital signal-processing (DSP) system.

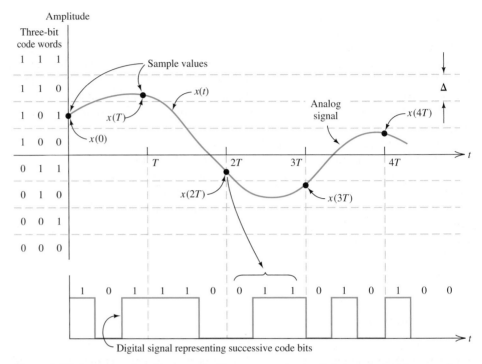

**Figure 6.45** An analog signal is converted to an approximate digital equivalent by sampling. Each sample value is represented by a three-bit code word. (Practical converters use longer code words, and the width Δ of each amplitude zone is much smaller.)

Of course, the interval between samples $T$ is the reciprocal of the sampling rate:

$$T = \frac{1}{f_s} \tag{6.49}$$

A second consideration important in converting analog signals to digital form is the number of amplitude zones to be used. Exact signal amplitudes cannot be represented, because all amplitudes falling into a given zone have the same code word. Thus, when a DAC converts the code words to recreate the original analog waveform, it is possible to reconstruct only an approximation to the original signal with the reconstructed voltage in the middle of each zone, which is illustrated in Figure 6.46. Thus, some **quantization error** exists between the original signal and the reconstruction. This error can be reduced by using a larger number of zones, which requires longer code words. The number $N$ of amplitude zones is related to the number of bits $k$ in a code word by

$$N = 2^k \tag{6.50}$$

Hence, if we are using an 8-bit ($k = 8$) ADC, there are $N = 2^8 = 256$ amplitude zones. In compact-disc technology, 16-bit words are used to represent sample values. With this number of bits, it is very difficult for a listener to detect the effects of quantization error on the reconstructed audio signal.

Often, in engineering instrumentation, we need to determine the DAC specifications needed for converting sensor signals to digital form. For example,

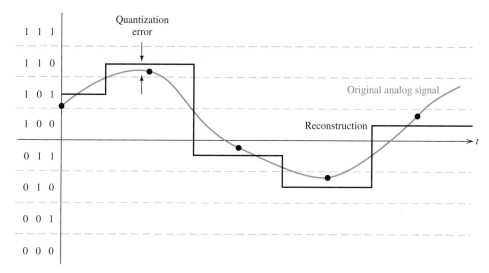

**Figure 6.46** Quantization error occurs when an analog signal is reconstructed from its digital form.

suppose that we need to digitize a signal that ranges from $-1$ to $+1$ V with a resolution of at most $\Delta = 0.5$ mV. ($\Delta$ is illustrated in the upper right-hand corner of Figure 6.45.) Then, the minimum number of zones is the total signal range (2 V) divided by $\Delta$, which yields $N = 4000$. However, $N$ must be an integer power of two. Thus, we require $k = 12$. (In other words, a 12-bit ADC is needed.)

In the remainder of this section, we will ignore quantization error and assume that the exact sample values are available to the digital computer.

## Digital Filters

We have seen that ADCs convert analog signals into sequences of code words that can accurately represent the amplitudes of the signals at the sampling instants. Although the computer actually manipulates code words that represent signal amplitudes, it is convenient to focus on the numbers that the code words represent. Conceptually, the signal $x(t)$ is converted into a list of values $x(nT)$ in which $T$ is the interval between samples and $n$ is a variable that takes on integer values. Often, we omit the sampling period from our notation and write the input and output samples simply as $x(n)$ and $y(n)$, respectively.

## Digital Lowpass Filter

Digital filters can be designed to mimic the $RLC$ filters that we discussed earlier in this chapter. For example, consider the first-order $RC$ lowpass filter shown in Figure 6.47, in which we have denoted the input voltage as $x(t)$ and the output voltage as $y(t)$. Writing a Kirchhoff's current equation at the top node of the capacitance, we have

$$\frac{y(t) - x(t)}{R} + C\frac{dy(t)}{dt} = 0 \qquad (6.51)$$

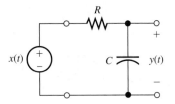

Figure 6.47 First-order *RC* lowpass filter.

Multiplying each term by $R$ and using the fact that the time constant is $\tau = RC$, we find that

$$y(t) - x(t) + \tau\frac{dy(t)}{dt} = 0 \qquad (6.52)$$

We can approximate the derivative as

$$\frac{dy(t)}{dt} \cong \frac{\Delta y}{\Delta t} = \frac{y(n) - y(n-1)}{T} \qquad (6.53)$$

and write the approximate equivalent to the differential equation

$$y(n) - x(n) + \tau\frac{y(n) - y(n-1)}{T} = 0 \qquad (6.54)$$

This type of equation is sometimes called a **difference equation** because it involves differences between successive samples. Solving for the $n$th output value, we have

$$y(n) = ay(n-1) + (1-a)x(n) \qquad (6.55)$$

in which we have defined the parameter

$$a = \frac{\tau/T}{1 + \tau/T} \qquad (6.56)$$

Equation 6.55 defines the calculations that need to be carried out to perform lowpass filtering of the input $x(n)$. For each sample point, the output is $a$ times the previous output value plus $(1-a)$ times the present input value. Usually, we have $\tau \gg T$ and $a$ is slightly less than unity.

---

### Example 6.11    Step Response of a First-Order Digital Lowpass Filter

Compute and plot the input and output samples for $n = 0$ to 20, given $a = 0.9$. The input is a step function defined by

$$x(n) = 0 \text{ for } n < 0$$
$$= 1 \text{ for } n \geq 0$$

Assume that $y(n) = 0$ for $n < 0$.

**Solution**  We have

$$y(0) = ay(-1) + (1-a)x(0) = 0.9 \times 0 + 0.1 \times 1 = 0.1$$
$$y(1) = ay(0) + (1-a)x(1) = 0.19$$
$$y(2) = ay(1) + (1-a)x(2) = 0.271$$

$$\cdots$$

$$y(20) = 0.8906$$

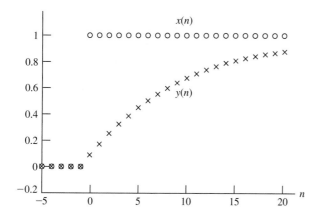

**Figure 6.48** Step input and corresponding output of a first-order digital lowpass filter.

Plots of $x(n)$ and $y(n)$ are shown in Figure 6.48. Notice that the response of the digital filter to a step input is very similar to that of the *RC* filter shown in Figure 4.4 on page 172.  ∎

**Exercise 6.23    a.** Determine the value of the time constant $\tau$, in terms of the sampling interval $T$ corresponding to $a = 0.9$. **b.** Recall that the time constant is the time required for the step response to reach $1 - \exp(-1) = 0.632$ times its final value. Estimate the value of the time constant for the response shown in Figure 6.48.
**Answer    a.** $\tau = 9T$; **b.** $\tau \cong 9T$.  □

## Other Digital Filters

We could develop digital bandpass, notch, or highpass filters that mimic the behavior of the *RLC* filters discussed earlier in this chapter. Furthermore, high-order digital filters are possible. In general, the equations defining such filters are of the form

$$y(n) = \sum_{\ell=1}^{N} a_\ell y(n - \ell) + \sum_{k=0}^{M} b_k x(n - k) \tag{6.57}$$

The type of filter and its performance depend on the values selected for the coefficients $a_\ell$ and $b_k$. For the first-order lowpass filter considered in Example 6.11, the coefficients are $a_1 = 0.9$, $b_0 = 0.1$, and all of the other coefficients are zero.

**Exercise 6.24** Consider the *RC* highpass filter shown in Figure 6.49. Apply the method that we used for the lowpass filter to find an equation having the form of Equation 6.57 for the highpass filter. Give expressions for the coefficients in terms of the time constant $\tau = RC$ and the sampling interval $T$.

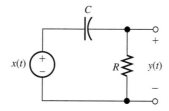

**Figure 6.49** *RC* highpass filter. See Exercise 6.24.

**Answer**    $y(n) = a_1 y(n-1) + b_0 x(n) + b_1 x(n-1)$ in which

$$a_1 = b_0 = -b_1 = \frac{\tau/T}{1 + \tau/T} \qquad \square$$

### A Simple Notch Filter

A simple way to obtain a notch filter is to select $a_\ell = 0$ for all $\ell$, $b_0 = 0.5$, $b_d = 0.5$, and to set the remaining $b_k$ coefficients to zero. Then, the output of the digital filter is given by

$$y(n) = 0.5x(n) + 0.5x(n-d) = 0.5[x(n) + x(n-d)]$$

Thus, each input sample is delayed in time by $Td$ and added to the current sample. Finally, the sum of the input and its delayed version is multiplied by 0.5. To see that this results in a notch filter, consider a sinewave delayed by an interval $Td$. We can write

$$A\cos[\omega(t - Td)] = A\cos(\omega t - \omega Td) = A\cos(\omega t - \theta)$$

Hence, a time delay of $Td$ amounts to a phase shift of $\omega Td$ radians or $fTd \times 360°$. (Keep in mind that, in this discussion, $T$ represents the interval between samples, *not the period of the sinewave*.) For low frequencies, the phase shift is small, so the low-frequency components of $x(n)$ add nearly in phase with those of $x(n-d)$. On the other hand, for the frequency

$$f_{notch} = \frac{1}{2Td} = \frac{f_s}{2d} \qquad (6.58)$$

the phase shift is 180°. Of course, when we phase shift a sinewave by 180° and add it to the original, the sum is zero. Thus, any input component having the frequency $f_{notch}$ does not appear in the output. The first-order lowpass filter and this simple notch filter are just two of many possible digital filters that can be realized by selection of the coefficient values in Equation 6.57.

**Exercise 6.25**    Suppose that the sampling frequency is $f_s = 10$ kHz, and we want to eliminate the 500-Hz component with a simple notch filter. **a.** Determine the value needed for $d$. **b.** What difficulty would be encountered if we wanted to eliminate the 300-Hz component?
**Answer**    **a.** $d = 10$; **b.** Equation 6.58 yields $d = 16.67$, but $d$ is required to be an integer value.    $\square$

### Digital Filter Demonstration

Next, we will use MATLAB to demonstrate the operation of a digital filter. First, we will create samples of a virtual signal including noise and interference. The signal of interest consists of a 1-Hz sinewave and is representative of many types of real world signals such as delta waves contained in the electroencephalogram (EEG) of an individual in deep sleep, or the output of a pressure sensor submerged in the ocean with waves passing over. Part of the interference consists of a 60-Hz sinewave, which is a common real-world problem due to coupling between the ac power line and the signal sensor. The other part of the interference is random noise, which is also common in real-world data.

The MATLAB code that we use to create our simulated data is

```
t = 0:1/6000:2;
signal = cos(2*pi*t);
interference = cos(120*pi*t);
white_noise = randn(size(t));
noise = zeros(size(t));
for n = 2:12001
noise(n) = 0.25*(white_noise(n) - white_noise(n - 1));
end
x = signal + interference + noise; % This is the simulated data.
```

The first command generates a 12,001-element row vector containing the sample times for a two-second interval with a sampling frequency of $f_s = 6000$ Hz. The second and third commands set up row matrices containing samples of the signal and the 60-Hz interference. In the next line, the random-number generator feature of MATLAB generates "white noise" that contains components of equal amplitudes (on average) at all frequencies up to half of the sampling frequency. The white noise is then manipulated by the commands in the for-end loop, producing noise with components from dc to 3000 Hz peaking around 1500 Hz. Then, the signal, interference and noise are added to produce the simulated data $x(n)$. (Of course, in a real-world application, the data are obtained by applying the outputs of sensors, such as EEG electrodes, to analog-to-digital converters.)

Next, we use MATLAB to plot the signal, interference, noise, and the simulated data.

```
subplot(2,2,1)
plot(t, signal)
axis([0 2 -2 2])
subplot(2,2,2)
plot(t, interference)
axis([0 2 -2 2])
subplot(2,2,3)
plot(t,noise)
axis([0 2 -2 2])
subplot(2,2,4)
plot(t,x)
axis([0 2 -3 3])
```

The resulting plots are shown in Figure 6.50. The simulated data is typical of what is often obtained from sensors in real-world experiments. In a biomedical setting, for example, an electrocardiograph produces data that is the sum of the heart signal, 60-Hz power-line interference, and noise from muscle contractions, especially when the subject is moving, as in a stress test.

Actually, the plot of the 60-Hz interference appears a little uneven in Figure 6.50(b) because of finite screen resolution for the display. This is a form of distortion called aliasing that occurs when the sampling rate is too low. If you run the commands on your own computer and use the zoom tool to expand the display horizontally, you will see a smooth plot of the 60-Hz sinewave interference. An m-file named DSPdemo that contains the commands used in this demonstration of a digital filter appears in the MATLAB folder.

What we need is a digital filter that processes the data $x(n)$ of Figure 6.50(d) and produces an output closely matching the signal in Figure 6.50(a). This filter should

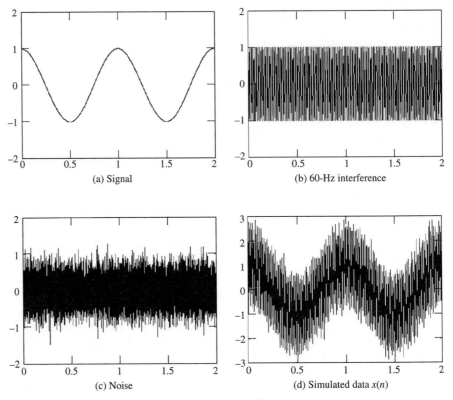

Figure 6.50  Simulated pressure-sensor output and its components.

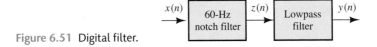

Figure 6.51  Digital filter.

pass the signal (1-Hz sinewave), reject the 60-Hz interference, and reject the noise, which has its largest components in the vicinity of 1500 Hz.

To achieve this, we will use a digital notch filter to remove the 60-Hz sinewave interference cascaded with a lowpass filter to remove most of the noise. The conceptual diagram of the digital filter is shown in Figure 6.51.

Equation 6.58 reveals that by using $d = 50$ and $f_s = 6000$ Hz, we can realize a notch filter with zero gain at precisely 60 Hz. (If 60-Hz interference is a problem, it is a good idea to pick the sampling frequency to be an even integer multiple of 60 Hz, which is one reason we picked the sampling frequency to be 6000 Hz.) The output $z(n)$ of the notch filter is given in terms of the input data $x(n)$ as

$$z(n) = \frac{1}{2}[x(n) + x(n-50)]$$

Also, we need a lowpass filter to eliminate the noise. We decide to use the first-order lowpass filter discussed earlier in this section. Because we do not want the lowpass filter to disturb the signal, we choose its break frequency to much higher than 1 Hz, say $f_B = 50$ Hz. For an $RC$ lowpass filter, the break frequency is

$$f_B = \frac{1}{2\pi RC}$$

Solving for the time constant and substituting values, we have

$$\tau = RC = \frac{1}{2\pi f_B} = \frac{1}{2\pi(50)} = 3.183 \text{ ms}$$

The gain constant for the (approximately) equivalent digital filter is given by Equation 6.56 in which $T = 1/f_s = 1/6000$ s is the sampling interval. We then have

$$a = \frac{\tau/T}{1 + \tau/T} = 0.9503$$

Substituting this value into Equation 6.55 yields the equation for the present $y(n)$ output of the lowpass filter in terms of its input $z(n)$ and previous output $y(n - 1)$.

$$y(n) = 0.9503y(n - 1) + 0.0497z(n)$$

The MATLAB commands to filter the simulated data $x(n)$ and plot the output $y(n)$ are:

```
for n = 51:12001
z(n) = (x(n) + x(n - 50))/2; % This is the notch filter.
end
y = zeros(size(z));
for n = 2:12001
y(n) = 0.9503*y(n-1) + 0.0497*z(n); % This is the lowpass filter.
end
figure
plot(t,y)
```

The resulting plot is shown in Figure 6.52. As desired, the output is nearly identical to the 1-Hz sinewave signal. This relatively simple digital filter has done a very good job of eliminating the noise and interference because most of the noise and the interference have frequencies much higher than does the signal. When the frequencies of the signal are nearer to those of the noise and interference, we would need to resort to higher-order filters.

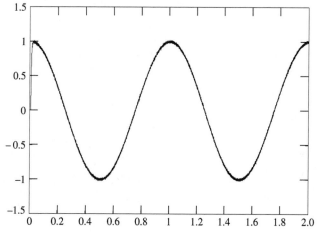

Figure 6.52 Output signal.

## Comparison of Filter Technologies

We have discussed two ways to filter signals: *RLC* circuits and digital filters. There are a number of other filter types, such as **active filters** that are composed of resistances, capacitances and **operational amplifiers**, or **op amps** (which we discuss in Chapter 14). Other filters are based on mechanical resonances in piezoelectric crystals, surface acoustic waves, the propagation of electric fields in wave guides, switched capacitor networks, and transmission lines.

In all cases, the objective of a filter is to separate a desired signal from noise and interference. Radio amateurs operating in the frequency band between 28 and 29.7 MHz often need to place a band reject filter between the transmitter and antenna to eliminate second-harmonic frequency components from reaching the antenna. If they are not removed, second-harmonic components can cause some very annoying interference on their neighbor's television screens. In this application, an *RLC* filter would be the technology of choice because of the large currents and voltages involved.

*The objective of a filter is to separate a desired signal from noise and interference.*

On the other hand, a sleep researcher may wish to filter brain waves to separate delta waves that appear at frequencies of 4 Hz or less from higher frequency brain waves. In this case, a digital filter is appropriate.

In summary, there are many applications for filters and many technologies for implementing filters. Most of the principles we have introduced in our discussion of *RLC* circuits and digital filters apply to filters based on other technologies.

## Summary

1. The fundamental concept of Fourier theory is that we can construct any signal by adding sinusoids with the proper amplitudes, frequencies, and phases.

2. In effect, a filter decomposes the input signal into its sinusoidal components, adjusts the amplitude and phase of each component, depending on its frequency, and sums the adjusted components to produce the output signal. Often, we need a filter that passes components in a given frequency range to the output, without change in amplitude or phase, and that rejects components at other frequencies.

3. The transfer function of a filter circuit is the phasor output divided by the phasor input as a function of frequency. The transfer function is a complex quantity that shows how the amplitudes and phases of input components are affected when passing through the filter.

4. We can use circuit analysis with phasors and complex impedances to determine the transfer function of a given circuit.

5. A first-order filter is characterized by its half-power frequency $f_B$.

6. A transfer-function magnitude is converted to decibels by taking 20 times the common logarithm of the magnitude.

7. Two-port filters are cascaded by connecting the output of the first to the input of the second. The overall transfer function of the cascade is the product of the transfer functions of the individual filters. If the transfer functions are converted to decibels, they are added for a cascade connection.

8. On a logarithmic frequency scale, frequency is multiplied by a given factor for equal increments of length along the axis. A decade is a range of frequencies for which the ratio of the highest frequency to the lowest is 10. An octave is a two-to-one change in frequency.

9. A Bode plot shows the magnitude of a network function in decibels versus frequency, using a logarithmic scale for frequency.

10. The Bode plots for first-order filters can be closely approximated by straight-line asymptotes. In the case of a first-order lowpass filter, the transfer-function magnitude slopes downward at 20 dB/decade for frequencies that are higher than the

half-power frequency. For a first-order highpass filter, the transfer-function magnitude slopes at 20 dB/decade below the break frequency.

11. At low frequencies, inductances behave as short circuits, and capacitances behave as open circuits. At high frequencies, inductances behave as open circuits, and capacitances behave as short circuits. Often, $RLC$ filters can be readily analyzed at low- or high-frequencies, providing checks on computer-generated Bode plots.

12. The key parameters of series and parallel resonant circuits are the resonant frequency and quality factor. The impedance of either type of circuit is purely resistive at the resonant frequency. High-quality-factor circuits can have responses that are much larger in magnitude than the driving source.

13. Filters may be classified as lowpass, highpass, bandpass, and band-reject filters. Ideal filters have constant (nonzero) gain (transfer-function magnitude) in the passband and zero gain in the stopband.

14. The series resonant circuit can be used to form any of the four filter types.

15. A second-order filter is characterized by its resonant frequency and quality factor.

16. MATLAB is useful in deriving and plotting network functions of complex $RLC$ filters.

17. In using DSP to filter a signal, the analog input signal $x(t)$ is converted to digital form (a sequence of numbers) by an ADC. A digital computer uses the digitized input signal to compute a sequence of values for the output signal, and, finally, (if desired) the computed values are converted to analog form by a DAC to produce the output signal $y(t)$.

18. If a signal contains no components with frequencies higher than $f_H$, the signal can be exactly reconstructed from its samples, provided that the sampling rate $f_s$ is selected to be more than twice $f_H$.

19. Approximately equivalent digital filters can be found for $RLC$ filters.

## Problems

### Section 6.1: Fourier Analysis, Filters, and Transfer Functions

**P6.1.** What is the fundamental concept of Fourier theory?

**P6.2.** The triangular waveform shown in Figure P6.2 can be written as the infinite sum

$$v_t(t) = 1 + \frac{8}{\pi^2}\cos(2000\pi t)$$

$$+ \frac{8}{(3\pi)^2}\cos(6000\pi t) + \cdots$$

$$+ \frac{8}{(n\pi)^2}\cos(2000n\pi t) + \cdots$$

in which $n$ takes odd integer values only. Use MATLAB to compute and plot the sum through $n = 19$ for $0 \le t \le 2$ ms. Compare your plot with the waveform shown in Figure P6.2.

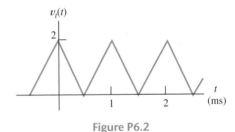

Figure P6.2

**P6.3.** The full-wave rectified cosine wave shown in Figure P6.3 can be written as

$$v_{fw} = \frac{2}{\pi} + \frac{4}{\pi(1)(3)}\cos(4000\pi t)$$

$$- \frac{4}{\pi(3)(5)}\cos(8000\pi t) + \cdots$$

$$+ \frac{4(-1)^{(n/2+1)}}{\pi(n-1)(n+1)}\cos(2000n\pi t) + \cdots$$

---

*Denotes that answers are contained in the Student Solutions files. See Appendix E for more information about accessing the Student Solutions.

in which $n$ assumes even integer values. Use MATLAB to compute and plot the sum through $n = 60$ for $0 \le t \le 2$ ms. Compare your plot with the waveform shown in Figure P6.3.

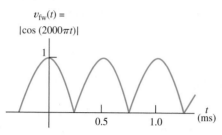

$v_{fw}(t) =$
$|\cos(2000\pi t)|$

Figure P6.3

**P6.4.** The Fourier series for the **half-wave rectified cosine** shown in Figure P6.4 is

$$v_{hw}(t) = \frac{1}{\pi} + \frac{1}{2}\cos(2\pi t) + \frac{2}{\pi(1)(3)}\cos(4\pi t)$$

$$- \frac{2}{\pi(3)(5)}\cos(8\pi t) + \cdots$$

$$+ \frac{2(-1)^{(n/2+1)}}{\pi(n-1)(n+1)}\cos(2n\pi t) + \cdots$$

in which $n = 2, 4, 6$, etc. Use MATLAB to compute and plot the sum through $n = 4$ for $-0.5 \le t \le 1.5$ s. Then plot the sum through $n = 50$. Compare your plots with the waveform in Figure P6.4.

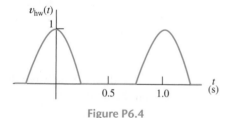

Figure P6.4

**P6.5.** Fourier analysis shows that the **sawtooth waveform** of Figure P6.5 can be written as

$$v_{st}(t) = 1 - \frac{2}{\pi}\sin(2000\pi t)$$

$$- \frac{2}{2\pi}\sin(4000\pi t) - \frac{2}{3\pi}\sin(6000\pi t)$$

$$- \cdots - \frac{2}{n\pi}\sin(2000 n\pi t) - \cdots$$

Use MATLAB to compute and plot the sum through $n = 3$ for $0 \le t \le 2$ ms. Repeat for the sum through $n = 50$.

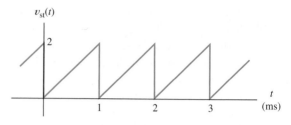

Figure P6.5

**P6.6.** What is the transfer function of a filter? Describe how the transfer function of a filter can be determined using laboratory methods.

**P6.7.** How does a filter process an input signal to produce the output signal in terms of sinusoidal components?

**\*P6.8.** The transfer function $H(f) = \mathbf{V}_{out}/\mathbf{V}_{in}$ of a filter is shown in Figure P6.8. The input signal is given by

$$v_{in}(t) = 5 + 2\cos(5000\pi t + 30°)$$

$$+ 2\cos(15000\pi t)$$

Find an expression (as a function of time) for the steady-state output of the filter.

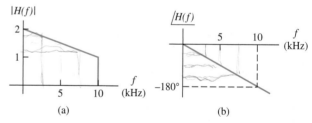

Figure P6.8

**P6.9.** Repeat Problem P6.8 for the input voltage given by

$$v_{in}(t) = 4 + 5\cos(10^4\pi t - 30°)$$

$$+ 2\sin(24000\pi t)$$

**P6.10.** Repeat Problem P6.8 for the input voltage given by

$$v_{in}(t) = 6 + 2\cos(6000\pi t) - 4\cos(12000\pi t)$$

**\*P6.11.** The input to a certain filter is given by

$$v_{in}(t) = 2\cos(10^4\pi t - 25°)$$

and the steady-state output is given by

$$v_{out}(t) = 2\cos(10^4\pi t + 20°)$$

Determine the (complex) value of the transfer function of the filter for $f = 5000$ Hz.

*P6.12. The input and output voltages of a filter operating under sinusoidal steady-state conditions are observed on an oscilloscope. The peak amplitude of the input is 5 V and the output is 15 V. The period of both signals is 4 ms. The input reaches a positive peak at $t = 1$ ms, and the output reaches its positive peak at $t = 1.5$ ms. Determine the frequency and the corresponding value of the transfer function.

*P6.13. The triangular waveform of Problem P6.2 is the input for a filter with the transfer function shown in Figure P6.13. Assume that the phase of the transfer function is zero for all frequencies. Determine the steady-state output of the filter.

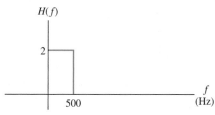

Figure P6.13

*P6.14. Consider a circuit for which the output voltage is the running-time integral of the input voltage, as illustrated in Figure P6.14. If the input voltage is given by $v_{in}(t) = V_{max} \cos(2\pi ft)$, find an expression for the output voltage as a function of time. Then, find an expression for the transfer function of the integrator. Plot the magnitude

and phase of the transfer function versus frequency.

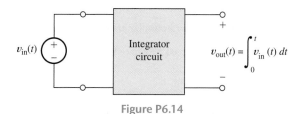

Figure P6.14

P6.15. The sawtooth waveform of Problem P6.5 is applied as the input to a filter with the transfer function shown in Figure P6.15. Assume that the phase of the transfer function is zero for all frequencies. Determine the steady-state output of the filter.

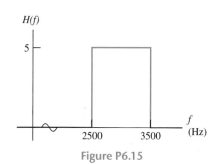

Figure P6.15

P6.16. Figure P6.16 shows the input and output voltages of a certain filter operating in steady state with a sinusoidal input. Determine the frequency and the corresponding value of the transfer function.

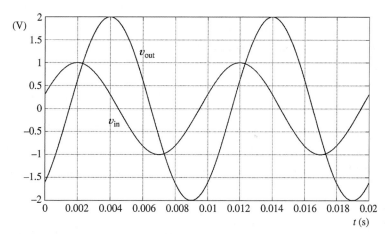

Figure P6.16

**P6.17.** List the frequencies in hertz for which the transfer function of a filter can be determined given that the input to the filter is

$$v_{in}(t) = 2 + 3\cos(1000\pi t) + 3\sin(2000\pi t)$$
$$+ \cos(3000\pi t) \text{ V}$$

and the output is

$$v_{out}(t) = 3 + 2\cos(1000\pi t + 30°)$$
$$+ 3\cos(3000\pi t) \text{ V}$$

Compute the transfer function for each of these frequencies.

**P6.18.** Consider a system for which the output voltage is $v_o(t) = v_{in}(t) + v_{in}(t - 10^{-3})$. (In other words, the output equals the input plus the input delayed by 1 ms.) Given that the input voltage is $v_{in}(t) = V_{max}\cos(2\pi ft)$, find an expression for the output voltage as a function of time. Then, find an expression for the transfer function of the system. Use MATLAB to plot the magnitude of the transfer function versus frequency for the range from 0 to 2000 Hz. Comment on the result.

**P6.19.** Suppose we have a system for which the output voltage is

$$v_o(t) = 1000 \int_{t-10^{-3}}^{t} v_{in}(t)dt$$

Given the input voltage $v_{in}(t) = V_{max}\cos(2\pi ft)$, find an expression for the output voltage as a function of time. Then, find an expression for the transfer function of the system. Use MATLAB to plot the magnitude of the transfer function versus frequency for the range from 0 to 2000 Hz. Comment on the result.

**P6.20.** Suppose we have a circuit for which the output voltage is the time derivative of the input voltage, as illustrated in Figure P6.20. For an input voltage given by $v_{in}(t) = V_{max}\cos(2\pi ft)$, find an expression for the output voltage as a function of time. Then, find an expression for the transfer function of the differentiator. Plot the magnitude and phase of the transfer function versus frequency.

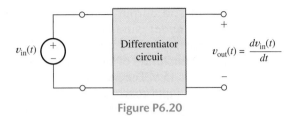

Figure P6.20

## Section 6.2: First-Order Lowpass Filters

**P6.21.** Draw the circuit diagram of a first-order $RC$ lowpass filter and give the expression for the half-power frequency in terms of the circuit components. Sketch the magnitude and phase of the transfer function versus frequency.

**P6.22.** Repeat Problem P6.21 for a first-order $RL$ filter.

**\*P6.23.** Consider a first-order $RC$ lowpass filter. At what frequency (in terms of $f_B$) is the phase shift equal to $-1°$? $-10°$? $-89°$?

**P6.24.** In Chapter 4, we used the time constant to characterize first-order RC circuits. Find the relationship between the half-power frequency and the time constant.

**\*P6.25.** An input signal given by

$$v_{in}(t) = 5\cos(500\pi t) + 5\cos(1000\pi t)$$
$$+ 5\cos(2000\pi t)$$

is applied to the lowpass $RC$ filter shown in Figure P6.25. Find an expression for the output signal.

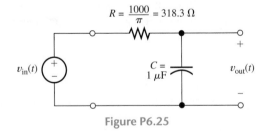

Figure P6.25

**P6.26.** The input signal of a first-order lowpass filter with the transfer function given by Equation 6.9 on page 297 and a half-power frequency of 200 Hz is

$$v_{in}(t) = 3 + 2\sin(800\pi t + 30°)$$
$$+ 5\cos(20 \times 10^3 \pi t)$$

Find an expression for the output voltage.

**P6.27.** Suppose that we need a first-order $RC$ lowpass filter with a half-power frequency of 1 kHz. Determine the value of the capacitance, given that the resistance is 5 k$\Omega$.

**P6.28.** The input signal to a filter contains components that range in frequency from 100 Hz to 50 kHz. We wish to reduce the amplitude of the 50-kHz component by a factor of 200 by passing the signal through a first-order lowpass filter. What half-power frequency is required for the filter? By what factor is a component at 2 kHz changed in amplitude in passing through this filter?

**P6.29.** Suppose we have a first-order lowpass filter that is operating in sinusoidal steady-state conditions at a frequency of 5 kHz. Using an oscilloscope, we observe that the positive-going zero crossing of the output is delayed by 30 $\mu$s compared with that of the input. Determine the break frequency of the filter.

**\*P6.30.** Sketch the magnitude of the transfer function $H(f) = \mathbf{V}_{out}/\mathbf{V}_{in}$ to scale versus frequency for the circuit shown in Figure P6.30. What is the value of the half-power frequency? [*Hint:* Start by finding the Thévenin equivalent circuit seen by the capacitance.]

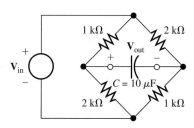

Figure P6.30

**P6.31.** In steady-state operation, a first-order $RC$ lowpass filter has the input signal $v_{in}(t) = 5\cos(20 \times 10^3 \pi t)$ and the output signal $v_{out}(t) = 0.2\cos(20 \times 10^3 \pi t - \theta)$. Determine the break frequency of the filter and the value of $\theta$.

**P6.32.** Consider the circuit shown in Figure P6.32(a). This circuit consists of a source having an internal resistance of $R_s$, an $RC$ lowpass filter, and a load resistance $R_L$. **a.** Show that the transfer function of this circuit is given by

$$H(f) = \frac{\mathbf{V}_{out}}{\mathbf{V}_s} = \frac{R_L}{R_s + R + R_L} \times \frac{1}{1 + j(f/f_B)}$$

in which the half-power frequency $f_B$ is given by

$$f_B = \frac{1}{2\pi R_t C} \quad \text{where} \quad R_t = \frac{R_L (R_s + R)}{R_L + R_s + R}$$

Notice that $R_t$ is the parallel combination of $R_L$ and $(R_s + R)$. [*Hint:* One way to make this problem easier is to rearrange the circuit as shown in Figure P6.32(b) and then to find the Thévenin equivalent for the source and resistances.] **b.** Given that $C = 0.2\ \mu$F, $R_s = 2$ k$\Omega$, $R = 47$ k$\Omega$, and $R_L = 1$ k$\Omega$, sketch (or use MATLAB to plot) the magnitude of $H(f)$ to scale versus $f/f_B$ from 0 to 3.

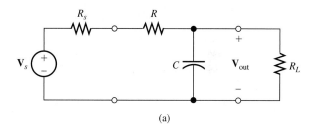

(a)

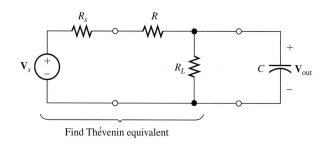

Find Thévenin equivalent

(b)

Figure P6.32

**P6.33.** **a.** Derive an expression for the transfer function $H(f) = \mathbf{V}_{out}/\mathbf{V}_{in}$ for the circuit shown in Figure P6.33. Find an expression for the half-power frequency. **b.** Given $R_1 = 50\ \Omega$, $R_2 = 50\ \Omega$, and $L = 15\ \mu$H, sketch (or use MATLAB to plot) the magnitude of the transfer function versus frequency.

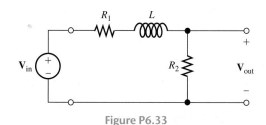

Figure P6.33

**P6.34.** We apply a 5-V-rms 20-kHz sinusoid to the input of a first-order $RC$ lowpass filter, and the output voltage in steady state is 0.5 V rms. Predict the steady-state rms output voltage after the frequency of the input signal is raised to 150 kHz and the amplitude remains constant.

**P6.35.** Perhaps surprisingly, we can apply the transfer-function concept to mechanical systems. Suppose we have a mass $m$ moving through a liquid with an applied force $f$ and velocity $v$. The motion of the mass is described by the first-order differential equation

$$f = m\frac{dv}{dt} + kv$$

in which $k$ is the coefficient of viscous friction. Find an expression for the transfer function

$$H(f) = \frac{\mathbf{V}}{\mathbf{F}}$$

Also, find the half-power frequency (defined as the frequency at which the transfer function magnitude is $1/\sqrt{2}$ times its dc value) in terms of $k$ and $m$. [*Hint:* To determine the transfer function, assume a steady-state sinusoidal velocity $v = V_m \cos(2\pi ft)$, solve for the force, and take the ratio of their phasors.]

**Section 6.3: Decibels, the Cascade Connection, and Logarithmic Frequency Scales**

**P6.36.** What is a logarithmic frequency scale? A linear frequency scale?

**P6.37.** What is a notch filter? What is one application?

**P6.38.** What is the main advantage of converting transfer function magnitudes to decibels before plotting?

**P6.39.** What is the passband of a filter?

**\*P6.40.** **a.** Given $|H(f)|_{dB} = -10$ dB, find $|H(f)|$. **b.** Repeat for $|H(f)|_{dB} = 10$ dB.

**\*P6.41.** **a.** What frequency is halfway between 100 and 3000 Hz on a logarithmic frequency scale? **b.** On a linear frequency scale?

**P6.42.** Find the decibel equivalent for $|H(f)| = 0.5$. Repeat for $|H(f)| = 2, |H(f)| = 1/\sqrt{2} \cong 0.7071$, and $|H(f)| = \sqrt{2}$.

**P6.43.** Find the frequency that is **a.** one octave higher than 800 Hz; **b.** two octaves lower; **c.** two decades lower; **d.** one decade higher.

**P6.44.** Explain what we mean when we say that two filters are cascaded.

**P6.45.** We have a list of successive frequencies 2, $f_1, f_2, f_3,$ 50 Hz. Determine the values of $f_1, f_2,$ and $f_3$ so that the frequencies are evenly spaced on: **a.** a linear frequency scale, and **b.** a logarithmic frequency scale.

**\*P6.46.** Two first-order lowpass filters are in cascade as shown in Figure P6.46. The transfer functions are

$$H_1(f) = H_2(f) = \frac{1}{1 + j(f/f_B)}$$

**a.** Write an expression for the overall transfer function. **b.** Find an expression for the half-power frequency for the overall transfer function in terms of $f_B$.

[Comment: This filter cannot be implemented by cascading two simple $RC$ lowpass filters like the one shown in Figure 6.7 on page 296 because the transfer function of the first circuit is changed when the second is connected. Instead, a buffer amplifier, such as the voltage follower discussed in Section 14.3, must be inserted between the $RC$ filters.]

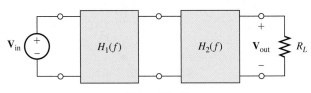

Figure P6.46

**P6.47.** How many decades are between $f_1 = 20$ Hz and $f_2 = 4.5$ kHz? **b.** How many octaves?

**P6.48.** We have two filters with transfer functions $H_1(f)$ and $H_2(f)$ cascaded in the order 1–2. Give the expression for the overall transfer function of the cascade. Repeat if the transfer function magnitudes are expressed in decibels denoted as $|H_1(f)|_{dB}$ and $|H_2(f)|_{dB}$. What caution concerning $H_1(f)$ must be considered?

**P6.49.** Two filters are in cascade. At a given frequency $f_1$, the transfer function values are

$|H_1(f_1)|_{dB} = -30$ and $|H_2(f_1)|_{dB} = +10$. Find the magnitude of the overall transfer function in decibels at $f = f_1$.

### Section 6.4: Bode Plots

**P6.50.** What is a Bode plot?

**P6.51.** What is the slope of the high-frequency asymptote for the Bode magnitude plot for a first-order lowpass filter? The low-frequency asymptote? At what frequency do the asymptotes meet?

**\*P6.52.** A transfer function is given by

$$H(f) = \frac{100}{1 + j(f/1000)}$$

Sketch the asymptotic magnitude and phase Bode plots to scale.

**P6.53.** Suppose that three filters, having identical first-order lowpass transfer functions, are cascaded, what will be the rate at which the overall transfer function magnitude declines above the break frequency? Explain.

**P6.54.** Solve for the transfer function $H(f) = \mathbf{V}_{out}/\mathbf{V}_{in}$ and sketch the asymptotic Bode magnitude and phase plots to scale for the circuit shown in Figure P6.54.

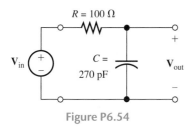

Figure P6.54

**P6.55.** A transfer function is given by

$$H(f) = \frac{10}{1 - j(f/500)}$$

Sketch the asymptotic magnitude and phase Bode plots to scale. What is the value of the half-power frequency?

**P6.56.** Consider a circuit for which

$$v_{out}(t) = v_{in}(t) - 200\pi \int_0^t v_{out}(t)dt$$

**a.** Assume that $v_{out}(t) = A\cos(2\pi ft)$, and find an expression for $v_{in}(t)$. **b.** Use the results of part (a) to find an expression for the transfer function $H(f) = \mathbf{V}_{out}/\mathbf{V}_{in}$ for the system. **c.** Draw the asymptotic Bode plot for the transfer function magnitude.

**P6.57.** Solve for the transfer function $H(f) = \mathbf{V}_{out}/\mathbf{V}_{in}$ and draw the asymptotic Bode magnitude and phase plots for the circuit shown in Figure P6.57.

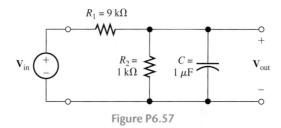

Figure P6.57

**P6.58.** Sketch the asymptotic magnitude and phase Bode plots to scale for the transfer function

$$H(f) = \frac{1 - j(f/100)}{1 + j(f/100)}$$

**P6.59.** Solve for the transfer function $H(f) = \mathbf{V}_{out}/\mathbf{V}_{in}$ and draw the Bode magnitude and phase plots for the circuit shown in Figure P6.59.

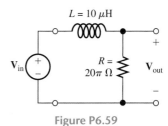

Figure P6.59

**\*P6.60.** In solving Problem P6.14, we find that the transfer function of an integrator circuit is given by $H(f) = 1/(j2\pi f)$. Sketch the Bode magnitude and phase plots to scale. What is the slope of the magnitude plot?

**P6.61.** In solving Problem P6.20, we find that the transfer function of a differentiator circuit is given by $H(f) = j2\pi f$. Sketch the Bode magnitude and phase plots to scale. What is the slope of the magnitude plot?

**Section 6.5:** First-Order Highpass Filters

**P6.62.** Draw the circuit diagram of a first-order $RC$ highpass filter and give the expression for the half-power frequency in terms of the circuit components.

**P6.63.** What is the slope of the high-frequency asymptote for the Bode magnitude plot for a first-order highpass filter? The low-frequency asymptote? At what frequency do the asymptotes meet?

**\*P6.64.** Consider the circuit shown in Figure P6.64. Sketch the asymptotic Bode magnitude and phase plots to scale for the transfer function $H(f) = \mathbf{V}_{out}/\mathbf{V}_{in}$.

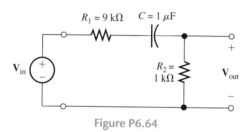

Figure P6.64

**\*P6.65.** Consider the first-order highpass filter shown in Figure P6.65. The input signal is given by

$$v_{in}(t) = 5 + 5\cos(2000\pi t)$$

Find an expression for the output $v_{out}(t)$ in steady-state conditions.

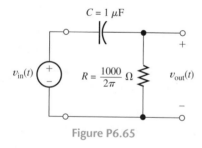

Figure P6.65

**P6.66.** Repeat Problem P6.65 for the input signal given by

$$v_{in}(t) = 10\cos(400\pi t) + 20\cos(4000\pi t)$$

**P6.67.** Suppose we need a first-order highpass filter (such as Figure 6.19 on page 309) to attenuate a 60-Hz input component by 60 dB. What value is required for the break frequency

of the filter? By how many dB is the 600-Hz component attenuated by this filter? If $R = 5\,k\Omega$, what is the value of $C$?

**P6.68.** Consider the circuit shown in Figure P6.68. Sketch the Bode magnitude and phase plots to scale for the transfer function $H(f) = \mathbf{V}_{out}/\mathbf{V}_{in}$.

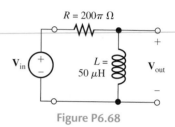

Figure P6.68

**P6.69.** Consider the circuit shown in Figure P6.69. Sketch the Bode magnitude and phase plots to scale for the transfer function $H(f) = \mathbf{V}_{out}/\mathbf{V}_{in}$.

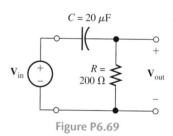

Figure P6.69

**Section 6.6:** Series Resonance

**P6.70.** What can you say about the impedance of a series $RLC$ circuit at the resonant frequency? How are the resonant frequency and the quality factor defined?

**P6.71.** What is a *bandpass filter*? How is its bandwidth defined?

**\*P6.72.** Consider the series resonant circuit shown in Figure P6.72, with $L = 20\,\mu H$, $R = 14.14\,\Omega$, and $C = 1000\,pF$. Compute the resonant frequency, the bandwidth, and the half-power frequencies. Assuming that the frequency of the source is the same as the resonant frequency, find the phasor voltages across the elements and sketch a phasor diagram.

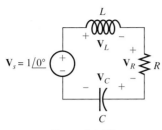

Figure P6.72

**P6.73.** Work Problem P6.72 for $L = 80 \, \mu\text{H}$, $R = 14.14 \, \Omega$, and $C = 1000 \, \text{pF}$.

**P6.74.** Suppose we have a series resonant circuit for which $B = 30 \, \text{kHz}$, $f_0 = 300 \, \text{kHz}$, and $R = 40 \, \Omega$. Determine the values of $L$ and $C$.

**\*P6.75.** At the resonant frequency $f_0 = 1 \, \text{MHz}$, a series resonant circuit with $R = 50 \, \Omega$ has $|\mathbf{V}_R| = 2 \, \text{V}$ and $|\mathbf{V}_L| = 20 \, \text{V}$. Determine the values of $L$ and $C$. What is the value of $|\mathbf{V}_C|$?

**P6.76.** Suppose we have a series resonant circuit for which $f_0 = 12 \, \text{MHz}$ and $B = 600 \, \text{kHz}$. Furthermore, the minimum value of the impedance magnitude is $20 \, \Omega$. Determine the values of $R$, $L$, and $C$.

**P6.77.** Derive an expression for the resonant frequency of the circuit shown in Figure P6.77. (Recall that we have defined the resonant frequency to be the frequency for which the impedance is purely resistive.)

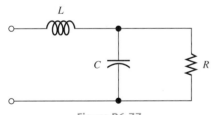

Figure P6.77

### Section 6.7: Parallel Resonance

**P6.78.** What can you say about the impedance of a parallel $RLC$ circuit at the resonant frequency? How is the resonant frequency defined? Compare the definition of quality factor for the parallel resonant circuit with that for the series resonant circuit.

**\*P6.79.** A parallel resonant circuit has $R = 5 \, \text{k}\Omega$, $L = 50 \, \mu\text{H}$, and $C = 200 \, \text{pF}$. Determine the resonant frequency, quality factor, and bandwidth.

**P6.80.** A parallel resonant circuit has $f_0 = 20 \, \text{MHz}$ and $B = 200 \, \text{kHz}$. The maximum value of $|Z_p|$ is $5 \, \text{k}\Omega$. Determine the values of $R$, $L$, and $C$.

**P6.81.** Consider the parallel resonant circuit shown in Figure 6.29 on page 319. Determine the $L$ and $C$ values, given $R = 1 \, \text{k}\Omega$, $f_0 = 10 \, \text{MHz}$, and $B = 500 \, \text{kHz}$. If $\mathbf{I} = 10^{-3} \angle 0°$, draw a phasor diagram showing the currents through each of the elements in the circuit at resonance.

**P6.82.** A parallel resonant circuit has $f_0 = 100 \, \text{MHz}$, $B = 5 \, \text{MHz}$, and $R = 2 \, \text{k}\Omega$. Determine the values of $L$ and $C$.

### Section 6.8: Ideal and Second-Order Filters

**P6.83.** Name four types of ideal filters and sketch their transfer functions.

**\*P6.84.** An ideal bandpass filter has cutoff frequencies of 9 and 11 kHz and a gain magnitude of two in the passband. Sketch the transfer-function magnitude to scale versus frequency. Repeat for an ideal band-reject filter.

**P6.85.** An ideal lowpass filter has a cutoff frequency of 10 kHz and a gain magnitude of two in the passband. Sketch the transfer-function magnitude to scale versus frequency. Repeat for an ideal highpass filter.

**P6.86.** Each AM radio signal has components ranging from 10 kHz below its carrier frequency to 10 kHz above its carrier frequency. Various radio stations in a given geographical region are assigned different carrier frequencies so that the frequency ranges of the signals do not overlap. Suppose that a certain AM radio transmitter has a carrier frequency of 980 kHz. What type of filter should be used if we want the filter to pass the components from this transmitter and reject the components of all other transmitters? What are the best values for the cutoff frequencies?

**P6.87.** In an electrocardiograph, the heart signals contain components with frequencies ranging from dc to 100 Hz. During exercise on a treadmill, the signal obtained from the electrodes also contains noise generated by muscle contractions. Most of the noise components have frequencies exceeding 100 Hz. What type of filter should be used to reduce the noise? What cutoff frequency is appropriate?

**\*P6.88.** Draw the circuit diagram of a second-order highpass filter. Suppose that $R = 1\ k\Omega$, $Q_s = 1$, and $f_0 = 100\ kHz$. Determine the values of $L$ and $C$.

**P6.89.** Draw the circuit diagram of a second-order highpass filter. Given that $R = 50\ \Omega$, $Q_s = 0.5$, and $f_0 = 30\ MHz$, determine the values of $L$ and $C$.

**P6.90.** Suppose that sinewave interference has been inadvertently added to an audio signal that has frequency components ranging from 20 Hz to 15 kHz. The frequency of the interference slowly varies in the range 950 to 1050 Hz. A filter that attenuates the interference by at least 20 dB and passes most of the audio components is desired. What type of filter is needed? Sketch the magnitude Bode plot of a suitable filter, labeling its specifications.

**Section 6.9: Transfer Functions and Bode Plots with MATLAB**

**P6.91.** Consider the filter shown in Figure P6.91. **a.** Derive an expression for the transfer function $H(f) = \mathbf{V}_{out}/\mathbf{V}_{in}$. **b.** Use MATLAB to obtain a Bode plot of the transfer-function magnitude for $R_1 = 9\ k\Omega$, $R_2 = 1\ k\Omega$, and $C = 0.01\ \mu F$. Allow frequency to range from 10 Hz to 1 MHz. **c.** At very low frequencies, the capacitance becomes an open circuit. In this case, determine an expression for the transfer function and evaluate for the circuit parameters of part (b). Does the result agree with the value plotted in part (b)? **d.** At very high frequencies, the capacitance becomes a short circuit. In this case, determine an

expression for the transfer function and evaluate for the circuit parameters of part (b). Does the result agree with the value plotted in part (b)?

**P6.92.** Repeat Problem P6.91 for the circuit of Figure P6.92.

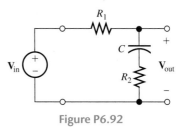

Figure P6.92

**P6.93.** Suppose that we need a filter with the Bode plot shown in Figure P6.93(a). We decide to cascade a highpass circuit and a lowpass circuit as shown in Figure P6.93(b). So that the second (i.e., right-hand) circuit looks like an approximate open circuit across the output of the first (i.e., left-hand) circuit, we choose $R_2 = 100R_1$. **a.** Which of the components form the lowpass filter? Which form the

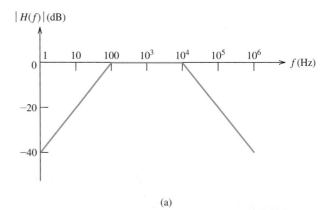

(a)

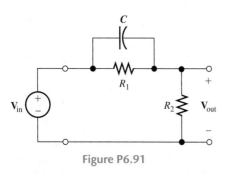

Figure P6.91

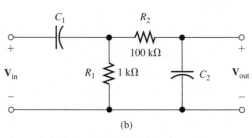

(b)

Figure P6.93

highpass filter? **b.** Compute the capacitances needed to achieve the desired break frequencies, making the approximation that the left-hand circuit has an open-circuit load. **c.** Write expressions that can be used to compute the exact transfer function $H(f) = \mathbf{V}_{out}/\mathbf{V}_{in}$ and use MATLAB to produce a Bode magnitude plot for $f$ ranging from 1 Hz to 1 MHz. The result should be a close approximation to the desired plot shown in Figure P6.93(a).

**P6.94.** Suppose that we need a filter with the Bode plot shown in Figure P6.93(a). We decide to cascade a highpass circuit and a lowpass circuit, as shown in Figure P6.94. So that the second (i.e., right-hand) circuit looks like an approximate open circuit across the output of the first (i.e., left-hand) circuit, we choose $C_2 = C_1/100$. **a.** Which of the components form the lowpass filter? Which form the highpass filter? **b.** Compute the resistances needed to achieve the desired break frequencies, making the approximation that the left-hand circuit has an open-circuit load. **c.** Write expressions that can be used to compute the exact transfer function $H(f) = \mathbf{V}_{out}/\mathbf{V}_{in}$ and use MATLAB to produce a Bode magnitude plot for $f$ ranging from 1 Hz to 1 MHz. The result should be a close approximation to the desired plot shown in Figure P6.93(a).

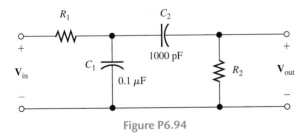

Figure P6.94

**P6.95.** Other combinations of $R$, $L$, and $C$ have behaviors similar to that of the series resonant circuit. For example, consider the circuit shown in Figure P6.95. **a.** Derive an expression for the resonant frequency of this circuit. (We have defined the resonant frequency to be the frequency for which the impedance is purely resistive.) **b.** Compute

the resonant frequency, given $L = 1$ mH, $R = 1000\ \Omega$, and $C = 0.25\ \mu\text{F}$. **c.** Use MATLAB to obtain a plot of the impedance magnitude of this circuit for $f$ ranging from 95 to 105 percent of the resonant frequency. Compare the result with that of a series $RLC$ circuit.

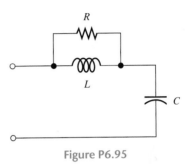

Figure P6.95

**P6.96.** Consider the circuit of Figure P6.77 with $R = 1\ \text{k}\Omega$, $L = 1$ mH, and $C = 0.25\ \mu\text{F}$. **a.** Using MATLAB, obtain a plot of the impedance magnitude of this circuit for $f$ ranging from 9 to 11 kHz. **b.** From the plot, determine the minimum impedance, the frequency at which the impedance is minimum, and the bandwidth (i.e., the band of frequencies for which the impedance is less than $\sqrt{2}$ times the minimum value). **c.** Determine the component values for a series $RLC$ circuit having the same parameters as those found in part (b). **d.** Plot the impedance magnitude of the series circuit on the same axes as the plot for part (a).

**P6.97.** Other combinations of $R$, $L$, and $C$ have behaviors similiar to that of the parallel circuit. For example, consider the circuit shown in Figure P6.97. **a.** Derive an expression for the resonant frequency of this circuit. (We have defined the resonant frequency to be the frequency for which the impedance is purely resistive. However, in this case you may find the algebra easier if you work with admittances.) **b.** Compute the resonant frequency, given $L = 1$ mH, $R = 1\ \Omega$, and $C = 0.25\ \mu\text{F}$. **c.** Use MATLAB to obtain a plot of the impedance magnitude of this circuit for $f$ ranging from 95 to 105 percent of the resonant frequency. Compare the result with that of a parallel $RLC$ circuit.

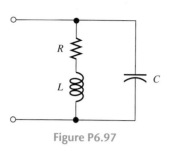

Figure P6.97

**P6.98.** Consider the filter shown in Figure P6.98. **a.** Derive an expression for the transfer function $H(f) = \mathbf{V}_{out}/\mathbf{V}_{in}$. **b.** Use MATLAB to obtain a Bode plot of the transfer function magnitude for $R = 10\ \Omega$, $L = 10\ \text{mH}$, and $C = 0.02533\ \mu\text{F}$. Allow frequency to range from 1 kHz to 100 kHz. **c.** At very low frequencies, the capacitance becomes an open circuit and the inductance becomes a short circuit. In this case, determine an expression for the transfer function and evaluate for the circuit parameters of part (b). Does the result agree with the value plotted in part (b)? **d.** At very high frequencies, the capacitance becomes a short circuit and the inductance becomes an open circuit. In this case, determine an expression for the transfer function and evaluate for the circuit parameters of part (b). Does the result agree with the value plotted in part (b)?

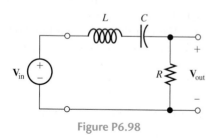

Figure P6.98

**P6.99.** Repeat Problem P6.98 for the circuit of Figure P6.99.

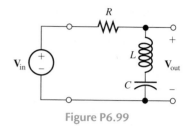

Figure P6.99

**Section 6.10:** **Digital Signal Processing**

**P6.100.** Develop a digital filter that mimics the action of the $RL$ filter shown in Figure P6.100. Determine expressions for the coefficients in terms of the time constant and sampling interval $T$. [*Hint:* If your circuit equation contains an integral, differentiate with respect to time to obtain a pure differential equation.] **b.** Given $R = 10\ \Omega$ and $L = 200\ \text{mH}$, sketch the step response of the circuit to scale. **c.** Use MATLAB to determine and plot the step response of the digital filter for several time constants. Use the time constant of part (b) and $f_s = 500\ \text{Hz}$. Compare the results of parts (b) and (c).

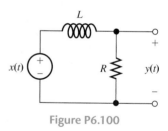

Figure P6.100

**P6.101.** Repeat Problem P6.100 for the filter shown in Figure P6.101.

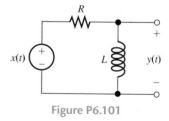

Figure P6.101

**\*P6.102.** Consider the second-order bandpass filter shown in Figure P6.102. **a.** Derive expressions for $L$ and $C$ in terms of the resonant frequency $\omega_0$ and quality factor $Q_s$. **b.** Write the KVL equation for the circuit and use it to develop a digital filter that mimics the action of the $RLC$ filter. Use the results of part (a) to write the coefficients in terms of the resonant frequency $\omega_0$, circuit quality factor $Q_s$, and sampling interval $T$. [*Hint:* The circuit equation contains an integral, so differentiate with respect to time to obtain a pure differential equation.]

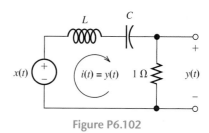

**Figure P6.102**

## Practice Test

Here is a practice test you can use to check your comprehension of the most important concepts in this chapter. Answers can be found in Appendix D and complete solutions are included in the Student Solutions files. See Appendix E for more information about the Student Solutions.

**T6.1.** What is the basic concept of Fourier theory as it relates to real-world signals? How does the transfer function of a filter relate to this concept?

**T6.2.** An input signal given by

$$v_{in}(t) = 3 + 4\cos(1000\pi t)$$
$$+ 5\cos(2000\pi t - 30°)$$

is applied to the $RL$ filter shown in Figure T6.2. Find the expression for the output signal $v_{out}(t)$.

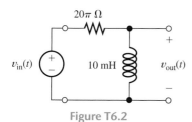

**Figure T6.2**

**T6.3.** Consider the Bode magnitude plot for the transfer function of a certain filter given by

$$H(f) = \frac{\mathbf{V}_{out}}{\mathbf{V}_{in}} = 50\frac{j(f/200)}{1 + j(f/200)}$$

**a.** What is the slope of the low-frequency asymptote?

**b.** What is the slope of the high-frequency asymptote?

**c.** What are the coordinates of the point at which the asymptotes meet?

**d.** What type of filter is this?

**e.** What is the value of the break frequency?

**T6.4.** A series resonant circuit has $R = 5\,\Omega$, $L = 20\,mH$, and $C = 1\,\mu F$. Determine the values of:

**a.** the resonant frequency in Hz.

**b.** $Q$.

**c.** bandwidth in Hz.

**d.** the impedance of the circuit at the resonant frequency.

**e.** the impedance of the circuit at dc.

**f.** the impedance of the circuit as the frequency approaches infinity.

**T6.5.** Repeat question T6.4 for a parallel resonant circuit with $R = 10\,k\Omega$, $L = 1\,mH$, and $C = 1000\,pF$.

**T6.6.** We have a signal consisting of voice conversations and music with frequency components from about 30 Hz to 8 kHz plus a loud sinusoidal tone of 800 Hz. Specify the type of ideal filter and cutoff frequencies if

**a.** we want nearly all of the voice and music components to pass through the filter with the 800 Hz tone eliminated, so we can monitor the conversations better.

**b.** we want to eliminate nearly all of the voice and music components and pass the 800 Hz tone through the filter so we can monitor slow variations in its amplitude, which can give information about movement of the persons speaking.

**T6.7.** Consider the transfer function $\mathbf{V}_{out}/\mathbf{V}_{in}$ for each of the circuits shown in Figure T6.7. Classify each circuit as a first-order lowpass filter, second-order bandpass filter, etc. Justify your answers.

**T6.8.** Give a list of MATLAB commands to produce the magnitude Bode plot for the transfer function of question T6.3 for frequency ranging from 10 Hz to 10 kHz.

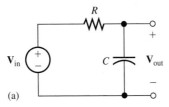

(a)

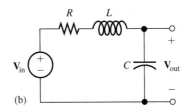

(b)

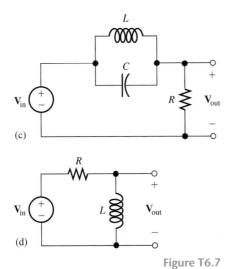

(c)

(d)

Figure T6.7

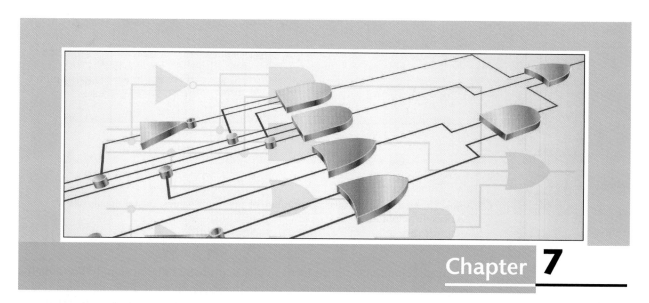

# Chapter 7

# Logic Circuits

## Study of this chapter will enable you to:

- State the advantages of digital technology over analog technology.
- Understand the terminology of digital circuits.
- Convert numbers between decimal, binary, and other forms.
- Use the Gray code for position and angular sensors.
- Understand the binary arithmetic operations used in computers and other digital systems.

- Interconnect logic gates of various types to implement a given logic function.
- Use Karnaugh maps to minimize the number of gates needed to implement a logic function.
- Understand how gates are connected together to form flip-flops and registers.

## Introduction to this chapter:

So far, we have considered circuits, such as filters, that process analog signals. For an **analog signal**, each amplitude in a continuous range has a unique significance. For example, a position sensor may produce an analog signal that is proportional to displacement. Each amplitude represents a different position. An analog signal is shown in Figure 7.1(a).

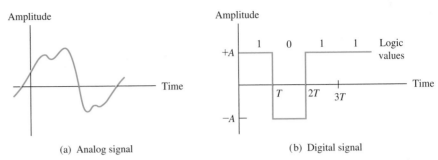

Figure 7.1 Analog signals take a continuum of amplitude values. Digital signals take a few discrete amplitudes.

In this chapter, we introduce circuits that process digital signals. For a **digital signal**, only a few restricted ranges of amplitude are allowed, and each amplitude in a given range has the same significance. Most common are **binary** signals that take on amplitudes in only two ranges, and the information associated with the ranges is represented by the **logic values** 1 or 0. An example of a digital signal is shown in Figure 7.1(b). Computers are examples of digital circuits. We will see that digital approaches have some important advantages over analog approaches.

## 7.1   BASIC LOGIC CIRCUIT CONCEPTS

We often encounter analog signals in instrumentation of physical systems. For example, a pressure transducer can yield a voltage that is proportional to pressure versus time in the cylinder of an internal combustion engine. In Section 6.10, we saw that analog signals can be converted to equivalent digital signals that contain virtually the same information. Then, computers or other digital circuits can be used to process this information. In many applications, we have a choice between digital and analog approaches.

### Advantages of the Digital Approach

Provided that the noise amplitude is not too large, the logic values represented by a digital signal can still be determined after noise is added.

Digital signals have several important advantages over analog signals. After noise is added to an analog signal, it is usually impossible to determine the precise amplitude of the original signal. On the other hand, after noise is added to a digital signal, we can still determine the logic values—provided that the noise amplitude is not too large. This is illustrated in Figure 7.2.

For a given type of logic circuit, one range of voltages represents logic 1, and another range of voltages represents logic 0. For proper operation, a logic circuit only needs to produce a voltage somewhere in the correct range. Thus, component values in digital circuits do not need to be as precise as in analog circuits.

With modern IC technology, it is possible to manufacture exceedingly complex digital circuits economically.

It turns out that with modern integrated-circuit (IC) manufacturing technology, very complex digital logic circuits (containing millions of components) can be produced economically. Analog circuits often call for large capacitances and precise component values that are impossible to manufacture as large-scale ICs. Thus, digital systems have become increasingly important in the past few decades, a trend that will continue.

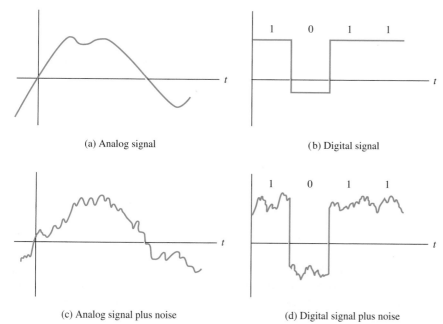

(a) Analog signal

(b) Digital signal

(c) Analog signal plus noise

(d) Digital signal plus noise

**Figure 7.2** The information (logic values) represented by a digital signal can still be determined precisely after noise is added. Noise obscures the information contained in an analog signal because the original amplitude cannot be determined exactly after noise is added.

## Positive versus Negative Logic

Usually, the higher amplitude in a binary system represents 1 and the lower-amplitude range represents 0. In this case, we say that we have **positive logic**. On the other hand, it is possible to represent 1 by the lower amplitude and 0 by the higher amplitude, resulting in **negative logic**. Unless stated otherwise, we assume positive logic throughout this book.

The logic value 1 is also called **high**, **true**, or **on**. Logic 0 is also called **low**, **false**, or **off**. Signals in logic systems switch between high and low as the information being represented changes. We denote these signals, or **logic variables**, by uppercase letters such as $A$, $B$, and $C$.

## Logic Ranges and Noise Margins

Logic circuits are typically designed so that a range of input voltages is accepted as logic 1 and another nonoverlapping range of voltages is accepted as logic 0. The input voltage accepted as logic 0, or low, is denoted as $V_{IL}$, and the smallest input voltage accepted as logic 1, or high, is denoted as $V_{IH}$. This is illustrated in Figure 7.3. No meaning is assigned to voltages between $V_{IL}$ and $V_{IH}$, which normally occur only during transitions.

Furthermore, the circuits are designed so that the output voltages fall into narrower ranges than the inputs (provided that the inputs are in the acceptable ranges). This is also illustrated in Figure 7.3. $V_{OL}$ is the highest logic-0 output voltage, and $V_{OH}$ is the lowest logic-1 output voltage.

For a logic signal, one range of amplitudes represents logic 1, a nonoverlapping range represents logic 0, and no meaning is assigned to the remaining amplitudes, which ordinarily do not occur or occur only during transitions.

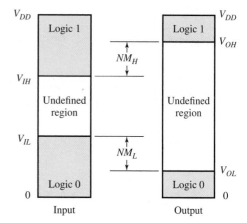

**Figure 7.3** Voltage ranges for logic-circuit inputs and outputs.

Because noise can be added to a logic signal in the interconnections between outputs and inputs, it is important that the outputs have narrower ranges than the acceptable inputs. The differences are called **noise margins** and are given by

$$NM_L = V_{IL} - V_{OL}$$
$$NM_H = V_{OH} - V_{IH}$$

Ideally, noise margins are as large as possible.

### Digital Words

A single binary digit, called a **bit**, represents a very small amount of information. For example, a logic variable $R$ could be used to represent whether or not it is raining in a particular location (say $R = 1$ if it is raining, and $R = 0$ if it is not raining).

To represent more information, we resort to using groups of logic variables called **digital words**. For example, the word $RWS$ could be formed, in which $R$ represents rain, $W$ is 1 if the wind velocity is greater than 15 miles per hour and 0 for less wind, and $S$ could be 1 for sunny conditions and 0 for cloudy. Then the digital word 110 would tell us that it is rainy, windy, and cloudy. A **byte** is a word consisting of eight bits, and a **nibble** is a four-bit word.

### Transmission of Digital Information

In **parallel transmission**, an $n$-bit word is transferred on $n + 1$ wires, one wire for each bit, plus a common or ground wire. On the other hand, in **serial transmission**, the successive bits of the word are transferred one after the other with a single pair of wires. At the receiving end, the bits are collected and combined into words. Parallel transmission is faster and often used for short distances, such as internal data transfer in a computer. Long-distance digital communication systems are usually serial.

### Examples of Digital Information-Processing Systems

By using a prearranged 100-bit word consisting of logic values and binary numbers, we could give a rather precise report of weather conditions at a given location. Computers, such as those used by the National Weather Bureau, process words received from various weather stations to produce contour maps of temperature, wind velocity, cloud state, precipitation, and so on. These maps are useful in understanding and predicting weather patterns.

Analog signals can be reconstructed from their periodic samples (i.e., measurements of instantaneous amplitude at uniformly spaced points in time), provided that the sampling rate is high enough. Each amplitude value can be represented as a digital word. Thus, an analog signal can be represented by a sequence of digital words. In playback, the digital words are converted to the corresponding analog amplitudes. This is the principle of the compact-disc recording technique.

Thus, electronic circuits can gather, store, transmit, and process information in digital form to produce results that are useful or pleasing.

## 7.2   REPRESENTATION OF NUMERICAL DATA IN BINARY FORM

### Binary Numbers

Digital words can represent numerical data. First, consider the decimal (base 10) number 743.2. We interpret this number as

$$7 \times 10^2 + 4 \times 10^1 + 3 \times 10^0 + 2 \times 10^{-1}$$

Similarly, the binary or base-two number 1101.1 is interpreted as

$$1 \times 2^3 + 1 \times 2^2 + 0 \times 2^1 + 1 \times 2^0 + 1 \times 2^{-1} = 13.5$$

Because digital circuits are (almost always) designed to operate with only two symbols, 0 or 1, it is necessary to represent numerical and other data as words composed of 0s and 1s.

Hence, the binary number 1101.1 is equivalent to the decimal number 13.5. Where confusion seems likely to occur, we use a subscript to distinguish binary numbers (such as $1101.1_2$) from decimal numbers (such as $13.5_{10}$).

With three bits, we can form $2^3$ distinct words. These words can represent the decimal integers 0 through 7 as shown:

| | |
|---|---|
| 000 | 0 |
| 001 | 1 |
| 010 | 2 |
| 011 | 3 |
| 100 | 4 |
| 101 | 5 |
| 110 | 6 |
| 111 | 7 |

Similarly, a four-bit word has 16 combinations that represent the integers 0 through 15. (Frequently, we include leading zeros in discussing binary numbers in a digital circuit, because circuits are usually designed to operate on fixed-length words and the circuit produces the leading zeros.)

### Conversion of Decimal Numbers to Binary Form

To convert a decimal integer to binary, we repeatedly divide by two until the quotient is zero. Then, the remainders read in reverse order give the binary form of the number.

|  | | Quotient | Remainder | |
|---|---|---|---|---|
| 343/2 | = | 171 | 1 | → $101010111_2$ |
| 171/2 | = | 85 | 1 | |
| 85/2 | = | 42 | 1 | |
| 42/2 | = | 21 | 0 | Read binary equivalent |
| 21/2 | = | 10 | 1 | in reverse order |
| 10/2 | = | 5 | 0 | |
| 5/2 | = | 2 | 1 | |
| 2/2 | = | 1 | 0 | |
| 1/2 | = | 0 | 1 | |

Stop when quotient equals zero

Figure 7.4  Conversion of $343_{10}$ to binary form.

### Example 7.1    Converting a Decimal Integer to Binary

Convert the decimal integer $343_{10}$ to binary.

Solution    The operations are shown in Figure 7.4. The decimal number is repeatedly divided by two. When the quotient reaches zero, we stop. Then, the binary equivalent is read as the remainders in reverse order. From the figure, we see that

$$343_{10} = 101010111_2$$

To convert decimal fractions to binary fractions, we repeatedly multiply the fractional part by two and retain the whole parts of the results as the successive bits of the binary fraction.

### Example 7.2    Converting a Decimal Fraction to Binary

Convert $0.392_{10}$ to its closest six-bit binary equivalent.

Solution    The conversion is illustrated in Figure 7.5. The fractional part of the number is repeatedly multiplied by two. The whole part of each product is retained as a bit of the binary equivalent. We stop when the desired degree of precision has been reached. Thus, from the figure, we have

$$0.392_{10} \cong 0.011001_2$$

To convert a decimal number having both whole and fractional parts, we convert each part separately and then combine the parts.

| | | | | |
|---|---|---|---|---|
| $2 \times 0.392$ | = | 0 | + | 0.784 |
| $2 \times 0.784$ | = | 1 | + | 0.568 |
| $2 \times 0.568$ | = | 1 | + | 0.136 |
| $2 \times 0.136$ | = | 0 | + | 0.272 |
| $2 \times 0.272$ | = | 0 | + | 0.544 |
| $2 \times 0.544$ | = | 1 | + | 0.088 |

$0.011001_2$    (approximate binary equivalent)

Figure 7.5  Conversion of $0.392_{10}$ to binary.

**Example 7.3** | **Converting Decimal Values to Binary**

Convert $343.392_{10}$ to binary.

**Solution**   From Examples 7.1 and 7.2, we have

$$343_{10} = 101010111_2$$

*Convert the whole and fractional parts of the number separately and combine the results.*

and

$$0.392_{10} \cong 0.011001_2$$

Combining these results, we get

$$343.392_{10} \cong 101010111.011001_2 \qquad ■$$

**Example 7.4** | **Converting Binary Numbers to Decimal**

Convert the binary number $10011.011_2$ to decimal.

**Solution**   We have

$$10011.011_2 = 1 \times 2^4 + 0 \times 2^3 + 0 \times 2^2 + 1 \times 2^1 + 1 \times 2^0 + 0 \times 2^{-1}$$
$$+ 1 \times 2^{-2} + 1 \times 2^{-3} = 19.375_{10} \qquad ■$$

**Exercise 7.1**   Convert the following numbers to binary form, stopping after you have found six bits (if necessary) for the fractional part: **a.** 23.75; **b.** 17.25; **c.** 4.3.
**Answer**   **a.** 10111.11; **b.** 10001.01; **c.** 100.010011.  □

**Exercise 7.2**   Convert the following to decimal equivalents: **a.** $1101.111_2$; **b.** $100.001_2$.
**Answer**   **a.** $13.875_{10}$; **b.** $4.125_{10}$.  □

## Binary Arithmetic

We add binary numbers in much the same way that we add decimal numbers, except that the rules of addition are different (and much simpler). The rules for binary addition are shown in Figure 7.6.

|  |  | Sum | Carry |
|---|---|---|---|
| $0 + 0$ | $=$ | 0 | 0 |
| $0 + 1$ | $=$ | 1 | 0 |
| $1 + 1$ | $=$ | 0 | 1 |
| $1 + 1 + 1$ | $=$ | 1 | 1 |

**Figure 7.6** Rules of binary addition.

**Example 7.5** | **Adding Binary Numbers**

Add the binary numbers 1000.111 and 1100.011.

**Solution**   See Figure 7.7.  ■

## Hexadecimal and Octal Numbers

Binary numbers are inconvenient for humans because it takes many bits to write large numbers (or fractions to a high degree of precision). Hexadecimal (base 16) and octal (base 8) numbers are easily converted to and from binary numbers. Furthermore, they are much more efficient than binary numbers in representing information.

```
0001 11  ←— Carries
  1000.111
+1100.011
 10101.010
```

**Figure 7.7** Addition of binary numbers.

**Table 7.1.** Symbols for Octal and Hexadecimal Numbers and Their Binary Equivalents

| Octal | | Hexadecimal | |
|---|---|---|---|
| 0 | 000 | 0 | 0000 |
| 1 | 001 | 1 | 0001 |
| 2 | 010 | 2 | 0010 |
| 3 | 011 | 3 | 0011 |
| 4 | 100 | 4 | 0100 |
| 5 | 101 | 5 | 0101 |
| 6 | 110 | 6 | 0110 |
| 7 | 111 | 7 | 0111 |
| | | 8 | 1000 |
| | | 9 | 1001 |
| | | A | 1010 |
| | | B | 1011 |
| | | C | 1100 |
| | | D | 1101 |
| | | E | 1110 |
| | | F | 1111 |

Table 7.1 shows the symbols used for hexadecimal and octal numbers and their binary equivalents. Notice that we need 16 symbols for the digits of a hexadecimal number. Customarily, the letters A through F are used to represent the digits for 10 through 15.

**Example 7.6**    Converting Octal Numbers to Decimal

Convert the octal number $173.21_8$ to decimal.

**Solution**    We have

$$173.21_8 = 1 \times 8^2 + 7 \times 8^1 + 3 \times 8^0 + 2 \times 8^{-1} + 1 \times 8^{-2} = 123.265625_{10}$$

■

**Example 7.7**    Converting Hexadecimal Numbers to Decimal

Convert the hexadecimal number $1FA.2A_{16}$ to decimal.

**Solution**    We have

$$1FA.2A_{16} = 1 \times 16^2 + 15 \times 16^1 + 10 \times 16^0 + 2 \times 16^{-1} + 10 \times 16^{-2}$$
$$= 506.1640625_{10}$$

■

We can convert an octal or hexadecimal number to binary simply by substituting the binary equivalents for each digit.

**Example 7.8**    Converting Octal and Hexadecimal Numbers to Binary

Convert the numbers $317.2_8$ and $F3A.2_{16}$ to binary.

**Solution**    We simply use Table 7.1 to replace each digit by its binary equivalent. Thus, we have

In converting an octal or hexadecimal number to binary, use Table 7.1 to replace each digit by its binary equivalent.

$$317.2_8 = 011\ 001\ 111.010_2$$
$$= 011001111.010_2$$

and

$$F3A.2_{16} = 1111\ 0011\ 1010.0010$$
$$= 111100111010.0010_2 \quad \blacksquare$$

In converting binary numbers to octal, we first arrange the bits in groups of three, starting from the binary point and working outward. If necessary, we insert leading or trailing zeros to complete the groups. Then, we convert each group of three bits to its octal equivalent. Conversion to hexadecimal uses the same approach, except that the binary number is arranged in groups of four bits.

| Example 7.9 | Converting Binary Numbers to Octal or Hexadecimal |
|---|---|

Convert $11110110.1_2$ to octal and to hexadecimal.

**Solution**    For conversion to octal, we first form three-bit groups, working outward from the binary point:

$$11110110.1_2 = 011\ 110\ 110.100$$

Notice that we have appended leading and trailing zeros so that each group contains three bits. Next, we write the octal digit for each group. Thus, we have

$$11110110.1_2 = 011\ 110\ 110.100 = 366.4_8$$

For conversion to hexadecimal, we form four-bit groups appending leading and trailing zeros as needed. Then, we convert each group to its equivalent hexadecimal integer, yielding

$$11110110.1_2 = 1111\ 0110.1000 = F6.8_{16} \quad \blacksquare$$

Working both directions from the binary point, group the bits into three-(octal) or four-(hexadecimal) bit words. Add leading or trailing zeros to complete the groups. Then, use Table 7.1 to replace each binary word by the corresponding symbol.

**Exercise 7.3**    Convert the following numbers to binary, octal, and hexadecimal forms: **a.** $97_{10}$; **b.** $229_{10}$.
**Answer    a.** $97_{10} = 1100001_2 = 141_8 = 61_{16}$; **b.** $229_{10} = 11100101_2 = 345_8 = E5_{16}$. □

**Exercise 7.4**    Convert the following numbers to binary form: **a.** $72_8$; **b.** $FA6_{16}$.
**Answer    a.** $111010_2$; **b.** $111110100110_2$. □

## Binary-Coded Decimal Format

Sometimes, decimal numbers are represented in binary form simply by writing the four-bit equivalents for each digit. The resulting numbers are said to be in **binary-coded decimal** (BCD) format. For example, 93.2 becomes

In converting a decimal number to BCD, each digit is replaced by its four-bit equivalent.

$$93.2 = 1001\ 0011.0010_{BCD}$$

Code groups 1010, 1011, 1100, 1101, 1110, and 1111 do not occur in BCD (unless an error has occurred). Calculators frequently represent numbers internally in BCD format. As each key is pressed, the BCD code group is stored. The operation

$$9 \times 3 = 27$$

would appear in BCD format as

$$1001 \times 0011 = 0010\ 0111$$

Even though binary code words are used to represent the decimal integers, the operations inside a calculator are partly decimal in nature. On the other hand, calculations are often carried out in true binary fashion in computers.

**Exercise 7.5**   Express $197_{10}$ in BCD form.
**Answer**   $197_{10} = 000110010111_{BCD}$.                                            □

## Gray Code

Consider a transducer for encoding the position of a robot arm in which black-and-white bands are placed on the arm as illustrated in Figure 7.8. In the figure, we assume that the bands are read by light-sensitive diodes in which a black band is converted to logic 1 and a white band is converted to logic 0.

A problem occurs if the successive positions are represented by the binary code shown in Figure 7.8(a). For example, when the arm moves from the position represented by 0011 to that of 0100, three bits of the code word must change. Suppose that because the photodiode sensors are not perfectly aligned, 0011 first changes to 0001, then to 0000, and finally to 0100. During this transition, the indicated position is far from the actual position.

A better scheme for coding the positions is to use the **Gray code** shown in Figure 7.8(b). *In a Gray code, each code word differs in only one bit from its neighboring code words.* Thus, erroneous position indications are avoided during transitions. Gray codes of any desired length can be constructed as shown in Figure 7.9. (Notice that successive code words differ in a single bit.)

> In a Gray code, each word differs in only one bit from each of its adjacent words.

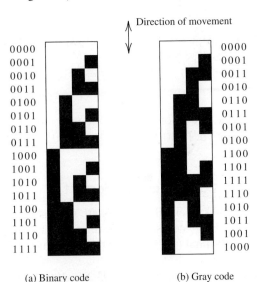

**Figure 7.8** Black and white bands that can be read by a photodiode array resulting in a digital word representing the position of a robot arm.

(a) Binary code          (b) Gray code

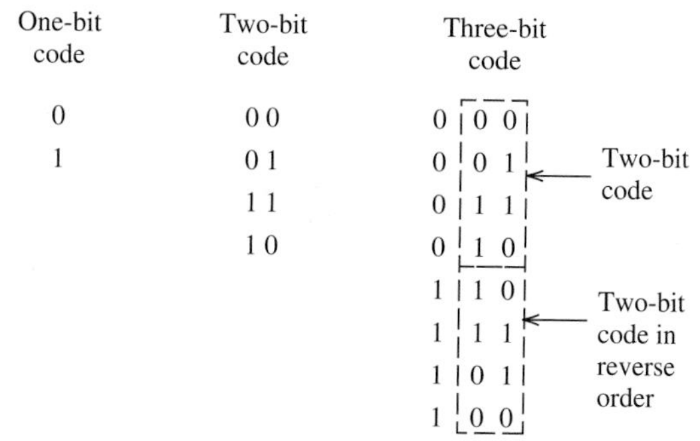

| One-bit code | Two-bit code | Three-bit code |
|---|---|---|
| 0 | 0 0 | 0 &#124; 0 0 &#124; |
| 1 | 0 1 | 0 &#124; 0 1 &#124;  ← Two-bit code |
|  | 1 1 | 0 &#124; 1 1 &#124; |
|  | 1 0 | 0 &#124; 1 0 &#124; |
|  |  | 1 &#124; 1 0 &#124; |
|  |  | 1 &#124; 1 1 &#124;  ← Two-bit code in reverse order |
|  |  | 1 &#124; 0 1 &#124; |
|  |  | 1 &#124; 0 0 &#124; |

**Figure 7.9** A one-bit Gray code simply consists of the two words 0 and 1. To create a Gray code of length $n$, repeat the code of length $n - 1$ in reverse order, place a 0 on the left-hand side of each word in the first half of the list, and place a 1 on the left-hand side of each word in the second half of the list.

Gray codes are also used to encode the angular position of rotating shafts. Then, the last word in the list wraps around and is adjacent to the first word. To represent angular position with a resolution better than 1 degree, we would need a nine-bit Gray code consisting of $2^9 = 512$ words. Each word would represent an angular sector of $360/512 = 0.703$ degree in width.

**Exercise 7.6**  We wish to represent the position of a robot arm with a resolution of 0.01 inch or better. The range of motion is 20 inches. How many bits are required for a Gray code that can represent the arm position?
**Answer**   11 bits. □

## Complement Arithmetic

The **one's complement** of a binary number is obtained by replacing 1s by 0s, and vice versa. For example, an eight-bit binary number and its one's complement are

$$01001101$$

$$10110010 \text{ (one's complement)}$$

The **two's complement** of a binary number is obtained by adding 1 to the one's complement, neglecting the carry (if any) out of the most significant bit (MSB). For example, to find the two's complement of

$$01001100$$

we first form the one's complement, which is

$$10110011$$

and then add 1. This is illustrated in Figure 7.10(a).

Another way to obtain the two's complement of a number is to copy the number—working from right to left—until after the first 1 is copied. Then, the remaining bits are inverted. An example of this process is shown in Figure 7.10(b).

Complements are useful for representing negative numbers and performing subtraction in computers. Furthermore, the use of complement arithmetic simplifies the design of digital computers. Most common is the **signed two's-complement** representation, in which the first bit is taken as the sign bit. If the number is positive, the first bit is 0, whereas if the number is negative, the first bit is 1. Negative numbers are represented as the two's complement of the corresponding positive number.

$$10110011 \quad \text{One's complement}$$
$$\underline{\phantom{1011001}+1}$$
$$10110100 \quad \text{Two's complement}$$

(a) First find the one's complement, then add 1; neglect the carry (if it occurs) out of the most significant bit

$$01001100 \quad \text{Number}$$
$$\underbrace{\phantom{0100}}_{\text{Invert}} \underbrace{\phantom{1100}}_{\text{Copy}}$$
$$10110100 \quad \text{Two's complement}$$

(b) Working from right to left, copy bits until after the first 1 is copied; then invert the remaining bits

**Figure 7.10** Two ways to find the two's complement of the binary number 01001100.

| | |
|---|---|
| +127 | 01111111 |
| | ... |
| +2 | 00000010 |
| +1 | 00000001 |
| 0 | 00000000 |
| −1 | 11111111 |
| −2 | 11111110 |
| | ... |
| −128 | 10000000 |

Sign bit

**Figure 7.11** Signed two's-complement representation using eight-bit words.

Figure 7.11 shows the signed two's-complement representation using eight bits. In this case, the range of numbers that can be represented runs from $-128$ to $+127$. Of course, if longer words are used, the range is extended.

Subtraction is performed by first finding the two's complement of the subtrahend and then adding in binary fashion and ignoring any carry out of the sign bit.

---

**Example 7.10**    Subtraction Using Two's-Complement Arithmetic

Perform the operation $29_{10} - 27_{10}$ by using eight-bit signed two's-complement arithmetic.

**Solution**    First, we convert $29_{10}$ and $27_{10}$ to binary form. This yields

$$29_{10} = 00011101$$

and

$$27_{10} = 00011011$$

Next, we find the two's complement of the subtrahend:

$$-27_{10} = 11100101$$

Finally, we add the numbers to find the result:

$$
\begin{array}{rr}
00011101 & 29 \\
+\ 11100101 & +(-27) \\
\hline
\text{ignore carry out of sign bit} \rightarrow 00000010 & 2
\end{array}
$$

∎

Of course, performing addition and subtraction in this manner is tedious for humans. However, computers excel at performing simple operations rapidly and accurately.

In performing two's-complement arithmetic, we must be aware of the possibility of **overflow** in which the result exceeds the maximum value that can be represented by the word length in use. For example, if we use eight-bit words to add

$$97_{10} = 01100001$$

and

$$63_{10} = 00111111$$

we obtain

$$01100001$$
$$+\underline{00111111}$$
$$10100000$$

The result is the signed two's-complement representation for $-96$, rather than the correct answer, which is $97 + 63 = 160$. This error occurs because the signed two's-complement representation has a maximum value of $+127$ (assuming eight-bit words).

Similarly, **underflow** occurs if the result of an arithmetic operation is less than $-128$. Overflow and underflow are not possible if the two numbers to be added have opposite signs. If the two numbers to be added have the same sign and the result has the opposite sign, underflow or overflow has occurred.

If the two numbers to be added have the same sign and the result has the opposite sign, overflow or underflow has occurred.

**Exercise 7.7**   Find the eight-bit signed two's-complement representation of **a.** $22_{10}$ and **b.** $-30_{10}$.
**Answer   a.** 00010110; **b.** 11100010. ☐

**Exercise 7.8**   Carry out $19_{10} - 4_{10}$ in eight-bit signed two's-complement form.
**Answer**

$$
\begin{array}{rl}
19 & 00010011 \\
+\underline{(-4)} & +\underline{11111100} \\
15 & 00001111
\end{array}
$$   ☐

## 7.3   COMBINATORIAL LOGIC CIRCUITS

In this section, we consider circuits called **logic gates** that combine several logic-variable inputs to produce a logic-variable output. We focus on the external behavior of logic gates. Later, in Chapter 11, we will see how gate circuits can be implemented with field-effect transistors.

The circuits that we are about to discuss are said to be **memoryless** because their output values at a given instant depend only on the input values at that instant. Later, we consider logic circuits that are said to possess **memory**, because their present output values depend on previous, as well as present, input values.

### AND Gate

An important logic function is called the AND operation. The AND operation on two logic variables, $A$ and $B$, is represented as $AB$, read as "$A$ and $B$." The AND operation is also called **logical multiplication**.

One way to specify a combinatorial logic system is to list all the possible combinations of the input variables and the corresponding output values. Such a listing is called a **truth table**. The truth table for the AND operation of two variables is shown in Figure 7.12(a). Notice that $AB$ is 1 if and only if $A$ and $B$ are both 1.

For the AND operation, we can write the following relations:

$$AA = A \qquad\qquad (7.1)$$

| A | B | C = AB |
|---|---|--------|
| 0 | 0 | 0 |
| 0 | 1 | 0 |
| 1 | 0 | 0 |
| 1 | 1 | 1 |

(a) Truth table

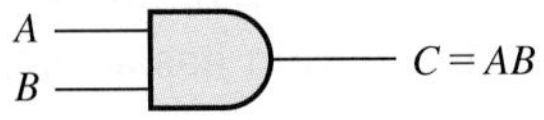

$C = AB$

(b) Symbol for two-input AND gate

**Figure 7.12** Two-input AND gate.

| A | B | C | D = ABC |
|---|---|---|---------|
| 0 | 0 | 0 | 0 |
| 0 | 0 | 1 | 0 |
| 0 | 1 | 0 | 0 |
| 0 | 1 | 1 | 0 |
| 1 | 0 | 0 | 0 |
| 1 | 0 | 1 | 0 |
| 1 | 1 | 0 | 0 |
| 1 | 1 | 1 | 1 |

(a) Truth table

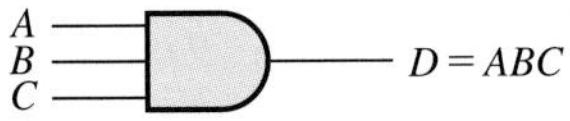

$D = ABC$

(b) Symbol for three-input AND gate

**Figure 7.13** Three-input AND gate.

$$A1 = A \tag{7.2}$$

$$A0 = 0 \tag{7.3}$$

$$AB = BA \tag{7.4}$$

$$A(BC) = (AB)C = ABC \tag{7.5}$$

The circuit symbol for a two-input AND gate (i.e., a circuit that produces an output equal to the AND operation of the inputs) is shown in Figure 7.12(b).

It is possible to have AND gates with more than two inputs. For example, the truth table and circuit symbol for a three-input AND gate are shown in Figure 7.13.

## Logic Inverter

The NOT operation on a logic variable is represented by placing a bar over the symbol for the logic variable. The symbol $\bar{A}$ is read as "not $A$" or as "$A$ inverse." If $A$ is 0, $\bar{A}$ is 1, and vice versa.

Circuits that perform the NOT operation are called **inverters**. The truth table and circuit symbol for an inverter are shown in Figure 7.14. The *bubble* placed at the output of the inverter symbol is used to indicate inversion.

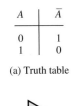

| A | $\bar{A}$ |
|---|-----------|
| 0 | 1 |
| 1 | 0 |

(a) Truth table

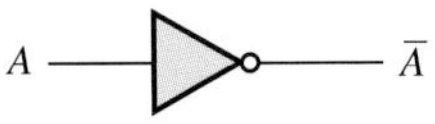

(b) Symbol for an inverter

**Figure 7.14** Logical inverter.

We can readily establish the following operations for the NOT operation:

$$A\overline{A} = 0 \qquad (7.6)$$

$$\overline{\overline{A}} = A \qquad (7.7)$$

## OR Gate

The OR operation of logic variables is written as $A + B$, which is read as "$A$ or $B$." The truth table and the circuit symbol for a two-input OR gate are shown in Figure 7.15. Notice that $A + B$ is 1 if $A$ or $B$ (or both) are 1. The OR operation is also called **logical addition**. The truth table and circuit symbol for a three-input OR gate are shown in Figure 7.16. For the OR operation, we can write

$$(A + B) + C = A + (B + C) = A + B + C \qquad (7.8)$$

$$A(B + C) = AB + AC \qquad (7.9)$$

$$A + 0 = A \qquad (7.10)$$

$$A + 1 = 1 \qquad (7.11)$$

$$A + \overline{A} = 1 \qquad (7.12)$$

$$A + A = A \qquad (7.13)$$

| A | B | C = A + B |
|---|---|---|
| 0 | 0 | 0 |
| 0 | 1 | 1 |
| 1 | 0 | 1 |
| 1 | 1 | 1 |

(a) Truth table

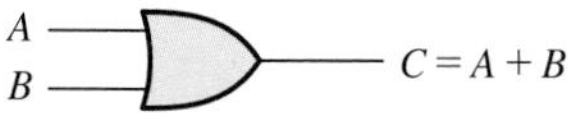

(b) Symbol for two-input OR gate

**Figure 7.15** Two-input OR gate.

| A | B | C | D = A + B + C |
|---|---|---|---|
| 0 | 0 | 0 | 0 |
| 0 | 0 | 1 | 1 |
| 0 | 1 | 0 | 1 |
| 0 | 1 | 1 | 1 |
| 1 | 0 | 0 | 1 |
| 1 | 0 | 1 | 1 |
| 1 | 1 | 0 | 1 |
| 1 | 1 | 1 | 1 |

(a) Truth table

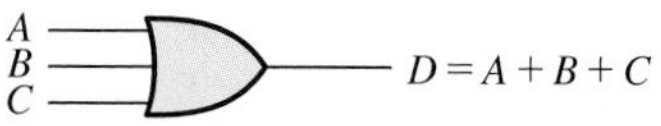

(b) Circuit symbol

**Figure 7.16** Three-input OR gate.

## Boolean Algebra

Equation Equation 7.13 illustrates that even though we use the addition sign (+) to represent the OR operation, manipulation of logic variables by the AND, OR, and NOT operations is different from ordinary algebra. The mathematical theory of logic variables is called **Boolean algebra**, named for mathematician George Boole.

One way to prove a Boolean algebra identity is to produce a truth table that lists all possible combinations of the variables and to show that both sides of the expression yield the same results.

---

**Example 7.11**    Using a Truth Table to Prove a Boolean Expression

Prove the associative law for the OR operation (Equation 7.8), which states that

$$(A + B) + C = A + (B + C)$$

Solution    The truth table listing all possible combinations of the variables and the values of both sides of Equation 7.8 is shown in Table 7.2. We can see from the truth table that $A + (B + C)$ and $(A + B) + C$ take the same logic values for all combinations of $A$, $B$, and $C$. Because both expressions yield the same results, the parentheses are not necessary, and we can write

$$A + (B + C) = (A + B) + C = A + B + C$$    ∎

---

**Exercise 7.9**    Use truth tables to prove Equations 7.5 and 7.9.
**Answer**    See Tables 7.3 and 7.4.    ☐

---

**Table 7.2** Truth Table Used to Prove the Associative Law for the OR Operation (Equation 7.8)

| $A$ | $B$ | $C$ | $(A + B)$ | $(B + C)$ | $A + (B + C)$ | $(A + B) + C$ | $A + B + C$ |
|---|---|---|---|---|---|---|---|
| 0 | 0 | 0 | 0 | 0 | 0 | 0 | 0 |
| 0 | 0 | 1 | 0 | 1 | 1 | 1 | 1 |
| 0 | 1 | 0 | 1 | 1 | 1 | 1 | 1 |
| 0 | 1 | 1 | 1 | 1 | 1 | 1 | 1 |
| 1 | 0 | 0 | 1 | 0 | 1 | 1 | 1 |
| 1 | 0 | 1 | 1 | 1 | 1 | 1 | 1 |
| 1 | 1 | 0 | 1 | 1 | 1 | 1 | 1 |
| 1 | 1 | 1 | 1 | 1 | 1 | 1 | 1 |

**Table 7.3** Truth Table Used to Prove That $A(BC) = (AB)C$ (Equation 7.5)

| $A$ | $B$ | $C$ | $(AB)$ | $(BC)$ | $(AB)C$ | $A(BC)$ |
|---|---|---|---|---|---|---|
| 0 | 0 | 0 | 0 | 0 | 0 | 0 |
| 0 | 0 | 1 | 0 | 0 | 0 | 0 |
| 0 | 1 | 0 | 0 | 0 | 0 | 0 |
| 0 | 1 | 1 | 0 | 1 | 0 | 0 |
| 1 | 0 | 0 | 0 | 0 | 0 | 0 |
| 1 | 0 | 1 | 0 | 0 | 0 | 0 |
| 1 | 1 | 0 | 1 | 0 | 0 | 0 |
| 1 | 1 | 1 | 1 | 1 | 1 | 1 |

**Table 7.4** Truth Table Used to Prove That $A(B + C) = AB + AC$ (Equation 7.9)

| A | B | C | (B + C) | AB | AC | AB + AC | A(B + C) |
|---|---|---|---------|----|----|---------|----------|
| 0 | 0 | 0 | 0 | 0 | 0 | 0 | 0 |
| 0 | 0 | 1 | 1 | 0 | 0 | 0 | 0 |
| 0 | 1 | 0 | 1 | 0 | 0 | 0 | 0 |
| 0 | 1 | 1 | 1 | 0 | 0 | 0 | 0 |
| 1 | 0 | 0 | 0 | 0 | 0 | 0 | 0 |
| 1 | 0 | 1 | 1 | 0 | 1 | 1 | 1 |
| 1 | 1 | 0 | 1 | 1 | 0 | 1 | 1 |
| 1 | 1 | 1 | 1 | 1 | 1 | 1 | 1 |

**Table 7.5** Truth Table for $D = AB + C$

| A | B | C | AB | D = AB + C |
|---|---|---|----|-----------|
| 0 | 0 | 0 | 0 | 0 |
| 0 | 0 | 1 | 0 | 1 |
| 0 | 1 | 0 | 0 | 0 |
| 0 | 1 | 1 | 0 | 1 |
| 1 | 0 | 0 | 0 | 0 |
| 1 | 0 | 1 | 0 | 1 |
| 1 | 1 | 0 | 1 | 1 |
| 1 | 1 | 1 | 1 | 1 |

**Exercise 7.10** Prepare a truth table for the logic expression $D = AB + C$.
**Answer** See Table 7.5. □

## Implementation of Boolean Expressions

Boolean algebra expressions can be implemented by interconnection of AND gates, OR gates, and inverters. For example, the logic expression

$$F = A\overline{B}C + ABC + (C + D)(\overline{D} + E) \qquad (7.14)$$

can be implemented by the logic circuit shown in Figure 7.17.

Boolean algebra expressions can be implemented by interconnection of AND gates, OR gates, and inverters.

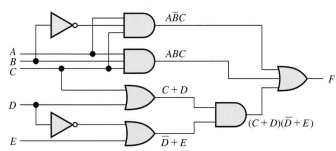

**Figure 7.17** A circuit that implements the logic expression $F = A\overline{B}C + ABC + (C + D)(\overline{D} + E)$.

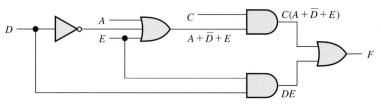

Figure 7.18  A simpler circuit equivalent to that of Figure 7.17.

Sometimes, we can manipulate a logic expression to find an equivalent expression that is simpler. For example, the last term on the right-hand side of Equation 7.14 can be expanded, resulting in

$$F = A\overline{B}C + ABC + C\overline{D} + CE + D\overline{D} + DE \tag{7.15}$$

But the term $D\overline{D}$ always has the logic value 0, so it can be dropped from the expression. Factoring the first two terms on the right-hand side of Equation 7.15 results in

$$F = AC(\overline{B} + B) + C\overline{D} + CE + DE \tag{7.16}$$

However, the quantity $\overline{B} + B$ always equals 1, so we can write

$$F = AC + C\overline{D} + CE + DE \tag{7.17}$$

Factoring $C$ from the first three terms on the right-hand side, we have

$$F = C(A + \overline{D} + E) + DE \tag{7.18}$$

This can be implemented as shown in Figure 7.18.

We can often find alternative implementations for a given logic function.

Thus, we can often find alternative implementations for a given. Later, we consider methods for finding the implementation using the fewest gates of a given type.

## De Morgan's Laws

Two important results in Boolean algebra are De Morgan's laws, which are given by

$$\overline{AB} = \overline{A} + \overline{B} \tag{7.19}$$

and

$$\overline{A + B} = \overline{A}\,\overline{B} \tag{7.20}$$

De Morgan's laws can be extended to three variables as follows:

$$\overline{ABC} = \overline{A} + \overline{B} + \overline{C} \quad \text{and} \quad \overline{A + B + C} = \overline{A}\,\overline{B}\,\overline{C}$$

Another way to state these laws is as follows: If the variables in a logic expression are replaced by their inverses, the AND operation is replaced by OR, the OR operation is replaced by AND, and the entire expression is inverted, the resulting logic expression yields the same values as before the changes.

Thus, De Morgan's laws can be used to change any (or all) of the AND operations in a logic expression to OR operations and vice versa. By changing various combinations of the operations, a variety of equivalent expressions can be found.

---

**Example 7.12**   **Applying De Morgan's Laws**

Apply De Morgan's laws, changing all of the ORs to ANDs and all of the ANDs to ORs, to the right-hand side of the logic expression:

$$D = AC + \overline{B}C + \overline{A}(\overline{B} + BC)$$

**Solution**   First, we replace each variable by its inverse, resulting in the expression

$$\overline{A}\,\overline{C} + B\overline{C} + A(B + \overline{B}\,\overline{C})$$

Then, we replace the AND operation by OR, and vice versa:

$$(\overline{A} + \overline{C})(B + \overline{C})[A + B(\overline{B} + \overline{C})]$$

Finally, inverting the expression, we can write

$$D = \overline{(\overline{A} + \overline{C})(B + \overline{C})[A + B(\overline{B} + \overline{C})]}$$

Therefore, De Morgan's laws give us alternative ways to write logic expressions. ∎

---

**Exercise 7.11**   Use De Morgan's laws to find alternative expressions for

$$D = AB + \overline{B}C$$

and

$$E = \overline{[F(G + \overline{H}) + F\overline{G}]}$$

**Answer**

$$D = \overline{(\overline{A} + \overline{B})(B + \overline{C})}$$

$$E = (\overline{F} + \overline{G}H)(\overline{F} + G)$$  □

An important implication of De Morgan's laws is that *we can implement any logic function by using AND gates and inverters.* This is true because Equation 7.20 can be employed to replace the OR operation by the AND operation (and logical inversions).

Similarly, *any logic function can be implemented with OR gates and inverters*, because Equation 7.19 can be used to replace the AND operation with the OR operation (and logical inversions). Consequently, to implement a logic function, we need inverters and either AND gates or OR gates, not both.

Any combinatorial logic function can be implemented solely with AND gates and inverters.

Any combinatorial logic function can be implemented solely with OR gates and inverters.

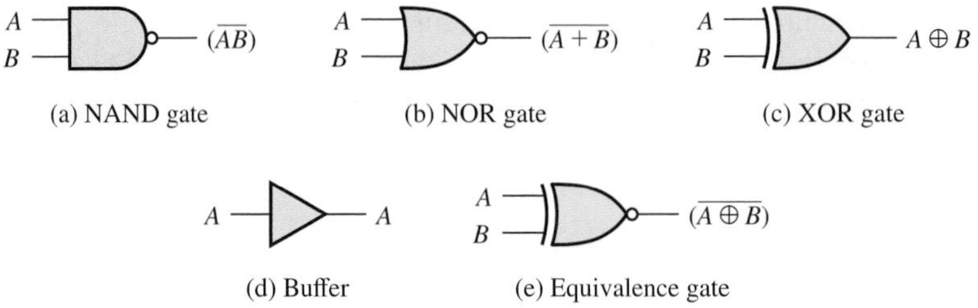

(a) NAND gate          (b) NOR gate          (c) XOR gate

(d) Buffer          (e) Equivalence gate

**Figure 7.19** Additional logic-gate symbols.

## NAND, NOR, and XOR Gates

Some additional logic gates are shown in Figure 7.19. The NAND gate is equivalent to an AND gate followed by an inverter. Notice that the symbol is the same as for an AND gate, with a bubble at the output terminal to indicate that the output has been inverted after the AND operation. Similarly, the NOR gate is equivalent to an OR gate followed by an inverter.

The exclusive-OR (XOR) operation for two logic variables $A$ and $B$ is represented by $A \oplus B$ and is defined by

$$0 \oplus 0 = 0$$
$$1 \oplus 0 = 1$$
$$0 \oplus 1 = 1$$
$$1 \oplus 1 = 0$$

Notice that the XOR operation yields 1 if $A$ is 1 or if $B$ is 1, but yields 0 if both $A$ and $B$ are 1. The XOR operation is also known as **modulo-two addition**.

A **buffer** has a single input and produces an output with the same value as the input. (Buffers are commonly used to provide large currents when a logic signal must be applied to a low-impedance load.)

The **equivalence gate** produces a high output only if both inputs have the same value. In effect, it is an XOR followed by an inverter as the symbol of Figure 7.19(e) implies.

## Logical Sufficiency of NAND Gates or of NOR Gates

As we have seen, several combinations of gates often can be found that perform the same function. For example, if the inputs to a NAND are tied together, an inverter results. This is true because

$$\overline{(AA)} = \overline{A}$$

which is illustrated in Figure 7.20(a).

Furthermore, as shown by De Morgan's laws, the OR operation can be realized by inverting the input variables and combining the results in a NAND gate. This is shown in Figure 7.20(b), in which the inverters are formed from NAND gates. Finally, a NAND followed by an inverter results in an AND gate. *Since the basic logic functions (AND, OR, and NOT) can be realized by using only NAND gates, we conclude that NAND gates are sufficient to realize any combinatorial logic function.*

Any combinatorial logic function can be implemented solely with NAND gates.

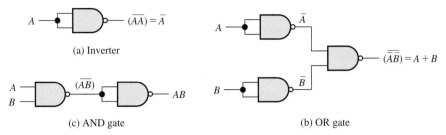

(a) Inverter

(c) AND gate

(b) OR gate

**Figure 7.20** Basic Boolean operations can be implemented with NAND gates. Therefore, any Boolean function can be implemented by the use of NAND gates alone.

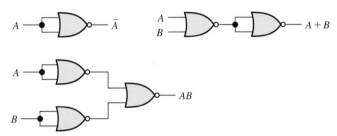

**Figure 7.21** The AND, OR, and NOT operations can be implemented with NOR gates. Thus, any combinatorial logic circuit can be designed by using only NOR gates. See Exercise 7.12.

**Exercise 7.12**   Show how to use only NOR gates to realize the AND, OR, and NOT functions.
**Answer**   See Figure 7.21.

Any combinatorial logic function can be implemented solely with NOR gates.

## 7.4 SYNTHESIS OF LOGIC CIRCUITS

In this section, we consider methods to implement logic circuits given the specification for the output in terms of the inputs. Often, the initial specification for a logic circuit is given in natural language. This is translated into a truth table or a Boolean logic expression that can be manipulated to find a practical implementation.

### Sum-of-Products Implementation

Consider the truth table shown in Table 7.6. $A$, $B$, and $C$ are input logic variables, and $D$ is the desired output. Notice that we have numbered the rows of the truth table with the decimal number corresponding to the binary number formed by $ABC$.

Suppose that we want to find a logic circuit that produces the output variable $D$. One way to write a logic expression for $D$ is to concentrate on the rows of the truth table for which $D$ is 1. In Table 7.6, these are the rows numbered 0, 2, 6, and 7. Then, we write a logical product of the input logic variables or their inverses that equals 1 for each of these rows. Each input variable or its inverse is included in each product. In writing the product for each row, we invert the logic variables that are 0 in that row. For example, the logical product $\overline{A}\,\overline{B}\,\overline{C}$ equals logic 1 only for row 0. Similarly, $\overline{A}B\overline{C}$ equals logic 1 only for row 2, $AB\overline{C}$ equals logic 1 only for row 6, and $ABC$ equals logic 1 only for row 7. Product terms that include all of the input variables (or their inverses) are called **minterms**.

**Table 7.6** Truth Table Used to Illustrate SOP and POS Logical Expressions

| Row | $A$ | $B$ | $C$ | $D$ |
| --- | --- | --- | --- | --- |
| 0 | 0 | 0 | 0 | 1 |
| 1 | 0 | 0 | 1 | 0 |
| 2 | 0 | 1 | 0 | 1 |
| 3 | 0 | 1 | 1 | 0 |
| 4 | 1 | 0 | 0 | 0 |
| 5 | 1 | 0 | 1 | 0 |
| 6 | 1 | 1 | 0 | 1 |
| 7 | 1 | 1 | 1 | 1 |

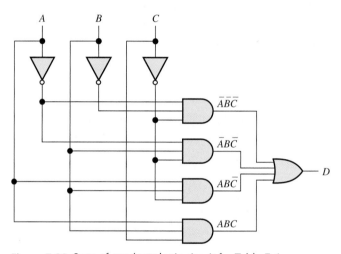

**Figure 7.22** Sum-of-products logic circuit for Table 7.6.

Finally, we write an expression for the output variable as a logical sum of minterms. For Table 7.6, it yields

$$D = \overline{A}\,\overline{B}\,\overline{C} + \overline{A}B\overline{C} + AB\overline{C} + ABC \tag{7.21}$$

In a sum-of-products expression, we form a product of all the input variables (or their inverses) for each row of the truth table for which the result is logic 1. The output is the sum of these products.

This type of expression is called a **sum of products** (SOP). Following this procedure, we can always find an SOP expression for a logic output given the truth table. A logic circuit that implements Equation 7.21 directly is shown in Figure 7.22.

A shorthand way to write an SOP is simply to list the row numbers of the truth table for which the output is logic 1. Thus, we can write

$$D = \sum m(0, 2, 6, 7) \tag{7.22}$$

in which $m$ indicates that we are summing the minterms corresponding to the rows enumerated.

## Product-of-Sums Implementation

Another way to write a logic expression for $D$ is to concentrate on the rows of the truth table for which $D$ is 0. For example, in Table 7.6, these are the rows numbered 1, 3, 4, and 5. Then, we write a logical sum that equals 0 for each of these rows.

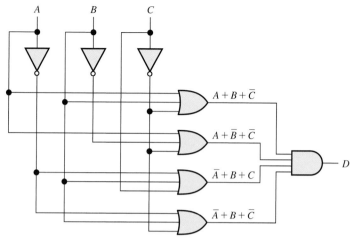

Figure 7.23  Product-of-sums logic circuit for Table 7.6.

Each input variable or its inverse in included in each sum. In writing the sum for each row, we invert the logic variables that are 1 in that row. For example, the logical sum $(A + B + \overline{C})$ equals logic 0 only for row 1. Similarly, $(A + \overline{B} + \overline{C})$ equals logic 0 only for row 3, $(\overline{A} + B + C)$ equals logic 0 only for row 4, and $(\overline{A} + B + \overline{C})$ equals 0 only for row 5. Sum terms that include all of the input variables (or their inverses) are called **maxterms**.

Finally, we write an expression for the output variable as the logical product of maxterms. For Table 7.6, it yields

$$D = (A + B + \overline{C})(A + \overline{B} + \overline{C})(\overline{A} + B + C)(\overline{A} + B + \overline{C}) \qquad (7.23)$$

This type of expression is called a **product of sums** (POS). We can always find a POS expression for a logic output given the truth table. A circuit that implements Equation 7.23 is shown in Figure 7.23.

A shorthand way to write a POS is simply to list the row numbers of the truth table for which the output is logic 1. Thus, we write

$$D = \prod M(1, 3, 4, 5) \qquad (7.24)$$

in which $M$ indicates the maxterms corresponding to the rows enumerated.

In a product-of-sums expression, we form a sum of all the input variables (or their inverses) for each row of the truth table for which the result is logic 0. The output is the product of these sums.

## Example 7.13   Combinatorial Logic Circuit Design

The control logic for a residential heating system is to operate as follows: During the daytime, heating is required only if the temperature falls below 68°F. At night, heating is required only for temperatures below 62°F. Assume that logic signals $D$, $L$, and $H$ are available. $D$ is high during the daytime and low at night. $H$ is high only if the temperature is above 68°F. $L$ is high only if the temperature is above 62°F. Design a logic circuit that produces an output signal $F$ that is high only when heating is required.

Solution   First, we translate the description of the desired operation into a truth table. This is shown in Figure 7.24(a). We have listed all combinations of the inputs.

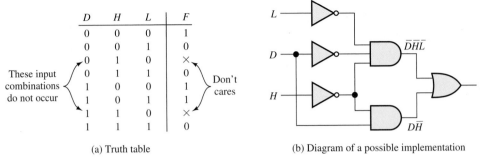

| D | H | L | F |
|---|---|---|---|
| 0 | 0 | 0 | 1 |
| 0 | 0 | 1 | 0 |
| 0 | 1 | 0 | × |
| 0 | 1 | 1 | 0 |
| 1 | 0 | 0 | 1 |
| 1 | 0 | 1 | 1 |
| 1 | 1 | 0 | × |
| 1 | 1 | 1 | 0 |

These input combinations do not occur

Don't cares

(a) Truth table

(b) Diagram of a possible implementation

**Figure 7.24**  See Example 7.10.

Some combinations of input variables may not occur; if so, the corresponding outputs are called "don't cares."

However, two combinations do not occur because temperature cannot be below 62°F ($L = 0$) and also be above 68°F ($H = 1$). The output listed for these combinations is ×, which is called a **don't care** because we don't care what the output of the logic circuit is for these input combinations.

As we have seen, one way to translate the truth table into a logic expression is to write the SOP in which there is a separate term for each high output. Applying this approach to Figure 7.24(a) yields

$$F = \overline{D}\,\overline{H}\,\overline{L} + D\overline{H}\,\overline{L} + D\overline{H}L \tag{7.25}$$

The first term on the right-hand side $\overline{D}\,\overline{H}\,\overline{L}$ is high only for row 0 of the truth table. Also, the second term $D\overline{H}\,\overline{L}$ is high only for row 4, and the third term $D\overline{H}L$ is high only for row 5. Thus, the shorthand way to write the logic expression is

$$F = \sum m(0, 4, 5) \tag{7.26}$$

Notice that in Equations 7.25 and 7.26, the don't cares turn out to be low.

The logic expression of Equation 7.25 can be manipulated into the form

$$F = D\overline{H} + \overline{D}\,\overline{H}\,\overline{L}$$

A logic diagram for this is shown in Figure 7.24(b).

An alternative approach is to write a POS with a separate sum term for each row of the truth table having an output value of 0. For the truth table of Figure 7.24(a), we have

$$F = (D + H + \overline{L})(D + \overline{H} + \overline{L})(\overline{D} + \overline{H} + \overline{L}) \tag{7.27}$$

The first term in the product $(D + H + \overline{L})$ is low only for row 1 of the truth table. (Recall that row 1 is actually the second row because we start numbering with 0.) The second term $(D + \overline{H} + \overline{L})$ is low only for row 3, and the last term in the product is low only for the last row of the truth table.

In short form, we can write Equation 7.27 as

$$F = \prod M(1, 3, 7) \tag{7.28}$$

For Equations 7.27 and 7.28, the don't cares are high. Because of the different outputs for the don't cares, the expressions we have given for $F$ in Equations 7.25 and 7.27 are not equivalent. ∎

**Exercise 7.13**   Show two ways to realize the XOR operation by using AND, OR, and NOT gates.
**Answer**   See Figure 7.25. ☐

**Exercise 7.14**   A traditional children's riddle concerns a farmer who is traveling with a sack of rye, a goose, and a mischievous dog. The farmer comes to a river that he must cross from east to west. A boat is available, but it only has room for the farmer and one of his possessions. If the farmer is not present, the goose will eat the rye or the dog will eat the goose.

   We wish to design a circuit to emulate the conditions of this riddle. A separate switch is provided for the farmer, the rye, the goose, and the dog. Each switch has two positions depending on whether the corresponding object is on the east bank or the west bank of the river. The rules of play stipulate that no more than two switches be moved at a time and that the farmer must move (to row the boat) each time switches are moved. The switch for the farmer provides logic signal $F$, which is high if the farmer is on the east bank and low if he is on the west bank. Similar logic signals ($G$ for the goose, $D$ for the dog, and $R$ for the rye) are high if the corresponding object is on the east bank and low if it is on the west bank.

   Find a Boolean logic expression based on the sum-of-products approach for a logic signal $A$ (alarm) that is high anytime the rye or the goose is in danger of being eaten. Repeat for the product-of-sums approach.
**Answer**   The truth table is shown in Table 7.7. The Boolean expressions are

$$A = \sum m(3, 6, 7, 8, 9, 12) = \overline{F}\,\overline{D}GR + \overline{F}DG\overline{R} + \overline{F}DGR$$
$$+ F\overline{D}\,\overline{G}\,\overline{R} + F\overline{D}\,\overline{G}R + FD\overline{G}\,\overline{R}$$

and

$$A = \prod M(0, 1, 2, 4, 5, 10, 11, 13, 14, 15)$$ ☐

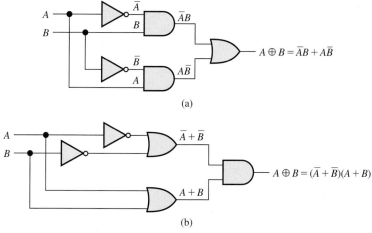

(a)

(b)

**Figure 7.25** Answer for Exercise 7.13.

**Table 7.7** Truth Table for Exercise 7.14

| F | D | G | R | A |
|---|---|---|---|---|
| 0 | 0 | 0 | 0 | 0 |
| 0 | 0 | 0 | 1 | 0 |
| 0 | 0 | 1 | 0 | 0 |
| 0 | 0 | 1 | 1 | 1 |
| 0 | 1 | 0 | 0 | 0 |
| 0 | 1 | 0 | 1 | 0 |
| 0 | 1 | 1 | 0 | 1 |
| 0 | 1 | 1 | 1 | 1 |
| 1 | 0 | 0 | 0 | 1 |
| 1 | 0 | 0 | 1 | 1 |
| 1 | 0 | 1 | 0 | 0 |
| 1 | 0 | 1 | 1 | 0 |
| 1 | 1 | 0 | 0 | 1 |
| 1 | 1 | 0 | 1 | 0 |
| 1 | 1 | 1 | 0 | 0 |
| 1 | 1 | 1 | 1 | 0 |

## Decoders, Encoders, and Translators

Many useful combinatorial circuits known as **decoders**, **encoders**, or **translators** are available as ICs. We discuss two examples. In a calculator or watch, we may represent information to be displayed in BCD form. Thus, 0000 is for 0, 0001 is for 1, 0010 is for 2, 0011 is for 3, and so on. Using four-bit words, 16 combinations are possible. However, only 10 combinations are used in BCD. Codes such as 1010 and 1011 do not occur in BCD.

The calculator display typically consists of liquid crystals with seven segments, as illustrated in Figure 7.26(a). The digits 0 through 9 are displayed by turning on appropriate segments as shown in Figure 7.26(b). Thus, a decoder is needed to translate the four-bit BCD words into seven-bit words of the form $ABCDEFG$, for which $A$ is high if segment $A$ of the display is required to be on, $B$ is high if segment $B$ is required to be on, and so on. Thus, 0000 is translated to 1111110 because all segments except $G$ are on to display the symbol for zero. Similarly, 0001 becomes 0110000, and 0010 becomes 1101101. Hence, the BCD-to-seven-segment decoder is a combinatorial circuit having four inputs and seven outputs.

Another example is the three-to-eight-line decoder that has a three-bit input and eight output lines. The three-bit input word selects one of the output lines and

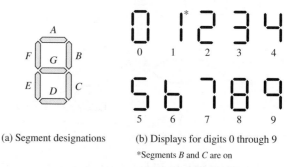

(a) Segment designations          (b) Displays for digits 0 through 9

*Segments $B$ and $C$ are on

Figure 7.26  Seven-segment display.

| C | B | A | $Y_0$ | $Y_1$ | $Y_2$ | $Y_3$ | $Y_4$ | $Y_5$ | $Y_6$ | $Y_7$ |
|---|---|---|---|---|---|---|---|---|---|---|
| 0 | 0 | 0 | 1 | 0 | 0 | 0 | 0 | 0 | 0 | 0 |
| 0 | 0 | 1 | 0 | 1 | 0 | 0 | 0 | 0 | 0 | 0 |
| 0 | 1 | 0 | 0 | 0 | 1 | 0 | 0 | 0 | 0 | 0 |
| 0 | 1 | 1 | 0 | 0 | 0 | 1 | 0 | 0 | 0 | 0 |
| 1 | 0 | 0 | 0 | 0 | 0 | 0 | 1 | 0 | 0 | 0 |
| 1 | 0 | 1 | 0 | 0 | 0 | 0 | 0 | 1 | 0 | 0 |
| 1 | 1 | 0 | 0 | 0 | 0 | 0 | 0 | 0 | 1 | 0 |
| 1 | 1 | 1 | 0 | 0 | 0 | 0 | 0 | 0 | 0 | 1 |

(a) Truth table

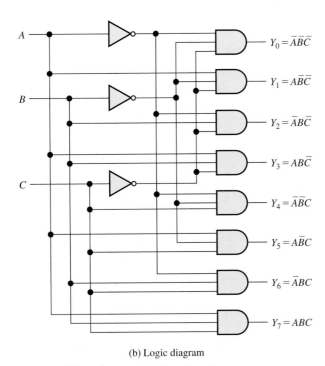

(b) Logic diagram

**Figure 7.27** Three-line-to-eight-line decoder.

that output becomes high. The truth table and a circuit implementation are shown in Figure 7.27.

Decoders are available that are used to convert binary numbers to BCD, or vice versa, test whether one number is larger or smaller than another, perform arithmetic operations on binary or BCD numbers, and many similar functions.

## 7.5   MINIMIZATION OF LOGIC CIRCUITS

We have seen that logic functions can be readily expressed either as a logical sum of minterms or as a logical product of maxterms. However, direct implementation of either of these expressions may not yield the best circuit in terms of minimizing the number of gates required. For example, consider the logical expression

$$F = \overline{A}\,\overline{B}D + \overline{A}BD + BCD + ABC \tag{7.29}$$

Implementations based on either a sum of minterms or a product of maxterms may not be optimum in minimizing the number of gates needed to realize a logic function.

**Table 7.8** Answer for Exercise 7.15

| A | B | C | D | F |
|---|---|---|---|---|
| 0 | 0 | 0 | 0 | 0 |
| 0 | 0 | 0 | 1 | 1 |
| 0 | 0 | 1 | 0 | 0 |
| 0 | 0 | 1 | 1 | 1 |
| 0 | 1 | 0 | 0 | 0 |
| 0 | 1 | 0 | 1 | 1 |
| 0 | 1 | 1 | 0 | 0 |
| 0 | 1 | 1 | 1 | 1 |
| 1 | 0 | 0 | 0 | 0 |
| 1 | 0 | 0 | 1 | 0 |
| 1 | 0 | 1 | 0 | 0 |
| 1 | 0 | 1 | 1 | 0 |
| 1 | 1 | 0 | 0 | 0 |
| 1 | 1 | 0 | 1 | 0 |
| 1 | 1 | 1 | 0 | 1 |
| 1 | 1 | 1 | 1 | 1 |

Implemented directly, this expression would require two inverters, four AND gates, and one OR gate.

Factoring the first pair of terms, we have

$$F = \overline{A}D(\overline{B} + B) + BCD + ABC$$

However, $\overline{B} + B = 1$, so we obtain

$$F = \overline{A}D + BCD + ABC$$

Of course, $BCD = 1$ only if $B = 1$, $C = 1$, and $D = 1$. In that case, either $\overline{A}D = 1$ or $ABC = 1$, because we must have either $\overline{A} = 1$ or $A = 1$. Thus, the term $BCD$ is redundant and can be dropped from the expression. Then, we get

$$F = \overline{A}D + ABC \qquad (7.30)$$

Only one inverter, two AND gates, and one OR gate are required to implement this expression.

**Exercise 7.15**   Create a truth table to verify that the right-hand sides of Equations 7.29 and 7.30 yield the same result.
**Answer**   See Table 7.8.    □

## Karnaugh Maps

As we have demonstrated, logic expressions can sometimes be simplified dramatically. However, the algebraic manipulations needed to simplify a given expression are often not readily apparent. By using a graphical approach known as the **Karnaugh map**, we will find it much easier to minimize the number of terms in a logic expression.

A Karnaugh map is an array of squares. Each square corresponds to one of the minterms of the logic variables or, equivalently, to one of the rows of the truth table.

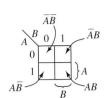

(a) Two-variable Karnaugh map

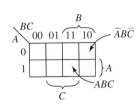

(b) Three-variable Karnaugh map

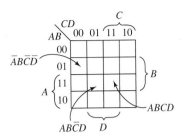

(c) Four-variable Karnaugh map

**Figure 7.28** Karnaugh maps showing the minterms corresponding to some of the squares.

Karnaugh maps for two, three, and four variables are shown in Figure 7.28. The two-variable map consists of four squares, one corresponding to each of the minterms. Similarly, the three-variable map has eight squares, and the four-variable map has 16 squares.

The minterms corresponding to some of the squares are shown in Figure 7.28. For example, for the three-variable map, the minterm $\overline{A}B\overline{C}$ corresponds to the upper right-hand square. Also, the bit combinations corresponding to the rows of the truth table are shown down the left-hand side and across the top of the map. For example, on the four-variable map, the row of the truth table for which the four-bit word $ABCD$ is 1101 corresponds to the square in the third row (i.e., the row labeled 11) and second column (i.e., the column labeled 01). Thus, we can readily find the square corresponding to any minterm or to any row of the truth table.

In case you are wondering about the order of the bit patterns along the side or top of the four-variable Karnaugh map (i.e., 00 01 11 10), notice that this is a two-bit Gray code. Thus, the patterns for squares with a common side differ in only one bit, so that similar minterms are grouped together. For example, the minterms containing $A$ (rather than $\overline{A}$) fall in the bottom half of each map. In the four-variable map, the minterms containing $B$ are in the middle two rows, the minterms containing $AB$ are in the third row, and so forth. This grouping of similar terms is the key to simplifying logic circuits.

> A Karnaugh map is a rectangular array of squares, where each square represents one of the minterms of the logic variables

**Exercise 7.16   a.** Write the minterm corresponding to the upper right-hand square in Figure 7.28(c). **b.** Write the minterm corresponding to the lower left-hand square.
**Answer   a.** $\overline{A}BC\overline{D}$; **b.** $A\overline{B}\,\overline{C}\,\overline{D}$.   □

We call two squares that have a common edge a **2-cube**. Similarly, four squares with common edges are called a **4-cube**. In locating cubes, the maps should be considered to fold around from top to bottom and from left to right. Therefore, the squares on the right-hand side are considered to be adjacent to those on the left-hand side, and the top of the map is adjacent to the bottom. Consequently, the four squares in the map corners form a 4-cube. Some cubes are illustrated in Figure 7.29.

To map a logic function, we place 1s in the squares for which the logic function takes a value of 1. Product terms map 1s into cubes. For example, some product terms are mapped in Figure 7.30.

In a four-variable map consisting of 16 squares, a single logic variable or its inverse covers (maps into) an 8-cube. A product of two variables (such as $AB$ or $A\overline{B}$) covers a 4-cube. A product of three variables maps into a 2-cube.

> The left and right (as well as top and bottom) edges of the Karnaugh map are considered to be adjacent.

> Rectangular arrays (known as cubes) of adjacent squares in a Karnaugh map represent products of logic variables or their inverses.

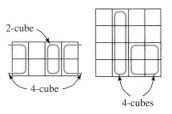

**Figure 7.29** Karnaugh maps illustrating cubes.

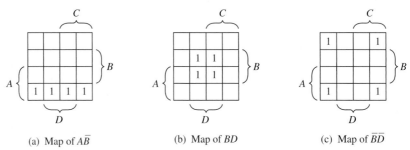

(a) Map of $A\overline{B}$    (b) Map of $BD$    (c) Map of $\overline{B}\overline{D}$

**Figure 7.30** Products of two variables map into 4-cubes on a 4-variable Karnaugh map.

By finding the fewest and largest (possibly overlapping) cubes for the region in which the logic expression is one, we obtain the minimum SOP for the logic expression.

The Karnaugh map of the logic function

$$F = \overline{A}\,\overline{B}\,\overline{C}D + \overline{A}\,\overline{B}CD + \overline{A}B\overline{C}D + \overline{A}BCD + ABC\overline{D} + ABCD \quad (7.31)$$

is shown in Figure 7.31. The squares containing 1s form a 4-cube corresponding to the product term $\overline{A}D$ plus a 2-cube corresponding to $ABC$. These are the largest cubes that cover the 1s in the map. Thus, the minimum SOP expression for $F$ is

$$F = \overline{A}D + ABC \quad (7.32)$$

Because it is relatively easy to spot the set of largest cubes that cover the 1s in a Karnaugh map, we can quickly minimize a logic function.

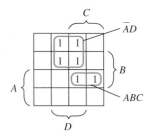

**Figure 7.31** Karnaugh map for the logic function of Equation 7.31. From the map, it is evident that $F = \overline{A}D + ABC$.

---

**Example 7.14    Finding the Minimum SOP Form for a Logic Function**

A logic circuit has inputs $A, B, C,$ and $D$. The output of the circuit is given by

$$E = \sum m(1, 3, 4, 5, 7, 10, 12, 13)$$

Find the minimum SOP form for $E$.

**Solution**    First, we construct the Karnaugh map. Because there are four input variables, the map contains 16 squares as shown in Figure 7.32. Converting the numbers of the minterms to binary numbers, we obtain $0001, 0011, 0100, 0101, 0111, 1010, 1100,$ and $1101$. Each of these locates a square on the map. For example, $1101$ locates the square in the third row and second column, $0011$ is the square in the first row and third column, and so forth. Placing a 1 in the square corresponding to each minterm results in the map shown in Figure 7.32.

Now we look for the smallest number of the largest size cubes that cover the ones in the map. To cover the ones in this map, we need two 4-cubes and a 1-cube (i.e., a single isolated square) as illustrated in the figure. Finally, the minimum SOP expression is

$$E = \overline{A}D + B\overline{C} + A\overline{B}C\overline{D}$$

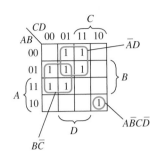

**Figure 7.32** Karnaugh map for Example 7.14.

## Minimum POS Forms

So far, we have concentrated on finding minimum SOP implementations for logic circuits. However, we can easily extend the methods to finding minimal POS circuits by following these steps:

1. Create the Karnaugh map for the desired output.
2. Invert the Karnaugh map, replacing 1s by 0s and vice versa.
3. Look for the least number of cubes of the largest sizes that cover the ones in the inverted map. Then, write the minimum SOP expression for the inverse of desired output.
4. Apply De Morgan's laws to convert the SOP expression to a POS expression. We illustrate with an example.

---

**Example 7.15** **Finding the Minimum POS Form for a Logic Function**

Find the minimum POS for the logic variable $E$ of Example 7.14.

**Solution**   The Karnaugh map for $E$ is shown in Figure 7.32. The map for $\overline{E}$ is obtained by replacing 1s with 0s (blank squares) and vice versa. The result is shown in Figure 7.33.

Now, we look for the smallest number of the largest size cubes that cover the ones in the map. Clearly, there are no 8-cubes or 4-cubes contained in Figure 7.33. A total of eight 1s appear in the map. Thus, the best we can do is to cover the map with four 2-cubes. One option is the grouping shown in the figure, which yields

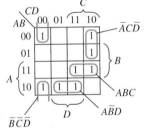

**Figure 7.33** Karnaugh map of Example 7.15. (This is the inverse of the map shown in Figure 7.32.)

$$\overline{E} = ABC + A\overline{B}D + \overline{A}C\overline{D} + \overline{B}C\overline{D}$$

Next, we apply De Morgan's laws to obtain a minimum POS form:

$$E = (\overline{A} + \overline{B} + \overline{C})(\overline{A} + B + \overline{D})(A + \overline{C} + D)(B + C + D)$$

Choosing a different grouping in Figure 7.33 produces another equally good form, which is

$$\overline{E} = A\overline{B}\,\overline{C} + \overline{A}\,B\overline{D} + ACD + BC\overline{D}$$

Then, applying De Morgan's laws gives another minimum POS form:

$$E = (\overline{A} + B + C)(A + B + D)(\overline{A} + \overline{C} + \overline{D})(\overline{B} + \overline{C} + D)$$

---

**Exercise 7.17**   Construct the Karnaugh maps and find the minimum SOP expressions for each of these logic functions:

**a.** $Z = \overline{W}\overline{X}\overline{Y} + \overline{W}X\overline{Y} + W\overline{X}Y + WXY$

**b.** $D = \overline{A}\,\overline{B}\,\overline{C} + A\overline{B}\,\overline{C} + \overline{A}\,\overline{B}C + A\overline{B}C + \overline{A}BC$

**c.** $E = \overline{A}BC\overline{D} + AB\overline{C}D + A\overline{B}\overline{C}D + ABC\overline{D}$

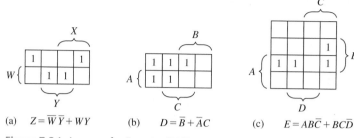

(a)    $Z = \overline{W}\,\overline{Y} + WY$    (b)    $D = \overline{B} + \overline{A}C$    (c)    $E = AB\overline{C} + BC\overline{D}$

**Figure 7.34**  Answers for Exercise 7.17.

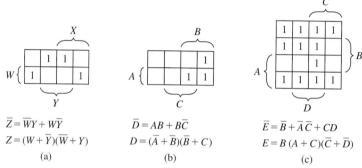

$\overline{Z} = \overline{W}Y + W\overline{Y}$            $\overline{D} = AB + B\overline{C}$            $\overline{E} = \overline{B} + \overline{A}\,\overline{C} + CD$
$Z = (W + \overline{Y})(\overline{W} + Y)$    $D = (\overline{A} + \overline{B})(\overline{B} + C)$    $E = B\,(A + C)(\overline{C} + \overline{D})$
(a)                                    (b)                                    (c)

**Figure 7.35**  Answers for Exercise 7.18.

**Answer**   See Figure 7.34.                                                                     □

**Exercise 7.18**   Construct the inverse maps and find the minimum POS expressions for each of the logic functions of Exercise 7.17.
**Answer**   See Figure 7.35.                                                                     □

## 7.6 SEQUENTIAL LOGIC CIRCUITS

So far, we have considered combinatorial logic circuits, such as gates, encoders, and decoders, for which the outputs at a given time depend only on the input values at that instant. In this section, we discuss **sequential logic circuits**, for which the outputs depend on past as well as present inputs. We say that such circuits have **memory** because they "remember" past input values.

Often, the operation of a sequential circuit is synchronized by a **clock signal** that consists of periodic logic-1 pulses, as shown in Figure 7.36. The clock signal regulates when the circuits respond to new inputs, so that operations occur in proper sequence. Sequential circuits that are regulated by a clock signal are said to be **synchronous**.

### Flip-Flops

One of the basic building blocks for sequential circuits is the **flip-flop**. A flip-flop has two stable operating states; therefore, it can store one bit of information. Many useful versions of flip-flops exist, differing in the manner that the clock signal and other input signals control the state of the flip-flop. We discuss several types shortly.

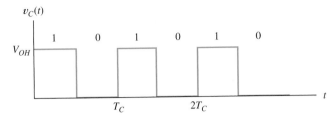

**Figure 7.36** The clock signal consists of periodic logic-1 pulses.

A simple flip-flop can be constructed by using two inverters, with the output of one connected to the input of the other, as shown in Figure 7.37. Two stable states are possible in the circuit. First, the output $Q$ of the top inverter can be high and then the output of the bottom inverter is low. Thus, the output of the bottom inverter is labeled as $\overline{Q}$. Notice that $Q$ high and $\overline{Q}$ low are consistent with the logic operation of the inverters, so the circuit can remain in that state. On the other hand, $Q$ low and $\overline{Q}$ high are also consistent. The circuit can remain in either state indefinitely.

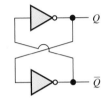

**Figure 7.37** Simple flip-flop.

**SR Flip-Flop.**   The simple two-inverter circuit of Figure 7.37 is not very useful because no provision exists for controlling its state. A more useful circuit is the **set—reset (SR) flip-flop**, consisting of two NOR gates, as shown in Figure 7.38. As long as the $S$ and $R$ inputs are low, the NOR gates act as inverters for the other input signal. Thus, with $S$ and $R$ both low, the $SR$ flip-flop behaves just as the two-inverter circuit of Figure 7.37 does.

If $S$ is high and $R$ is low, $\overline{Q}$ is forced low and $Q$ is high (or set). When $S$ returns low, the flip-flop remains in the **set state** (i.e., $Q$ stays high). On the other hand, if $R$ becomes high and $S$ low, $Q$ is forced low. When $R$ returns low, the flip-flop remains in the **reset state** (i.e., $Q$ stays low). In normal operation, $R$ and $S$ are not allowed to be high at the same time. Thus, with $R$ and $S$ low, the $SR$ flip-flop *remembers* which input ($R$ or $S$) was high most recently.

We use subscripts on logic variables to indicate a sequence of states. For example, the flip-flop output state $Q_{n-1}$ occurs before $Q_n$, which occurs before $Q_{n+1}$, and so on. The truth table for the $SR$ flip-flop is shown in Figure 7.39(a). In the first row of the truth table, we see that if both $R$ and $S$ are logic 0, the output remains in the previous state ($Q_n = Q_{n-1}$). The symbol for the $SR$ flip-flop is shown in Figure 7.39(b).

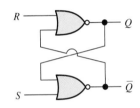

**Figure 7.38** An $SR$ flip-flop can be implemented by cross coupling two NOR gates.

| $R$ | $S$ | $Q_n$ |
|---|---|---|
| 0 | 0 | $Q_{n-1}$ |
| 0 | 1 | 1 |
| 1 | 0 | 0 |
| 1 | 1 | Not allowed |

(a) Truth table

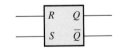

(b) Circuit symbol

**Figure 7.39** The truth table and symbol for the $SR$ flip-flop.

**Using an *SR* Flip-Flop to Debounce a Switch.**   One application for the $SR$ flip-flop is to *debounce* a switch. Consider the single-pole double-throw switch shown in Figure 7.40(a). When the switch is moved from position $A$ to position $B$, the waveforms shown in Figure 7.40(b) typically result. At first, $V_A$ is high because the switch is in position $A$. Then, the switch breaks contact, and $V_A$ drops to zero. Next, the switch makes initial contact with $B$ and $V_B$ goes high. Contact bounce at $B$ again causes $V_B$ to drop to zero, then back high several times, until finally it ends up high. Later, when the switch is returned to $A$, contact bounce occurs again.

This kind of behavior can be troublesome. For example, a computer keyboard consists of switches that are depressed to select a character. Contact bounce could cause several characters to be accepted by the computer or calculator each time a key is depressed.

An $SR$ flip-flop can eliminate the effects of contact bounce. The switch voltages $V_A$ and $V_B$ are connected to the $S$ and $R$ inputs as shown in Figure 7.40(a). At first, when the switch is at position $A$, the flip-flop is in the set state, and $Q$ is high. When

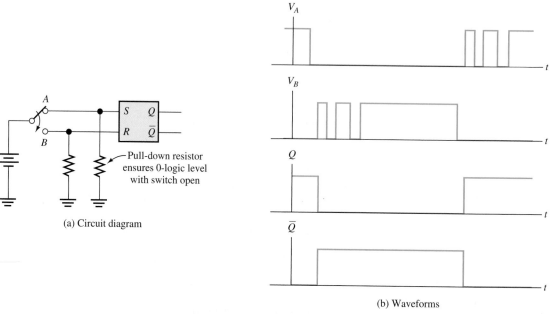

(a) Circuit diagram

(b) Waveforms

Figure 7.40 An *SR* flip-flop can be used to eliminate the effects of switch bounce.

contact is broken with $A$, $V_A$ drops to zero, but the flip-flop does not change state until the first time $V_B$ goes high. As contact bounce occurs, the flip-flop stays in the reset state with $Q$ low. The waveforms for the flip-flop outputs $Q$ and $\overline{Q}$ are shown in Figure 7.40(b).

**Exercise 7.19**   The waveforms present at the input terminals of an *SR* flip-flop are shown in Figure 7.41. Sketch the waveforms for $Q$ versus time.
**Answer**   See Figure 7.42.                                          □

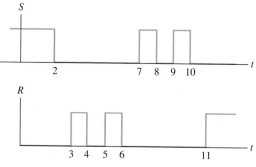

Figure 7.41  See Exercise 7.19.

Figure 7.42  Answer for Exercise 7.19.

**Exercise 7.20**   Prepare a truth table similar to that of Figure 7.39(a) for the circuit of Figure 7.43.

**Answer**   See Table 7.9.   □

Figure 7.43   A flip-flop implemented with NAND gates. See Exercise 7.20.

**Clocked *SR* Flip-Flop.**   Often, it is advantageous to control the point in time that a flip-flop responds to its inputs. This is accomplished with the **clocked *SR* flip-flop** shown in Figure 7.44. Two AND gates have been added at the inputs of an *SR* flip-flop. If the clock signal *C* is low, the inputs to the *SR* flip-flop are both low, and the state cannot change. The clock signal must be high for the *R* and *S* signals to be transmitted to the input of the *SR* flip-flop.

The truth table for the clocked *SR* flip-flop is shown in Figure 7.44(b), and the circuit symbol is shown in Figure 7.44(c). We say that a high clock level **enables** the inputs to the flip-flop. On the other hand, the low clock level **disables** the inputs.

Usually, we design digital systems so that *R*, *S*, and *C* are not all high at the same time. If all three signals are high and then *C* goes low, the state of the flip-flop settles either to $Q = 1$ or to $Q = 0$ unpredictably. Usually, systems that behave in an unpredictable manner are not useful.

Sometimes, a clocked *SR* flip-flop is needed, but it is also necessary to be able to set or clear the flip-flop state independent of the clock. A circuit having this feature is shown in Figure 7.45(a). If the **preset input** *Pr* is high, *Q* becomes high even if the clock is low. Similarly, the **clear input** *Cl* can force *Q* low. The *Pr* and *Cl* inputs are called **asynchronous inputs** because their effect is not synchronized by the clock signal. On the other hand, the *R* and *S* inputs are recognized only if the clock signal is high, and are therefore called **synchronous inputs**.

**Edge-Triggered *D* Flip-Flop.**   So far, we have considered circuits for which the level of the clock signal *enables* or *disables* other input signals. On the other hand, **edge-triggered** circuits respond to their inputs only at a transition in the clock signal. If the clock signal is steady, either high or low, the inputs are disabled. At the clock transition, the flip-flop responds to the inputs present just prior to the transition. **Positive-edge-triggered** circuits respond when the clock signal switches from low

**Table 7.9**   Truth Table for Exercise 7.20

| $A$ | $B$ | $C_n$ | $D_n$ |
|---|---|---|---|
| 0 | 0 | 1 | 1 |
| 0 | 1 | 1 | 0 |
| 1 | 0 | 0 | 1 |
| 1 | 1 | $C_{n-1}$ | $D_{n-1}$ |

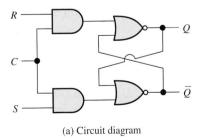

| $R$ | $S$ | $C$ | $Q_n$ |
|---|---|---|---|
| 0 | 0 | × | $Q_{n-1}$ |
| 0 | 1 | 1 | 1 |
| 1 | 0 | 1 | 0 |
| 1 | 1 | 1 | Not allowed |
| × | × | 0 | $Q_{n-1}$ |

(a) Circuit diagram               (b) Truth table               (c) Circuit symbol

**Figure 7.44**   A clocked *SR* flip-flop.

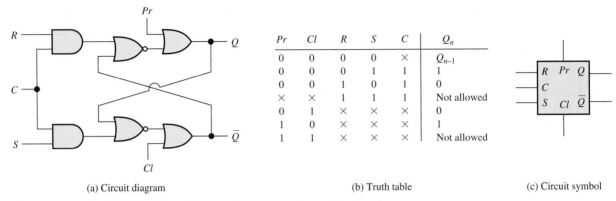

| Pr | Cl | R | S | C | $Q_n$ |
|----|----|----|----|----|----|
| 0 | 0 | 0 | 0 | × | $Q_{n-1}$ |
| 0 | 0 | 0 | 1 | 1 | 1 |
| 0 | 0 | 1 | 0 | 1 | 0 |
| × | × | 1 | 1 | 1 | Not allowed |
| 0 | 1 | × | × | × | 0 |
| 1 | 0 | × | × | × | 1 |
| 1 | 1 | × | × | × | Not allowed |

(a) Circuit diagram                (b) Truth table                (c) Circuit symbol

**Figure 7.45** A clocked *SR* flip-flop with asynchronous preset and clear inputs.

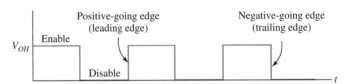

**Figure 7.46** Clock signal.

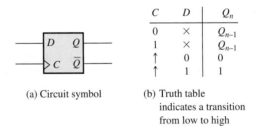

| C | D | $Q_n$ |
|----|----|----|
| 0 | × | $Q_{n-1}$ |
| 1 | × | $Q_{n-1}$ |
| ↑ | 0 | 0 |
| ↑ | 1 | 1 |

(a) Circuit symbol

(b) Truth table
indicates a transition
from low to high

**Figure 7.47** A positive-edge-triggered *D*
flip-flop.

to high. Conversely, **negative-edge-triggered** circuits respond on the transition from high to low. The positive-going edge of the clock is also called the **leading edge**, and the negative-going edge is called the **trailing edge**. A clock signal illustrating these points is shown in Figure 7.46. Thus, clocked flip-flops can be sensitive either to the level of the clock or to transitions.

An example of an edge-triggered circuit is the *D* **flip-flop**, which is also known as the **delay flip-flop**. Its output takes the value of the input that was present just prior to the triggering clock transition. The circuit symbol for the edge-triggered *D* flip-flop is shown in Figure 7.47(a). The "knife edge" symbol at the *C* input indicates that the flip-flop is edge triggered. The truth table for a positive-edge-triggered version is shown in Figure 7.47(b). Notice the symbols in the clock column of the truth table, indicating transitions of the clock signal from low to high.

**Exercise 7.21**    The input signals to a positive-edge-triggered *D* flip-flop are shown in Figure 7.48. Sketch the output *Q* to scale versus time. (Assume that *Q* is low prior to $t = 2$.)
**Answer**    See Figure 7.49.                                                                    ☐

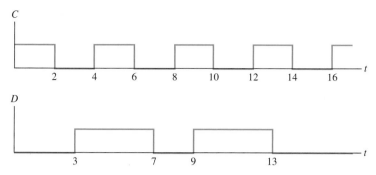

**Figure 7.48** See Exercise 7.21.

**Figure 7.49** Answer for Exercise 7.21.

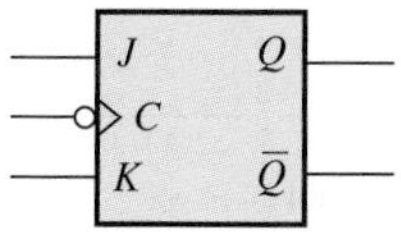

| $C$ | $J$ | $K$ | $Q_n$ | Comment |
|---|---|---|---|---|
| 0 | × | × | $Q_{n-1}$ | Memory |
| 1 | × | × | $Q_{n-1}$ | Memory |
| ↓ | 0 | 0 | $Q_{n-1}$ | Memory |
| ↓ | 0 | 1 | 0 | Reset |
| ↓ | 1 | 0 | 1 | Set |
| ↓ | 1 | 1 | $\overline{Q}_{n-1}$ | Toggle |

(a) Circuit symbol

(b) Truth table indicates a transition from low to high

**Figure 7.50** Negative-edge-triggered $JK$ flip-flop.

***JK* Flip-Flop.**   The circuit symbol and truth table for a negative-edge-triggered ***JK* flip-flop** are shown in Figure 7.50. Its operation is very similar to that of an *SR* flip-flop except that if both control inputs ($J$ and $K$) are high, the state changes on the next negative-going clock edge. Thus when both $J$ and $K$ are high, the output of the flip-flop **toggles** on each cycle of the clock–switching from high to low on one negative-going clock transition, back to high on the next negative transition, and so on.

## Serial-In Parallel-Out Shift Register

A **register** is an array of flip-flops that is used to store or manipulate the bits of a digital word. For example, if we connect several positive-edge-triggered $D$ flip-flops as shown in Figure 7.51, a **serial-in parallel-out shift register** results. As the name implies, the digital input word is shifted through the register moving one stage for each clock pulse.

The waveforms shown in Figure 7.51 illustrate the operation of the shift register. We assume that the flip-flops are initially ($t = 0$) all in the reset state ($Q_0 = Q_1 = Q_2 = Q_3 = 0$). The input data are applied to the input of the first stage serially (i.e., one bit after another). On the leading edge of the first clock pulse,

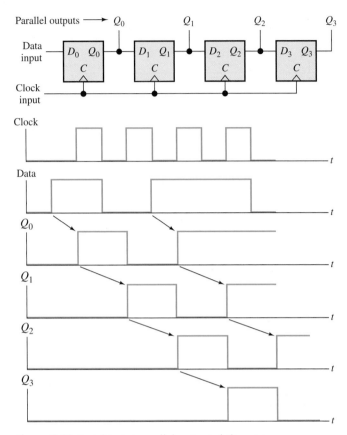

Figure 7.51  Serial-input parallel-output shift register.

the first data bit is transferred into the first stage. On the second clock pulse, the first bit is transferred to the second stage, and the second bit is transferred into the first stage. After four clock pulses, four bits of input data have been transferred into the shift register. Thus, serial data applied to the input are converted to parallel form available at the outputs of the stages of the shift register.

### Parallel-In Serial-Out Shift Register

Sometimes, we have parallel data that we wish to transmit serially. Then, the **parallel-in serial-out shift register** shown in Figure 7.52 is useful. This register consists of four positive-edge-triggered $D$ flip-flops with asynchronous preset and clear inputs. First, the register is cleared by applying a high pulse to the clear input. (The clear input is asynchronous, so a clock pulse is not necessary to clear the register.) Parallel data are applied to the $A, B, C$, and $D$ inputs. Then, a high pulse is applied to the parallel enable (PE) input. The result is to set each flip-flop for which the corresponding data line is high. Thus, four parallel bits are loaded into the stages of the register. Then, application of clock pulses produces the data in serial form at the output of the last stage.

### Counters

Counters are used to count the pulses of an input signal. An example is the **ripple counter** shown in Figure 7.53. It consists of a cascade of $JK$ flip-flops. Reference to Figure 7.50 shows that with the $J$ and $K$ inputs high, the $Q$-output of the flip-flop

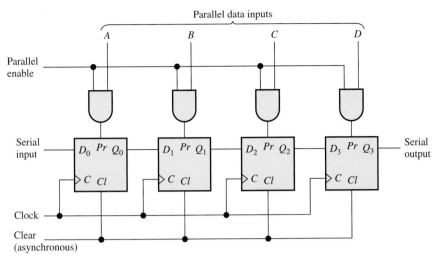

Figure 7.52 Parallel-input serial-output shift register.

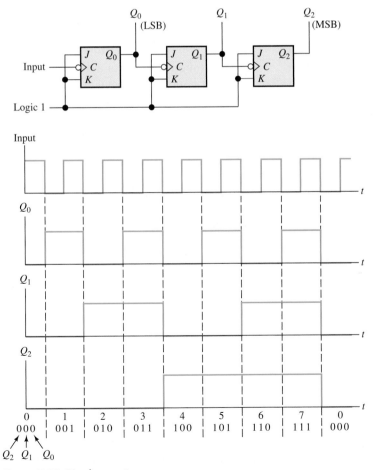

Figure 7.53 Ripple counter.

toggles on each falling edge of the clock input. The input pulses to be counted are connected to the clock input of the first stage, and the output of the first stage is connected to the clock input of the second stage.

Assume that the flip-flops are initially all in the reset state ($Q = 0$). When the falling edge of the first input pulse occurs, $Q_0$ changes to logic 1. On the falling edge of the second pulse, $Q_0$ toggles back to logic 0, and the resulting falling input to the second stage causes $Q_1$ to become high. As shown by the waveforms in Figure 7.53, after seven pulses, the shift register is in the 111 state. On the eighth pulse, the counter returns to the 000 state. Thus, we say that this is a modulo-8 or mod-8 counter.

## PRACTICAL APPLICATION    7.1

### Biomedical Engineering Application of Electronics: Cardiac Pacemaker

In certain types of heart disease, the biological signals that should stimulate the heart to beat are blocked from reaching the heart muscle. When this blockage occurs, the heart muscle may spontaneously beat at a very low rate, so death does not occur. However, the afflicted person is not able to function at a normal level of activity because of the low heart rate. The application of electrical pacemaker pulses to force beating at a higher rate is dramatically helpful in many of these cases.

Sometimes, the blockage of the natural pacemaking is not complete. In this case, the heart beats normally part of the time but experiences missed beats sporadically. A demand pacemaker can be useful for the patient with partial blockage. The demand pacemaker contains circuits that sense natural heartbeats and apply an electrical pulse to the heart muscle only if a beat does not occur within a predetermined interval. If natural beats are detected, no pulses are applied. This type of circuit is called a *demand* pacemaker because pulses are issued only when needed.

It turns out to be advantageous to the patient for the heart to beat naturally, provided that its natural rate is above some limit. On the other hand, if artificial pulses are required, a slightly higher rate is better. Typical values are a natural limit of 66.7 beats per minute (corresponding to 0.9 s between beats) and 75 beats per minute (corresponding to 0.8 s between beats) for forced pacing. Thus, in a typical situation, the circuit waits 0.9 s after a natural beat before applying a pacing pulse but waits only 0.8 s after an artificial pulse before applying another artificial pulse.

Another feature of the pacemaker is that it should ignore signals from the heart for a short period (about 0.4 s) after detection of a natural beat or after issuing a pacemaker pulse. This is because natural signals occur during the contraction and relaxation of the heart muscle. These signals should not cause the timing functions of the pacemaker to be reset. Thus, when the start of a contraction is sensed (or is stimulated by the circuit), the timing circuits are reset, but cannot be reset again until the contraction and relaxation is over.

The electrical signals present at the terminals of the pacemaker are shown for a typical case in Figure PA7.1. At the left side of the tracing, the

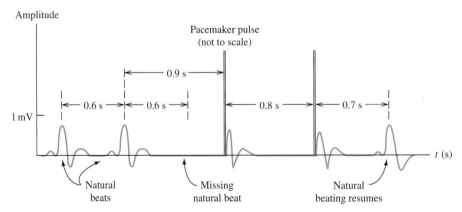

**Figure PA7.1** Typical electrical signals at the terminals of a demand cardiac pacemaker.

signals occurring during natural beating are shown. Then, blockage of the natural beating occurs, and a pacemaker pulse is issued 0.9 s after the last natural beat. The amplitude of these pulses is typically 5 V and their durations are 0.7 ms. After the pacemaker pulse, natural signals occur from the contraction and relaxation of the heart. These are ignored by the circuit. After two forced cycles, the heart again begins natural beating.

The pacemaker circuitry and battery are enclosed in a metal case. This is implanted under the skin on the chest of the patient. A wire (enclosed in an insulating tube known as a catheter) leads from the pacemaker through an artery into the interior of the heart. The electrical terminals of the pacemaker are the metal case and the tip of the catheter. A pacemaker and catheter are shown in Figure PA7.2.

The block diagram of a typical demand pacemaker is shown in Figure PA7.3. Notice that the electrical terminals serve both as the input to the amplifier and the output terminals for the pulse generator. The input amplifier increases the amplitude of the natural signals. Amplification is necessary because the natural heart signals have a very small amplitude (on the order of 1 mV), which must be increased before a comparator circuit can be employed to decide on the presence or absence of a natural heartbeat. Filtering to eliminate certain frequency components is employed in the amplifier to enhance the detectability of the heartbeats. Furthermore, proper filtering eliminates the possibility that radio or power-line signals will interfere with the pacemaker. Thus, the important specifications of the amplifier are its gain and frequency response.

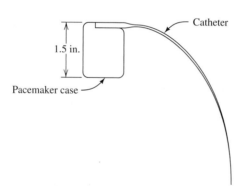

Figure PA7.2  A cardiac pacemaker and catheter.

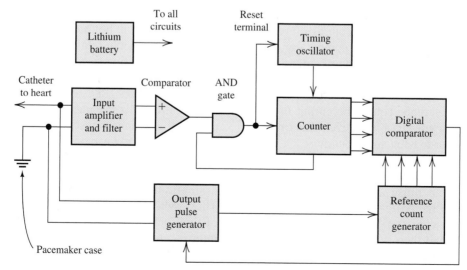

Figure PA7.3  Block diagram of a demand cardiac pacemaker.

Amplifiers and comparators are discussed at length in Part III of this book.

The output of the amplifier is applied to an analog comparator circuit that compares the amplified and filtered signal to a threshold value. If the input signal becomes higher than the threshold, the output of the comparator becomes high. Thus, the comparator output is a digital signal that indicates the detection of either a natural heartbeat or an output-pacing pulse.

This detection decision is passed through an AND gate to the counting and timer circuitry. The second input to the AND gate comes from the counter circuit and is low for 0.4 s after a detected beat. Thus, the AND gate prevents another decision from passing through for 0.4 s after the first. In this manner, the pacemaker ignores input signals for 0.4 s after the start of a natural or forced beat.

The timing functions are accomplished by counting the output cycles of a timing oscillator. The timing oscillator generates a square wave with a period of 0.1 s. When a heartbeat is detected, the timing oscillator is reset to the beginning of a cycle, so the completion of each oscillator cycle occurs exactly at an integer multiple of 0.1 s after the heartbeat. The timing oscillator must maintain a precise period because the proper operation of the circuit depends on accurate timing. Thus, the frequency stability of the timing oscillator is its primary specification.

The counter is a digital circuit that counts the output cycles of the timing oscillator. The counter is also capable of being reset to zero when a heartbeat occurs or when a pacemaker pulse is issued. The digital signals produced by the counter are applied to a digital comparator. Signals from a reference circuit are also applied to the digital comparator. The reference count is nine if the last beat was natural, but the reference count is eight if the last beat was forced. When the counter input to the digital comparator agrees with the reference count, the output of the digital comparator goes to a high level. This causes the pulse generator to issue an output pulse.

The pulse generator must produce output pulses of a specified amplitude and duration. In some designs, the output pulse amplitude is required to be higher than the battery voltage. This can be accomplished by charging capacitors in parallel with the battery and then switching them to series to generate the higher voltage.

What we have described so far is a relatively simple demand pacemaker readily implemented with registers, counters, and gates. By substituting a **microcontroller** and **software** (discussed in Chapter 8) for the digital functions of the pacemaker, many additional useful features can be accommodated. For example, an accelerometer can sense the physical activity of the patient and the software can use this information to adjust heart rate. Communicating through magnetic fields linked to a coil in the pacemaker, the physician can instruct the software to alter the operating characteristics appropriately for each patient.

Extremely low power consumption is an important requirement for all pacemaker circuits. This is because the circuit must operate from a small battery for many years. After all, replacement of the battery requires a surgical procedure. When pacing pulses are not needed, a typical circuit can function with a few microamperes from a 2.5-V battery. When pacing pulses are required, the average current drain increases to a few tens of microamperes. This higher current consumption is unavoidable because of the output power required in the form of pacing pulses.

High reliability is very important because malfunction can be life-threatening. A very detailed failure-mode analysis must be performed for every component in the circuit. This is necessary because some failures are much more threatening than others. For example, if the pacemaker fails to issue pacemaking pulses, the person may survive because of the natural (low-rate) pacing of the heart muscle. On the other hand, if the timing generator fails in such a manner that it runs too fast, the unfortunate person's heart will be forced to beat much too fast. This can be quickly fatal, especially for those in a weakened condition from heart disease.

Clearly, circuit design is not the total solution to this problem. Physicians must provide the specifications for the pacemaker. Mechanical and chemical engineers must be involved in selecting the materials and form of the catheter and the case. By working in teams, engineers and physicians have designed electronic pacemakers that provide very dramatic health improvements for many people. Those who have contributed can be most proud of their achievements. Nevertheless, many further improvements are possible and may be achieved by some of the students of this book.

## CONCLUSIONS

In this chapter, we have seen that complex combinatorial logic functions can be achieved simply by interconnecting NAND gates (or NOR gates). Furthermore, logic gates can be interconnected to form flip-flops. Interconnections of flip-flops form registers. A complex digital system, such as a computer, consists of many gates, flip-flops, and registers. Thus, logic gates are the basic building blocks for complex digital systems.

## Summary

1. Digital signals are more immune to the effects of noise than analog signals. The logic levels of a digital signal can be determined after noise is added, provided that the noise amplitude is not too high.

2. Component values in digital circuits do not need to be as precise as in analog circuits.

3. Digital circuits are more amenable than analog circuits to implementation as large-scale ICs.

4. In positive logic, the higher voltage represents logic 1.

5. Numerical data can be represented in decimal, binary, octal, hexadecimal, or BCD forms.

6. In the Gray code, each word differs from adjacent words in only a single bit. The Gray code is useful for representing position or angular displacement.

7. In computers, numbers are frequently represented in signed two's-complement form. (See Figure 7.11 on page 366.)

8. Logic variables take two values, logic 1 or logic 0. Logic variables may be combined by the AND, OR, and inversion operations according to the rules of Boolean algebra. A truth table lists all combinations of input variables and the corresponding output.

9. De Morgan's laws state that

$$AB = \overline{\overline{A} + \overline{B}}$$

and

$$A + B = \overline{\overline{A}\,\overline{B}}$$

10. NAND (or NOR) gates are sufficient to realize any combinatorial logic function.

11. Any combinatorial logic function can be written as a Boolean expression consisting of a logical SOP. Each product is a minterm corresponding to a line of the truth table for which the output variable is logic 1.

12. Any combinatorial logic function can be written as a Boolean expression consisting of a logical POS. Each sum is a maxterm corresponding to a line of the truth table for which the output variable is logic 0.

13. Many useful combinatorial circuits, known as decoders, encoders, or translators, are available as ICs.

14. Karnaugh maps can be used to minimize the number of gates needed to implement a given logic function.

15. Sequential logic circuits are said to have memory because their outputs depend on past as well as present inputs. Synchronous or clocked sequential circuits are regulated by a clock signal.

16. Various types of flip-flops are the $SR$ flip-flop, the clocked flip-flop, the $D$ flip-flop, and the $JK$ flip-flop.

17. Flip-flops can be combined to form registers that are used to store or manipulate digital words.

18. Logic gates can be interconnected to form flip-flops. Interconnections of flip-flops form registers. A complex digital system, such as a computer, consists of many gates, flip-flops, and registers. Thus, logic gates are the basic building blocks for complex digital systems.

## Problems

### Section 7.1: Basic Logic Circuit Concepts

**\*P7.1.** State three advantages of digital technology compared with analog technology.

**P7.2.** Define these terms: *bit, byte,* and *nibble.*

**P7.3.** Explain the difference between positive logic and negative logic.

**P7.4.** What are noise margins? Why are they important?

**P7.5.** How is serial transmission of a digital word different from parallel transmission?

### Section 7.2: Representation of Numerical Data in Binary Form

**P7.6.** Convert the following binary numbers to decimal form: **a.**\* 101.101; **b.** 0111.11; **c.** 1010.01; **d.** 111.111; **e.** 1000.0101; **f.**\* 10101.011.

**P7.7.** Express the following decimal numbers in binary form and in BCD form: **a.** 17; **b.** 8.5; **c.**\* 9.75; **d.** 73.03125; **e.** 67.375.

**P7.8.** How many bits per word are needed to represent the decimal integers 0 through 100? 0 through 1000? 0 through $10^6$?

**P7.9.** Add these pairs of binary numbers: **a.**\* 1101.11 and 101.111; **b.** 1011 and 101; **c.** 10001.111 and 0101.001.

**P7.10.** Find the result (in BCD format) of adding the BCD numbers: **a.**\* 10010011.0101 and 00110111.0001; **b.** 01011000.1000 and 10001001.1001.

**P7.11.** Express the following decimal numbers in binary, octal, and hexadecimal forms: **a.** 173; **b.** 299.5; **c.** 735.75; **d.**\* 313.0625; **e.** 112.25.

**P7.12.** Write each of the following decimal numbers as an eight-bit signed two's-complement number: **a.** 19; **b.** −19; **c.**\* 75; **d.**\* −87; **e.** −95; **f.** 99.

**P7.13.** Express each of the following hexadecimal numbers in binary, octal, and decimal forms: **a.** $FA.F_{16}$; **b.** $2A.1_{16}$; **c.** $777.7_{16}$.

**P7.14.** Express each of the following octal numbers in binary, hexadecimal, and decimal forms: **a.** $777.7_8$; **b.** $123.5_8$; **c.** $24.4_8$.

**P7.15.** What number follows 777 when counting in **a.** decimal; **b.** octal; **c.** hexadecimal?

**P7.16.** What range of decimal integers can be represented by **a.** three-bit binary numbers; **b.** three-digit octal numbers; **c.** three-digit hexadecimal numbers?

**\*P7.17.** Starting with the three-bit Gray code listed in Figure 7.9, construct a four-bit Gray code. For what applications is a Gray code advantageous? Why?

**P7.18.** Convert the following numbers to decimal form: **a.**\* $FA5.6_{16}$; **b.**\* $725.3_8$; **c.** $3F4.8_{16}$; **d.** $73.25_8$; **e.** $FF.F0_{16}$.

**P7.19.** Find the one's and two's complements of the binary numbers: **a.**\* 11101000; **b.** 00000000; **c.** 10101010; **d.** 11111100; **e.** 11000000.

**P7.20.** Perform these operations by using eight-bit signed two's-complement arithmetic: **a.** $17_{10} + 15_{10}$; **b.** $17_{10} - 15_{10}$; **c.**\* $33_{10} - 37_{10}$; **d.** $15_{10} - 63_{10}$; **e.** $49_{10} - 44_{10}$.

**P7.21.** Describe how to test whether overflow or underflow has occurred in adding signed two's-complement numbers.

### Section 7.3: Combinatorial Logic Circuits

**P7.22.** What is a truth table?

**\*P7.23.** State De Morgan's laws.

**P7.24.** Draw the circuit symbol and list the truth table for the following: an AND gate, an OR gate, an inverter, a NAND gate, a NOR gate, and an XOR gate. Assume two inputs for each gate (except the inverter).

**P7.25.** Describe a method for proving the validity of a Boolean algebra identity.

**P7.26.** Write the truth table for each of these Boolean expressions:

    **a.** $D = ABC + A\overline{B}$

    **b.** \*$E = AB + A\overline{B}C + \overline{C}D$

    **c.** $Z = WX + \overline{(W + Y)}$

    **d.** $D = A + \overline{A}B + C$

    **e.** $D = \overline{(A + BC)}$

---

\*Denotes that answers are contained in the Student Solutions files. See Appendix E for more information about accessing the Student Solutions.

**P7.27.** Write a Boolean expression for the output of each of the logic circuits shown in Figure P7.27.

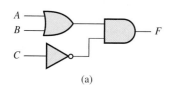

(a)

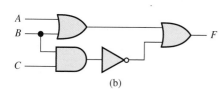

(b)

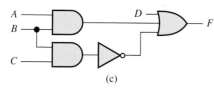

(c)

**Figure P7.27**

**\*P7.28.** Use a truth table to prove the identity

$$(A + B)(A + C) = A + BC$$

**P7.29.** Use a truth table to prove the identity

$$(A + B)(\overline{A} + AB) = B$$

**P7.30.** Use a truth table to prove the identity

$$A + \overline{A}B = A + B$$

**P7.31.** Use a truth table to prove the identity

$$ABC + AB\overline{C} + A\overline{B}\,\overline{C} + A\overline{B}C = A$$

**P7.32.** Draw a circuit to realize each of the following expressions using AND gates, OR gates, and inverters:

   **a.** $F = A + \overline{B}C$
   **b.** $F = A\overline{B}C + AB\overline{C} + \overline{A}BC$
   **c.** \* $F = (\overline{A} + B + C)(A + B + \overline{C})$
        $(A + \overline{B} + C)$

**P7.33.** Replace the AND operations by ORs and vice versa by applying De Morgan's laws to each of these expressions:

   **a.** $F = AB + (\overline{C} + A)\overline{D}$
   **b.** $F = A(\overline{B} + C) + D$
   **c.** $F = A\overline{B}C + A(B + C)$

   **d.** \* $F = (A + B + C)(A + \overline{B} + C)$
        $(\overline{A} + B + \overline{C})$
   **e.** \* $F = ABC + A\overline{B}C + \overline{A}B\overline{C}$

**P7.34.** Why are NAND gates said to be *sufficient* for combinatorial logic? What other type of gate is sufficient?

**P7.35.** Consider the circuit shown in Figure P7.35. The switches are controlled by logic variables such that, if $A$ is high, switch $A$ is closed, and if $A$ is low, switch $A$ is open. Conversely, if $B$ is high, the switch labeled $\overline{B}$ is open, and if $B$ is low, the switch labeled $\overline{B}$ is closed. The output variable is high if the output voltage is 5 V, and the output variable is low if the output voltage is zero. Write a logic expression for the output variable. Construct the truth table for the circuit.

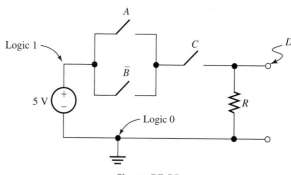

**Figure P7.35**

**P7.36.** Repeat Problem P7.35 for the circuit shown in Figure P7.36.

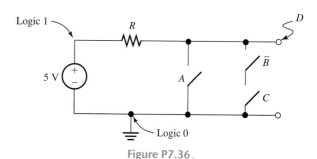

**Figure P7.36**

**P7.37.** Sometimes "bubbles" are used to indicate inverters on the input lines to a gate, as illustrated in Figure P7.37. What are the equivalent gates for those of Figure P7.37? Justify your answers.

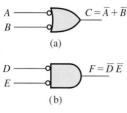

(a)

(b)

**Figure P7.37**

### Section 7.4: Synthesis of Logic Circuits

**P7.38.** Using the sum-of-products approach, describe the synthesis of a logic expression from a truth table. Repeat for the product-of-sums approach.

**P7.39.** Give an example of a decoder.

**\*P7.40.** Consider Table P7.40. $A$, $B$, and $C$ represent logic-variable input signals; $F$ through $K$ are outputs. Using the product-of-sums approach, write a Boolean expression for $F$ in terms of the inputs. Repeat by using the sum-of-products approach.

Table P7.40

| $A$ | $B$ | $C$ | $F$ | $G$ | $H$ | $I$ | $J$ | $K$ |
|---|---|---|---|---|---|---|---|---|
| 0 | 0 | 0 | 1 | 1 | 1 | 0 | 0 | 1 |
| 0 | 0 | 1 | 0 | 0 | 1 | 0 | 1 | 1 |
| 0 | 1 | 0 | 1 | 0 | 1 | 0 | 0 | 0 |
| 0 | 1 | 1 | 0 | 1 | 0 | 1 | 1 | 0 |
| 1 | 0 | 0 | 0 | 0 | 1 | 0 | 0 | 0 |
| 1 | 0 | 1 | 1 | 0 | 1 | 0 | 1 | 0 |
| 1 | 1 | 0 | 0 | 0 | 1 | 1 | 1 | 1 |
| 1 | 1 | 1 | 1 | 0 | 1 | 1 | 1 | 1 |

**P7.41.** Repeat Problem P7.40 for $G$.

**P7.42.** Repeat Problem P7.40 for $H$.

**P7.43.** Repeat Problem P7.40 for $I$.

**P7.44.** Repeat Problem P7.40 for $J$.

**P7.45.** Repeat Problem P7.40 for $K$.

**P7.46.** Show how to implement the sum-of-products circuit shown in Figure P7.46 by using only NAND gates.

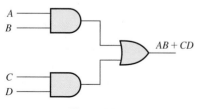

**Figure P7.46**

**P7.47.** Show how to implement the product-of-sums circuit shown in Figure P7.47 by using only NOR gates.

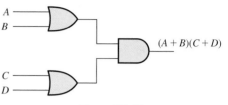

**Figure P7.47**

**P7.48.** Design a logic circuit to control electrical power to the engine ignition of a speed boat. Logic output $I$ is to become high if ignition power is to be applied and is to remain low otherwise. Gasoline fumes in the engine compartment present a serious hazard of explosion. A sensor provides a logic input $F$ that is high if fumes are present. Ignition power should not be applied if fumes are present. To help prevent accidents, ignition power should not be applied while the outdrive is in gear. Logic signal $G$ is high if the outdrive is in gear and is low otherwise. A blower is provided to clear fumes from the engine compartment and is to be operated for 5 minutes before applying ignition power. Logic signal $B$ becomes high after the blower has been in operation for 5 minutes. Finally, an emergency override signal $E$ is provided so that the operator can choose to apply ignition power even if the blower has not operated for 5 minutes and if the outdrive is in gear, but not if gasoline fumes are present.
**a.** Prepare a truth table listing all combinations of the input signals $B$, $E$, $F$, and $G$. Also, show the desired value of $I$ for each row in the table.
**b.** Using the sum-of-products approach, write a Boolean expression for $I$. **c.** Using the product-of-sums approach, write a Boolean expression for $I$. **d.** Try to manipulate the expressions of parts (b) and (c) to obtain a logic circuit having the least number of gates and inverters. Use AND gates, OR gates, and inverters.

**\*P7.49.** Use only NAND gates to find a way to implement the XOR function for two inputs, $A$ and $B$. [*Hint:* The inputs of a two-input NAND can be wired together to obtain an inverter. List the truth table and write the SOP expression. Then, apply De Morgan's laws to convert the OR operation to AND.]

**P7.50.** Use only two-input NOR gates to find a way to implement the XOR function for two inputs, $A$ and $B$. [*Hint:* The inputs of a two-input NOR can be wired together to obtain an inverter. List the truth table and write the POS expression. Then, apply De Morgan's laws to convert the AND operation to OR.]

**P7.51.** Consider the BCD-to-seven-segment decoder discussed in conjunction with Figure 7.26 on page 380. Suppose that the BCD data are represented by the logic variables $B_8$, $B_4$, $B_2$, and $B_1$. For example, the decimal number 7 is represented in BCD by the word 0111 in which the leftmost bit is $B_8 = 0$, the second bit is $B_4 = 1$, and so forth. **a.** Find a logic circuit based on the product of maxterms having output $A$ that is high only if segment $A$ of the display is to be on. **b.** Repeat for segment $B$.

**\*P7.52.** Suppose that two numbers in signed two's-complement form have been added. $S_1$ is the sign bit of the first number, $S_2$ is the sign bit of the second number, and $S_T$ is the sign bit of the total. Suppose that we want a logic circuit with output $E$ that is high if either overflow or underflow has occurred; otherwise, $E$ is to remain low. **a.** Write the truth table. **b.** Find an SOP expression composed of minterms for $E$. **c.** Draw a circuit that yields $E$, using AND, OR, and NOT gates.

**Section 7.5: Minimization of Logic Circuits**

**\*P7.53. a.** Construct a Karnaugh map for the logic function

$$F = \overline{A}\,\overline{B}\,\overline{C}\,D + AB\overline{C}\,\overline{D} + \overline{A}\,BC\overline{D}$$
$$+ AB\overline{C}D + \overline{A}BC\overline{D} + \overline{A}\,\overline{B}C\overline{D}$$

**b.** Find the minimum SOP expression. **c.** Find the minimum POS expression.

**P7.54.** A logic circuit has inputs $A$, $B$, and $C$. The output of the circuit is given by

$$D = \sum m(0, 3, 4)$$

  **a.** Construct the Karnaugh map for $D$.
  **b.** Find the minimum SOP expression.
  **c.** Find two equally good minimum POS expressions.

**P7.55.** A logic circuit has inputs $A$, $B$, and $C$. The output of the circuit is given by

$$D = \prod M(1, 3, 4, 6)$$

  **a.** Construct the Karnaugh map for $D$.
  **b.** Find the minimum SOP expression.
  **c.** Find the minimum POS expression.

**P7.56. a.** Construct a Karnaugh map for the logic function

$$D = ABC + \overline{A}BC + AB\overline{C} + BC$$

**b.** Find the minimum SOP expression and realize the function, using AND, OR, and NOT gates.
**c.** Find the minimum POS expression and realize the function, using AND, OR, and NOT gates.

**P7.57. a.** Construct a Karnaugh map for the logic function

$$F = AB\overline{C}\,\overline{D} + ABCD + ABC\overline{D} + \overline{A}BCD$$

**b.** Find the minimum SOP expression. **c.** Realize the minimum SOP function, using AND, OR, and NOT gates. **d.** Find the minimum POS expression.

**\*P7.58.** Consider Table P7.58 in which $A$, $B$, $C$, and $D$ are input variables. $F$, $G$, $H$, and $I$ are the output variables. **a.** Construct a Karnaugh map for the output variable $F$. **b.** Find the minimum SOP expression for this logic function. **c.** Use AND, OR, and NOT gates to realize the minimum SOP function. **d.** Find the minimum POS expression.

**P7.59.** Repeat Problem P7.58 for output variable $G$.

Table P7.58

| $A$ | $B$ | $C$ | $D$ | $F$ | $G$ | $H$ | $I$ |
|---|---|---|---|---|---|---|---|
| 0 | 0 | 0 | 0 | 0 | 0 | 0 | 1 |
| 0 | 0 | 0 | 1 | 1 | 0 | 0 | 1 |
| 0 | 0 | 1 | 0 | 0 | 0 | 0 | 1 |
| 0 | 0 | 1 | 1 | 0 | 0 | 0 | 0 |
| 0 | 1 | 0 | 0 | 0 | 0 | 0 | 1 |
| 0 | 1 | 0 | 1 | 1 | 0 | 0 | 1 |
| 0 | 1 | 1 | 0 | 0 | 0 | 1 | 1 |
| 0 | 1 | 1 | 1 | 0 | 1 | 1 | 0 |
| 1 | 0 | 0 | 0 | 0 | 0 | 0 | 0 |
| 1 | 0 | 0 | 1 | 1 | 0 | 1 | 0 |
| 1 | 0 | 1 | 0 | 0 | 0 | 0 | 0 |
| 1 | 0 | 1 | 1 | 0 | 0 | 1 | 0 |
| 1 | 1 | 0 | 0 | 1 | 0 | 0 | 0 |
| 1 | 1 | 0 | 1 | 1 | 0 | 1 | 0 |
| 1 | 1 | 1 | 0 | 1 | 1 | 1 | 0 |
| 1 | 1 | 1 | 1 | 1 | 1 | 1 | 0 |

**P7.60.** Repeat Problem P7.58 for output variable $H$.

**P7.61.** Repeat Problem P7.58 for output variable $I$.

**P7.62.** We need a logic circuit that gives an output $X$ that is high only if a given hexadecimal digit is even (including 0) and less than 7. The inputs to the logic circuit are the bits $B_8, B_4, B_2,$ and $B_1$ of the binary equivalent for the hexadecimal digit. (The MSB is $B_8$, and the LSB is $B_1$.) Construct a truth table and the Karnaugh map; then, write the minimized SOP expression for $X$.

**P7.63.** We need a logic circuit that gives an output $X$ that is high when an error in the form of an unused code occurs in a given BCD codeword. The inputs to the logic circuit are the bits $B_8, B_4, B_2,$ and $B_1$ of the BCD codeword. (The MSB is $B_8$, and the LSB is $B_1$.) Construct the Karnaugh map and write the minimized SOP and POS expressions for $X$.

**P7.64.** We need a logic circuit that gives a high output if a given hexadecimal digit is 4, 6, C, or E. The inputs to the logic circuit are the bits $B_8, B_4, B_2,$ and $B_1$ of the binary equivalent for the hexadecimal digit. (The MSB is $B_8$, and the LSB is $B_1$.) Construct the Karnaugh map and write the minimized SOP and POS expressions for $X$.

**P7.65.** We need to design a logic circuit for interchanging two logic signals. The system has three inputs $I_1, I_2,$ and $S$ as well as two outputs $O_1$ and $O_2$. When $S$ is low, we should have $O_1 = I_1$ and $O_2 = I_2$. On the other hand, when $S$ is high, we should have $O_1 = I_2$ and $O_2 = I_1$. Thus, $S$ acts as the control input for a reversing switch. Use Karnaugh maps to obtain a minimal SOP design. Draw the circuit.

**P7.66.** A city council has three members, $A, B,$ and $C$. Each member votes on a proposition (1 for yes, 0 for no). Find a minimized SOP logic expression having inputs $A, B,$ and $C$ and output $X$ that is high when the majority vote is yes and low otherwise. Show that the minimized logic circuit checks to see if any pair of the three board members have voted yes. Repeat for a council with five members. [*Hint*: In this case, the circuit checks to see if any group of three has all voted yes.]

**P7.67.** A city council has four members, $A, B, C,$ and $D$. Each member votes on a proposition (1 for yes, 0 for no). Find a minimized SOP logic expression having inputs $A, B, C,$ and $D$ and output $X$ that is high when the vote is tied and low otherwise.

**P7.68.** One way to help ensure that data are communicated correctly is to append a parity bit to each data word such that the number of 1s in the transmitted word is even. Then, if an odd number are found in the received result, we know that at least one error has occurred. **a.** Show that the circuit in Figure P7.68 produces the correct parity bit $P$ for the nibble (four-bit data word) $ABCD$. In other words, show that the transmitted word $ABCDP$ contains an even number of 1s for all combinations of data. **b.** Determine the minimum SOP expression for $P$ in terms of the data bits. **c.** If the received word contains a single bit error, the number of ones in the word will be odd. Draw a circuit using four XOR gates that outputs a 1 if the received word $ABCDP$ contains an odd number of 1s and outputs a 0 otherwise.

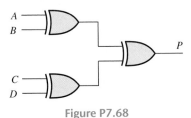

Figure P7.68

**P7.69.** Suppose we want circuits to convert the binary codes into the three-bit Gray codes shown in Table P7.69. Find the minimum SOP expressions for $X, Y,$ and $Z$ in terms of $A, B,$ and $C$.

Table P7.69

| Binary Code $ABC$ | Gray Code $XYZ$ |
| --- | --- |
| 000 | 000 |
| 001 | 001 |
| 010 | 011 |
| 011 | 010 |
| 100 | 110 |
| 101 | 111 |
| 110 | 101 |
| 111 | 100 |

**P7.70.** Find the minimum SOP expressions for $A$, $B$, and $C$ in terms of $X$, $Y$, and $Z$ for the codes of Table P7.69.

**\*P7.71.** We have discussed BCD numbers in which the bits have weights of 8, 4, 2, and 1. Another way to represent decimal integers is the 4221 code in which the weights of the bits are 4, 2, 2, and 1. The decimal integers, the BCD equivalents, and the 4221 equivalents are shown in Table P7.71. We want to design logic circuits to convert BCD codewords to 4221 codewords. **a.** Fill in the Karnaugh map for $F$, placing $x$'s (don't cares) in the squares for BCD codes that do not occur in the table. Find the minimum SOP expression allowing the various $x$'s to be either 1s or 0s to make the expression as simple as possible. **b.** Repeat (a) for $G$. **c.** Repeat (a) for $H$. **d.** Repeat (a) for $I$.

Table P7.71 **BCD, 4221, and excess-3 codewords for the decimal integers.**

| Decimal Integer | BCD Codeword $ABCD$ | 4221 Codeword $FGHI$ | Excess-3 Codeword $WXYZ$ |
|---|---|---|---|
| 0 | 0000 | 0000 | 0011 |
| 1 | 0001 | 0001 | 0100 |
| 2 | 0010 | 0010 | 0101 |
| 3 | 0011 | 0011 | 0110 |
| 4 | 0100 | 1000 | 0111 |
| 5 | 0101 | 0111 | 1000 |
| 6 | 0110 | 1100 | 1001 |
| 7 | 0111 | 1101 | 1010 |
| 8 | 1000 | 1110 | 1011 |
| 9 | 1001 | 1111 | 1100 |

**P7.72.** We want to design logic circuits to convert the 4221 codewords of Problem P7.71 to BCD codewords. **a.** Fill in the Karnaugh map for $A$, placing $x$'s (don't cares) in the squares for 4221 codes that do not occur in the table. Find the minimum SOP expression allowing the various $x$'s to be either 1s or 0s to make the expression as simple as possible. **b.** Repeat (a) for $B$. **c.** Repeat (a) for $C$. **d.** Repeat (a) for $D$.

**P7.73.** Another code that is sometimes used to represent decimal digits is the excess-3 code.

To convert a decimal digit to excess-3, we add 3 to the digit and express the sum as a four-bit binary number. For example, to convert the decimal digit 9 to excess-3 code, we have

$$9_{10} + 3_{10} = 12_{10} = 1100_2$$

Thus, 1100 is the excess-3 codeword for 9. The excess-3 codewords for the other decimal digits are shown in Table P7.71.

We want to design logic circuits to convert BCD codewords to excess-3 codewords. **a.** Fill in the Karnaugh map for $W$, placing $x$'s (don't cares) in the squares for BCD codes that do not occur in the table. Find the minimum SOP expression allowing the various $x$'s to be either 1s or 0s to make the expression as simple as possible. **b.** Repeat (a) for $X$. **c.** Repeat (a) for $Y$. **d.** Repeat (a) for $Z$.

**P7.74.** We want to design logic circuits to convert the excess-3 codewords of Problem P7.73 to BCD codewords. **a.** Fill in the Karnaugh map for $A$, placing $x$'s (don't cares) in the squares for excess-3 codes that do not occur in the table. Find the minimum SOP expression allowing the various $x$'s to be either 1s or 0s to make the expression as simple as possible. **b.** Repeat (a) for $B$. **c.** Repeat (a) for $C$. **d.** Repeat (a) for $D$.

### Section 7.6: Sequential Logic Circuits

**P7.75.** Use NOR gates to draw the diagram of an $SR$ flip-flop. Repeat using NAND gates.

**P7.76.** Draw the circuit symbol and give the truth table for an $SR$ flip-flop.

**P7.77.** Draw the circuit symbol and give the truth table for a clocked $SR$ flip-flop.

**P7.78.** Explain the distinction between synchronous and asynchronous inputs to a flip-flop.

**P7.79.** What is edge triggering?

**P7.80.** Draw the circuit symbol and give the truth table for a positive-edge-triggered $D$ flip-flop.

**\*P7.81.** Assuming that the initial state of the shift register shown in Figure P7.81 is 100 (i.e., $Q_0 = 1$, $Q_1 = 0$, and $Q_2 = 0$), find the successive states. After how many shifts does the register return to the starting state?

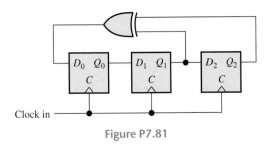

Figure P7.81

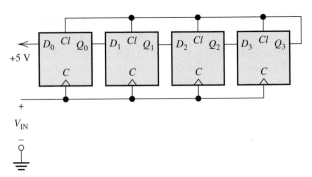

Figure P7.84

**P7.82.** Repeat Problem P7.81 if the XOR gate is replaced with **a.** an OR gate; **b.** an AND gate.

**P7.83.** The $D$ flip-flops of Figure P7.83 are positive-edge triggered. Assuming that prior to $t = 0$, the states are $Q_0 = Q_1 = 0$, sketch the voltage waveforms at $Q_0$ and $Q_1$ versus time. Assume logic levels of 0 V and 5 V.

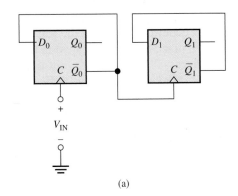

(a)

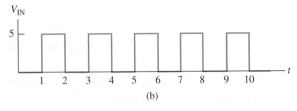

(b)

Figure P7.83

**P7.84.** The $D$ flip-flops of Figure P7.84 are positive-edge triggered, and the $Cl$ input is an asynchronous clear. Assume that the states are $Q_0 = Q_1 = Q_2 = Q_3 = 0$ at $t = 0$. The clock input $V_{IN}$ is shown in Figure P7.83. Sketch the voltage waveforms at $Q_0$, $Q_1$, $Q_2$, and $Q_3$ versus time. Assume logic levels of 0 V and 5 V.

**\*P7.85.** Use AND gates, OR gates, inverters, and a negative-edge-triggered $D$ flip-flop to show how to construct the $JK$ flip-flop of Figure 7.50 on page 391.

**P7.86.** Consider the ripple counter of Figure 7.53 on page 393. Suppose that the flip-flops have asynchronous clear inputs. Show how to add gates so that the count resets to zero immediately when the count reaches six. This results in a modulo-six counter.

**P7.87.** Figure P7.87 shows the functional diagram of an electronic die that can be used in games of chance. The system contains a high-speed clock, a push-button momentary contact switch that returns to the upper (logic 1) position when released, and a counter that counts through the cycle of states: 001, 010, 011, 100, 101, 110 (i.e., the binary equivalents of the number of spots on the various sides of the die). $Q_3$ is the MSB, and $Q_1$ is the LSB. The system has a display consisting of seven light-emitting diodes

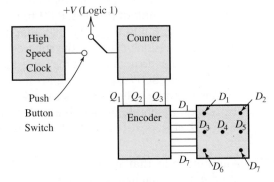

Figure P7.87 Electronic die.

(LED), each of which lights when logic 1 is applied to it. The encoder is a combinatorial logic circuit that translates the state of the counter into the logic signals needed by the display. Each time the switch is depressed, the counter operates, stopping in a random state when the switch is released. **a.** Use *JK* flip-flops having asynchronous preset and clear inputs to draw the detailed diagram of the counter. **b.** Design the encoder, using Karnaugh maps to minimize the logic elements needed to produce each of the seven output signals.

**P7.88.** Four LED are arranged at the corners of a diamond, as illustrated in Figure P7.88. When logic 1 is applied to an LED, it lights. Only one diode is to be on at a time. The on state should move from diode to diode either clockwise or counterclockwise, depending on whether *S* is high or low, respectively. One complete revolution should be completed in each two-second

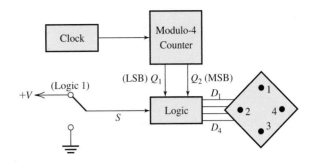

$S = 0$: Counterclockwise (1, 2, 3, 4, 1,...)
$S = 1$: Clockwise (1, 4, 3, 2, 1,...)

**Figure P7.88**

interval. **a.** What is the frequency of the clock? **b.** Draw a suitable logic circuit for the counter. **c.** Construct the truth table and use Karnaugh maps to determine the minimum SOP expressions for $D_1$ through $D_4$ in terms of $S$, $Q_1$, and $Q_2$.

## Practice Test

Here is a practice test you can use to check your comprehension of the most important concepts in this chapter. Answers can be found in Appendix D and complete solutions are included in the Student Solutions files.

See Appendix E for more information about the Student Solutions.

**T7.1.** First, think of one or more correct ways to complete each statement in Table T7.1(a).

## Table T7.1

| Item | Best Match |
|---|---|
| **(a)** | |

**a.** The truth table for a logic expression contains . . .
**b.** De Morgan's laws imply that . . .
**c.** If the higher voltage level represents logic 1 and the lower level represents logic 0, we have . . .
**d.** For a Gray code . . .
**e.** If we invert each bit of a binary number and then add 1 to the result, we have . . .
**f.** If we add two negative signed two's complement numbers, and the left-hand bit of the result is zero, we have had . . .
**g.** Squares on the top and bottom of a Karnaugh map are considered to be . . .
**h.** If the $\overline{Q}$ output of a positive-edge-triggered *D* flip-flop is connected to the *D* input, at each positive clock edge, the state of the flip-flop . . .

| Item | Best Match |
|------|-----------|
| **(a)** | |

**i.** Provided that it is not too large, noise can be completely eliminated from . . .
**j.** Registers are composed of . . .
**k.** An SOP logic circuit is composed of . . .
**l.** Working out from the decimal point and converting four-bit groups of a BCD number to their hexadecimal equivalents produces . . .

| **(b)** | |
|------|-----------|

1. the decimal equivalent
2. old recordings of music
3. inverse logic
4. congruent
5. OR gates
6. AND gates and one OR gate
7. digital signals
8. inverters, AND gates, and one OR gate
9. inverses
10. flip-flops
11. overflow
12. a listing of all combinations of inputs and the corresponding outputs
13. a table of ones and zeros
14. code words appear in numerical order
15. analog signals
16. adjacent
17. each code word is rotated to form the next word
18. if AND operations are changed to OR and vice versa, and the result is inverted, the result is equivalent to the original logic expression
19. NAND gates are sufficient to implement any logic expression
20. positive logic
21. the two's complement
22. negative logic
23. adjacent words differ in a single bit
24. underflow
25. toggles

Then, select the best choice from the list given in Table T7.1(b). [Items in Table T7.1(b) may be used more than once or not at all.]

**T7.2.** Convert the decimal integer, $353.875_{10}$ to each of these forms: **a.** binary; **b.** octal; **c.** hexadecimal; **d.** BCD.

**T7.3.** Find the octal equivalent of $FA.7_{16}$.

**T7.4.** Find the decimal equivalent for each of these eight-bit signed two's complement integers: **a.** 01100001; **b.** 10111010.

**T7.5.** For the logic circuit of Figure T7.5: **a.** write the logic expression for $D$ in terms of $A, B,$ and $C$ directly from the logic diagram; **b.** construct the truth table and the Karnaugh map;

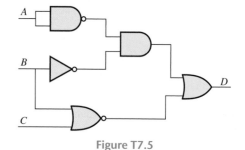

Figure T7.5

**c.** determine the minimum SOP expression for $D$; **d.** determine the minimum POS expression for $D$.

**T7.6.** Suppose we need a logic circuit with a logic output $G$ that is high only if a certain hexadecimal digit is 1, 5, $B$, or $F$. The inputs to the logic circuit are the bits $B_8$, $B_4$, $B_2$, and $B_1$ of the binary equivalent for the hexadecimal digit. (The MSB is $B_8$, and the LSB is $B_1$.) **a.** Fill in the Karnaugh map shown in Figure

T7.6. **b.** Determine the minimized SOP expression for $G$. **c.** Determine the minimum POS expression for $G$.

**T7.7.** Consider the shift register shown in Figure T7.7. Assuming that the initial shift-register state is 100 (i.e., $Q_0 = 1$, $Q_1 = 0$, and $Q_2 = 0$), list the next six states. After how many shifts does the register return to its initial state?

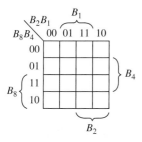

Figure T7.6 Karnaugh map to be filled in for G.

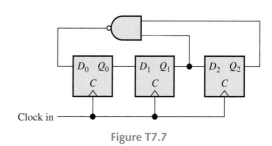

Figure T7.7

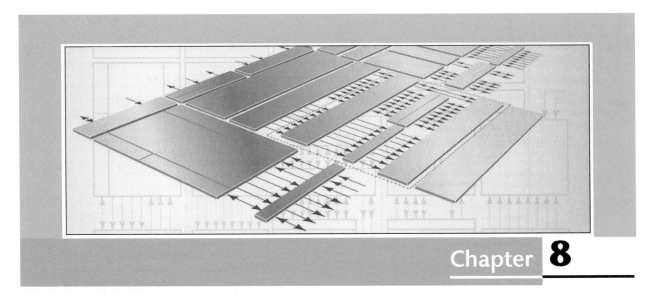

# Chapter 8

# Computers, Microcontrollers and Computer-Based Instrumentation Systems

## Study of this chapter will enable you to:

- Identify and describe the functional blocks of a computer.

- Define the terms *microprocessor*, *microcomputer*, and *microcontroller*.

- Select the type of memory needed for a given application.

- Understand how microcontrollers can be applied in your field of specialization.

- Identify the registers in the programmer's model and their functions for the HCS12/9S12 microcontroller family from Freescale Semiconductor, Inc.

- List some of the instructions and addressing modes of HCS12/9S12 microcontrollers.

- Write simple assembly language programs, using the CPU12 instruction set.

- Describe the operation of the elements of a computer-based instrumentation system.

- Identify the types of errors that may be encountered in instrumentation systems.

- Avoid common pitfalls such as ground loops, noise coupling, and loading when using sensors.

- Determine specifications for the elements of computer-based instrumentation systems such as data-acquisition boards.

## Introduction to this chapter:

Certainly you are familiar with general-purpose electronic computers that are used for business, engineering design, word processing, and other applications. Although it is sometimes not readily apparent, special-purpose computers can be found in many products such as automobiles, appliances, cameras, fax machines, garage-door openers, and instrumentation. An **embedded computer** is part of a product that is not called a computer. Virtually any recently manufactured device that is partly electrical in nature is almost certain to contain one or more embedded computers. Typical automobiles contain over 100 embedded computers. The emphasis of this chapter is on embedded computers.

Relatively simple computers for embedded control applications can be completely implemented on a single silicon chip costing less than a dollar. This type of computer is often called a **microcontroller** (MCU) and is useful for problems such as control of a washing machine, a printer, or a toaster.

In this chapter, we give an overview of MCU organization and instruction sets using the Freescale Semiconductor HCS12/9S12 family as an example. Hundreds of types of MCUs and their variations are in use, but most of the underlying concepts are similar from one to another. The primary intent is to give you an understanding of these basic concepts. Space is not available in this book for the intensive coverage of a particular MCU needed to prepare you to design complex mechatronic systems.

Instrumentation concepts are important in systems with embedded microcontrollers. We discuss the concepts related to computer-based instrumentation in the last several sections of this chapter.

Computer capability has advanced rapidly and costs have fallen dramatically, a trend that will continue for the foreseeable future. For the past several decades, the price for a given computer capability has been cut in half about every 18 months. You should view embedded MCUs as powerful, but inexpensive, resources that are appropriate for solving virtually any control or instrumentation problem in your field of engineering, no matter how complex or mundane the problem may be.

> An embedded computer is part of a product, such as an automobile, printer, or bread machine, that is not called a computer.

## 8.1  COMPUTER ORGANIZATION

Figure 8.1 shows the system-level diagram of a computer. The **central processing unit** (CPU) is composed of the **arithmetic/logic unit** (ALU) and the **control unit**.

The ALU carries out arithmetic and logic operations on data such as addition, subtraction, comparison, or multiplication. Basically, the ALU is a logic circuit similar to those discussed in Chapter 7 (but much more complex).

The control unit supervises the operation of the computer, such as determining the location of the next instruction to be retrieved from memory and setting up the ALU to carry out operations on data. The ALU and control unit contain various **registers** that hold operands, results, and control signals. (Recall from Section 7.6 that a register is simply an array of flip-flops that can store a word composed of binary digits.) Later in this chapter, we will discuss the functions of various CPU registers for the CPU12 which is used in the Freescale HCS12/9S12 family.

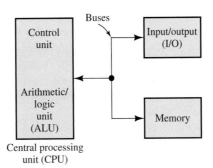

Figure 8.1  A computer consists of a central processing unit, memory, buses, and input output devices.

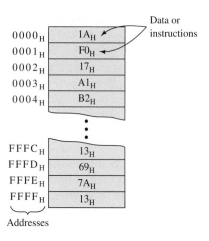

Figure 8.2 A 64-Kbyte memory that has $2^{16} = 65{,}536$ memory locations, each of which contains one byte (eight bits) of data. Each location has a 16-bit address. It is convenient to represent the addresses and data in hexadecimal form as shown. The addresses range from $0000_H$ to $FFFF_H$. For example, the memory location $0004_H$ contains the byte $B2_H = 10110010_2$.

## Memory

Several notations are used for hexadecimal numbers, including the subscript 16, the subscript H, and the prefix $. Thus, $F2_{16}$, $F2_H$, and $F2 are alternative ways to indicate that F2 is a hexadecimal number.

**Memory** can be thought of as a sequence of locations that store data and instructions. Each memory location has a unique address and typically stores one byte of data, which can conveniently be represented by two hexadecimal digits. (Of course, within the computer circuits, data appear in binary form.)

Usually, memory capacity is expressed in Kbytes, where $1\text{ K} = 2^{10} = 1024$. Similarly, 1 Mbyte is $2^{20} = 1.048{,}576$ bytes. A 64-Kbyte memory is illustrated in Figure 8.2. Under the direction of the control unit, information can either be written to or read from each memory location. (We are assuming that we have the read/write type of memory. Later, we will see that another type, known as read-only memory, can also be very useful.)

## Programs

Programs are sequences of instructions stored in memory. Typically, the controller fetches (i.e., retrieves) an instruction, determines what operation is called for by the instruction, fetches data from memory as required, causes the ALU to perform the operation, and writes results back to memory. Then, the next instruction is fetched, and the process is repeated. We will see that the CPU12 can execute a rich variety of instruction types.

## Buses

The various elements of a computer are connected by **buses**, which are sets of conductors that transfer multiple bits at a time. For example, the **data bus** transfers data (and instructions) between the CPU and memory (or I/O devices). In small computers, the width of the data bus (i.e., the number of bits that can be transferred at a time) is typically eight bits. Then, one byte can be transferred between the CPU and memory (or I/O) at a time. (The bus is wider in more powerful general-purpose CPUs such as those found in personal computers, which typically have data-bus widths of 64 bits.)

Several **control buses** are used to direct the operations of the computer. For example, one control bus sends the addresses for memory locations (or I/O devices) as well as signals that direct whether data are to be read or written. With an address bus width of 16 bits, $2^{16} = 64\text{K}$ of memory locations (and I/O devices) can

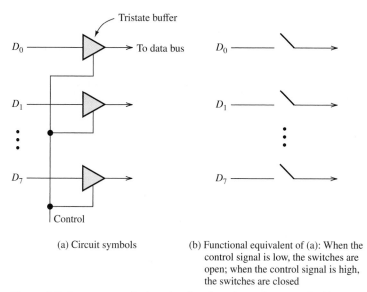

(a) Circuit symbols

(b) Functional equivalent of (a): When the control signal is low, the switches are open; when the control signal is high, the switches are closed

**Figure 8.3** Data are applied to the data bus through tristate buffers, which can function either as closed or as open switches.

be addressed. Another control bus internal to the CPU transfers signals from the control unit to the ALU. These control signals direct the ALU to perform a particular operation, such as addition.

Buses can be bidirectional. In other words, they can transfer data in either direction. Let us consider the data bus connecting the CPU and memory. Of course, the memory and the CPU cannot apply conflicting data signals to the bus at the same time. Conflict is avoided by transferring data to the bus through **tristate buffers**, as illustrated in Figure 8.3. Depending on the control signal, the tristate buffers function either as open or as closed switches. When a byte of data is to be transferred from the CPU to memory, the tristate buffers are enabled (switches closed) on the CPU and disabled (switches open) in the memory. The data inputs of both the CPU and the memory are connected to the bus at all times, so data can be accepted from the bus as desired. Thus, the data from the CPU appear on the bus and can be stored by the memory. When data are to be transferred from memory to the CPU, the conditions of the tristate buffers are reversed.

## Input Output

Some examples of I/O devices are keyboards, display devices, and printers. An important category of input devices in control applications are **sensors**, which convert temperatures, pressures, displacements, flow rates, and other physical values to digital form that can be read by the computer. **Actuators** are output devices such as valves, motors, and switches that allow the computer to affect the system being controlled.

Some computers are said to have **memory-mapped I/O**, in which I/O devices are addressed by the same bus as memory locations. Then, the same instructions used for storing and reading data from memory can be used for I/O. Other computers have a separate address bus and instructions for I/O. We discuss primarily systems that use memory-mapped I/O.

A **microprocessor** is a CPU contained on a single integrated-circuit chip. The first microprocessor was the Intel 4004, which appeared in 1971 and cost several thousand dollars each. Subsequently, microprocessors have dramatically fallen in price and

A microcontroller is a complete computer containing the CPU, memory, and I/O on a single silicon chip.

increased in performance. A **microcomputer**, such as a PC or a laptop, combines a microprocessor with memory and I/O chips. A MCU combines CPU, memory, buses, and I/O on a single chip and is optimized for embedded control applications. We give a more detailed overview of the HCS12/9S12 MCU family later in this chapter.

There are several variations of computer organization. In computers with **Harvard architecture**, there are separate memories for data and instructions. If the same memory contains both data and instructions, we have **von Neumann architecture**. The HCS12/9S12 MCU family uses von Neumann architecture.

**Exercise 8.1**    Suppose that a microprocessor has an address bus width of 20 bits. How many memory locations can it access?
**Answer**    $2^{20} = 1,048,576 = 1024$ K $= 1$ M.    ☐

**Exercise 8.2**    How many bits can be stored in a 64-Kbyte memory?
**Answer**    524,288.    ☐

## 8.2  MEMORY TYPES

Several types of memory are used in computers: (1) Read-and-write memory (RAM), (2) Read-only memory (ROM), and (3) Mass storage. We discuss each type in turn. Then, we consider how to select the best type of memory for various applications.

### RAM

**Read-and-write memory** (RAM) is used for storing data, instructions, and results during execution of a program. Semiconductor RAM consists of one or more silicon integrated circuits (each of which has many storage cells) and control logic so that information can be transferred into or out of the cell specified by the address.

RAM and ROM do not incur any loss of speed when the memory locations are accessed in random order. In fact, RAM originally meant random-access memory.

Usually, the information that is stored in RAM is lost when power is removed. Thus, we say that RAM is **volatile**. Originally, the acronym RAM meant random-access memory, but the term has changed its meaning over time. As the term is used now, RAM means volatile semiconductor memory. (Actually, RAM is also available with small batteries that maintain information in the absence of other power.)

The time required to access data in RAM is the same for all memory locations. The fastest RAM is capable of access times of a few nanoseconds. No time penalty is incurred by accessing locations in random order.

There are two types of RAM in common use. In **static RAM**, the storage cells are *SR* flip-flops that can store data indefinitely, provided that power is applied continuously. In **dynamic RAM**, information is stored in each cell as charge (or lack of charge) on a capacitor. Because the charge leaks off the capacitors, it is necessary to refresh the information periodically. This makes the use of dynamic RAM more complex than the use of static RAM. The advantage of dynamic RAM is that the basic storage cell is smaller, so that chips with larger capacities are available. A relatively small amount of RAM is needed in most control applications, and it is simpler to use static RAM.

An 8K-word by 8-bit static RAM chip is illustrated in Figure 8.4. The chip has 13 address lines, eight data lines, and three control lines. The "bubbles" on the control input lines indicate that they are active when low. Unless the chip select line is low, the chip neither stores data nor places data on the data bus. If both the output-enable and the chip-select inputs are low, the data stored in the location specified by the address appear on the data lines. If both the write-enable and the chip-select lines

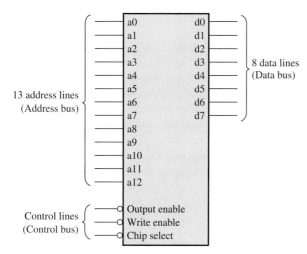

**Figure 8.4** A generic 8K-word by 8-bit word RAM.

are low, the data appearing on the data bus are stored in the location specified by the address signals. In normal operation, both the output enable and write enable lines are never low at the same time.

## ROM

In normal operation, **read-only memory** (ROM) can be read, but not written to. The chief advantages of ROM are that data can be read quickly in random order and that information is not lost when power is turned off. Thus, we say that ROM is **non-volatile** (i.e., permanent). ROM is useful for storing programs such as the *boot program*, which is executed automatically when power is applied to a computer. In simple dedicated applications such as the controller for a clothes washer, all of the programs are stored in ROM.

Several types of ROM exist. For example, in **mask-programmable ROM**, the data are written when the chip is manufactured. A substantial cost is incurred in preparing the mask that is used to write the data while manufacturing this type of ROM. However, mask-programmable ROM is the least expensive form of ROM when the mask cost is spread over a sufficiently large number of units. Mask-programmable ROM is not a good choice if frequent changes in the information stored are necessary, as in initial system development.

In **programmable read-only memory** (PROM), data can be written by special circuits that blow tiny fuses or leave them unblown, depending on whether the data bits are zeros or ones. Thus, with PROM, we write data once and can read it as many times as desired. PROM is an economical choice if a small number of units are needed.

**Erasable PROM** (EPROM) is another type that can be erased by exposure to ultraviolet light (through a window in the chip package) and rewritten by using special circuits. **Electrically erasable PROMs** (EEPROMs) can be erased by applying proper voltages to the chip. Although we can write data to an EEPROM, the process is much slower than for RAM.

**Flash memory** is a non-volatile technology in which data can be erased and rewritten relatively quickly in blocks of locations, ranging in size from 512 bytes up to 512 Kbytes. Flash memory has a limited lifetime, typically on the order of 10 thousand to 100 thousand read/write cycles. Flash is a rapidly advancing technology and may eventually replace hard drives for mass storage in general purpose computers.

The primary advantage of ROM is that it is non-volatile. Information stored in RAM is lost when power is interrupted.

## Mass Storage

Mass-storage units include hard disks and flash memory, both of which are read/write memory. Another type is CD-ROM and DVD-ROM disks, which are used for storing large amounts of data. Mass storage is the least expensive type of memory per unit of capacity. With all forms of mass storage except flash, a relatively long time is required to access a particular location. Initial access times for mass storage range upward from several milliseconds, compared with fractions of a microsecond for RAM or ROM. However, if mass-storage locations are accessed sequentially, the transfer rate is considerably higher (but still lower than for RAM or ROM). Usually, data and instructions need to be accessed quickly in random order during execution of a program. Thus, programs are stored in RAM or ROM during execution.

> Hard disks, CD-ROMs, and DVDs are examples of sequential memories in which access is faster if memory locations are accessed in order.

## Selection of Memory

The main considerations in choosing the type of memory to be used are:

1. The trade-off between speed and cost.
2. Whether the information is to be stored permanently or must be changed frequently.
3. Whether data are to be accessed in random order or in sequence.

In general-purpose computers, programs and data are read into RAM before execution from mass-storage devices such as hard disks. Because many different programs are used, it is not practical to store programs in semiconductor ROM, which would be too expensive for the large memory space required. Furthermore, information stored in ROM is more difficult to modify compared to data stored on a hard disk. We often find a small amount of ROM used for the startup or boot program in general-purpose computers, but most of the memory is RAM and mass storage.

On the other hand, in embedded MCUs, programs are usually stored in semiconductor ROM, and only a small amount of RAM is needed to store temporary results. For example, in a controller for a television receiver, the programs for operating the TV are stored in ROM, but time and channel information entered by the user is stored in RAM. In this application, power is applied to the RAM even when the TV is "turned off." However, during a power failure, the data stored in RAM are lost (unless the TV has a battery backup for its RAM). Usually, we do not find mass-storage devices used in embedded computers.

## 8.3 DIGITAL PROCESS CONTROL

Figure 8.5 shows the general block diagram of a control scheme for a physical process such as an internal combustion engine. Various physical inputs such as power and material flow are regulated by actuators that are in turn controlled by the MCU.

Some actuators are analog and some are digital. Examples of digital actuators are switches or valves that are either on or off, depending on the logic value of their control signals. Digital actuators can be controlled directly by digital control lines. Analog actuators require an analog input. For example, the rudder of an airplane may deflect in proportion to an analog input signal. Then, **digital-to-analog** (D/A) **converters** are needed to convert the digital signals to analog form before they are applied to analog actuators.

> D/A and DAC are both acronyms for digital-to-analog converter.

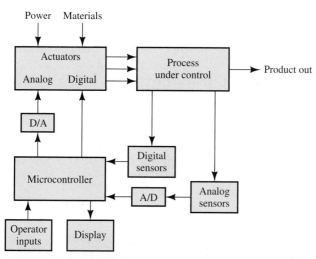

**Figure 8.5** Microcontroller-based control of a physical process.

Various sensors produce electrical signals related to process parameters (such as temperature, pressure, pH, velocity, or displacement) of the process under control. Some sensors are digital and some are analog. For example, a pressure sensor may consist of a switch that closes producing a high output signal when pressure exceeds a particular value. On the other hand, an analog pressure sensor produces an output voltage that is proportional to pressure. **Analog-to-digital** (A/D) **converters** are used to convert the analog sensor signals to digital form.

Often, a display is provided so that information about the process can be accessed by the operator. A keyboard or other input device enables the operator to direct operation of the control process.

Many variations of the system shown in Figure 8.5 are possible. For example, sometimes we simply want to instrument a process and present information to the operator. This is the situation for automotive instrumentation in which sensors provide signals for speed, fuel reserve, oil pressure, engine temperature, battery voltage, and so on. These data are presented to the driver by one or more displays.

Actuators, sensors, and I/O tend to be unique to each application and do not lend themselves to integration with the MCU. A/D and D/A converters often are included within the MCU. Thus, a typical system consists of an MCU, sensors, actuators, and I/O devices. Systems may not contain all of these elements. Within a given MCU family, variations are usually available with respect to the amount and type of memory, the number of A/D channels, and so forth, that are included on the chip.

Virtually any system can be controlled or monitored by an MCU. Here is a short list: traffic signals, engines, chemical plants, antiskid brakes, manufacturing processes, stress measurement in structures, machine tools, aircraft instrumentation, monitoring of patients in a cardiac-care unit, nuclear reactors, and laboratory experiments.

### Interrupts versus Polling

In many control applications, the MCU must be able to respond to certain input signals very quickly. For example, an overpressure indication in a nuclear power plant may require immediate attention. When such an event occurs, the MCU must **interrupt** what it is doing and start a program known as an **interrupt handler** that determines the source of the interrupt and takes appropriate action. Many MCUs have hardware capability and instructions for handling these interrupts.

Instead of using interrupts, an MCU can use **polling** to determine if any parts of the system need attention. The processor checks each sensor in turn and takes appropriate actions as needed. However, continuous polling is wasteful of processor time. In complex applications, the processor may be required to carry out extensive, but lower-priority activities, much of the time. In this case, interrupts provide faster response to critical events than polling does. For a breadmaker, polling would be acceptable because no activity ties up the MCU for more than a few milliseconds. Furthermore, any of the actions required could be delayed by a few tens of milliseconds without undue consequences.

## PRACTICAL APPLICATION    8.1

### Fresh Bread Anyone?

Let us consider a relatively simple MCU application: a breadmaker. Possibly, you have had experience with this popular appliance. The chef measures and adds the ingredients (flour, water, dried milk, sugar, salt, yeast, and butter) to the bread pan, makes selections from the menu by using a keypad, and takes out a finished loaf of bread after about four hours.

The diagram of a bread machine is shown in Figure PA8.1. There are three digital actuators in the bread machine: a switch to control the heating element, a switch for the mixing and kneading motor, and a switch for the fan used to cool the loaf

after baking is finished. Analog actuators are not needed in this application.

An analog sensor is used to measure temperature. The sensor output is converted to digital form by an A/D converter.

A timer circuit is part of the MCU that is initially loaded with the time needed to complete the loaf. The timer is a digital circuit that counts down, similar to the counter circuits discussed in Section 7.6.

The timer indicates the number of hours and minutes remaining in the process. The MCU can

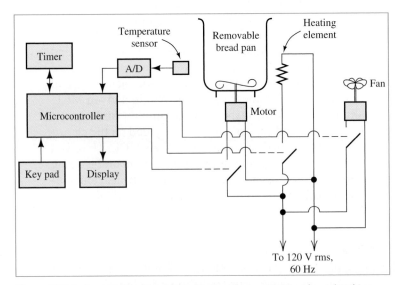

**Figure PA8.1** A relatively simple application for an MCU—a breadmaking machine.

read the time remaining and use it to make decisions. Time remaining to completion of the loaf is also displayed for the convenience of the chef.

The control programs are stored in ROM. The parameters entered by the chef are written into RAM (for example, whether the bread crust is to be light, medium, or dark). The MCU continually checks the time remaining and the temperature. By executing the program stored in ROM, the computer determines when the machine should mix the ingredients, turn on the heating element to warm the dough and cause it to rise, knead the dough, bake, or cool down. The duration and temperature of the various parts of the cycle depend on the initial selections made by the chef.

First, the machine mixes the ingredients for several minutes, and the heating element is turned on to warm the yeast that makes the dough rise. While the dough is rising, a warm temperature is required, say 90°F. Thus, the heating element is turned on and

the MCU reads the temperature frequently. When the temperature reaches the desired value, the heating element is turned off. If the temperature falls too low, the element is again turned on.

The MCU continues to check the time remaining and the temperature. According to the programs stored in ROM and the parameters entered by the chef (which are saved in RAM), the motors and heating element are turned on and off.

In this application, about 100 bytes of RAM would be needed to store information entered by the operator and temporary data. Also, about 16 Kbytes of ROM would be needed to store the programs. Compared to the total price of the appliance, the cost of this amount of ROM is very small. Therefore, many variations of the program can be stored in ROM, and bread machines can be very versatile. In addition to finished loaves of bread, they can also bake cakes, cook rice, make jam, or prepare dough for other purposes, such as cinnamon rolls.

## 8.4 PROGRAMMING MODEL FOR THE HCS12/9S12 FAMILY

Earlier, we discussed a generic computer shown in Figure 8.1 on page 409. In this section, we give a more detailed internal description of the HCS12/9S12 MCU family from Freescale Semiconductor. Space does not allow us to discuss all of the features, instructions, and programming techniques for these MCUs. However, we will describe the programming model, selected instructions, and a few simple programs to give you a better understanding of how MCUs can be used for embedded applications that you will encounter in your field.

### The HCS12/9S12 Programming Model

The ALU and the control unit contain various registers that are used to hold operands, the address of the next instruction to be executed, addresses of data, and results. For example, the programmer's model for the CPU12 is illustrated in Figure 8.6. (Actually, the MCUs contain many other registers–only the registers of concern to the programmer are shown in the figure; thus, Figure 8.6 is often called the programming model.)

The **accumulators** are general-purpose registers that hold one of the arguments and the result of all arithmetic and logical operations. Registers A and B each contain 8 bits with the least significant bit on the right (bit 0 in Figure 8.6) and the most significant bit on the left. Sometimes A and B are used as separate registers, and other times they are used in combination as a single 16-bit register, denoted as register D. It is important to remember that D is not separate from A and B.

It is important to remember that register D is not separate from registers A and B.

The **program counter** (PC) is a 16-bit register that contains the address of the first byte of the next instruction to be fetched (read) from memory by the control

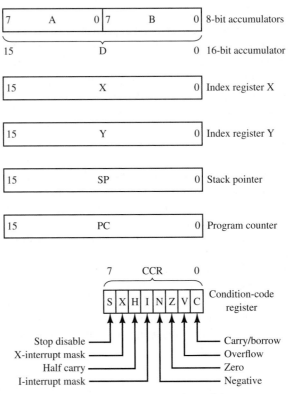

Figure 8.6 The CPU12 programmer's model.

unit. The size of the PC is the same as the size of memory addresses; thus, the memory potentially contains up to $2^{16} = 64$ K locations, each of which contains one byte of data or instructions as illustrated in Figure 8.2 on page 410.

The **index registers** X and Y are mainly used for a type of addressing (of data) known as indexed addressing, which we will discuss later.

The **condition-code register** (CCR) is an 8-bit register in which each bit depends either on a condition of the processor or on the result of the preceding logic or arithmetic operation. The details of the CCR are shown in Figure 8.6. For example, the carry bit C (bit 0 of the CCR) is set (to logic 1) if a carry (or borrow) occurred in the preceding arithmetic operation. Bit 1 (overflow or V) is set if the result of the preceding operation resulted in overflow or underflow. Bit 2 (zero or Z) is set to 1 if the result of the preceding operation was zero. Bit 3 (negative or N) is set if the result was negative. The meaning and use of the remaining bits will be discussed as the need arises.

## Stacks and the Stack Pointer Register

A **stack** is a sequence of locations in memory used to store information such as the contents of the program counter and other registers when a subroutine is executed or when an interrupt occurs. (We discuss subroutines shortly.) As the name implies, information is added to (pushed onto) the top of the stack and later read out (pulled off) in the reverse order that it was written. This is similar to adding plates to the top of a stack when clearing a dinner table and then taking the plates off the top of the stack when loading a dishwasher. After data are pulled off the stack, they are considered to no longer exist in memory and are written over by later push commands. The first word pushed onto the stack is the last to be pulled off, and stacks are called **last-in first-out memories** (LIFOs).

The **stack pointer** is a CPU register that keeps track of the address of the top of the stack. Each time the content of a register is pushed onto the stack, the content of the stack pointer is decremented by one if the register contained one byte. If the register contained two bytes, the stack-pointer content is decreased by two. (Addresses are smaller in value as we progress upward in the stack.) Conversely, when data is pulled from the stack and transferred to a register, the stack-pointer content is increased by one or two (depending on the length of the register). When the content of one of the 8-bit registers (A, B, or CCR) is pushed onto the stack (by the commands PSHA, PSHB, or PSHC, respectively), these operations take place:

1. The content of the stack pointer is reduced by one.
2. The content of the 8-bit register is stored at the address corresponding to the content of the stack pointer.

When the content of one of the 16-bit registers D, X, or Y is pushed onto the stack (by the commands PSHD, PSHX, or PSHY), the following operations take place:

1. The content of the stack pointer is decremented by one and the least significant byte (bits 8 through 15) of the content of the 16-bit register is stored at the address corresponding to the content of the stack pointer.
2. The content of the stack pointer is again decremented by one, and the most significant byte of the content of the 16-bit register is stored at the address corresponding to the content of the stack pointer.

In pulling data off of the stack, the operations are reversed. For an 8-bit register (commands PULA, PULB, or PULC):

1. The data in the memory location pointed to by the stack pointer is stored in the register.
2. The content of the stack pointer is incremented by one.

For a 16-bit register (commands PULD, PULX, or PULY):

1. The data in the memory location to which the stack pointer points is stored in the high byte of the register, and the content of the stack pointer is incremented by one.
2. The data in the memory location to which the stack pointer points is stored in the low byte of the register, and the content of the stack pointer is again incremented by one.

Figure 8.7 illustrates the effects of the command sequence PSHA, PSHB, PULX. Figure 8.7(a) shows the original contents of pertinent registers and memory locations. (Memory locations always contain something; they are never blank. However, when the content of a memory location is unknown or does not matter, we have left the location blank.) Figure 8.7(b) shows the new contents after the command PSHA has been executed. Notice that the initial content of register A has been stored in location 090A and that the content of SP has been decremented by one. (Furthermore, the initial contents of A and B are unchanged.) Figure 8.7(c) shows the contents after the command PSHB has been executed. Notice that the content of register B has been stored in location 0909 and that the content of SP has

Stacks are last-in first-out memories. Information is added to (pushed onto) the top of the stack and eventually read out (pulled off) in the reverse order that it was written.

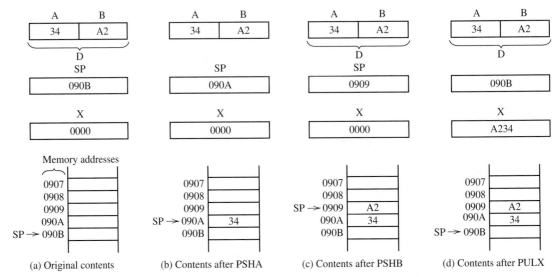

**Figure 8.7** Register and memory contents for the command sequence: PSHA, PSHB, PULX.

been decremented by one. Finally, Figure 8.7(d) shows the new contents after the command PULX has been executed. Notice that the content of memory location 0909 has been stored in the first byte of register X and the content of memory location 090A has been stored in the second byte of register X. Also, the content of SP has been increased by two.

**Exercise 8.3**  Starting from the initial contents shown in Figure 8.7(a), determine the content of register X after execution of the command sequence PSHB, PSHA, PULX.
**Answer**  The content of the X register is 34A2.  ☐

**Exercise 8.4**  Starting from the initial contents shown in Figure 8.7(a), determine the content of register X after the command sequence PSHX, PSHA, PULX.
**Answer**  The content of the X register is 3400.  ☐

**Exercise 8.5**  Suppose that initially the contents of all memory locations in the stack are 00, and the stack pointer register contains 0806. Then, the following operations occur in sequence:

1. The data byte $A7_H$ is pushed onto the stack.
2. $78_H$ is pushed onto the stack.
3. One byte is pulled from the stack.
4. FF is pushed onto the stack.

List the contents of memory locations 0800 through 0805 after each step. Also, give the content of the stack pointer (SP).
**Answer**  After step 1, we have:

```
0800: 00    SP: 0805
0801: 00
0802: 00
0803: 00
0804: 00
0805: A7
```

After step 2, we have:
```
0800: 00   SP: 0804
0801: 00
0802: 00
0803: 00
0804: 78
0805: A7
```

After step 3, we have:
```
0800: 00   SP: 0805
0801: 00
0802: 00
0803: 00
0804: 78
0805: A7
```

After step 4, we have:
```
0800: 00   SP: 0804
0801: 00
0802: 00
0803: 00
0804: FF (New data replaces old.)
0805: A7
```

□

## 8.5 THE INSTRUCTION SET AND ADDRESSING MODES FOR THE CPU12

Computers excel at executing simple instructions, such as quickly and accurately adding a number stored in a given memory location to the content of a specified register. Computers are capable of highly sophisticated and seemingly intelligent behavior by following instruction sequences called **programs** or **software**. These are prepared by a human programmer.

Unfortunately, even the smallest oversight on the part of the programmer can render a program useless until the error is corrected. A substantial part of the effort in designing an MCU-based controller is in writing software. To be effective, the programmer must be fully knowledgeable about the fine details of the instruction set for the MCU in use. Our objective in this and the next section is simply to give you a brief overview, not to make you an expert programmer.

*A substantial part of the effort in designing an MCU-based controller is in writing software.*

In general, instruction sets are similar between different MCU types, but details differ. Once one has mastered programming of a given machine, it is much easier to learn and make good use of the instruction set of another processor. Here again, we take the CPU12 as an example. More details about it can be readily found on the Web.

### Instructions for the CPU12

A selected set of instructions for the CPU12 is listed in Table 8.1. The first column in the table gives the mnemonic for each instruction, the second column is a brief description and an equivalent Boolean expression for the instruction. For example, the ABA instruction adds the content of register B to the content of A with the result residing in A. We can indicate this operation as

*(A) represents the content of register A.*

$$(A) + (B) \rightarrow A$$

as shown in the second column of the table.

**Table 8.1.** Selected Instructions for the CPU12

| Source Form | Operation | Addr. Mode | Machine Code | S | X | H | I | N | Z | V | C |
|---|---|---|---|---|---|---|---|---|---|---|---|
| ABA | Add Accumulators $(A) + (B) \rightarrow A$ | INH | 18  06 | - | - | ↕ | - | ↕ | ↕ | ↕ | ↕ |
| ADDA (opr) | Add Memory to A $(A) + (M) \rightarrow A$ | IMM<br>DIR<br>EXT<br>IDX | 8B  ii<br>9B  dd<br>BB  hh  ll<br>AB  * | - | - | ↕ | - | ↕ | ↕ | ↕ | ↕ |
| ADDB (opr) | Add Memory to B $(B) + (M) \rightarrow B$ | IMM<br>DIR<br>EXT<br>IDX | CB  ii<br>DB  dd<br>FB  hh  ll<br>EB  * | - | - | ↕ | - | ↕ | ↕ | ↕ | ↕ |
| ADDD (opr) | Add Memory to D $(D) + (M: M + 1) \rightarrow D$ | IMM<br>DIR<br>EXT<br>IDX | C3  jj  kk<br>D3  dd<br>F3  hh  ll<br>E3  * | - | - | - | - | ↕ | ↕ | ↕ | ↕ |
| BCS (rel) | Branch if Carry Set   (if C = 1) | REL | 25  rr | - | - | - | - | - | - | - | - |
| BEQ (rel) | Branch if Equal   (if Z = 1) | REL | 27  rr | - | - | - | - | - | - | - | - |
| BLO (rel) | Branch if Lower[U]   (if C = 1) | REL | 25  rr | - | - | - | - | - | - | - | - |
| BMI (rel) | Branch if Minus[s]   (if N = 1) | REL | 2B  rr | - | - | - | - | - | - | - | - |
| BNE (rel) | Branch if Not Equal   (if Z = 0) | REL | 26  rr | - | - | - | - | - | - | - | - |
| BPL (rel) | Branch if Plus[s]   (if N = 0) | REL | 2A  rr | - | - | - | - | - | - | - | - |
| BRA (rel) | Branch Always | REL | 20  rr | - | - | - | - | - | - | - | - |
| CLRA | Clear Accumulator A   $00 \rightarrow A$ | INH | 87 | - | - | - | - | 0 | 1 | 0 | 0 |
| CLRB | Clear Accumulator B   $00 \rightarrow B$ | INH | C7 | - | - | - | - | 0 | 1 | 0 | 0 |
| COMA | Complement Accumulator A   $(\overline{A}) \rightarrow A$ | INH | 41 | - | - | - | - | ↕ | ↕ | 0 | 1 |
| INCA | Increment Accumulator A   $(A) + \$01 \rightarrow A$ | INH | 42 | - | - | - | - | ↕ | ↕ | ↕ | - |
| INCB | Increment Accumulator B   $(B) + \$01 \rightarrow B$ | INH | 52 | - | - | - | - | ↕ | ↕ | ↕ | - |
| INX | Increment Index Register X $(X) + \$0001 \rightarrow X$ | INH | 08 | - | - | - | - | - | ↕ | - | - |
| JMP (opr) | Jump Routine Address $\rightarrow$ PC | EXT<br>IDX | 06  hh  ll<br>05  * | - | - | - | - | - | - | - | - |
| JSR (opr) | Jump to Subroutine (See Text) | DIR<br>EXT<br>IDX | 17  dd<br>16  hh  ll<br>15  * | | | | | | | | |
| LDAA (opr) | Load Accumulator A $(M) \rightarrow A$ | IMM<br>DIR<br>EXT<br>IDX | 86  ii<br>96  dd<br>B6  hh  ll<br>A6  * | - | - | - | - | ↕ | ↕ | 0 | - |
| LDAB (opr) | Load Accumulator B $(M) \rightarrow B$ | IMM<br>DIR<br>EXT<br>IDX | C6  ii<br>D6  dd<br>F6  hh  ll<br>E6  * | - | - | - | - | ↕ | ↕ | 0 | - |

**Table 8.1.** Selected Instructions for the CPU12 (*Cont.*)

| Source Form | Operation | Addr. Mode | Machine Code | S | X | H | I | N | Z | V | C |
|---|---|---|---|---|---|---|---|---|---|---|---|
| LDD (opr) | Load Accumulator D $(M):(M+1) \rightarrow D$ | IMM<br>DIR<br>EXT<br>IDX | CC jj kk<br>DC dd<br>FC hh ll<br>EC * | - | - | - | - | ↕ | ↕ | 0 | - |
| LDX (opr) | Load Index Register X $(M):(M+1) \rightarrow X$ | IMM<br>DIR<br>EXT<br>IDX | CE jj kk<br>DE dd<br>FE hh ll<br>EE * | - | - | - | - | ↕ | ↕ | 0 | - |
| LDY (opr) | Load Index Register Y $(M):(M+1) \rightarrow Y$ | IMM<br>DIR<br>EXT<br>IDX | CD jj kk<br>DD dd<br>FD hh ll<br>ED * | - | - | - | - | ↕ | ↕ | 0 | - |
| MUL | Multiply A by BU $(A) \times (B) \rightarrow D$ | INH | 12 | - | - | - | - | - | - | - | ↕ |
| NOP | No Operation | INH | A7 | - | - | - | - | - | - | - | - |
| PSHA | Push A onto Stack $(SP) - 1 \Rightarrow SP; (A) \Rightarrow M_{(SP)}$ | INH | 36 | - | - | - | - | - | - | - | - |
| PSHB | Push B onto Stack $(SP) - 1 \Rightarrow SP; (B) \Rightarrow M_{(SP)}$ | INH | 37 | - | - | - | - | - | - | - | - |
| PSHX | Push X onto Stack $(SP) - 2 \Rightarrow SP; (X_H : X_L) \Rightarrow M_{(SP)}: M_{(SP+1)}$ | INH | 34 | - | - | - | - | - | - | - | - |
| PSHY | Push Y onto Stack $(SP) - 2 \Rightarrow SP; (Y_H : Y_L) \Rightarrow M_{(SP)}: M_{(SP+1)}$ | INH | 35 | - | - | - | - | - | - | - | - |
| PULA | Pull A from Stack $(M_{(SP)}) \Rightarrow A; (SP) + 1 \Rightarrow SP$ | INH | 32 | - | - | - | - | - | - | - | - |
| PULB | Pull B from Stack $(M_{(SP)}) \Rightarrow B; (SP) + 1 \Rightarrow SP$ | INH | 33 | - | - | - | - | - | - | - | - |
| PULX | Pull X from Stack $(M_{(SP)}: M_{(SP+1)}) \Rightarrow X_H : X_L; (SP) + 2 \Rightarrow SP$ | INH | 30 | - | - | - | - | - | - | - | - |
| PULY | Pull Y from Stack $(M_{(SP)}: M_{(SP+1)}) \Rightarrow Y_H : Y_L; (SP) + 2 \Rightarrow SP$ | INH | 31 | - | - | - | - | - | - | - | - |
| RTS | Return from Subroutine $(M_{(SP)}: M_{(SP+1)}) \Rightarrow PC; (SP) + 2 \Rightarrow SP$ | INH | 3D | - | - | - | - | - | - | - | - |
| STAA (opr) | Store Accumulator A $(A) \rightarrow M$ | DIR<br>EXT<br>IDX | 5A dd<br>7A hh ll<br>6A * | - | - | - | - | ↕ | ↕ | 0 | - |
| STAB (opr) | Store Accumulator B $(B) \rightarrow M$ | DIR<br>EXT<br>IDX | 5B dd<br>7B hh ll<br>6B * | - | - | - | - | ↕ | ↕ | 0 | - |
| STD (opr) | Store Accumulator D $(A) \rightarrow M; (B) \rightarrow M + 1$ | DIR<br>EXT<br>IDX | 5C dd<br>7C hh ll<br>6C * | - | - | - | - | ↕ | ↕ | 0 | - |

**Table 8.1.** Selected Instructions for the CPU12 (*Cont.*)

| Source Form | Operation | Addr. Mode | Machine Code | Condition Codes S | X | H | I | N | Z | V | C |
|---|---|---|---|---|---|---|---|---|---|---|---|
| STOP | Stop Internal Clocks. If S control bit = 1, the STOP instruction is disabled and acts like a NOP | INH | 18   3E | - | - | - | - | - | - | - | - |
| TSTA | Test Accumulator A;   (A) − 00 | INH | 97 | - | - | - | - | $\updownarrow$ | $\updownarrow$ | 0 | 0 |
| TSTB | Test Accumulator B;   (B) − 00 | INH | D7 | - | - | - | - | $\updownarrow$ | $\updownarrow$ | 0 | 0 |

S      indicates instructions that are intended for two's complement signed numbers
U      indicates instructions that are intended for unsigned numbers
*      For indexed addressing (IDX). Only the first byte of the machine coded is given. An additional one to three bytes are needed. The details are beyond the scope of our discussion.
ii      8-bit immediate data
dd     low byte of a direct address
hh ll   high and low bytes of an extended address
jj kk   high and low bytes of 16-bit immediate data
rr      signed 8-bit offset in branch instruction

Mnemonics are easy for humans to remember. However, in the microcomputer memory, the instructions are stored as machine codes (or op codes) consisting of one or more 8-bit numbers, each of which is represented in the table as a two-digit hexadecimal number. For example, in the row for the ABA instruction, we see that the op code is 1806. Thus, the ABA instruction appears in memory as the binary numbers 00011000 and 00000110.

Look at the row for the ADDA(opr) instruction in which (opr) stands for the address of a memory location. The effect of the instruction is to add the content of a memory location to the content of accumulator A with the result residing in A. This is represented by the expression

$$(A) + (M) \rightarrow A$$

in which (M) represents the content of a memory location. Several **addressing modes** can be used to select the memory location to be accessed by some instructions. For example, the ADDA instruction can use any of several addressing modes. We will discuss the CPU12 addressing modes shortly.

Table 8.1 also shows the effect of each instruction on the contents of the CCR. The meanings of the symbols shown for each bit of the condition code are

-      the bit is unchanged by this instruction
0      the bit is always cleared by this instruction
1      the bit is always set by this instruction
$\updownarrow$      the bit is set or cleared depending on the result

The CPU12 has many more instructions than those listed in the table; we have just given a sample of various kinds. Next, we briefly describe each of the addressing modes used by the CPU12.

## Extended (EXT) Addressing

Recall that the CPU12 uses 16 bits (usually written as four hexadecimal digits) for memory addresses. In extended addressing, the complete address of the operand is included in the instruction. Thus, the instruction

ADDA $CA01

adds the content of memory location CA01 to the content of register A. (Later, we will see that a program called an assembler is used to convert the mnemonics to op codes. The $ sign indicates to the assembler that the address is given in hexadecimal form.) The op codes appear in three successive memory locations as

In CPU12 assembly language, a prefix of $ indicates that the number is hexadecimal.

BB      (op code for ADDA with extended addressing)
CA      (high byte of address)
01      (low byte of address)

Notice that the high byte of the address is given first followed by the low byte.

## Direct (DIR) Addressing

In **direct addressing**, only the least significant two (hexadecimal) digits of the address are given, and the most significant two digits are assumed to be zero. Therefore, the effective address falls between 0000 and 00FF. For example, the instruction

ADDA $A9

adds the content of memory location 00A9 to the content of register A. The instruction appears in two successive memory locations as

9B      (the op code for ADDA with direct addressing)
A9      (the low byte of the address)

Notice that the same result could be obtained by using extended addressing, in which case the instruction would appear as

ADDA $00A9

However, the extended addressing form of the instruction would occupy three bytes of memory, rather than two with direct addressing. Furthermore, the direct addressing form is completed more quickly.

## Inherent (INH) Addressing

Some instructions, such as ABA, access only the MCU registers. We say that this instruction uses **inherent addressing**. An instruction sequence that adds the numbers in locations 23A9 and 00AA, then stores the result in location 23AB is

LDAA $23A9      (extended addressing, load A from location 23A9)
LDAB $AA        (direct addressing, load B from location 00AA)
ABA             (inherent addressing, add B to A)
STAA $23AB      (extended addressing, store result in 23AB)

## Immediate (IMM) Addressing

In **immediate addressing**, which is denoted by the symbol#, the address of the operand is the address immediately following the instruction. For example, the instruction ADDA #$83 adds the hexadecimal number 83 to the contents of A. It is stored in two successive memory locations as

In CPU12 assembly language, the symbol # indicates immediate addressing.

8B      (op code for ADDA with immediate addressing)
83      (operand)

Because A is a single-byte register, only one byte of memory is needed to store the operand.

On the other hand, D is a double-byte (16-bit) register, and its operand is assumed to occupy two memory bytes. For example, the instruction ADDD #$A276 adds the two-byte hexadecimal number A276 to the contents of D. It is stored in three successive memory locations as

C3   (op code for ADDD with immediate addressing)
A2   (high byte of operand)
76   (low byte of operand)

### Indexed (IDX) Addressing

Indexed addressing is useful when we want to access a list of items either one after another in order, or perhaps skipping forward or backward through the list by twos, threes, etc. The CPU12 has a variety of indexed addressing options. (In Table 8.1, we have grouped all of these options under the IDX label. Only the first bite of the machine code, of two to four in all, is given in the table. This has been done because of space limitations in this chapter.)

**Constant-Offset Indexed Addressing.** In constant-offset indexed addressing, the effective address is formed by adding a signed offset to the content of a selected CPU register (X, Y, SP, or PC). The contents of X, Y, SP, or PC are not changed in this type of addressing. Suppose that X contains $1005 and Y contains $200A. Then, some examples of source code using this type of addressing and their effects are

STAA 5,X       store the content of A in location $100A
STD − 3,Y      store the content of D in locations $2007 and $2008
ADDB $A,X      add the content of location $100F to register B

**Accumulator-Offset Indexed Addressing.** In this form of addressing, the content of one of the accumulators (A, B, or D) is added as an unsigned number to the content of a designated register (X, Y, SP, or PC) to obtain the effective address. For example, if X contains $2000 and A contains $FF, the command

LDAB A,X

loads the content of memory location $20FF into B. The contents of A and X remain the same after the command is completed as they were before the command.

Next, we discuss four types of indexed addressing that increment or decrement the contents of the selected CPU register (X, Y, or SP) either before or after the instruction is carried out. The amount of the increment or decrement can range from 1 to 8 and the selected CPU register contains the incremented or decremented value after the instruction is completed.

**Auto Pre-Incremented Indexed Addressing.** Suppose that X contains $1005. Then, the instruction

STAA 5,+X

pre-increments X so the content of X becomes $100A and the content of A is stored in location $100A. X contains $100A after the completion of the instruction.

**Auto Pre-Decremented Indexed Addressing.** Again, suppose that X contains $1005. Then, the instruction

STD 5,-X

pre-decrements X so the content of X becomes $1000 and the content of D is stored in locations $1000 and $1001. X contains $1000 after the completion of the instruction.

Notice that the sign preceding X determines whether we have a pre-increment or a pre-decrement. On the other hand, if the algebraic sign comes after the register name, we have either a post-increment or post-decrement. Here again, the increment or decrement can range from 1 to 8.

**Auto Post-Incremented Indexed Addressing.** Suppose that X contains $1005. Then, the instruction

STAA 5,X+

stores the content of A in location 1005 and then increments X so the content of X becomes $100A. As before, the increment can vary from 1 to 8.

**Auto Post-Decremented Indexed Addressing.** Again, suppose that X contains $1005. Then, the instruction

STAA 3,X-

stores the content of A in location $1005 and then decrements X so the content of X becomes $1002.

**Indirect Indexed Addressing.** In this type of addressing, a 16-bit (or equivalently, four-digit hexadecimal) constant given in the command is added to the content of a selected CPU register (X, Y, SP, or PC). This results in a pointer to a location containing the address of the operand. To illustrate an example, first assume that the contents of the CPU registers and some of the memory locations are as shown in Figure 8.8. Then, if the command

LDY [$1002, X]

is executed, $1002 is added to the content of X resulting in $2003. The content of locations $2003 and $2004 contain the high byte and low byte for the starting address of the operand. Thus, the starting address of the operand is $3003. Finally, the content of locations $3003 and $3004 is written to register Y. Thus, after the execution of this command Y contains $A3F6.

Instead of specifying an offset value in the command, the content of register D can be used. Thus, in Figure 8.8, the command

LDY[D,X]

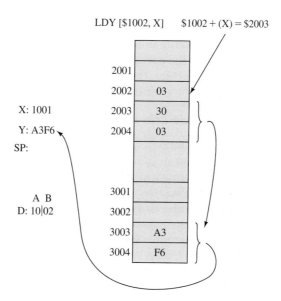

**Figure 8.8** Illustration of the command LDY [$1002, X].

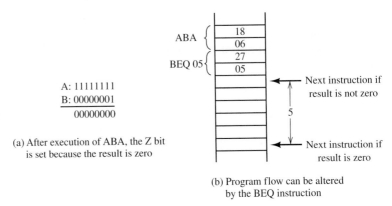

A: 11111111
B: 00000001
00000000

(a) After execution of ABA, the Z bit
is set because the result is zero

(b) Program flow can be altered
by the BEQ instruction

Figure 8.9  Using the BEQ instruction.

adds the content of D to the content of X again resulting in $2003. The content of locations $2003 and $2004 is $3003. This is the (initial) address of the operand. So $A3F6 is loaded into register Y as it was for the command LDY[$1002,X].

To recap, the steps followed by indexed indirect addressing are:

1. Add either the offset or the content of D to the content of the CPU register (X, Y, SP, or PC) specified in the command.

2. Go to the address resulting from step 1. The content of this location and the next is the address of the operand.

### Relative Addressing

Branch instructions are used to alter the sequence of program flow. Recall that if we add two numbers, the Z bit of the condition code register is clear if the result was not zero and is set if the result was zero. The BEQ (branch if result equals zero) command can be used to change the program flow depending on the value of the Z bit.

In the CPU12, branch instructions use only relative addressing. Conversely, branches are the only instructions that use relative addressing.

For example, suppose that initially the A register contains FF and the B register contains 01. Then, if the instruction ABA is executed, the binary addition shown in Figure 8.9(a) is performed. Because the result is zero, the Z bit of the condition code register is set. If the branch instruction BEQ $05 is executed, the next instruction is the content of the program counter plus the offset (which is 05 in this case). If the Z bit had been clear, the instruction immediately following the branch instruction would have been executed. This is illustrated in Figure 8.9(b).

### Machine Code and Assemblers

We have seen that ADDA is a mnemonic for an instruction executed by the processor. It turns out that in the CPU12, the instruction ADDA with extended addressing is stored in memory as BB = $10111011_2$. We say that BB is the **machine code** for the instruction ADDA with extended addressing. Machine codes are also known as **operation codes** or simply **op codes**. In extended addressing, the address of the operand is stored in the two memory locations immediately following the instruction code. The instruction ADDA $070A appears in three successive memory locations as

BB
07
0A

| | | | | |
|---|---|---|---|---|
| 0200 | 10 | | 0000 | EF |
| 0201 | 04 | | 0001 | 01 |
| 0202 | 1A | | 0002 | AE |
| 0203 | 10 | | 0003 | F1 |
| 0204 | 07 | | 0004 | 78 |
| 0205 | FF | | 0005 | 96 |
| 0206 | A3 | | 0006 | 13 |
| 0207 | 16 | | 0007 | A4 |

**Figure 8.10** Contents of memory for Exercise 8.6.

It would be a daunting task for a human to make conversions from instruction mnemonics to machine codes, whereas computers excel at this type of task. Furthermore, mnemonics are much easier for us to remember than machine codes. Thus, we generally start writing a program by using mnemonics. A computer program called an **assembler** is then employed to convert the mnemonics into machine code. Assemblers also help in other chores associated with programming, such as converting decimal numbers to hexadecimal and keeping track of branching addresses and operand addresses. We will have more to say about assembly language in the next section.

**Exercise 8.6**   Suppose that the contents of certain memory locations are as shown in Figure 8.10. Furthermore, the content of register A is zero and the X register contains 0200 prior to execution of each of these instructions:

**a.** LDAA $0202
**b.** LDAA #$43
**c.** LDAA $05,X
**d.** LDAA $06
**e.** LDAA $07,X-
**f.** LDAA $05,+X

Find the contents of registers A and X after each instruction.
**Answer**    **a.** Register A contains 1A and X contains 0200. **b.** Register A contains 43 and X contains 0200. **c.** Register A contains FF and X contains 0200. **d.** Register A contains 13 and X contains 0200 **e.** Register A contains 10 and X contains 01F9. **f.** Register A contains FF and X contains 0205. ☐

**Exercise 8.7**   Suppose that starting in location 0200 successive memory locations contain op codes for the instructions

```
CLRA
BEQ $15
```

**a.** Show the memory addresses and contents (in hexadecimal form) for these instructions.
**b.** What is the address of the instruction executed immediately after the branch instruction?

**Answer    a.** The memory addresses and contents are:

    0200:87    (op code for CLRA)

    0201:27    (op code for BEQ)

    0202:15    (offset for branch instruction)

**b.** The address of the next instruction is 0218.    □

## 8.6    ASSEMBLY-LANGUAGE PROGRAMMING

A program consists of a sequence of instructions used to accomplish some task. No doubt, you have been introduced to programming that uses high-level languages such as BASIC, C, Java, MATLAB, or Pascal. When using high-level languages, a **compiler** or **interpreter** converts program statements into machine code before they are executed. It would be much too tedious to write machine-language programs for sophisticated engineering analysis. In application-oriented software, such as computer-aided design packages, even greater emphasis is placed on making the programs easy to use. However, in writing programs for embedded computers in control applications, we often need to keep the number of instructions relatively small and to minimize the time required to execute the various operations; quick response to events in the system being controlled can be highly important.

Though we may program MCUs for control applications in machine language, it is possible to relieve much of the drudgery by using an **assembler**. This provides many conveniences, such as allowing us to write instructions with mnemonics, using labels for memory addresses, and including user in the source program file.

In practice, we write the program as **source code** using a text editor on a general-purpose computer, called the **host computer**. The source code is then converted to **object code** (machine code) by the assembler program. Finally, the machine code is loaded into the memory of the MCU, which is called the **target system**. Sometimes, we say that the source code is written in assembly language. Assembly language code, nevertheless, is very close to the actual machine code executed by the computer.

In general, CPU12 assembly language statements take the following form:

```
LABEL    INSTRUCTION/DIRECTIVE    OPERAND    COMMENT
```

Typically each line of source code is converted into one machine instruction. Some of the source code statements, called **directives**, however, are used to give commands to the assembler. One of these is the origin directive ORG. For example,

```
ORG    $0100
```

instructs the assembler to place the first instruction following the directive in memory location 0100 of the target system.

In CPU12 assembly language, labels must begin in the first column. The various fields are separated by spaces. Thus, when we want ORG to be treated as a directive, rather than as a label, we need to place one or more spaces ahead of it. If the first character of a line is a semicolon, the line is ignored by the assembler. Such lines are useful for comments and line spaces that make the source code more understandable to humans.

*Usually, there are many ways to write a program to accomplish a given task.*

In writing a program, we start by describing the algorithm for accomplishing the task. We then create a sequence of instructions to carry out the algorithm. Usually, there are many ways to write a program to accomplish a given task.

## Example 8.1   An Assembly-Language Program

Suppose that we want a program starting in memory location 0400 that retrieves the number stored in location 0500, adds 5 to the number, writes the result to location 0500, and then stops. (We will use only the instructions listed in Table 8.1, even though the CPU12 has many additional instructions, which often could make our programs shorter.)

**Solution**   The source code is:

```
; SOURCE CODE FOR EXAMPLE  8.1
; THIS LINE IS A COMMENT THAT IS IGNORED BY THE ASSEMBLER
;
        ORG     $0400       ;ORIGIN DIRECTIVE
BEGIN   LDAA    $0500       ;LOAD NUMBER INTO A
        ADDA    #$05        ;ADD 5, IMMEDIATE ADDRESSING
        STAA    $0500       ;STORE RESULT
        STOP
        END                 ;END DIRECTIVE
```

Comments have been included to explain the purpose of each line. BEGIN is a label that identifies the address of the LDAA instruction. (In this case, BEGIN has a value of 0400.) If we wanted to reference this location somewhere in a more complex program, the label would be useful. STOP is the mnemonic for the instruction that halts further action by the MCU. END is a directive that informs the assembler that there are no further instructions. ∎

## Example 8.2   Absolute Value Assembly Program

Write the source code for a program starting in location $0300 that loads register A with the signed two's-complement number in location $0200, computes its absolute value, returns the result to location $0200, clears the A register, and then stops. Use the instructions listed in Table 8.1. (Assume that the initial content of location $0200 is never $1000000 = -128_{10}$ which does not have a positive equivalent in 8-bit two's complement form.)

**Solution**   Recall that branch instructions (also known as conditional instructions) allow different sets of instructions to be executed depending on the values of certain bits in the condition code register. For example, in Table 8.1, we see that the branch on plus instruction (BPL) causes a branch if the N bit of the condition code register is clear (i.e., logic 0).

Testing occurs automatically in many instructions. For example, in the load A instruction LDAA, the N and Z bits of the condition code register are set if the value loaded is negative or zero, respectively.

Our plan is to load the number, compute its two's complement if it is negative, store the result, clear the A register, and then stop. Recall that one way to find the two's complement is to first find the one's complement and add one. If the number is positive, no calculations are needed. The source code is:

```
; SOURCE CODE FOR EXAMPLE  8.2
;
        ORG     $0300       ;ORIGIN DIRECTIVE
        LDAA    $0200       ;LOAD NUMBER INTO REGISTER A
        BPL     PLUS        ;BRANCH IF A IS POSITIVE
        COMA                ;ONES'S COMPLEMENT
```

```
            INCA                    ;ADD ONE TO FORM TWO'S COMPLEMENT
            STAA        $0200       ;RETURN THE RESULT TO MEMORY
    PLUS    CLRA                    ;CLEAR A
            STOP
            END                     ;DIRECTIVE
```

In this program, the number is first loaded into register A from memory location $0200. If the number is negative (i.e., if the most significant bit is 1), the N bit of the CCR is set (logic 1); otherwise, it is not set. If the N bit is zero, the BPL PLUS instruction causes the next instruction executed to be the one starting in the location labeled PLUS. On the other hand, if the N bit is one, the next instruction is the one immediately following the branch instruction. Thus, if the content of the memory location is negative, the two's complement is computed to change its sign. Then, the result is written to the original location and A is cleared.    ■

Next, to illustrate some of the chores performed by the assembler, we manually convert the source code of the previous example to machine code.

### Example 8.3    Manual Conversion of Source Code to Machine Code

Manually determine the machine code for each memory location produced by the source code of Example 8.2. What is the value of the label PLUS? (*Hint:* Use Table 8.1 to determine the op codes for each instruction.)

**Solution**    The assembler ignores the title and other comments. Because of the ORG directive, the machine code is placed in memory starting at location 0300. The memory addresses and their contents are:

```
0300: B6 Op code for LDAA with extended addressing.
0301: 02 High byte of address.
0302: 00 Low byte of address.
0303: 2A Op code for BPL which uses relative addressing.
0304: 05 Offset (On the first pass this value is unknown.)
0305: 41 Op code for COMA which computes one's complement.
0306: 42 Op code for INCA.
0307: 7A Op code for STAA with extended addressing.
0308: 02 High byte of address.
0309: 00 Low byte of address.
030A: 87 Clear A.
030B: 18 Halt processor.
030C: 3E
```

A comment has been added to explain each line; however, the assembler does not produce these comments.

Recall that branch instructions use relative addressing. The required offset for the BPL command is not known on the first pass through the source code. However, after the first pass, we see that the location corresponding to the label PLUS is 030A and that an offset of 05 is needed for the BPL command. The END directive does not produce object code.    ■

### Subroutines

Sometimes, certain sequences of instructions are used over and over in many different places. Memory is saved if these sequences are stored once and used wherever needed. A sequence of instructions such as this is called a **subroutine**. At

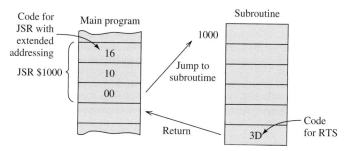

**Figure 8.11** Illustration of the jump-to-subroutine command with extended addressing. (Other addressing modes are allowed with the JSR instruction.)

any point in the main program that the subroutine needs to be executed, we place the *J*ump to *S*ub*R*outine instruction:

```
JSR address
```

in which address is a direct, extended, or indexed address of the first instruction of the subroutine. At the end of the subroutine, we place the *Re T*urn from *S*ubroutine command

```
RTS
```

which causes the next instruction to be taken from the location following the JSR instruction in the main program. This is illustrated in Figure 8.11.

The stack is used to keep track of where to return after the subroutine is finished. The address of the instruction following the JSR is pushed onto the system stack when the jump is executed. This address is pulled off the stack and loaded into the program counter when the return instruction is executed. After the subroutine is completed, the next instruction executed is the one following the JSR.

One of the chores that the assembler can perform is to keep track of the starting addresses of the subroutines. We simply label the first instruction of the subroutine in the source code. This is convenient because we usually don't know where subroutines will eventually be located when writing programs. After all of the source code is written, the assembler can calculate the amount of memory needed for each portion of the program and determine the subroutine starting addresses, which are then substituted for the labels.

---

**Example 8.4**   **Subroutine Source Code**

Assume that the content of register A is a signed two's-complement number *n*. Using the instructions of Table 8.1, write a subroutine called SGN that replaces the content of A with +1 (in signed two's-complement form) if *n* is positive, replaces it with −1 if *n* is negative, and does not change the content of A if *n* is zero.

**Solution**   The source code for the subroutine is:

```
;  SOURCE CODE FOR SUBROUTINE OF  EXAMPLE 8.4
;
SGN        TSTA                    ;TEST CONTENT OF A
           BEQ      END            ;BRANCH IF A IS ZERO
           BPL      PLUS           ;BRANCH IF A IS POSITIVE
           LDAA     #$FF           ;LOAD -1, IMMEDIATE ADDRESSING
           JMP      END            ;JUMP TO END OF SUBROUTINE
```

```
PLUS      LDAA    #$01        ;LOAD +1, IMMEDIATE ADDRESSING
END       RTS                 ;RETURN FROM SUBROUTINE
```

First, the number is tested. If it is zero, the Z flag is set. If it is negative, the N flag is set. Next, if the number is zero, the BEQ instruction compels a branch to END, causing a return from the subroutine. If the number is positive, the subroutine branches to PLUS, loads the hexadecimal code for the signed two's-complement representation of +1, and returns. If the number is negative, the LDA #FF instruction is executed, followed by the return. [Notice that in this subroutine, END is a label (rather than a directive) because it begins in column 1.] ■

**Exercise 8.8**    Write a program starting in location $0100 that adds $52_{10}$ to the content of location $0500, stores the result in location $0501, and then stops. Assume that all values are represented in signed two's-complement form.

**Answer**    The source code is:

```
; SOLUTION FOR  EXERCISE 8.8
;
          ORG       $0100
          LDAA      #$34        ;LOAD HEX EQUIVALENT OF 52 BASE TEN
          ADDA      $0500       ;ADD CONTENT OF 0500
          STAA      $0501       ;STORE RESULT IN 0501
          STOP
          END                   ;DIRECTIVE
```

**Exercise 8.9**    Write a subroutine named MOVE that tests the content of register A. If A is not zero, the subroutine should move the content of location 0100 to 0200. Otherwise, no move is made. The contents of A, B, X, and Y must be the same on return as before the subroutine is called.

**Answer**    One version of the desired subroutine is:

```
; SUBROUTINE FOR  EXERCISE 8.9.
;
MOVE      TSTA              ;TEST CONTENT OF A
BEQ       END               ;BRANCH IF CONTENT OF A IS ZERO
          PSHA              ;SAVE CONTENT OF A ON STACK
          LDAA $0100        ;LOAD CONTENT OF MEMORY LOCATION 0100
          STAA $0200        ;STORE CONTENT OF A IN LOCATION 0200
          PULA              ;RETRIEVE ORIGINAL CONTENT OF A
END       RTS               ;RETURN FROM SUBROUTINE
```

## Resources for Additional Study

In this chapter, we have given a very brief overview of MCUs focusing on the HCS12/9S12 family of devices from Freescale Semiconductor. Much more detail about the HCS12/9S12 can be found at www.freescale.com.

We have emphasized assembly language programming because it gives a clear picture of the inner workings of MCUs. As MCUs have gained higher speed and complexity, the trend has been away from assembly language and toward higher-level languages such as C.

If you are interested in applying MCUs to a project of your own, you should take a course devoted to MCUs exclusively or a course in mechatronics. Hands-on work is very important in learning about MCUs. Typically, one starts with a training board containing the MCU of interest, provisions for prototyping, LEDs, a small display, switches, and other components. Such boards are often equipped with an interface so programs can be downloaded to the MCU from a host computer through a USB link. An example for the HCS12 MCU is the Dragon12-Plus-USB board from EVBplus.com.

Design and construction of robots, model railroads, remotely controlled model airplanes equipped with video cameras, and so forth can become an engrossing hobby for those with an interest in combining MCUs, mechanical systems, and electronic elements. For numerous examples, look at *Nuts and Volts Magazine* at www.nutsvolts. com or *Make Magazine* at www.makezine.com. These may give you some good ideas for a senior project.

The Arduino MCU boards are very popular with artists, do-it-yourself enthusiasts, and students. You will find many articles related to these boards in *Nuts and Volts* and *Make Magazine*. Also, look at http://spectrum.ieee.org/geek-life/ hands-on/the-making-of-arduino/0.

## 8.7  MEASUREMENT CONCEPTS AND SENSORS

### Overview of Computer-Based Instrumentation

Figure 8.12 shows a computer-based system for instrumentation of a physical system such as an automobile or chemical process. Physical phenomena such as temperatures, angular speeds, displacements, and pressures produce changes in the voltages, currents, resistances, capacitances, or inductances of the **sensors**. If the sensor output is not already a voltage, **signal conditioners** provide an **excitation source** that transforms the changes in electrical parameters to voltages. Furthermore, the signal conditioner amplifies and filters these voltages. The conditioned signals are input to a **data-acquisition** (DAQ) board. On the DAQ board, each of the conditioned signals is sent to a **sample-and-hold circuit** (S/H) that periodically samples the signal and holds the value steady while the **multiplexer** (MUX) connects it to the (A/D or ADC) that converts the values to digital words. The words are read by the computer, which then processes the data further before storing and displaying the results. For example, the signals derived from a force sensor and a velocity sensor could be multiplied to obtain a plot of power versus time. Furthermore, the power could be integrated to show energy expended versus time. Long-term statistical analysis of a process can be carried out to facilitate quality control.

Computer-based instrumentation systems consist of four main elements: sensors, a DAQ board, software, and a general-purpose computer.

In this section, we consider sensors. Then, in the next several sections, we discuss other aspects of computer-based DAQ systems.

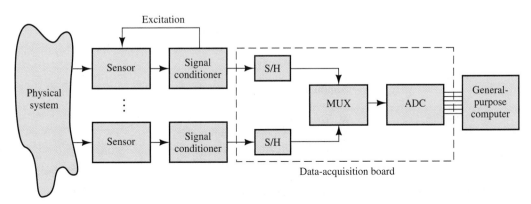

Data-acquisition board

**Figure 8.12** Computer-based DAQ system.

### Sensors

We emphasize sensors or signal conditioners that produce electrical signals (usually voltages) which are analogous to the physical quantity, or **measurand**, to be measured. Often, the voltage is proportional to the measurand. Then, the sensor voltage is given by

$$V_{\text{sensor}} = Km \tag{8.1}$$

in which $V_{\text{sensor}}$ is the voltage produced by the sensor, $K$ is the **sensitivity constant**, and $m$ is the measurand. For example, a load cell is a sensor consisting of four strain-gauge elements (see page 30) connected in a Wheatstone bridge (see page 107) and bonded to a load-bearing element. As force is applied to the load cell, a proportional voltage appears across two terminals of the bridge. **Excitation** in the form of a constant voltage is applied to the other two terminals of the bridge. For a given excitation voltage, the sensitivity constant has units of V/N or V/lbf.

Some examples of measurands and sensor types are shown in Table 8.2. These are only a few of the many types of sensors that are available.

A good source of detailed information about computer-based instrumentation and control, including sensors, is the National Instruments, website: www.ni.com.

**Table 8.2** Measurands and Sensor Types

| Measurands | Sensor types |
| --- | --- |
| Acceleration | Seismic mass accelerometers |
| | Piezoelectric accelerometers |
| Angular displacement | Rotary potentiometers |
| | Optical shaft encoders |
| | Tachometric generators |
| Light | Photoconductive sensors |
| | Photovoltaic cells |
| | Photodiodes |
| Liquid level | Capacitance probes |
| | Electrical conductance probes |
| | Ultrasonic level sensors |
| | Pressure sensors |
| Linear displacement | Linear variable differential transformers (LVDTs) (see page 152) |
| | Strain gauges (see page 30) |
| | Potentiometers |
| | Piezoelectric devices |
| | Variable-area capacitance sensors |
| Force/torque | Load cells |
| | Strain gauges |
| Fluid flow | Magnetic flowmeters (see page 758) |
| | Paddle wheel sensors |
| | Constriction-effect pressure sensors |
| | Ultrasonic flow sensors |
| Gas flow | Hot-wire anemometers |
| Pressure | Bourdon tube/linear variable differential transformer combinations |
| | Capacitive pressure sensors |
| Proximity | Microswitches |
| | Variable-reluctance proximity sensors |
| | Hall-effect proximity sensors |
| | Optical proximity sensors |
| | Reed-switch sensors |
| Temperature | Diode thermometers |
| | Thermistors |
| | Thermocouples |

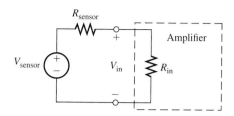

**Figure 8.13** Model for a sensor connected to the input of an amplifier.

## Equivalent Circuits and Loading

An equivalent circuit that applies to many sensors is shown in Figure 8.13; the source voltage $V_{sensor}$ is analogous to the measurand, and $R_{sensor}$ is the Thévenin resistance. Frequently, as part of the signal conditioning, the sensor voltage must be amplified. Figure 8.13 shows the sensor connected to the input terminals of an amplifier. (Amplifiers are discussed in Chapter 10.) Looking into the input terminals of any amplifier, we see a finite impedance, which is represented as $R_{in}$ in Figure 8.13. Using the voltage-division principle, we have

$$V_{in} = V_{sensor}\frac{R_{in}}{R_{in} + R_{sensor}} \tag{8.2}$$

Because of the current flowing through the circuit, the amplifier input voltage is less than the internal voltage of the sensor. This effect is known as **loading**. Loading is unpredictable and, therefore, undesirable. Provided that $R_{in}$ is very large compared with $R_{sensor}$, the amplifier input voltage is nearly equal to the internal sensor voltage. Thus, when we need to measure the internal voltage of the sensor, we should specify a signal-conditioning amplifier having an input impedance that is much larger in magnitude than the Thévenin impedance of the sensor.

When we need to measure the internal voltage of the sensor, we should specify a signal-conditioning amplifier having an input impedance that is much larger in magnitude than the Thévenin impedance of the sensor.

---

| Example 8.5 | Sensor Loading |
|---|---|

Suppose that we have a temperature sensor for which the open-circuit voltage is proportional to temperature. What is the minimum input resistance required for the amplifier so that the system sensitivity constant changes by less than 0.1 percent when the Thévenin resistance of the sensor changes from 15 kΩ to 5 kΩ?

**Solution**   The sensitivity constant is proportional to the voltage division ratio between the input resistance and the Thévenin resistance of the sensor. We require that this ratio changes by 0.1% (or less) when the Thévenin resistance changes. Thus, with resistances in kΩ, we have

$$V_{sensor}\frac{R_{in}}{15 + R_{in}} \geq 0.999V_{sensor}\frac{R_{in}}{5 + R_{in}}$$

Solving, we determine that $R_{in}$ is required to be greater than 9985 kΩ.   ∎

## Sensors with Electrical Current Output

Some types of sensors produce electrical current that is proportional to the measurand. For example, with suitable applied voltages, photodiodes produce currents that are proportional to the light intensities falling on the diodes. A photodiode is shown

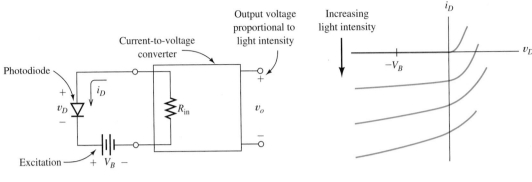

(a) Photodiode light sensor connected to a
current-to-voltage converter

(b) Volt–ampere characteristic of photodiode

**Figure 8.14** Photodiode light-sensing system. Because the diode voltage should be constant, $R_{in}$ should ideally equal zero.

in Figure 8.14(a). Like the load cell, the photodiode requires a constant-voltage excitation source. Figure 8.14(b) shows diode current versus diode voltage for various light intensities. If we want the current to depend only on light intensity, the diode voltage must be held nearly constant.

Usually a **current-to-voltage converter** (also known as a transresistance amplifier) is used to produce an output voltage that is proportional to the photodiode current. As in the case of amplifiers, we see an impedance looking into the input terminals of the current-to-voltage converter. This is shown as $R_{in}$ in Figure 8.14(a). In order for the diode voltage to remain constant as the current varies, $R_{in}$ must be very small (so the voltage across it is negligible). Thus, when we want to sense the current produced by a sensor, we use a current-to-voltage converter having a very small (ideally zero) input impedance magnitude.

> When we want to sense the current produced by a sensor, we need a current-to-voltage converter having a very small (ideally zero) input impedance magnitude.

### Variable-Resistance Sensors

Other sensors produce a changing resistance in response to changes in the measurand. For example, the resistance of a thermistor changes with temperature. Changes in resistance can be converted to changes in voltage by driving the sensor with a constant current source. To avoid loading effects, the voltage is applied to a high-input impedance amplifier as illustrated in Figure 8.15. Similar circuits that use ac excitation can convert changes in capacitance or inductance into voltage changes.

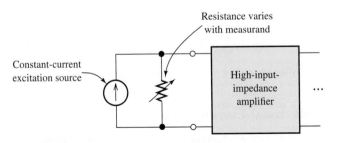

**Figure 8.15** Variable-resistance sensor.

## Errors in Measurement Systems

Many types of errors can occur in making measurements. We define the error of a measurement as

$$\text{Error} = x_m - x_{\text{true}} \tag{8.3}$$

in which $x_m$ is the measured value and $x_{\text{true}}$ is the actual or true value of the measurand. Often, error is expressed as a percentage of the full-scale value $x_{\text{full}}$ (i.e., the maximum value that the system is designed to measure).

$$\text{Percentage error} = \frac{x_m - x_{\text{true}}}{x_{\text{full}}} \times 100\% \tag{8.4}$$

There are many possible sources of error, some of which are specific to particular measurands and measurement systems. However, it is useful to classify the types of errors that can occur. Some are **bias errors**, also called **systematic errors**, that are the same each time a measurement is repeated under the same conditions. Sometimes, bias errors can be quantified by comparing the measurements with more accurate standards. For example, we could calibrate a weight scale by using it to measure the weights of highly accurate standards of mass. Then, the calibration data could be used to correct subsequent weight measurements.

Bias errors include **offset**, **scale error**, **nonlinearity**, and **hysteresis**, which are illustrated in Figure 8.16. Offset consists of a constant that is added to, or subtracted from, the true value. Scale error produces measurement errors that are proportional to the true value of the measurand. Nonlinearity can result from improper design or overdriving an electronic amplifier. When hysteresis

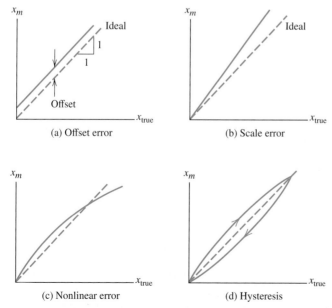

**Figure 8.16** Illustration of some types of instrumentation error. $x_m$ represents the value of the measurand reported by the measurement system, and $x_{\text{true}}$ represents the true value.

error is present, the error depends on the direction and distance from which the measurand arrived at its current value. For example, hysteresis can result from static friction in measuring displacement or from materials effects in sensors that involve magnetic fields. All types of bias error are potentially subject to slow **drift** due to aging and changes in environmental factors such as temperature or humidity.

Although bias errors are the same for each measurement made under apparently identical conditions (except for drift), **random errors** are different for each instance and have zero average value. For example, in measuring a given distance, friction combined with vibration may cause repeated measurements to vary. We can sometimes reduce the effect of random errors by making repeated measurements and averaging the results.

Some additional terms used in rating instrumentation performance are as follows:

1. **Accuracy:** The maximum expected difference in magnitude between measured and true values (often expressed as a percentage of the full-scale value).

2. **Precision:** The ability of the instrument to repeat the measurement of a constant measurand. More precise measurements have less random error.

3. **Resolution:** The smallest possible increment discernible between measured values. As the term is used, higher resolution means smaller increments. Thus, an instrument with a five-digit display (e.g., 0.0000 to 9.9999) is said to have higher resolution than an otherwise identical instrument with a three-digit display (e.g., 0.00 to 9.99).

---

**Exercise 8.10**   Suppose that a given magnetic flow sensor has an internal resistance (say, with variations in the electrical conductivity of the fluid) that varies from $5\,k\Omega$ to $10\,k\Omega$. The internal (open-circuit) voltage of the sensor is proportional to the flow rate. Suppose that we want the changes in the sensitivity constant of the measurement system (including loading effects) to vary by less than 0.5 percent with changes in sensor resistance. What specification is required for the input resistance of the amplifier in this system?

**Answer** The input resistance of the amplifier must be greater than $990\,k\Omega$.   □

---

**Exercise 8.11   a.** Can a very precise instrument be very inaccurate? **b.** Can a very accurate instrument be very imprecise?

**Answer   a.** Yes. Precision implies that the measurements are repeatable; however they could have large bias errors. **b.** No. If repeated measurements vary a great deal under apparently identical conditions, some of the measurements must have large errors, and therefore must be inaccurate.   □

---

## 8.8  SIGNAL CONDITIONING

Some functions of signal conditioners are amplification of the sensor signals, conversion of currents to voltages, supply of (ac or dc) excitations to the sensors so that changes in resistance, inductance, or capacitance are converted to changes in voltage, and filtering to eliminate noise or other unwanted signal components. Signal conditioners are often specific to particular applications. For example, a signal conditioner for a diode thermometer may not be appropriate for use with a thermocouple.

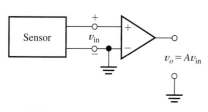

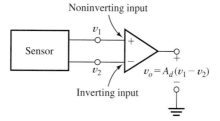

(a) Single-ended input: one
amplifier input terminal is grounded

(b) Differential input: neither amplifier input
is grounded and the output is gain $A_d$ times
the difference between the input voltages

Figure 8.17 Amplifiers with single-ended and differential input terminals.

## Single-Ended versus Differential Amplifiers

Often, the signal from the sensor is very small (one millivolt or less) and an important step in signal conditioning is amplification. Thus, the sensor is often connected to the input terminals of an amplifier. In an amplifier with a **single-ended input**, one of its input terminals is grounded as shown in Figure 8.17(a), and the output voltage is a gain constant $A$ times the input voltage.

An amplifier with a **differential input** is shown in Figure 8.17(b). Differential amplifiers have non-inverting and inverting input terminals as indicated in the figure, and ideally, the output is the differential gain $A_d$ times the difference between the input voltages.

A model for the voltages produced by a typical sensor connected to a differential amplifier is shown in Figure 8.18. The difference between the amplifier input voltages is the **differential signal**:

$$v_d = v_1 - v_2 \tag{8.5}$$

Sometimes, a large **common-mode signal** is also present, which is given by

$$v_{cm} = \frac{1}{2}(v_1 + v_2) \tag{8.6}$$

Almost always, the differential signal is of interest, and the common-mode signal represents unwanted noise. Thus, it is often very important for the differential amplifier to respond only to the differential signal. Great care must be taken in

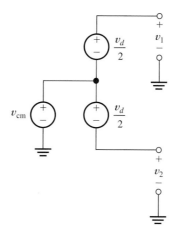

Figure 8.18 Model for a sensor with differential and common-mode components.

When large common-mode
signals are present, it
is important to select a
differential amplifier having
a large CMRR specification.

designing amplifiers that reject large common-mode signals sufficiently. A measure of how well a differential amplifier rejects the common-mode signal is the **common-mode rejection ratio** (CMMR). When large common-mode signals are present, it is important to select a differential amplifier with a large CMRR. **Instrumentation amplifiers** are very good in this respect. Differential amplifiers, CMRR, and instrumentation amplifiers are discussed at length in Chapters 10 and 13.

## Ground Loops

Often, the sensor and the signal-conditioning unit (such as an amplifier or current-to-voltage converter) are located some distance apart and are connected by a cable. Furthermore, the voltage (or current) produced by the sensor may be very small (less than one millivolt or one microampere). Then, several problems can occur that reduce accuracy or, in extreme cases, totally obscure the desired signal.

One of these problems is known as **ground loops**. If we have a single-ended amplifier, one of its input terminals is connected to a ground wire of the electrical distribution system. These ground wires eventually lead to the ground bus in the electrical distribution panel, which in turn is connected to a cold-water pipe or to a conducting rod driven into the earth. In instrumentation systems, we often have several pieces of equipment that are connected to ground through different wires. Ideally, the ground wires would have zero impedance, and all of the ground points would be at the same voltage. In reality, because of currents flowing through small but nonzero resistances of the various ground wires, small but significant voltages exist between various ground points.

Consider Figure 8.19, in which we have a sensor, an amplifier with a single-ended input, and a cable connecting the sensor to the amplifier. The cable wires have small resistances denoted by $R_{cable}$. Several ground wires are shown with their resistances $R_{g1}$ and $R_{g2}$. The current source $I_g$ represents current flowing to ground. Typically, $I_g$ originates from the 60-Hz line voltage through power-supply circuits of the instruments. If we connect both the sensor and the amplifier input to ground, part of $I_g$ flows through the connecting cable, and the input voltage is the sensor voltage minus the drop across $R_{cable}$:

$$V_{in} = V_{sensor} - I_{g1}R_{cable} \qquad (8.7)$$

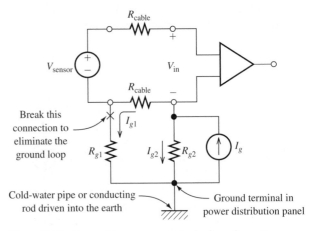

**Figure 8.19** Ground loops are created when the system is grounded at several points.

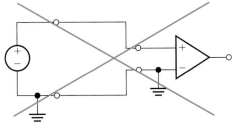

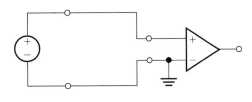

(a) Grounded sensor with single-ended amplifier
Avoid to help prevent ground-loop noise

(b) Floating sensor with single-ended amplifier

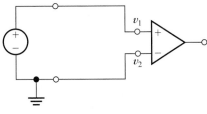

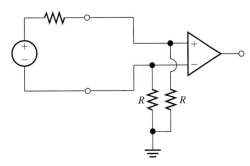

(c) Grounded sensor with differential amplifier

(d) Floating sensor with differential amplifier including
resistors to provide a path for the input bias current

**Figure 8.20** Four sensor amplifier combinations.

When the sensor voltage is very small, it can be totally obscured by the drop across $R_{\text{cable}}$.

On the other hand, if we break the sensor ground connection so only the amplifier is grounded, $I_{g1}$ becomes zero and the input voltage is the sensor voltage as desired. Thus, in connecting a sensor to an amplifier with a single-ended input, we should select an ungrounded or **floating** sensor.

*In connecting a sensor to an amplifier with a single-ended input, we should select a floating sensor.*

If you have connected several audiovisual components, such as VCRs, TVs, radio tuners, CD players, stereo amplifiers, and so forth, you have probably encountered ground loops, which cause an annoying 60-Hz hum to be produced by loudspeakers.

## Alternative Connections

Figure 8.20 shows four combinations of sensors and amplifiers. As we have seen, we need to avoid the combination of grounded sensor and single-ended input shown in part (a) of the figure because of ground loops. Any of the other three connections can be used. However, for a floating sensor with a differential amplifier as shown in part (d), it is often necessary to include two high-valued (much greater than the internal impedance of the sensor to avoid loading effects) resistors to provide a path for the input bias current of the amplifier. (Input bias current is discussed further in Section 10.12.) If the resistors are not included, the common-mode voltage of the source can become so large that the amplifier does not function properly. In part (c) of the figure, the ground connection to the sensor provides a path for the bias current.

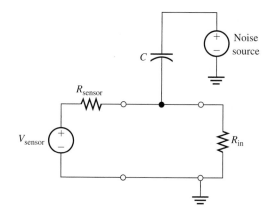

**Figure 8.21** Noise can be coupled into the sensor circuit by electric fields. This effect is modeled by small capacitances between the noise source and the sensor cable.

## Noise

Another problem that can occur in connecting sensors to signal-conditioning units is the inadvertent addition of noise produced by the electric or magnetic fields generated by nearby circuits. (Computers are infamous for creating high-frequency electrical noise.) Electric field coupling can be modeled as small capacitances that are connected between nearby circuits and the cables, as shown in Figure 8.21. Currents are injected into the cable through these capacitances. This is particularly a problem with unshielded cables and when the sensor impedance is large. A shielded cable, in which an outer conductor in the form of metallic foil or braided wire encases the signal conductors, can eliminate much of the noise caused by electric fields. The shield is connected to ground, providing a low-resistance path for the capacitive currents. This is illustrated in Figure 8.22.

Noise problems can also occur due to magnetic coupling. Many circuits, particularly power-supply transformers, produce time-varying magnetic fields. When these fields pass through the region bounded by the cable conductors, voltages are induced in the cable. Magnetically coupled noise can be greatly diminished by reducing the effective area bounded by the conductors. Twisted-pair and coaxial cables (see Figure 8.23) are two good ways to accomplish this. Because the center lines of the conductors in coaxial cable are coincident, the effective bounded area is very small.

> Electric field coupling of noise can be reduced by using shielded cables.

> Magnetically coupled noise is reduced by using coaxial or twisted-pair cables.

**Exercise 8.12** The voltages produced by a sensor are $v_1 = 5.7$ V and $v_2 = 5.5$ V. Determine the differential and common-mode components of the sensor signal.
**Answer** $v_d = 0.2$ V; $v_{cm} = 5.6$ V.    □

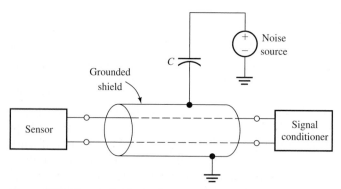

**Figure 8.22** Electric field coupling can be greatly reduced by using shielded cables.

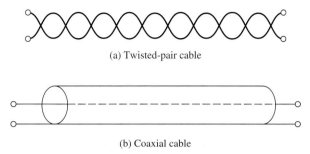

(a) Twisted-pair cable

(b) Coaxial cable

**Figure 8.23** Magnetic field coupling can be greatly reduced by using twisted-pair or coaxial cables.

## PRACTICAL APPLICATION    8.2

### The Virtual First-Down Line

In American football, the team on offense must gain 10 yards in a series of four plays to retain possession of the ball. Thus, a line marking the needed advance is of constant interest to fans viewing a game.

On September 27, 1998, in a Sunday Night Football game on ESPN, Sportvision introduced their "1st and Ten" system for electronically drawing the first-down line on television images. The system has been enthusiastically accepted and even won an Emmy award for technical innovation. In the 2003 season, 18 crews covered about 300 NCAA and NFL games.

While the concept of drawing a virtual line on a television image sounds simple, there are some formidable problems that need to be overcome to produce a result that appears to be painted on the field. Typically, three main cameras are situated above and back from the 50-yard line and both of the 25-yard lines. Each camera pans, tilts, zooms, and changes focus rapidly during the game. The virtual line needs to change its position, orientation, and width as the camera view changes. In addition, some football fields are not flat—they are crowned to ensure drainage and the yard lines are not exactly straight. If the virtual line does not closely match the curvature of the lines on the field, it will not look natural. Furthermore, the line needs to be drawn thirty times a second, once for each video frame. Of course, for realism, part of the line needs to disappear when a player, an official, or the ball moves across it. An impressive array of sophisticated electronic and computer technology has been employed by Sportvision engineers to meet these demands.

To set up the system on a given field, Sportvision starts by using laser surveying instruments to measure the elevation at a number of points along each 10-yard line. A computer uses this data to produce a virtual three-dimensional model of the field.

Sensors attached to each of the cameras measure pan, tilt, zoom, and focus. This data is fed into a computer that alters the model to match the perspective for a given camera, and a virtual map is drawn in blue lines over the image of the field seen by the camera. Finally, the virtual map is tweaked to match the real image for many combinations of pan, tilt, and zoom as illustrated in Figure PA8.2. The resulting calibration data is saved for use by the system during an actual game.

A technique known as "chroma keying" has been around for a long time and is widely used in televised weather reports. A meteorologist forecasting weather stands in front of a light blue wall. Computers substitute weather maps and graphics for all the pixels (i.e., picture elements) that are light blue. Thus, the forecaster seems to stand in front of the weather map. The same kind of technology is used to allow officials, players, and the ball to seem to move over the virtual first-down line. However, discerning which pixels are players and which are part of the field is much more difficult than separating a forecaster from a blue wall. Weather forecasters usually avoid wearing clothing that matches the color of the wall, and the wall is all the same color. On the other hand, the field can be many different shades of white (painted yard lines on the field), green (grass or artificial turf), or brown (grass or mud). Part of the field may be sunlit while

Tilt, pan, zoom, and focus sensors on the camera provide information about which part of the field is in view.

During system calibration, computers use the field survey data to draw the virtual map (in gray) of the field on the TV screen.

Then the virtual map is tweaked to match the actual field. This is repeated for many combinations of tilt, pan, zoom, and focus for each of the three main cameras.

**Figure PA8.2** Computers use a virtual map based on surveying the football field to draw the 1st and Ten line on the television screen in real time.

other parts are in shadow. Football teams, such as the Green Bay Packers, have uniforms that are partly green, which is especially difficult to distinguish from sunlit artificial turf. Other colors, such as brown, can also be difficult.

By constant recalibrating, the 1st and Ten system can keep track of which colors are part of the field and which are not, so the virtual down line is not drawn over players.

During a game, a team of four people operates the system, which contains five computers. The "spotter" is in the stadium and radios the location of the down line to a truck containing the equipment and two of the other team members. The "line position technician" enters the location data into the computers, monitors line position, and makes any needed adjustments. Another operator monitors

changes in field colors so the chroma keying is properly accomplished. Finally, a troubleshooter looks for problems and solves them.

One of the computers receives and processes the pan, tilt, and zoom data from the cameras. Another keeps track of which camera is "on air." A third displays the on-air video and superimposes the virtual map of the field including the current down line. Another computer discerns which parts of the image are field and which are players or officials. Finally, the fifth computer places the virtual down line on the broadcast image while avoiding any superimposed graphics that the network may place on the screen.

More information about the 1st and Ten system as well as similar technology for other sports can be found at www.sportvision.com.

## 8.9   ANALOG-TO-DIGITAL CONVERSION

As discussed starting on page 331, analog signals are converted to digital form by a two-step process. First, the analog signal is sampled (i.e., measured) at periodic points in time. A code word is then assigned to represent the approximate value of each sample. The sampling rate and the number of bits used to represent each sample are two very important considerations in the selection of a DAQ system.

### Sampling Rate

The rate at which a signal must be sampled depends on the frequencies of the signal's components. (All signals can be considered to be sums of sinusoidal components that have various frequencies, amplitudes, and phases.) If a signal contains no components with frequencies higher than $f_H$, all of the information contained in the signal is present in its samples, provided that the sampling rate is selected to be more than twice $f_H$.

> If a signal contains no components with frequencies higher than $f_H$, all of the information contained in the signal is present in its samples, provided that the sampling rate is selected to be more than twice $f_H$.

### Aliasing

Sometimes, we may only be interested in the components with frequencies up to $f_H$, but the signal may contain noise or other components with frequencies higher than $f_H$. Then, if the sampling rate is too low, a phenomenon called **aliasing** can occur. In aliasing, the samples of a high-frequency component appear to be those of a lower frequency component and may obscure the components of interest. For example, Figure 8.24 shows a 7-kHz sinusoid sampled at 10 kHz. As illustrated by the dashed line, the sample values appear to be those of a 3-kHz sinusoid. Because the sampling rate (10 kHz) is less than twice the signal frequency (7 kHz), the samples appear to be those of an alias frequency (3 kHz). (Notice that from the samples it is impossible to determine whether a 3-kHz or a 7-kHz signal was sampled.)

Figure 8.25 shows the alias frequency as a function of the signal frequency $f$. When the signal frequency $f$ exceeds one-half of the sampling frequency $f_s$, the apparent frequency of the samples is different from the true signal frequency.

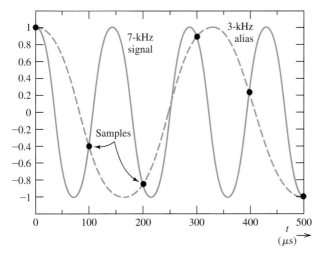

**Figure 8.24**  When a 7-kHz sinusoid is sampled at 10 kHz, the sample values appear to be those of a 3-kHz sinusoid.

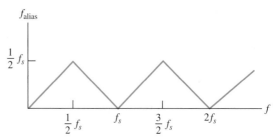

**Figure 8.25** Alias or apparent frequency versus true signal frequency.

One way to avoid aliasing is to pick the sampling frequency high enough so that the alias frequencies are higher than the frequencies of interest. Then, the computer can remove the unwanted components by processing the samples with digital-filtering software. However, when high-frequency noise is present, it can result in a sampling rate that exceeds the ability of the computer to process the resulting data. Then, it is better to use an analog **antialias filter** ahead of the ADC to remove the noise above the highest frequency of interest. Typically, this is a high-order Butterworth filter implemented with operational amplifiers, such as those discussed in Section 13.10 starting on page 681. Because real filters are unable to sufficiently reject components slightly above their cutoff frequencies, it is usually necessary to select the sampling frequency at least three times the highest frequency of interest. For example, the highest audible frequency is about 15 kHz, but in CD technology a sampling rate of 44.1 kHz is used.

## Quantization Noise

A second consideration important in converting analog signals to digital form is the number of amplitude zones to be used. Exact signal amplitudes cannot be represented, since all amplitudes falling into a given zone have the same codeword. Thus, when a DAC converts the codewords to form the original analog waveform, it is possible to reconstruct only an approximation to the original signal the reconstructed voltage is in the middle of each zone, which was illustrated in Figure 6.46 on page 333. Hence, some **quantization error** exists between the original signal and the reconstruction. This error can be reduced by using a larger number of zones, which requires a longer codeword for each sample. The number $N$ of amplitude zones is related to the number of bits $k$ in a codeword by

Analog-to-digital conversion is a two-step process. First, the signal is sampled at uniformly spaced points in time. Second, the sample values are quantized so they can be represented by words of finite length.

$$N = 2^k \tag{8.8}$$

Therefore, if we are using an 8-bit ($k = 8$) ADC, we find that there are $N = 2^8 = 256$ amplitude zones. The resolution of a computer-based measurement system is limited by the word length of the ADC. In compact-disc technology, 16-bit words are used to represent sample values. With that number of bits, it is very difficult for a listener to detect the effects of quantization error on the reconstructed audio signal. In the telephone system, 8-bit words are used and the fidelity of the reconstructed signal is relatively poor.

The effect of finite word length can be modeled as adding quantization noise to the reconstructed signal. It can be shown that the rms value of the quantization noise is approximately

The effect of finite word length can be modeled as adding quantization noise to the reconstructed signal.

$$N_{qrms} = \frac{\Delta}{2\sqrt{3}}$$   (8.9)

where $\Delta$ is the width of a quantization zone.

In the example that follows, we illustrate how the various factors we have discussed are used in selecting the components of a computer-based measurement system.

---

### Example 8.6   Specifications for a Computer-Based Measurement System

Suppose that we have a single-ended (i.e., one terminal of the sensor is connected to power-system ground) piezoelectric vibration sensor that produces a signal of interest having peak values of $\pm 25$ mV, an rms value of 3 mV, and components with frequencies up to 5 kHz. The internal impedance of the sensor is 1 k$\Omega$. We want the system resolution to be 2 $\mu$V or better (i.e., smaller) and the accuracy to be $\pm 0.2$ percent of the peak signal or better. (Note that the desired resolution is considerably better than the accuracy. This allows the system to discern changes in the signal that are smaller than the error.) The probe wiring is likely to be exposed to electric and magnetic field noise having components at frequencies higher than 5 kHz. An ADC having an input range from $-5$ V to $+5$ V is to be used. Draw the block diagram of the measurement system and give key specifications for each block.

**Solution**   Because one end of the sensor is grounded, we need to use an instrumentation amplifier with a differential input to avoid ground-loop problems. (See Figure 8.20 on pg 443.)

To help reduce capacitively and inductively coupled noise, we should select a shielded twisted-pair or coaxial cable to connect the sensor to the system. To avoid ground loops, the shield should be grounded at the sensor end only. Furthermore, we should use an antialias filter to reduce noise above 5 kHz. The block diagram of the system is shown in Figure 8.26.

The voltage gain of the instrumentation amplifier/antialias filter combination should be (5 V)/(25 mV) = 200; so the sensor signal is amplified to match the range of the ADC. The input impedance of the amplifier should be very large compared

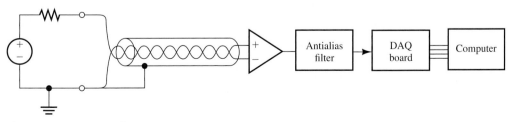

Figure 8.26  See Example 8.6.

with the internal sensor impedance; therefore, loading effects are insignificant. If we specify a minimum input impedance of 1 MΩ, loading effects will reduce the signal by 0.1 percent, which is within the desired accuracy. (We need to allow for errors from other unknown sources.)

To achieve a resolution of 2 $\mu$V for a $\pm$25-mV signal, we need an ADC with at least (50 mV)/(2 $\mu$V) = 25,000 amplitude levels. This implies an ADC word length of at least $k = \log_2(25,000) = 14.6$. Since word length must be an integer, $k = 15$ is the smallest word length that will meet the desired specifications. It turns out that DAQ boards with 16-bit ADCs are readily available, and to provide some design margin, that is the word length we should specify.

Because the highest frequency of interest is 5 kHz, we need a sampling frequency of at least 10 kHz. However, the antialias filter will not effectively remove the components that are only slightly above 5 kHz, so a greater sampling rate (20 kHz or greater) should be chosen.    ∎

**Exercise 8.13** A certain 8-bit ADC accepts signals ranging from −5 V to +5 V. Determine the width of each quantization zone and the approximate rms value of the quantization noise.
**Answer** $\Delta = 39.1$ mV; $N_{qrms} = 11.3$ mV. ☐

**Exercise 8.14** A 25-kHz sinewave is sampled at 30 kHz. Determine the value of the alias frequency.
**Answer** $f_{alias} = 5$ kHz. ☐

## Summary

1. A computer is composed of a central processing unit (CPU), memory, and input output (I/O) devices. These are connected together with bidirectional data and control buses. The CPU contains the control unit, the ALU, and various registers.

2. Memory is used to store programs and data. Three types of memory are RAM, ROM, and mass storage.

3. In von Neumann computer architecture, data and instructions are stored in the same memory. In Harvard architecture, separate memories are used for data and instructions.

4. Sensors are input devices that convert physical values to electrical signals. Actuators are output devices that allow the MCU to affect the system being controlled.

5. Figure 8.5 (on pg 415) shows the elements of a typical microcomputer used for process control.

6. Analog-to-digital converters (A/D) transform analog voltages into digital words. Digital-to-analog converters (D/A) transform digital words into analog voltages. Converters are needed to interface analog sensors and actuators with an MCU.

7. Figure 8.6 (on pg 418) shows the register set for the CPU12.

8. In a stack memory, data is added to or read from the top of the stack. It is a last-in first-out (LIFO) memory. The stack pointer is a register that contains the address of the top of the stack.

9. Table 8.1 (on pg 422) contains some of the instructions for the CPU12.

10. Six addressing modes are supported by the CPU12: extended addressing, direct addressing, inherent addressing, immediate addressing, indexed addressing, and program-relative addressing.

11. In writing programs for embedded microcontrollers, we often start by writing a source program using labels and mnemonics. An assembler converts the source program into an object program consisting of machine code that is loaded into the target system.

12. High costs are incurred in software development for MCUs. However, when the cost can be spread over many units, assembly language programming can be the best solution.

13. The block diagram of a typical computer-based instrumentation system is shown in Figure 8.12 on page 435.

14. When we need to sense the internal (open-circuit) voltage of a sensor, we should specify an amplifier having an input impedance that is much larger in magnitude than the Thévenin impedance of the sensor.

15. When we want to sense the current produced by a sensor, we need a current-to-voltage converter having a very small (ideally zero) input impedance magnitude compared to the Thévenin impedance of the sensor.

16. Bias errors are the same each time a measurement is repeated under the same apparent conditions. Bias errors include offset, scale error, nonlinearity, and hysteresis.

17. Random errors are different for each measurement and have zero average value.

18. The accuracy of an instrument is the maximum expected difference in magnitude between measured and true values (often expressed as a percentage of the full-scale value).

19. Precision is the ability of the instrument to repeat the measurement of a constant measurand. More precise measurements have less random error.

20. The resolution of an instrument is the smallest increment discernible between measured values.

21. Some of the functions of signal conditioners are amplification, conversion of current signals to voltage signals, supply of excitation to the sensor, and filtering to eliminate noise or other unwanted signal components.

22. One of the input terminals of a single-ended amplifier is grounded. Neither input terminal of a differential amplifier is grounded. The output of an ideal differential amplifier is its differential gain times the difference between the input voltages.

23. If the input voltages to a differential amplifier are $v_1$ and $v_2$, the differential input signal is $v_d = v_1 - v_2$ and the common-mode signal is $v_{cm} = \frac{1}{2}(v_1 + v_2)$. Usually in instrumentation systems, the differential signal is of interest and the common-mode signal is unwanted noise.

24. When large unwanted common-mode signals are present, it is important to select a differential amplifier having a large CMRR (common-mode rejection ratio) specification.

25. Ground loops are created in an instrumentation system when several points are connected to ground. Currents flowing through the ground conductors can produce noise that makes the measurements less accurate and less precise.

26. To avoid ground-loop noise when we must connect a sensor to an amplifier with a single-ended input, we should select a floating sensor (i.e., neither terminal of the sensor should be grounded).

27. Shielded cables reduce the noise coupled by electric fields.

28. Coaxial or twisted-pair cables reduce magnetically coupled noise.

29. If a signal contains no components with frequencies higher than $f_H$, all of the information contained in the signal is present in its samples, provided that the sampling rate is selected to be more than twice $f_H$.

30. Analog-to-digital conversion is a two-step process. First, the signal is sampled at uniformly spaced points in time. Second, the sample values are quantized so they can be represented by words of finite length.

31. We model the effect of finite word length as the addition of quantization noise to the signal.

32. If a sinusoidal signal component is sampled at a rate $f_s$ that is less than twice the signal frequency $f$, the samples appear to be from a component with a different frequency known as the alias frequency $f_{alias}$. Alias frequency is plotted versus the true signal frequency in Figure 8.25 on pg 448.

33. If a sensor signal contains high-frequency components that are not of interest, we often use an analog antialias filter to remove them so a lower sampling frequency can be used without aliasing.

## Problems

### Section 8.1: Computer Organization

**P8.1.** List the functional parts of a computer.

**P8.2.** What are tristate buffers? What are they used for?

**P8.3.** Give several examples of I/O devices.

**P8.4.** What is memory-mapped I/O?

**P8.5.** What is a bus? What is the function of the data bus? Of the address bus?

**P8.6.** What is an embedded computer?

**\*P8.7.** The address bus of a computer is 16 bits wide and the data bus is 32 bits wide. How many bytes does the memory potentially contain?

**P8.8.** Define the terms *microprocessor, microcomputer*, and *microcontroller*.

**P8.9.** Explain the difference between Harvard computer architecture and von Neumann computer architecture.

### Section 8.2: Memory Types

**P8.10.** What is RAM? List two types. Is it useful for storing programs in embedded computers? Explain.

**\*P8.11.** What is ROM? List four types. Is it useful for storing programs in embedded computers? Explain.

**P8.12.** List three examples of mass-storage devices.

**P8.13.** Which type of memory is least expensive per unit of storage? (Assume that many megabytes of capacity are required.)

**P8.14.** How many memory locations can be addressed if the address bus has a width of 32 bits?

**\*P8.15.** Which type of memory would be best in the controller for an ignition system for automobiles?

**P8.16.** When might we choose EEPROM rather than mask-programmed ROM?

**P8.17.** What types of memory are volatile? Non-volatile?

### Section 8.3: Digital Process Control

**P8.18.** List the elements that are potentially found in a microcomputer-based control application.

**P8.19.** What is a sensor? Give three examples.

**P8.20.** What is an actuator? Give three examples.

**\*P8.21.** Explain the difference between a digital sensor and an analog sensor. Give an example of each.

**P8.22.** List five common household products that potentially include an MCU.

**P8.23.** List two potential applications of MCU-based control or instrumentation in your field of specialization.

**P8.24.** What is an A/D? Why might one be needed in an MCU-based controller?

**\*P8.25.** What is a D/A? Why might one be needed in an MCU-based controller?

**P8.26.** What is polling? What is an interrupt? What is the main potential advantage of interrupts versus polling?

### Section 8.4: Programming Model for the HCS12/9S12 Family

**P8.27.** What is the function of the A, B, and D registers of the CPU12?

**P8.28.** What is the function of the program counter register? Of the MCU?

**\*P8.29.** What is a stack? What is the stack pointer used for?

**P8.30.** What is a LIFO memory?

**\*P8.31.** Suppose that initially the contents of the registers are

A:07    B:A9    SP:004F    X:34BF

and that memory locations 0048 through 004F initially contain all zeros. The commands PSHA, PSHB, PULA, PULB, PSHX are then executed in sequence. List the contents of the registers A, B, SP, and X, and the memory locations 0048 through 004F after each command is executed.

---

*Denotes that answers are contained in the Student Solutions files. See Appendix E for more information about accessing the Student Solutions.

**P8.32.** Suppose that initially the contents of the registers are

A:A7   B:69   SP:004E   Y:B804

and that memory locations 0048 through 004F initially contain all zeros. The commands PSHY, PSHB, PULY, PSHA are then executed in sequence. List the contents of the registers A, B, SP, and X and the memory locations 0048 through 004F after each command is executed.

**\*P8.33.** Write a sequence of push and pull commands to swap the high byte and low byte of the X register. After the sequence of commands is executed, the contents of the other registers should be the same as before.

## Section 8.5: The Instruction Set and Addressing Modes for the CPU12

**P8.34.** For each part of this problem, assume that the X register contains 2000 and the A register initially contains 01. Name the type of addressing and give the content of A after each instruction listed next. The contents of memory are shown in Figure P8.34.

  **a.** \*LDDA $2002

  **b.** LDDA #$43

  **c.** \*LDDA $04

  **d.** LDDA 6,X

  **e.** \*INCA

  **f.** CLRA

  **g.** \*LDAA $2007

  **h.** INX

**P8.35.** Suppose that the contents of certain memory locations are as shown in Figure P8.34. Furthermore, for each part of this problem, the initial contents of the CPU registers are: (D) = $0003, (X) = $1 FFF, and (Y) = $1000. Name the type of addressing and determine the contents of registers D, X, and Y after each of these instructions is executed:

  **a.** ADDB $1002,Y

  **b.** LDAA B,X

  **c.** LDAB 7,+X

  **d.** LDX [$1004,Y]

  **e.** LDAA [D,X]

**P8.36. a.** \*Assume that the A register initially contains FF and that the program counter is 2000. What is the address of the instruction executed immediately after the branch command? The content of memory and the corresponding instruction mnemonics are shown in Figure P8.36(a).

  **b.** Repeat for Figure P8.36(b).

  **c.** Repeat for Figure P8.36(c).

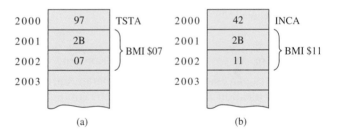

(a)                    (b)

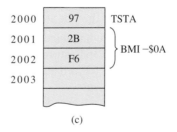

(c)

Figure P8.36

**P8.37.** Give the machine code for each of the following instructions:

  **a.** \*CLRA

  **b.** \*ADDA $4A

  **c.** ADDA $02FF

| 0000 | 01 |  | 2000 | 37 |
| 0001 | FA |  | 2001 | AF |
| 0002 | 9B |  | 2002 | 20 |
| 0003 | 61 |  | 2003 | 07 |
| 0004 | 9A |  | 2004 | 20 |
| 0005 | B6 |  | 2005 | 00 |
| 0006 | 73 |  | 2006 | FF |
| 0007 | 41 |  | 2007 | F3 |

(a)                    (b)

Figure P8.34

**d.** BNE −$06

**e.** ADDA #$0D

How many memory locations are occupied by each instruction?

**P8.38.** Find the content of the A register after each instruction as the following sequence of instructions is executed:

```
LDAA    #$01
ADDA    #$F1
CLRA
```

Is the N bit of the condition-code register set or clear? Is the Z bit of the condition-code register set or clear?

**\*P8.39. a.** Suppose that the content of A is 43. Find the content of A after the instruction ADDA #$05 is executed. **b.** Suppose that the content of A is FA. Find the content of A after the instruction ADDA #$0F is executed. (In this case, overflow occurs.)

**\*P8.40.** Assume that the content of A is $A7 and that the content of B is $20. Find the contents of A and B after the MUL instruction is executed. [*Hint:* The MUL instruction assumes that the contents of A and B are unsigned integer values.]

### Section 8.6: Assembly-Language Programming

**\*P8.41.** Write an assembly-language program starting in location 200 that multiplies the content of the A register by $11_{10}$ and stores the result in memory locations $FF00 and $FF01, with the most significant byte in location $FF00. The processor should then be stopped.

**P8.42.** Write an assembly-language program starting in location 0400 that stores 00 in location 0800, 01 in 0801, 02 in 0802, 03 in 0803, and then halts the processor.

**\*P8.43.** Write a subroutine called DIV3 that divides the content of A by three. Assume that the initial content of A is a positive integer in two's-complement form. On return from the subroutine, the quotient should reside in B and the remainder in A.

**P8.44.** Consider the following assembly-language code for the CPU12:

```
;    PROBLEM 8.44
;
```

```
            ORG     $0600
START       LDAA    #$07
            LDAB    #$AF
            STD     $0609
            STOP
            END
```

Describe the effect of each line of this code and list the contents of memory locations 0600 through 060A after the code has been assembled and executed.

**P8.45.** Write a subroutine called MUL3 that rounds the content of A to its nearest integer multiple of 3. Assume that the initial content of A is a positive integer in two's-complement form. Memory location $0A can be used for temporary storage. Include comments in your source code to explain the program and its operation to human readers. [*Hint:* Repeatedly subtract 3 until the result becomes negative. If the result is −3, the original content of A was a multiple of 3 and should not be changed. If the result is −2, the original content of A was one plus an integer multiple of 3, and we should subtract one from the original number to obtain the nearest multiple of 3. If the result is −1, the original content of A is 2 plus an integer multiple of 3, and we should add 1 to the original number to obtain the nearest multiple of 3.]

**P8.46.** Suppose that register B contains a two-decimal-digit BCD number $n$. Write a subroutine called CONVERT that replaces the content of register B by its binary equivalent. The content of the other registers (except the program counter) should be unchanged at the completion of the subroutine. Memory locations $1A, $1B, and $1C can be used for temporary storage. [*Hint:* We need to separate the upper nibble (four bits) of $n$ from its lower nibble. This can be achieved by shifting $n$ four bits to the left, with the result appearing in the D register. Shifting by four bits to the left is accomplished by multiplying by $2^4$.]

### Section 8.7: Measurement Concepts and Sensors

**P8.47.** Name the elements of a computer-based instrumentation system.

*P8.48. Draw the equivalent circuit of a sensor in which the open-circuit sensor voltage is proportional to the measurand. What are loading effects? How do we avoid them when we need to measure the Thévenin (i.e., open-circuit) sensor voltage?

P8.49. What signal-conditioner input impedance is best if the sensor produces short-circuit (Norton) current that is proportional to the measurand?

*P8.50. A load cell produces an open-circuit voltage of 200 $\mu$V for a full-scale applied force of $10^4$ N, and the Thévenin resistance is 1 k$\Omega$. The sensor terminals are connected to the input terminals of an amplifier. What is the minimum input resistance of the amplifier so the overall system sensitivity is reduced by less than 1 percent by loading?

*P8.51. A certain liquid-level sensor has a Thévenin (or Norton) resistance that varies randomly from 10 k$\Omega$ to 1 M$\Omega$. The short-circuit current of the sensor is proportional to the measurand. What type of signal-conditioning unit is required? Suppose we allow for up to 1 percent change in overall sensitivity due to changes in the sensor resistance. Determine the specification for the input resistance of the signal conditioner.

P8.52. How are bias errors different from random errors?

P8.53. List four types of bias error.

*P8.54. An instrumentation system measures distances ranging from 0 to 1 m full scale. The system accuracy is specified as $\pm0.5$ percent of full scale. If the measured value is 70 cm, what is the potential range of the true value?

P8.55. Explain how precision, accuracy, and resolution of an instrument are different.

*P8.56. Three instruments each make 10 repeated measurements of a flow rate known to be 1.500 m$^3$/s with the results given in Table P8.56.

a. Which instrument is most precise? Least precise? Explain.

b. Which instrument has the best accuracy? Worst accuracy? Explain.

c. Which instrument has the best resolution? Worst resolution? Explain.

**Table P8.56**

| Trial | Instrument A | Instrument B | Instrument C |
|---|---|---|---|
| 1 | 1.5 | 1.73 | 1.552 |
| 2 | 1.3 | 1.73 | 1.531 |
| 3 | 1.4 | 1.73 | 1.497 |
| 4 | 1.6 | 1.73 | 1.491 |
| 5 | 1.3 | 1.73 | 1.500 |
| 6 | 1.7 | 1.73 | 1.550 |
| 7 | 1.5 | 1.73 | 1.456 |
| 8 | 1.7 | 1.73 | 1.469 |
| 9 | 1.6 | 1.73 | 1.503 |
| 10 | 1.5 | 1.73 | 1.493 |

**Section 8.8: Signal Conditioning**

P8.57. List four or more functions of signal conditioners.

P8.58. How is a single-ended amplifier different from a differential amplifier?

P8.59. Suppose that the input voltages to an ideal differential amplifier are equal. Determine the output voltage.

*P8.60. The input voltages to a differential amplifier are

$$v_1(t) = 0.002 + 5\cos(\omega t)$$

and

$$v_2(t) = -0.002 + 5\cos(\omega t)$$

Determine the differential input voltage and the common-mode input voltage. Assuming that the differential amplifier is ideal with a differential gain $A_d = 1000$, determine the output voltage of the amplifier.

P8.61. A sensor produces a differential signal of 6 mV dc and a 2-V-rms 60-Hz ac common-mode signal. Write expressions for the voltages between the sensor output terminals and ground.

*P8.62. Suppose we have a sensor that has one terminal grounded. The sensor is to be connected to a DAQ board in a computer 5 meters away. What type of amplifier should we select? To mitigate noise from electric and magnetic fields, what type of cable should we use? Draw the schematic diagram of the sensor, cable, and amplifier.

**P8.63.** What is a floating sensor? When would we want to use a floating sensor?

**\*P8.64.** Suppose that the data collected from a sensor is found to contain an objectionable 60-Hz ac component. What potential causes for this interference would you look for? What are potential solutions for each cause of the interference?

### Section 8.9:    Analog-to-Digital Conversion

**P8.65.** In principle, analog-to-digital conversion involves two operations. What are they?

**P8.66.** What is aliasing? Under what conditions does it occur?

**P8.67.** What causes quantization noise?

**\*P8.68.** We need to use the signal from a piezoelectric vibration sensor in a computer-aided instrumentation system. The signal is known to contain components with frequencies up to 30 kHz. What is the minimum sampling rate that should be specified? Suppose that we want the resolution of the sampled values to be 0.1 percent (or better) of the full range of the ADC. What is the fewest number of bits that should be specified for the ADC?

**\*P8.69.** A 2-V-peak sinewave signal is converted to digital form by a 12-bit ADC that has been designed to accept signals ranging from −5 V to +5 V. (In other words, codewords are assigned for equal increments of amplitude for amplitudes between −5 V and +5 V.)

    **a.** Determine the width $\Delta$ of each quantization zone.

    **b.** Determine the rms value of the quantization noise and the power that the quantization noise would deliver to a resistance $R$.

    **c.** Determine the power that the 2-V sinewave signal would deliver to a resistance $R$.

    **d.** Divide the signal power found in part (c) by the noise power determined in part (b). This ratio is called the signal-to-noise ratio (SNR). Express the SNR in decibels, using the formula $\text{SNR}_{dB} = 10\log(P_{signal}/P_{noise})$.

**\*P8.70.** We need an ADC that can accept input voltages ranging from 0 to 5 V and have a resolution of 0.02 V. How many bits must the codewords have?

**\*P8.71.** A 10-kHz sinewave is sampled. Determine the apparent frequency of the samples. Has aliasing occurred? The sampling frequency is **a.** 11 kHz; **b.** 8 kHz; **c.** 40 kHz.

**P8.72.** A 60-Hz sinewave $x(t) = A\cos(120\pi t + \phi)$ is sampled at a rate of 360 Hz. Thus, the sample values are $x(n) = A\cos(120\pi n T_s + \phi)$, in which $n$ assumes integer values and $T_s = 1/360$ is the time interval between samples. A new signal is computed by the equation

$$y(n) = \frac{1}{2}[x(n) + x(n-3)]$$

    **a.** Show that $y(n) = 0$ for all $n$.

    **b.** Now suppose that $x(t) = V_{signal} + A\cos(120\pi t + \phi)$, in which $V_{signal}$ is constant with time and again find an expression for $y(n)$.

    **c.** When we use the samples of an input $x(n)$ to compute the samples for a new signal $y(n)$, we have a **digital filter**. Describe a situation in which the filter of parts (a) and (b) could be useful.

---

### Practice Test

Here is a practice test you can use to check your comprehension of the most important concepts in this chapter. Answers can be found in Appendix D and complete solutions are included in the Student Solutions files. See Appendix E for more information about the Student Solutions.

**T8.1.** First, think of one or more correct ways to complete each statement in Table T8.1(a). Then, select the best choice from the list given in Table T8.1(b). [Items in Table T8.1(b) may be used more than once or not at all.]

## Table T8.1

(a)

**a.** Tristate buffers . . .

**b.** When I/O devices are accessed by the same address
and data buses as data memory locations, we have . . .

**c.** In an microcontroller, programs are usually stored in . . .

**d.** The type of memory most likely to be volatile is . . .

**e.** The type of memory most likely to be used for temporary
data in an microcontroller is . . .

**f.** Arithmetic operations are carried out in the . . .

**g.** An microcontroller may be designed to respond to external events by . . .

**h.** The registers of a CPU12 that hold one of the arguments and
the results of arithmetic and logical operations are . . .

**i.** The registers of a CPU12 that are used mainly for indexed addressing are . . .

**j.** The TSTA instruction may change the content of . . .

**k.** Stacks are . . .

**l.** The type of addressing used by the ABA instruction is . . .

**m.** The type of addressing used by the ADDA $0AF2 instruction is . . .

**n.** The type of addressing used by the ADDA #$0A instruction is . . .

**o.** The register that holds the address of the next instruction
to be retrieved from memory is . . .

**p.** The type of addressing used by the BEQ instruction is . . .

(b)

1. are composed of switches that are always closed
2. mass storage
3. B and Y
4. I/O devices have their own addresses and data buses
5. are composed of open switches
6. A and X
7. control unit
8. extended
9. A, B, and D
10. contain switches that open and close under the direction of the program counter
11. facilitate the ability to transfer data in either direction over a bus
12. the condition code register
13. ALU
14. polling
15. LIFO memories
16. inherent
17. memory-mapped I/O
18. A and D
19. interrupts
20. X and Y
21. ROM
22. direct
23. the program counter
24. dynamic RAM
25. the stack pointer
26. either interrupts or polling
27. static RAM
28. indexed
29. immediate
30. relative

**T8.2.** We have a CPU12 MCU. For each part of this problem, assume that the initial content of A is 00, the initial content of B is FF, the initial content of Y is 2004, and the initial content of selected memory locations is as shown in Figure P8.34 on page 453. Name the type of addressing used and give the content of A (in hexadecimal form) after execution of the instruction: **a.** LDAA $03; **b.** LDAA $03,Y; **c.** COMA; **d.** INCA; **e.** LDAA #$05; **f.** ADDD #A001.

**T8.3.** Suppose that initially the contents of the registers of a CPU12 microcontroller are

A:A6    B:32    SP:1039    X:1958

and that memory locations 1034 through 103C initially contain all zeros. List the contents of the registers A, B, SP, and X, and the contents of the memory locations 1034 through 103C after the sequence of commands, PSHX, PSHB, PULA, PSHX, has been executed.

**T8.4.** Name the four elements of a computer-based instrumentation system.

**T8.5.** Name four types of systematic errors in measurement systems.

**T8.6.** How are bias errors different from random errors?

**T8.7.** What causes ground loops in an instrumentation system? What are the effects of a ground loop?

**T8.8.** If a sensor must have one end connected to ground, what type of amplifier should we choose? Why?

**T8.9.** What types of cable are best for connecting a sensor to an instrumentation amplifier to avoid coupling of noise by electrical and magnetic fields?

**T8.10.** If we need to sense the open-circuit voltage of a sensor, what specification is important for the instrumentation amplifier?

**T8.11.** How do we choose the sampling rate for an ADC? Why?

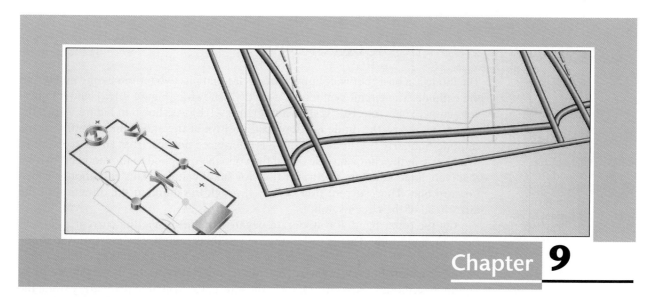

# Chapter 9

# Diodes

## Study of this chapter will enable you to:

- Understand diode operation and select diodes for various applications.

- Use the graphical load-line technique to analyze nonlinear circuits.

- Analyze and design simple voltage-regulator circuits.

- Use the ideal-diode model and piecewise-linear models to solve circuits.

- Understand various rectifier and wave-shaping circuits.

- Understand small-signal equivalent circuits.

## Introduction to this chapter:

Electronic circuits are useful for processing information and controlling energy. Some applications of electronic circuits are computers, radio, television, navigation systems, light dimmers, calculators, appliances, controls for machines, motion sensors, and surveying equipment. A basic understanding of electronic circuits will help you in working with instrumentation in any field of engineering. In the next several chapters, we introduce the most important electronic devices, their basic circuit applications, and several important analysis techniques. In this chapter, we discuss the diode.

## 9.1 BASIC DIODE CONCEPTS

The diode is a basic but very important device that has two terminals, the **anode** and the **cathode**. The circuit symbol for a diode is shown in Figure 9.1(a), and a typical volt–ampere characteristic is shown in Figure 9.1(b). As shown in Figure 9.1(a), the voltage $v_D$ across the diode is referenced positive at the anode and negative at the cathode. Similarly, the diode current $i_D$ is referenced positive from anode to cathode.

Notice in the characteristic that if the voltage $v_D$ applied to the diode is positive, relatively large amounts of current flow for small voltages. This condition is called **forward bias**. Thus, current flows easily through the diode in the direction of the arrowhead of the circuit symbol.

On the other hand, for moderate negative values of $v_D$, the current $i_D$ is very small in magnitude. This is called the **reverse-bias region**, as shown on the diode characteristic. In many applications, the ability of the diode to conduct current easily in one direction, but not in the reverse direction, is very useful. For example, in an automobile, diodes allow current from the alternator to charge the battery when the engine is running. However, when the engine stops, the diodes prevent the battery from discharging through the alternator. In these applications, the diode is analogous to a one-way valve in a fluid-flow system, as illustrated in Figure 9.1(d).

If a sufficiently large reverse-bias voltage is applied to the diode, operation enters the **reverse-breakdown region** of the characteristic, and currents of large magnitude flow. Provided that the power dissipated in the diode does not raise its temperature too high, operation in reverse breakdown is not destructive to the

*Diodes readily conduct current from anode to cathode (in the direction of the arrow), but do not readily allow current to flow in the opposite direction.*

*If a sufficiently large reverse-bias voltage is applied to the diode, operation enters the reverse-breakdown region of the characteristic, and currents of large magnitude flow.*

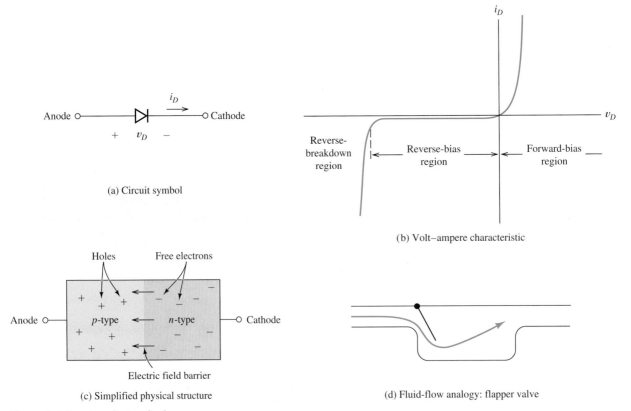

(a) Circuit symbol

(b) Volt–ampere characteristic

(c) Simplified physical structure

(d) Fluid-flow analogy: flapper valve

**Figure 9.1** Semiconductor diode.

device. In fact, we will see that diodes are sometimes deliberately operated in the reverse-breakdown region.

## Brief Sketch of Diode Physics

We concentrate our discussion on the external behavior of diodes and some of their circuit applications. However, at this point, we give a thumbnail sketch of the internal physics of the diode.

The diodes that we consider consist of a junction between two types of semiconducting material (usually, silicon with carefully selected impurities). On one side of the junction, the impurities create *n*-**type material**, in which large numbers of electrons move freely. On the other side of the junction, different impurities are employed to create (in effect) positively charged particles known as **holes**. Semiconductor material in which holes predominate is called *p*-**type material**. Most diodes consist of a junction between *n*-type material and *p*-type material, as shown in Figure 9.1(c).

Even with no external applied voltage, an electric-field **barrier** appears naturally at the *pn* junction. This barrier holds the free electrons on the *n*-side and the holes on the *p*-side of the junction. If an external voltage is applied with positive polarity on the *n*-side, the barrier is enhanced and the charge carriers cannot cross the junction. Thus, virtually no current flows. On the other hand, if a voltage is applied with positive polarity on the *p*-side, the barrier is reduced and large currents cross the junction. Thus, the diode conducts very little current for one polarity and large current for the other polarity of applied voltage. The anode corresponds to the *p*-type material and the cathode is the *n*-side.

## Small-Signal Diodes

Various materials and structures are used to fabricate diodes. For now, we confine our discussion to small-signal silicon diodes, which are the most common type found in low- and medium-power electronic circuits.

The characteristic curve of a typical small-signal silicon diode operated at a temperature of 300 K is shown in Figure 9.2. Notice that the voltage and current

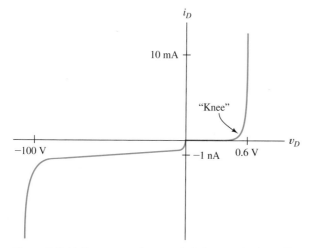

**Figure 9.2** Volt–ampere characteristic for a typical small-signal silicon diode at a temperature of 300 K. Notice the change of scale for negative current and voltage.

scales for the forward-bias region are different than for the reverse-bias region. This is necessary for displaying details of the characteristic, because the current magnitudes are much smaller in the reverse-bias region than in the forward-bias region. Furthermore, the forward-bias voltage magnitudes are much less than typical breakdown voltages.

In the forward-bias region, small-signal silicon diodes conduct very little current (much less than 1 mA) until a forward voltage of about 0.6 V is applied (assuming that the diode is at a temperature of about 300 K). Then, current increases very rapidly as the voltage is increased. We say that the forward-bias characteristic displays a *knee* in the forward bias characteristic at about 0.6 V. (The exact value of the knee voltage depends on the device, its temperature, and the current magnitude. Typical values are 0.6 or 0.7 V.) As temperature is increased, the knee voltage decreases by about 2 mV/K. (Because of the linear change in voltage with temperature, diodes are useful as temperature sensors. The diode is operated at a fixed current, and the voltage across the diode depends on its temperature. Electronic thermometers used by physicians contain a diode sensor, amplifiers, and other electronic circuits that drive the liquid-crystal temperature display.)

In the reverse-bias region, a typical current is about 1 nA for small-signal silicon diodes at room temperature. As temperature increases, reverse current increases in magnitude. A rule of thumb is that the reverse current doubles for each 10-K increase in temperature.

When reverse breakdown is reached, current increases in magnitude very rapidly. The voltage for which this occurs is called the **breakdown voltage**. For example, the breakdown voltage of the diode characteristic shown in Figure 9.2 is approximately −100 V. Breakdown-voltage magnitudes range from several volts to several hundred volts. Some applications call for diodes that operate in the forward-bias and nonconducting reverse-bias regions without entering the breakdown region. Diodes intended for these applications have a specification for the minimum magnitude of the breakdown voltage.

### Shockley Equation

Under certain simplifying assumptions, theoretical considerations result in the following relationship between current and voltage for a junction diode:

$$i_D = I_s \left[ \exp\left(\frac{v_D}{nV_T}\right) - 1 \right] \tag{9.1}$$

This is known as the **Shockley equation**. The **saturation current** $I_s$, has a value on the order of $10^{-14}$ A for small-signal junction diodes at 300 K. ($I_s$ depends on temperature, doubling for each 5-K increase in temperature for silicon devices.) The parameter $n$, known as the **emission coefficient**, takes values between 1 and 2, depending on details of the device structure. The voltage $V_T$ is given by

$$V_T = \frac{kT}{q} \tag{9.2}$$

and is called the **thermal voltage**. The temperature of the junction in kelvin is represented by $T$. Furthermore, $k = 1.38 \times 10^{-23}$ J/K is Boltzmann's constant, and

$q = 1.60 \times 10^{-19}$ C is the magnitude of the electrical charge of an electron. At a temperature of 300 K, we have $V_T \cong 0.026$ V.

If we solve the Shockley equation for the diode voltage, we find that

$$v_D = nV_T \ln\left[\left(\frac{i_D}{I_s}\right) + 1\right] \tag{9.3}$$

For small-signal junction diodes operated at forward currents between 0.01 $\mu$A and 10 mA, the Shockley equation with $n$ taken as unity is usually very accurate. Because the derivation of the Shockley equation ignores several phenomena, the equation is not accurate for smaller or larger currents. For example, under reverse bias, the Shockley equation predicts $i_D \cong -I_s$, but we usually find that the reverse current is much larger in magnitude than $I_s$ (although still small). Furthermore, the Shockley equation does not account for reverse breakdown.

With forward bias of at least several tenths of a volt, the exponential in the Shockley equation is much larger than unity; with good accuracy, we have

$$i_D \cong I_s \exp\left(\frac{v_D}{nV_T}\right) \tag{9.4}$$

This approximate form of the equation is often easier to use.

Occasionally, we are able to derive useful analytical results for electronic circuits by use of the Shockley equation, but much simpler models for diodes are usually more useful.

## Zener Diodes

Diodes that are intended to operate in the breakdown region are called **Zener diodes**. Zener diodes are useful in applications for which a constant voltage in breakdown is desirable. Therefore, manufacturers try to optimize Zener diodes for a nearly vertical characteristic in the breakdown region. The modified diode symbol shown in Figure 9.3 is used for Zener diodes. Zener diodes are available with breakdown voltages that are specified to a tolerance of ±5%.

Figure 9.3 Zener-diode symbol.

**Exercise 9.1**    At a temperature of 300 K, a certain junction diode has $i_D = 0.1$ mA for $v_D = 0.6$ V. Assume that $n$ is unity and use $V_T = 0.026$ V. Find the value of the saturation current $I_s$. Then, compute the diode current at $v_D = 0.65$ V and at 0.70 V.
**Answer**    $I_s = 9.50 \times 10^{-15}$ A, $i_D = 0.684$ mA, $i_D = 4.68$ mA.    ☐

**Exercise 9.2**    Consider a diode under forward bias so that the approximate form of the Shockley equation (Equation 9.4) applies. Assume that $V_T = 0.026$ V and $n = 1$. **a.** By what increment must $v_D$ increase to double the current? **b.** To increase the current by a factor of 10?
**Answer**    **a.** $\Delta v_D = 18$ mV; **b.** $\Delta v_D = 59.9$ mV.    ☐

## 9.2  LOAD-LINE ANALYSIS OF DIODE CIRCUITS

In Section 9.1, we learned that the volt–ampere characteristics of diodes are nonlinear. We will see shortly that other electronic devices are also nonlinear. On the other hand, resistors have linear volt–ampere characteristics, as shown in Figure 9.4. Because of this nonlinearity, many of the techniques that we have studied

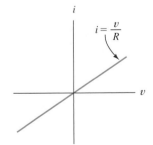

Figure 9.4  In contrast to diodes, resistors have linear volt–ampere characteristics.

for linear circuits in Chapters 1 through 6 do not apply to circuits involving diodes. In fact, much of the study of electronics is concerned with techniques for analysis of circuits containing nonlinear elements.

Graphical methods provide one approach to analysis of nonlinear circuits. For example, consider the circuit shown in Figure 9.5. By application of Kirchhoff's voltage law, we can write the equation

$$V_{SS} = Ri_D + v_D \qquad (9.5)$$

We assume that the values of $V_{SS}$ and $R$ are known and that we wish to find $i_D$ and $v_D$. Thus, Equation 9.5 has two unknowns, and another equation (or its equivalent) is needed before a solution can be found. This is available in graphical form in Figure 9.6, which shows the volt–ampere characteristic of the diode.

We can obtain a solution by plotting Equation 9.5 on the same set of axes used for the diode characteristic. Since Equation 9.5 is linear, it plots as a straight line, which can be drawn if two points satisfying the equation are located. A simple method is to assume that $i_D = 0$, and then Equation 9.5 yields $v_D = V_{SS}$. This pair of values is shown as point $A$ in Figure 9.6. A second point results if we assume that $v_D = 0$, for which the equation yields $i_D = V_{SS}/R$. The pair of values is shown as point $B$ in Figure 9.6. Then, connecting points $A$ and $B$ result in a plot called the **load line**. The **operating point** is the intersection of the load line and the diode characteristic. This point represents the simultaneous solution of Equation 9.5 and the diode characteristic.

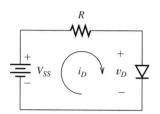

**Figure 9.5** Circuit for load-line analysis.

---

### Example 9.1    Load-Line Analysis

If the circuit of Figure 9.5 has $V_{SS} = 2$ V, $R = 1$ kΩ, and a diode with the characteristic shown in Figure 9.7, find the diode voltage and current at the operating point.

**Solution**    First, we locate the ends of the load line. Substituting $v_D = 0$ and the values given for $V_{SS}$ and $R$ into Equation 9.5 yields $i_D = 2$ mA. These values plot as point $B$ in Figure 9.7. Substitution of $i_D = 0$ and circuit values results in

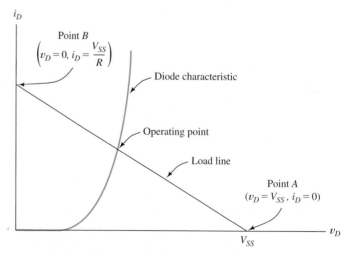

**Figure 9.6** Load-line analysis of the circuit of Figure 9.5.

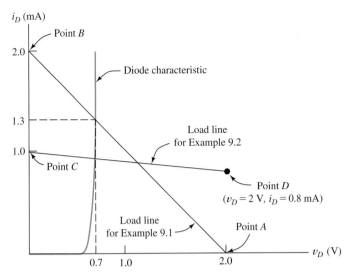

**Figure 9.7** Load-line analysis for Examples 9.1 and 9.2.

$v_D = 2$ V. These values plot as point $A$ in the figure. Constructing the load line results in an operating point of $V_{DQ} \cong 0.7$ V and $I_{DQ} \cong 1.3$ mA, as shown in the figure.  ■

---

### Example 9.2    Load-Line Analysis

Repeat Example 9.1 if $V_{SS} = 10$ V and $R = 10$ kΩ.

**Solution**    If we let $v_D = 0$ and substitute values into Equation 9.5, we find that $i_D = 1$ mA. This is plotted as point $C$ in Figure 9.7.

When an intercept of the load line falls off the page, we select a point at the edge of the page.

If we proceed as before by assuming that $i_D = 0$, we find that $v_D = 10$ V. This is a perfectly valid point on the load line, but it plots at a point far off the page. Of course, we can use any other point satisfying Equation 9.5 to locate the load line. Since we already have point $C$ on the $i_D$ axis, a good point to use would be on the

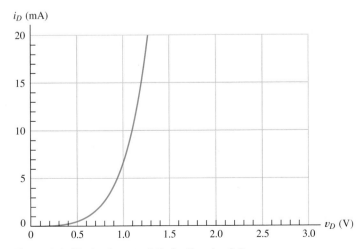

**Figure 9.8** Diode characteristic for Exercise 9.3.

right-hand edge of Figure 9.7. Thus, we assume that $v_D = 2$ V and substitute values into Equation 9.5, resulting in $i_D = 0.8$ mA. These values plot as point $D$. Then, we can draw the load line and find that the operating-point values are $V_{DQ} \cong 0.68$ V and $I_{DQ} \cong 0.93$ mA.  ∎

**Exercise 9.3**   Find the operating point for the circuit of Figure 9.5 if the diode characteristic is shown in Figure 9.8 and: **a.** $V_{SS} = 2$ V and $R = 100$ $\Omega$; **b.** $V_{SS} = 15$ V and $R = 1$ k$\Omega$; **c.** $V_{SS} = 1.0$ V and $R = 20$ $\Omega$.
**Answer    a.** $V_{DQ} \cong 1.1$ V, $I_{DQ} \cong 9.0$ mA;   **b.** $V_{DQ} \cong 1.2$ V, $I_{DQ} \cong 13.8$ mA; **c.** $V_{DQ} \cong 0.91$ V, $I_{DQ} \cong 4.5$ mA.   □

## 9.3   ZENER-DIODE VOLTAGE-REGULATOR CIRCUITS

Sometimes, a circuit that produces constant output voltage while operating from a variable supply voltage is needed. Such circuits are called **voltage regulators**. For example, if we wanted to operate computer circuits from the battery in an automobile, a voltage regulator would be needed. Automobile battery voltage typically varies between about 10 and 14 V (depending on the state of the battery and whether or not the engine is running). Many computer circuits require a nearly constant voltage of 5 V. Thus, a regulator is needed that operates from the 10 to 14 V supply and produces a nearly constant 5-V output.

A voltage regulator circuit provides a nearly constant voltage to a load from a variable source.

In this section, we use the load-line technique that we introduced in Section 9.2 to analyze a simple regulator circuit. The regulator circuit is shown in Figure 9.9. (For proper operation, it is necessary for the minimum value of the variable source voltage to be somewhat larger than the desired output voltage.) The Zener diode has a breakdown voltage equal to the desired output voltage. The resistor $R$ limits the diode current to a safe value so that the Zener diode does not overheat.

Assuming that the characteristic for the diode is available, we can construct a load line to analyze the operation of the circuit. As before, we use Kirchhoff's voltage law to write an equation relating $v_D$ and $i_D$. (In this circuit, the diode operates in the breakdown region with negative values for $v_D$ and $i_D$.) For the circuit of Figure 9.9, we obtain

$$V_{SS} + Ri_D + v_D = 0 \qquad (9.6)$$

Once again, this is the equation of a straight line, so location of any two points is sufficient to construct the load line. The intersection of the load line with the diode characteristic yields the operating point.

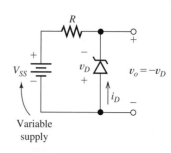

**Figure 9.9** A simple regulator circuit that provides a nearly constant output voltage $v_o$ from a variable supply voltage.

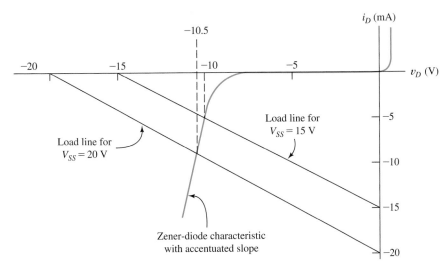

Figure 9.10  See Example 9.3.

---

Example 9.3   **Load-Line Analysis of a Zener-Diode Voltage Regulator**

The voltage-regulator circuit of Figure 9.9 has $R = 1 \text{ k}\Omega$ and uses a Zener diode having the characteristic shown in Figure 9.10. Find the output voltage for $V_{SS} = 15$ V. Repeat for $V_{SS} = 20$ V.

**Solution**   The load lines for both values of $V_{SS}$ are shown in Figure 9.10. The output voltages are determined from the points where the load lines intersect the diode characteristic. The output voltages are found to be $v_o = 10.0$ V for $V_{SS} = 15$ V and $v_o = 10.5$ V for $V_{SS} = 20$ V. Thus, a 5-V change in the supply voltage results in only a 0.5-V change in the regulated output voltage.

Actual Zener diodes are capable of much better performance than this. The slope of the characteristic has been accentuated in Figure 9.10 for clarity–actual Zener diodes have a more nearly vertical slope in breakdown.   ∎

## Slope of the Load Line

Notice that the two load lines shown in Figure 9.10 are parallel. Inspection of Equation 9.5 or Equation 9.6 shows that the slope of the load line is $-1/R$. Thus, a change of the supply voltage changes the position, but not the slope of the load line.

Load lines for different source voltages (but the same resistance) are parallel.

## Load-Line Analysis of Complex Circuits

Any circuit that contains resistors, voltage sources, current sources, and a single two-terminal nonlinear element can be analyzed by the load-line technique. First, the Thévenin equivalent is found for the linear portion of the circuit as illustrated in Figure 9.11. Then, a load line is constructed to find the operating point on the characteristic of the nonlinear device. Once the operating point of the nonlinear element is known, voltages and currents can be determined in the original circuit.

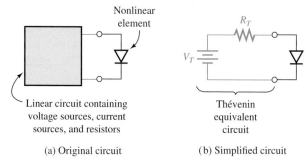

(a) Original circuit

(b) Simplified circuit

**Figure 9.11** Analysis of a circuit containing a single nonlinear element can be accomplished by load-line analysis of a simplified circuit.

**Example 9.4    Analysis of a Zener-Diode Regulator with a Load**

Consider the Zener-diode regulator circuit shown in Figure 9.12(a). The diode characteristic is shown in Figure 9.13. Find the load voltage $v_L$ and source current $I_S$ if $V_{SS} = 24$ V, $R = 1.2$ kΩ, and $R_L = 6$ kΩ.

**Solution**    First, consider the circuit as redrawn in Figure 9.12(b), in which we have grouped the linear elements together on the left-hand side of the diode. Next, we find the Thévenin equivalent for the linear portion of the circuit. The Thévenin voltage is the open-circuit voltage (i.e., the voltage across $R_L$ with the diode replaced by an open circuit), which is given by

$$V_T = V_{SS}\frac{R_L}{R + R_L} = 20 \text{ V}$$

The Thévenin resistance can be found by zeroing the voltage source and looking back into the circuit from the diode terminals. This is accomplished by reducing

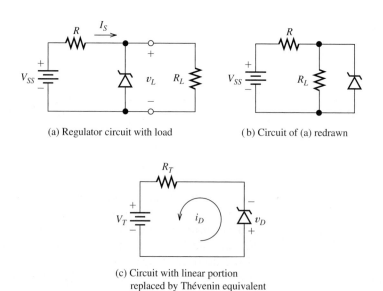

(a) Regulator circuit with load

(b) Circuit of (a) redrawn

(c) Circuit with linear portion replaced by Thévenin equivalent

**Figure 9.12**  See Example 9.4.

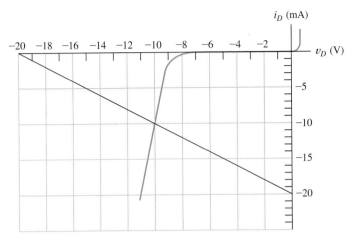

Figure 9.13  Zener-diode characteristic for Example 9.4 and Exercise 9.4.

$V_{SS}$ to zero so that the voltage source becomes a short circuit. Then, we have $R$ and $R_L$ in parallel, so the Thévenin resistance is

$$R_T = \frac{RR_L}{R + R_L} = 1 \text{ k}\Omega$$

The resulting equivalent circuit is shown in Figure 9.12(c).

Now, we can use Kirchhoff's voltage law to write the load-line equation from the equivalent circuit as

$$V_T + R_T i_D + v_D = 0$$

Using the values found for $V_T$ and $R_T$, we can construct the load line shown in Figure 9.13 and locate the operating point. This yields $v_L = -v_D = 10.0$ V.

Once $v_L$ is known, we can find the voltages and currents in the original circuit. For example, using the output voltage value of 10.0 V in the original circuit of Figure 9.12(a), we find that $I_S = (V_{SS} - v_L)/R = 11.67$ mA.  ■

**Exercise 9.4**  Find the voltage across the load in Example 9.4 if: **a.** $R_L = 1.2$ k$\Omega$; **b.** $R_L = 400$ $\Omega$.
**Answer  a.** $v_L \cong 9.4$ V; **b.** $v_L \cong 6.0$ V. (Notice that this regulator is not perfect because the load voltage varies as the load current changes.)  □

**Exercise 9.5**  Consider the circuit of Figure 9.14(a). Assume that the breakdown characteristic is vertical, as shown in Figure 9.14(b). Find the output voltage $v_o$ for: **a.** $i_L = 0$; **b.** $i_L = 20$ mA; **c.** $i_L = 100$ mA. [*Hint:* Applying Kirchhoff's voltage law to the circuit, we have

$$15 = 100(i_L - i_D) - v_D$$

Construct a different load line for each value of $i_L$.]
**Answer  a.** $v_o = 10.0$ V; **b.** $v_o = 10.0$ V; **c.** $v_o = 5.0$ V. (Notice that the regulator is not effective for large load currents.)  □

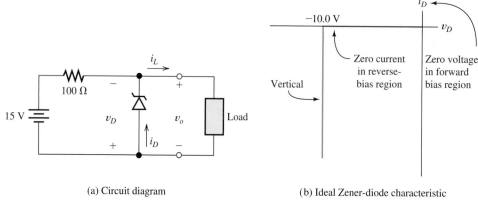

(a) Circuit diagram

(b) Ideal Zener-diode characteristic

**Figure 9.14  See Exercise 9.5.**

## 9.4  IDEAL-DIODE MODEL

Graphical load-line analysis is useful for some circuits, such as the voltage regulator studied in Section 9.3. However, it is too cumbersome for more complex circuits. Instead, we often use simpler models to approximate diode behavior.

> The ideal diode acts as a short circuit for forward currents and as an open circuit with reverse voltage applied.

One model for a diode is the **ideal diode**, which is a perfect conductor with zero voltage drop in the forward direction. In the reverse direction, the ideal diode is an open circuit. We use the ideal-diode assumption if our judgment tells us that the forward diode voltage drop and reverse current are negligible, or if we want a basic understanding of a circuit rather than an exact analysis.

The volt–ampere characteristic for the ideal diode is shown in Figure 9.15. If $i_D$ is positive, $v_D$ is zero, and we say that the diode is in the *on* state. On the other hand, if $v_D$ is negative, $i_D$ is zero, and we say that the diode is in the *off* state.

### Assumed States for Analysis of Ideal-Diode Circuits

In analysis of a circuit containing ideal diodes, we may not know in advance which diodes are on and which are off. Thus, we are forced to make a considered guess. Then, we analyze the circuit to find the currents in the diodes assumed to be on and the voltages across the diodes assumed to be off. If $i_D$ is positive for the diodes assumed to be on and if $v_D$ is negative for the diodes assumed to be off, our assumptions are correct, and we have solved the circuit. (We are assuming that $i_D$ is referenced positive in the forward direction and that $v_D$ is referenced positive at the anode.)

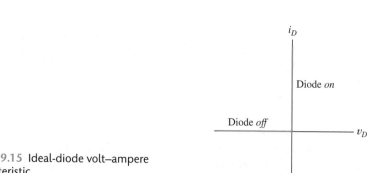

**Figure 9.15  Ideal-diode volt–ampere characteristic.**

Otherwise, we must make another assumption about the diodes and try again. After a little practice, our first guess is usually correct, at least for simple circuits.

A step-by-step procedure for analyzing circuits that contain ideal diodes is to:

1. Assume a state for each diode, either on (i.e., a short circuit) or off (i.e., an open circuit). For $n$ diodes there are $2^n$ possible combinations of diode states.

2. Analyze the circuit to determine the current through the diodes assumed to be on and the voltage across the diodes assumed to be off.

3. Check to see if the result is consistent with the assumed state for each diode. Current must flow in the forward direction for diodes assumed to be on. Furthermore, the voltage across the diodes assumed to be off must be positive at the cathode (i.e., reverse bias).

4. If the results are consistent with the assumed states, the analysis is finished. Otherwise, return to step 1 and choose a different combination of diode states.

---

### Example 9.5    Analysis by Assumed Diode States

Use the ideal-diode model to analyze the circuit shown in Figure 9.16(a). Start by assuming that $D_1$ is off and $D_2$ is on.

**Solution**   With $D_1$ off and $D_2$ on, the equivalent circuit is shown in Figure 9.16(b). Solving results in $i_{D2} = 0.5$ mA. Since the current in $D_2$ is positive, our assumption that $D_2$ is on seems to be correct. However, continuing the solution of the circuit of Figure 9.16(b), we find that $v_{D1} = +7$ V. This is not consistent with the assumption that $D_1$ is off. Therefore, we must try another assumption.

This time, we assume that $D_1$ is on and $D_2$ is off. The equivalent circuit for these assumptions is shown in Figure 9.16(c). We can solve this circuit to find that $i_{D1} = 1$ mA and $v_{D2} = -3$ V. These values are consistent with the assumptions about the diodes ($D_1$ on and $D_2$ off) and, therefore, are correct.  ∎

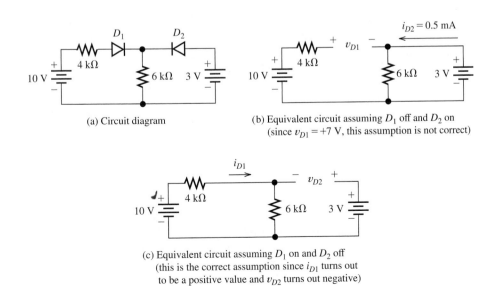

(a) Circuit diagram

(b) Equivalent circuit assuming $D_1$ off and $D_2$ on (since $v_{D1} = +7$ V, this assumption is not correct)

(c) Equivalent circuit assuming $D_1$ on and $D_2$ off (this is the correct assumption since $i_{D1}$ turns out to be a positive value and $v_{D2}$ turns out negative)

**Figure 9.16** Analysis of a diode circuit, using the ideal-diode model. See Example 9.5.

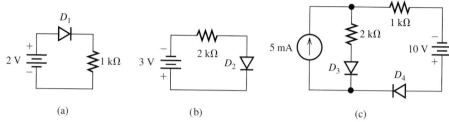

**Figure 9.17** Circuits for Exercise 9.8.

Notice in Example 9.5 that even though current flows in the forward direction of $D_2$ for our first guess about diode states ($D_1$ off and $D_2$ on), the correct solution is that $D_2$ is off. Thus, in general, we cannot decide on the state of a particular diode until we have found a combination of states that works for all the diodes in the circuit.

In general, we cannot decide on the state of a particular diode until we have found a combination of states that works for all of the diodes in the circuit.

For a circuit containing $n$ diodes, there are $2^n$ possible states. Thus, an exhaustive search eventually yields the solution for each circuit.

**Exercise 9.6**   Show that the condition $D_1$ off and $D_2$ off is not valid for the circuit of Figure 9.16(a).    □

**Exercise 9.7**   Show that the condition $D_1$ on and $D_2$ on is not valid for the circuit of Figure 9.16(a).    □

**Exercise 9.8**   Find the diode states for the circuits shown in Figure 9.17. Assume ideal diodes.
**Answer    a.** $D_1$ is on; **b.** $D_2$ is off; **c.** $D_3$ is off; and $D_4$ is on.    □

## 9.5  PIECEWISE-LINEAR DIODE MODELS

Sometimes, we want a more accurate model than the ideal-diode assumption, but do not want to resort to nonlinear equations or graphical techniques. Then, we can use **piecewise-linear models** for the diodes. First, we approximate the actual volt–ampere characteristic by straight-line segments. Then, we model each section of the diode characteristic with a resistance in series with a constant-voltage source. Different resistance and voltage values are used in the various sections of the characteristic.

Consider the resistance $R_a$ in series with a voltage source $V_a$ shown in Figure 9.18(a). We can write the following equation, relating the voltage and current of the series combination:

$$v = R_a i + V_a \tag{9.7}$$

The current $i$ is plotted versus $v$ in Figure 9.18(b). Notice that the intercept on the voltage axis is at $v = V_a$ and that the slope of the line is $1/R_a$.

Given a straight-line volt–ampere characteristic, we can work backward to find the corresponding series voltage and resistance. Thus, after a nonlinear volt–ampere characteristic has been approximated by several straight-line segments, a circuit model consisting of a voltage source and series resistance can be found for each segment.

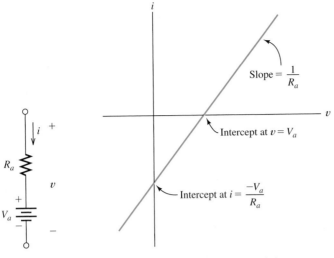

(a) Circuit diagram                    (b) Volt–ampere characteristic

**Figure 9.18** Circuit and volt–ampere characteristic for piecewise-linear models.

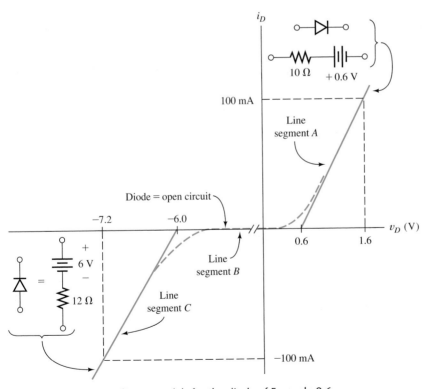

**Figure 9.19** Piecewise-linear models for the diode of Example 9.6.

---

| Example 9.6 | Piecewise-Linear Model for a Zener Diode |
|---|---|

Find circuit models for the Zener-diode volt–ampere characteristic shown in Figure 9.19. Use the straight-line segments shown.

**Solution**   For line segment $A$ of Figure 9.19, the intercept on the voltage axis is 0.6 V and the reciprocal of the slope is 10 $\Omega$. Hence, the circuit model for the diode on this segment is a 10-$\Omega$ resistance in series with a 0.6-V source, as shown in the figure. Line segment $B$ has zero current, and therefore, the equivalent circuit for segment $B$ is an open circuit, as illustrated in the figure. Finally, line segment $C$ has an intercept of $-6$ V and a reciprocal slope of 12 $\Omega$, resulting in the equivalent circuit shown. Thus, this diode can be approximated by one of these linear circuits, depending on where the operating point is located.  ∎

| Example 9.7 | Analysis Using a Piecewise-Linear Model |
|---|---|

Use the circuit models found in Example 9.6 to solve for the current in the circuit of Figure 9.20(a).

**Solution**   Since the 3-V source has a polarity that results in forward bias of the diode, we assume that the operating point is on line segment $A$ of Figure 9.19. Consequently, the equivalent circuit for the diode is the one for segment $A$. Using this equivalent circuit, we have the circuit of Figure 9.20(b). Solving, we find that $i_D = 80$ mA.  ∎

**Exercise 9.9**   Use the appropriate circuit model from Figure 9.19 to solve for $v_o$ in the circuit of Figure 9.21 if: **a.** $R_L = 10$ k$\Omega$; and **b.** $R_L = 1$ k$\Omega$. (*Hint:* Be sure that your answers are consistent with your choice of equivalent circuit for the diode–the various equivalent circuits are valid only for specific ranges of diode voltage and current. The answer must fall into the valid range for the equivalent circuit used.)
**Answer**   **a.** $v_o = 6.017$ V; **b.** $v_o = 3.333$ V.  ☐

**Exercise 9.10**   Find a circuit model for each line segment shown in Figure 9.22(a). Draw the circuit models identifying terminals $a$ and $b$ for each equivalent circuit.
**Answer**   See Figure 9.22(b). Notice the polarity of the voltage sources with respect to terminals $a$ and $b$.  ☐

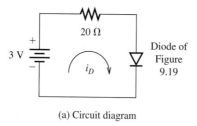

(a) Circuit diagram

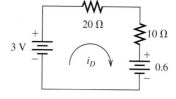

(b) Circuit with diode modeled by the equivalent circuit for the forward-bias region

**Figure 9.20**  Circuit for Example 9.7.

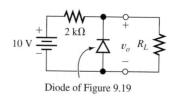

**Figure 9.21**  Circuit for Exercise 9.9.

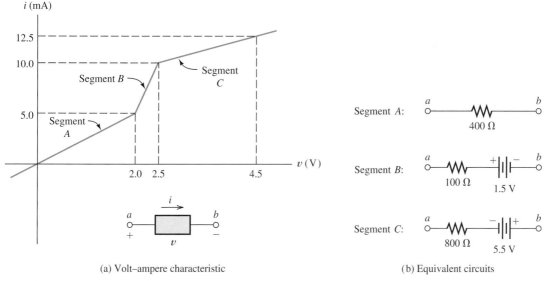

(a) Volt–ampere characteristic

(b) Equivalent circuits

**Figure 9.22** Hypothetical nonlinear device for Exercise 9.10.

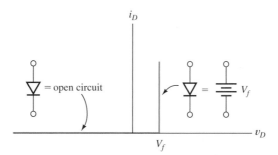

**Figure 9.23** Simple piecewise-linear equivalent for the diode.

## Simple Piecewise-Linear Diode Equivalent Circuit

Figure 9.23 shows a simple piecewise-linear equivalent circuit for diodes that is often sufficiently accurate. It is an open circuit in the reverse-bias region and a constant voltage drop in the forward direction. This model is equivalent to a battery in series with an ideal diode.

## 9.6  RECTIFIER CIRCUITS

Now that we have introduced the diode and some methods for analysis of diode circuits, we consider some additional practical circuits. First, we consider several types of **rectifiers**, which convert ac power into dc power. These rectifiers form the basis for electronic **power supplies** and battery-charging circuits. Typically, a power supply takes power from a raw source, which is often the 60-Hz ac power line, and delivers steady dc voltages to a load such as computer circuits or television circuits. Other applications for rectifiers are in signal processing, such as demodulation of a radio signal. (*Demodulation* is the process of retrieving the message, such as a voice or video signal.) Another application is precision conversion of an ac voltage to dc in an electronic voltmeter.

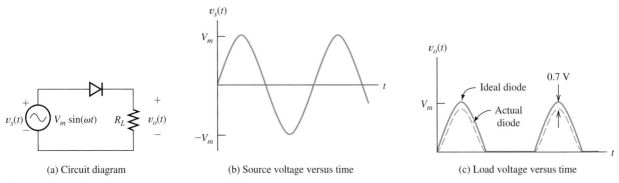

**Figure 9.24**  Half-wave rectifier with resistive load.

## Half-Wave Rectifier Circuits

A **half-wave rectifier** with a sinusoidal source and resistive load $R_L$ is shown in Figure 9.24. When the source voltage $v_s(t)$ is positive, the diode is in the forward-bias region. If an ideal diode is assumed, the source voltage appears across the load. For a typical real diode, the output voltage is less than the source voltage by an amount equal to the drop across the diode, which is approximately 0.7 V for silicon diodes at room temperature. When the source voltage is negative, the diode is reverse biased and no current flows through the load. Even for typical real diodes, only a very small reverse current flows. Thus, only the positive half-cycles of the source voltage appear across the load.

**Battery-Charging Circuit.**   We can use a half-wave rectifier to charge a battery as shown in Figure 9.25. Current flows whenever the instantaneous ac source voltage is higher than the battery voltage. As shown in the figure, it is necessary to add resistance to the circuit to limit the magnitude of the current. When the ac source voltage is less than the battery voltage, the diode is reverse biased and the current is zero. Hence, the current flows only in the direction that charges the battery.

**Half-Wave Rectifier with Smoothing Capacitor.**   Often, we want to convert an ac voltage into a nearly constant dc voltage to be used as a power supply for electronic circuits. One approach to smoothing the rectifier output voltage is to place a large

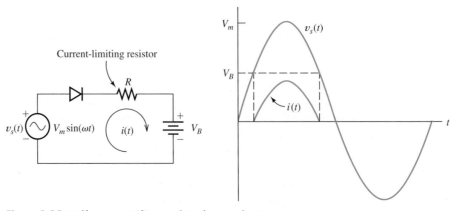

**Figure 9.25**  Half-wave rectifier used to charge a battery.

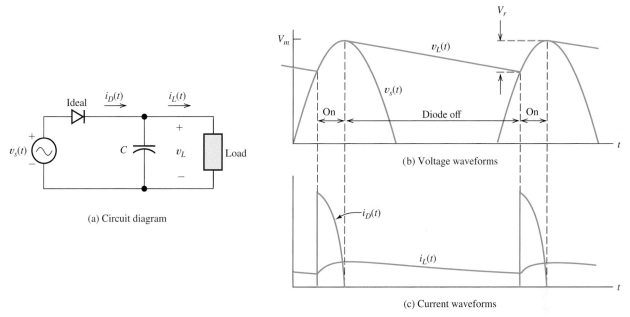

**Figure 9.26** Half-wave rectifier with smoothing capacitor.

capacitance across the output terminals of the rectifier. The circuit and waveforms of current and voltage are shown in Figure 9.26. When the ac source reaches a positive peak, the capacitor is charged to the peak voltage (assuming an ideal diode). When the source voltage drops below the voltage stored on the capacitor, the diode is reverse biased and no current flows through the diode. The capacitor continues to supply current to the load, slowly discharging until the next positive peak of the ac input. As shown in the figure, current flows through the diode in pulses that recharge the capacitor.

Because of the charge and discharge cycle, the load voltage contains a small ac component called **ripple**. Usually, it is desirable to minimize the amplitude of the ripple, so we choose the largest capacitance value that is practical. In this case, the capacitor discharges for nearly the entire cycle, and the charge removed from the capacitor during one discharge cycle is

$$Q \cong I_L T \qquad (9.8)$$

where $I_L$ is the average load current and $T$ is the period of the ac voltage. Since the charge removed from the capacitor is the product of the change in voltage and the capacitance, we can also write

$$Q = V_r C \qquad (9.9)$$

where $V_r$ is the peak-to-peak ripple voltage and $C$ is the capacitance. Equating the right-hand sides of Equations 9.8 and 9.9 allows us to solve for $C$:

$$C = \frac{I_L T}{V_r} \qquad (9.10)$$

In practice, Equation 9.10 is approximate because the load current varies and because the capacitor does not discharge for a complete cycle. However, it gives a good starting value for calculating the capacitance required in the design of power-supply circuits.

The average voltage supplied to the load if a smoothing capacitor is used is approximately midway between the minimum and maximum voltages. Thus, referring to Figure 9.26, the average load voltage is

$$V_L \cong V_m - \frac{V_r}{2} \tag{9.11}$$

### Peak Inverse Voltage

An important aspect of rectifier circuits is the **peak inverse voltage** (PIV) across the diodes. Of course, the breakdown specification of the diodes should be greater in magnitude than the PIV. For example, in the half-wave circuit with a resistive load, shown in Figure 9.24, the PIV is $V_m$.

The addition of a smoothing capacitor in parallel with the load increases the PIV to (approximately) $2V_m$. Referring to Figure 9.26, for the negative peak of the ac input, we see that the reverse bias of the diode is the sum of the source voltage and the voltage stored on the capacitor.

### Full-Wave Rectifier Circuits

A wire or other conductor, not shown explicitly in the diagram, connects all of the points that are connected to ground symbols.

Several **full-wave rectifier** circuits are in common use. One approach uses two ac sources and two diodes, as shown in Figure 9.27(a). One feature of this diagram is the **ground symbol**. Usually in electronic circuits, many components are connected to a common point known as *ground*. Often, the chassis containing the circuit is the electrical ground. Therefore, in Figure 9.27(a), the lower end of $R_L$ and the point between the voltage sources are connected together.

When the upper source applies a positive voltage to the left-hand end of diode $A$, the lower source applies a negative voltage to the left-hand end of diode $B$, and vice versa. We say that the sources are **out of phase**. Thus, the circuit consists of two half-wave rectifiers with out-of-phase source voltages and a common load. The diodes conduct on alternate half-cycles.

Usually, the two out-of-phase ac voltages are provided by a **transformer**. (Transformers are discussed in Chapter 15.) Besides providing the out-of-phase ac voltages, the transformer also allows the designer to adjust $V_m$ by selection of

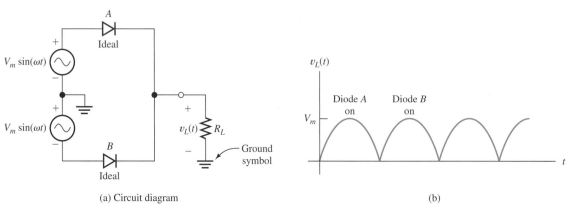

(a) Circuit diagram                                    (b)

**Figure 9.27**  Full-wave rectifier.

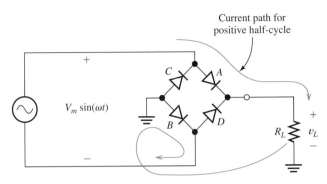

Figure 9.28 Diode-bridge full-wave rectifier.

the turns ratio. This is important, because the ac voltage available is often not of a suitable amplitude for direct rectification–usually either a higher or lower dc voltage is required.

A second type of full-wave rectifier uses the **diode bridge** shown in Figure 9.28. When the ac voltage, $V_m \sin(\omega t)$, is positive, current flows through diode $A$, then through the load, and returns through diode $B$, as shown in the figure. For the opposite polarity, current flows through diodes $C$ and $D$. Notice that in either case, current flows in the same direction through the load.

Usually, neither of the ac source terminals is connected to ground. This is necessary if one side of the load is to be connected to ground, as shown in the figure. (If both the ac source and the load have a common ground connection, part of the circuit is shorted.)

If we wish to smooth the voltage across the load, a capacitor can be placed in parallel with the load, similar to the half-wave circuit discussed earlier. In the full-wave circuits, the capacitor discharges for only a half-cycle before being recharged. Hence, the capacitance required is only half as much in the full-wave circuit as for the half-wave circuit. Therefore, we modify Equation 9.10 to obtain

$$C = \frac{I_L T}{2V_r} \tag{9.12}$$

for the full-wave rectifier with a capacitive filter.

**Exercise 9.11**   Consider the battery-charging circuit of Figure 9.25 with $V_m = 20$ V, $R = 10$ Ω, and $V_B = 14$ V. **a.** Find the peak current assuming an ideal diode. **b.** Find the percentage of each cycle for which the diode is in the on state.
**Answer**   **a.** $I_{peak} = 600$ mA; **b.** the diode is on for 25.3 percent of each cycle.   □

**Exercise 9.12**   A power-supply circuit is needed to deliver 0.1 A and 15 V (average) to a load. The ac source has a frequency of 60 Hz. Assume that the circuit of Figure 9.26 is to be used. The peak-to-peak ripple voltage is to be 0.4 V. Instead of assuming an ideal diode, allow 0.7 V for forward diode drop. Find the peak ac voltage $V_m$ needed and the approximate value of the smoothing capacitor. (*Hint:* To achieve an average load voltage of 15 V with a ripple of 0.4 V, design for a peak load voltage of 15.2 V.)
**Answer**   $V_m = 15.9$, $C = 4166$ μF.   □

## 9.7 WAVE-SHAPING CIRCUITS

A wide variety of **wave-shaping circuits** are used in electronic systems. These circuits are used to transform one waveform into another. Numerous examples of wave-shaping circuits can be found in transmitters and receivers for television or radar. In this section, we discuss a few examples of wave-shaping circuits that can be constructed with diodes.

### Clipper Circuits

A clipper circuit "clips off" part of the input waveform to produce the output waveform.

Diodes can be used to form **clipper circuits**, in which a portion of an input signal waveform is "clipped" off. For example, the circuit of Figure 9.29 clips off any part of the input waveform above 6 V or less than −9 V. (We are assuming ideal diodes.) When the input voltage is between −9 and +6 V, both diodes are off and no current flows. Then, there is no drop across $R$ and the output voltage $v_o$ is equal to the input voltage $v_{in}$. On the other hand, when $v_{in}$ is larger than 6 V, diode $A$ is on and the output voltage is 6 V, because the diode connects the 6-V battery to the output terminals. Similarly, when $v_{in}$ is less than −9 V, diode $B$ is on and the output voltage is −9 V. The output waveform resulting from a 15-V-peak sinusoidal input

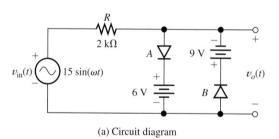

(a) Circuit diagram

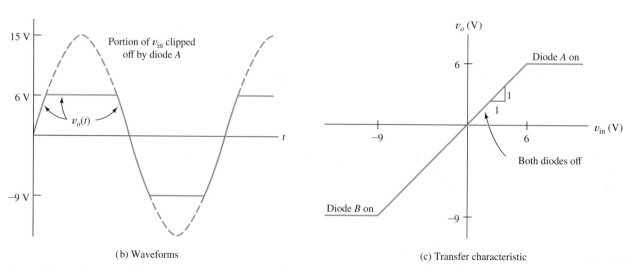

(b) Waveforms

(c) Transfer characteristic

**Figure 9.29** Clipper circuit.

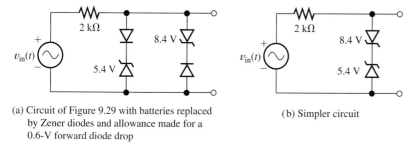

(a) Circuit of Figure 9.29 with batteries replaced by Zener diodes and allowance made for a 0.6-V forward diode drop

(b) Simpler circuit

**Figure 9.30** Circuits with nearly the same performance as the circuit of Figure 9.29.

is shown in Figure 9.29(b), and the transfer characteristic of the circuit is shown in Figure 9.29(c).

The resistance $R$ is selected large enough so that the forward diode current is within reasonable bounds (usually, a few milliamperes), but small enough so that the reverse diode current results in a negligible voltage drop. Often, we find that a wide range of resistance values provides satisfactory performance in a given circuit.

In Figure 9.29, we have assumed ideal diodes. If small-signal silicon diodes are used, we expect a forward drop of 0.6 or 0.7 V, so we should reduce the battery voltages to compensate. Furthermore, batteries are not desirable for use in circuits if they can be avoided, because they may need periodic replacement. Thus, a better design uses Zener diodes instead of batteries. Practical circuits equivalent to Figure 9.29 are shown in Figure 9.30. The Zener diodes are labeled with their breakdown voltages.

**Exercise 9.14    a.** Sketch the transfer characteristics to scale for the circuits of Figure 9.31(a) and (b). Allow a 0.6-V forward drop for the diodes. **b.** Sketch the output waveform to scale if $v_{in}(t) = 15 \sin(\omega t)$.
**Answer    a.** See Figure 9.31(c); **b.** see Figure 9.31(d).    □

**Exercise 9.15**    Design clipper circuits that have the transfer characteristics shown in **a.** Figure 9.32(a) and **b.** Figure 9.32(b). Allow for a 0.6-V drop in the forward direction for the diodes. [*Hint for part (b):* Include a resistor in series with the diode that begins to conduct at $v_{in} = 3$ V to achieve the slope required for the section between $v_{in} = 3$ V and 6 V.]
**Answer    a.** See Figure 9.32(c); **b.** see Figure 9.32(d).    □

## Clamp Circuits

Another diode wave-shaping circuit is the **clamp circuit**, which is used to add a dc component to an ac input waveform so that the positive (or negative) peaks are forced to take a specified value. In other words, the peaks of the waveform are "clamped" to a specified voltage value. An example circuit is shown in Figure 9.33. In this circuit, the positive peaks are clamped to −5 V.

The capacitance is a large value, so it discharges only very slowly and we can consider the voltage across the capacitor to be constant. Because the capacitance is large, it has a very small impedance for the ac input signal. The output voltage of the circuit is given by

$$v_o(t) = v_{in}(t) - V_C \qquad (9.13)$$

If a positive swing of the input signal attempts to force the output voltage to become greater than −5 V, the diode conducts, increasing the value of $V_C$. Thus, the capacitor

> In a clamp circuit, a variable dc voltage is added to the input waveform so that one of the peaks of the output is clamped to a specified value.

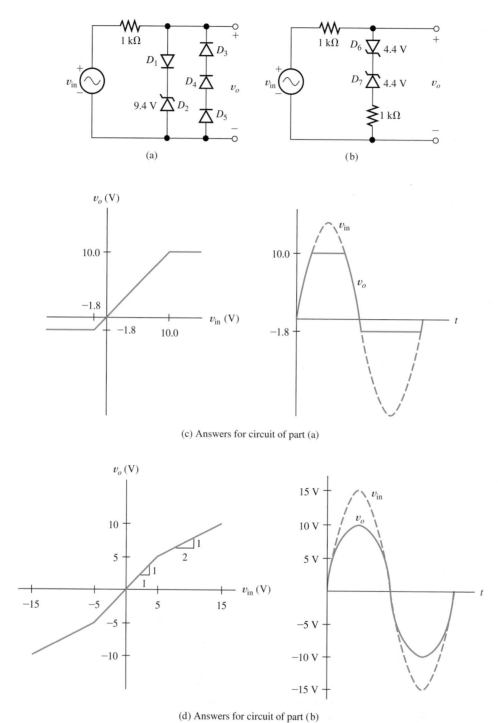

(a)

(b)

(c) Answers for circuit of part (a)

(d) Answers for circuit of part (b)

**Figure 9.31** See Exercise 9.14.

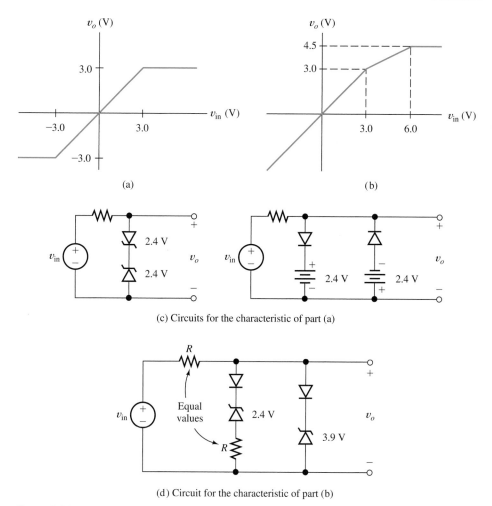

(c) Circuits for the characteristic of part (a)

(d) Circuit for the characteristic of part (b)

**Figure 9.32** See Exercise 9.15.

is charged to a value that adjusts the maximum value of the output voltage to $-5$ V. A large resistance $R$ is provided so that the capacitor can discharge slowly. This is necessary so the circuit can adjust if the input waveform changes to a smaller peak amplitude.

Of course, we can change the voltage to which the circuit clamps by changing the battery voltage. Reversing the direction of the diode causes the negative peak to be clamped instead of the positive peak. If the desired clamp voltage requires the diode to be reverse biased, it is necessary to return the discharge resistor to a suitable dc supply voltage to ensure that the diode conducts and performs the clamping operation. Furthermore, it is often more convenient to use Zener diodes rather than batteries. A circuit including these features is shown in Figure 9.34.

**Exercise 9.16**   Consider the circuit of Figure 9.34(a). Assume that the capacitance is large enough so that the voltage across it does not discharge through $R$ appreciably during one cycle of input. **a.** What is the steady-state output voltage if $v_{in}(t) = 0$? **b.** Sketch the steady-state output to scale versus time if $v_{in}(t) = 2\sin(\omega t)$. **c.** Suppose that the resistor is returned directly to ground instead of $-15$ V (i.e., replace the 15-V source by a short circuit). In this case, sketch the steady-state output versus time if $v_{in}(t) = 2\sin(\omega t)$.

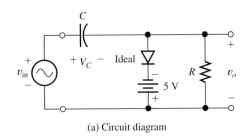

(a) Circuit diagram

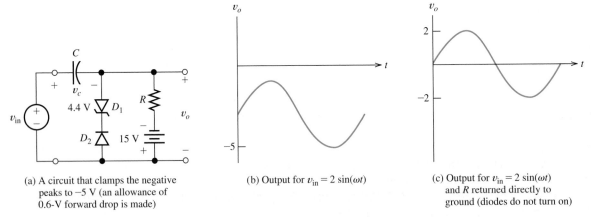

(b) Output waveform for $v_{in} = 5 \sin(\omega t)$

**Figure 9.33** Example clamp circuit.

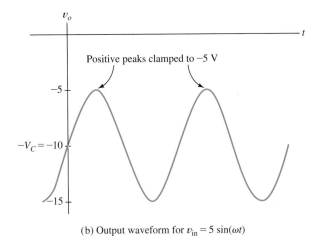

(a) A circuit that clamps the negative
peaks to −5 V (an allowance of
0.6-V forward drop is made)

(b) Output for $v_{in} = 2 \sin(\omega t)$

(c) Output for $v_{in} = 2 \sin(\omega t)$
and R returned directly to
ground (diodes do not turn on)

**Figure 9.34** See Exercise 9.16.

**Answer**   **a.** For $v_{in}(t) = 0$, we have $v_o = -5$ V; **b.** see Figure 9.34(b); **c.** see Figure 9.34(c).  □

**Exercise 9.17**   Design a circuit that clamps the negative peaks of an ac signal to +6 V. You can use batteries, resistors, and capacitors of any value desired in addition to Zener or conventional diodes. Allow 0.6 V for the forward drop.
**Answer**   A solution is shown in Figure 9.35. Other solutions are possible.  □

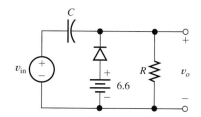

**Figure 9.35** Answer for Exercise 9.17.

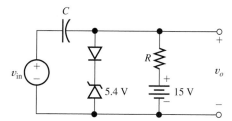

**Figure 9.36** Answer for Exercise 9.18.

**Exercise 9.18**   Repeat Exercise 9.17 for a circuit that clamps the positive peaks to +6 V.

**Answer**   A solution is shown in Figure 9.36. Other solutions are possible.   □

## 9.8   LINEAR SMALL-SIGNAL EQUIVALENT CIRCUITS

We will encounter many examples of electronic circuits in which dc supply voltages are used to **bias** a nonlinear device at an operating point, and a small ac signal is injected into the circuit. We often split the analysis of such circuits into two parts. First, we analyze the dc circuit to find the operating point. In this analysis of bias conditions, we must deal with the nonlinear aspects of the device. In the second part of the analysis, we consider the small ac signal. Since virtually any nonlinear characteristic is approximately linear (straight) if we consider a sufficiently small portion, we can find a **linear small-signal equivalent circuit** for the nonlinear device to use in the ac analysis.

Often, the main concern in the design of such circuits is what happens to the ac signal. The dc supply voltages simply bias the device at a suitable operating point. For example, in a portable radio, the main interest is the signal being received, demodulated, amplified, and delivered to the speaker. The dc currents supplied by the battery are required for the devices to perform their intended function on the ac signals. However, most of our design time is spent in consideration of the small ac signals to be processed.

The small-signal linear equivalent circuit is an important analysis approach that applies to many types of electronic circuits. In this section, we demonstrate the principles with a simple diode circuit. In Chapters 11 and 12, we use similar techniques for transistor amplifier circuits.

Now, we show that in the case of a diode, the small-signal equivalent circuit consists simply of a resistance. Consider the diode characteristic shown in Figure 9.37. Assume that the dc supply voltage results in operation at the **quiescent point**, or $Q$ **point**, indicated on the characteristic. Then, a small ac signal injected into the circuit swings the instantaneous point of operation slightly above and below the $Q$ point.

The small-signal equivalent circuit for a diode is a resistance.

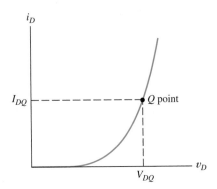

**Figure 9.37** Diode characteristic, illustrating the Q point.

For a sufficiently small ac signal, the characteristic is straight. Thus, we can write

$$\Delta i_D \cong \left(\frac{di_D}{dv_D}\right)_Q \Delta v_D \qquad (9.14)$$

where $\Delta i_D$ is the small change in diode current from the $Q$-point current caused by the ac signal, $\Delta v_D$ is the change in the diode voltage from the $Q$-point value, and $(di_D/dv_D)_Q$ is the slope of the diode characteristic evaluated at the $Q$ point. Notice that the slope has the units of inverse resistance.

Hence, we define the **dynamic resistance** of the diode as

$$r_d = \left[\left(\frac{di_D}{dv_D}\right)_Q\right]^{-1} \qquad (9.15)$$

and Equation 9.14 becomes

$$\Delta i_D \cong \frac{\Delta v_D}{r_d} \qquad (9.16)$$

We find it convenient to drop the $\Delta$ notation and denote changes of current and voltage from the $Q$-point values as $v_d$ and $i_d$. (Notice that lowercase subscripts are used for the small changes in current and voltage.) Therefore, for these small ac signals, we write

$$i_d = \frac{v_d}{r_d} \qquad (9.17)$$

As shown by Equation 9.15, we can find the equivalent resistance of the diode for the small ac signal as the reciprocal of the slope of the characteristic curve. The current of a junction diode is given by the Shockley equation (Equation 9.1), repeated here for convenience:

$$i_D = I_s\left[\exp\left(\frac{v_D}{nV_T}\right) - 1\right]$$

The slope of the characteristic can be found by differentiating the Shockley equation, resulting in

$$\frac{di_D}{dv_D} = I_s\frac{1}{nV_T}\exp\left(\frac{v_D}{nV_T}\right) \qquad (9.18)$$

Substituting the voltage at the $Q$ point, we have

$$\left(\frac{di_D}{dv_D}\right)_Q = I_s\frac{1}{nV_T}\exp\left(\frac{V_{DQ}}{nV_T}\right) \tag{9.19}$$

For forward-bias conditions with $V_{DQ}$ at least several times as large as $V_T$, the $-1$ inside the brackets of the Shockley equation is negligible. Thus, we can write

$$I_{DQ} \cong I_s\exp\left(\frac{V_{DQ}}{nV_T}\right) \tag{9.20}$$

Substituting this into Equation 9.19, we have

$$\left(\frac{di_D}{dv_D}\right)_Q = \frac{I_{DQ}}{nV_T} \tag{9.21}$$

Taking the reciprocal and substituting into Equation 9.15, we have the dynamic small-signal resistance of the diode at the $Q$ point:

$$r_d = \frac{nV_T}{I_{DQ}} \tag{9.22}$$

To summarize, for signals that cause small changes from the $Q$ point, we can treat the diode simply as a linear resistance. The value of the resistance is given by Equation 9.22 (provided that the diode is forward biased). As the $Q$-point current $I_{DQ}$ increases, the resistance becomes smaller. Thus, an ac voltage of fixed amplitude produces an ac current that has higher amplitude as the $Q$ point moves higher. This is illustrated in Figure 9.38.

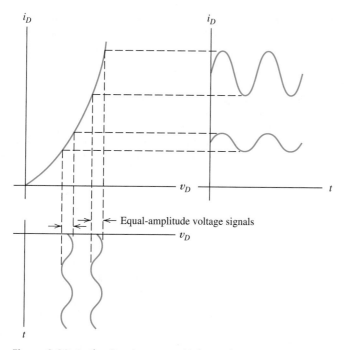

**Figure 9.38** As the $Q$ point moves higher, a fixed-amplitude ac voltage produces an ac current of larger amplitude.

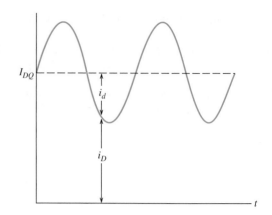

**Figure 9.39** Illustration of diode currents.

## Notation for Currents and Voltages in Electronic Circuits

Perhaps we should review the notation we have used for the diode currents and voltages, because we use similar notation throughout this book:

- $v_D$ and $i_D$ represent the total instantaneous diode voltage and current. At times, we may wish to emphasize the time-varying nature of these quantities, and then we use $v_D(t)$ and $i_D(t)$.
- $V_{DQ}$ and $I_{DQ}$ represent the dc diode current and voltage at the quiescent point.
- $v_d$ and $i_d$ represent the (small) ac signals. If we wish to emphasize their time-varying nature, we use $v_d(t)$ and $i_d(t)$.

This notation is illustrated for the waveform shown in Figure 9.39.

**Exercise 9.19**   Compute the dynamic resistance of a junction diode having $n = 1$ at a temperature of 300 K for $I_{DQ} =$ **a.** 0.1 mA; **b.** 1 mA; **c.** 10 mA.
**Answer**   **a.** 260 $\Omega$; **b.** 26 $\Omega$; **c.** 2.6 $\Omega$. □

## Voltage-Controlled Attenuator

Now, we consider an example of linear-equivalent-circuit analysis for the relatively simple, but useful circuit shown in Figure 9.40. The function of this circuit is to produce an output signal $v_o(t)$ that is a variable fraction of the ac input signal $v_{\text{in}}(t)$.

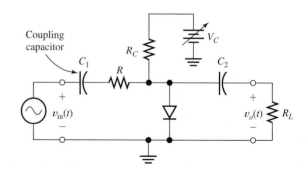

**Figure 9.40** Variable attenuator using a diode as a controlled resistance.

It is similar to the resistive voltage divider (see Section 2.3), except that in this case, we want the division ratio to depend on another voltage $V_C$ called the **control signal**. We refer to the process of reduction of the amplitude of a signal as **attenuation**. Thus, the circuit to be studied is called a **voltage-controlled attenuator**. The degree of attenuation depends on the value of the dc control voltage $V_C$.

Notice that the ac signal to be attenuated is connected to the circuit by a **coupling capacitor**. The output voltage is connected to the load $R_L$ by a second coupling capacitor. Recall that the impedance of a capacitance is given by

$$Z_C = \frac{1}{j\omega C}$$

in which $\omega$ is the angular frequency of the ac signal. We select the capacitance values large enough so that they are effectively short circuits for the ac signal. However, the coupling capacitors are open circuits for dc. Thus, the quiescent operating point ($Q$ point) of the diode is unaffected by the signal source or the load. This can be important for a circuit that must work for various sources and loads that could affect the $Q$ point. Furthermore, the coupling capacitors prevent (sometimes undesirable) dc currents from flowing in the source or the load.

Because of the coupling capacitors, we only need to consider $V_C$, $R_C$, and the diode to perform the bias analysis to find the $Q$ point. Hence, the dc circuit is shown in Figure 9.41. We can use any of the techniques discussed earlier in this chapter to find the $Q$ point. Once it is known, the $Q$-point value of the diode current $I_{DQ}$ can be substituted into Equation 9.22 to determine the dynamic resistance of the diode.

Now, we turn our attention to the ac signal. The dc control source should be considered as a short circuit for ac signals. The signal source causes an ac current to flow through the $V_C$ source. However, $V_C$ is a dc voltage source, and by definition, the voltage across it is constant. *Since the dc voltage source has an ac component of current, but no ac voltage, the dc voltage source is equivalent to a short circuit for ac signals.* This is an important concept that we will use many times in drawing ac equivalent circuits.

The equivalent circuit for ac signals is shown in Figure 9.42. The control source and the capacitors have been replaced by short circuits, and the diode has been replaced by its dynamic resistance. This circuit is a voltage divider and can be analyzed by ordinary linear-circuit analysis. The parallel combination of $R_C$, $R_L$, and $r_d$ is denoted as $R_p$, given by

$$R_p = \frac{1}{1/R_C + 1/R_L + 1/r_d} \tag{9.23}$$

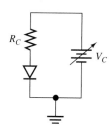

**Figure 9.41** Dc circuit equivalent to Figure 9.40 for $Q$-point analysis.

Dc sources and coupling capacitors are replaced by short circuits in small-signal ac equivalent circuits. Diodes are replaced with their dynamic resistances.

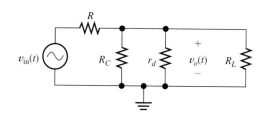

**Figure 9.42** Small-signal ac equivalent circuit for Figure 9.40.

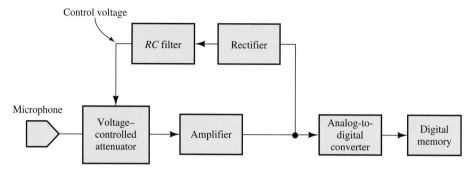

**Figure 9.43** The voltage-controlled attenuator is useful in maintaining a suitable signal amplitude at the recording head.

Then, the **voltage gain** of the circuit is

$$A_v = \frac{v_o}{v_{\text{in}}} = \frac{R_p}{R + R_p} \tag{9.24}$$

(Of course, $A_v$ is less than unity.)

---

**Exercise 9.20**   Suppose that the circuit of Figure 9.40 has $R = 100\ \Omega$, $R_C = 2\ \text{k}\Omega$, and $R_L = 2\ \text{k}\Omega$. The diode has $n = 1$ and is at a temperature of 300 K. For purposes of $Q$-point analysis, assume a constant diode voltage of 0.6 V. Find the $Q$-point value of the diode current and $A_v$ for $V_C =$ **a.** 1.6 V; **b.** 10.6 V.
**Answer**   **a.** $I_{DQ} = 0.5\ \text{mA}$ and $A_v = 0.331$; **b.** $I_{DQ} = 5\ \text{mA}$ and $A_v = 0.0492$.   ◻

---

An application for voltage-controlled attenuators occurs in digital voice recorders in which the audio signal from a microphone is amplified to a suitable level, converted to digital form in an analog-to-digital converter (ADC), and stored in a digital memory. (Analog-to-digital conversion is discussed in Section 6.10, starting on page 331.) A problem frequently encountered in recording audio is that some persons speak quietly, while others speak loudly. Furthermore, some may be far from the microphone, while others are close. If an amplifier with fixed gain is used between the microphone and the ADC, either the weak signals are small compared with the quantization error or the strong signals exceed the maximum limits of the ADC so that severe distortion occurs.

A solution is to use a voltage-controlled attenuator in a system such as the one shown in Figure 9.43. The attenuator is placed between the microphone and a highgain amplifier. When the signal being recorded is weak, the control voltage is small and very little attenuation occurs. On the other hand, when the signal is strong, the control voltage is large so that the signal is attenuated, preventing distortion. The control voltage is generated by rectifying the output of the amplifier. The rectified signal is filtered by a long-time constant $RC$ filter so that the attenuation responds to the average signal amplitude rather than adjusting too rapidly. With proper design, this system can provide an acceptable signal at the converter for a wide range of input signal amplitudes.

While the diode circuit we have discussed is convenient for illustrating principles, integrated-circuit transistor amplifiers in which gain is controlled by changing the $Q$-points of the transistors offer better performance. Examples are the AN-934 from Analog Devices and the MAX9814 from Maxim Integrated Products.

## Summary

1. A *pn*-junction diode is a two-terminal device that conducts current easily in one direction (from anode to cathode), but not in the opposite direction. The volt–ampere characteristic has three regions: forward bias, reverse bias, and reverse breakdown.

2. The Shockley equation relates current and voltage in a *pn*-junction diode.

3. Nonlinear circuits, such as those containing a diode, can be analyzed by using the load-line technique.

4. Zener diodes are intended to be operated in the reverse-breakdown region as constant-voltage references.

5. Voltage regulators are circuits that produce a nearly constant output voltage while operating from a variable source.

6. The ideal-diode model is a short circuit (on) if current flows in the forward direction and an open circuit (off) if voltage is applied in the reverse direction.

7. In the method of assumed states, we assume a state for each diode (on or off), analyze the circuit, and check to see if the assumed states are consistent with the current directions and voltage polarities. This process is repeated until a valid set of states is found.

8. In a piecewise-linear model for a nonlinear device, the volt–ampere characteristic is approximated by straight-line segments. On each segment, the device is modeled as a voltage source in series with a resistance.

9. Rectifier circuits can be used to charge batteries and to convert ac voltages into constant dc voltages. Half-wave rectifiers conduct current only for one polarity of the ac input, whereas full-wave circuits conduct for both polarities.

10. Wave-shaping circuits change the waveform of an input signal and deliver the modified waveform to the output terminals. Clipper circuits remove that portion of the input waveform above (or below) a given level. Clamp circuits add or subtract a dc voltage, so that the positive (or negative) peaks have a specified voltage.

11. The small-signal (incremental) equivalent circuit of a diode consists of a resistance. The value of the resistance depends on the operating point (*Q* point).

12. Dc sources and coupling capacitors are replaced by short circuits in small-signal ac equivalent circuits. Diodes are replaced with their dynamic resistances.

## Problems

### Section 9.1: Basic Diode Concepts

**P9.1.** Draw the circuit symbol for a diode, labeling the anode and cathode.

**P9.2.** Draw the volt–ampere characteristic of a typical diode and label the various regions.

**P9.3.** Describe a fluid-flow analogy for a diode.

**P9.4.** Write the Shockley equation and define all of the terms.

**P9.5.** Compute the values of $V_T$ for temperatures of 20°C and 150°C.

**\*P9.6.** Sketch $i$ versus $v$ to scale for the circuits shown in Figure P9.6. The reverse-breakdown voltages of the Zener diodes are shown. Assume voltages of 0.6 V for all diodes including the Zener diodes when current flows in the forward direction.

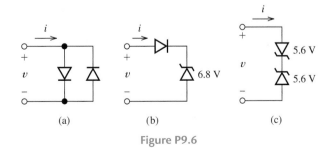

Figure P9.6

**P9.7.** Repeat Problem P9.6 for the circuits shown in Figure P9.7.

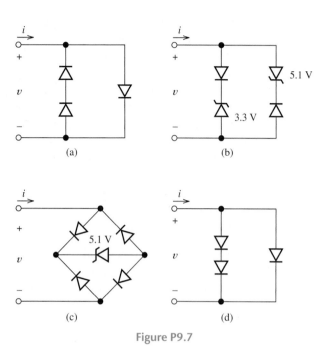

(a)

(b)

(c)

(d)

Figure P9.7

**\*P9.8.** A diode operates in forward bias and is described by Equation 9.4, with $V_T = 0.026$ V. For $v_{D1} = 0.600$ V, the current is $i_{D1} = 1$ mA. For $v_{D2} = 0.680$ V, the current is $i_{D2} = 10$ mA. Determine the values of $I_s$ and $n$.

**P9.9.** With constant current flowing in the forward direction in a small-signal silicon diode, the voltage across the diode decreases with temperature by about 2 mV/K. Such a diode has a voltage of 0.650 V, with a current of 1 mA at a temperature of 25°C. Find the diode voltage at 1 mA and a temperature of 175°C.

**P9.10.** We have a junction diode that has $i_D = 0.2$ mA for $v_D = 0.6$ V. Assume that $n = 2$ and $V_T = 0.026$ V. Use the Shockley equation to compute the diode current at $v_D = 0.65$ V and at $v_D = 0.70$ V.

**P9.11.** We have a diode with $n = 1$, $I_s = 10^{-14}$ A, and $V_T = 26$ mV. **a.** Using a computer program of your choice, obtain a plot of $i_D$ versus $v_D$ for $i_D$ ranging from 10 $\mu$A to 10 mA.

Choose a logarithmic scale for $i_D$ and a linear scale for $v_D$. What type of curve results? **b.** Place a 100-$\Omega$ resistance in series with the diode, and plot current versus voltage across the series combination on the same axes used for part (a). Compare the two curves. When is the added series resistance significant?

**P9.12.** A silicon diode described by the Shockley equation has $n = 2$ and operates at 150°C with a current of 1 mA and voltage of 0.25 V. Determine the current after the voltage is increased to 0.30 V.

**\*P9.13.** The diodes shown in Figure P9.13 are identical and have $n = 1$. The temperature of the diodes is constant at 300 K. Before the switch is closed, the voltage $v$ is 600 mV. Find $v$ after the switch is closed. Repeat for $n = 2$.

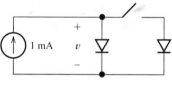

Figure P9.13

**P9.14.** Suppose we have a junction diode operating at a constant temperature of 300 K. With a forward current of 1 mA, the voltage is 600 mV. Furthermore, with a current of 10 mA, the voltage is 700 mV. Find the value of $n$ for this diode.

**\*P9.15.** **Current hogging.** The diodes shown in Figure P9.15 are identical and have $n = 1$. For each diode, a forward current of 100 mA results in a voltage of 700 mV at a temperature of 300 K. **a.** If both diodes are at 300 K, what are the values of $I_A$ and $I_B$? **b.** If diode $A$ is at 300 K and diode $B$ is at 305 K, again find $I_A$ and $I_B$, given that $I_s$ doubles in value for a 5-K increase in temperature. [*Hint:* Answer part (a) by use of symmetry. For part (b), a transcendental equation for the voltage across the diodes can be found. Solve by trial and error. An important observation to be made from this problem is that, starting at the same temperature, the diodes should theoretically each conduct half of the total current. However, if one diode conducts slightly more, it becomes warmer, resulting in even more current. Eventually, one of the diodes "hogs" most of the current. This is particularly noticeable for devices that are thermally isolated from one another with large currents, for which significant heating occurs.]

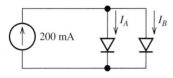

Figure P9.15

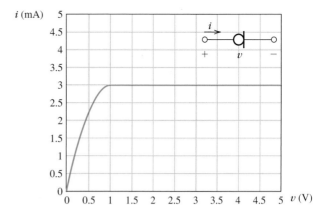

(a)  Volt–ampere characteristic of a constant-current diode

## Section 9.2: Load-Line Analysis of Diode Circuits

*P9.16. The nonlinear circuit element shown in Figure P9.16 has $i_x = [\exp(v_x) - 1]/10$. Also, we have $V_s = 3$ V and $R_s = 1\ \Omega$. Use graphical load-line techniques to solve for $i_x$ and $v_x$. (You may prefer to use a computer program to plot the characteristic and the load line.)

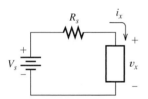

Figure P9.16

P9.17. Repeat Problem P9.16 for $V_s = 20$ V, $R_s = 5\ \text{k}\Omega$, and $i_x = 0.01/(1 - v_x/5)^3$ mA.

P9.18. Repeat Problem P9.16 for $V_s = 6$ V, $R_s = 3\ \Omega$, and $i_x = v_x^3/8$.

P9.19. Repeat Problem P9.16 for $V_s = 3$ V, $R_s = 1\ \Omega$, and $i_x = v_x + v_x^2$.

P9.20. Several types of special-purpose diodes exist. One is the constant-current diode for which the current is constant over a wide range of voltage. The circuit symbol and volt–ampere characteristic for a constant-current diode are shown in Figure P9.20(a). Another special type is the light-emitting diode (LED) for which the circuit symbol and a typical volt–ampere characteristic are shown in Figure P9.20(b). Sometimes, the series combination of these two devices is used to

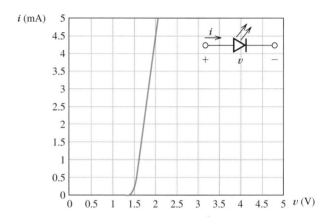

(b)  Volt–ampere characteristic of a light-emitting diode (LED).

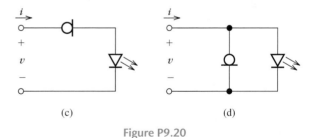

(c)                              (d)

Figure P9.20

provide constant current to the LED from a variable voltage shown in Figure P9.20(c).
**b.** Sketch the overall volt–ampere characteristic to scale for the parallel combination shown in Figure P9.20(d).

P9.21. Determine the values for $i$ and $v$ for the circuit of Figure P9.21. The diode is the LED having the characteristic shown in Figure P9.20(b).

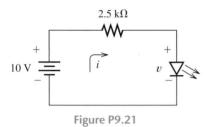

Figure P9.21

**P9.22.** Determine the values for $i_1$ and $i_2$ for the circuit of Figure P9.22. The device is the constant-current diode having the characteristic shown in Figure P9.20(a).

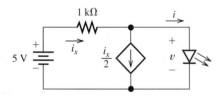

Figure P9.22

**P9.23.** Determine the values for $i$ and $v$ for the circuit of Figure P9.23. The diode is the LED having the characteristic shown in Figure P9.20(b).

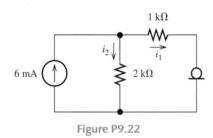

Figure P9.23

**P9.24.** Repeat Problem P9.23 for the circuit of Figure P9.24.

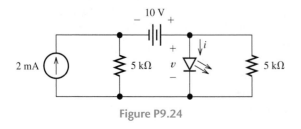

Figure P9.24

**Section 9.3:   Zener-Diode Voltage-Regulator Circuits**

**P9.25.** What is a Zener diode? For what is it typically used? Draw the volt–ampere characteristic of an ideal 5.8-V Zener diode.

*****P9.26.** Draw the circuit diagram of a simple voltage regulator.

**P9.27.** Consider the Zener-diode regulator shown in Figure 9.14 on page 470. What is the minimum load resistance for which $v_o$ is 10 V?

**P9.28.** Consider the voltage regulator shown in Figure P9.28. The source voltage $V_s$ varies from 10 to 14 V, and the load current $i_L$ varies from 50 to 100 mA. Assume that the Zener diode is ideal. Determine the largest value allowed for the resistance $R_s$ so that the load voltage $v_L$ remains constant with variations in load current and source voltage. Determine the maximum power dissipation in $R_s$.

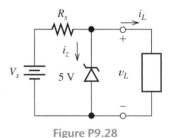

Figure P9.28

**P9.29.** Design a voltage-regulator circuit to provide a constant voltage of 5 V to a load from a variable supply voltage. The load current varies from 0 to 100 mA, and the source voltage varies from 8 to 10 V. You may assume that ideal Zener diodes are available. Resistors of any value may be specified. Draw the circuit diagram of your regulator, and specify the value of each component. Also, find the worst case (maximum) power dissipated in each component in your regulator. Try to use good judgment in your design.

**P9.30.** Repeat Problem P9.29 if the supply voltage ranges from 6 to 10 V.

**P9.31.** Repeat Problem P9.29 if the load current varies from 0 to 1 A.

**P9.32.** Outline a method for solving a circuit that contains a single nonlinear element plus resistors, dc voltage sources, and dc current sources, given the volt–ampere characteristic of the nonlinear device.

**\*P9.33.** A certain linear two-terminal circuit has terminals $a$ and $b$. Under open-circuit conditions, we have $v_{ab} = 10$ V. A short circuit is connected across the terminals, and a current of 2 A flows from $a$ to $b$ through the short circuit. Determine the value of $v_{ab}$ when a nonlinear element that has $i_{ab} = \sqrt[3]{v_{ab}}$ is connected across the terminals.

### Section 9.4: Ideal-Diode Model

**P9.34.** What is an ideal diode? Draw its volt–ampere characteristic. After solving a circuit with ideal diodes, what check is necessary for diodes initially assumed to be on? Off?

**P9.35.** Two ideal diodes are placed in series, pointing in opposite directions. What is the equivalent circuit for the combination? What is the equivalent circuit if the diodes are in parallel and pointing in opposite directions?

**P9.36.** Find the values of $I$ and $V$ for the circuits of Figure P9.36, assuming that the diodes are ideal.

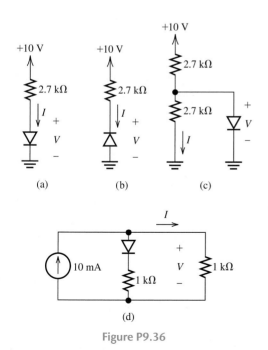

Figure P9.36

**\*P9.37.** Find the values of $I$ and $V$ for the circuits of Figure P9.37, assuming that the diodes are ideal.

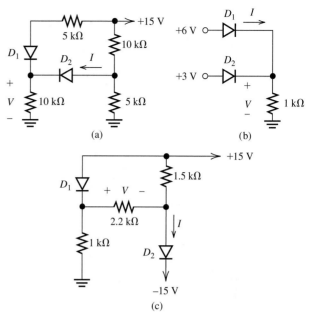

Figure P9.37

**P9.38.** Find the values of $I$ and $V$ for the circuits of Figure P9.38, assuming that the diodes are ideal. For part (b), consider $V_{in} = 0, 2, 6,$ and 10 V. Also, for part (b) of the figure, plot $V$ versus $V_{in}$ for $V_{in}$ ranging from $-10$ V to 10 V.

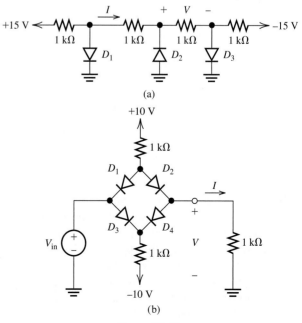

Figure P9.38

**P9.39.** Sketch $i$ versus $v$ to scale for each of the circuits shown in Figure P9.39. Assume that the diodes are ideal and allow $v$ to range from $-10$ V to $+10$ V.

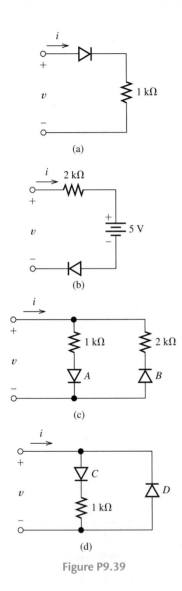

Figure P9.39

**P9.40.** The circuit shown in Figure P9.40(a) is a type of logic gate. Assume that the diodes are ideal. The voltages $V_A$ and $V_B$ independently have values of either 0 V (for logic 0, or low) or 5 V (for logic 1, or high). For which of the four combinations of input voltages is the output high (i.e., $V_o = 5$ V)? What type of logic gate is this? Repeat for the circuit of Figure P9.40(b).

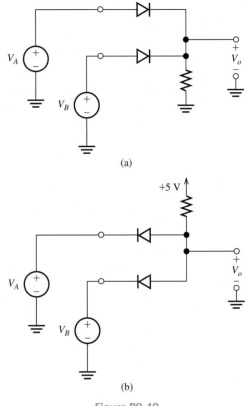

Figure P9.40

**P9.41.** Sketch $v_o(t)$ to scale versus time for the circuit shown in Figure P9.41. Assume that the diodes are ideal.

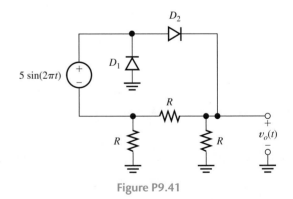

Figure P9.41

**Section 9.5: Piecewise-Linear Diode Models**

**P9.42.** If a nonlinear two-terminal device is modeled by the piecewise-linear approach, what is the equivalent circuit of the device for each linear segment?

**P9.43.** A resistor $R_a$ is in series with a voltage source $V_a$. Draw the circuit. Label the voltage across the combination as $v$ and the current as $i$. Draw and label the volt–ampere characteristic ($i$ versus $v$).

**P9.44.** The volt–ampere characteristic of a certain two-terminal device is a straight line that passes through the points (2 V, 5 mA) and (3 V, 15 mA). The current reference points into the positive reference for the voltage. Determine the equivalent circuit for this device.

**P9.45.** Consider the volt–ampere characteristic of an ideal 10-V Zener diode shown in Figure 9.14 on page 470. Determine the piecewise-linear equivalent circuit for each segment of the characteristic.

**\*P9.46.** Assume that we have approximated a nonlinear volt–ampere characteristic by the straight-line segments shown in Figure P9.46(c). Find the equivalent circuit for each segment. Use these equivalent circuits to find $v$ in the circuits shown in Figure P9.46(a) and (b).

**\*P9.47.** The Zener diode shown in Figure P9.47 has a piecewise-linear model shown in Figure 9.19 on page 473. Plot load voltage $v_L$ versus load current $i_L$ for $i_L$ ranging from 0 to 100 mA.

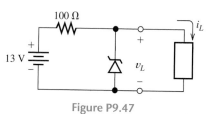

Figure P9.47

**P9.48.** The diode shown in Figure P9.48 can be represented by the model of Figure 9.23 on page 475, with $V_f = 0.7$ V. **a.** Assume that the diode operates as an open circuit and solve for the node voltages $v_1$ and $v_2$. Are the results consistent with the model? Why or why not? **b.** Repeat part (a), assuming that the diode operates as a 0.7-V voltage source.

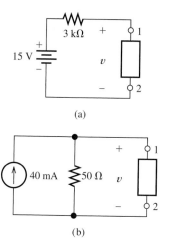

(a)

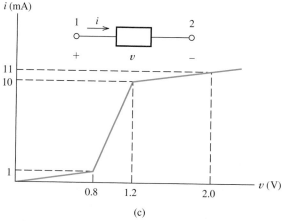

(b)

(c)

Figure P9.46

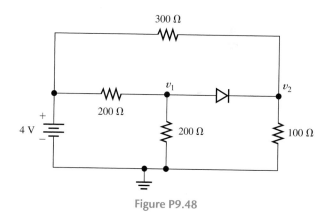

Figure P9.48

**Section 9.6: Rectifier Circuits**

**P9.49.** Draw the circuit diagram of a half-wave rectifier for producing a nearly steady dc voltage from an ac source. Draw two different full-wave circuits.

**P9.50.** A 20-V-rms 60-Hz ac source is in series with an ideal diode and a 100-$\Omega$ resistance. Determine the peak current and PIV for the diode.

**P9.51.** Consider the battery charging circuit shown in Figure 9.25 on page 476. The ac source has a peak value of 24 V and a frequency of 60 Hz. The resistance is 2 Ω, the diode is ideal, and $V_B = 12$ V. Determine the average current (i.e., the value of the charge that passes through the battery in 1 second). Suppose that the battery starts from a totally discharged state and has a capacity of 100 ampere hours. How long does it take to fully charge the battery?

**P9.52.** Consider the half-wave rectifier shown in Figure 9.26 on page 477. The ac source has an rms value of 20 V and a frequency of 60 Hz. The diodes are ideal, and the capacitance is very large, so the ripple voltage $V_r$ is very small. The load is a 100-Ω resistance. Determine the PIV across the diode and the charge that passes through the diode per cycle.

**P9.53.** Most dc voltmeters produce a reading equal to the average value of the voltage measured. The mathematical definition of the average value of a periodic waveform is

$$V_{avg} = \frac{1}{T} \int_0^T v(t)\,dt$$

in which $T$ is the period of the voltage $v(t)$ applied to the meter.

a. What does a dc voltmeter read if the applied voltage is $v(t) = V_m \sin(\omega t)$?

b. What does the meter read if the applied voltage is a half-wave rectified version of the sinewave?

c. What does the meter read if the applied voltage is a full-wave rectified version of the sinewave?

**∗P9.54.** Design a half-wave rectifier power supply to deliver an average voltage of 9 V with a peak-to-peak ripple of 2 V to a load. The average load current is 100 mA. Assume that ideal diodes and 60-Hz ac voltage sources of any amplitudes needed are available. Draw the circuit diagram for your design. Specify the values of all components used.

**P9.55.** Repeat Problem P9.54 with a full-wave bridge rectifier.

**P9.56.** Repeat Problem P9.54 with two diodes and out-of-phase voltage sources to form a full-wave rectifier.

**P9.57.** Repeat Problem P9.54, assuming that the diodes have forward drops of 0.8 V.

**∗P9.58.** A half-wave rectifier is needed to supply 15-V dc to a load that draws an average current of 250 mA. The peak-to-peak ripple is required to be 0.2 V or less. What is the minimum value allowed for the smoothing capacitance? If a full-wave rectifier is needed?

**P9.59.** Consider the battery-charging circuit shown in Figure 9.25 on page 476, in which $v_s(t) = 20 \sin(200\pi t)$, $R = 80$ Ω, $V_B = 12$ V, and the diode is ideal.

a. Sketch the current $i(t)$ to scale versus time.

b. Determine the average charging current for the battery.

[*Hint:* The average current is the charge that flows through the battery in one cycle, divided by the period.]

**P9.60. a.** Consider the full-wave rectifier shown in Figure 9.27 on page 478, with a large smoothing capacitance placed in parallel with the load $R_L$ and $V_m = 12$ V. Assuming that the diodes are ideal, what is the approximate value of the load voltage? What PIV appears across the diodes? **b.** Repeat for the full-wave bridge shown in Figure 9.28 on page 479.

**P9.61.** Figure P9.61 shows the equivalent circuit for a typical automotive battery charging system. The three-phase delta-connected source represents the stator coils of the alternator. (Three-phase ac sources are discussed in Section 5.7. Actually, the alternator stator is usually wye connected, but the terminal voltages are the same as for the equivalent delta.) Not shown in the figure is a voltage regulator that controls the current applied to the rotor coil of the alternator and, consequently, $V_m$ and the charging current to the battery. **a.** Sketch the load voltage $v_L(t)$ to scale versus time. Assume ideal diodes and that $V_m$ is large enough that current flows into the battery at all times. [*Hint:* Each source and four of the diodes form a full-wave bridge rectifier.] **b.** Determine the peak-to-peak ripple and the average load voltage in terms of $V_m$. **c.** Determine the value of $V_m$ needed to provide an average charging current of 30 A. **d.** What additional factors would need to be considered in a realistic computation of $V_m$?

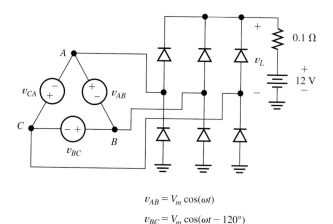

$$v_{AB} = V_m \cos(\omega t)$$

$$v_{BC} = V_m \cos(\omega t - 120°)$$

$$v_{CA} = V_m \cos(\omega t + 120°)$$

**Figure P9.61** Idealized model of an automotive battery-charging system.

## Section 9.7: Wave-Shaping Circuits

**P9.62.** What is a clipper circuit? Draw an example circuit diagram, including component values, an input waveform, and the corresponding output waveform.

**P9.63.** Sketch to scale the output waveform for the circuit shown in Figure P9.63. Assume that the diodes are ideal.

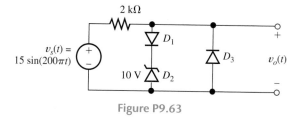

**Figure P9.63**

**P9.64.** Sketch the transfer characteristic ($v_o$ versus $v_{in}$) to scale for the circuit shown in Figure P9.64. Assume that the diode is ideal.

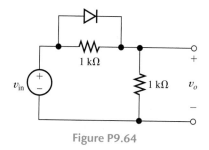

**Figure P9.64**

**P9.65.** Sketch the transfer characteristic ($v_o$ versus $v_{in}$) to scale for the circuit shown in Figure P9.65. Assume that the diodes are ideal.

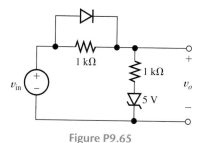

**Figure P9.65**

**P9.66.** Sketch the transfer characteristic ($v_o$ versus $v_{in}$) to scale for the circuit shown in Figure P9.66. Allow $v_{in}$ to range from $-5$ V to $+5$ V and assume that the diodes are ideal.

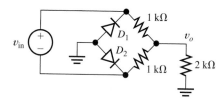

**Figure P9.66**

**P9.67.** Sketch the transfer characteristic ($v_o$ versus $v_{in}$) for the circuit shown in Figure P9.67, carefully labeling the breakpoint and slopes. Allow $v_{in}$ to range from $-5$ V to $+5$ V and assume that the diodes are ideal.

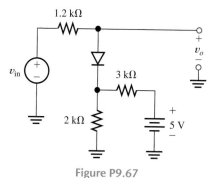

**Figure P9.67**

**P9.68.** What is a clamp circuit? Draw an example circuit diagram, including component values, an input waveform, and the corresponding output waveform.

**P9.69.** Consider the circuit shown in Figure P9.69, in which the $RC$ time constant is very long compared with the period of the input and in which the diode is ideal. Sketch $v_o(t)$ to scale versus time.

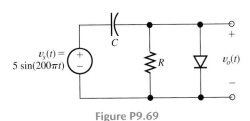

**Figure P9.69**

**\*P9.70.** Sketch to scale the steady-state output waveform for the circuit shown in Figure P9.70. Assume that $RC$ is much larger than the period of the input voltage and that the diodes are ideal.

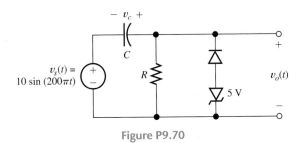

**Figure P9.70**

**P9.71. Voltage-doubler circuit.** Consider the circuit of Figure P9.71. The capacitors are very large, so they discharge only a very small amount per cycle. (Thus, no ac voltage appears across the capacitors, and the ac input plus the dc voltage of $C_1$ must appear at point $A$.) Sketch the voltage at point $A$ versus time. Find the voltage across the load. Why is this called a voltage doubler? What is the PIV across each diode?

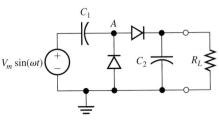

**Figure P9.71**

**\*P9.72.** Design a clipper circuit to clip off the portions of an input voltage that fall above 3 V or below −5 V. Assume that diodes having a constant forward drop of 0.7 V are available. Ideal Zener diodes of any breakdown voltage required are available. Dc voltage sources of any value needed are available.

**P9.73.** Repeat Problem P9.72, with clipping levels of +2 V and +5 V (i.e., every part of the input waveform below +2 or above +5 is clipped off).

**P9.74.** Design circuits that have the transfer characteristics shown in Figure P9.74. Assume that $v_{in}$ ranges from −10 to +10 V. Use diodes, Zener diodes, and resistors of any values needed. Assume a 0.6-V forward drop for all diodes and that the Zener diodes have an ideal characteristic in the breakdown region. Power-supply voltages of ±15 V are available.

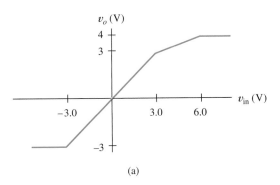

(a)

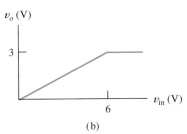

(b)

**Figure P9.74**

**\*P9.75.** Design a clamp circuit to clamp the negative extreme of a periodic input waveform to −5 V. Use diodes, Zener diodes, and resistors of any values required. Assume a 0.6-V forward drop for all diodes and that the Zener diodes have an ideal characteristic in the breakdown region. Power-supply voltages of ±15 V are available.

**P9.76.** Repeat Problem P9.75 for a clamp voltage of +5 V.

## Section 9.8: Linear Small-Signal Equivalent Circuits

**P9.77.** A certain diode has $I_{DQ} = 4$ mA and $i_d(t) = 0.5 \cos(200\pi t)$ mA. Find an expression for $i_D(t)$, and sketch it to scale versus time.

**P9.78.** Of what does the small-signal equivalent circuit of a diode consist? How is the dynamic resistance of a nonlinear circuit element determined at a given operating point?

**P9.79.** With what are dc voltage sources replaced in a small-signal ac equivalent circuit? Why?

**P9.80.** With what should we replace a dc current source in a small-signal ac equivalent circuit? Justify your answer.

**\*P9.81.** A certain nonlinear device has $i_D = v_D^3/8$. Sketch $i_D$ versus $v_D$ to scale for $v_D$ ranging from $-2$ V to $+2$ V. Is this device a diode? Determine the dynamic resistance of the device and sketch it versus $v_D$ to scale for $v_D$ ranging from $-2$ V to $+2$ V.

**P9.82.** A breakdown diode has

$$i_D = \frac{-10^{-6}}{(1 + v_D/5)^3} \quad \text{for } -5 \text{ V} < v_D < 0$$

where $i_D$ is in amperes. Plot $i_D$ versus $v_D$ in the reverse-bias region. Find the dynamic resistance of this diode at $I_{DQ} = -1$ mA and at $I_{DQ} = -10$ mA.

**P9.83.** A certain nonlinear device is operating with an applied voltage given by

$$v_D(t) = 5 + 0.01 \cos(\omega t) \text{ V}$$

The current is given by

$$i_D(t) = 3 + 0.2 \cos(\omega t) \text{ mA}$$

Determine the dynamic resistance and $Q$ point of the device under the conditions given.

**P9.84.** Ideally, we want the voltage for a Zener diode to be constant in the breakdown region. What does this imply about the dynamic resistance in the breakdown region for an ideal Zener diode?

**\*P9.85.** Consider the voltage-regulator circuit shown in Figure P9.85. The ac ripple voltage is 1 V peak to peak. The dc (average) load voltage is 5 V. What is the $Q$-point current in the Zener diode? What is the maximum dynamic resistance allowed for the Zener diode if the output ripple is to be less than 10 mV peak to peak?

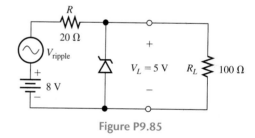

Figure P9.85

## Practice Test

Here is a practice test you can use to check your comprehension of the most important concepts in this chapter. Answers can be found in Appendix D and complete solutions are included in the Student Solutions files. See Appendix E for more information about the Student Solutions.

**T9.1.** Determine the value of $i_D$ for each of the circuits shown in Figure T9.1. The characteristic for the diode is shown in Figure 9.8 on page 465.

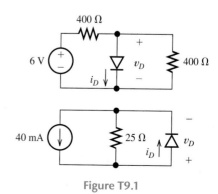

Figure T9.1

**T9.2.** The diode shown in Figure T9.2 is ideal. Determine the state of the diode and the values of $v_x$ and $i_x$.

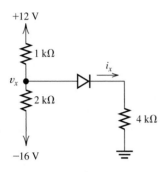

**Figure T9.2**

**T9.3.** The current versus voltage characteristic of a certain two-terminal device passes through the points (5 V, 2 mA) and (10 V, 7 mA). The reference for the current points into the positive reference for the voltage. Determine the values for the resistance and voltage source for the piecewise linear equivalent circuit for this device between the two points given.

**T9.4.** Draw the circuit diagram for a full-wave bridge rectifier with a resistance as the load.

**T9.5.** Suppose we have a 10-V-peak sinusoidal voltage source. Draw the diagram of a circuit that clips off the part of the sinusoid above 5 V and below −4 V. The circuit should be composed of ideal diodes, dc voltage sources, and other components as needed. Be sure to label the terminals across which the clipped output waveform $v_o(t)$ appears.

**T9.6.** Suppose we have a 10-Hz sinusoidal voltage source, $v_{in}(t)$. Draw the diagram of a circuit that clamps the positive peaks to −4 V. The circuit should be composed of ideal diodes, dc voltage sources, and other components as needed. List any constraints that should be observed in selecting component values. Be sure to label the terminals across which the clamped output waveform $v_o(t)$ appears.

**T9.7.** Suppose we have a silicon diode operating with a bias current of 5 mA at a temperature of 300 K. The diode current is given by the Shockley equation with $n = 2$. Draw the small-signal equivalent circuit for the diode including numerical values for the components.

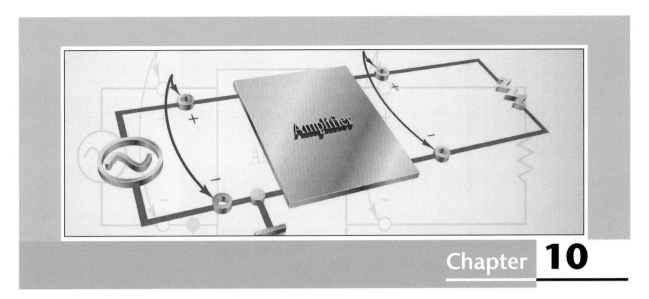

# Chapter 10

# Amplifiers: Specifications and External Characteristics

## Study of this chapter will enable you to:

- Use various amplifier models to calculate amplifier performance for given sources and loads.

- Compute amplifier efficiency.

- Understand the importance of input and output impedances of amplifiers.

- Determine the best type of ideal amplifier for various applications.

- Specify the frequency-response requirements for various amplifier applications.

- Understand linear and nonlinear distortion in amplifiers.

- Specify the pulse-response parameters of amplifiers.

- Work with differential amplifiers and specify common-mode rejection requirements.

- Understand the various sources of dc offsets and design balancing circuits.

## Introduction to this chapter:

The most important analog blocks found in electronic systems are amplifiers. Basically, amplifiers are used to increase the amplitudes of electrical signals. For example, the signals from most sensors (such as strain gauges used in mechanical engineering or flow meters used in chemical processes) are small in amplitude and need to be amplified before they can be utilized.

In this chapter, we consider the external characteristics of amplifiers that are important in selecting them for instrumentation applications. After introducing the basic concepts of amplifiers, we consider several nonideal properties of real amplifiers. To avoid errors when working with electronic instrumentation in your field, you need to be familiar with these amplifier imperfections. The internal operation of amplifier circuits is treated in Chapters 11, 12, and 13.

## 10.1    BASIC AMPLIFIER CONCEPTS

Ideally, an amplifier produces an output signal with identical waveshape as the input signal, but with a larger amplitude.

Ideally, an amplifier produces an output signal with identical waveshape as the input signal, but with a larger amplitude. This concept is illustrated in Figure 10.1. The signal source produces a voltage $v_i(t)$ that is applied to the input terminals of the amplifier, which generates an output voltage

$$v_o(t) = A_v v_i(t) \tag{10.1}$$

across a **load resistance** $R_L$ connected to the output terminals. The constant $A_v$ is called the **voltage gain** of the amplifier. Often, the voltage gain is much larger in magnitude than unity, but we will see later that useful amplification can take place even if $A_v$ is less than unity.

An example of a signal source is a microphone that typically produces a signal of 1-mV peak as we speak into it. This small signal can be used as the input to an amplifier with a voltage gain of 10,000 to produce an output signal with a peak value of 10 V. If this larger output voltage is applied to a loudspeaker, a much louder version of the sound entering the microphone results the principle of operation for the electronic megaphone.

Sometimes, $A_v$ is a negative number, so the output voltage is an inverted version of the input, and the amplifier is then called an **inverting amplifier**. On the other hand, if $A_v$ is a positive number, we have a **noninverting amplifier**. A typical input waveform and the corresponding output waveforms for a noninverting amplifier and for an inverting amplifier are shown in Figure 10.2.

Inverting amplifiers have negative voltage gain, and the output waveform is an inverted version of the input waveform. Noninverting amplifiers have positive voltage gain.

For monaural audio signals, it does not matter whether the amplifier is inverting or noninverting because the sounds produced by the loudspeaker are perceived the same either way. However, in a stereo system, it is important that the amplifiers for the left and right channels are the same (i.e., either both inverting or both noninverting), so that the signals applied to the two loudspeakers have the proper phase relationship. If video signals are inverted, a negative image with black and

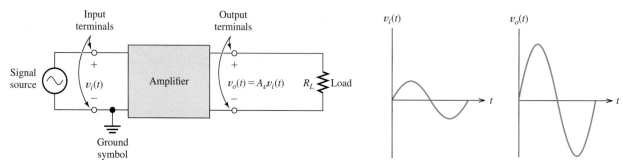

**Figure 10.1** Electronic amplifier.

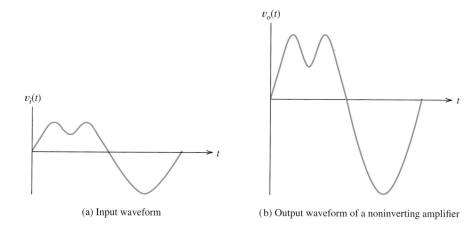

(a) Input waveform

(b) Output waveform of a noninverting amplifier

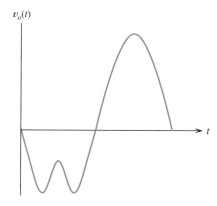

(c) Output waveform of an inverting amplifier

**Figure 10.2** Input waveform and corresponding output waveforms.

white interchanges results; hence, it is important whether video amplifiers are inverting or noninverting.

## Common Ground Node

Often, one of the amplifier input terminals and one of the output terminals are connected to a common **ground**. Notice the ground symbol shown in Figure 10.1. Typically, the ground terminal consists of the metal chassis that contains the circuit as well as circuit-board conductors. This common ground serves as the return path for signal currents and, as we will see later, the dc power-supply currents in electronic circuits.

You may be familiar with the concept of electrical grounds in automobile wiring. Here, the ground conductor consists of the frame, fenders, and other conductive parts of the car. For example, current is carried to the taillights by a wire, but may return through the ground conductors, consisting of the fenders and frame. Similarly, residential 60-Hz power distribution systems are grounded, often to a cold-water pipe. However, in this case, return currents are not intended to flow through the ground conductors because that could pose safety hazards.

Sometimes, *but not always*, the chassis ground is connected through the line cord to the 60-Hz power-system ground. *Always be careful in working with electrical circuits.* In some types of electronic circuits, the chassis ground can be at 120 V ac with respect to the power-system ground. Touching the chassis while in contact with the power-system ground (through a water pipe or a damp concrete floor for example) *can be fatal.*

---

**Exercise 10.1**    A certain noninverting amplifier has a voltage-gain magnitude of 50. The input voltage is $v_i(t) = 0.1 \sin(2000\pi t)$. **a.** Find an expression for the output voltage $v_o(t)$. **b.** Repeat for an inverting amplifier.
**Answer    a.** $5 \sin(2000\pi t)$; **b.** $-5 \sin(2000\pi t)$.                                      ☐

---

## Voltage-Amplifier Model

Amplification can be modeled by a controlled source as illustrated in Figure 10.3. Because real amplifiers draw some current from the signal source, a realistic model of an amplifier must include a resistance $R_i$ across the input terminals. Furthermore, a resistance $R_o$ must be included in series with the output terminals to account for the fact that the output voltage of a real amplifier is reduced when load current flows. The complete amplifier model shown in Figure 10.3 is called the **voltage-amplifier model**. Later, we will see that other models can be used for amplifiers.

The **input resistance** $R_i$ of the amplifier is the equivalent resistance seen when looking into the input terminals. As we will find later, the input circuitry can sometimes include capacitive or inductive effects, and we would then refer to the **input impedance**. For example, the input amplifiers of typical oscilloscopes have an input impedance consisting of a 1-M$\Omega$ resistance in parallel with a 47-pF capacitance. In this chapter, we assume that the input impedance is purely resistive, unless stated otherwise.

The resistance $R_o$ in series with the output terminals is known as the **output resistance**. Real amplifiers are not able to deliver a fixed voltage to an arbitrary load resistance. Instead, the output voltage becomes smaller as the load resistance becomes smaller, and the output resistance accounts for this reduction. When the load draws current, a voltage drop occurs across the output resistance, resulting in a reduction of the output voltage.

The voltage-controlled voltage source models the amplification properties of the amplifier. Notice that the voltage produced by this source is simply a constant $A_{voc}$ times the input voltage $v_i$. If the load is an open circuit, there is no drop across

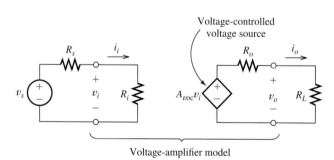

Figure 10.3  Model of an electronic amplifier, including input resistance $R_i$ and output resistance $R_o$.

the output resistance, and then, $v_o = A_{voc}v_i$. For this reason, $A_{voc}$ is called the **open-circuit voltage gain**.

In sum, the voltage-amplifier model includes the input impedance, the output impedance, and the open-circuit voltage gain in an equivalent circuit for the amplifier.

*The voltage-amplifier model includes the input impedance, the output impedance, and the open-circuit voltage gain in an equivalent circuit for the amplifier.*

## Current Gain

As shown in Figure 10.3, the input current $i_i$ is the current delivered to the input terminals of the amplifier, and the output current $i_o$ is the current flowing through the load. The **current gain** $A_i$ of an amplifier is the ratio of the output current to the input current:

$$A_i = \frac{i_0}{i_i} \qquad (10.2)$$

The input current can be expressed as the input voltage divided by the input resistance, and the output current is the output voltage divided by the load resistance. Thus, we can find the current gain in terms of the voltage gain and the resistances as

$$A_i = \frac{i_o}{i_i} = \frac{v_o/R_L}{v_i/R_i} = A_v\frac{R_i}{R_L} \qquad (10.3)$$

in which

$$A_v = \frac{v_o}{v_i}$$

*$A_v$ is the voltage gain with the load attached, whereas $A_{voc}$ is the voltage gain with the output terminals open circuited.*

is the voltage gain with the load resistance connected. Usually, $A_v$ is smaller in magnitude than the open-circuit voltage gain $A_{voc}$ because of the voltage drop across the output resistance.

## Power Gain

The power delivered to the input terminals by the signal source is called the input power $P_i$, and the power delivered to the load by the amplifier is the output power $P_o$. The **power gain** $G$ of an amplifier is the ratio of the output power to the input power:

$$G = \frac{P_o}{P_i} \qquad (10.4)$$

Because we are assuming that the input impedance and load are purely resistive, the average power at either set of terminals is simply the product of the root-mean-square (rms) current and rms voltage. Thus, we can write

$$G = \frac{P_o}{P_i} = \frac{V_oI_o}{V_iI_i} = A_vA_i = (A_v)^2\frac{R_i}{R_L} \qquad (10.5)$$

Notice that we have used uppercase symbols, such as $V_o$ and $I_o$, for the rms values of the currents and voltages. We use lowercase symbols, such as $v_o$ and $i_o$, for the instantaneous values. Of course, since we have assumed so far that the instantaneous output is a constant times the instantaneous input, the ratio of the rms voltages is the same as the ratio of the instantaneous voltages, and both ratios are equal to the voltage gain of the amplifier.

**Example 10.1    Calculating Amplifier Performance**

A source with an internal voltage of $V_s = 1$ mV rms and an internal resistance of $R_s = 1$ MΩ is connected to the input terminals of an amplifier having an open-circuit voltage gain of $A_{voc} = 10^4$, an input resistance of $R_i = 2$ MΩ, and an output resistance of $R_o = 2$ Ω. The load resistance is $R_L = 8$ Ω. Find the voltage gains $A_{vs} = V_o/V_s$ and $A_v = V_o/V_i$. Also, find the current gain and power gain.

**Solution**    First, we draw the circuit containing the source, amplifier, and load as shown in Figure 10.4. We can apply the voltage-divider principle to the input circuit to write

$$V_i = \frac{R_i}{R_i + R_s} V_s = 0.667 \text{ mV rms}$$

The voltage produced by the voltage-controlled source is given by

$$A_{voc}V_i = 10^4 V_i = 6.67 \text{ V rms}$$

Next, the output voltage can be found by using the voltage-divider principle, resulting in

$$V_o = A_{voc} V_i \frac{R_L}{R_L + R_o} = 5.33 \text{ V rms}$$

Now, we can find the required voltage gains:

$$A_v = \frac{V_o}{V_i} = A_{voc} \frac{R_L}{R_o + R_L} = 8000$$

and

$$A_{vs} = \frac{V_o}{V_s} = A_{voc} \frac{R_i}{R_i + R_s} \frac{R_L}{R_o + R_L} = 5333$$

Using Equations 10.3 and 10.5, we find that the current gain and power gain are

$$A_i = A_v \frac{R_i}{R_L} = 2 \times 10^9$$

$$G = A_v A_i = 16 \times 10^{12}$$

Notice that the current gain is very large, because the high input resistance allows only a small amount of input current to flow, whereas the relatively small load resistance allows the output current to be relatively large.    ∎

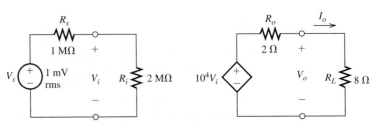

**Figure 10.4  Source, amplifier, and load for Example 10.1.**

## Loading Effects

Notice that not all of the internal voltage of the source appears at the input terminals of the amplifier in Example 10.1. This is because the finite input resistance of the amplifier allows current to flow into the input terminals, resulting in a voltage drop across the internal resistance $R_s$ of the source. Similarly, the voltage produced by the controlled source does not all appear across the load. These reductions in voltage are called **loading effects**. Because of loading effects, the voltage gains ($A_v$ or $A_{vs}$) realized are less than the internal gain $A_{voc}$ of the amplifier.

Because of loading effects, the voltage gains realized ($A_v$ or $A_{vs}$) are smaller in magnitude than the internal gain $A_{voc}$ of the amplifier.

**Exercise 10.2**   An amplifier has an input resistance of 2000 Ω, an output resistance of 25 Ω, and an open-circuit voltage gain of 500. The source has an internal voltage of $V_s = 20$ mV rms and a resistance of $R_s = 500$ Ω. The load resistance is $R_L = 75$ Ω. Find the voltage gains $A_v = V_o/V_i$ and $A_{vs} = V_o/V_s$. Find the current gain and the power gain.
**Answer**   $A_v = 375, A_{vs} = 300, A_i = 10^4, G = 3.75 \times 10^6$. ☐

**Exercise 10.3**   Assume that we can change the load resistance in Exercise 10.2. What value of load resistance maximizes the power gain? What is the power gain for this load resistance?
**Answer**   $R_L = 25$ Ω, $G = 5 \times 10^6$. ☐

## 10.2  CASCADED AMPLIFIERS

Sometimes, we connect the output of one amplifier to the input of another as shown in Figure 10.5. This is called a **cascade connection** of the amplifiers. The overall voltage gain of the cascade connection is given by

$$A_v = \frac{v_{o2}}{v_{i1}}$$

When the output of one amplifier is connected to the input of another amplifier, we say that the amplifiers are cascaded.

By multiplying and dividing by $v_{o1}$, this becomes

$$A_v = \frac{v_{o1}}{v_{i1}} \times \frac{v_{o2}}{v_{o1}}$$

Moreover, referring to Figure 10.5, we see that $v_{i2} = v_{o1}$. Therefore, we can write

$$A_v = \frac{v_{o1}}{v_{i1}} \times \frac{v_{o2}}{v_{i2}}$$

However, $A_{v1} = v_{o1}/v_{i1}$ is the gain of the first stage, and $A_{v2} = v_{o2}/v_{i2}$ is the gain of the second stage, so we have

$$A_v = A_{v1}A_{v2} \qquad\qquad (10.6)$$

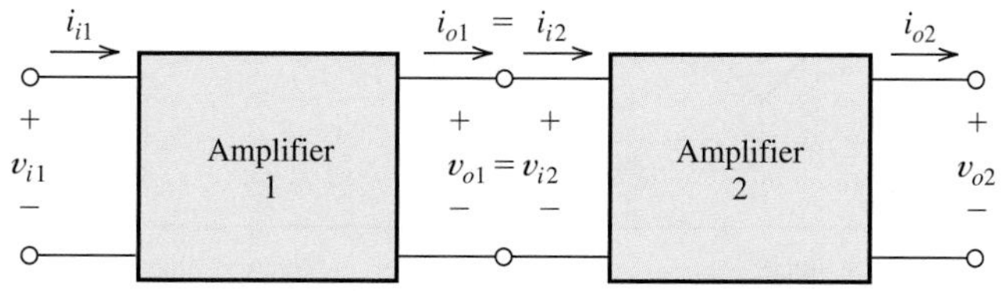

**Figure 10.5**  Cascade connection of two amplifiers.

Thus, the overall voltage gain of cascaded amplifier stages is the product of the voltage gains of the individual stages. (Of course, it is necessary to include loading effects in computing the gain of each stage. Notice that the input resistance of the second stage loads the first stage.)

Similarly, the overall current gain of a cascade connection of amplifiers is the product of the current gains of the individual stages. Furthermore, the overall power gain is the product of the individual power gains.

It is necessary to include loading effects in computing the gain of each stage.

### Example 10.2    Calculating Performance of Cascaded Amplifiers

Consider the cascade connection of the two amplifiers shown in Figure 10.6. Find the current gain, voltage gain, and power gain of each stage and for the overall cascade connection.

**Solution**    Considering loading by the input resistance of the second stage, the voltage gain of the first stage is

$$A_{v1} = \frac{v_{o1}}{v_{i1}} = \frac{v_{i2}}{v_{i1}} = A_{voc1}\frac{R_{i2}}{R_{i2} + R_{o1}} = 150$$

where we have used the fact that $A_{voc1} = 200$, as indicated in Figure 10.6. Similarly,

$$A_{v2} = \frac{v_{o2}}{v_{i2}} = A_{voc2}\frac{R_L}{R_L + R_{o2}} = 50$$

The overall voltage gain is

$$A_v = A_{v1}A_{v2} = 7500$$

Because $R_{i2}$ is the load resistance for the first stage, we can find the current gain of the first stage by use of Equation 10.3:

$$A_{i1} = A_{v1}\frac{R_{i1}}{R_{i2}} = 10^5$$

Similarly, the current gain of the second stage is found as

$$A_{i2} = A_{v2}\frac{R_{i2}}{R_L} = 750$$

The overall current gain is

$$A_i = A_{i1}A_{i2} = 75 \times 10^6$$

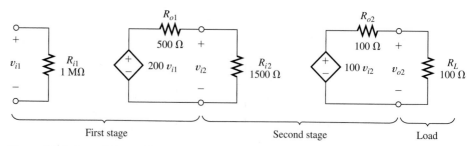

Figure 10.6 Cascaded amplifiers of Examples 10.2 and 10.3.

Now, the power gains can be found as

$$G_1 = A_{v1}A_{i1} = 1.5 \times 10^7$$

$$G_2 = A_{v2}A_{i2} = 3.75 \times 10^4$$

and

$$G = G_1G_2 = 5.625 \times 10^{11}$$ ■

## Simplified Models for Cascaded Amplifier Stages

Sometimes, we will want to find a simplified model for a cascaded amplifier. The input resistance of the cascade is the input resistance of the first stage, and the output resistance of the cascade is the output resistance of the last stage. The open-circuit voltage gain of the cascade is computed with an open-circuit load on the last stage. However, loading effects of each stage on the preceding stage must be considered. Once the open-circuit voltage gain of the overall cascade connection is found, a simplified model can be drawn.

Simplified models can be found for cascaded amplifiers.

---

**Example 10.3**    Simplified Model for an Amplifier Cascade

Find the overall simplified model for the cascade connection of Figure 10.6.

**Solution**    The voltage gain of the first stage, accounting for the loading of the second stage, is

$$A_{v1} = A_{voc1}\frac{R_{i2}}{R_{i2} + R_{o1}} = 150$$

First, determine the voltage gain of the first stage accounting for loading by the second stage.

With an open-circuit load, the gain of the second stage is

$$A_{v2} = A_{voc2} = 100$$

The overall open-circuit voltage gain is

$$A_{voc} = A_{v1}A_{v2} = 15 \times 10^3$$

The overall voltage gain is the product of the gains of the separate stages.

The input resistance of the cascade amplifier is

$$R_i = R_{i1} = 1\,\text{M}\Omega$$

and the output resistance is

The input impedance is that of the first stage, and the output impedance is that of the last stage.

$$R_o = R_{o2} = 100\,\Omega$$

The simplified model for the cascade is shown in Figure 10.7 ■

Figure 10.7  Simplified model for the cascaded amplifiers of Figure 10.6. See Example 10.3.

**Exercise 10.4** Three amplifiers with the following characteristics are cascaded:

$$\text{Amplifier 1:} \quad A_{voc1} = 10, R_{i1} = 1 \text{ K}\Omega, R_{o1} = 100 \text{ }\Omega$$

$$\text{Amplifier 2:} \quad A_{voc2} = 20, R_{i2} = 2 \text{ K}\Omega, R_{o2} = 200 \text{ }\Omega$$

$$\text{Amplifier 3:} \quad A_{voc3} = 30, R_{i3} = 3 \text{ K}\Omega, R_{o3} = 300 \text{ }\Omega$$

Find the parameters for the simplified model of the cascaded amplifier. Assume that the amplifiers are cascaded in the order 1, 2, 3.
**Answer** $R_i = 1 \text{ k}\Omega, R_o = 300 \text{ }\Omega, A_{voc} = 5357.$

**Exercise 10.5** Repeat Exercise 10.4 if the order of the amplifiers is 3, 2, 1.
**Answer** $R_i = 3\text{k }\Omega, R_o = 100 \text{ }\Omega, A_{voc} = 4348.$

## 10.3 POWER SUPPLIES AND EFFICIENCY

Power is supplied to the internal circuitry of amplifiers from **power supplies**. The power supply typically delivers current from several dc voltages to the amplifier; an example configuration is shown in Figure 10.8. The average power supplied to the amplifier by each voltage source is the product of the average current and the voltage. The total power supplied is the sum of the powers supplied by each voltage source. For example, the total average power supplied to the amplifier of Figure 10.8 is

> The number of terms in this equation depends on the number of supply voltages applied to the amplifier.

$$P_s = V_{AA}I_A + V_{BB}I_B \tag{10.7}$$

Notice that we have assumed that the current directions in the supply voltages are such that both sources deliver power to the amplifier. Rarely, a condition occurs for which some of the power taken from one supply source is returned to another source. We may only have a single supply voltage or there can be several, so the number of terms in a supply-power calculation such as Equation 10.7 is variable. It is customary

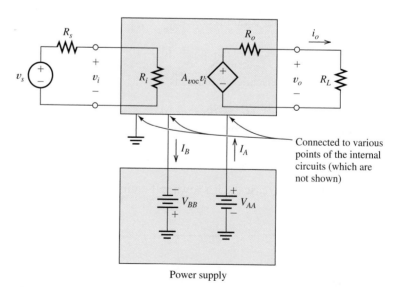

Power supply

**Figure 10.8** The power supply delivers power to the amplifier from several dc voltage sources.

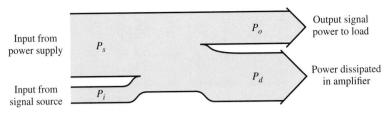

**Figure 10.9** Illustration of power flow.

to use uppercase symbols with repeated uppercase subscripts, such as $V_{CC}$, for dc supply voltages in electronic circuits.

We have seen that the power gain of typical amplifiers can be very large. Thus, the output power delivered to the load is much greater than the power taken from the signal source. This additional power is taken from the power supply. Power taken from the power supply can also be **dissipated** as heat in the internal circuits of the amplifier. Such dissipation is an undesirable effect that we usually try to minimize when designing the internal circuitry of an amplifier.

The sum of the power entering the amplifier from the signal source $P_i$ and the power from the power supply $P_s$ must be equal to the sum of the output power $P_o$ and the power dissipated $P_d$:

> Power flows into an amplifier from dc power supplies and from the signal source. Part of this power is delivered to the load as a useful signal, and part is dissipated as heat.

$$P_i + P_s = P_o + P_d \tag{10.8}$$

This is illustrated in Figure 10.9. Often, the input power $P_i$ from the signal source is insignificant compared with the other terms in this equation.

To summarize, we can view an amplifier as a system that takes power from the dc power supply and converts part of this power into output signal power. For example, a stereo audio system converts part of the power taken from the power supply into signal power that is finally converted to sound by the loudspeakers.

## Efficiency

The **efficiency** $\eta$ of an amplifier is the percentage of the power supplied that is converted into output power, or

$$\eta = \frac{P_o}{P_s} \times 100 \text{ percent} \tag{10.9}$$

### Example 10.4   Amplifier Efficiency

Find the input power, output power, supply power, and power dissipated in the amplifier shown in Figure 10.10. Also, find the efficiency of the amplifier. (The values given in this example are typical of one channel of a stereo amplifier under high-output test conditions.)

**Solution**   The average signal power delivered to the amplifier is given by

$$P_i = \frac{V_i^2}{R_i} = 10^{-11} \text{ W} = 10 \text{ pW}$$

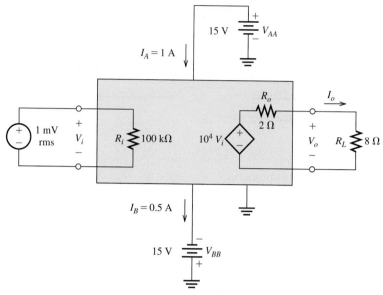

Figure 10.10 Amplifier of Example 10.4.

(Recall that 1 pW = 1 picowatt = $10^{-12}$ W.) The output voltage is

$$V_o = A_{voc} V_i \frac{R_L}{R_L + R_o} = 8 \text{ V rms}$$

Then, we find the average output power as

$$P_o = \frac{V_o^2}{R_L} = 8 \text{ W}$$

The supply power is given by

$$P_s = V_{AA} I_A + V_{BB} I_B = 15 + 7.5 = 22.5 \text{ W}$$

Notice that (as often happens) the power of the input signal is insignificant compared with the output and supply powers. The power dissipated as heat in the amplifier is

$$P_d = P_s + P_i - P_o = 14.5 \text{ W}$$

and the efficiency of the amplifier is

$$\eta = \frac{P_o}{P_s} \times 100 \text{ percent} = 35.6 \text{ percent}$$

■

**Exercise 10.6**   A certain amplifier is supplied with 1.5 A from a 15-V supply. The output signal power is 2.5 W, and the input signal power is 0.5 W. Find the power dissipated in the amplifier and the efficiency.
**Answer**   $P_d$ = 20.5 W, $\eta$ = 11.1 percent.                                                  □

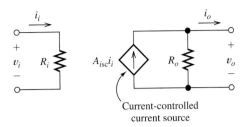

**Figure 10.11**  Current-amplifier model.

## 10.4  ADDITIONAL AMPLIFIER MODELS

### Current-Amplifier Model

Until now, we have modeled amplifiers as shown in Figure 10.3, in which the gain property of the amplifier is represented by a voltage-controlled voltage source. An alternative model known as a **current amplifier** is shown in Figure 10.11. In this model, the gain property is modeled by a current-controlled current source. As before, the input resistance accounts for the current that the amplifier draws from the signal source. The output resistance is now in parallel with the controlled source and accounts for the fact that the amplifier cannot supply a fixed current to an arbitrarily high-load resistance.

If the load is a short circuit, no current flows through $R_o$, and the ratio of output current to input current is $A_{isc}$. For this reason, $A_{isc}$ is known as the **short-circuit current gain**. An amplifier, initially modeled as a voltage amplifier, can also be modeled as a current amplifier. The input resistance and output resistance are the same for both models. The short-circuit current gain can be found from the voltage-amplifier model by connecting a short circuit to the output and computing the current gain.

$A_{isc}$ is the current gain of the amplifier with the output short circuited.

Notice that we have converted the Thévenin circuit of the voltage-amplifier model to a Norton circuit in the current-amplifier model.

---

**Example 10.5**  Determining the Current-Amplifier Model from the Voltage-Amplifier Model

A certain amplifier is modeled by the voltage-amplifier model shown in Figure 10.12. Find the current-amplifier model.

**Solution**  To find the short-circuit current gain, we connect a short circuit to the output terminals of the amplifier as shown in Figure 10.12. Then, we find that

Connect a short circuit across the output terminals and analyze the circuit to determine $A_{isc}$.

$$i_i = \frac{v_i}{R_i} \quad \text{and} \quad i_{osc} = \frac{A_{voc}v_i}{R_o}$$

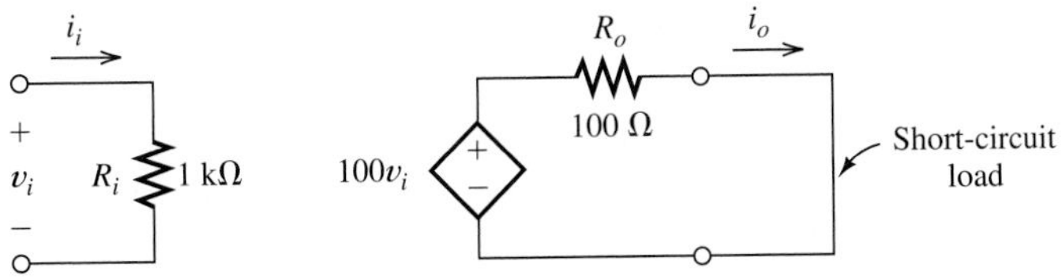

**Figure 10.12**  Voltage amplifier of Examples 10.5, 10.6, and 10.7.

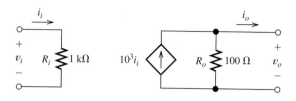

**Figure 10.13** Current-amplifier model equivalent to the voltage-amplifier model of Figure 10.12.

The short-circuit current gain is

$$A_{isc} = \frac{i_{osc}}{i_i} = A_{voc}\frac{R_i}{R_o} = 10^3$$

The resulting current-amplifier model is shown in Figure 10.13. ∎

**Exercise 10.7**   A certain amplifier modeled as a current amplifier has an input resistance of 1 kΩ, an output resistance of 20 Ω, and a short-circuit current gain of 200. Find the parameters for the voltage-amplifier model.
**Answer**   $A_{voc} = 4$, $R_i = 1$ kΩ, $R_o = 20$ Ω.   □

## Transconductance-Amplifier Model

Another model for an amplifier, known as a **transconductance amplifier**, is shown in Figure 10.14. In this case, the gain is modeled by a voltage-controlled current source, and the gain parameter $G_{msc}$ is called the **short-circuit transconductance gain**. $G_{msc}$ is the ratio of the short-circuit output current $i_{osc}$ to the input voltage $v_i$:

$$G_{msc} = \frac{i_{osc}}{v_i}$$

The units of transconductance gain are siemens. The input resistance and output resistance model the same effects as they do in the voltage-amplifier and current-amplifier models. A given amplifier can be modeled as a transconductance amplifier if the input resistance, output resistance, and short-circuit transconductance gain can be found.

The input resistance is the resistance seen looking into the input terminals. It has the same value for all models of a given amplifier. Similarly, the output resistance is the Thévenin resistance seen looking back into the output terminals and is the same for all the models.

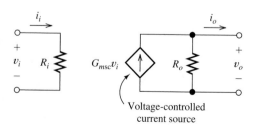

**Figure 10.14** Transconductance-amplifier model.

Voltage-controlled current source

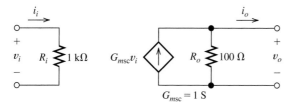

**Figure 10.15** Transconductance-amplifier equivalent to the voltage amplifier of Figure 10.12. See Example 10.6.

---

**Example 10.6    Determining the Transconductance-Amplifier Model**

Find the transconductance model for the amplifier of Figure 10.12.

**Solution**    The short-circuit transconductance gain is given by

$$G_{msc} = \frac{i_{osc}}{v_i}$$

The output current for a short-circuit load is

$$i_{osc} = \frac{A_{voc}v_i}{R_o}$$

Thus, we find that

$$G_{msc} = \frac{A_{voc}}{R_o} = 1.0 \text{ S}$$

The resulting amplifier model is shown in Figure 10.15.  ■

Connect a short circuit across the output terminals, and analyze the circuit to determine $G_{msc}$.

**Exercise 10.8** A current amplifier has an input resistance of 500 Ω, an output resistance of 50 Ω, and a short-circuit current gain of 100. Find the parameters for the transconductance-amplifier model.
**Answer**    $G_{msc} = 0.2$ S, $R_i = 500$ Ω, and $R_o = 50$ Ω.  □

## Transconductance-Amplifier Model

Finally, we can model an amplifier as a **transresistance amplifier** as shown in Figure 10.16. In this case, the gain property is modeled by a current-controlled voltage source. The gain parameter $R_{moc}$ is called the **open-circuit transresistance gain** and has units of ohms. It is the ratio of the open-circuit output voltage $v_{ooc}$ to the input current $i_i$:

$$R_{moc} = \frac{v_{ooc}}{i_i}$$

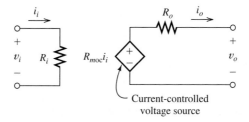

**Figure 10.16** Transresistance-amplifier model.

Current-controlled voltage source

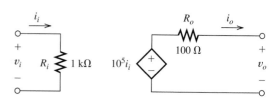

**Figure 10.17** Transresistance amplifier that is equivalent to the voltage amplifier of Figure 10.12. See Example 10.7.

The values of the input resistance and output resistance are the same as in any of the other amplifier models.

---

### Example 10.7    Determining the Transresistance-Amplifier Model

Find the transresistance-amplifier model for the amplifier shown in Figure 10.12.

**Solution**    With an open-circuit load, the output voltage is

$$v_{ooc} = A_{voc}v_i$$

and the input current is

$$i_i = \frac{v_i}{R_i}$$

Thus, we find the transresistance gain as

$$R_{moc} = \frac{v_{ooc}}{i_i} = A_{voc}R_i = 100 \text{ k}\Omega$$

The resulting transresistance model of the amplifier is shown in Figure 10.17. ■

Open circuit the output terminals, and analyze the circuit to determine $R_{moc}$.

**Exercise 10.9**    An amplifier has an input resistance of 1 M$\Omega$, an output resistance of 10 $\Omega$, and $G_{msc} = 0.05$ S. Find $R_{moc}$ for this amplifier.
**Answer**    $R_{moc} = 500 \text{ k}\Omega$.    □

We have seen that an amplifier can be modeled by any of the four models: voltage amplifier, current amplifier, transconductance amplifier, or transresistance amplifier. However, in cases for which either of the resistances (input or output) is zero or infinity, it is not possible to make conversions to all the models because the gain parameter is not defined for all the models. For example, if $R_i = 0$, then $v_i = 0$, and the voltage gain $A_{voc} = v_o/v_i$ is not defined.

## 10.5  IMPORTANCE OF AMPLIFIER IMPEDANCES IN VARIOUS APPLICATIONS

### Applications Calling for High or Low Input Impedance

Sometimes, we have an application for an amplifier that calls for the internal voltage produced by the source to be amplified. For example, an electrocardiograph amplifies and records the small voltages generated by a person's heart. These voltages are detected by placing electrodes on the person's skin. The impedance of the electrodes is variable from person to person and can be quite high. If the input impedance of the electrocardiograph is low, a variable reduction in voltage occurs because of loading. Thus, the amplitude of the signal can be affected by the contact resistance of the

Some applications call for amplifiers with high input impedance, while others call for low input impedance.

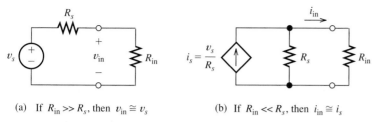

(a)  If $R_{in} \gg R_s$, then $v_{in} \cong v_s$        (b)  If $R_{in} \ll R_s$, then $i_{in} \cong i_s$

**Figure 10.18** If we want to sense the open-circuit voltage of a source, the amplifier should have a high input resistance, as in (a). To sense the short-circuit current of the source, low input resistance is called for, as in (b).

electrodes with the skin and therefore does not truly represent the electrical activity of the heart. On the other hand, if the input impedance of the electrocardiograph is much higher than the source impedance, the actual voltage produced by the heart appears at the input terminals. Thus, the input impedance of an electrocardiograph amplifier should be very high.

Other applications call for the amplifier to respond to the short-circuit current of a source. Then a very low input impedance is needed. An example is an electronic ammeter inserted in series with a circuit to measure current. Usually, we do not want the ammeter to change the current that is being measured. This is accomplished by designing the ammeter to have a low enough input impedance so that it does not change the impedance of the circuit significantly.

To summarize, if the input impedance of an amplifier is much higher than the internal impedance of the source, the voltage produced across the input terminals is nearly the same as the internal source voltage. This is illustrated in Figure 10.18(a). On the other hand, if the input impedance is very low, the input current is nearly equal to the short-circuit current of the source. This is illustrated in Figure 10.18(b).

## Applications Calling for High or Low Output Impedance

Diverse requirements for output impedance also occur. For example, we could have an audio amplifier that supplies background music to loudspeakers in many rooms of an office building, as shown in Figure 10.19. A switch is provided so that each loudspeaker can be turned off independent of the others (by opening its switch). Therefore, the load impedance presented to the amplifier is quite variable, depending on the number of loudspeakers turned on. If the amplifier output impedance is high compared with the load, the voltage supplied depends on the load impedance. Thus, as loudspeakers are turned off, the voltage applied to the others becomes higher, resulting in louder music. This effect could be undesirable. On the other hand, if the output impedance of the amplifier is very low compared with the load impedance,

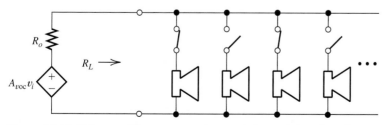

**Figure 10.19** If the amplifier output resistance $R_o$ is much less than the (lowest) load resistance, the load voltage is nearly independent of the number of switches closed.

the output voltage is nearly independent of the load. Thus, in this situation, a low output impedance is better.

Another example occurs in optical communication systems, where a *light-emitting diode* (LED) can be used to produce a light wave with intensity proportional to a message signal such as a voice waveform. Over a certain range of operation, the intensity of the light output of an LED is proportional to the current through it. Because LEDs have a nonlinear relationship between voltage and current, light intensity is *not* proportional to the voltage across the LED. Thus, it is desirable to force a current proportional to the message waveform to flow through the diode. This can be achieved by designing an amplifier with a very high output impedance to drive the LED. (On the other hand, if a very low output impedance were used, the voltage supplied to the diode would be proportional to the input signal to the amplifier, but because of the nonlinear relationship between current and voltage for the diode, the light output would no longer be proportional to the message.)

To summarize, we can force a desired voltage waveform to appear across a variable load by designing the amplifier to have a very low output impedance compared with the load impedance. On the other hand, we can force a given current waveform through a variable load by designing the amplifier to have a very high output impedance compared with the load impedance.

## Applications Calling for a Particular Impedance

Not all applications call for amplifiers with either very small or very large impedances. For example, consider an amplifier whose input is connected to a source by a **transmission line** as shown in Figure 10.20. An example of a transmission line with which you may be familiar is *coaxial cable*, commonly used to connect TV sets to cable systems or to digital television (DTV) antennas. Each type of transmission line has a **characteristic impedance**, which is typically 75 $\Omega$ for the coaxial line used in television applications. Unless the transmission line is terminated in its characteristic impedance, a signal traveling along a transmission line is partially reflected and travels back toward the source. This is illustrated in Figure 10.20. When connecting a signal source, such as a set-top box or antenna, to a television set, the reflections from the TV set can be reflected again at the source, so the signal arrives a second time at the set. These extra signals are delayed because of the round-trip travel along the transmission line and can cause degradation of picture quality. In the case of analog signals, the effect of the reflection is a faint image known as a "ghost" slightly to the right-hand side of the main image. Digital signals can become so corrupted by

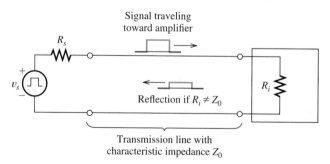

**Figure 10.20** To avoid reflections, the amplifier input resistance $R_i$ should equal the characteristic resistance $Z_0$ of the transmission line.

these reflections that the set cannot decode them. Therefore, it is important for the input impedance of the television set and the output impedance of the source to be nearly equal to the characteristic impedance of the transmission line so that significant reflections do not occur.

The output impedance of an audio amplifier is another situation that sometimes calls for an intermediate value. The frequency response of a loudspeaker depends on the output impedance of the amplifier driving it. Thus, if high fidelity is a primary consideration, the amplifier should be designed so that it has the output impedance that gives the most nearly constant response versus frequency.

## 10.6  IDEAL AMPLIFIERS

We have seen in Section 10.5 that certain applications call for amplifiers with very high or very low input impedance (compared with the source impedance) and very high or very low output impedance (compared with the load). Such amplifiers can be classified as follows:

1. An **ideal voltage amplifier** senses the open-circuit voltage of the source and produces an amplified voltage across the load independent of the load impedance. Thus, the ideal voltage amplifier has infinite input impedance (so the open-circuit voltage appears across the input terminals) and zero output impedance (so the output voltage is independent of the load impedance).

2. An **ideal current amplifier** senses the short-circuit current of the source and forces an amplified version of this current to flow through the load. Thus, the ideal current amplifier has zero input impedance and infinite output impedance.

3. An **ideal transconductance amplifier** senses the open-circuit voltage of the source and forces a current proportional to this voltage to flow through the load. Thus, the ideal transconductance amplifier has infinite input impedance and infinite output impedance.

4. An **ideal transresistance amplifier** senses the short-circuit current of the source and causes a voltage proportional to this current to appear across the load. Thus, the ideal transresistance amplifier has zero input impedance and zero output impedance. Table 10.1 shows the input impedance, output impedance, and gain parameter for each type of ideal amplifier.

> According to their input and output impedances, ideal amplifiers can be classified into four types. These are the ideal voltage amplifier, the ideal current amplifier, the ideal transconductance amplifier, and the ideal transresistance amplifier. The best amplifier type to select depends on the application.

### Classifying Real Amplifiers

In practice, amplifiers do not have either zero or infinite impedances. However, real amplifiers can often be classified as approximately ideal amplifiers. For example, if

**Table 10.1** Characteristics of Ideal Amplifiers

| Amplifier Type | Input Impedance | Output Impedance | Gain Parameter |
|---|---|---|---|
| Voltage | $\infty$ | 0 | $A_{voc}$ |
| Current | 0 | $\infty$ | $A_{isc}$ |
| Transconductance | $\infty$ | $\infty$ | $G_{msc}$ |
| Transresistance | 0 | 0 | $R_{moc}$ |

the input impedance is very large (compared with the source impedance) and the output impedance is very small (compared with the load), we have an approximate ideal voltage amplifier.

Notice that a given amplifier cannot be classed as an approximate ideal amplifier unless the source and load impedances to be encountered are known in advance. For example, an amplifier with an input impedance of 1000 $\Omega$ and an output impedance of 100 $\Omega$ would be classed as an approximate ideal voltage amplifier if the source impedances to be encountered are much less than 1000 $\Omega$ and the load impedances are much greater than 100 $\Omega$. On the other hand, if the source impedances are on the order of 1 M$\Omega$ and the load impedances are on the order of 1 $\Omega$, the same amplifier would be properly classed as an approximate ideal current amplifier.

In general, the "middle range" of impedances in low-power electronic circuits runs from 1 k$\Omega$ to 100 k$\Omega$. Impedances less than 100 $\Omega$ are usually considered to be "small," and impedances greater than 1 M$\Omega$ are classed as "large." Thus, we would usually be inclined to classify an amplifier with an input impedance of 10 $\Omega$ and an output impedance of 2 M$\Omega$ as an approximate ideal current amplifier. However, we might want to change this classification, depending on the actual load and source impedances.

> The proper classification of a given amplifier depends on the ranges of source and load impedances with which the amplifier is used.

**Exercise 10.10**    A certain amplifier has an input resistance of $R_i = 1$ k$\Omega$ and an output resistance of $R_o = 1$ k$\Omega$. $R_s$ is the source resistance, and $R_L$ is the load. Classify the amplifier if: **a.** $R_s$ is less than 10 $\Omega$ and $R_L$ is greater than 100 k$\Omega$; **b.** $R_s$ is greater than 100 k$\Omega$ and $R_L$ is less than 10 $\Omega$; **c.** $R_s$ is less than 10 $\Omega$ and $R_L$ is less than 10 $\Omega$; **d.** $R_s$ is greater than 100 k$\Omega$ and $R_L$ is greater than 100 k$\Omega$; **e.** $R_s$ is approximately 1 k$\Omega$ and $R_L$ is less than 10 $\Omega$.

**Answer**    **a.** Approximate ideal voltage amplifier; **b.** approximate ideal current amplifier; **c.** approximate ideal transconductance amplifier; **d.** approximate ideal transresistance amplifier; **e.** for this source resistance, the amplifier does not fit into any ideal amplifier category. ☐

**Exercise 10.11**    A particular transducer is to be used in measuring liquid level in a chemical process. The short-circuit current of the transducer is proportional to the liquid level. (However, the open-circuit voltage of the transducer is nearly independent of the level.) An amplifier is needed to deliver a voltage signal proportional to liquid level to a resistive load that may vary in value between 1 k$\Omega$ and 10 k$\Omega$. What type of ideal amplifier is needed?

**Answer**    A transresistance amplifier is needed, because we want the input impedance of the amplifier to be small enough so it responds to the short-circuit transducer current. Furthermore, to deliver an output voltage independent of the load, the output impedance of the amplifier must be very small compared with the load impedance. ☐

## 10.7 FREQUENCY RESPONSE

So far, we have considered the gain parameter of an amplifier to be a constant. However, if we apply a variable-frequency sinusoidal input signal to an amplifier, we will find that gain is a function of frequency. Moreover, the amplifier affects the phase as well as the amplitude of the sinusoid. Therefore, we now give a more general definition of amplifier gain. We define complex gain to be the ratio of the phasor for the output signal to the phasor for the input signal:

$$A_v = \frac{\mathbf{V}_o}{\mathbf{V}_i} \qquad (10.10)$$

We use uppercase bold symbols to stand for the phasors of the input and output voltages. Similarly, we define complex current gain, transconductance gain, and transresistance gain as the ratio of the appropriate phasor quantities. We have used the term *complex gain* to emphasize the fact that these gains have both magnitude and phase. Shortly, we will drop the word *complex*, for simplicity. Actually, complex gain is the same concept as that of a transfer function, which we discussed in Chapter 6. Recall that to express a transfer function in decibels, we take 20 times the common logarithm of its magnitude.

> Complex voltage gain is the output-voltage phasor divided by the input-voltage phasor. Complex gain has magnitude and phase.

---

**Example 10.8**   Determining Complex Gain

The input voltage to a certain amplifier is

$$v_i(t) = 0.1 \cos(2000\pi t - 30°)$$

and the output voltage is

$$v_o(t) = 10 \cos(2000\pi t + 15°)$$

Find the complex voltage gain of the amplifier and express the magnitude of the gain in decibels.

**Solution**   Recall that the phasor for the input voltage is a complex number whose magnitude is the peak value of the sinusoidal signal and whose angle is the phase angle of the sinusoidal signal. Thus,

$$\mathbf{V}_i = 0.1\underline{/-30°}$$

Similarly,

$$\mathbf{V}_o = 10\underline{/15°}$$

Now, we can find the complex voltage gain as

$$A_v = \frac{\mathbf{V}_o}{\mathbf{V}_i} = \frac{10\underline{/15°}}{0.1\underline{/-30°}}$$

$$= 100\underline{/45°}$$

The meaning of this complex voltage gain is that the output signal is 100 times larger in amplitude than the input signal. Furthermore, the output signal is phase shifted by 45° relative to the input signal.

To express gain in decibels, we first find the magnitude of the gain by dropping the angle and then compute decibel gain:

$$|A_v|_{dB} = 20 \log |A_v| = 20 \log(100) = 40 \text{ dB} \qquad ∎$$

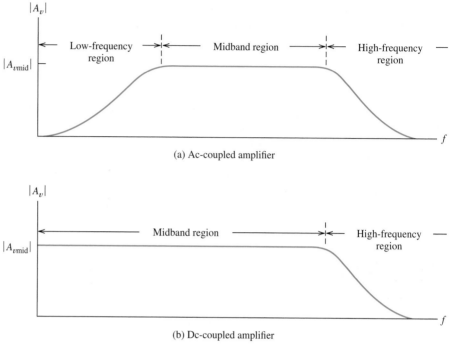

Figure 10.21  Gain magnitude versus frequency.

## Gain as a Function of Frequency

Many amplifiers have a midband range in which the gain magnitude is constant.

If we plot the magnitude of the gain of a typical amplifier versus frequency, a plot such as the one shown in Figure 10.21 results. Notice that the gain magnitude is constant over a wide range of frequencies known as the **midband region**.

## AC Coupling versus Direct Coupling

In some cases, such as the one shown in Figure 10.21(a), the gain drops to zero at dc (zero frequency). Such amplifiers are said to be **ac coupled** because only ac signals are amplified. These amplifiers are often constructed by cascading several amplifier circuits or stages that are connected together by **coupling capacitors** so that the dc voltages of the amplifier circuits do not affect the signal source, adjacent stages, or the load. This is illustrated in Figure 10.22. (Sometimes, transformers are used to couple individual stages together, which also leads to an ac-coupled amplifier with zero gain at dc. Transformers are discussed in Chapter 14.)

Amplifiers may be ac coupled or dc coupled.

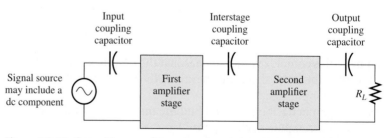

Figure 10.22  Capacitive coupling prevents a dc input component from affecting the first stage, dc voltages in the first stage from reaching the second stage, and dc voltages in the second stage from reaching the load.

Other amplifiers have constant gain all the way down to dc, as shown in Figure 10.21(b). They are said to be **dc coupled** or **direct coupled**. Amplifiers that are realized as integrated circuits are often dc coupled because the capacitors or transformers needed for ac coupling cannot be fabricated in integrated form.

Audio amplifiers are almost always ac coupled because audible sounds span the frequency range from about 20 Hz to 15 kHz. Therefore, there is no need to provide gain down to dc, and furthermore, it is not desirable to apply dc voltages to the loudspeakers.

Electrocardiograph amplifiers are deliberately ac coupled because a dc voltage of nearly a volt often occurs at the input due to electrochemical potentials developed by the electrodes. The ac signal generated by the heart is on the order of 1 mV, and therefore, the gain of the amplifier is high typically, 1000 or more. A 1-V dc input would cause the amplifier to try to produce an output of 1000 V. It would be difficult (and highly undesirable) to design an amplifier capable of such large outputs. Therefore, it is necessary to ac couple the input circuit of an electrocardiograph to prevent the dc component from overloading the amplifier.

Amplifiers for (analog) video signals need to be dc coupled because video signals have frequency components from dc into the MHz range. Dark pictures result in a different dc component than bright pictures. To obtain pictures with the proper brightness, it is necessary to use a dc-coupled amplifier to preserve the dc component.

## High-Frequency Region

As indicated in Figure 10.21(a) and (b), the gain of an amplifier always drops off at high frequencies. This is caused either by small amounts of capacitance in parallel with the signal path or by small inductances in series with the signal path in the amplifier circuitry, as illustrated in Figure 10.23. Recall that the impedance of a capacitor is inversely proportional to frequency, resulting in an effective short circuit at sufficiently high frequencies. The impedance of an inductor is proportional to frequency, so it becomes an open circuit at very high frequencies.

Some of these small capacitances occur because of stray wiring capacitance between signal-carrying conductors and ground. Other capacitances are integral parts of the active devices (transistors) necessary for amplification. Small inductances result from the magnetic fields surrounding the conductors in the circuit. For example, a critically placed piece of wire one-half inch long can have enough inductance to limit severely the frequency response of an amplifier intended to operate at several gigahertz.

*Gain magnitude declines for all amplifiers at sufficiently high frequencies.*

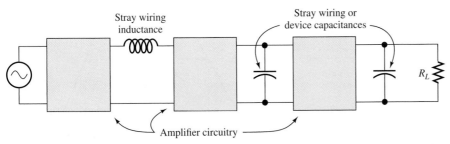

**Figure 10.23** Capacitance in parallel with the signal path and inductance in series with the signal path reduce the gain in the high-frequency region.

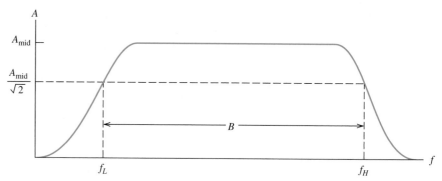

**Figure 10.24** Gain versus frequency for a typical amplifier showing the upper and lower half-power (3-dB) frequencies ($f_H$ and $f_L$) and the half-power bandwidth $B$.

### Half-Power Frequencies and Bandwidth

Usually, we specify the approximate useful frequency range of an amplifier by giving the frequencies for which the voltage (or current) gain magnitude is $1/\sqrt{2}$ times the midband gain magnitude. These are known as the **half-power frequencies** because the output power level is half the value for the midband region if a constant-amplitude variable-frequency input test signal is used. Expressing the factor $1/\sqrt{2}$ in decibels, we have $20 \log(1/\sqrt{2}) = -3.01$ dB. Thus, at the half-power frequencies, the voltage (or current) gain is approximately 3 dB lower than the midband gain. The bandwidth $B$ of an amplifier is the distance between the half-power frequencies. These definitions are illustrated in Figure 10.24.

### Wideband versus Narrowband Amplifiers

Amplifiers that are either dc coupled or have a lower half-power frequency that is a small fraction of the upper half-power frequency are called **wideband** or **baseband amplifiers**. Wideband amplifiers are used for signals that occupy a wide range of frequencies, such as audio signals (20 Hz to 15 kHz) or video signals (dc into the MHz range).

On the other hand, the frequency response of an amplifier is sometimes deliberately limited to a small bandwidth compared with the center frequency. Such an amplifier is called a **narrowband** or **bandpass amplifier**. The gain versus frequency response of a bandpass amplifier is shown in Figure 10.25. Bandpass amplifiers are

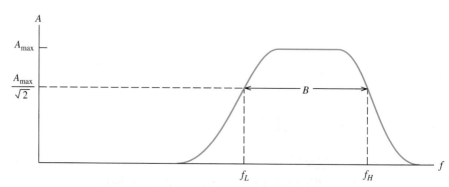

**Figure 10.25** Gain magnitude versus frequency for a bandpass amplifier.

used in radio receivers, where it is desired to amplify the signal from one transmitter and reject the signals from other transmitters in adjacent frequency ranges.

## 10.8  LINEAR WAVEFORM DISTORTION

### Amplitude Distortion

If the gain of an amplifier has a different magnitude for the various frequency components of the input signal, a form of distortion known as **amplitude distortion** occurs.

Audio systems often suffer from amplitude distortion because the amplifier, and particularly the loudspeakers, tend to reduce the amplitudes of the high-pitched and low-pitched components relative to the midband components. This is especially true for telephone systems. Hence, the music we hear while on hold is of poor quality.

---

**Example 10.9** | **Amplitude Distortion**

The input signal to a certain amplifier contains two frequency components and is given by

$$v_i(t) = 3\cos(2000\pi t) - 2\cos(6000\pi t)$$

The gain of the amplifier at 1000 Hz is $10\underline{/0°}$, and the gain at 3000 Hz is $2.5\underline{/0°}$. Plot the input waveform and output waveform to scale versus time.

**Solution**    The first term of the input signal is at a frequency of 1000 Hz, so it experiences a gain of $10\underline{/0°}$, whereas the second term of the input is at a frequency of 3000 Hz, so the gain for it is $2.5\underline{/0°}$. Applying these gains and phase shifts to the terms of the input signal, we find the output:

$$v_o(t) = 30\cos(2000\pi t) - 5\cos(6000\pi t)$$

Plots of the input and output waveforms are shown in Figure 10.26. Notice that the output waveform has a different shape than the input waveform because of amplitude distortion. ∎

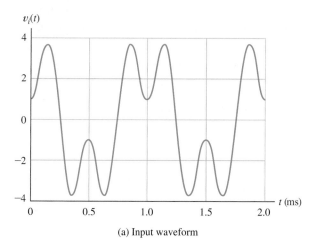

(a) Input waveform

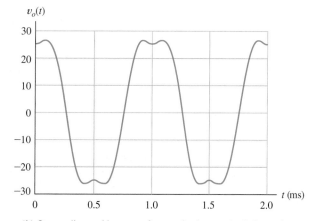

(b) Output distorted because of unequal gain magnitude for various frequency components

**Figure 10.26** Linear amplitude distortion. See Example 10.9.

## Phase Distortion

If the phase shift of an amplifier is not proportional to frequency, **phase distortion** occurs. Zero phase at all frequencies results in an output waveform identical to the input. On the other hand, if the phase shift of the amplifier is proportional to frequency, the output waveform is a time-shifted version of the input. However, we do not say that distortion has occurred because the shape of the waveform is unchanged. If phase is not proportional to frequency, the waveform shape is changed in passing through the amplifier and phase distortion has occurred.

---

**Example 10.10**   **Phase Distortion**

Suppose that the input signal given by

$$v_i(t) = 3\cos(2000\pi t) - \cos(6000\pi t)$$

is applied to the inputs of three amplifiers having the gains shown in Table 10.2. Find and plot the output of each amplifier.

**Solution**   Applying the gains and phase shifts to the input signal, we find the output signals for the amplifiers to be

$$v_{oA}(t) = 30\cos(2000\pi t) - 10\cos(6000\pi t)$$

$$v_{oB}(t) = 30\cos(2000\pi t - 45°) - 10\cos(6000\pi t - 135°)$$

$$v_{oC}(t) = 30\cos(2000\pi t - 45°) - 10\cos(6000\pi t - 45°)$$

Plots of the output waveforms are shown in Figure 10.27. Amplifier $A$ produces an output waveform identical to the input, and amplifier $B$ produces an output waveform identical to the input, except for a time delay. For amplifier $A$, the phase shift is zero for both frequency components, whereas the phase shift of amplifier $B$ is proportional to frequency. (The phase shift for the 3000-Hz component is three times the phase shift for the 1000-Hz component.) Amplifier $C$ produces a distorted output waveform because its phase response is not proportional to frequency.   ∎

Amplitude and phase distortion are sometimes called **linear distortion** because they occur even though the amplifier is linear (i.e., obeys superposition). Later, we will see that another type of distortion, known as nonlinear distortion, can also occur in amplifiers.

Let us briefly consider the meaning of superposition, which we discussed in the context of resistive circuits in Section 2.7 starting on page 103. For a given amplifier, suppose that the input $v_{inA}$ results in the output $v_{oA}$ and input $v_{inB}$ results in the output $v_{oB}$. Then, if the input $v_{inA} + v_{inB}$ always produces the output $v_{oA} + v_{oB}$, we say that the amplifier obeys superposition or is linear. In other words, if adding input

> If adding input signals always corresponds to adding the output signals, an amplifier is said to be linear.

**Table 10.2** Complex Gains of the Amplifiers Considered in Example 10.10

| Amplifier | Gain at 1000 Hz | Gain at 3000 Hz |
|-----------|-----------------|-----------------|
| A | $10\underline{/0°}$ | $10\underline{/0°}$ |
| B | $10\underline{/-45°}$ | $10\underline{/-135°}$ |
| C | $10\underline{/-45°}$ | $10\underline{/-45°}$ |

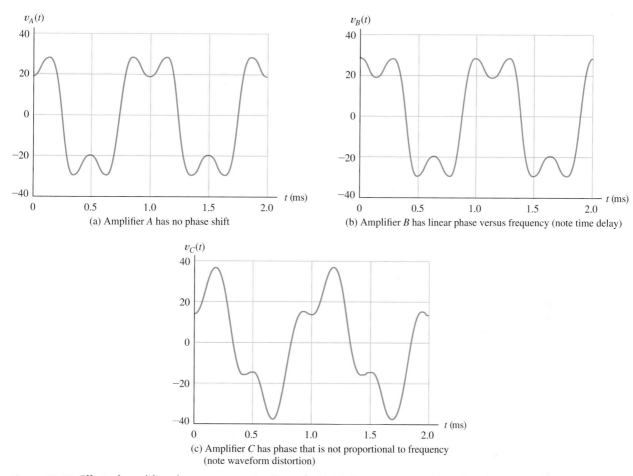

Figure 10.27  Effect of amplifier phase response. See Example 10.10. [*Note:* Input waveform has the same shape as $v_A(t)$.]

signals always corresponds to adding the output signals, an amplifier is said to be linear. For example, an amplifier for which $v_o(t) = 10v_{in}(t)$ is linear because $10(v_{inA} + v_{inB}) = 10v_{inA} + 10v_{inB}$. However, an amplifier for which $v_o(t) = [v_{in}(t)]^2$ is not linear because $(v_{inA} + v_{inB})^2 \neq v_{inA}^2 + v_{inB}^2$.

## Requirements for Distortionless Amplification

To avoid linear waveform distortion, an amplifier should have constant gain magnitude and a phase response that is linear versus frequency for the range of frequencies contained in the input signal. Of course, departure from these requirements outside the frequency range of the input signal components does not result in distortion. These requirements for distortionless amplification are illustrated in Figure 10.28.

In the examples we have given, the input signals contained only a few components at specific frequencies. However, most signals of interest in electronic systems contain components spread over a continuous range of frequencies. For example, audio signals contain components from about 20 Hz to about 15 kHz. Thus, we require an audio amplifier to have nearly constant gain magnitude over that range. (However, since it turns out that the ear is not sensitive to phase distortion, we would not require the phase response of an audio amplifier to be proportional to frequency.)

To avoid linear waveform distortion, an amplifier should have constant gain magnitude and a phase response that is linear versus frequency for the range of frequencies contained in the input signal.

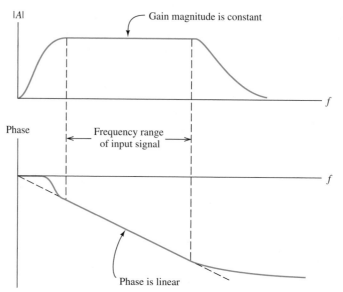

**Figure 10.28** Linear distortion does not occur if the gain magnitude is constant and the phase is proportional to frequency over the frequency range of the input signal.

Analog video signals contain significant frequency components from dc to several MHz. Since the shape of the waveform ultimately determines the brightness of various points in the picture, either phase distortion or amplitude distortion would affect the image. Therefore, we require the gain magnitude to be constant and the phase response to be proportional to frequency over the stated range for a video amplifier.

### Definition of Gain Revisited

As a final comment, recall that we originally defined the gain of an amplifier to be the ratio of the output signal to the input:

$$A_v = \frac{v_o(t)}{v_i(t)}$$

However, if linear waveform distortion occurs (or even a time delay), the ratio of output to input is a function of time, rather than a constant. Thus, we should not try to find the gain of an amplifier by taking the ratio of the instantaneous output and input. Instead, we recognize that gain is a function of frequency and take the ratio of the phasors for a sinusoidal input signal to find the (complex) gain for each frequency.

**Exercise 10.12** Suppose that an input signal is given by

$$v_i(t) = \sin(1000\pi t) + \cos(2000\pi t) + 2\cos(3000\pi t)$$

and the gain of an amplifier at 1000 Hz is $5\underline{/30°}$. What are the required amplifier gain and phase shift at the frequencies of the other components if both types of linear waveform distortion are to be avoided?

**Answer**   $5\underline{/15°}$ for the 500-Hz component and $5\underline{/45°}$ for the 1500-Hz component. ▫

**Exercise 10.13** The output of a certain amplifier is given by $v_o(t) = 10v_{\text{in}}(t - 0.01)$. Consider a sinusoidal input signal $v_{\text{in}}(t) = V_m \cos(\omega t)$, and find the complex gain (magnitude and phase) as a function of $\omega$.
**Answer**   $10\underline{/-0.01\omega}$.                                                    ▫

## 10.9   PULSE RESPONSE

Often, we need to amplify a pulse signal such as the one shown in Figure 10.29(a). Pulses contain components spread over a wide range of frequencies; therefore, amplification of pulses calls for a wideband amplifier. A typical amplified output pulse is shown in Figure 10.29(b). The output waveform differs from the input in several important respects: The pulse displays **overshoot** and **ringing**, the leading and trailing edges are gradual rather than abrupt, and if the amplifier is ac coupled, the top of the output pulse is **tilted**.

### Rise Time

The gradual rise of the leading edge of the amplifier response is often quantified by giving the **rise time** $t_r$, which is the time interval between the point $t_{10}$ at which the amplifier achieves 10 percent of the eventual output amplitude and the point $t_{90}$ at which the output is 90 percent of the final value. This is illustrated in Figure 10.30.

The rounding of the leading edge can be attributed to the roll-off of gain in the high-frequency region. A rule-of-thumb relationship between the half-power bandwidth $B$ and the rise time $t_r$ of a wideband amplifier is

This approximate relationship is very useful in estimating bandwidth given rise time or vice versa.

$$t_r \cong \frac{0.35}{B} \tag{10.11}$$

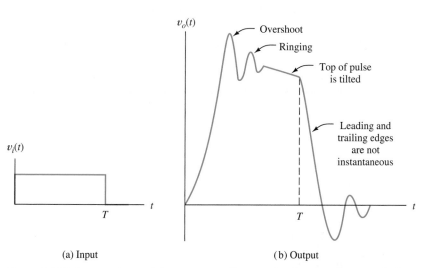

(a) Input

(b) Output

**Figure 10.29** Input pulse and the corresponding output of a typical ac-coupled broadband amplifier.

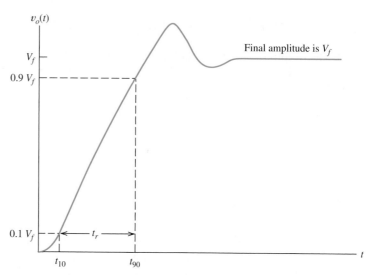

**Figure 10.30** Rise time of the output pulse. (*Note:* No tilt is shown. When tilt is present, some judgment is necessary to estimate the amplitude $V_f$.)

This relationship is not exact for all types of wideband amplifiers, but it is a useful guide for estimating performance. (It is accurate for first-order circuits; see Problem P10.80.)

Since pulse amplifiers are broadband, the bandwidth is almost equal to the upper half-power frequency. Thus, it is mainly the high-frequency characteristics of the amplifier that restrict rise time.

## Overshoot and Ringing

Another aspect of the output pulse shown in Figure 10.29 is overshoot and ringing, which are also related to the way the gain of the amplifier behaves in the high-frequency region. An amplifier that displays pronounced overshoot and ringing usually has a peak in its gain characteristic, as shown in Figure 10.31. The frequency of maximum gain approximately matches the ringing frequency.

Because both rise time and overshoot are related to the high-frequency response, there is usually some trade-off between these specifications. In a particular design, component values that reduce rise time tend to produce more overshoot and ringing. However, more than about 10 percent overshoot is usually undesirable.

**Figure 10.31** Gain versus frequency for an amplifier that displays pronounced ringing in its pulse response. The frequency of the ringing is approximately $f_r$.

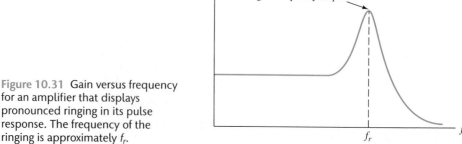

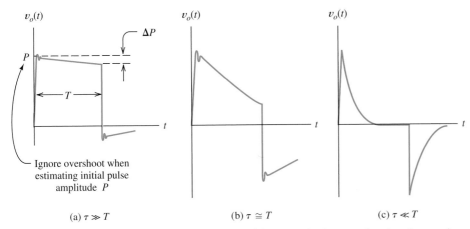

Figure 10.32  Pulse responses of ac-coupled amplifiers. $T$ is the input pulse duration, and $\tau$ represents the shortest time constant of the coupling circuits.

## Tilt

The tilt of the top of the output pulse, shown in Figure 10.32(a), occurs if the amplifier is ac coupled and arises from charging of coupling capacitors during the pulse. (After all, if the pulse lasted indefinitely, it would be the same as a new dc level at the input, and eventually the output voltage of an ac-coupled amplifier would return to zero.) Tilt is specified as a percentage of the initial pulse amplitude,

The pulse response of an amplifier may contain overshoot, ringing, and tilt. Rise time is always nonzero.

$$\text{percentage tilt} = \frac{\Delta P}{P} \times 100 \text{ percent} \qquad (10.12)$$

where $\Delta P$ and $P$ are defined in Figure 10.32(a). As the duration of the pulse is increased (or as the lower half-power frequency of the amplifier is raised by changing the coupling circuits to have shorter time constants), output waveforms such as those in Figure 10.32(b) and (c) result.

For small amounts of tilt, the percentage tilt is related to the lower half-power frequency by the approximate relation

$$\text{percentage tilt} \cong 200\pi f_L T \qquad (10.13)$$

where $T$ is the duration of the pulse and $f_L$ is the lower half-power frequency of the amplifier. (See Problem P10.81 for a derivation of this formula for percentage tilt.)

---

**Exercise 10.14**  In a radar system, pulses of radio waves are transmitted and objects are detected by their reflected signals. After conversion of the reflected signals to baseband, they appear as pulses, and the time interval between pulses indicates the distance between objects. To distinguish objects a given distance apart, the maximum rise time allowed for the amplifiers is approximately equal to the time separation of the reflections. For example, if it is desired to distinguish objects that are 10 m apart on a line from the radar transmitter, the time separation of the echoes is 20 m (because the waves must make a round trip) divided by the speed of light. This gives a required maximum rise time of approximately 66.7 ns. Estimate the minimum bandwidth required for the amplifier.

**Answer**   $B \cong 5.25$ MHz.

> **Exercise 10.15**  An amplifier is needed to amplify pulses with a duration of 100 $\mu$s with a sag (tilt) of not more than 1 percent. Estimate the highest value allowed for the lower half-power frequency of the amplifier.
> **Answer**   $f_L = 15.9$ Hz.                                                             □

## 10.10  TRANSFER CHARACTERISTIC AND NONLINEAR DISTORTION

The transfer characteristic is a plot of instantaneous output amplitude versus instantaneous input amplitude.

The **transfer characteristic** of an amplifier is a plot of the instantaneous output amplitude versus the instantaneous input amplitude. For an ideal amplifier, the output is simply a larger version of the input waveform, and the transfer characteristic is a straight line whose slope is the gain. Real amplifiers have transfer characteristics that depart from straight lines, particularly at large amplitudes. This is shown in Figure 10.33. Curvature of the transfer characteristic results in an undesirable effect known as **nonlinear distortion**.

Curvature of the transfer characteristic results in nonlinear distortion.

Sometimes, the departure from a straight characteristic can be very abrupt. Then the result of applying a high-amplitude input signal is **clipping** of the output waveform, as shown in Figure 10.34. However, even small departures from a straight characteristic can be very serious in some applications.

### Harmonic Distortion

The input output relationship of a nonlinear amplifier can be written as

$$v_o = A_1 v_i + A_2(v_i)^2 + A_3(v_i)^3 + \cdots \tag{10.14}$$

where $A_1$, $A_2$, $A_3$, and so on, are constants selected so that the equation matches the curvature of the nonlinear transfer characteristic.

Consider the case for which the input signal is a sinusoid given by

$$v_i(t) = V_a \cos(\omega_a t) \tag{10.15}$$

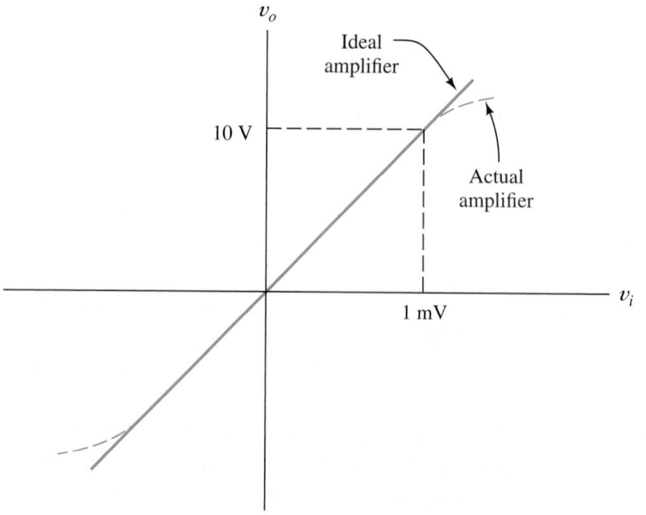

**Figure 10.33**  Transfer characteristics. $A_v = 10{,}000$.

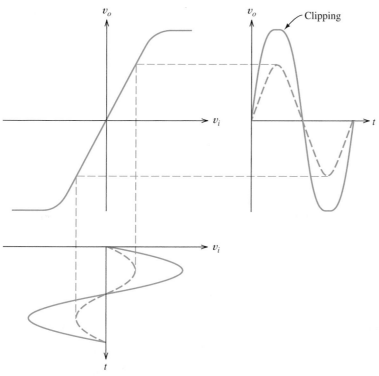

**Figure 10.34** Illustration of input signal, amplifier transfer characteristic, and output signal, showing clipping for large signal amplitude.

Let us find an expression for the corresponding output signal. Substituting Equation 10.15 into 10.14, applying trigonometric identities for $[\cos(\omega_a t)]^n$, collecting terms, and defining $V_0$ to be equal to the sum of all of the constant terms, $V_1$ to be the sum of the coefficients of the terms with frequency $\omega_a$, and so on, we find that

$$v_o(t) = V_0 + V_1 \cos(\omega_a t) + V_2 \cos(2\omega_a t) + V_3 \cos(3\omega_a t) + \cdots \quad (10.16)$$

The desired output is the $V_1 \cos(\omega_a t)$ term, which we call the **fundamental** component. The $V_0$ term represents a shift in the dc level (which does not appear at the load if it is ac coupled). In addition, terms at multiples of the input frequency have resulted from the second and higher power terms of the transfer characteristic. These terms are called **harmonic distortion**. The $2\omega_a$ term is called the **second harmonic**, the $3\omega_a$ term is the **third harmonic**, and so on. The higher order terms in the transfer characteristic given by Equation 10.14 produce the higher order harmonics. For example, the squared term produces the second harmonic. Similarly, the cubic term generates the third harmonic.

*For a sinewave input, nonlinear distortion produces output components having frequencies that are integer multiples of the input frequency.*

Harmonic distortion is objectionable in a wideband amplifier because the harmonics can fall in the frequency range of the desired signal. In an audio amplifier, harmonic distortion degrades the aesthetic qualities of the sound produced by the loudspeakers.

The **second-harmonic distortion factor** $D_2$ is defined as the ratio of the amplitude of the second harmonic to the amplitude of the fundamental. In equation form, we have

$$D_2 = \frac{V_2}{V_1} \quad (10.17)$$

where $V_1$ is the amplitude of the fundamental term of Equation 10.16 and $V_2$ is the amplitude of the second harmonic. Similarly, the third-harmonic distortion factor, and so on, are defined as

$$D_3 = \frac{V_3}{V_1} \quad D_4 = \frac{V_4}{V_1} \quad \cdots \tag{10.18}$$

Total harmonic distortion is a specification that indicates the degree of nonlinear distortion produced by an amplifier.

The **total harmonic distortion** (THD), denoted by $D$, is the ratio of the rms value of the sum of all the harmonic distortion terms to the rms value of the fundamental. The total harmonic distortion can be found from

$$D = \sqrt{D_2^2 + D_3^2 + D_4^2 + D_5^2 + \cdots} \tag{10.19}$$

We will often find THD expressed as a percentage. A well-designed audio amplifier might have a THD specification of 0.01 percent (i.e., $D = 0.0001$) at rated power output. (Some years ago, THD of 5 percent was typical for amplifiers found in inexpensive radios or phonographs.)

Notice that the THD specification of an amplifier depends on the amplitude of the output signal because the degree of nonlinearity of the transfer characteristic is amplitude dependent. Certainly, any amplifier eventually clips the output signal if the input signal becomes large enough. When severe clipping occurs, THD becomes large.

**Exercise 10.16**   A certain amplifier has a transfer characteristic given by

$$v_o = 100v_i + v_i^2$$

**a.** Find the THD rating of the amplifier for a sinusoidal input voltage $v_i(t) = \cos(\omega t)$.
**b.** Repeat for $v_i(t) = 5\cos(\omega t)$. [*Hint:* Use the fact that $\cos^2 x = 1/2 + (1/2)\cos(2x)$. This amplifier produces no third or higher harmonic distortion. Thus, $D_3 = 0$, $D_4 = 0$, etc.]
**Answer**   **a.** $D = 0.005$; **b.** $D = 0.025$. Notice that the THD is larger for the larger input amplitude.                                                                                    □

## 10.11   DIFFERENTIAL AMPLIFIERS

A differential amplifier has two input terminals: an inverting input and a noninverting input.

Until now, we have considered amplifiers that have only one input source. Now we consider **differential amplifiers**, which have two input sources, as shown in Figure 10.35. An ideal differential amplifier produces an output voltage proportional to the difference between the input voltages. This is demonstrated by

$$v_o(t) = A_d[v_{i1}(t) - v_{i2}(t)]$$
$$= A_d v_{i1}(t) - A_d v_{i2}(t) \tag{10.20}$$

Ideally, a differential amplifier produces an output that is proportional to the difference between two input signals.

Notice that gain is positive for the voltage applied to terminal 1 and negative for the voltage applied to terminal 2. Therefore, terminal 2 is called an **inverting input**, and terminal 1 is called a **noninverting input**. Inverting input terminals are marked with a − (minus) sign, and noninverting input terminals are marked with a + (plus) sign, as indicated in Figure 10.35.

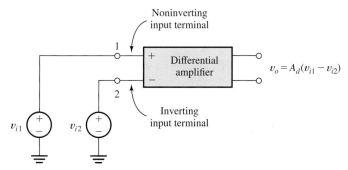

Figure 10.35 Differential amplifier with input sources.

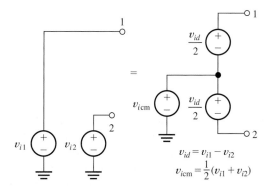

Figure 10.36 The input sources $v_{i1}$ and $v_{i2}$ can be replaced by the equivalent sources $v_{icm}$ and $v_{id}$.

The difference between the input voltages, known as the **differential signal**, is given by

$$v_{id} = v_{i1} - v_{i2} \tag{10.21}$$

We refer to the gain $A_d$ as the **differential gain**. Thus, we can write the output of the ideal differential amplifier as

$$v_o = A_d v_{id} \tag{10.22}$$

The **common-mode signal** $v_{icm}$ is the average of the input voltages, given by

$$v_{icm} = \frac{1}{2}(v_{i1} + v_{i2}) \tag{10.23}$$

The original input sources $v_{i1}$ and $v_{i2}$ can be replaced by the equivalent system of sources shown in Figure 10.36. Thus, we can consider the inputs to the differential amplifier to be the differential signal $v_{id}$ and the common-mode signal $v_{icm}$.

Sometimes, we have a small differential signal that we wish to amplify, but a large common-mode signal is also present that is of no interest. A good example of this is in recording the electrocardiogram (ECG) of a patient. Imagine a patient lying on a bed insulated from electrical ground as shown in Figure 10.37. If electrodes are placed in contact with each of the patient's arms, a differential signal generated by the patient's heart appears between the electrodes, which is the signal of interest to the cardiologist. Also, we often find a large 60-Hz common-mode signal present between each electrode and the local power-system ground. This occurs because patients are connected to the 60-Hz power line by very small

Many applications exist in which a small differential signal is important and a strong interfering common-mode signal is present.

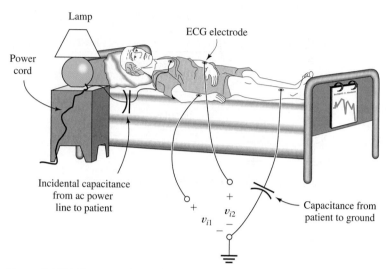

**Figure 10.37** Electrocardiographs encounter large 60-Hz common-mode signals.

incidental capacitances between their bodies and the power line. Similar small capacitances connect the patient to ground. That network of capacitances forms a voltage-divider network, so the patient's body is at a significant fraction of the power-line voltage with respect to ground. (You may have observed this 60-Hz common-mode signal in the laboratory if you have touched the input terminals of a high-input-impedance ac meter or oscilloscope.) Thus, at the input to the electrocardiograph amplifier, there exists a differential signal of about 1 mV and a 60-Hz common-mode signal of several tens of volts. Ideally, the electrocardiograph should respond only to the differential signal.

Interfacing sensors to computers is pervasive in all fields of science and engineering. The problem of large common-mode interference from the power line is also very common. Hence, it is important for you to have a good grasp of these concepts.

### Common-Mode Rejection Ratio

Unfortunately, real differential amplifiers respond to both the common-mode signal and the differential signal. Recall that the gain for the differential signal is denoted as $A_d$. If we denote the gain for the common-mode signal as $A_{cm}$, the output voltage of a real differential amplifier is given by

$$v_o = A_d v_{id} + A_{cm} v_{icm} \tag{10.24}$$

For well-designed differential amplifiers, the differential gain $A_d$ is much larger than the common-mode gain $A_{cm}$. A quantitative specification is the **common-mode rejection ratio** (CMRR), which is defined as the ratio of the magnitude of the differential gain to the magnitude of the common-mode gain. Often, CMRR is expressed in decibels as

Common-mode rejection ratio (CMRR) is a specification that indicates how well the common-mode signal is rejected relative to the differential signal.

$$\text{CMRR} = 20 \log \frac{|A_d|}{|A_{cm}|} \tag{10.25}$$

The CMRR of an amplifier is generally a function of frequency, becoming lower as frequency is raised. At 60 Hz, a CMRR of 120 dB is considered good.

## Example 10.11  Determination of the Minimum CMRR Specification

Find the minimum CMRR for an electrocardiograph amplifier if the differential gain is 1000, the desired differential input signal has a peak amplitude of 1 mV, the common-mode signal is a 100-V-peak 60-Hz sine wave, and it is desired that the output contain a peak common-mode contribution that is 1 percent or less of the peak output caused by the differential signal.

**Solution**   Since the peak differential input is 1 mV and the differential gain is 1000, the peak output of the desired signal is 1 V. To meet the specification required, the common-mode output signal must have a peak value of 0.01 V or less. Thus, the common-mode gain is

$$A_{cm} = \frac{0.01 \text{ V}}{100 \text{ V}} = 10^{-4}$$

(Therefore, the common-mode gain actually amounts to attenuation.) Now, we can find the CMRR by application of Equation 10.25:

$$CMRR = 20 \log \frac{|A_d|}{|A_{cm}|} = 20 \log \frac{1000}{10^{-4}} = 140 \text{ dB}$$

Hence, an electrocardiograph requires a very good CMRR specification.  ■

Perhaps, we should note in passing that another simpler—but dangerous—approach exists to solving the common-mode problem for the electrocardiograph: It is to short out the common-mode signal by attaching another electrode to the patient and connecting it to the power-system ground. This would reduce the 60-Hz interference to a very low level, so an amplifier with a much less stringent CMRR specification could be used. However, once the patient is in good electrical contact with the power-system ground, any contact with power-line voltages is potentially fatal. That is particularly true if the patient is too ill to protest. Even small currents imperceptible under ordinary circumstances can be fatal if conducted directly to the patient's heart. Such small currents can be conducted through other medical instrumentation or even through a surgeon's hands. *Keeping the patient isolated from ground provides some measure of protection from such problems.*

## Measurement of CMRR

Measurements to find the CMRR of an amplifier are fairly straightforward. We must find both the differential and common-mode gains. The common-mode gain is found by connecting the input terminals of the amplifier together and attaching a test source, as shown in Figure 10.38. Notice that with the input terminals of the amplifier connected together, the differential signal $v_{id}$ is zero, and any output is caused by the common-mode signal applied to both input terminals by the test source. Thus, we measure both the input voltage and output voltage, and then we compute their ratio to find the common-mode gain.

In theory, to apply a pure differential signal, we must provide two sources out of phase with each other at the amplifier input terminals, as shown in Figure 10.39(a). However, since the common-mode gain is usually much smaller than the differential gain, only a small error results if a single source is used, as shown in Figure 10.39(b). [In Figure 10.39(b), the input contains both a differential signal $v_{id}$ and a common-mode signal $v_{icm} = v_{id}/2$.] In any case, the differential gain is found by taking the

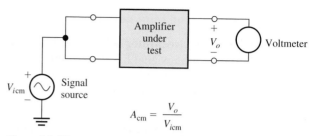

Figure 10.38 Setup for measurement of common-mode gain.

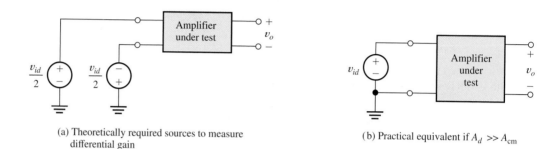

(a) Theoretically required sources to measure differential gain

(b) Practical equivalent if $A_d \gg A_{cm}$

Figure 10.39 Setup for measuring differential gain. $A_d = v_o/v_{id}$.

ratio of the output voltage to the input voltage when the common-mode voltage is zero or negligible. Finally, the CMRR is found by taking the ratio of the gains.

**Exercise 10.17**　A certain amplifier has a differential gain $A_d = 50{,}000$. If the input terminals are connected together and a 1-V signal is applied to them, an output signal of 0.1 V results. What is the common-mode gain of the amplifier and the CMRR, both expressed in dB?

**Answer**　$A_{cm} = -20$ dB, CMRR $= 114$ dB.　　□

**Exercise 10.18**　A certain amplifier has $v_o = A_1 v_{i1} - A_2 v_{i2}$. **a.** Assume that $v_{i1} = 1/2$ and $v_{i2} = -1/2$. Find $v_{id}$ and $v_{icm}$. Find $v_o$ and $A_d$ in terms of $A_1$ and $A_2$. **b.** Assume that $v_{i1} = 1$ and $v_{i2} = 1$. Find $v_{id}$ and $v_{icm}$. Find $v_o$ and $A_{cm}$ in terms of $A_1$ and $A_2$. **c.** Use the results of parts (a) and (b) to find an expression for the CMRR in terms of $A_1$ and $A_2$. Evaluate the CMRR if $A_1 = 100$ and $A_2 = 101$.

**Answer**　**a.** $v_{id} = 1$, $v_{icm} = 0$, $v_o = A_d = (1/2)A_1 + (1/2)A_2$; **b.** $v_{id} = 0$, $v_{icm} = 1$, $v_o = A_{cm} = A_1 - A_2$; **c.** CMRR $= 20 \log |(A_1 + A_2)/2(A_1 - A_2)| = 40.0$ dB.　□

## 10.12　OFFSET VOLTAGE, BIAS CURRENT, AND OFFSET CURRENT

Until now, we have assumed that the output of an amplifier is zero if the input sources are zero, but in real direct-coupled amplifiers that is not true. A dc output voltage is usually observed even if the input sources are zero. This is caused by undesired imbalances in the internal component values of the amplifier and because, in some types of amplifier circuits, it is necessary for the external input circuits to supply small dc currents to the amplifier input terminals. Assuming a differential

## PRACTICAL APPLICATION    10.1

### Electronic Stud Finder

When we want to hang a heavy picture or shelf on a wall, we often need to locate a wood stud capable of bearing the weight of the picture or shelf. Usually, this can be accomplished with an electronic stud finder.

A simple electronic stud finder, illustrated in Figures PA10.1 and PA10.2, can be designed by using several of the electrical-engineering concepts discussed in this book. First, as discussed in Section 3.3, the capacitance between metal plates depends on the dielectric constants of the materials surrounding the plates. For the configuration shown in Figure PA10.1, the capacitance between plates $A$ and $B$ is less than the capacitance between $B$ and $C$ because the dielectric constant of the wood stud is higher than that of air. As the stud finder moves to the right and becomes centered on the stud, the capacitances become equal. Then, as the stud finder moves slightly past center, the capacitance between $A$ and $B$ becomes higher.

The second concept used in the stud finder is an ac bridge circuit similar to the Wheatstone bridge of Section 2.8. As shown in Figure PA10.2, the variable capacitances are connected in the bridge with two equal resistances and an ac source. Recall that the voltage between nodes $A$ and $C$ becomes zero when the bridge is balanced, which, in this case, occurs when the capacitances are equal.

The third concept is the use of a high-gain differential amplifier (such as those discussed in Section 10.11) and a beeper (which is a simple loudspeaker) to form a sensitive detector for the bridge circuit.

When the stud finder is over the stud, but not centered, the capacitances are unequal and the bridge is not balanced. Then, an ac voltage appears as the input to the differential amplifier, and a sound is emitted from the beeper. When the stud finder is centered, the bridge becomes balanced, and the sound disappears. Thus, by moving the stud finder over the surface of a wall, we can easily locate the center lines of the studs.

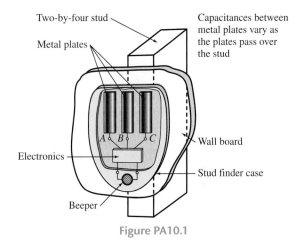

**Figure PA10.1**

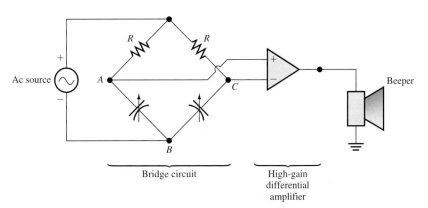

**Figure PA10.2**

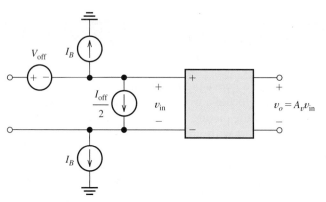

**Figure 10.40** Differential amplifier, including dc sources to account for the dc output that exists even when the input signals are zero.

amplifier, these effects can be modeled by the addition of three current sources and one voltage source to the input terminals of an otherwise ideal amplifier. These sources are shown in Figure 10.40.

The two current sources labeled $I_B$ are known as **bias-current sources**. These sources account for the small dc currents drawn by the internal amplifier circuitry through the input terminals. The bias currents have the same value and direction (either both flow toward the amplifier input terminals or both flow toward ground). The value of the bias current $I_B$ is a function of temperature, and it varies from unit to unit of a given amplifier type.

The current $I_{off}$ is called **offset current**. Offset current arises from incidental imbalances in the internal components of the amplifier. The offset current value is usually somewhat smaller than the bias current. The direction of the offset current is unpredictable—it can flow toward either input terminal. The direction of flow may be different from unit to unit of a given amplifier model. Notice that the offset current source (Figure 10.40) has a value of $I_{off}/2$.

The voltage source $V_{off}$ in series with the input terminals is called an **offset voltage**. Like the offset current, it is caused by internal circuit imbalances. The value of the offset voltage is usually a function of temperature. Furthermore, it changes in value and polarity from unit to unit. The offset voltage source can be placed in series with either input terminal.

Real differential amplifiers suffer from imperfections that can be modeled by several dc sources: two bias-current sources, an offset current source, and an offset voltage source. The effect of these sources is to add a (usually undesirable) dc term to the ideal output.

## Minimizing the Effect of Bias Current

The effects of bias current can be mitigated by ensuring that the Thévenin imped-ances of the circuits connected to the input terminals are the same. (Recall from Section 2.6 that to find the Thévenin impedance of a network, we zero the indepen-dent sources and then compute the impedance of the network. Independent volt-age sources are zeroed by replacing them with short circuits, whereas independent current sources are replaced by open circuits.) Figure 10.41(a) shows a differen-tial amplifier with source resistances and bias-current sources. Each current source can be converted to a voltage source in series with the corresponding resistance as shown in Figure 10.41(b). If the source resistances are equal, these voltages are equal, so there is no differential signal supplied to the amplifier. Assuming that the com-mon-mode gain is zero, the resulting output voltage is zero.

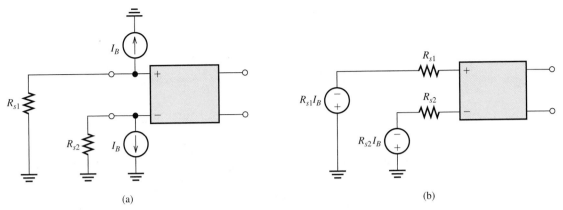

(a)                                           (b)

**Figure 10.41** The effects of the bias-current sources cancel if $R_{s1} = R_{s2}$.

---

**Example 10.12**   **Calculation of Worst-Case DC Output Voltage**

A certain direct-coupled differential amplifier has a differential voltage gain of 100, an input impedance of 1 M$\Omega$, an input bias current of 200 nA, a maximum offset current of 80 nA, and a maximum offset voltage of 5 mV. Compute the worst-case output voltage if the amplifier input terminals are connected to ground through 100-k$\Omega$ source resistances.

**Solution**   The circuit, including the source resistances, is shown in Figure 10.42(a). Since the circuit is linear, we can use superposition, considering each source separately. Because the impedances for the two inputs are the same, the effects of the bias currents balance and, therefore, can be ignored.

> Because the circuit is linear, we use superposition, thereby dividing the problem into several relatively simple problems.

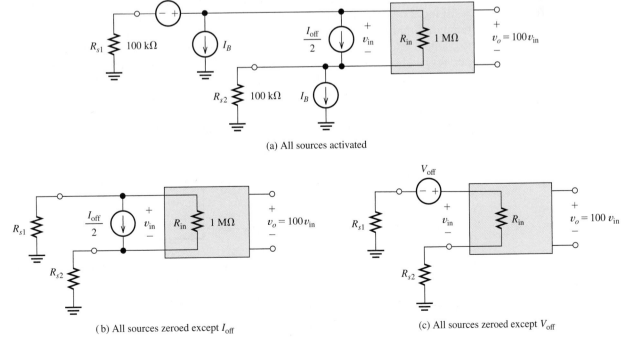

(a) All sources activated

(b) All sources zeroed except $I_{off}$

(c) All sources zeroed except $V_{off}$

**Figure 10.42** Amplifier of Example 10.12.

The offset current flows through the parallel combination of $R_{in}$ and the sum of the source resistances as shown in Figure 10.42(b). Thus, the differential input voltage arising from the offset current has a maximum value given by

$$V_{Ioff} = \frac{I_{off}}{2} \frac{R_{in}(R_{s1} + R_{s2})}{R_{in} + R_{s1} + R_{s2}} = 6.67 \text{ mV}$$

The circuit with only the offset voltage source activated is shown in Figure 10.42(c). The differential input voltage resulting from the offset voltage is found by noting that part of the input offset source voltage appears across the input terminals and the rest appears across $R_{s1}$ and $R_{s2}$. The portion across the input terminals can be computed by use of the voltage-divider principle as

$$V_{Voff} = V_{off} \frac{R_{in}}{R_{in} + R_{s1} + R_{s2}} = 4.17 \text{ mV}$$

Multiplying by the amplifier gain, we find that the maximum output voltage caused by the offset current source is 0.667 V and the maximum output voltage caused by the offset voltage source is 0.417 V. These voltages are maximum values and they can have either polarity, so the total output voltage can range between $-1.084$ and $+1.084$ V. ∎

## Balancing Circuits

Balancing circuits can be used to cancel the dc offset added to the output signal by amplifier imperfections.

The effects of the offset current and voltage can be canceled by the use of a balancing circuit such as that shown in Figure 10.43. The resistors $R_1$ and $R_2$ on each side of the potentiometer form voltage dividers that supply small voltages to opposite ends of the potentiometer positive on one end and negative on the other. In use, the potentiometer is simply adjusted so that the amplifier output is zero if the input from the signal source is zero.

Even if such a balancing circuit is used, it is good practice to maintain equal resistances from both input terminals to ground, because bias current varies with temperature. Equal resistances provide balancing for bias current, independent of its value. Unfortunately, the offset current and voltage vary with temperature, so perfect balance at all temperatures is not possible with a fixed circuit.

In principle, the dividers ($R_1$ and $R_2$) could be left out of the circuit of Figure 10.43, and the ends of the potentiometer could be connected directly to the power-supply voltages. However, the range of adjustment would then be much larger than necessary, and the correct adjustment would be very difficult to achieve.

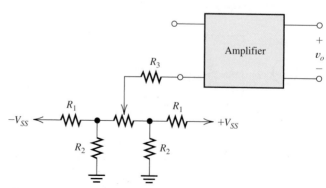

**Figure 10.43** Network that can be adjusted to cancel the effects of offset and bias sources.

Some amplifiers provide separate terminals for attachment of balancing circuits so that the signal input terminals are not encumbered.

**Exercise 10.19**  A certain direct-coupled differential amplifier has a differential voltage gain of 500, an input impedance of 100 kΩ, an input bias current of 400 nA, a maximum offset current of 100 nA, and a maximum offset voltage of 10 mV. Compute the worst-case output voltages if the amplifier input terminals are connected to ground through 50-kΩ resistances.
**Answer**  $v_o$ ranges from −3.75 to +3.75 V.  □

**Exercise 10.20**  Repeat Exercise 10.19 if the inverting input terminal is grounded directly and the noninverting input is connected to ground through a 50-kΩ resistance.
**Answer**  $v_o$ ranges from +2.5 V to +10.84 V.  □

## Summary

1. The purpose of an amplifier is to deliver a larger signal to a load than is available from the signal source.

2. Amplifiers are characterized by their input impedance, output impedance, and a gain parameter.

3. Inverting amplifiers have negative voltage gain, so the output waveform is an inverted version of the input waveform. Noninverting amplifiers have positive voltage gain.

4. Loading effects result from voltage drops across the internal source impedance and across the output impedance of the amplifier.

5. In a cascade connection, the output of each amplifier is connected to the input of the next amplifier.

6. The efficiency of an amplifier is the percentage of the supply power that is converted into output signal power.

7. Several models are useful in characterizing amplifiers. They are the voltage-amplifier model, the current-amplifier model, the transconductance-amplifier model, and the transresistance-amplifier model.

8. According to their input and output impedances, ideal amplifiers can be classified into four types: the ideal voltage amplifier, the ideal current amplifier, the ideal transconductance amplifier, and the ideal transresistance amplifier. The best amplifier type to select depends on the application.

9. Amplifiers may be direct coupled, in which case constant gain extends to dc. On the other hand, amplifiers may be ac coupled, in which case the gain falls off at low frequencies, reaching zero gain at dc. Gain magnitude falls to zero at sufficiently high frequencies for all amplifiers.

10. Linear distortion can be either amplitude distortion or phase distortion. Amplitude distortion occurs if the gain magnitude is different for various components of the input signal. Phase distortion occurs if amplifier phase shift is not proportional to frequency.

11. Amplifier pulse response is characterized by rise time, overshoot, ringing, and tilt.

12. Nonlinear distortion occurs if the transfer characteristic of an amplifier is not straight. Assuming a sinusoidal input signal, nonlinear distortion causes harmonics to appear in the output. The total harmonic distortion rating of an amplifier indicates the degree of nonlinear distortion.

13. A differential amplifier ideally responds only to the difference between its two input signals (i.e., the differential input signal).

14. The common-mode input is the average of the two inputs to a differential amplifier. CMRR is the ratio of the differential gain to the common-mode gain. CMRR is an important specification for many instrumentation applications.

15. Dc offset is the addition of a dc term to the signal being amplified. It is the result of bias current, offset current, and offset voltage, and it can be canceled by use of a properly designed balance circuit.

## Problems

### Section10.1: Basic Amplifier Concepts

**P10.1.** What are two causes of "loading effects" in an amplifier circuit?

**P10.2.** Explain how an inverting amplifier differs from a noninverting amplifier.

**P10.3.** Draw the voltage-amplifier model and label its elements.

**\*P10.4.** The output voltage $v_o$ of the circuit of Figure P10.4 is 100 mV with the switch closed. With the switch open, the output voltage is 50 mV. Find the input resistance of the amplifier.

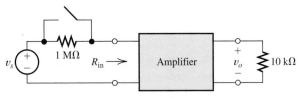

**Figure P10.4**

**\*P10.5.** A certain amplifier operating with a 100-$\Omega$ load has a voltage gain of 50 and a power gain of 5000. Determine the current gain and input resistance of the amplifier.

**\*P10.6.** A signal source with an open-circuit voltage of $V_s = 2$ mV rms and an internal resistance of 50 k$\Omega$ is connected to the input terminals of an amplifier having an open-circuit voltage gain of 100, an input resistance of 100 k$\Omega$, and an output resistance of 4 $\Omega$. A 4-$\Omega$ load is connected to the output terminals. Find the voltage gains $A_{vs} = V_o/V_s$ and $A_v = V_o/V_i$. Also, find the power gain and current gain.

**\*P10.7.** An ideal ac current source is applied to the input terminals of an amplifier, and the amplifier output voltage is 2 V rms. Then, a 2-k$\Omega$ resistance is placed in parallel with the current source and the amplifier input terminals, and the output voltage is 1.5 V rms. Determine the input resistance of the amplifier.

**P10.8.** An amplifier with $R_i = 12$ k$\Omega$, $R_o = 1$ k$\Omega$, and $A_{voc} = -10$ is operated with a 1-k$\Omega$ load. A source having a Thévenin resistance of 4 k$\Omega$ and a short-circuit current of $2\cos(200\pi t)$ mA is connected to the input

terminals. Determine the output voltage as a function of time and the power gain.

**P10.9.** An amplifier has an open-circuit voltage gain of 100. With a 10-k$\Omega$ load connected, the voltage gain is found to be only 80. Find the output resistance of the amplifier.

**P10.10.** An amplifier having $R_i = 1$ M$\Omega$, $R_o = 1$ k$\Omega$, and $A_{voc} = -10^4$ is operated with a 1-k$\Omega$ load. A source having a Thévenin resistance of 2 M$\Omega$ and an open-circuit voltage of $3\cos(200\pi t)$ mV is connected to the input terminals. Determine the output voltage as a function of time and the power gain.

**P10.11.** The current gain of an amplifier is 500, the load resistance is 100 $\Omega$, and the input resistance of the amplifier is 1 M$\Omega$. Determine the voltage gain and power gain under these conditions.

**P10.12.** A certain amplifier has an open-circuit voltage gain of unity, an input resistance of 1 M$\Omega$, and an output resistance of 100 $\Omega$. The signal source has an internal voltage of 5 V rms and an internal resistance of 100 k$\Omega$. The load resistance is 50 $\Omega$. If the signal source is connected to the amplifier input terminals and the load is connected to the output terminals, find the voltage across the load and the power delivered to the load. Next, consider connecting the load directly across the signal source without the amplifier, and again find the load voltage and power. Compare the results. What do you conclude about the usefulness of a unity-gain amplifier in delivering signal power to a load?

**P10.13.** Suppose we have a resistive load that varies from 5 k$\Omega$ to 10 k$\Omega$. We connect this load to an amplifier, and we need the voltage across the load to vary by less than 1 percent with variations in the load resistance. What parameter of the amplifier is important in this situation? What range of values is allowed for the parameter?

**P10.14.** A certain amplifier has a voltage gain of 0.1. However, the power gain is 10. How is this possible? What is the value of the current gain? How does the load resistance compare with the input resistance of the amplifier?

**P10.15.** An amplifier has an open-circuit voltage gain of 1000, an input resistance of 20 k$\Omega$, and an output resistance of 2 $\Omega$. A signal source with an internal resistance of 10 k$\Omega$ is connected to the input terminals of the amplifier. An 8-$\Omega$ load is connected to the output terminals. Find the voltage gains $A_{vs} = V_o/V_s$ and $A_v = V_o/V_i$. Also, find the power gain and current gain.

**P10.16.** Suppose we have a sensor, with a Thévenin resistance that varies from zero to 10 k$\Omega$, connected to the input of an amplifier. We want the output voltage of the amplifier to vary by less than 2 percent with changes in the Thévenin resistance of the sensor. What parameter of the amplifier is important in this situation? What range of values is allowed for the parameter?

**P10.17.** A certain amplifier operates with a resistive load. The current gain and the voltage gain are equal. What can you say about the input resistance and the load resistance?

**Section10.2:** Cascaded Amplifiers

*__P10.18.__ Amplifiers having $A_{voc} = 10$, $R_i = 2$ k$\Omega$, and $R_o = 2$ k$\Omega$ are available. How many of these amplifiers must be cascaded to attain a voltage gain of at least 1000 when operating with a 1-k$\Omega$ load?

*__P10.19.__ Three amplifiers with the following characteristics are cascaded in the order 1, 2, 3.

Amplifier 1: $A_{voc1} = 100$, $R_{i1} = 2$ k$\Omega$,
$R_{o1} = 1$ k$\Omega$
Amplifier 2: $A_{voc2} = 200$, $R_{i2} = 4$ k$\Omega$,
$R_{o2} = 2$ k$\Omega$
Amplifier 3: $A_{voc3} = 300$, $R_{i3} = 6$ k$\Omega$,
$R_{o3} = 3$ k$\Omega$

Find the parameters for the simplified model of the cascaded amplifier.

**P10.20.** Three identical amplifiers having $A_{voc} = 25$, $R_i = 2$ k$\Omega$, and $R_o = 3$ k$\Omega$ are cascaded. Determine the input resistance, the open-circuit voltage gain, and the output resistance of the cascade.

**P10.21.** Given that the amplifiers having the characteristics shown in Table P10.21 are cascaded in the order $A - B$, find the input impedance, output impedance, and open-circuit voltage gain of the cascade. Repeat when the order is $B - A$.

**Table P10.21** Amplifier Characteristics

| Amplifier | Open-Circuit Voltage Gain | Input Resistance | Output Resistance |
|---|---|---|---|
| A | 100 | 3 k$\Omega$ | 400 $\Omega$ |
| B | 500 | 1 M$\Omega$ | 2 k$\Omega$ |

**P10.22.** Draw the cascade connection of two amplifiers. Write an expression for the open-circuit voltage gain of the cascade connection in terms of the open-circuit voltage gains and impedances of the individual amplifiers.

**P10.23.** Three amplifiers with the following characteristics are cascaded in the order 1, 2, 3.

Amplifier 1: $A_{voc1} = 300$, $R_{i1} = 6$ k$\Omega$,
$R_{o1} = 3$ k$\Omega$
Amplifier 2: $A_{voc2} = 200$, $R_{i2} = 4$ k$\Omega$,
$R_{o2} = 2$ k$\Omega$
Amplifier 3: $A_{voc3} = 100$, $R_{i3} = 2$ k$\Omega$,
$R_{o3} = 1$ k$\Omega$

Find the parameters for the simplified model of the cascaded amplifier.

**Section10.3:** Power Supplies and Efficiency

*__P10.24.__ Find the net power delivered to the amplifier by the three dc supply voltages shown in Figure P10.24.

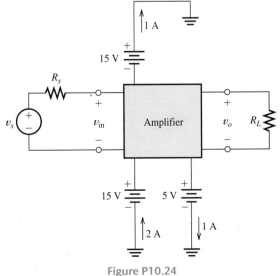

Figure P10.24

**P10.25.** Under high-signal test conditions, a certain audio amplifier supplies a 24 V rms 1-kHz sinusoidal voltage to an 8-$\Omega$ load. The power supply delivers 4 A at a voltage of 50 V to the amplifier. The signal power supplied by the input source is negligible. Determine the efficiency and the power dissipated in the amplifier.

**P10.26.** An amplifier operates from a 12-V power supply that supplies a current of 1.5 A. The input signal current is 1 $\mu$A rms, and the input resistance is 100 k$\Omega$. The amplifier delivers 10 V rms to a 10-$\Omega$ load. Determine the power dissipated in the amplifier and the efficiency of the amplifier.

**P10.27.** A certain amplifier has an input voltage of 100 mV rms, an input resistance of 100 k$\Omega$, and produces an output of 10 V rms across an 8-$\Omega$ load resistance. The power supply has a voltage of 15 V and delivers an average current of 2 A. Find the power dissipated in the amplifier and the efficiency of the amplifier.

**P10.28.** Define the efficiency of a power amplifier. What is dissipated power in an amplifier? What form does dissipated power take?

**P10.29.** Two amplifiers are cascaded. The first has supply power of 2 W, an input resistance of 1 M$\Omega$, and an input voltage of 2 mV rms. The second has a supply power of 22 W, a load resistance of 8 $\Omega$, and output voltage of 12 V rms. Determine the overall power gain, dissipated power, and efficiency.

**Section10.4:** Additional Amplifier Models

**\*P10.30.** An amplifier has an input resistance of 1 k$\Omega$, an output resistance of 200 $\Omega$, and a short-circuit transconductance gain of 0.5 S. Determine the open-circuit voltage gain, the short-circuit current gain, and the open-circuit transresistance gain.

**\*P10.31.** Amplifier $A$ has an input resistance of 1 M$\Omega$, an output resistance of 200 $\Omega$, and an open-circuit transresistance gain of 100 M$\Omega$. Amplifier $B$ has an input resistance of 50 $\Omega$, an output impedance of 500 k$\Omega$, and a short-circuit current gain of 100. Find the voltage amplifier model for the cascade of $A$ followed by $B$. Then, determine the corresponding transconductance amplifier model.

**\*P10.32.** An amplifier has an input resistance of 10 k$\Omega$, an output resistance of 2 k$\Omega$, and an open-circuit transresistance gain of 200 k$\Omega$. Determine the open-circuit voltage gain, the short-circuit current gain, and the short-circuit transconductance gain.

**\*P10.33.** An amplifier has an input resistance of 20 $\Omega$, an output resistance of 10 $\Omega$, and a short-circuit current gain of 3000. The signal source has an internal voltage of 100 mV rms and an internal impedance of 200 $\Omega$. The load is a 5-$\Omega$ resistance. Find the current gain, voltage gain, and power gain of the amplifier. If the power supply has a voltage of 12 V and supplies an average current of 2 A, find the power dissipated in the amplifier and the efficiency.

**\*P10.34.** An amplifier has $R_i = 100$ $\Omega$, $R_o = 1$ k$\Omega$, and $R_{moc} = 10$ k$\Omega$. Determine the values (including units) of $A_{voc}$, $G_{msc}$, and $A_{isc}$ for this amplifier.

**P10.35.** Draw a voltage-amplifier model. Is the gain parameter measured under open-circuit or short-circuit conditions? Repeat for a current amplifier model, a transresistance-amplifier model, and a transconductance-amplifier model.

**P10.36.** **a.** Which amplifier model contains a current-controlled voltage source? **b.** A current-controlled current source? **c.** A voltage-controlled current source?

**P10.37.** Amplifier $A$ has an input resistance of 50 $\Omega$, an output impedance of 500 k$\Omega$, and a short-circuit current gain of 100. Amplifier $B$ has an input resistance of 1 M$\Omega$, an output resistance of 200 $\Omega$, and an open-circuit transresistance gain of 100 M$\Omega$. Find the voltage amplifier model for the cascade of $A$ followed by $B$. Then, determine the corresponding transconductance amplifier model.

**P10.38.** An amplifier has an input resistance of 100 $\Omega$, an output resistance of 10 $\Omega$, and a short-circuit current gain of 500. Draw the voltage amplifier model for the amplifier, including numerical values for all parameters. Repeat for the transresistance and transconductance models.

**P10.39.** An amplifier has $R_i = 10$ k$\Omega$, $R_o = 100$ $\Omega$, and $G_{msc} = 0.5$ S. Determine the values (including units) of $A_{voc}$, $R_{moc}$, and $A_{isc}$ for this amplifier.

**P10.40.** An amplifier with $R_i = 2\,k\Omega$, $R_o = 500\,\Omega$, and $R_{moc} = -10^7\,\Omega$ is operated with a 1-k$\Omega$ load. A source having a Thévenin resistance of $1\,k\Omega$ and an open-circuit voltage of $2\cos(200\pi t)$ mV is connected to the input terminals. Determine the output voltage as a function of time and the power gain.

**P10.41.** An amplifier has an open-circuit transresistance gain of 200 $\Omega$, a short-circuit transconductance gain of 0.5 S, and a short-circuit current gain of 50. Determine the input resistance, the output resistance, and the open-circuit voltage gain.

**P10.42.** An amplifier has a short-circuit current gain of 10. When operated with a 50-$\Omega$ load, the current gain is 8. Find the output resistance of the amplifier.

**P10.43.** An amplifier has $R_i = 2\,k\Omega$, $R_o = 300\,\Omega$, and $A_{isc} = 200$. Determine the values (including units) of $A_{voc}$, $R_{moc}$, and $G_{msc}$ for this amplifier.

**P10.44.** An amplifier has an open-circuit voltage gain of 100, a short-circuit transconductance gain of 0.2 S, and a short-circuit current gain of 50. Determine the input resistance, the output resistance, and the open-circuit transresistance gain.

**Section10.5: Importance of Amplifier Impedances in Various Applications**

**P10.45.** Give an example of a situation in which a specific input impedance is needed for an amplifier.

**P10.46.** Suppose we have a voltage source $v(t) = V_{dc} + V_m\cos(\omega t)$ connected to the input terminals of an amplifier. The load is a non-linear device such as an LED. **a.** What output impedance is needed for the amplifier if we need the current through the load to be proportional to $v(t)$? **b.** If we need the voltage across the load to be proportional to $v(t)$?

**P10.47.** We need an amplifier to supply a constant signal to each of a variable number of loads connected in parallel. What output impedance is needed in this situation? Why? What if the loads are connected in series?

**P10.48.** Give an application in which an amplifier with very low input impedance is needed.

**P10.49.** Describe an application in which an amplifier with very high input impedance is needed.

**Section10.6: Ideal Amplifiers**

*__P10.50.__ The amplifier shown in Figure P10.50 has an input resistance of 1000 $\Omega$, an output impedance of 20 $\Omega$, and an open-circuit transresistance gain of 10 k$\Omega$. Find the resistance $R_x = v_x/i_x$ seen from the input terminals.

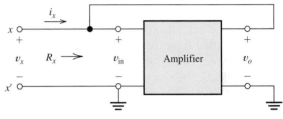

Figure P10.50

*__P10.51.__ Suppose we have a two-stage cascaded amplifier with an ideal transconductance amplifier as the first stage and an ideal transresistance amplifier as the second stage. What type of amplifier results and what is its gain in terms of the gains of the two stages? Repeat for the amplifiers cascaded in the opposite order.

*__P10.52.__ In instrumenting a physics experiment, we need to record the open-circuit voltage of a certain sensor. The voltage needs to be amplified by a factor of 1000 and applied to a variable load resistance. What type of ideal amplifier is needed? Justify your answer.

*__P10.53.__ The output terminals of an ideal transresistance amplifier are connected to the input terminals of an ideal transconductance amplifier. What type of ideal amplifier results? Determine its gain parameter in terms of the gain parameters of the separate stages.

**P10.54.** Give the input and output impedances for an ideal-voltage amplifier. Repeat for each of the other ideal amplifier types.

**P10.55.** An ideal transconductance amplifier having a short-circuit transconductance gain of 0.1 S is connected as shown in Figure P10.55. Find the resistance $R_x = v_x/i_x$ seen from the input terminals.

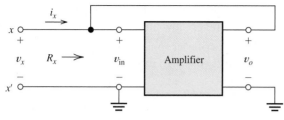

**Figure 10.55**

**P10.56.** The output terminals of an ideal voltage amplifier are connected to the input terminals of an ideal transconductance amplifier. What type of ideal amplifier results? Determine its gain parameter in terms of the gain parameters of the separate stages.

**P10.57.** In a certain application, an amplifier is needed to sense the open-circuit voltage of a source and force current to flow through a load. The source resistance and load resistance are variable. The current delivered to the load is to be nearly independent of both the source resistance and load resistance. What type of ideal amplifier is needed? If the source resistance increases from $1\,k\Omega$ to $2\,k\Omega$ and it causes a 1 percent decrease in load current, what is the value of the input resistance? If the load resistance increases from 100 to $300\,\Omega$ and this causes a 1 percent decrease in load current, what is the value of the output resistance?

**P10.58.** An amplifier has an input resistance of $1\,\Omega$, an output resistance of $1\,\Omega$, and an open-circuit voltage gain of 10. Classify this amplifier as an approximate ideal type and find the corresponding gain parameter. In deciding on an amplifier classification, assume that the source and load impedances are on the order of $1\,k\Omega$.

**P10.59.** In recording automotive emissions, we need to sense the short-circuit current of a chemical sensor that has a variable Thévenin impedance. A voltage that is proportional to the current must be applied to the input of a data-acquisition module. What type of ideal amplifier is needed? Justify your answer.

**P10.60.** What type of ideal amplifier is needed if we need to sense the short-circuit current of a sensor and drive a proportional current through a variable load? Explain your answer.

**P10.61.** An amplifier is needed as a part of a system for documentation of voltages in the earth created by an electrical power distribution system. Voltage waveforms occurring between probes to be placed in the earth are to be amplified before being applied to the analog-to-digital converter (ADC) inputs of laptop computers. The internal impedance of the probe can be as high as $10\,k\Omega$ in dry sand or as low as $10\,\Omega$ in muck. Because several different models of ADCs are to be used in the project, the load impedance for the amplifier varies from $10\,k\Omega$ to $1\,M\Omega$. Nominally, the voltage applied to the ADC is required to be 1000 times the open-circuit voltage of the probe $\pm 3$ percent. What type of ideal amplifier is best suited for this application? Using your best judgment, find the specifications for the impedances and gain parameter of this amplifier.

**P10.62.** We need to design an amplifier for use in recording the short-circuit current of experimental electrochemical cells versus time. (For this purpose, a short circuit is any resistance less than $10\,\Omega$.) The amplifier output is to be applied to a strip-chart recorder that deflects $1\,cm\ \pm 1$ percent for each volt applied. The input resistance of the recorder is unknown and likely to be variable, but it is greater than $10\,k\Omega$. A deflection of $1\,cm$ per milliampere of cell current with an accuracy of about $\pm 3$ percent is desired. What type of ideal amplifier is best suited for this application? Using your best judgment, find specifications for the amplifier's input impedance, output impedance, and gain parameter.

**P10.63.** An amplifier has an input resistance of $1\,M\Omega$, an output resistance of $1\,M\Omega$, and an open-circuit voltage gain of 100. Classify this amplifier as an approximate ideal type and find the corresponding gain parameter. In deciding on an amplifier classification, assume that the source and load impedances are on the order of $1\,k\Omega$.

**P10.64.** An amplifier is needed as a part of a system for documentation of voltages in the earth created by thunderstorms. Voltages occurring between probes to be placed in the earth are to be amplified before being applied to a strip chart recorder. The internal impedance of the probe pair can be as high as $10\,k\Omega$ in dry sand or as low as $10\,\Omega$ in muck. The strip chart recorder has an unknown impedance of less than 100 $\Omega$ and deflects 1cm $\pm1$ percent per milliampere of applied current. It is desired that the amplifier be designed so that the recorder deflects 1 cm for each 0.1 V of probe voltage. What type of ideal amplifier is best suited for this application? Using your best judgment, find the specifications for the impedances and gain parameter of this amplifier.

## Section10.7: Frequency Response

**\*P10.65.** The gain of an amplifier is given by

$$A = \frac{1000}{[1 + j(f/f_B)]^2}$$

Determine the upper half-power frequency in terms of $f_B$.

**P10.66.** Sketch the gain magnitude of a typical de-coupled amplifier versus frequency. Repeat for an ac-coupled amplifier.

**\*P10.67.** Consider the amplifier of Problem P10.62. Should this amplifier be ac coupled or dc coupled? Explain your answer.

**\*P10.68.** The input to a certain amplifier is

$$v_{in}(t) = 0.1 \cos(2000\pi t)$$
$$+ 0.2 \cos(4000\pi t + 30°)$$

and the corresponding output voltage is

$$v_o(t) = 10 \cos(2000\pi t - 20°)$$
$$+ 15 \cos(4000\pi t + 20°)$$

Determine the values of the complex gain at $f = 1000$ Hz and at $f = 2000$ Hz.

**P10.69.** How is a wideband amplifier different from a narrowband amplifier?

**P10.70.** Consider Figure P10.70, in which block $A$ is an ideal voltage amplifier and block $B$ is an ideal transresistance amplifier. **a.** Derive an expression for $v_o(t)$ in terms of the amplifier gains, $v_{in}(t)$, and the capacitance $C$. **b.** Derive an expression for the overall voltage gain of the system as a function of frequency. [*Hint:* Assume that $v_{in}(t) = V_m \cos(2\pi ft)$, determine the expression for $v_o(t)$, and then determine the complex voltage gain by taking the ratio

of the phasors for the input and output.] **c.** Given $R_{moc} = 10^3 \, \Omega$, $A_{voc} = 50/\pi$, and $C = 1 \, \mu F$, sketch Bode plots of the voltage-gain magnitude and phase to scale for the range from 1 Hz to 1 kHz.

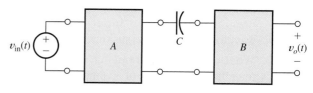

Figure P10.70

**P10.71.** The output signal produced by a certain electret microphone consists of a 2-V dc term plus an ac audio signal that has an rms value of 10 mV. The frequencies of the components of the audio signal range from 20 Hz to 10 kHz. We need to amplify the audio signal to 10 V rms, which is to be applied to a loudspeaker. Should this amplifier be ac coupled or dc coupled? Explain your answer. What midband voltage gain is needed? What values are appropriate for the half-power frequencies?

**P10.72.** Consider Figure P10.72, in which block $A$ is an ideal transconductance amplifier and block $B$ is an ideal voltage amplifier. The capacitance is initially uncharged. **a.** Derive an expression for $v_o(t)$ for $t \geq 0$ in terms of the amplifier gains, $v_{in}(t)$, and the capacitance $C$. **b.** Derive an expression for the overall voltage gain of the system as a function of frequency. [*Hint:* Assume that $v_{in}(t) = V_m \cos(2\pi ft)$, determine the expression for $v_o(t)$, and then determine the complex voltage gain by taking the ratio of the phasors for the input and output.] **c.** Given $G_{msc} = 10^{-6}$ S, $A_{voc} = 200\pi$, and $C = 1 \, \mu F$, sketch Bode plots of the voltage-gain magnitude and phase to scale for the range from 1 Hz to 1 kHz.

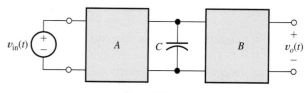

Figure P10.72

**Section10.8:** **Linear Waveform Distortion**

**\*P10.73.** The input signal to an amplifier is $v_{in}(t) = 0.01 \cos(2000\pi t) + 0.02 \cos(4000\pi t)$. The complex gain of the amplifier at 1000 Hz is $100 \underline{/-45°}$. What complex value must the gain have at 2000 Hz for distortionless amplification? Sketch or write a computer program to plot the input and output waveforms to scale versus time.

**P10.74.** What are the requirements for the gain magnitude and phase of an amplifier so that linear distortion does not occur?

**P10.75.** The output of an amplifier used to create special effects for audio signals is given by

$$v_o(t) = v_{in}(t) + Kv_{in}(t - t_d)$$

in which $k$ and $t_d$ are constants. **a.** Is this amplifier linear? Explain carefully. **b.** Determine the complex voltage gain as a function of frequency. [*Hint:* Assume that $v_{in}(t) = V_m \cos(2\pi f t)$, determine the corresponding output, and divide the phasor output by the phasor input.] **c.** Given $K = 0.5$ and $t_d = 1$ ms, use a computer to plot the gain magnitude and phase versus frequency for $0 \le f \le 10$ kHz. **d.** Does this amplifier produce amplitude distortion? Phase distortion? Explain carefully.

**P10.76.** The input signal to an amplifier is $v_i(t) = 0.01 \cos(2000\pi t) + 0.02 \cos(4000\pi t)$. The gain of the amplifier as a function of frequency is given by

$$A = \frac{1000}{1 + j(f/1000)}$$

Find an expression for the output signal of the amplifier as a function of time.

**P10.77.** The output of a certain amplifier in terms of the input is $v_o(t) = Kv_{in}(t - t_d)$. **a.** Is this amplifier linear? Explain carefully. **b.** Determine the complex voltage gain as a function of frequency. [*Hint:* Assume that $v_{in}(t) = V_m \cos(2\pi f t)$, determine the corresponding output, and divide the phasor output by the phasor input.] **c.** Given $K = 100$ and $t_d = 0.1$ ms, plot the gain magnitude and phase versus frequency for $0 \le f \le 10$ kHz. **d.** Does this amplifier produce amplitude distortion? Phase distortion? Explain carefully.

**P10.78.** The output of an amplifier used to create special effects for audio signals is given by

$$v_o(t) = v_{in}(t) + K\frac{d}{dt}v_{in}(t)$$

in which $K$ is a constant. **a.** Is this amplifier linear? Explain carefully. **b.** Determine the complex voltage gain as a function of frequency. [*Hint:* Assume that $v_{in}(t) = V_m \cos(2\pi f t)$, determine the corresponding output, and divide the phasor output by the phasor input.] **c.** Given $K = 1/(2000\pi)$, use a computer to plot the gain magnitude and phase versus frequency for $0 \le f \le 10$ kHz. **d.** Does this amplifier produce amplitude distortion? Phase distortion? Explain carefully.

**Section10.9:** **Pulse Response**

**\*P10.79.** The gain magnitudes of several amplifiers are shown versus frequency in Figure P10.79. If the input to the amplifiers is

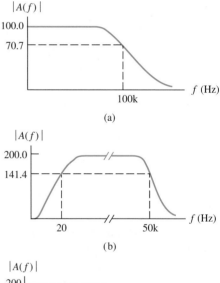

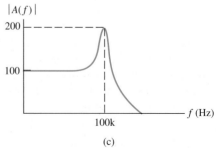

**Figure P10.79**

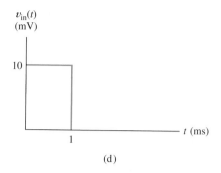

Figure P10.79  (Cont)

the pulse shown in the figure, sketch the output of each amplifier versus time. Give quantitative estimates of as many features on each waveform sketch as you can.

*P10.80. An audio amplifier is specified to have half-power frequencies of 15 Hz and 15 kHz. The amplifier is to be used to amplify the pulse shown in Figure P10.82(b). Estimate the rise time and tilt of the amplifier output. The pulse width $T$ is 2 ms.

P10.81. Sketch the pulse response of an amplifier, showing the rise time, overshoot, ringing, and tilt. Give an approximate relationship between rise time and the upper half-power frequency of a broadband amplifier. Give an approximate relationship between percentage tilt and the lower half-power frequency.

P10.82. Consider the simple highpass filter shown in Figure P10.82(a).
   **a.** Find the complex gain $A = V_2/V_1$ as a function of frequency.
   **b.** What is the magnitude of the gain at dc? At very high frequencies? Find the half-power frequency in terms of $R$ and $C$.
   **c.** Consider the input pulse shown in Figure P10.82(b). Assuming that the capacitor is initially uncharged, find an expression for the output voltage $v_2(t)$ for $t$ between 0 and $T$. Assuming that $RC$ is much greater than $T$, find an approximate expression for percentage tilt.
   **d.** Combine the results of parts (b) and (c) to find a relationship between percentage tilt and the half-power frequency.

P10.83. The input signal and corresponding output signals are shown for several amplifiers in Figure P10.83. Sketch the gain magnitude

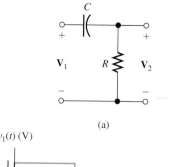

(a)

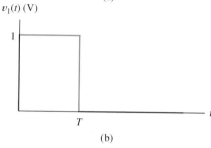

(b)

Figure P10.82

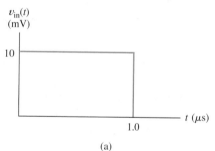

(a)

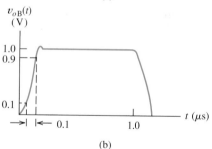

(b)

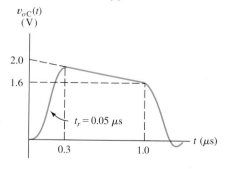

(c)

Figure P10.83

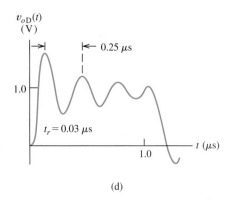

(d)

**Figure P10.83** (*Cont.*)

of each amplifier versus frequency. Give quantitative estimates of as many features on the gain sketches as you can.

**P10.84.** Consider the simple lowpass filter shown in Figure P10.84. **a.** Find the complex gain $A = \mathbf{V}_2/\mathbf{V}_1$ as a function of frequency. What are the magnitudes of $A$ at dc and at very high frequencies? Find the half-power bandwidth $B$ of the circuit in terms of $R$ and $C$. **b.** Consider the case for which the capacitor is initially uncharged and $v_1(t)$ is a unit-step function. Find $v_2(t)$ and an expression for the rise time $t_r$ of the circuit in terms of $R$ and $C$. **c.** Combine the results found in parts (a) and (b) to obtain a relationship between bandwidth and rise time for this circuit. Compare your result with Equation 10.11on page 531.

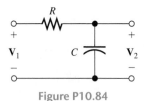

**Figure P10.84**

**Section10.10:** **Transfer Characteristic and Nonlinear Distortion**

**P10.85.** What is harmonic distortion? What causes it?

**\*P10.86.** The input to an amplifier is

$$v_{in}(t) = 0.1 \cos(2000\pi t)$$

and the corresponding output is

$$v_o(t) = 10 \cos(2000\pi t) + 0.2 \cos(4000\pi t)$$
$$+ \ 0.1 \cos(6000\pi t)$$

Determine the distortion factors $D_2$, $D_3$, and $D_4$. Also, determine the percentage of total harmonic distortion.

**P10.87.** The transfer characteristic of an amplifier is described by the equation

$$v_o(t) = 10v_{in}(t) + 0.6v_{in}^2(t) + 0.4v_{in}^3(t)$$

For the input $v_{in}(t) = 2 \cos(200\pi t)$, determine the distortion factors $D_2$, $D_3$, and $D_4$. Also, compute the total harmonic distortion. You may find the following trigonometric identities useful:

$$\cos^2(A) = \frac{1}{2} + \frac{1}{2}\cos(2A)$$

$$\cos^3(A) = \frac{3}{4}\cos(A) + \frac{1}{4}\cos(3A)$$

**P10.88.** The transfer characteristic of an amplifier is described by the equation

$$v_o(t) = v_{in}(t) + 0.1v_{in}^2(t)$$

For the input $v_{in}(t) = \cos(\omega_1 t) + \cos(\omega_2 t)$, determine the frequency and amplitude of each component of the output. You may find the following trigonometric identities useful:

$$\cos^2(A) = \frac{1}{2} + \frac{1}{2}\cos(2A)$$

$$\cos(A)\cos(B) = \frac{1}{2}\cos(A - B)$$

$$+ \ \frac{1}{2}\cos(A + B)$$

**Section10.11:** **Differential Amplifiers**

**P10.89.** What is a differential amplifier? Define the differential input voltage and the common-mode input voltage. Write an expression for the output in terms of the differential and common-mode input components.

**\*P10.90.** A certain amplifier has a differential gain of 500. If the two input terminals are tied together and a 10-mV-rms input signal is applied, the output signal is 20 mV rms. Find the CMRR for this amplifier.

**P10.91.** The input signals $v_{i1}$ and $v_{i2}$ shown in Figure P10.91 are the inputs to a differential amplifier with a gain of $A_d = 10$. (Assume

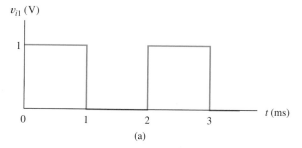

(a)

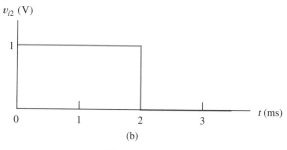

(b)

Figure P10.91

that the common-mode gain is zero.) Sketch the output of the amplifier to scale versus time. Sketch the common-mode input signal to scale versus time.

**P10.92.** In your own words, describe a situation in which a small differential signal is of interest and a large common-mode signal is also present.

**P10.93.** Define the common-mode rejection ratio of a differential amplifier.

**P10.94.** The output of a certain instrumentation amplifier in terms of the inputs is $v_o(t) = 1000v_{i1}(t) - 1001v_{i2}(t)$. Determine the CMRR of this amplifier in decibels.

**P10.95.** In a certain instrumentation amplifier, the input signal consists of a 20-mV-rms differential signal and a 5-V-rms 60-Hz interfering common-mode signal. It is desired that the common-mode contribution to the output signal be at least 60 dB lower than the contribution from the differential signal. What is the minimum CMRR allowed for the amplifier in decibels?

**Section10.12: Offset Voltage, Bias Current, and Offset Current**

**P10.96.** Sketch the circuit diagram of a balancing circuit for a differential amplifier.

**\*P10.97.** A differential amplifier has a differential gain of 500 and negligible common-mode gain. The input terminals are tied to ground through 1-k$\Omega$ resistors having tolerances of $\pm 5$ percent. What are the extreme values of the output voltage caused by a bias current of 100 nA? What is the output voltage if the resistors are exactly equal?

**P10.98.** Draw the differential amplifier symbol and show sources for the offset voltage, bias current, and offset current. What effect do these sources have on the output signal of the amplifier?

**\*P10.99.** A differential amplifier has a bias current of 100 nA, a maximum offset current of 20 nA, a maximum offset voltage of 2 mV, an input resistance of 1 M$\Omega$, and a differential gain of 1000. The input terminals are tied to ground through (exactly equal) 100-k$\Omega$ resistors. Find the extreme values of the output voltage if the common-mode gain is assumed to be zero.

**P10.100.** Repeat Problem P10.99 if the CMRR of the amplifier is 60 dB. By what percentage is the extreme output voltage increased in this case, compared with zero common-mode gain?

**P10.101.** A differential amplifier, including sources to model its dc imperfections, is shown under three different test conditions in Figure P10.101. The amplifier has a differential voltage gain of 100, a common-mode voltage gain of zero, and infinite input impedance. Determine the values of $V_{\text{off}}$, $I_B$, and $I_{\text{off}}$.

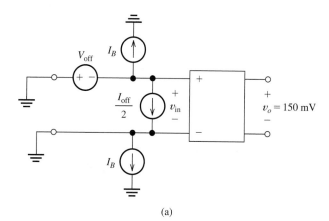

(a)

Figure P10.101

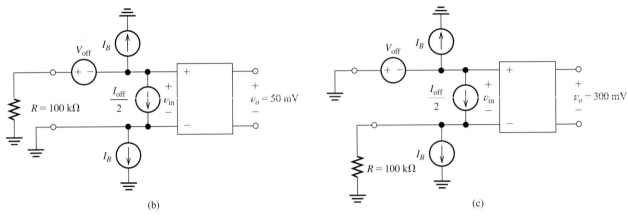

(b)

(c)

Figure P10.101  (Cont)

## Practice Test

Here is a practice test you can use to check your comprehension of the most important concepts in this chapter. Answers can be found in Appendix D and complete solutions are included in the Student Solutions files. See Appendix E for more information about the Student Solutions.

**T10.1.** Suppose we have two identical amplifiers each of which has $A_{voc} = 50$, $R_i = 60\ \Omega$, and $R_o = 40\ \Omega$ connected in cascade. Determine the open-circuit voltage gain, the input resistance, and the output resistance of the cascade.

**T10.2.** Prepare a table showing the gain parameter, the input impedance, and the output impedance for the four types of ideal amplifiers.

**T10.3.** Suppose we have a sensor with a variable internal impedance and a load of variable impedance. What type of ideal amplifier would be best if we need: **a.** the load current to be proportional to the Thévenin voltage of the source; **b.** the load current to be proportional to the short-circuit current of the source; **c.** the load voltage to be proportional to the open-circuit voltage of the source; **d.** the load voltage to be proportional to the short-circuit current of the source?

**T10.4.** Suppose we have an amplifier with $R_i = 200\ \Omega$, $R_o = 1\ \mathrm{k}\Omega$, and $A_{isc} = 50$. Determine the values (including units) of $A_{voc}$, $R_{moc}$, and $G_{msc}$ for this amplifier. Then draw each of the four models for the amplifier showing the value of each parameter.

**T10.5.** We have an amplifier that draws 2 A from a 15-V dc power supply. The input signal current is 1 mA rms, and the input resistance is 2 kΩ. The amplifier delivers 12 V rms to an 8-Ω load. Determine the power dissipated in the amplifier and the efficiency of the amplifier.

**T10.6.** Suppose that the input voltage to an amplifier has a peak amplitude of 100 mV and contains components with frequencies ranging from 1 to 10 kHz. We want the output voltage waveform to be nearly identical to that of the input except with larger amplitude (by a factor of 100) and possibly delayed in time. Based on this information, what specifications can you make for the amplifier?

**T10.7.** What is the principal effect of offset current, bias current, and offset voltage of an amplifier on the signal being amplified?

**T10.8.** What is harmonic distortion? What causes it?

**T10.9.** What is the CMRR? For what types of applications is it important?

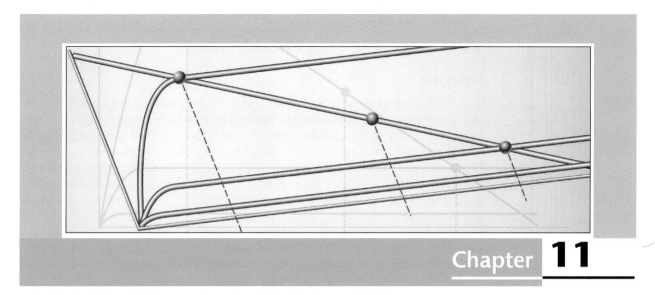

# Chapter 11

# Field-Effect Transistors

## Study of this chapter will enable you to:

- Understand MOSFET operation.
- Use the load-line technique to analyze basic FET amplifiers.
- Analyze bias circuits.
- Use small-signal equivalent circuits to analyze FET amplifiers.

- Compute the performance parameters of several FET amplifier configurations.
- Select a FET amplifier configuration that is appropriate for a given application.
- Understand the basic operation of CMOS logic gates.

## Introduction to this chapter:

Field-effect transistors (FETs) are important devices that are used in amplifiers and logic gates. In this chapter, we discuss the enhancement-mode **metal–oxide–semiconductor field-effect transistor** (MOSFET), which is the primary device that underlies the rapid advances in digital electronics over the past several decades. (Several other types of FETs exist, but to simplify our discussion, we only discuss enhancement-mode MOSFETs, which are presently the most important type.)

In the next chapter, we will discuss another device, the bipolar junction transistor, which is also used in amplifiers and logic gates. Compared to BJTs, MOSFETs can occupy less chip area and can be fabricated with fewer processing steps. Complex digital circuits such as memories and microprocessors are often implemented solely with MOSFETs. On the other hand, BJTs are capable of producing large output currents that are needed for fast switching of a capacitive load such as a circuit-board trace interconnecting digital chips. Each type of device has some applications in which it performs better than the others.

## 11.1 NMOS AND PMOS TRANSISTORS

### Overview

The physical structure of an *n*-channel enhancement-mode MOSFET (also known as an NMOS transistor) is shown in Figure 11.1. It is a chip of silicon crystal with impurities added to the various regions to produce *n*-type and *p*-type material. In *n*-type material, conduction is due mainly to negatively charged electrons, whereas in *p*-type material, conduction is due mainly to positively charged holes. (Although positively charged particles called holes are a helpful concept for understanding conduction in semiconductors, they are not fundamental subatomic particles, as electrons are.)

The device terminals are the **drain** (*D*), **gate** (*G*), **source** (*S*), and **body** (*B*). (Another commonly used term for the body is substrate.) In normal operation, negligible current flows through the body terminal. Sometimes, the body is connected to the source so that we have a three-terminal device. The gate is insulated from the substrate by a thin layer of silicon dioxide, and negligible current flows through the gate terminal. When a sufficiently large (positive) voltage is applied to the gate relative to the source, electrons are attracted to the region under the gate, and a channel of *n*-type material is induced between the drain and the source. Then, if voltage is applied between the drain and the source, current flows into the drain through the channel and out the source. Drain current is controlled by the voltage applied to the gate.

Although the acronym MOS stands for metal–oxide–semiconductor, the gates of modern MOSFETs are actually composed of polysilicon.

The channel length *L* and width *W* are illustrated in Figure 11.1. The trend over the past four decades has been to reduce both *L* and *W* in order to fit an ever-increasing number of transistors into a given chip area. In 1971, Intel introduced the first microprocessor, the 4004, which had a minimum feature size of 10 $\mu$m and contained 2300 transistors. By 2016, processor chips had channel lengths of 7 nm, oxide thickness on the order of 1 nm, and in excess of 10 billion transistors. This has

> Drain current is controlled by the voltage applied to the gate.

**Figure 11.1** *n*-channel enhancement MOSFET showing channel length *L* and channel width *W*.

led to the remarkable performance improvements that we have come to expect for computers and other electronic products. However, this trend is expected to slow in the near feature because of fundamental physical limitations such as the size of the atoms involved. Certainly, oxide thickness or gate width cannot have dimensions smaller than that of an atom.

Device characteristics depend on $L$, $W$, and process parameters such as doping levels and oxide thickness. Usually, the process parameters are predetermined, but the circuit designer can adjust $L$ and $W$ to obtain the device best suited for a given application.

The circuit symbol for the *n*-channel enhancement MOSFET is shown in Figure 11.2. Next, we discuss the basic operation of this device.

The circuit designer can adjust *L* and *W* to obtain the device best suited for a given application.

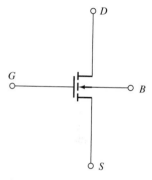

Figure 11.2  Circuit symbol for an enhancement-mode *n*-channel MOSFET.

## Operation in the Cutoff Region

Consider the situation shown in Figure 11.3. Suppose that positive voltage $v_{DS}$ is applied to the drain relative to the source and that we start with $v_{GS} = 0$. Notice that *pn* junctions (i.e., diodes) appear at the drain/body and at the source/body interfaces. Under forward bias (positive on the *p*-side), electrons flow easily across a *pn* junction, but under reverse bias (positive on the *n*-side), virtually no current flows. Thus, virtually no current flows into the drain terminal because the drain/body junction is reverse biased by the $v_{DS}$ source. This is called the **cutoff region** of operation. As $v_{GS}$ is increased, the device remains in cutoff until $v_{GS}$ reaches a particular value called the **threshold voltage** $V_{to}$. Typically, the threshold voltage ranges from a fraction of a volt to one volt. Thus, in cutoff, we have

$$i_D = 0 \text{ for } v_{GS} \leq V_{to} \tag{11.1}$$

Key equation for enhancement NMOS in cutoff.

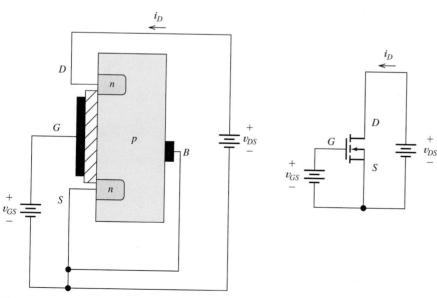

**Figure 11.3** For $v_{GS} < V_{to}$, the *pn* junction between drain and body is reverse biased and $i_D = 0$.

## Operation in the Triode Region

The triode region is also called the linear region of operation.

For $v_{DS} < v_{GS} - V_{to}$ and $v_{GS} \geq V_{to}$, we say that the NMOS is operating in the **triode region**. Consider the situation shown in Figure 11.4, in which $v_{GS}$ is greater than the threshold voltage. The electric field resulting from the applied gate voltage has repelled holes from the region under the gate and attracted electrons that can easily flow in the forward direction across the source/body junction. This results in an $n$-type channel between the drain and the source. Then, when $v_{DS}$ is increased, current flows into the drain, through the channel, and out the source. For small values of $v_{DS}$, the drain current is proportional to $v_{DS}$. Furthermore, for a given (small) value of $v_{DS}$, drain current is also proportional to the **excess gate voltage** $v_{GS} - V_{to}$.

In the triode region, the NMOS device behaves as a resistor connected between drain and source, but the resistance decreases as $v_{GS}$ increases.

Plots of $i_D$ versus $v_{DS}$ are shown in Figure 11.4 for several values of gate voltage. In the triode region, the NMOS device behaves as a resistor connected between drain and source, but the resistance decreases as $v_{GS}$ increases.

Now, consider what happens if we continue to increase $v_{DS}$. Because of the current flow, the voltages between points along the channel and the source become greater as we move toward the drain. Thus, the voltage between gate and channel becomes smaller as we move toward the drain, resulting in tapering of the channel thickness as illustrated in Figure 11.5. Because of the tapering of the channel, its resistance becomes larger with increasing $v_{DS}$, resulting in a lower rate of increase of $i_D$.

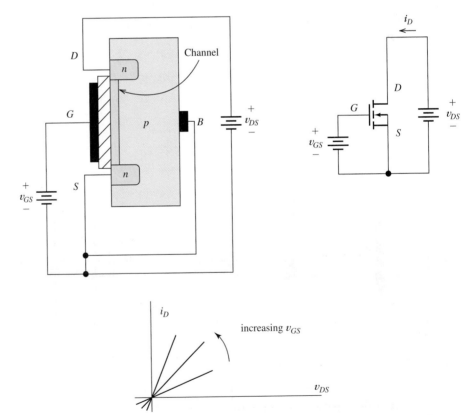

**Figure 11.4** For $v_{GS} > V_{to}$, a channel of $n$-type material is induced in the region under the gate. As $v_{GS}$ increases, the channel becomes thicker. For small values of $v_{DS}$, $i_D$ is proportional to $v_{DS}$. The device behaves as a resistance whose value depends on $v_{GS}$.

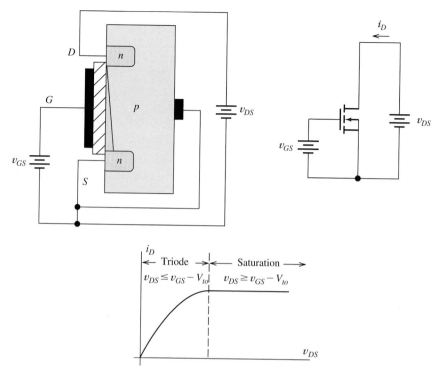

**Figure 11.5** As $v_{DS}$ increases, the channel pinches down at the drain end and $i_D$ increases more slowly. Finally, for $v_{DS} > v_{GS} - V_{to}$, $i_D$ becomes constant.

For $v_{DS} < v_{GS} - V_{to}$ and $v_{GS} \geq V_{to}$, the device is operating in the triode region, and the drain current is given by

$$i_D = K \left[ 2(V_{GS} - V_{to})v_{DS} - v_{DS}^2 \right] \qquad (11.2)$$

in which $K$ is given by

$$K = \left( \frac{W}{L} \right) \frac{KP}{2} \qquad (11.3)$$

Key equation for enhancement NMOS in the triode region.

As illustrated in Figure 11.1, $W$ is the width of the channel and $L$ is its length. The device parameter $KP$ depends on the thickness of the oxide layer and certain properties of the channel material. A typical value of $KP$ for $n$-channel enhancement devices is 50 $\mu$A/V$^2$.

A typical value of $KP$ for $n$-channel enhancement devices is 50 $\mu$A/V$^2$.

Usually, $KP$ is determined by the fabrication process. However, in designing a circuit, we can vary the ratio $W/L$ to obtain transistors best suited to various parts of the circuit. The condition $v_{DS} \leq v_{GS} - V_{to}$ is equivalent to $v_{GD} \geq V_{to}$. Thus, the device is in the triode region if both $v_{GD}$ and $v_{GS}$ are greater than the threshold voltage $V_{to}$.

## Operation in the Saturation Region

We have seen that as $v_{DS}$ is increased the voltage between the gate and the drain end of the channel decreases. When the gate-to-drain voltage $v_{GD}$ equals the threshold

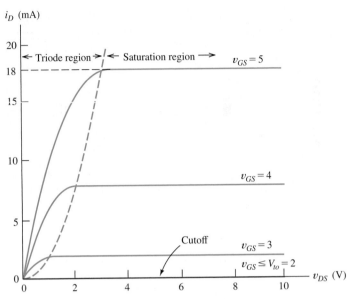

**Figure 11.6** Characteristic curves for an NMOS transistor.

voltage $V_{to}$, the channel thickness at the drain end becomes zero. For further increases in $v_{DS}$, $i_D$ is constant, as illustrated in Figure 11.5. This is called the **saturation region**, in which we have $v_{GS} \geq V_{to}$ and $v_{DS} \geq v_{GS} - V_{to}$, and the current is given by

$$i_D = K(v_{GS} - V_{to})^2 \tag{11.4}$$

Keep in mind that in the saturation region, $v_{GS}$ is greater than the threshold, but $v_{GD}$ is less than the threshold. Figure 11.6 shows the drain characteristics of an NMOS transistor.

## Boundary between the Triode and Saturation Regions

Next, we derive the equation for the boundary between the triode region and the saturation region in the $i_D$–$v_{DS}$ plane. At this boundary, the channel thickness at the drain is zero, which occurs when $v_{GD} = V_{to}$. Thus, we want to find $i_D$ in terms of $v_{DS}$ under the condition that $v_{GD} = V_{to}$. Since $v_{GD} = v_{GS} - v_{DS}$, the condition at the boundary is given by

$$v_{GS} - v_{DS} = V_{to} \tag{11.5}$$

Solving this for $v_{GS}$, substituting into Equation 11.4, and reducing, we have the desired boundary equation given by

$$i_D = Kv_{DS}^2 \tag{11.6}$$

Notice that the boundary between the triode region and the saturation region is a parabola.

Solving Equation 11.5 for $v_{GS}$ and substituting into Equation 11.2 also produces Equation 11.6. (Equations 11.2 and 11.4 give the same values for $i_D$ on the boundary.)

Given the values for $KP$, $L$, $W$, and $V_{to}$, we can plot the static characteristics of an NMOS transistor.

---

### Example 11.1   Plotting the Characteristics of an NMOS Transistor

A certain enhancement-mode NMOS transistor has $W = 160\ \mu m$, $L = 2\ \mu m$, $KP = 50\ \mu A/V^2$, and $V_{to} = 2$ V. Plot the drain characteristic curves to scale for $v_{GS} = 0, 1, 2, 3, 4$, and 5 V.

**Solution**   First, we use Equation 11.3 to compute the device constant:

$$K = \left(\frac{W}{L}\right)\frac{KP}{2} = 2\ \text{mA/V}^2$$

Equation 11.6 gives the boundary between the triode region and the saturation region. Thus, we have

$$i_D = Kv_{DS}^2 = 2v_{DS}^2$$

where $i_D$ is in mA and $v_{DS}$ is in volts. The plot of this equation is the dashed line shown in Figure 11.6.

Next, we use Equation 11.4 to compute the drain current in the saturation region for each of the $v_{GS}$ values of interest. Hence, we get

$$i_D = K(v_{GS} - V_{to})^2 = 2(v_{GS} - 2)^2$$

where again the current is in mA. Substituting values, we find that

$$i_D = 18\ \text{mA for } v_{GS} = 5\ \text{V}$$
$$i_D = \phantom{0}8\ \text{mA for } v_{GS} = 4\ \text{V}$$
$$i_D = \phantom{0}2\ \text{mA for } v_{GS} = 3\ \text{V}$$
$$i_D = \phantom{0}0\ \text{mA for } v_{GS} = 2\ \text{V}$$

For $v_{GS} = 0$ and 1 V, the device is in cutoff and $i_D = 0$. These values are plotted in the saturation region as shown in Figure 11.6.

Finally, Equation 11.2 is used to plot the characteristics in the triode region. For each value of $v_{GS}$, this equation plots as a parabola that passes through the origin ($i_D = 0$ and $v_{DS} = 0$). The apex of each parabola is on the boundary between the triode region and the saturation region. ∎

*Summary:* In an NMOS transistor, when a sufficiently large (positive) voltage is applied to the gate relative to the source, electrons are attracted to the region under the gate, and a channel of *n*-type material is induced between the drain and the source. Then, if positive voltage is applied to the drain relative to the source, current flows into the drain through the channel and out the source. Drain current is controlled by the voltage applied to the gate.

**Exercise 11.1**   Consider an NMOS having $V_{to} = 2$ V. What is the region of operation (triode, saturation, or cutoff) if: **a.** $v_{GS} = 1$ V and $v_{DS} = 5$ V; **b.** $v_{GS} = 3$ V and $v_{DS} = 0.5$ V; **c.** $v_{GS} = 3$ V and $v_{DS} = 6$ V; **d.** $v_{GS} = 5$ V and $v_{DS} = 6$ V?
**Answer**   **a.** cutoff; **b.** triode; **c.** saturation; **d.** saturation. ☐

**Exercise 11.2**   Suppose that we have an NMOS transistor with $KP = 50\ \mu A/V^2$, $V_{to} = 1$ V, $L = 2\ \mu m$, and $W = 80\ \mu m$. Sketch the drain characteristics for $v_{DS}$ ranging from 0 to 10 V and $v_{GS} = 0, 1, 2, 3$, and 4 V.
**Answer**   The plots are shown in Figure 11.7. ☐

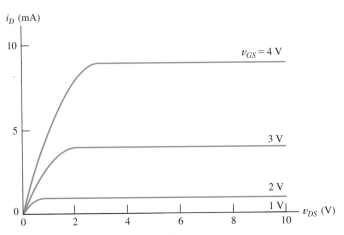

Figure 11.7 Answer for Exercise 11.2.

Figure 11.8 Circuit symbol for PMOS transistor.

## PMOS Transistors

MOSFETs can also be constructed by interchanging the $n$ and $p$ regions of $n$-channel devices, resulting in $p$-channel devices. The circuit symbol for the $p$-channel MOSFET is shown in Figure 11.8. As indicated in the figure, we usually orient the $p$-channel FETs with the source at the top and reference the current out of the drain. The PMOS symbol is the same as the NMOS symbol, except for orientation and the direction of the arrowhead.

The characteristics of a PMOS transistor are very similar to those for the NMOS transistors, except that voltage polarities are inverted. Because we reference the drain current into the drain for $n$-channel devices and out of the drain for $p$-channel devices, the drain current assumes positive values in both devices. Thus, the characteristic curves of a $p$-channel device are like those of an $n$-channel device, except that the algebraic signs of the voltages must be inverted.

Table 11.1 gives the equations of operation for both NMOS and PMOS enhancement-mode transistors. A typical value of $KP$ for PMOS transistors is $25\ \mu\text{A/V}^2$, which is about half the value for NMOS transistors. This is due to differences in the conduction properties of electrons and holes in silicon. Notice that the threshold voltage for an enhancement-mode PMOS transistor assumes a negative value.

**Exercise 11.3**   Suppose that we have a PMOS transistor with $KP = 25\ \mu\text{A/V}^2$, $V_{to} = -1$ V, $L = 2\ \mu\text{m}$, and $W = 200\ \mu\text{m}$. Sketch the drain characteristics for $v_{DS}$ ranging from 0 to $-10$ V and $v_{GS} = 0, -1, -2, -3$, and $-4$ V.
**Answer**   The plots are shown in Figure 11.9.                                                     □

## Channel-Length Modulation and Charge-Carrier-Velocity Saturation

The description of MOSFETs that we have given up to this point is reasonably accurate for devices with channel lengths in excess of about 10 $\mu$m. As the channel lengths become shorter, several effects that modify MOSFET behavior eventually come into play. One of these is channel-length modulation, which is caused because the effective length of the channel is reduced as $v_{DS}$ increases in magnitude. Channel-length modulation causes the characteristics to slope upward in the saturation region

**Table 11.1** MOSFET Summary

|  | NMOS | PMOS |
|---|---|---|
| Circuit symbol | 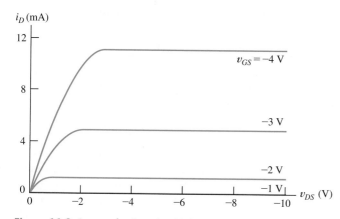 | |
| $KP$ (typical value) | $50\ \mu\text{A/V}^2$ | $25\ \mu\text{A/V}^2$ |
| $K$ | $(1/2)\ KP\ (W/L)$ | $(1/2)\ KP\ (W/L)$ |
| $V_{to}$ (typical value) | $+1$ V | $-1$ V |
| Cutoff region | $v_{GS} \leq V_{to}$ <br> $i_D = 0$ | $v_{GS} \geq V_{to}$ <br> $i_D = 0$ |
| Triode region | $v_{GS} \geq V_{to}$ and $0 \leq v_{DS} \leq v_{GS} - V_{to}$ <br> $i_D = K\,[2(v_{GS} - V_{to})\,v_{DS} - v_{DS}^2]$ | $v_{GS} \leq V_{to}$ and $0 \geq v_{DS} \geq v_{GS} - V_{to}$ <br> $i_D = K\,[2(v_{GS} - V_{to})\,v_{DS} - v_{DS}^2]$ |
| Saturation region | $v_{GS} \geq V_{to}$ and $v_{DS} \geq v_{GS} - V_{to}$ <br> $i_D = K\,(v_{GS} - V_{to})^2$ | $v_{GS} \leq V_{to}$ and $v_{DS} \leq v_{GS} - V_{to}$ <br> $i_D = K\,(v_{GS} - V_{to})^2$ |
| $v_{DS}$ and $v_{GS}$ | Normally assume positive values | Normally assume negative values |

**Figure 11.9** Answer for Exercise 11.3.

and can be taken into account by including the factor $(1 + \lambda|v_{DS}|)$ in the expressions for $i_D$ given in Table 11.1. The channel-length modulation factor $\lambda$ is typically equal to $0.1/L$, in which $L$ is the channel length in microns.

A second effect is caused by velocity saturation of the charge carriers. The equations we have given for $i_D$ are based on the assumption that charge carrier velocity is proportional to electric field strength in the channel. However, at the higher field strengths encountered in devices with channels shorter than about $2\ \mu\text{m}$, carrier velocity tends to become more nearly constant as the field strength is increased. The result is that the characteristic curves enter saturation at smaller values of $|v_{DS}|$. Furthermore, $i_D$ is more nearly linear with variations in $v_{GS}$, and the characteristic curves are more uniformly spaced than we have shown.

Although these and other secondary effects are important to advanced MOS designers, we can gain a basic understanding of MOSFET amplifiers and logic circuits by using the equations presented earlier in this section, except when otherwise noted.

## 11.2  LOAD-LINE ANALYSIS OF A SIMPLE NMOS AMPLIFIER

In this section, we analyze the NMOS amplifier circuit shown in Figure 11.10 by using a graphical load-line approach similar to the analysis we carried out for diode circuits in Section 9.2. The dc sources **bias** the MOSFET at a suitable operating point so that amplification of the input signal $v_{in}(t)$ can take place. We will see that the input voltage $v_{in}(t)$ causes $v_{GS}$ to vary with time, which in turn causes $i_D$ to vary. The changing voltage drop across $R_D$ causes an amplified version of the input signal to appear at the drain terminal.

> The input voltage $v_{in}(t)$ causes $v_{GS}$ to vary with time, which in turn causes $i_D$ to vary. The changing voltage drop across $R_D$ causes an amplified version of the input signal to appear at the drain terminal.

Applying Kirchhoff's voltage law to the input loop, we obtain the following expression:

$$v_{GS}(t) = v_{in}(t) + V_{GG} \tag{11.7}$$

For our example, we assume that the input signal is a 1-V-peak, 1-kHz sinusoid and that $V_{GG}$ is 4 V. Then, we have

$$v_{GS}(t) = \sin(2000\pi t) + 4 \tag{11.8}$$

Writing a voltage equation around the drain circuit, we obtain

> Load-line equation.

$$V_{DD} = R_D i_D(t) + v_{DS}(t) \tag{11.9}$$

For our example, we assume that $R_D = 1\ k\Omega$ and $V_{DD} = 20$ V, so Equation 11.9 becomes

$$20 = i_D(t) + v_{DS}(t) \tag{11.10}$$

where we have assumed that $i_D(t)$ is in milliamperes. A plot of this equation on the drain characteristics of the transistor is a straight line called the **load line**.

> To establish the load line, we first locate two points on it.

To establish the load line, we first locate two points on it. Assuming that $i_D = 0$ in Equation 11.10, we find that $v_{DS} = 20$ V. These values plot as the lower right-hand end of the load line shown in Figure 11.11. For a second point, we assume that $v_{DS} = 0$, which yields $i_D = 20$ mA when substituted into Equation 11.10. This pair of values ($v_{DS} = 0$ and $i_D = 20$ mA) plots as the upper left-hand end of the load line.

> The term "quiescent" implies that the input signal is zero.

The operating point of an amplifier for zero input signal is called the **quiescent operating point** or **Q point**. For $v_{in}(t) = 0$, Equation 11.8 yields $v_{GS} = V_{GG} = 4$ V. Therefore, the intersection of the curve for $v_{GS} = 4$ V with the load line is the $Q$ point. The quiescent values are $I_{DQ} = 9$ mA and $V_{DSQ} = 11$ V.

The maximum and minimum values of the gate-to-source voltage are $V_{GSmax} = 5$ V and $V_{GSmin} = 3$ V (see Equation 11.8). The intersections of the corresponding curves with the load line are labeled as points $A$ and $B$, respectively, in Figure 11.11.

At point $A$, we find that $V_{DSmin} = 4$ V and $I_{Dmax} = 16$ mA. At point $B$, we find that $V_{DSmax} = 16$ V and $I_{Dmin} = 4$ mA.

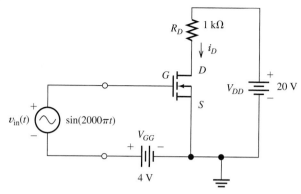

Figure 11.10  Simple NMOS amplifier circuit.

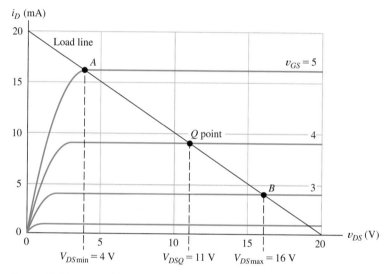

Figure 11.11  Drain characteristics and load line for the circuit of Figure 11.10.

Plots of $v_{GS}(t)$ and $v_{DS}(t)$ versus time are shown in Figure 11.12. Notice that the peak-to-peak swing of $v_{DS}(t)$ is 12 V, whereas the peak-to-peak swing of the input signal is 2 V. Furthermore, the ac voltage at the drain is inverted compared to the input signal. (In other words, the positive peak of the input occurs at the same time as the minimum value of $v_{DS}$.) Therefore, this is an inverting amplifier. Apparently, the circuit has a voltage gain $A_v = -12V/2V = -6$, where the minus sign is due to the inversion.

Notice, however, that the output waveform shown in Figure 11.12(b) is not a symmetrical sinusoid like the input. For illustration, we see that starting from the $Q$ point at $V_{DSQ} = 11$ V, the output voltage swings down to $V_{CEmin} = 4$ V for a change of 7 V. On the other hand, the output swings up to 16 V for a change of only 5 V from the $Q$ point on the positive-going half cycle of the output. We cannot properly define gain for the circuit, because the ac output signal is not proportional to the ac input. Nevertheless, the output signal is larger than the input even if it is distorted. (This is an example of nonlinear distortion, which we discuss in Section 11.10.)

In this circuit, distortion is due to the fact that the characteristic curves for the FET are not uniformly spaced. If a much smaller input amplitude was applied, we would

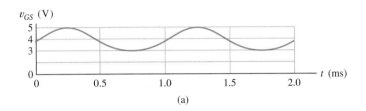

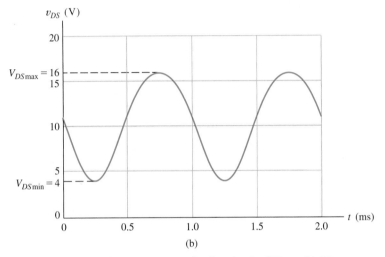

**Figure 11.12** $v_{GS}$ and $v_{DS}$ versus time for the circuit of Figure 11.10.

Distortion is due to the fact that the characteristic curves for the FET are not uniformly spaced. If a much smaller input amplitude was applied, we would have amplification without appreciable distortion.

have amplification without appreciable distortion. This is true because the curves are more uniformly spaced if a very restricted region of the characteristics is considered. If we plotted the curves for 0.1-V increments in $v_{GS}$, this would be apparent.

The amplifier circuit we have analyzed in this section is fairly simple. Practical amplifier circuits are more difficult to analyze by graphical methods. Later in the chapter, we develop a linear small-signal equivalent circuit for the FET, and then we can use mathematical circuit-analysis techniques instead of graphical analysis. Usually, the equivalent-circuit approach is more useful for practical amplifier circuits. However, graphical analysis of simple circuits provides an excellent way to understand the basic concepts of amplifiers.

**Exercise 11.4**   Find $V_{DSQ}$, $V_{DSmin}$, and $V_{DSmax}$ for the circuit of Figure 11.10 if the circuit values are changed to $V_{DD} = 15$ V, $V_{GG} = 3$ V, $R_D = 1$ k$\Omega$, and $v_{in}(t) = \sin(2000\pi t)$. The characteristics for the MOSFET are shown in Figure 11.11.
**Answer**   $V_{DSQ} \cong 11$ V, $V_{DSmin} \cong 6$ V, $V_{DSmax} \cong 14$ V.   □

## 11.3   BIAS CIRCUITS

Amplifier analysis has two steps: 1. Determine the Q point. 2. Use a small-signal equivalent circuit to determine impedances and gains.

Analysis of amplifier circuits is often undertaken in two steps. First, we analyze the dc circuit to determine the $Q$ point. In this analysis, the nonlinear device equations or the characteristic curves must be used. Then, after the bias analysis is completed, we use a linear small-signal equivalent circuit to determine the input resistance, voltage gain, and so on.

The bias circuits that we discuss are suitable for discrete-component designs in which large capacitances (for isolating the amplifier bias circuit from the source, load, and adjacent amplifier stages) and relatively tight tolerance (±5 percent or better) resistances are practical. Close-tolerance resistances and large capacitances are not practical in integrated-circuit amplifiers; and their design is more complicated due to interaction between source, amplifier stages, and the load. Integrated-circuit design is mostly beyond the scope of this book.

The two-battery bias circuit used in the amplifier of Figure 11.10 is not practical. Usually, only one dc voltage is readily available instead of two. However, a more significant problem is that FET parameters vary considerably from device to device. In general, we want to establish a $Q$ point near the middle of the load line so the output signal can swing in both directions without clipping. When the FET parameters vary from unit to unit, the two-battery circuit can wind up with some circuits biased near one end or the other.

### The Fixed- plus Self-Bias Circuit

The **fixed- plus self-bias circuit** shown in Figure 11.13(a) is a good circuit for establishing $Q$ points that are relatively independent of device parameters.

For purposes of analysis, we replace the gate circuit with its Thévenin equivalent as shown in Figure 11.13(b). The Thévenin voltage is

$$V_G = V_{DD}\frac{R_2}{R_1 + R_2} \tag{11.11}$$

and the Thévenin resistance $R_G$ is the parallel combination of $R_1$ and $R_2$. Writing a voltage equation around the gate loop of Figure 11.13(b), we obtain

$$V_G = v_{GS} + R_S i_D \tag{11.12}$$

(We have assumed that the voltage drop across $R_G$ is zero because the gate current of an NMOS transistor is extremely small.)

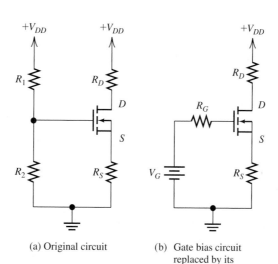

(a) Original circuit

(b) Gate bias circuit replaced by its Thévenin equivalent

**Figure 11.13** Fixed- plus self-bias circuit.

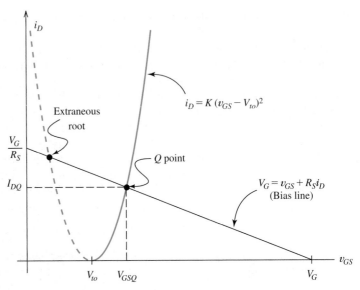

**Figure 11.14** Graphical solution of Equations 11.12 and 11.13.

Usually, we want to bias the transistor in its saturation region, so we have

$$i_D = K(v_{GS} - V_{to})^2 \tag{11.13}$$

Simultaneous solution of Equations 11.12 and 11.13 yields the operating point (provided that it falls in the saturation region). Plots of these two equations are shown in Figure 11.14. Equation 11.12 plots as a straight line called the **bias line**. Notice that two roots will be found because Equation 11.13 plots as the dashed curve for $v_{GS} < V_{to}$. Thus, the smaller root found for $v_{GS}$ is extraneous and should be discarded. The larger root found for $v_{GS}$ and the smaller root for $i_D$ are the true operating point.

Finally, writing a voltage equation around the drain loop of Figure 11.13 gives us

$$v_{DS} = V_{DD} - (R_D + R_S)i_D \tag{11.14}$$

Simultaneous solution of Equations 11.12 and 11.13 yields the operating point (provided that it falls in the saturation region). The larger root found for $v_{GS}$ and the smaller root for $i_D$ are the true operating point.

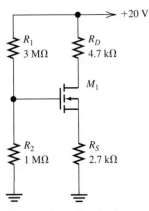

**Figure 11.15** Fixed- plus self-bias circuit of Example 11.2.

**Example 11.2    Determination of Q Point for the Fixed- plus Self-Bias Circuit**

Analyze the fixed- plus self-bias circuit shown in Figure 11.15. The transistor has $KP = 50\ \mu A/V^2$, $V_{to} = 2$ V, $L = 10\ \mu m$, and $W = 400\ \mu m$.

**Solution**    First, we use Equation 11.3 to compute the device constant, which yields

$$K = \left(\frac{W}{L}\right)\frac{KP}{2} = 1\ mA/V^2$$

Substituting values into Equation 11.11, we have

$$V_G = V_{DD}\frac{R_2}{R_1 + R_2} = 20\frac{1}{(3+1)} = 5\ V$$

The $Q$-point values must satisfy Equations 11.12 and 11.13. Thus, we need to find the solution to the following pair of equations:

$$V_G = V_{GSQ} + R_S I_{DQ}$$
$$I_{DQ} = K(V_{GSQ} - V_{to})^2$$

Using the last equation to substitute for $I_{DQ}$ in the expression for $V_G$, we have

$$V_G = V_{GSQ} + R_S K (V_{GSQ} - V_{to})^2$$

Rearranging, we have

$$V_{GSQ}^2 + \left(\frac{1}{R_S K} - 2V_{to}\right) V_{GSQ} + V_{to}^2 - \frac{V_G}{R_S K} = 0$$

After values are substituted, we have

$$V_{GSQ}^2 - 3.630 V_{GSQ} + 2.148 = 0$$

Solving, we find $V_{GSQ} = 2.886$ V and $V_{GSQ} = 0.744$ V. The second root is extraneous and should be discarded. Then, we have

$$I_{DQ} = K (V_{GSQ} - V_{to})^2 = 0.784 \text{ mA}$$

Solving for the drain-to-source voltage, we get

$$V_{DSQ} = V_{DD} - (R_D + R_S) I_{DQ} = 14.2 \text{ V}$$

which is high enough to ensure that operation is in saturation as assumed in the solution.    ■

**Exercise 11.5**   Determine $I_{DQ}$ and $V_{DSQ}$ for the circuit shown in Figure 11.16. The transistor has $KP = 50 \ \mu A/V^2$, $V_{to} = 1$ V, $L = 10 \ \mu m$, and $W = 200 \ \mu m$.
**Answer**   $I_{DQ} = 2$ mA; $V_{DSQ} = 16$ V.    ☐

**Exercise 11.6**   Determine $I_{DQ}$ and $V_{DSQ}$ for the PMOS circuit shown in Figure 11.17. The transistor has $KP = 25 \ \mu A/V^2$, $V_{to} = -1$ V, $L = 10 \ \mu m$, and $W = 400 \ \mu m$.
**Answer**   $I_{DQ} = 4.5$ mA; $V_{DSQ} = -11$ V.    ☐

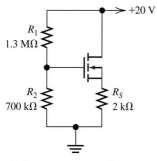

**Figure 11.16**  Circuit for Exercise 11.5.

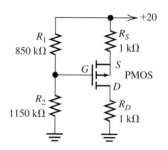

**Figure 11.17** Circuit for Exercise 11.6.

## 11.4 SMALL-SIGNAL EQUIVALENT CIRCUITS

In the preceding section, we considered discrete-component dc bias circuits for FET amplifiers. Now, we consider the relationships between signal currents and voltages resulting in small changes from the $Q$ point. As in Section 9.8, we denote total quantities by lowercase letters with uppercase subscripts, such as $i_D(t)$ and $v_{GS}(t)$. The dc $Q$ point values are denoted by uppercase letters with an additional $Q$ subscript, such as $I_{DQ}$ and $V_{GSQ}$. The signals are denoted by lowercase letters with lowercase subscripts, such as $i_d(t)$ and $v_{gs}(t)$. The total current or voltage is the sum of the $Q$ point value and the signal. Thus, we can write

$$i_D(t) = I_{DQ} + i_d(t) \tag{11.15}$$

and

$$v_{GS}(t) = V_{GSQ} + v_{gs}(t) \tag{11.16}$$

Figure 11.18 illustrates the terms in Equation 11.15.

In the discussion that follows, we assume that the FETs are biased in the saturation region, which is usually the case for amplifier circuits. Equation 11.4, repeated here for convenience,

$$i_D = K \left(v_{GS} - V_{to}\right)^2$$

gives the total drain current in terms of the total gate-to-source voltage. Using Equations 11.15 and 11.16 to substitute into 11.4, we get

$$I_{DQ} + i_d(t) = K \left[V_{GSQ} + v_{gs}(t) - V_{to}\right]^2 \tag{11.17}$$

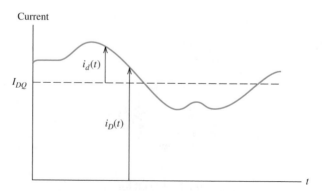

**Figure 11.18** Illustration of the terms in Equation 11.15.

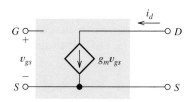

**Figure 11.19** Small-signal equivalent circuit for FETs.

The right-hand side of Equation 11.17 can be expanded to obtain

$$I_{DQ} + i_d(t) = K(V_{GSQ} - V_{to})^2 + 2K(V_{GSQ} - V_{to})v_{gs}(t) + Kv_{gs}^2(t) \quad (11.18)$$

However, the $Q$ point values are also related by Equation 11.4, so we have

$$I_{DQ} = K(V_{GSQ} - V_{to})^2 \qquad (11.19)$$

Therefore, the first term on each side of Equation 11.18 can be canceled. Furthermore, we are interested in small-signal conditions for which the last term on the right-hand side of Equation 11.18 is negligible and can be dropped [i.e., we assume that $|v_{gs}(t)|$ is much smaller than $|(V_{GSQ} - V_{to})|$ for all values of time]. With these changes, Equation 11.18 becomes

$$i_d(t) = 2K(V_{GSQ} - V_{to})v_{gs}(t) \qquad (11.20)$$

If we define the **transconductance** of the FET as

$$g_m = 2K(V_{GSQ} - V_{to}) \qquad (11.21)$$

Equation 11.20 can be written as

$$i_d(t) = g_m v_{gs}(t) \qquad (11.22)$$

The gate current for the FET is negligible, so we obtain

$$i_g(t) = 0 \qquad (11.23)$$

Equations 11.22 and 11.23 are represented by the **small-signal equivalent circuit** shown in Figure 11.19. Thus for small signals, the FET is modeled by a voltage-controlled current source connected between the drain and source terminals. The model has an open circuit between gate and source.

> For small signals, the FET is modeled by a voltage-controlled current source connected between the drain and source terminals. The model has an open circuit between gate and source.

### Dependence of Transconductance on Q Point and Device Parameters

We will see that transconductance $g_m$ is an important parameter in the analysis of amplifier circuits. In general, better performance is obtained with higher values of $g_m$. Thus, it is important to know how $Q$ point and device parameters influence transconductance.

Solving Equation 11.19 for the quantity $(V_{GSQ} - V_{to})$ and substituting into Equation 11.21, we obtain

$$g_m = 2\sqrt{KI_{DQ}} \qquad (11.24)$$

We can increase $g_m$ by choosing a higher value of $I_{DQ}$.

An important point to notice is that $g_m$ is proportional to the square root of the $Q$ point drain current. Thus, we can increase $g_m$ by choosing a higher value of $I_{DQ}$.

If we use Equation 11.3 to substitute for $K$ in Equation 11.24, we obtain

$$g_m = \sqrt{2KP}\sqrt{W/L}\sqrt{I_{DQ}} \tag{11.25}$$

Thus, we can obtain higher values of $g_m$ for a given value of $I_{DQ}$ by increasing the width-to-length ratio of the MOSFET.

We can obtain higher values of $g_m$ for a given value of $I_{DQ}$ by increasing the width-to-length ratio of the MOSFET.

## More Complex Equivalent Circuits

Sometimes, additions to the equivalent circuit are needed to accurately model FETs. For example, we would need to include small capacitances between the device terminals if we considered the high-frequency response of FET amplifiers. The device equations and the equivalent circuit that we have derived from them describe only the static behavior of the device. For an accurate model with rapidly changing currents and voltages, capacitances must be considered.

Furthermore, the first-order equations we have used to obtain the equivalent circuit for the FET did not account for the effect of $v_{DS}$ on the drain current. We have assumed that the drain characteristics are horizontal in the saturation region, but this is not exactly true—the drain characteristics of real devices slope slightly upward with increasing $v_{DS}$. If we wish to account for the effect of $v_{DS}$ in the small-signal equivalent circuit, we must add a resistance $r_d$ called the **drain resistance** between drain and source as shown in Figure 11.20. In this case, Equation 11.22 becomes

$$i_d = g_m v_{gs} + v_{ds}/r_d \tag{11.26}$$

## Transconductance and Drain Resistance as Partial Derivatives

An alternative definition of $g_m$ can be found by examination of Equation 11.26. Notice that if $v_{ds} = 0$, $g_m$ is the ratio of $i_d$ and $v_{gs}$. In equation form, we have

$$g_m = \left.\frac{i_d}{v_{gs}}\right|_{v_{ds}=0} \tag{11.27}$$

However, $i_d$, $v_{gs}$, and $v_{ds}$ represent small changes from the $Q$ point. Therefore, the condition $v_{ds} = 0$ is equivalent to requiring $v_{DS}$ to remain constant at the $Q$ point value, namely $V_{DSQ}$. Thus, we can write

$$g_m \cong \left.\frac{\Delta i_D}{\Delta v_{GS}}\right|_{v_{DS}=V_{DSQ}} \tag{11.28}$$

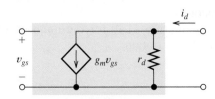

**Figure 11.20** FET small-signal equivalent circuit that accounts for the dependence of $i_D$ on $v_{DS}$.

where $\Delta i_D$ is an increment of drain current centered at the $Q$ point. Similarly, $\Delta v_{GS}$ is an increment of gate-to-source voltage centered at the $Q$ point.

Equation 11.28 is an approximation to a partial derivative. Therefore, $g_m$ is the partial derivative of $i_D$ with respect to $v_{GS}$, evaluated at the $Q$ point:

$$g_m = \left. \frac{\partial i_D}{\partial v_{GS}} \right|_{Q \text{ point}} \tag{11.29}$$

Similarly, the reciprocal of the drain resistance is

$$\frac{1}{r_d} \cong \left. \frac{\Delta i_D}{\Delta v_{DS}} \right|_{v_{GS}=V_{GSQ}} \tag{11.30}$$

Therefore, we can write

$$\frac{1}{r_d} \cong \left. \frac{\partial i_D}{\partial v_{DS}} \right|_{Q \text{ point}} \tag{11.31}$$

Given the drain characteristics, we can determine approximate values of the partial derivatives for a given $Q$ point. Then, we can model the FET by its small-signal equivalent in analysis of an amplifier circuit and use the values found for $g_m$ and $r_d$ to compute amplifier gains and impedances. In the next several sections, we show examples of this process. First, we show how to determine the values of $g_m$ and $r_d$ starting from the characteristic curves.

---

**Example 11.3**    Determination of $g_m$ and $r_d$ from the Characteristic Curves

Determine the values of $g_m$ and $r_d$ for the MOSFET having the characteristics shown in Figure 11. 21 at a $Q$ point defined by $V_{GSQ} = 3.5$ and $V_{DSQ} = 10$ V.

**Solution**    First, we locate the $Q$ point as shown in the figure. Then, we use Equation 11.28 to find $g_m$:

$$g_m \cong \left. \frac{\Delta i_D}{\Delta v_{GS}} \right|_{v_{DS}=V_{DSQ}=10 \text{ V}}$$

We must make changes around the $Q$ point while holding $v_{DS}$ constant at 10 V. Thus, the incremental changes are made along a vertical line through the $Q$ point. To obtain a representative value for $g_m$, we consider an increment centered on the $Q$ point (rather than making the changes in one direction from the $Q$ point). Taking the changes starting from the curve below the $Q$ point and ending at the curve above the $Q$ point, we have $\Delta i_D \cong 10.7 - 4.7 = 6$ mA and $\Delta v_{GS} = 1$ V. The $\Delta i_D$ increment is labeled in the figure. Thus, we have

$$g_m = \frac{\Delta i_D}{\Delta v_{GS}} = \frac{6 \text{ mA}}{1 \text{ V}} = 6 \text{ mS}$$

The drain resistance is found by applying Equation 11.30:

$$\frac{1}{r_d} = \left. \frac{\Delta i_D}{\Delta v_{DS}} \right|_{v_{GS}=V_{GSQ}}$$

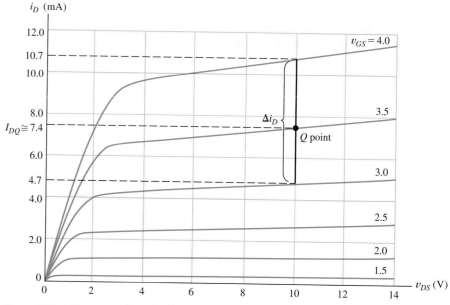

**Figure 11.21** Determination of $g_m$ and $r_d$. See Example 11.3.

Because the incremental changes are to be made while holding $v_{GS}$ constant, the changes are made along the characteristic curve through the $Q$ point. Thus, $1/r_d$ is the slope of the curve through the $Q$ point. For $v_{GS} = V_{GSQ} = 3.5$ V, we obtain $i_D \cong 6.7$ mA at $v_{DS} = 4$ V, and $i_D \cong 8.0$ mA at $v_{DS} = 14$ V. Thus, we get

$$\frac{1}{r_d} = \frac{\Delta i_D}{\Delta v_{DS}} \cong \frac{(8.0 - 6.7)\,\text{mA}}{(14 - 4)\,\text{V}} = 0.13 \times 10^{-3}$$

Taking the reciprocal, we find $r_d = 7.7$ k$\Omega$.    ■

**Exercise 11.7**    Find the values of $g_m$ and $r_d$ for the characteristics of Figure 11.21 at a $Q$ point of $V_{GSQ} = 2.5$ V and $V_{DSQ} = 6$ V.
**Answer**    $g_m \cong 3.3$ mS, $r_d \cong 20$ k$\Omega$.    ☐

**Exercise 11.8**    Show that Equation 11.21 results from application of Equation 11.29 to Equation 11.4.    ☐

## 11.5    COMMON-SOURCE AMPLIFIERS

The circuit diagram of a **common-source amplifier** is shown in Figure 11.22. The ac signal to be amplified is $v(t)$. The **coupling capacitors** $C_1$ and $C_2$, as well as the **bypass capacitor** $C_S$, are intended to have very small impedances for the ac signal. In this section, we carry out a midband analysis in which we assume that these capacitors are short circuits for the signal. The resistors $R_1$, $R_2$, $R_S$, and $R_D$ form the bias network, and their values are selected to obtain a suitable $Q$ point. The amplified output signal is applied to the load $R_L$.

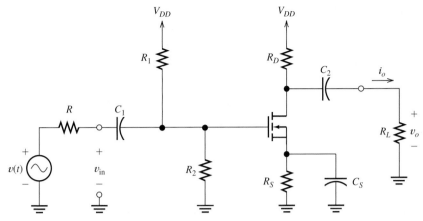

**Figure 11.22** Common-source amplifier.

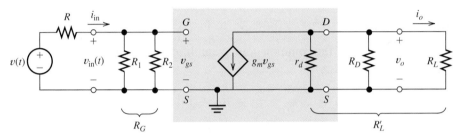

**Figure 11.23** Small-signal equivalent circuit for the common-source amplifier.

## The Small-Signal Equivalent Circuit

The small-signal equivalent circuit for the amplifier is shown in Figure 11.23. The input coupling capacitor $C_1$ has been replaced by a short circuit. The MOSFET has been replaced by its small-signal equivalent. Because the bypass capacitor $C_S$ is assumed to be a short circuit, the source terminal of the FET is connected directly to ground—which is why the circuit is called a *common-source amplifier*.

The dc supply voltage source acts as a short circuit for the ac signal. (Even if ac current flows through the dc source, the ac voltage across it is zero. Thus, for ac currents, the dc voltage source is a short.) Consequently, both $R_1$ and $R_2$ are connected from gate to ground in the equivalent circuit. Similarly, $R_D$ is connected from drain to ground.

The dc supply voltage source acts as a short circuit for ac current.

## Voltage Gain

Next, we derive an expression for the voltage gain of the common-source amplifier. Refer to the small-signal equivalent circuit, and notice that the resistances $r_d$, $R_D$, and $R_L$ are in parallel. We denote the equivalent resistance by

$$R'_L = \frac{1}{1/r_d + 1/R_D + 1/R_L} \qquad (11.32)$$

The output voltage is the product of the current from the controlled source and the equivalent resistance, given by

$$v_o = -(g_m v_{gs})\, R'_L \qquad (11.33)$$

The minus sign is needed because of the reference directions selected (i.e., the current $g_m v_{gs}$ flows out of the positive end of the voltage reference for $v_o$). Furthermore, the input voltage and the gate-to-source voltage are equal:

$$v_{in} = v_{gs} \tag{11.34}$$

Now if we divide the respective sides of Equation 11.33 by those of Equation 11.34, we obtain the voltage gain:

$$A_v = \frac{v_o}{v_{in}} = -g_m R_L' \tag{11.35}$$

The minus sign in the expression for the voltage gain shows that the common-source amplifier is inverting. Notice that the voltage gain is proportional to $g_m$.

In small-signal midband analysis of FET amplifiers, the coupling capacitors, bypass capacitors, and dc voltage sources are replaced by short circuits. The FET is replaced with its small-signal equivalent circuit. Then, we write circuit equations and derive useful expressions for gains, input impedance, and output impedance.

## Input Resistance

The input resistance of the common-source amplifier is given by

$$R_{in} = \frac{v_{in}}{i_{in}} = R_G = R_1 \| R_2 \tag{11.36}$$

in which $R_1 \| R_2$ denotes the parallel combination of $R_1$ and $R_2$. The resistances $R_1$ and $R_2$ form part of the bias network, but their values are not critical. (See Section 11.3 for a discussion of the bias circuit.) Practical resistance values range from 0 to perhaps 10 M$\Omega$ in discrete-component circuits. Thus, we have a great deal of freedom in design of the input resistance of a common-source amplifier. (We will see in the next chapter that this is not true for BJT amplifier circuits.)

We have a great deal of freedom in design of the input resistance of a common-source amplifier.

## Output Resistance

To find the output resistance of an amplifier, we disconnect the load, replace the signal source by its internal resistance, and then find the resistance looking into the output terminals. The equivalent circuit with these changes is shown in Figure 11.24.

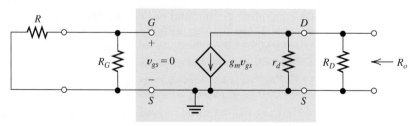

**Figure 11.24** Circuit used to find $R_o$.

Because there is no source connected to the input side of the circuit, we conclude that $v_{gs} = 0$. Therefore, the controlled current source $g_m v_{gs}$ produces zero current and appears as an open circuit. Consequently, the output resistance is the parallel combination of $R_D$ and $r_d$:

$$R_o = \frac{1}{1/R_D + 1/r_d} \qquad (11.37)$$

To find the output resistance of an amplifier, we disconnect the load, replace the signal source by its internal resistance, and then find the resistance by looking into the output terminals.

---

**Example 11.4** **Gain and Impedance Calculations for a Common-Source Amplifier**

Consider the common-source amplifier shown in Figure 11.25. The NMOS transistor has $KP = 50\ \mu\text{A/V}^2$, $V_{to} = 2$ V, $L = 10\ \mu\text{m}$, and $W = 400\ \mu\text{m}$. Find the midband voltage gain, input resistance, and output resistance of the amplifier. Then, assuming that the input source is given by

$$v(t) = 100 \sin(2000\pi t)\ \text{mV}$$

compute the output voltage. Also, assume that the frequency of the source (which is 1000 Hz) is in the midband region.

**Solution** First, we need to find the $Q$ point so we can determine the value of $g_m$ for the MOSFET. The bias circuit consists of $R_1$, $R_2$, $R_D$, $R_S$, and the MOSFET. This circuit was analyzed in Example 11.2, where we determined that $I_{DQ} = 0.784$ mA.

Next, we use Equation 11.25 to find the transconductance of the device:

$$g_m = \sqrt{2KP}\sqrt{W/L}\sqrt{I_{DQ}} = 1.77\ \text{mS}$$

Because the drain characteristics are horizontal in the saturation region, we have $r_d = \infty$.

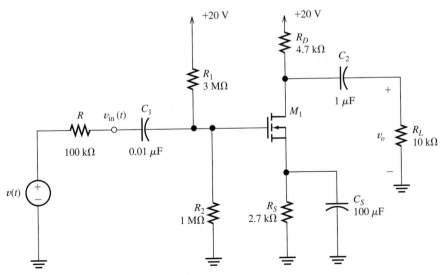

**Figure 11.25** Common-source amplifier.

Now, we use Equations 11.32, 11.35, 11.36, and 11.37 to find

$$R'_L = \frac{1}{1/r_d + 1/R_D + 1/R_L} = 3197 \ \Omega$$

$$A_v = \frac{v_o}{v_{in}} = -g_m R'_L = -5.66$$

$$R_{in} = \frac{v_{in}}{i_{in}} = R_G = R_1 \| R_2 = 750 \ k\Omega$$

$$R_o = \frac{1}{1/R_D + 1/r_d} = 4.7 \ k\Omega$$

The signal voltage divides between the internal source resistance and the input resistance of the amplifier. Thus, we have

$$v_{in} = v(t) \frac{R_{in}}{R + R_{in}} = 88.23 \ \sin(2000\pi t) \ \text{mV}$$

Then, the output voltage can be found as

$$v_o(t) = A_v v_{in}(t) = -500 \ \sin(2000\pi t) \ \text{mV}$$

Notice the phase inversion of $v_o(t)$ compared to $v_{in}(t)$.                      ■

**Exercise 11.9**   Find the voltage gain of the amplifier of Example 11.4 with $R_L$ replaced by an open circuit.
**Answer**   $A_{voc} = -8.32$.                                                    □

**Exercise 11.10**   Consider the circuit of Figure 11.22 with the bypass capacitor $C_S$ replaced by an open circuit. Draw the small-signal equivalent circuit. Then assuming that $r_d$ is an open circuit for simplicity, derive an expression for the voltage gain in terms of $g_m$ and the resistances.
**Answer**   $A_v = -g_m R'_L/(1 + g_m R_S)$.                                       □

**Exercise 11.11**   Evaluate the gain expression found in Exercise 11.10 by using the values given in Example 11.4. Compare the result with the voltage gain found in the example.
**Answer**   $A_v = -0.979$ without the bypass capacitor compared to $A_v = -5.66$ with the bypass capacitor in place. Notice that unbypassed impedance between the FET source terminal and ground strongly reduces the gain of a common-source amplifier.                                                        □

Unbypassed impedance between the FET source terminal and ground strongly reduces the gain of a common-source amplifier.

## 11.6  SOURCE FOLLOWERS

Another amplifier circuit known as a **source follower** is shown in Figure 11.26. The signal to be amplified is $v(t)$, and $R$ is the internal (Thévenin) resistance of the signal source. The coupling capacitor $C_1$ causes the ac input signal to appear at the gate of the FET. The capacitor $C_2$ connects the load to the source terminal of the MOSFET. (In the midband analysis of the amplifier, we assume that the coupling capacitors behave as short circuits.) The resistors $R_S$, $R_1$, and $R_2$ form the bias circuit.

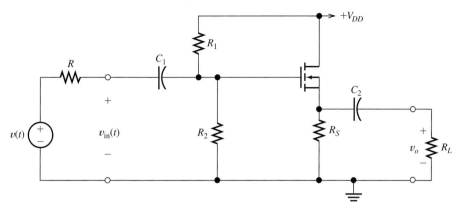

**Figure 11.26**  Source follower.

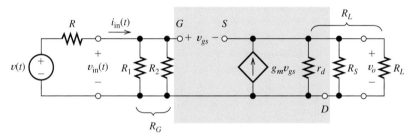

**Figure 11.27**  Small-signal ac equivalent circuit for the source follower.

## The Small-Signal Equivalent Circuit

The small-signal equivalent circuit is shown in Figure 11.27. The coupling capacitors have been replaced by short circuits, and the FET has been replaced by its small-signal equivalent. Notice that the drain terminal is connected directly to ground because the dc supply becomes a short (for ac currents) in the small-signal equivalent. Here the FET equivalent circuit is drawn in a different configuration (i.e., with the drain at the bottom) from that shown earlier, but it is the same electrically.

The ability to draw the small-signal equivalent for an amplifier circuit is important. Test yourself to see if you can obtain the small-signal circuit starting from Figure 11.26.

## Voltage Gain

Now, we derive an expression for the voltage gain of the source follower. Notice that $r_d$, $R_S$, and $R_L$ are in parallel. We denote the parallel combination by

$$R'_L = \frac{1}{1/r_d + 1/R_s + 1/R_L} \tag{11.38}$$

The output voltage is given by

$$v_o = g_m v_{gs} R'_L \tag{11.39}$$

Furthermore, we can write the following voltage equation:

$$v_{in} = v_{gs} + v_o \tag{11.40}$$

Using Equation 11.39 to substitute for $v_o$ in Equation 11.40, we have

$$v_{in} = v_{gs} + g_m v_{gs} R'_L \tag{11.41}$$

Dividing the respective sides of Equations 11.39 and 11.41, we obtain the following expression for the voltage gain:

$$A_v = \frac{v_o}{v_{in}} = \frac{g_m R'_L}{1 + g_m R'_L} \tag{11.42}$$

Notice that the voltage gain given in Equation 11.42 is positive and is less than unity. Thus, the source follower is a noninverting amplifier with voltage gain less than unity.

The source follower is a noninverting amplifier with voltage gain less than unity.

## Input Resistance

The input resistance is the resistance seen looking into the input terminals of the equivalent circuit. Thus, we have

$$R_{in} = \frac{v_{in}}{i_{in}} = R_G = R_1 \| R_2 \tag{11.43}$$

in which $R_1 \| R_2$ denotes the parallel combination of $R_1$ and $R_2$.

## Output Resistance

To find the output resistance, we remove the load resistance, replace the signal source with its internal resistance, and look back into the output terminals. It is helpful to attach a test source $v_x$ to the output terminals as shown in Figure 11.28. Then, the output resistance is found as

$$R_o = \frac{v_x}{i_x} \tag{11.44}$$

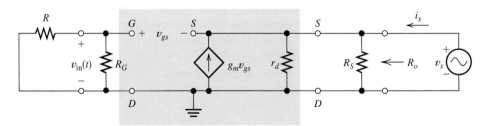

Figure 11.28 Equivalent circuit used to find the output resistance of the source follower.

where $i_x$ is the current supplied by the test source as shown in the figure. It can be shown that the output resistance is given by

$$R_o = \frac{1}{g_m + 1/R_S + 1/r_d} \qquad (11.45)$$

This can be quite low in value, and one reason for using a source follower is to obtain low output resistance.

One reason for using a source follower is to obtain low output resistance.

## Example 11.5   Gain and Impedance Calculations for a Source Follower

Consider the source follower shown in Figure 11.26 given $R_L = 1\,\text{k}\Omega$ and $R_1 = R_2 = 2\,\text{M}\Omega$. The NMOS transistor has $KP = 50\,\mu\text{A/V}^2$, $L = 2\,\mu\text{m}$, $W = 160\,\mu\text{m}$, and $V_{to} = 1\,\text{V}$. Find the value for $R_S$ to achieve $I_{DQ} = 10\,\text{mA}$. Then compute the voltage gain, input resistance, and output resistance.

**Solution**   From Equations 11.3 and 11.4, we have

$$K = \left(\frac{W}{L}\right)\frac{KP}{2} = 2\,\text{mA/V}^2$$

and

$$I_{DQ} = K\,(V_{GSQ} - V_{to})^2$$

Solving for $V_{GSQ}$ and substituting values, we get

$$V_{GSQ} = \sqrt{I_{DQ}/K} + V_{to} = 3.236\,\text{V}$$

The dc voltage at the gate terminal (with respect to ground) is given by

$$V_G = V_{DD} \times \frac{R_2}{R_1 + R_2} = 7.5\,\text{V}$$

The dc voltage at the source terminal of the NMOS is

$$V_S = V_G - V_{GSQ} = 4.264\,\text{V}$$

Finally, we find the source resistance as

$$R_S = \frac{V_S}{I_{DQ}} = 426.4\,\Omega$$

(Of course, in a discrete circuit, we would choose a standard nominal value for $R_S$. However, we will continue this example by using the exact value computed for $R_S$.)
    Next, we use Equation 11.25 to find the transconductance of the device:

$$g_m = \sqrt{2KP}\sqrt{W/L}\sqrt{I_{DQ}} = 8.944\,\text{mS}$$

Because the drain characteristics are horizontal in the saturation region, we have $r_d = \infty$.
    Next, we substitute values into Equation 11.38 to obtain

$$R'_L = \frac{1}{1/r_d + 1/R_S + 1/R_L} = 298.9\,\Omega$$

Then, the voltage gain, given by Equation 11.42, is

$$A_v = \frac{v_o}{v_{in}} = \frac{g_m R'_L}{1 + g_m R'_L} = 0.7272$$

The input resistance is

$$R_{in} = R_1 \| R_2 = 1 \text{ M}\Omega$$

The output resistance, given by Equation 11.45, is

$$R_o = \frac{1}{g_m + 1/R_S + 1/r_d} = 88.58 \ \Omega$$

This is a fairly low output resistance compared with that of other single-FET amplifier configurations.

The current gain is shown by the use of Equation 11.3 to be

$$A_i = A_v \frac{R_{in}}{R_L} = 727.2$$

The power gain is given by

$$G = A_v A_i = 528.8$$

The source follower has voltage gain slightly less than unity, high input impedance, and low output impedance. Current gain and power gain can be larger than unity.

Even though the voltage gain is less than unity, the output power is much greater than the input power because of the high input resistance. ■

**Exercise 11.12**   Derive Equation 11.45.    ☐

**Exercise 11.13**   Derive expressions for the voltage gain, input resistance, and output resistance of the **common-gate amplifier** shown in Figure 11.29, assuming that $r_d$ is an open circuit.
**Answer**   The small-signal equivalent circuit is shown in Figure 11.30. $A_v = g_m R'_L$; $R_{in} = 1/(g_m + 1/R_S)$; $R_o = R_D$.    ☐

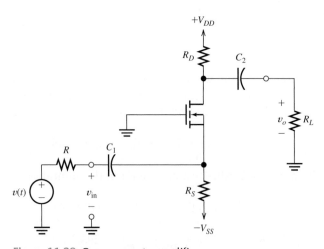

**Figure 11.29** Common-gate amplifier.

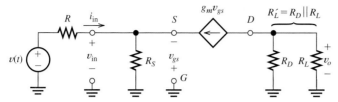

Figure 11.30  See Exercise 11.13.

## 11.7  CMOS LOGIC GATES

In **complementary metal–oxide–semiconductor** (CMOS) technology, both NMOS and PMOS transistors are fabricated on the same chip. Next, we describe how the basic building blocks of digital systems (NAND gates or NOR gates) are constructed using CMOS technology.

### CMOS Inverter

A CMOS inverter is shown in Figure 11.31. The NMOS and PMOS FETs are constructed by adding impurities to a silicon crystal, forming regions of $n$-type and $p$-type semiconductor as shown in part (a) of the figure. Notice that the gates $G$ are insulated from the rest of the circuit by layers of silicon dioxide ($SiO_2$). Thus, the input behaves as an open circuit (except for a small amount of capacitance).

**PRACTICAL APPLICATION   11.1**

**Where *Did* Those Trout Go?**

Fish biologists often need to learn about the migratory behavior of various kinds of fish to properly regulate fishing and habitat changes. One example is the effort to reestablish the coaster brook trout in various streams on the south shore of Lake Superior. Coaster brook trout migrate from their natal streams into Lake Superior, where they grow much larger than their stream-dwelling cousins. As adults, highly colored coasters return to their native streams each fall to spawn. Originally, coasters were found in nearly all of the rivers entering Lake Superior; however, over-fishing and habitat changes due mainly to logging have caused them to disappear from much of their original range. Currently, several projects are underway to attempt to reestablish these beautiful trout in south shore streams.

A powerful approach to gaining accurate information about fish migration is to implant radio frequency identification (RFID) tags in the fish. Then, antennas located in the streams can monitor movement of the fish. A wide variety of RFID systems are in use. We will describe a single representative system.

A typical RFID tag used in fisheries research consists of a coil of enameled copper wire wound on a ferrite core, a CMOS integrated circuit chip, and two capacitors, all hermetically sealed in a glass tube comparable in size to a large grain of rice. These tags are implanted in fish through hypodermic needles. The tags do not contain an internal power source, so they are sometimes called passive identification tags (PIT).

A typical streamside fish monitoring station is illustrated in Figure PA11.1. Because the important locations are often far from power lines, power for the station is provided by deep-cycle storage

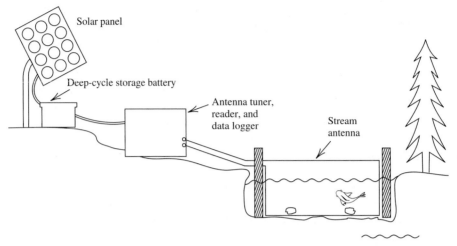

**Figure PA11.1  Typical monitoring station.**

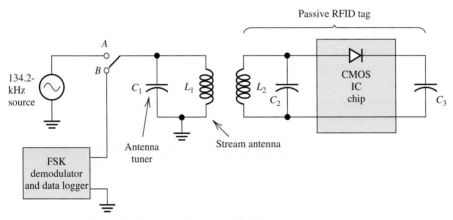

**Figure PA11.2  Electrical schematic diagram of a fish monitoring system.**

batteries recharged by solar panels. The antenna often consists of a loop of wire suspended above the stream on poles and weighted to the stream bottom with rocks.

A schematic diagram of a representative system is shown in Figure PA11.2. The stream antenna appears as the inductance $L_1$, typically 10 to 100 $\mu$H in value. A capacitance $C_1$, called an antenna tuner, forms a parallel resonant circuit with $L_1$ having a resonant frequency of 134.2 kHz. The coil in the RFID tag also acts as an antenna and is represented by the inductance $L_2$, which forms a parallel resonant circuit with $C_2$.

In operation, the switch periodically moves to contact A, applying a 134.2-kHz sine wave to the stream antenna, which creates an ac magnetic field in the vicinity of the antenna. When a tagged fish is present, part of the magnetic flux links $L_2$, resulting in a 134.2-kHz voltage at the input to the CMOS IC chip. This voltage is rectified by the diode contained in the chip charging $C_3$. After about 50 ms, the switch moves to contact $B$ so that the voltage applied to the stream antenna becomes zero. Then, power is supplied to the CMOS chip by the charge stored on $C_3$. When the CMOS chip senses the end of the pulse, it transmits a 64-bit codeword that identifies the particular tag and, hence, the fish into which it was implanted.

Frequency shift keying (FSK) is used to encode the bits. For a 1-bit, the chip applies 16 cycles of a 123.2-kHz signal to $L_2$; and for a 0-bit, it applies 16 cycles of a 134.2-kHz signal to $L_2$. The resulting magnetic field partially links $L_1$, inducing a voltage that is applied to the FSK demodulator and data logger.

The FSK demodulator determines the frequency of each 16-cycle segment and the resulting bit value. The resulting code words are saved by the data logger, which is read periodically by the fish biologist. The data logger can also save additional data, such as the time of day that the fish passed through the station, the stream temperature, and the stream flow rate, if additional sensors are placed in the stream.

Additional information about coaster brook-trout programs and fish monitoring systems can be found on the web.

The circuit diagram is shown in Figure 11.31(b). The dc supply voltage $V_{DD}$ is applied at the top of the circuit. When the input voltage is high ($V_{in} = V_{DD}$), a conducting channel is induced between the drain $D$ and the source $S$ of the NMOS transistor. Thus, the NMOS transistor becomes a low resistance and ideally behaves as a closed switch, as shown in Figure 11.31(c). Furthermore, the PMOS transistor

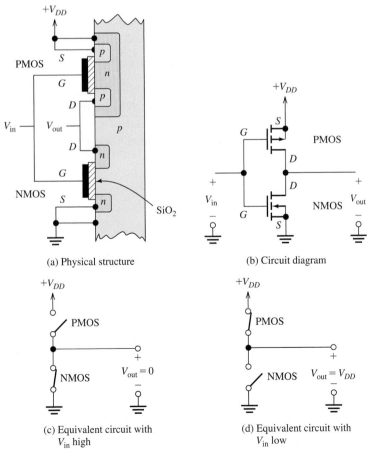

(a) Physical structure

(b) Circuit diagram

(c) Equivalent circuit with $V_{in}$ high

(d) Equivalent circuit with $V_{in}$ low

Figure 11.31  CMOS inverter.

is off, and it behaves as an open switch. Thus, with $V_{in}$ high, the output voltage $V_{out}$ becomes low (i.e., $V_{out} = 0$).

On the other hand, with $V_{in}$ low, a conducting layer is induced under the gate in the PMOS transistor, but not in the NMOS transistor. Therefore, the PMOS transistor is on and the NMOS transistor is off. This is illustrated in Figure 11.31(d). Then the output voltage $V_{out}$ is high ($V_{out} = V_{DD}$).

Because of the switching action of the transistors, the output is low when the input is high, and vice versa. This is exactly how a logic inverter is supposed to behave.

With the input low, the PMOS is on, the NMOS is off, and the output is high. With the input high, the situation is reversed.

## CMOS NAND Gate

By adding transistors to the inverter circuit, we can construct a NAND gate. The circuit for a two-input NAND gate is shown in Figure 11.32(a). Notice that we have two PMOS transistors in parallel and two NMOS transistors in series.

When their gates are high, the NMOS devices are on, and when their gates are low, the NMOS devices are off. The opposite is true for the PMOS transistors (i.e., the PMOS transistors are on when their gates are low, and they are off when their gates are high).

The equivalent circuit with $A$ high and $B$ low is shown in Figure 11.32(b). Furthermore, the equivalent circuit with both $A$ and $B$ high is shown in Figure 11.32(c). Notice that because of the switching action of the transistors, the output is low only if both $A$ and $B$ are high. This is exactly the way that a NAND gate is supposed to behave. By adding more transistors, we could produce a three-input NAND gate.

Placing $N$ PMOS transistors in parallel and $N$ NMOS transistors in series produces an $N$-input NAND gate.

## CMOS NOR Gate

The circuit diagram of a two-input NOR gate is shown in Figure 11.33. In this case, we have two PMOS devices in series and two NMOS devices in parallel. The operation of the NOR gate is very similar to that of the NAND gate discussed previously. Here again, the NMOS devices are on when their gates are high and off when their gates are low. The situation is reversed for the PMOS devices.

Placing $N$ PMOS transistors in series and $N$ NMOS transistors in parallel produces an $N$-input NOR gate.

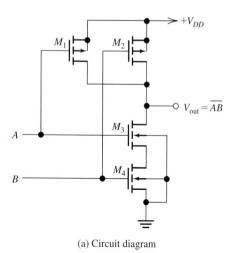
(a) Circuit diagram

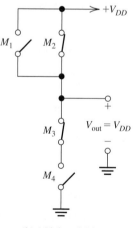
(b) $A$ high and $B$ low

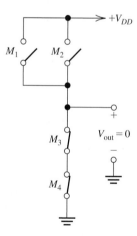
(c) $A$ and $B$ both high

**Figure 11.32** Two-input CMOS NAND gate.

**Exercise 11.14** Draw the equivalent circuits [similar to Figure 11.32(b) and (c)] of the NOR gate shown in Figure 11.33 for: **a.** $A$ high and $B$ high; **b.** $A$ high and $B$ low; **c.** $A$ low and $B$ low. Then give a truth table for the gate.

**Answer** See Figure 11.34. □

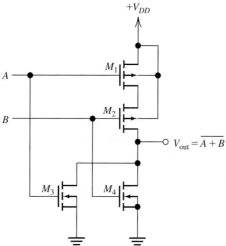

**Figure 11.33** Two-input CMOS NOR gate.

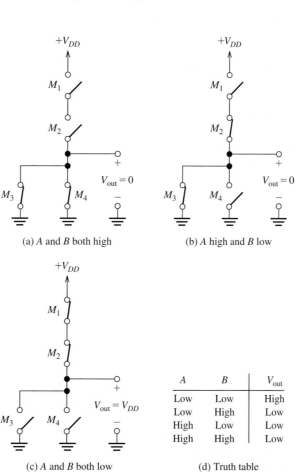

**Figure 11.34** Answers for Exercise 11.14.

| A | B | $V_{out}$ |
|------|------|------|
| Low | Low | High |
| Low | High | Low |
| High | Low | Low |
| High | High | Low |

(c) $A$ and $B$ both low          (d) Truth table

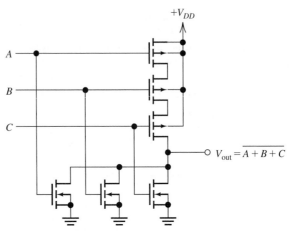

**Figure 11.35** Three-input CMOS NOR gate. (Answer for Exercise 11.15.)

**Exercise 11.15**    Draw the circuit diagram of a three-input NOR gate.
**Answer**    See Figure 11.35.                                    □

## Conclusions

In Chapter 7, we saw that complex combinatorial logic functions can be achieved simply by interconnecting NAND gates (or NOR gates). Furthermore, logic gates can be interconnected to form flip-flops. Interconnections of flip-flops form registers. A complex digital system, such as a computer, consists of many gates, flip-flops, and registers. Thus, logic gates are the basic building blocks for complex digital systems.

Complex digital systems can be constructed by interconnecting billions of NMOS and PMOS transistors, all of which are fabricated on a single silicon chip by a relatively small number of processing steps.

Modern technology can construct billions of CMOS gates on a silicon wafer by adding impurities, oxide layers, and metal interconnections. Relatively few (perhaps 20) steps are needed in the manufacturing process. This results in the production of powerful computers at low cost.

## Summary

1. The structure of an enhancement n-channel MOS transistor is shown in Figure 11.1 on page 558.

2. The MOSFET is the primary device that underlies the rapid advances in digital electronics over the past several decades.

3. In an NMOS transistor, when a sufficiently large (positive) voltage is applied to the gate relative to the source, electrons are attracted to the region under the gate, and a channel of $n$-type material is induced between the drain and the source. Then if voltage is applied between the drain and the source, current flows into the drain through the channel and out the source. Drain current is controlled by the voltage applied to the gate.

4. MOSFETs operate in cutoff, in the linear region, or in saturation.

5. Usually, $KP$ is determined by the fabrication process. However, in designing a circuit, we can vary the ratio $W/L$ to obtain transistors best suited to various parts of the circuit.

6. Simple amplifier circuits can be analyzed by using graphical (load-line) techniques.

7. Nonlinear distortion occurs in FET amplifiers because of the nonuniform spacing of the drain characteristics. Distortion is less pronounced for smaller signal amplitudes.

8. For use as amplifiers, FETs are usually biased in the saturation region.

9. In a small-signal midband analysis, a FET can be modeled by the equivalent circuit shown in Figure 11.20 on page 574.

10. Transconductance of a FET is defined as

$$g_m = \left.\frac{\partial i_D}{\partial v_{GS}}\right|_{Q \text{ point}}$$

11. Small-signal drain resistance of a FET is defined as

$$\frac{1}{r_d} = \left.\frac{\partial i_D}{\partial v_{DS}}\right|_{Q \text{ point}}$$

12. In small-signal midband analysis of FET amplifiers, the coupling capacitors, bypass capacitors, and dc voltage sources are replaced by short circuits. The FET is replaced with its small-signal equivalent circuit. Then, we write circuit equations and derive useful expressions for gains, input impedance, and output impedance.

13. To find the output resistance of an amplifier, we disconnect the load, replace the signal source by its internal resistance, and then find the resistance looking into the output terminals.

14. The common-source amplifier is inverting and can have voltage-gain magnitude larger than unity.

15. Unbypassed impedance between the FET source terminal and ground strongly reduces the gain of a common-source amplifier.

16. The source follower has voltage gain slightly less than unity, high current gain, and relatively low output impedance. It is noninverting.

17. Complex digital systems can be constructed by interconnecting millions of NMOS and PMOS transistors, all of which are fabricated on a single chip by a relatively small number of processing steps.

## Problems

### Section 11.1: NMOS and PMOS Transistors

**P11.1.** Sketch the physical structure of an $n$-channel enhancement MOSFET. Label the channel length $L$, the width $W$, the terminals, and the channel region. Draw the corresponding circuit symbol.

**P11.2.** Give the equations for the drain current and the ranges of $v_{GS}$, $v_{DS}$, and $v_{GD}$ in terms of the threshold voltage $V_{to}$ for each region (cutoff, saturation, and triode) of an $n$-channel MOSFET.

**\*P11.3.** A certain NMOS transistor has $V_{to} = 1$ V, $KP = 50\ \mu\text{A/V}^2$, $L = 5\ \mu\text{m}$, and $W = 50\ \mu\text{m}$. For each set of voltages, state the region of operation and compute the drain current. **a.** $v_{GS} = 4$ V and $v_{DS} = 10$ V; **b.** $v_{GS} = 4$ V and $v_{DS} = 2$ V; **c.** $v_{GS} = 0$ V and $v_{DS} = 10$ V.

**\*P11.4.** Suppose that we have an NMOS transistor with $KP = 50\ \mu\text{A/V}^2$, $V_{to} = 1$ V, $L = 10\ \mu\text{m}$, and $W = 200\ \mu\text{m}$. Sketch the drain characteristics

for $v_{DS}$ ranging from 0 to 10 V and $v_{GS} = 0, 1, 2, 3,$ and 4 V.

**P11.5.** We have an $n$-channel enhancement MOSFET with $V_{to} = 1$ V and $K = 0.1\ \text{mA/V}^2$. Given that $v_{GS} = 4$ V, for what range of $v_{DS}$ is the device in the saturation region? In the triode region? Plot $i_D$ versus $v_{GS}$ for operation in the saturation region.

**P11.6.** Suppose we have an NMOS transistor that has $V_{to} = 1$ V. What is the region of operation (linear, saturation, or cutoff) if **a.** $v_{GS} = 5$ V and $v_{DS} = 10$ V; **b.** $v_{GS} = 3$ V and $v_{DS} = 1$ V; **c.** $v_{GS} = 3$ V and $v_{DS} = 6$ V; **d.** $v_{GS} = 0$ V and $v_{DS} = 5$ V?

**P11.7.** What is the region of operation of an enhancement NMOS device if the gate is connected to the drain and a positive voltage greater than the threshold is applied to the drain with respect to the source? If the applied voltage is less than the threshold?

---

\*Denotes that answers are contained in the Student Solutions files. See Appendix E for more information about accessing the Student Solutions.

**P11.8.** Determine the region of operation for each of the enhancement transistors and the currents shown in Figure P11.8. The transistors have $|V_{to}| = 1$ V and $K = 0.1$ mA/V$^2$.

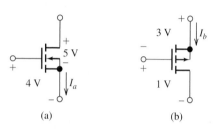

(a)                    (b)

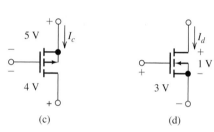

(c)                    (d)

**Figure P11.8**

**P11.9.** Suppose we need an NMOS transistor for which $i_D = 2$ mA when $v_{GS} = v_{DS} = 5$ V. Process constraints result in $KP = 50$ $\mu$A/V$^2$ and $V_{to} = 1$ V. Determine the width-to-length ratio needed for the transistor. If $L = 2$ $\mu$m, what is the value of $W$?

**P11.10.** Because of process constraints, $L$ and $W$ are required to be at least 0.25 $\mu$m. Furthermore, to save chip area, we do not want $L$ or $W$ to exceed 2 $\mu$m. How should we select $L$ and $W$ to obtain the least drain current for a given transistor? The greatest drain current? Assuming operation with identical voltages, what ratio of drain currents (between different transistors) can be achieved?

**\*P11.11.** Two points in the saturation region of a certain NMOS transistor are $(v_{GS} = 2$ V, $i_D = 0.2$ mA) and $(v_{GS} = 3$ V, $i_D = 1.8$ mA). Determine the values of $V_{to}$ and $K$ for this transistor.

**P11.12.** Suppose we have an NMOS transistor operating as a voltage-controlled resistance, as shown in Figure 11.4 on page 560, with $v_{DS} \ll v_{GS} - V_{to}$. Find an approximate expression for the resistance of the

channel in terms of the device parameters and voltages. Given $V_{to} = 0.5$ V and $K = 0.1$ mA/V$^2$, compute the resistance for $v_{GS} = 0.5, 1, 1.5$, and 2 V.

**P11.13.** Find the currents and the region of operation for each of the enhancement transistors shown in Figure P11.13 for $V_{in} = 0$ and for $V_{in} = 5$ V. The transistors have $|V_{to}| = 1$ V and $K = 0.2$ mA/V$^2$.

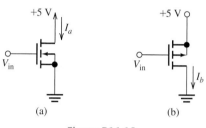

(a)                    (b)

**Figure P11.13**

**P11.14.** Given that the enhancement transistor shown in Figure P11.14 has $V_{to} = 1$ V and $K = 0.5$ mA/V$^2$, find the value of the resistance $R$.

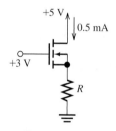

**Figure P11.14**

**\*P11.15.** A $p$-channel enhancement MOSFET has $V_{to} = -0.5$ V and $K = 0.2$ mA/V$^2$. Assuming operation in the saturation region, what value of $v_{GS}$ is required for $i_D = 0.8$ mA?

**Section 11.2:    Load-Line Analysis of a Simple NMOS Amplifier**

**P11.16.** What is the principal cause of distortion in FET amplifiers?

**\*P11.17.** Draw the load lines on the $i_D$–$v_{DS}$ axes for the circuit of Figure 11.10 on page 567 for

**a.** $R_D = 1$ k$\Omega$ and $V_{DD} = 20$ V

**b.** $R_D = 2$ k$\Omega$ and $V_{DD} = 20$ V

**c.** $R_D = 3$ k$\Omega$ and $V_{DD} = 20$ V

How does the position of the load line change as $R_D$ increases in value?

**P11.18.** Draw the load lines on the $i_D-v_{DS}$ axes for the circuit of Figure 11.10 on page 567 for

   **a.** $R_D = 1\ k\Omega$ and $V_{DD} = 5\ V$

   **b.** $R_D = 1\ k\Omega$ and $V_{DD} = 10\ V$

   **c.** $R_D = 1\ k\Omega$ and $V_{DD} = 15\ V$

How does the position of the load line change as $V_{DD}$ increases in value?

**\*P11.19.** Consider the circuit shown in Figure 11.10 on page 567. The transistor characteristics are shown in Figure 11.11. Suppose that $V_{GG}$ is changed to 0 V. Determine the values of $V_{DSQ}$, $V_{DSmin}$, and $V_{DSmax}$. Find the gain of the amplifier.

**P11.20.** Consider the amplifier shown in Figure P11.20.

   **a.** Find $v_{GS}(t)$, assuming that the coupling capacitor is a short circuit for the ac signal and an open circuit for dc. [*Hint:* Apply the superposition principle for the ac and dc sources.]

   **b.** If the FET has $V_{to} = 1\ V$ and $K = 0.5\ mA/V^2$, sketch its drain characteristics to scale for $v_{GS} = 1, 2, 3,$ and $4\ V$.

   **c.** Draw the load line for the amplifier on the characteristics.

   **d.** Find the values of $V_{DSQ}$, $V_{DSmin}$, and $V_{DSmax}$.

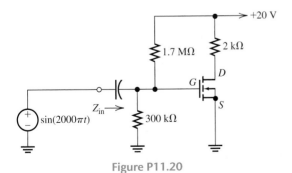

**Figure P11.20**

**\*P11.21.** What is the largest value of $R_D$ allowed in the circuit of Problem P11.20 if the instantaneous operating point is required to remain in the saturation region at all times?

**P11.22.** Use a load-line analysis of the circuit shown in Figure P11.22 to determine the values of $V_{DSQ}$, $V_{DSmin}$, and $V_{DSmax}$. The characteristics of the FET are shown in Figure 11.21 on page 576. [*Hint:* First, replace the 15-V source and the resistances by their Thévenin equivalent circuit.]

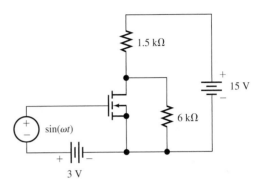

**Figure P11.22**

**P11.23.** Suppose that the resistance $R_D$ in Figure 11.10 (page 567) is replaced with an unusual two-terminal nonlinear device for which $v = 0.1i_D^2$, where $i_D$ is the current through the device in mA and v is the voltage across the device in volts (referenced positive at the end connected to $V_{DD}$). Carefully sketch the load line on Figure 11.11 (page 567). What shape is this load line?

**P11.24.** Use a load-line analysis for the PMOS amplifier shown in Figure P11.24 to determine the maximum, minimum, and Q-point values of $v_o(t)$. The characteristics of the transistor are shown in Figure 11.9 on page 565.

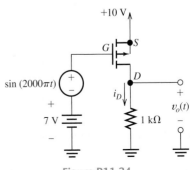

**Figure P11.24**

**P11.25.** The distorted signal shown in Figure 11.12(b) on page 568 can be written as

$$v_{DS}(t) = V_{DC} + V_{1m}\sin(2000\pi t)$$
$$+ V_{2m}\cos(4000\pi t)$$

The term $V_{1m}\sin(2000\pi t)$ is the desired signal. The term $V_{2m}\cos(4000\pi t)$ is distortion, which in this case has twice the frequency of the input signal, and is called second-harmonic distortion. Determine the values of $V_{1m}$, $V_{2m}$, and the percentage second-harmonic distortion, which is defined as $|V_{2m}/V_{1m}| \times 100\%$. (A high-quality audio amplifier has a distortion percentage less than 0.1 percent.)

### Section 11.3:  Bias Circuits

**P11.26.** In an amplifier circuit, why do we need to bias the MOSFET at an operating point? What would happen if the signal peak amplitude was smaller than 1 V, the transistor had $V_{to} = 1$ V, and we biased the transistor at $V_{GSQ} = 0$?

**\*P11.27.** Find $I_{DQ}$ and $V_{DSQ}$ for the circuit shown in Figure P11.27. The MOSFET has $V_{to} = 1$ V and $K = 0.25$ mA/V$^2$.

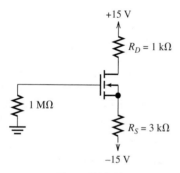

Figure P11.27

**\*P11.28.** We need a fixed- plus self-bias circuit for an NMOS source follower with $V_{DD} = 12$ V, $R_D = 0$, and $R_1 = 1$ MΩ. The transistor has $KP = 50\,\mu$A/V$^2$, $W = 800\,\mu$m, $L = 10\,\mu$m, and $V_{to} = 1$ V. The circuit is to have $V_{DSQ} = 6$ V and $I_{DQ} \cong 2$ mA. Determine the values of $R_2$ and $R_S$.

**\*P11.29.** The transistor of Figure P11.29 has $KP = 50\,\mu$A/V$^2$, $W = 600\,\mu$m, $L = 20\,\mu$m, and

$V_{to} = 1$ V. Determine the values of $R_1$ and $R_s$.

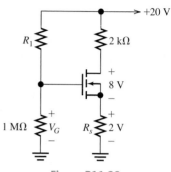

Figure P11.29

**P11.30.** The fixed- plus self-bias circuit of Figure 11.13 on page 569 has $V_{DD} = 15$ V, $R_1 = 2$ MΩ, $R_2 = 1$ MΩ, $R_S = 4.7$ kΩ, and $R_D = 4.7$ kΩ. The MOSFET has $V_{to} = 1$ V and $K = 0.25$ mA/V$^2$. Determine the $Q$ point.

**P11.31.** **a.** Find the value of $I_{DQ}$ for the circuit shown in Figure P11.31. Assume that $V_{to} = 4$ V and $K = 1$ mA/V$^2$. **b.** Repeat for $V_{to} = 2$ V and $K = 2$ mA/V$^2$.

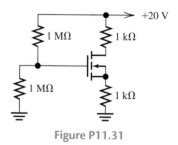

Figure P11.31

**P11.32.** Consider the fixed- plus self-bias circuit of Figure 11.13(a) on page 569, with $V_{DD} = 12$ V, $R_1 = 1$ MΩ, and $R_D = 3$ kΩ. Nominally, the transistor has $KP = 50\,\mu$A/V$^2$, $W = 80\,\mu$m, $L = 10\,\mu$m, and $V_{to} = 1$ V. The circuit is to have $V_{DSQ} = 6$ V and $I_{DQ} \cong 1$ mA. Determine the values needed for $R_2$ and for $R_S$.

**P11.33.** Find $I_{DQ}$ and $V_{DSQ}$ for the circuit shown in Figure P11.33. The MOSFET has $V_{to} = 1$ V and $K = 0.25$ mA/V$^2$.

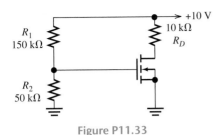

Figure P11.33

**\*P11.34.** Find $I_{DQ}$ and $V_{DSQ}$ for the circuit shown in Figure P11.34. The MOSFET has $V_{to} = 1$ V and $K = 0.25$ mA/V$^2$.

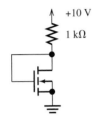

Figure P11.34

**P11.35.** Both transistors shown in Figure P11.35 have $KP = 100 \ \mu\text{A/V}^2$ and $V_{to} = 0.5$ V. Determine the value of $R$ needed so that $i_{D1} = 0.2$ mA. For what range of $V_x$ is the second transistor operating in the saturation region? What is the resulting value of $i_{D2}$? Provided that $V_x$ is large enough so that the second transistor operates in saturation, to what ideal circuit element is the transistor equivalent?

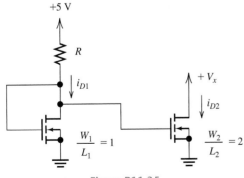

Figure P11.35

## Section 11.4: Small-Signal Equivalent Circuits

**P11.36.** Draw the small-signal equivalent circuit for a FET, including $r_d$.

**P11.37.** Give definitions of $g_m$ and $r_d$ as partial derivatives.

**P11.38.** The characteristic curves of a certain NMOS transistor have constant values for $i_D$ in the saturation region. What is the value of $r_d$, assuming operation in the saturation region?

**P11.39.** What is the value of $g_m$ for $V_{DSQ} = 0$? Draw the small-signal equivalent circuit at this bias point. For what applications could the FET be used at this bias point?

**\*P11.40.** Derive an expression for $g_m$ in terms of $K$, $V_{to}$, $V_{GSQ}$, and $I_{DQ}$ for an NMOS transistor operating in the *triode* region.

**\*P11.41.** Derive an expression for $r_d$ in terms of $K$, $V_{to}$, $V_{GSQ}$, and $I_{DQ}$ for an NMOS transistor operating in the *triode* region.

**P11.42.** A certain NMOS transistor has the characteristics shown in Figure P11.42. Graphically determine the values of $g_m$ and $r_d$ at the operating point defined by $V_{DSQ} = 6$ V and $V_{GSQ} = 2.5$ V.

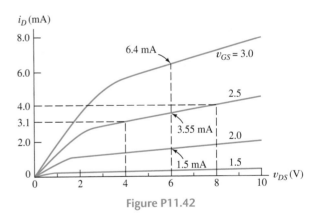

Figure P11.42

**P11.43.** Suppose that we have an unusual type of FET for which

$$i_D = 3v_{GS}^3 + 0.1v_{DS}$$

Here, $i_D$ is in mA, $v_{GS}$ is in volts, and $v_{DS}$ is in volts. Determine the values of $g_m$ and $r_d$ for a Q point of $V_{GSQ} = 1$ V and $V_{DSQ} = 10$ V.

**P11.44.** Suppose that we have an unusual type of FET for which

$$i_D = 3 \exp(v_{GS}) + 0.01v_{DS}^2$$

Here, $i_D$ is in mA, $v_{GS}$ is in volts, and $v_{DS}$ is in volts. Determine the values of $g_m$ and $r_d$ for a $Q$ point of $V_{GSQ} = 1$ V and $V_{DSQ} = 10$ V.

**P11.45.** Suppose we have an NMOS transistor that has $g_m = 2$ mS and $r_d = 5$ kΩ for a $Q$ point of $V_{GSQ} = 2$ V, $I_{DQ} = 4$ mA, and $V_{DSQ} = 10$ V. Sketch the drain characteristics to scale for a small region around the $Q$ point, say, for $v_{GS} = 1.8, 2.0,$ and $2.2$ V and for $9.0 < v_{DS} < 11.0$ V.

**P11.46.** A certain NMOS transistor has

$$v_{GS}(t) = 1 + 0.2 \sin(\omega t) \text{ V}$$

$$v_{DS}(t) = 4 \text{ V}$$

$$i_D(t) = 2 + 0.1 \sin(\omega t) \text{ mA}$$

Which small-signal parameter ($g_m$ or $r_d$) can be determined from this information? What is its value? For what $Q$ point ($V_{GSQ}, I_{DQ},$ and $V_{DSQ}$) does this parameter apply?

**P11.47.** A certain NMOS transistor has

$$v_{GS}(t) = 2 \text{ V}$$

$$v_{DS}(t) = 5 + 2 \sin(\omega t) \text{ V}$$

$$i_D(t) = 3 + 0.01 \sin(\omega t) \text{ mA}$$

Which small-signal parameter ($g_m$ or $r_d$) can be determined from this information? What is its value? For what $Q$ point ($V_{GSQ}, I_{DQ},$ and $V_{DSQ}$) does this parameter apply?

### Section 11.5: Common-Source Amplifiers

**P11.48.** What is the function of coupling capacitors? Assuming that they are performing their intended function, how do they appear in the ac equivalent circuit? In general, what effect do coupling capacitors have on the gain of an amplifier as a function of frequency?

**P11.49.** Draw the circuit diagram of a resistance–capacitance coupled common-source amplifier.

**\*P11.50.** Consider the common-source amplifier shown in Figure P11.50. The NMOS transistor has $KP = 50 \, \mu\text{A/V}^2$, $L = 5 \, \mu\text{m}$, $W = 500 \, \mu\text{m}$, $V_{to} = 1$ V, and $r_d = \infty$.
  **a.** Determine the values of $I_{DQ}, V_{DSQ},$ and $g_m$.

**b.** Compute the voltage gain, input resistance, and output resistance, assuming that the coupling capacitors are short circuits for the ac signal.

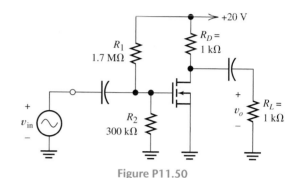

**Figure P11.50**

**P11.51.** Repeat Problem P11.50 for an NMOS transistor having $KP = 50 \, \mu\text{A/V}^2$, $W = 600 \, \mu\text{m}$, $L = 20 \, \mu\text{m}$, $V_{to} = 2$ V, and $r_d = \infty$. Compare the gain with that attained in Problem P11.50.

**P11.52.** Consider the amplifier shown in Figure P11.52.
  **a.** Draw the small-signal equivalent circuit, assuming that the capacitors are short circuits for the signal.
  **b.** Assume that $r_d = \infty$, and derive expressions for the voltage gain, input resistance, and output resistance.
  **c.** Find $I_{DQ}$ if $R = 100$ kΩ, $R_f = 100$ kΩ, $R_D = 3$ kΩ, $R_L = 10$ kΩ, $V_{DD} = 20$ V, $V_{to} = 5$ V, and $K = 1$ mA/V$^2$. Determine the value of $g_m$ at the $Q$ point.
  **d.** Evaluate the expressions found in part (b) by using the values given in part (c).
  **e.** Find $v_o(t)$ if $v(t) = 0.2 \sin(2000\pi t)$.
  **f.** Is this amplifier inverting or noninverting?

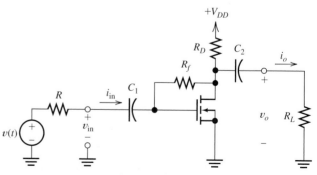

**Figure P11.52**

**\*P11.53.** Find $V_{DSQ}$ and $I_{DQ}$ for the FET shown in Figure P11.53, given $V_{to} = 3$ V and $K = 0.5$ mA/V$^2$. Find the value of $g_m$ at the operating point. Draw the small-signal equivalent circuit, assuming that $r_d = \infty$. Derive an expression for the resistance $R_o$ in terms of $R_D$ and $g_m$. Evaluate the expression for the values given.

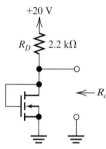

+20 V

$R_D$  2.2 kΩ

$\leftarrow R_o$

**Figure P11.53**

**Section 11.6:** Source Followers

**P11.54.** Draw the circuit diagram of a resistance–capacitance coupled source follower.

**P11.55.** Consider the common-source amplifier and the source follower. Which amplifier would be used if a voltage-gain magnitude larger than unity is needed? Which would be used to obtain low output resistance?

**\*P11.56.** Consider the source follower shown in Figure 11.26 on page 581, given $V_{DD} = 15$ V, $R_L = 2$ kΩ, $R_1 = 1$ MΩ, and $R_2 = 2$ MΩ.

The NMOS transistor has $KP = 50\ \mu$A/V$^2$, $L = 10\ \mu$m, $W = 160\ \mu$m, $r_d = \infty$, and $V_{to} = 1$ V. Find the value for $R_S$ to achieve $I_{DQ} = 2$ mA. Then, compute the voltage gain, input resistance, and output resistance.

**P11.57.** Consider the common-gate amplifier of Figure 11.29 on page 584, which was analyzed in Exercise 11.13 on page 584. The MOSFET has $KP = 50\ \mu$A/V$^2$, $W = 600\ \mu$m, $L = 10\ \mu$m, $V_{to} = 1$ V, and $r_d = \infty$. The supply voltages are $V_{DD} = 15$ V and $V_{SS} = 15$ V. The resistances are $R_S = 3$ kΩ, $R_L = 10$ kΩ, and $R_D = 3$ kΩ.
**a.** Determine the Q point and the value of $g_m$.
**b.** Determine the input resistance and the voltage gain.

**Section 11.7:** CMOS Logic Gates

**P11.58.** Draw the circuit diagram of a CMOS inverter. Draw its equivalent circuit (open and closed switches) if the input is high. Repeat if the input is low.

**P11.59.** Draw the circuit diagram of a two-input CMOS AND gate. [*Hint:* Use a two-input NAND followed by an inverter.]

**P11.60. a.** Draw the circuit diagram of a three-input CMOS NAND gate. **b.** Draw its equivalent circuit (open and closed switches) if all three inputs are high. **c.** Repeat if all three inputs are low.

**Practice Test**

Here is a practice test you can use to check your comprehension of the most important concepts in this chapter. Answers can be found in Appendix D and complete solutions are included in the Student Solutions files. See Appendix E for more information about the Student Solutions.

**T11.1.** An NMOS transistor has $KP = 80\ \mu$A/V$^2$, $V_{to} = 1$V, $L = 4\ \mu$m, and $W = 100\ \mu$m. Carefully sketch to scale the drain characteristics

for $v_{DS}$ ranging from 0 to 10V and $v_{GS} = 0.5$ and 4V.

**T11.2.** We have an amplifier identical to that of Figure 11.10 on page 567, except that $R_D$ is changed to 2 kΩ and the dc sources are changed to $V_{DD} = 10$ V and $V_{GG} = 3$ V. The drain characteristics for the transistor are shown in Figure 11.7 on page 564. Use load-line analysis to determine the maximum, minimum, and Q-point values of $v_{DS}$.

**T11.3.** Consider the biasing circuit shown in Figure T11.3. The transistor has $KP = 80\,\mu A/V^2$, $V_{to} = 1\,V$, $L = 4\,\mu m$, and $W = 100\,\mu m$. What value is required for $R_S$ so the operating current is $I_{DQ} = 0.5\,mA$?

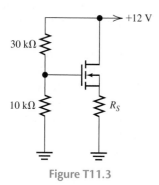

+12 V

30 kΩ

10 kΩ

$R_S$

**Figure T11.3**

**T11.4.** A certain NMOS transistor has

$$v_{GS}(t) = 2 + 0.02\sin(\omega t)\,V$$

$$v_{DS}(t) = 5\,V$$

$$i_D(t) = 0.5 + 0.05\sin(\omega t)\,mA$$

Which small-signal parameter ($g_m$ or $r_d$) can be determined from this information? What is its value? For what $Q$ point ($V_{GSQ}$, $I_{DQ}$, and $V_{DSQ}$) does this parameter apply?

**T11.5.** What replaces each of the following elements when we draw the mid-band small-signal equivalent circuit for an amplifier: **a.** a dc voltage source; **b.** a coupling capacitor; **c.** a dc current source?

**T11.6.** Draw the circuit diagram of a CMOS inverter. When the input is high, which transistor is on? Which is off?

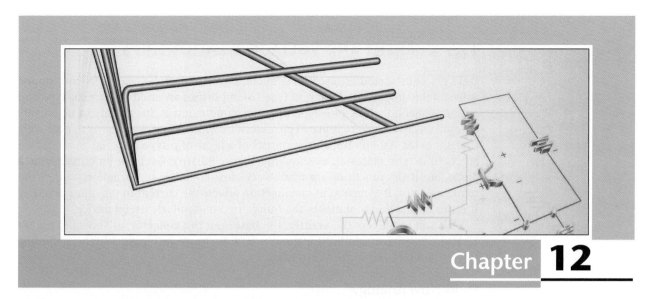

# Chapter 12

# Bipolar Junction Transistors

## Study of this chapter will enable you to:

- Understand bipolar junction transistor operation in amplifier circuits.

- Use the load-line technique to analyze simple amplifiers and understand the causes of nonlinear distortion.

- Use large-signal equivalent circuits to analyze BJT circuits.

- Analyze bias circuits.

- Use small-signal equivalent circuits to analyze BJT amplifiers.

- Compute performance of several important amplifier configurations.

- Select an amplifier configuration appropriate for a given application.

## Introduction to this chapter:

In Chapter 12, we considered the field-effect transistor, which is one of the important devices in modern electronics. Now, we turn our attention to another important device, the **bipolar junction transistor** (BJT), which is also very useful in amplifiers and digital logic circuits. These two devices are the key building blocks for modern electronics.

First, we discuss the device parameters and equations that relate the currents and voltages in the *npn* BJT. Next, we discuss the BJT common-emitter characteristics, which show the device operation graphically. Then, we use the graphical load-line technique to analyze a simple amplifier circuit. In Section 12.4, the *pnp* BJT is introduced. In the next several sections, we discuss large-signal models for the three regions of BJT operation (the active, saturation, and cutoff regions) and use the models to analyze bias circuits. Then, we develop a small-signal equivalent circuit for the BJT and use it to analyze two important amplifier configurations (common-emitter amplifiers and emitter followers).

## 12.1 CURRENT AND VOLTAGE RELATIONSHIPS

BJTs are constructed as layers of semiconductor materials (usually silicon) **doped** with suitable impurities. Different types of impurities are used to create *n*-type and *p*-type semiconductors. In *n*-type material, conduction is due mainly to negatively charged electrons, whereas in *p*-type material, conduction is due mainly to positively charged holes. An *npn* transistor consists of a layer of *p*-type material between two layers of *n*-type material, as shown in Figure 12.1(a). Each *pn* junction forms a diode, but if the junctions are made very close together in a single crystal of the semiconductor, the current in one junction affects the current in the other junction. It is this interaction that makes the transistor a particularly useful device.

We call the layers the **emitter**, the **base**, and the **collector**, as shown in Figure 12.1(a). The circuit symbol for an *npn* BJT is shown in Figure 12.1(b), including reference directions for the currents.

An *npn* BJT consists of a layer of *p*-type semiconductor, called the base, between two layers of *n*-type semiconductor, called the collector and the emitter. Assuming that proper voltages are applied, small amounts of current flowing into the base terminal cause much larger currents to flow from the collector to the emitter.

### Fluid-Flow Analogy

We will see that the BJT is somewhat analogous to a valve in a fluid-flow system. In suitable circuits, if a small current is made to flow into the base terminal, a much larger current flows into the collector and out of the emitter terminal. We can imagine that the base current opens a valve between collector and emitter. Larger base currents open the valve wider. When a signal to be amplified is applied as a current to the base, the valve between collector and emitter opens and closes in response to changes in the signal. Thus, a current with magnified fluctuations flows between collector and emitter.

### Equations of Operation

A *pn* junction is **forward biased** by applying voltage with the positive polarity on the *p*-side. On the other hand, **reverse bias** occurs if the positive polarity is applied to the *n*-side. This is illustrated in Figure 12.2.

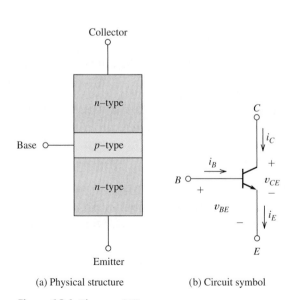

(a) Physical structure            (b) Circuit symbol

**Figure 12.1** The *npn* BJT.

In normal operation of a BJT as an amplifier, the base–collector junction is reverse biased and the base–emitter junction is forward biased. In the upcoming discussion, we assume that the junctions are biased in this fashion unless stated otherwise.

The Shockley equation gives the emitter current $i_E$ in terms of the base-to-emitter voltage $v_{BE}$:

$$i_E = I_{ES}\left[\exp\left(\frac{v_{BE}}{V_T}\right) - 1\right] \tag{12.1}$$

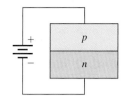

(a) Forward bias

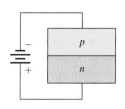

(b) Reverse bias

**Figure 12.2** Bias conditions for *pn* junctions.

This is exactly the same equation as that for the current in a junction diode given in Equation 10.1, except for changes in notation. (We have let the emission coefficient $n$ equal unity since that is the appropriate value for most junction transistors.) Typical values for the saturation current $I_{ES}$ range from $10^{-12}$ to $10^{-16}$ A, depending on the physical size of the device and other factors. (Recall that at a temperature of 300 K, $V_T$ is approximately 26 mV.)

Of course, Kirchhoff's current law requires that the current flowing out of the BJT is equal to the sum of the currents flowing into it. Thus, referring to Figure 12.1(b), we have

$$i_E = i_C + i_B \tag{12.2}$$

In normal operation as an amplifier, the base–collector junction is reverse biased and the base–emitter junction is forward biased.

(This equation is true regardless of the bias conditions of the junctions.)

We define the parameter $\alpha$ as the ratio of the collector current to the emitter current, displayed as

$$\alpha = \frac{i_C}{i_E} \tag{12.3}$$

Values for $\alpha$ range from 0.9 to 0.999, with 0.99 being very typical. Equation 12.2 indicates that the emitter current is supplied partly through the base terminal and partly through the collector terminal. However, since $\alpha$ is nearly unity, most of the emitter current is supplied through the collector.

Normally, $\alpha$ is slightly less than unity.

Substituting Equation 12.1 into 12.3 and rearranging, we have

$$i_C = \alpha I_{ES}\left[\exp\left(\frac{v_{BE}}{V_T}\right) - 1\right] \tag{12.4}$$

For $v_{BE}$ greater than a few tenths of a volt, the exponential term inside the brackets is much larger than unity. Then, the 1 inside the brackets can be dropped. Also, we define the **scale current** as

$$I_s = \alpha I_{ES} \tag{12.5}$$

and Equation 12.4 becomes

$$i_C \cong I_s \exp\left(\frac{v_{BE}}{V_T}\right) \tag{12.6}$$

Solving Equation 12.3 for $i_C$, substituting into Equation 12.2, and solving for the base current, we obtain

$$i_B = (1 - \alpha)i_E \tag{12.7}$$

Since $\alpha$ is slightly less than unity, only a very small fraction of the emitter current is supplied through the base. Using Equation 12.1 to substitute for $i_E$, we obtain

$$i_B = (1 - \alpha) I_{ES} \left[ \exp\left(\frac{v_{BE}}{V_T}\right) - 1 \right] \tag{12.8}$$

We define the parameter $\beta$ as the ratio of the collector current to the base current. Taking the ratio of Equations 12.4 and 12.8 results in

$$\beta = \frac{i_C}{i_B} = \frac{\alpha}{1 - \alpha} \tag{12.9}$$

Values for $\beta$ range from about 10 to 1000, and a very common value is $\beta = 100$. We can write

$$i_C = \beta i_B \tag{12.10}$$

Because $\beta$ is usually large compared to unity, the collector current is an amplified version of the base current.

Note that since $\beta$ is usually large compared to unity, *the collector current is an amplified version of the base current.* Current flow in an *npn* BJT is illustrated in Figure 12.3.

**Exercise 12.1**    A certain transistor has $\beta = 50$, $I_{ES} = 10^{-14}$ A, $v_{CE} = 5$ V, and $i_E = 10$ mA. Assume that $V_T = 0.026$ V. Find $v_{BE}$, $v_{BC}$, $i_B$, $i_C$, and $\alpha$.
**Answer**    $v_{BE} = 0.718$ V, $v_{BC} = -4.28$ V, $i_B = 0.196$ mA, $i_C = 9.80$ mA, $\alpha = 0.980$. ☐

**Exercise 12.2**    Compute the corresponding values of $\beta$ if $\alpha = 0.9, 0.99$, and $0.999$.
**Answer**    $\beta = 9, 99$, and $999$, respectively. ☐

**Exercise 12.3**    A certain transistor operated with forward bias of the base–emitter junction and reverse bias of the base–collector junction has $i_C = 9.5$ mA and $i_E = 10$ mA. Find the values of $i_B$, $\alpha$, and $\beta$.
**Answer**    $i_B = 0.5$ mA, $\alpha = 0.95$, $\beta = 19$. ☐

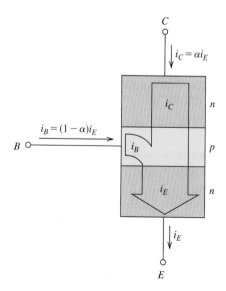

**Figure 12.3** Only a small fraction of the emitter current flows into the base (provided that the collector–base junction is reverse biased and the base–emitter junction is forward biased).

## 12.2  COMMON-EMITTER CHARACTERISTICS

The common-emitter configuration for an *npn* BJT is shown in Figure 12.4. The voltage source connected between the base and emitter supplies a positive voltage $v_{BE}$ that forward biases the base–emitter junction. The $v_{CE}$ voltage source produces a positive voltage at the collector with respect to the emitter. Notice that the voltage across the base–collector junction is given by

$$v_{BC} = v_{BE} - v_{CE} \qquad (12.11)$$

Thus, if $v_{CE}$ is greater than $v_{BE}$, the base-to-collector voltage $v_{BC}$ is negative (which is reverse bias).

The **common-emitter characteristics** of the transistor are plots of the currents $i_B$ and $i_C$ versus the voltages $v_{BE}$ and $v_{CE}$. Representative characteristics for a small-signal silicon device are shown in Figure 12.5.

The **common-emitter input characteristic** shown in Figure 12.5(a) is a plot of $i_B$ versus $v_{BE}$, which are related by Equation 12.8. Notice that the input characteristic takes the same form as the forward-bias characteristic of a junction diode. Thus, for appreciable current to flow at room temperature, the base-to-emitter voltage must be approximately 0.6 to 0.7 V. (The base-to-emitter voltage for a given current decreases with temperature by about 2 mV/K.)

The **common-emitter output characteristics** shown in Figure 12.5(b) are plots of $i_C$ versus $v_{CE}$ for constant values of $i_B$. The transistor illustrated has $\beta = 100$. As long as the collector–base junction is reverse biased ($v_{BC} < 0$ or equivalently, $v_{CE} > v_{BE}$),

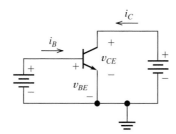

Figure 12.4  Common-emitter circuit configuration for the *npn* BJT.

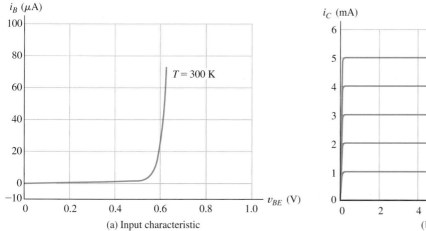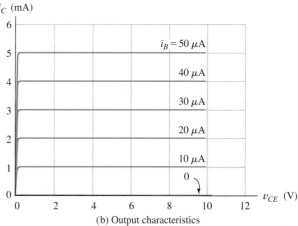

(a) Input characteristic

(b) Output characteristics

Figure 12.5  Common-emitter characteristics of a typical *npn* BJT.

we have

$$i_C = \beta i_B = 100 i_B$$

As $v_{CE}$ becomes less than $v_{BE}$, the base–collector junction becomes forward biased, and eventually the collector current falls as shown at the left-hand edge of the output characteristics.

## Amplification by the BJT

Refer to Figure 12.5(a), and notice that a very small change in the base-to-emitter voltage $v_{BE}$ can result in an appreciable change in the base current $i_B$, particularly if the base–emitter junction is forward biased so that some current (say 40 $\mu$A) is flowing before the change in $v_{BE}$ is made. Provided that $v_{CE}$ is more than a few tenths of a volt, this change in base current causes a much larger change in the collector current $i_C$ (because $i_C = \beta i_B$). In suitable circuits, the change in collector current is converted into a much larger voltage change than the initial change in $v_{BE}$. Thus, the BJT can amplify a signal applied to the base–emitter junction.

---

**Example 12.1**   Determining $\beta$ from the Characteristic Curves

Verify that the value of $\beta$ is 100 for the transistor with the characteristics shown in Figure 12.5.

**Solution**   The value of $\beta$ can be found by taking the ratio of collector current to base current provided that $v_{CE}$ is high enough so that the collector–base junction is reverse biased. For example, at $v_{CE} = 4$ V and $i_B = 30$ $\mu$A, the output characteristics yield $i_C = 3$ mA. Thus, the value of $\beta$ is

$$\beta = \frac{i_C}{i_B} = \frac{3 \text{ mA}}{30 \text{ } \mu\text{A}} = 100$$

(For most devices, slightly different values of $\beta$ result from different points on the output characteristics.) ∎

---

**Exercise 12.4**   Plot the common-emitter characteristics of an *npn* small-signal silicon transistor at a temperature of 300 K if $I_{ES} = 10^{-14}$ A and $\beta = 50$. Allow $i_B$ to range from 0 to 50 $\mu$A in 10-$\mu$A steps for the output characteristics. [*Hints:* For the input characteristic, use Equation 12.8 to calculate values of $v_{BE}$ for $i_B = 10$ $\mu$A, 20 $\mu$A, and so on. The output characteristic is identical to Figure 12.5(b), except for a change in scale for the $i_C$ axis.]
**Answer**   See Figure 12.6. ◻

## 12.3 LOAD-LINE ANALYSIS OF A COMMON-EMITTER AMPLIFIER

A simple BJT amplifier circuit is shown in Figure 12.7. The dc power-supply voltages $V_{BB}$ and $V_{CC}$ **bias** the device at an operating point for which amplification of the ac input signal $v_{in}(t)$ is possible. Next, we demonstrate that an amplified version of the input signal voltage appears between the collector and ground.

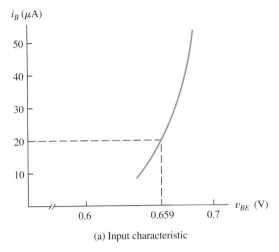

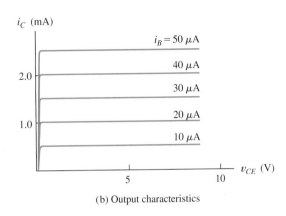

(a) Input characteristic

(b) Output characteristics

**Figure 12.6** See Exercise 12.4.

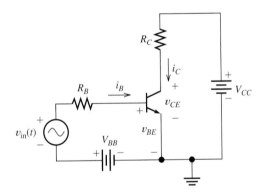

**Figure 12.7** A simple common-emitter amplifier that can be analyzed by load-line techniques.

## Analysis of the Input Circuit

We can analyze this circuit by use of load-line techniques similar to those that we used for diode circuits in Chapter 9. For example, if we apply Kirchhoff's voltage law to the loop consisting of $V_{BB}$, $v_{in}(t)$, and the base–emitter junction, we obtain

$$V_{BB} + v_{in}(t) = R_B i_B(t) + v_{BE}(t) \tag{12.12}$$

Equation for the input load line.

A plot of Equation 12.12 is shown as the **load line** on the input characteristics of the transistor in Figure 12.8(a). To establish this load line, we must locate two points. If we assume that $i_B = 0$, Equation 12.12 yields $v_{BE} = V_{BB} + v_{in}$. This establishes the point where the load line intersects the voltage axis. Similarly, assuming that $v_{BE} = 0$ results in $i_B = (V_{BB} + v_{in})/R_B$, which establishes the load-line intercept on the current axis. The load line is shown as the solid line in Figure 12.8(a).

Equation 12.12 represents the constraint placed on the values of $i_B$ and $v_{BE}$ by the external circuit. In addition, $i_B$ and $v_{BE}$ must fall on the device characteristic. The values that satisfy both constraints are the values at the intersection of the load line and the device characteristic.

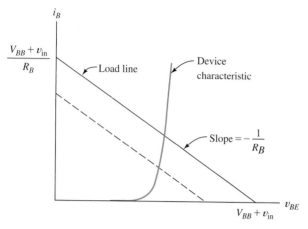

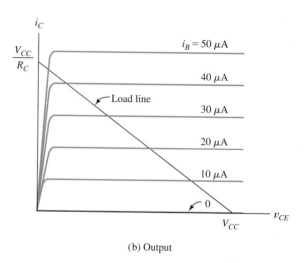

(a) Input load line (shifts to dashed line for a smaller value of $v_{in}$)

(b) Output

**Figure 12.8** Load-line analysis of the amplifier of Figure 12.7.

The input load line shifts position but maintains constant slope as $v_{in}(t)$ changes.

The slope of the load line is $-1/R_B$. Thus, the load line shifts position but maintains a constant slope as $v_{in}$ changes in value. For example, the dashed load line in Figure 12.8(a) is for a smaller value of $v_{in}$ than that for the solid load line.

The **quiescent operating point**, or **Q point**, corresponds to $v_{in}(t) = 0$. Thus, as the ac input signal $v_{in}(t)$ changes in value with time, the instantaneous operating point swings above and below the Q-point value. Values of $i_B$ can be found from the intersection of the load line with the input characteristic for each value of $v_{in}$.

### Analysis of the Output Circuit

After the input circuit has been analyzed to find values of $i_B$, a load-line analysis of the output circuit is possible. Referring to Figure 12.7, we can write a voltage equation for the loop through $V_{CC}$, $R_C$, and the transistor from collector to emitter. Thus, we have

Equation for the output load line.

$$V_{CC} = R_C i_C + v_{CE} \qquad (12.13)$$

This is plotted on the output characteristics of the transistor in Figure 12.8(b).

Now, with the values of $i_B$ found by prior analysis of the input circuit, we can locate the intersection of the corresponding output curve with the load line to find values for $i_C$ and $v_{CE}$. Thus, as $v_{in}$ swings through a range of values, $i_B$ changes, and the instantaneous operating point swings up and down the load line on the output characteristics. Usually, the ac component of $v_{CE}$ is much larger than the input voltage, and amplification has taken place.

Examination of Figure 12.8(a) shows that as $v_{in}(t)$ swings positive, the input load line moves upward and to the right, and the value of $i_B$ increases (i.e., the intersection of the load line with the input characteristic moves upward). This in turn causes the instantaneous operating point to move upward on the output load line, and $v_{CE}$ decreases in value. Thus, a swing in the positive direction for $v_{in}$ results in a (much larger) swing in the negative direction for $v_{CE}$. Therefore, as well as being amplified, the signal is inverted. Thus, the common-emitter circuit is an **inverting amplifier**.

As $v_{in}$ swings through a range of values, $i_B$ changes, and the instantaneous operating point swings up and down the load line on the output characteristics.

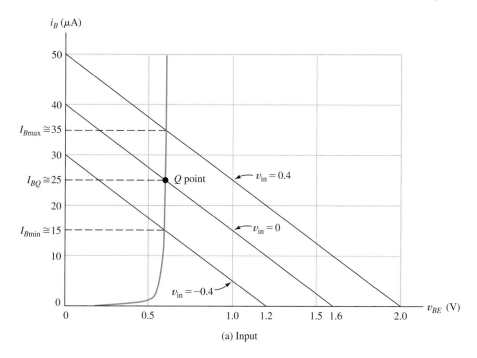

(a) Input

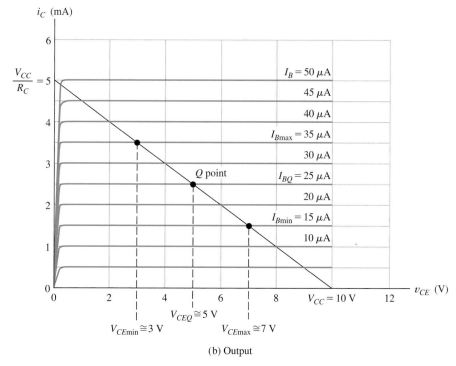

(b) Output

Figure 12.9  Load-line analysis for Example 12.2.

## Example 12.2   Load-Line Analysis of a BJT Amplifier

Assume that the circuit of Figure 12.7 has $V_{CC} = 10$ V, $V_{BB} = 1.6$ V, $R_B = 40$ k$\Omega$, and $R_C = 2$ k$\Omega$. The input signal is a 0.4-V-peak 1-kHz sinusoid given by $v_{in}(t) = 0.4 \sin(2000\pi t)$. The common-emitter characteristics for the transistor are shown in Figure 12.9. Find the maximum, minimum, and $Q$-point values for $v_{CE}$.

**Solution**    First, we must find values for $i_B$. The load lines for $v_{in} = 0$ (to find the $Q$ point), $v_{in} = 0.4$ (positive extreme), and $v_{in} = -0.4$ (negative extreme) are shown in Figure 12.9(a). The values for the base current are found at the intersection of the load lines with the input characteristic. The (approximate) values are $I_{B\,max} \cong 35\ \mu A$, $I_{BQ} \cong 25\ \mu A$, and $I_{B\,min} \cong 15\ \mu A$.

Next, the load line is constructed on the output characteristic as shown in Figure 12.9(b). The intersection of the output load line with the characteristic for $I_{BQ} = 25\ \mu A$ establishes the $Q$ point on the output characteristics. The values are $I_{CQ} = 2.5$ mA and $V_{CEQ} = 5$ V. Similarly, the intersection of the load line with the characteristic for $I_{B\,max} = 35\ \mu A$ yields $V_{CE\,min} \cong 3$ V. The opposite extreme is $I_{B\,min} \cong 15\ \mu A$ resulting in $V_{CE\,max} \cong 7$ V.

If more points are found as $v_{in}$ varies with time, we can eventually plot the $v_{CE}$ waveform versus time. The waveforms for $v_{in}(t)$ and $v_{CE}(t)$ are shown in Figure 12.10. Notice that the ac component of $v_{CE}(t)$ is inverted compared to the input signal [i.e., the minimum of $v_{CE}(t)$ occurs at the same instant as the maximum of $v_{in}(t)$, and vice versa].

The peak-to-peak value of the input voltage is 0.8 V and the peak-to-peak value of the ac component of $v_{CE}$ is 4 V. Thus, the voltage-gain magnitude is 5 (i.e., the ac component of $v_{CE}$ is five times larger in amplitude than $v_{in}$). Usually, we would state the gain as $-5$ to emphasize the fact that the amplifier inverts the input signal.    ■

### Nonlinear Distortion

It is not apparent in the waveforms of Figure 12.10, but unlike the input, the output signal is not a precise sine wave. The amplifier is slightly nonlinear because of the curvature of the characteristics of the transistor. Therefore, as well as being amplified and inverted, the signal is distorted. Of course, distortion is not usually desirable. Figure 12.11 shows the output of the amplifier of Example 12.2 if the input signal is increased in amplitude to 1.2-V peak. The distortion is obvious.

Notice that the positive peak of $v_{CE}$ has been "clipped" at $V_{CC} = 10$ V. This occurs when $i_B$ and $i_C$ have been reduced to zero by the negative peaks of the input signal, and the instantaneous operating point moves down to the voltage-axis intercept of the output load line. When this happens, we say that the transistor has been driven into **cutoff**.

*When $i_C$ becomes zero, we say that the transistor is cut off.*

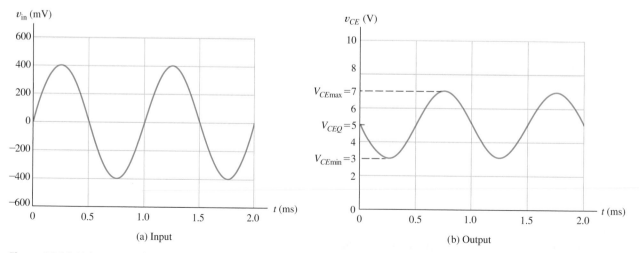

(a) Input

(b) Output

**Figure 12.10** Voltage waveforms for the amplifier of Figure 12.7. See Example 12.2.

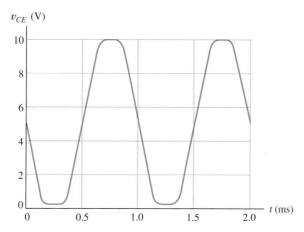

**Figure 12.11**  Output of the amplifier of Example 12.2 for $v_{in}(t) = 1.2 \sin(2000\pi t)$ showing gross nonlinear distortion.

The negative-going peak of the output waveform in Figure 12.11 is clipped at $v_{CE} \cong 0.2$ V. This occurs because $i_B$ becomes large enough so that operation is driven into the region at the upper end of the output load line, where the characteristic curves are crowded together. We call this the **saturation region**.

When $v_{CE} \cong 0.2$ V, we say that the transistor is in saturation.

Reasonably linear amplification occurs only if the signal swing remains in the **active region** between saturation and cutoff on the load line. An output load line is shown in Figure 12.12, including labels for the cutoff, saturation, and active regions.

**Exercise 12.5**   Repeat Example 12.2 if $v_{in}(t) = 0.8 \sin(2000\pi t)$. Find the values of $V_{CEmax}$, $V_{CEQ}$, and $V_{CEmin}$.
**Answer**   $V_{CEmax} \cong 9.0$ V, $V_{CEQ} \cong 5.0$ V, $V_{CEmin} \cong 1.0$ V.   □

**Exercise 12.6**   Repeat Example 12.2 if $v_{in}(t) = 0.8 \sin(2000\pi t)$ and $V_{BB} = 1.2$ V. Find the values of $V_{CEmax}$, $V_{CEQ}$, and $V_{CEmin}$.
**Answer**   $V_{CEmax} \cong 9.8$ V, $V_{CEQ} \cong 7.0$ V, $V_{CEmin} \cong 3.0$ V.   □

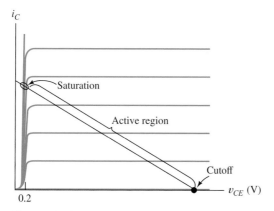

**Figure 12.12**  Amplification occurs in the active region. Clipping occurs when the instantaneous operating point enters saturation or cutoff. In saturation, $v_{CE} \cong 0.2$ V.

**PRACTICAL APPLICATION   12.1**

**Soup Up Your Automobile by Changing Its Software?**

Early automobiles contained no electronics and very little in the way of electrical circuits. Perhaps the most important initial application was electrical ignition, an early version of which is described briefly on page 198. The first step in modernizing the ignition system was to replace the "points" (a mechanically operated switch) with a BJT. The transistor cycles between saturation (in which it behaves as a closed switch) and cutoff (in which it behaves as an open switch). The ignition spark is produced as the result of rapidly switching off the current flowing through the coil.

A significant advantage of electronic switching is that the BJT does not wear out as points do. However, the use of an electronic switch in the place of mechanically operated points has paved the way for numerous additional improvements in ignition control. Optimum ignition timing relative to engine rotation varies with engine speed, throttle setting, air temperature, engine temperature, fuel quality, and load on the engine, as well as the design goals (good fuel economy, long engine life, or highest racing performance). Early ignition systems used mechanical and pneumatic systems to adjust timing, but such systems cannot achieve optimum performance under all conditions. Modern engine-control systems employ electrical sensors to determine operating conditions, various electronic circuits to process the sensor signals, and a special-purpose computer (including software) to compute the optimum firing instant for each cylinder.

The computer switches the BJT from saturation to cutoff, creating the ignition spark.

In the 1950s, souping up an engine involved boring out cylinders and milling heads. Today, an equally important factor is to modify the engine-control software. Hot-rod magazines abound with advertisements offering ROMs (read-only memories, discussed in Chapter 8) loaded with engine control software optimized for high performance (as opposed to fuel economy or long engine life).

A milestone event underscores the importance of electronics in what was originally an almost purely mechanical system. To commemorate the 100th anniversary of the first mass production of the automobile, *Automotive Engineering* magazine surveyed the Society of Automotive Engineers Fellows Committee to ascertain the top 10 significant events in automotive history. What came out at the top of the list? The answer is as follows: "automotive electronics, including applications in engine controls, brakes, steering, and stability control" (*Automotive Engineering*, February 1996, p. 4). Of course, advances in electronics applications for automobiles did not stop in 1996; instead, reports of automotive electronic innovations appear at an ever increasing rate. A few more recent examples are collision avoidance systems, automatic parallel parking, night vision systems, and even cars that drive themselves. Certainly, today's mechanical engineers need to be familiar with the capabilities and limitations of the electronics, as well as with mechanical design and materials issues.

## 12.4 *pnp* BIPOLAR JUNCTION TRANSISTORS

So far, we have only considered the *npn* BJT, but an equally useful device results if the base is a layer of *n*-type material between *p*-type emitter and collector regions. For proper operation as an amplifier, the polarities of the dc voltages applied to the *pnp* device must be opposite to those of the *npn* device. Furthermore, currents flow in opposite directions. Aside from the differences in voltage polarity and current directions, the two types of devices are nearly identical.

A diagram of the structure of a *pnp* BJT and the circuit symbol are shown in Figure 12.13. Notice that the arrow on the emitter of the *pnp* transistor symbol points into the device, which is the normal direction of the emitter current. Furthermore, we have reversed the reference directions for the currents to agree with the actual direction of current flow for the *pnp* in the active region.

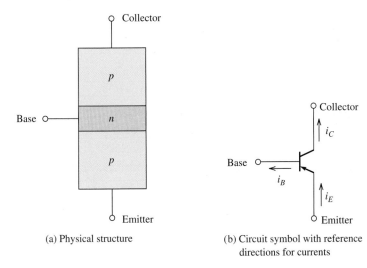

**Figure 12.13** The *pnp* BJT.

For the *pnp* transistor, we can write the following equations, which are exactly the same as for the *npn* transistor:

$$i_C = \alpha i_E \tag{12.14}$$

$$i_B = (1 - \alpha)i_E \tag{12.15}$$

$$i_C = \beta i_B \tag{12.16}$$

and

$$i_E = i_C + i_B \tag{12.17}$$

Equations 12.14 through 12.16 are valid only if the base–emitter junction is forward biased ($v_{BE}$ negative for a *pnp*) and the collector–base junction is reverse biased ($v_{BC}$ positive for a *pnp*). As for the *npn* transistor, typical values are $\alpha \cong 0.99$ and $\beta \cong 100$.

For the *pnp* transistor in the active region, we have

> Except for reversal of current directions and voltage polarities, the *pnp* BJT is almost identical to the *npn* BJT.

$$i_E = I_{ES}\left[\exp\left(\frac{-v_{BE}}{V_T}\right) - 1\right] \tag{12.18}$$

and

$$i_B = (1 - \alpha)\, I_{ES}\left[\exp\left(\frac{-v_{BE}}{V_T}\right) - 1\right] \tag{12.19}$$

These equations are identical to Equations 12.1 and 12.8 for the *npn* transistor, except that $-v_{BE}$ has been substituted for $v_{BE}$ (because $v_{BE}$ takes negative values for the *pnp* device). As for the *npn* device, typical values for $I_{ES}$ range from $10^{-12}$ to $10^{-16}$ A, and at 300 K, we have $V_T \cong 0.026$ V.

The common-emitter characteristics of a *pnp* transistor are exactly the same as for the *npn* transistor, except that the values on the voltage axes are negative. A typical set of characteristics is shown in Figure 12.14.

**Exercise 12.7**   Find the values of $\alpha$ and $\beta$ for the transistor having the characteristics shown in Figure 12.14.
**Answer**   $\alpha = 0.980, \beta = 50$. □

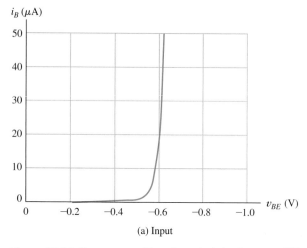

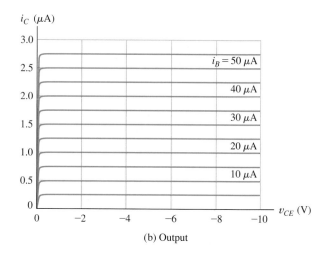

(a) Input    (b) Output

**Figure 12.14** Common-emitter characteristics for a *pnp* BJT.

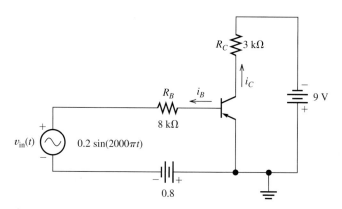

**Figure 12.15** Common-emitter amplifier for Exercise 12.8.

**Exercise 12.8    a.** Use load-line analysis to find the maximum, minimum, and $Q$-point values of $i_B$ and $v_{CE}$ for the amplifier circuit shown in Figure 12.15. Use the characteristics shown in Figure 12.14. **b.** Does this *pnp* common-emitter amplifier invert the signal?
**Answer    a.** $I_{B\max} \cong 48\ \mu A$, $I_{BQ} \cong 24\ \mu A$, $I_{B\min} \cong 5\ \mu A$, $V_{CE\max} \cong -1.8\ V$, $V_{CEQ} \cong -5.3\ V$, $V_{CE\min} \cong -8.3\ V$. **b.** Yes, the output signal is inverted. [If you are in doubt about this, try sketching the $v_{in}(t)$, $i_B(t)$, and $v_{CE}(t)$ waveforms to scale versus time.]    ☐

## 12.5 LARGE-SIGNAL DC CIRCUIT MODELS

In the analysis or design of BJT amplifier circuits, we often consider the dc operating point ($Q$ point) separately from the analysis of the signals. (This was illustrated for a voltage-controlled attenuator in Section 9.8.) Usually, we consider the dc operating point first. Then, we turn our attention to the signal to be amplified. In this section, we present models for large-signal dc analysis of BJT circuits. Then, in the next section, we show how to use these models to analyze bias circuits for BJT amplifiers. Later, we consider small-signal models used to analyze circuits for the signals being amplified.

It is customary to use uppercase symbols with uppercase subscripts to represent large-signal dc currents and voltages in transistor circuits. Thus, $I_C$ and $V_{CE}$ represent the dc collector current and collector-to-emitter voltage, respectively. Similar notation is used for the other currents and voltages.

As we have seen, BJTs can operate in the active region, in saturation, or in cutoff. In the active region, the base–emitter junction is forward biased, and the base–collector junction is reverse biased.

## Active-Region Model

Circuit models for BJTs in the active region are shown in Figure 12.16(a). A current-controlled current source models the dependence of the collector current on the base

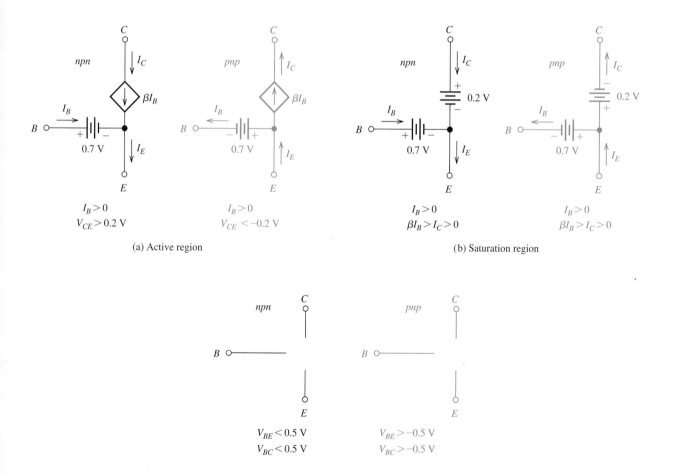

(a) Active region

(b) Saturation region

(c) Cutoff region

**Figure 12.16** BJT large-signal models. (*Note:* Values shown are appropriate for typical small-signal silicon devices at a temperature of 300 K.)

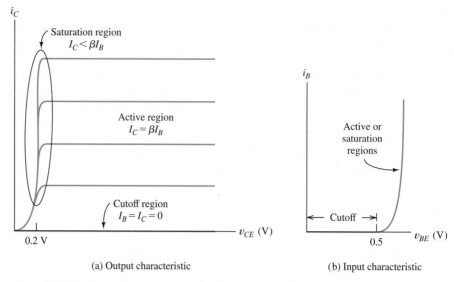

(a) Output characteristic                              (b) Input characteristic

**Figure 12.17** Regions of operation on the characteristics of an *npn* BJT.

current. The constraints given in the figure for $I_B$ and $V_{CE}$ must be satisfied to ensure validity of the active-region model.

Let us relate the active-region model to the device characteristics. Figure 12.17 shows the characteristic curves of an *npn* transistor. The base current $I_B$ is positive and $V_{BE} \cong 0.7$ V for forward bias of the base-to-emitter junction, as shown in Figure 12.17(b). Also notice in Figure 12.17(a) that $V_{CE}$ must be greater than about 0.2 V to ensure that operation is in the active region (i.e., above the *knees* of the characteristic curves).

Similarly, for the *pnp* BJT, we must have $I_B > 0$ and $V_{CE} < -0.2$ V for validity of the active-region model. (As usual, we assume that $I_B$ is referenced positive out of the base for the *pnp* BJT.)

### Saturation-Region Model

The BJT models for the saturation region are shown in Figure 12.16(b). In the saturation region, both junctions are forward biased. Examination of the collector characteristics in Figure 12.17(a) shows that $V_{CE} \cong 0.2$ V for an *npn* transistor in saturation. Thus, the model for the saturation region includes a 0.2-V source between collector and emitter. As in the active region, $I_B$ is positive. Also, we see in Figure 12.17(a) that for operation below the knee of the collector characteristic, the constraint is $\beta I_B > I_C > 0$.

### Cutoff-Region Model

In cutoff, both junctions are reverse biased and no currents flow in the device. Thus, the model consists of open circuits among all three terminals, as shown in Figure 12.16(c). (Actually, if small forward-bias voltages of up to about 0.5 V are applied, the currents are often negligible and we still use the cutoff-region model.) The constraints on the voltages for the BJT to be in the cutoff region are shown in the figure.

**Example 12.3**   **Determining the Operating Region of a BJT**

A given *npn* transistor has $\beta = 100$. Determine the region of operation if
**a.** $I_B = 50\ \mu\text{A}$ and $I_C = 3\ \text{mA}$; **b.** $I_B = 50\ \mu\text{A}$ and $V_{CE} = 5\ \text{V}$; **c.** $V_{BE} = -2\ \text{V}$ and
$V_{CE} = -1\ \text{V}$.

**Solution**

**a.** Since $I_B$ and $I_C$ are positive, the transistor is either in the active or saturation
region. The constraint for saturation

$$\beta I_B > I_C$$

is met, so the device is in saturation.

**b.** Since we have $I_B > 0$ and $V_{CE} > 0.2$, the transistor is in the active region.

**c.** We have $V_{BE} < 0$ and $V_{BC} = V_{BE} - V_{CE} = -1 < 0$. Therefore, both junctions
are reverse biased, and operation is in the cutoff region.                               ∎

---

**Exercise 12.9**   A certain *npn* transistor has $\beta = 100$. Determine the region of
operation if **a.** $V_{BE} = -0.2\ \text{V}$ and $V_{CE} = 5\ \text{V}$; **b.** $I_B = 50\ \mu\text{A}$ and $I_C = 2\ \text{mA}$;
**c.** $V_{CE} = 5\ \text{V}$ and $I_B = 50\ \mu\text{A}$.
**Answer**   **a.** cutoff; **b.** saturation; **c.** active.                   ☐

---

## 12.6 LARGE-SIGNAL DC ANALYSIS OF BJT CIRCUITS

In Section 12.5, we presented large-signal dc models for the BJT. In this section, we
use those models to analyze circuits. In dc analysis of BJT circuits, we first assume
that the operation of the transistor is in a particular region (i.e., active, cutoff, or
saturation). Then, we use the appropriate model for the device and solve the circuit.
Next, we check to see if the solution satisfies the constraints for the region assumed.
If so, the analysis is complete. If not, we assume operation in a different region and
repeat until a valid solution is found. This is very similar to the analysis of diode
circuits using the ideal-diode model or a piecewise-linear model. The step-by-step
procedure is as follows:

**1.** Choose one of the three BJT operating regions: saturation, cutoff, or active.

**2.** Analyze the circuit to determine $I_C$, $I_B$, $V_{BE}$, and $V_{CE}$, by using the transistor
   model for the region chosen.

**3.** Check to see if the constraints for the chosen region are met. If so, the analysis
   is completed. If not, return to Step 1, and choose a different region.

This approach is particularly useful in the analysis and design of bias circuits for
BJT amplifiers. The objective of the bias circuit is to place the operating point in the
active region so that signals can be amplified. Because transistors show considerable
variation of parameters, such as $\beta$, from unit to unit and with temperature, it is
important for the bias point to be independent of these variations.

The next several examples illustrate the technique and provide some observations
that are useful in bias-circuit design.

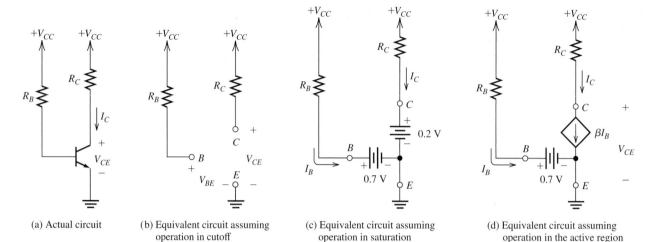

(a) Actual circuit  (b) Equivalent circuit assuming operation in cutoff  (c) Equivalent circuit assuming operation in saturation  (d) Equivalent circuit assuming operation in the active region

**Figure 12.18** Bias circuit of Examples 12.4 and 12.5.

### Example 12.4 Analysis of the Fixed Base Bias Circuit

The dc bias circuit shown in Figure 12.18(a) has $R_B = 200\text{ k}\Omega$, $R_C = 1\text{ k}\Omega$, and $V_{CC} = 15$ V. The transistor has $\beta = 100$. Solve for $I_C$ and $V_{CE}$.

**Solution** We will eventually see that the transistor is in the active region, but we start by assuming that the transistor is cut off (to illustrate how to test the initial guess of operating region). Since we assume operation in cutoff, the model for the transistor is shown in Figure 12.16(c), and the equivalent circuit is shown in Figure 12.18(b). We reason that $I_B = 0$ and that there is no voltage drop across $R_B$. Hence, we conclude that $V_{BE} = 15$ V. However, in cutoff, we must have $V_{BE} < 0.5$ for an *npn* transistor. Therefore, we conclude that the cutoff assumption is invalid.

Next, let us assume that the transistor is in saturation. The transistor model is shown in Figure 12.16(b). Then, the equivalent circuit is shown in Figure 12.18(c). Solving, we find that

and

$$I_C = \frac{V_{CC} - 0.2}{R_C} = 14.8\text{ mA}$$

$$I_B = \frac{V_{CC} - 0.7}{R_B} = 71.5\ \mu\text{A}$$

Checking the conditions required for saturation, we find that $I_B > 0$ is met, but $\beta I_B > I_C$ is not met. Therefore, we conclude that the transistor is not in saturation.

Finally, if we assume that the transistor operates in the active region, we use the BJT model of Figure 12.16(a), and the equivalent circuit is shown in Figure 12.18(d). Solving, we find that

$$I_B = \frac{V_{CC} - 0.7}{R_B} = 71.5\ \mu\text{A}$$

(We have assumed a forward bias of 0.7 V for the base–emitter junction. Some authors assume 0.6 V for small-signal silicon devices at room temperature; others assume 0.7 V. In reality, the value depends on the particular device and the current level. Usually, the difference is not significant.) Now, we have

$$I_C = \beta I_B = 7.15 \text{ mA}$$

Finally,

$$V_{CE} = V_{CC} - R_C I_C = 7.85 \text{ V}$$

The requirements for the active region are $V_{CE} > 0.2$ V and $I_B > 0$, which are met. Thus, the transistor operates in the active region. ∎

---

**Example 12.5**   **Analysis of the Fixed Base Bias Circuit**

Repeat Example 12.4 with $\beta = 300$.

**Solution**   First, we assume operation in the active region. This leads to

$$I_B = \frac{V_{CC} - 0.7}{R_B} = 71.5 \ \mu\text{A}$$

$$I_C = \beta I_B = 21.45 \text{ mA}$$

$$V_{CE} = V_{CC} - R_C I_C = -6.45 \text{ V}$$

The requirements for the active region are $V_{CE} > 0.2$ V and $I_B > 0$, which are not met. Thus, the transistor is not operating in the active region.

Next, we assume that the transistor is in saturation. This leads to

$$I_C = \frac{V_{CC} - 0.2}{R_C} = 14.8 \text{ mA}$$

and

$$I_B = \frac{V_{CC} - 0.7}{R_B} = 71.5 \ \mu\text{A}$$

Now, we find that the conditions for saturation ($I_B > 0$ and $\beta I_B > I_C$) are met. Thus, we have solved the circuit, and $V_{CE} = 0.2$ V. ∎

## Implications for Bias-Circuit Design

It is instructive to consider the load-line constructions that are shown in Figure 12.19 for Examples 12.4 and 12.5. For $\beta = 100$, the operating point ($Q$ point) is approximately in the center of the load line. On the other hand, for $\beta = 300$, the operating point has moved up into saturation.

To use this circuit as an amplifier, we would want a $Q$ point in the active region, where changes in base current cause the instantaneous operating point to move up and down the load line. In saturation, the operating point does not move significantly for small changes in base current, and amplification is not achieved. Thus, a suitable $Q$ point is achieved for $\beta = 100$ but not for $\beta = 300$. Since we often find unit-to-unit variations in $\beta$ of this magnitude, this circuit is not suitable as an amplifier bias circuit for mass production. (We could consider adjusting $R_B$ to compensate for unit-to-unit variations in $\beta$, but this is usually not practical.)

Sometimes this circuit [Figure 12.18(a)] is called a **fixed base bias circuit** because the base current is fixed by $V_{CC}$ and $R_B$ and does not adjust for changes in $\beta$. (Notice

The amplifiers that we discuss in this book need to be biased near the center of the active region.

The fixed base bias circuit shown in Figure 12.18(a) is not suitable for amplifiers because unit-to-unit variations in $\beta$ cause some of the circuits to operate in saturation or close to cutoff.

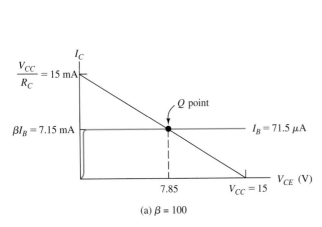

Figure 12.19 Load lines for Examples 12.4 and 12.5.

that if we want a circuit that achieves a particular operating point on the collector load line, the base current must change when $\beta$ changes.)

**Exercise 12.10** Repeat Example 12.4 for **a.** $\beta = 50$; **b.** $\beta = 250$.
**Answer** **a.** $I_C = 3.575$ mA, $V_{CE} = 11.43$ V; **b.** $I_C = 14.8$ mA, $V_{CE} = 0.2$ V. ☐

**Exercise 12.11** Assume that $R_C = 5$ k$\Omega$, $V_{BE} = 0.7$ V, and $V_{CC} = 20$ V in the circuit of Figure 12.18(a). Solve for the value of $R_B$ needed to place the operating point exactly in the middle of the output load line (i.e., the $Q$ point should fall at $V_{CE} = V_{CC}/2 = 10$ V) for: **a.** $\beta = 100$; **b.** $\beta = 300$.
**Answer** **a.** $R_B = 965$ k$\Omega$; **b.** $R_B = 2.90$ M$\Omega$. ☐

**Exercise 12.12** Solve the circuit shown in Figure 12.20 to find $I_C$ and $V_{CE}$ if:
**a.** $\beta = 50$; **b.** $\beta = 150$.
**Answer** **a.** $I_C = 0.965$ mA, $V_{CE} = -10.35$ V; **b.** $I_C = 1.98$ mA, $V_{CE} = -0.2$ V (transistor in saturation). ☐

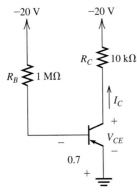

Figure 12.20 Circuit for Exercise 12.12.

In the next example, we consider a circuit that achieves an emitter current that is relatively independent of $\beta$.

---

**Example 12.6** **Analysis of a BJT Bias Circuit**

Solve for $I_C$ and $V_{CE}$ in the circuit of Figure 12.21(a) given that $V_{CC} = 15$ V, $V_{BB} = 5$ V, $R_C = 2$ k$\Omega$, $R_E = 2$ k$\Omega$, and $\beta = 100$. Repeat for $\beta = 300$.

**Solution** We assume that the transistor is in the active region and use the equivalent circuit shown in Figure 12.21(b). Writing a voltage equation through $V_{BB}$, the base–emitter junction, and $R_E$, we have

$$V_{BB} = 0.7 + I_E R_E$$

This can be solved for the emitter current:

$$I_E = \frac{V_{BB} - 0.7}{R_E} = 2.15 \text{ mA}$$

Notice that the emitter current does not depend on the value of $\beta$.

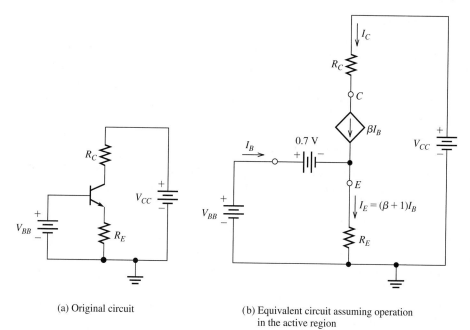

(a) Original circuit

(b) Equivalent circuit assuming operation
in the active region

**Figure 12.21**  Circuit for Example 12.6.

Next, we can compute the base current and collector current using Equations 12.2 and 12.10.

$$I_C = \beta I_B$$

$$I_E = I_B + I_C$$

Using the first equation to substitute for $I_C$ in the second equation, we have

$$I_E = I_B + \beta I_B = (\beta + 1)I_B$$

Solving for the base current, we obtain

$$I_B = \frac{I_E}{\beta + 1}$$

Substituting values, we obtain the results given in Table 12.1. Notice that $I_B$ is lower for the higher $\beta$ transistor, and $I_C$ is nearly constant.

Now, we can write a voltage equation around the collector loop to find $V_{CE}$:

$$V_{CC} = R_C I_C + V_{CE} + R_E I_E$$

Substituting values found previously, we find that $V_{CE} = 6.44$ V for $\beta = 100$ and $V_{CE} = 6.42$ V for $\beta = 300$.   ∎

**Table 12.1**  Results for the Circuit of Example 12.6

| $\beta$ | $I_B\,(\mu A)$ | $I_C\,(mA)$ | $V_{CE}\,(V)$ |
|-----|------|------|------|
| 100 | 21.3 | 2.13 | 6.44 |
| 300 | 7.14 | 2.14 | 6.42 |

The $Q$ point for the circuit of Figure 12.21(a) is almost independent of $\beta$. However, the circuit is not practical for use in most amplifier circuits. First, it requires two voltage sources $V_{CC}$ and $V_{BB}$, but often only one source is readily available. Second, we may want to inject the signal into the base (through a coupling capacitor), but the base voltage is fixed with respect to ground by the $V_{BB}$ source. Because the $V_{BB}$ source is constant, it acts as a short circuit to ground for ac signal currents (i.e., the $V_{BB}$ source does not allow an ac voltage to appear at the base).

### Analysis of the Four-Resistor Bias Circuit

The four-resistor BJT bias circuit is practical for amplifiers composed of discrete components. However, it is not practical in integrated circuits.

A circuit that avoids these objections is shown in Figure 12.22(a). We call this the **four-resistor BJT bias circuit**. The resistors $R_1$ and $R_2$ form a voltage divider that is intended to provide a nearly constant voltage at the base of the transistor (independent of transistor $\beta$). As we saw in Example 12.6, constant base voltage results in nearly constant values for $I_C$ and $V_{CE}$. Because the base is not directly connected to the supply or ground in the four-resistor bias circuit, it is possible to couple an ac signal to the base through a coupling capacitor.

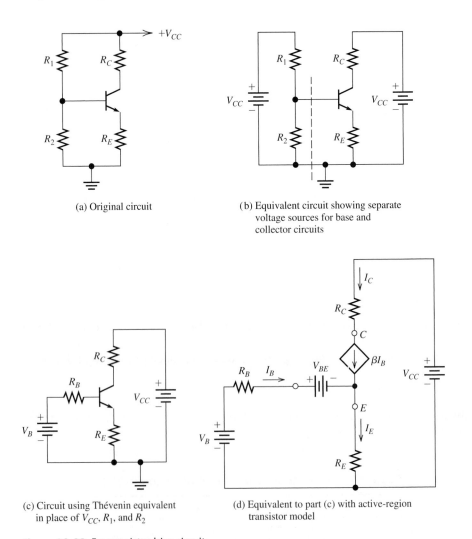

(a) Original circuit

(b) Equivalent circuit showing separate voltage sources for base and collector circuits

(c) Circuit using Thévenin equivalent in place of $V_{CC}$, $R_1$, and $R_2$

(d) Equivalent to part (c) with active-region transistor model

**Figure 12.22** Four-resistor bias circuit.

The circuit can be analyzed as follows. First, the circuit is redrawn as shown in Figure 12.22(b). Two separate voltage supplies are shown as an aid in the analysis to follow, but otherwise, the circuits in parts (a) and (b) of the figure are identical. Next, we find the Thévenin equivalent for the circuit to the left of the dashed line in Figure 12.22(b). The Thévenin resistance $R_B$ is the parallel combination of $R_1$ and $R_2$ given by

$$R_B = \frac{1}{1/R_1 + 1/R_2} = R_1 \| R_2 \qquad (12.20)$$

The Thévenin voltage $V_B$ is

$$V_B = V_{CC} \frac{R_2}{R_1 + R_2} \qquad (12.21)$$

The circuit with the Thévenin equivalent replacement is shown in Figure 12.22(c). Finally, the transistor is replaced by its active-region model, as shown in Figure 12.22(d).

Now, we can write a voltage equation around the base loop of Figure 12.22(d), resulting in

$$V_B = R_B I_B + V_{BE} + R_E I_E \qquad (12.22)$$

Of course, for small-signal silicon transistors at room temperature, we have $V_{BE} \cong 0.7$ V. Now, we can substitute

$$I_E = (\beta + 1) I_B$$

and solve to find that

$$I_B = \frac{V_B - V_{BE}}{R_B + (\beta + 1) R_E} \qquad (12.23)$$

Once $I_B$ is known, $I_C$ and $I_E$ can easily be found. Then, we can write a voltage equation around the collector loop of Figure 12.22(d) and solve for $V_{CE}$. This yields

$$V_{CE} = V_{CC} - R_C I_C - R_E I_E \qquad (12.24)$$

**Example 12.7**   Analysis of the Four-Resistor Bias Circuit

Find the values of $I_C$ and $V_{CE}$ in the circuit of Figure 12.23 for $\beta = 100$ and $\beta = 300$. Assume that $V_{BE} = 0.7$ V.

**Solution**   Substituting into Equations 12.20 and 12.21, we find that

$$R_B = \frac{1}{1/R_1 + 1/R_2} = 3.33 \text{ k}\Omega$$

$$V_B = V_{CC} \frac{R_2}{R_1 + R_2} = 5 \text{ V}$$

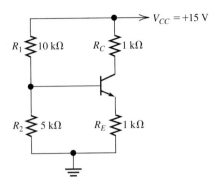

**Figure 12.23** Circuit for Example 12.7.

Then, substituting into Equation 12.23 and using $\beta = 100$, we have

$$I_B = \frac{V_B - V_{BE}}{R_B + (\beta + 1)R_E} = 41.2\ \mu\text{A}$$

For $\beta = 300$, we find that $I_B = 14.1\ \mu\text{A}$. Notice that the base current is significantly smaller for the higher $\beta$.

Now, we can compute the collector current by using $I_C = \beta I_B$. For $\beta = 100$, we find that $I_C = 4.12$ mA, and for $\beta = 300$, we have $I_C = 4.24$ mA. For a 3:1 change in $\beta$, the collector current changes by less than 3 percent. The emitter current is given by $I_E = I_C + I_B$. The results are $I_E = 4.16$ mA for $\beta = 100$ and $I_E = 4.25$ mA for $\beta = 300$.

Finally, Equation 12.24 can be used to find $V_{CE}$. The results are $V_{CE} = 6.72$ for $\beta = 100$ and $V_{CE} = 6.51$ for $\beta = 300$. ∎

Thus, we have seen that for proper resistance values, the four-resistor bias circuit can achieve a $Q$ point that is nearly independent of $\beta$. Because of this fact, the circuit is commonly used for biasing BJT amplifiers (except in integrated circuits, for which resistors tend to be impractical).

**Exercise 12.13**    Repeat Example 12.7 for $R_1 = 100$ k$\Omega$ and $R_2 = 50$ k$\Omega$. Compute the ratio of $I_C$ for $\beta = 300$ to $I_C$ for $\beta = 100$, and compare to the ratio of the currents found in Example 12.7. Comment.
**Answer**    For $\beta = 100$, $I_C = 3.20$ mA, and $V_{CE} = 8.57$ V. For $\beta = 300$, $I_C = 3.86$ mA, and $V_{CE} = 7.27$ V. The ratio of the collector currents is 1.21. On the other hand, in the example, the ratio of the collector currents is only 1.03. Larger values of $R_1$ and $R_2$ lead to larger changes in $I_C$ with changes in $\beta$. ☐

## 12.7 SMALL-SIGNAL EQUIVALENT CIRCUITS

*In this section, we develop a BJT equivalent circuit that is useful in analyzing amplifier circuits.*

Now, we turn our attention to the signal currents and voltages in the BJT. First, we establish the notation used in amplifier circuits. We denote the total currents and voltages by lowercase symbols with uppercase subscripts. Thus, $i_B(t)$ is the total base current as a function of time.

The dc $Q$-point currents and voltages are denoted by uppercase symbols with uppercase subscripts. Thus, $I_{BQ}$ is the dc base current if the input signal is set to zero.

Finally, we denote the changes in currents and voltages from the $Q$ point (due to the input signal being amplified) by lowercase symbols with lowercase subscripts.

Thus, $i_b(t)$ denotes the signal component of the base current. Since the total base current is the sum of the $Q$-point value and the signal component, we can write

$$i_B(t) = I_{BQ} + i_b(t) \qquad (12.25)$$

These quantities are illustrated in Figure 12.24. Similarly, we can write

$$v_{BE}(t) = V_{BEQ} + v_{be}(t) \qquad (12.26)$$

The $Q$ point is established by the bias circuit, as discussed in Section 12.6. Now, we consider how the (small) signal components are related in the BJT. The total base current is given in terms of the total base-to-emitter voltage by Equation 12.8, repeated here for convenience:

$$i_B = (1 - \alpha)I_{ES}\left[\exp\left(\frac{v_{BE}}{V_T}\right) - 1\right]$$

We are concerned with operation in the active region, for which the 1 inside the brackets is negligible and can be dropped.

We substitute Equations 12.25 and 12.26 into 12.8 to obtain

$$I_{BQ} + i_b(t) = (1 - \alpha)I_{ES}\exp\left[\frac{V_{BEQ} + v_{be}(t)}{V_T}\right] \qquad (12.27)$$

This can be written as

$$I_{BQ} + i_b(t) = (1 - \alpha)I_{ES}\exp\left(\frac{V_{BEQ}}{V_T}\right)\exp\left[\frac{v_{be}(t)}{V_T}\right] \qquad (12.28)$$

Equation 12.8 also relates the $Q$-point values, so we can write

$$I_{BQ} = (1 - \alpha)I_{ES}\exp\left(\frac{V_{BEQ}}{V_T}\right) \qquad (12.29)$$

Substituting into Equation 12.28, we have

$$I_{BQ} + i_b(t) = I_{BQ}\exp\left(\frac{v_{be}(t)}{V_T}\right) \qquad (12.30)$$

$i_b(t)$ denotes the signal current flowing into the base, $I_{BQ}$ is the dc current that flows when the signal is absent, and $i_B(t)$ is the total base current. Similar notation is used for the other currents and voltages.

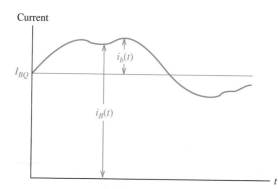

Current

$I_{BQ}$

$i_b(t)$

$i_B(t)$

$t$

**Figure 12.24**  Illustration of the $Q$-point base current $I_{BQ}$, signal current $i_b(t)$, and total current $i_B(t)$.

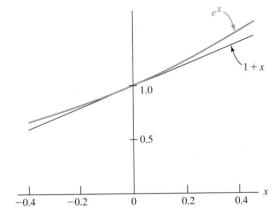

**Figure 12.25** Comparison of $e^x$ and its approximation $1 + x$.

We are interested in small signals for which the magnitude of $v_{be}(t)$ is much smaller than $V_T$ at all times. [Thus, the magnitude of $v_{be}(t)$ is confined to be a few millivolts or less.]

For $|x| << 1$, the following approximation holds:

$$\exp(x) \cong 1 + x \qquad (12.31)$$

This approximation is illustrated in Figure 12.25. Hence, we can write Equation 12.30 as

$$I_{BQ} + i_b(t) \cong I_{BQ}\left[1 + \frac{v_{be}(t)}{V_T}\right] \qquad (12.32)$$

If we subtract $I_{BQ}$ from both sides and define $r_\pi = V_T/I_{BQ}$, we have

$$i_b(t) = \frac{v_{be}(t)}{r_\pi} \qquad (12.33)$$

Therefore, for small-signal variations around the $Q$ point, the base-to-emitter junction of the transistor appears to be a resistance $r_\pi$, given by

$$r_\pi = \frac{V_T}{I_{BQ}} \qquad (12.34)$$

Substituting $I_{BQ} = I_{CQ}/\beta$, we have an alternative formula:

$$r_\pi = \frac{\beta V_T}{I_{CQ}} \qquad (12.35)$$

At room temperature, $V_T \cong 0.026$ V. A typical value of $\beta$ is 100, and a typical bias current for a small-signal amplifier is $I_{CQ} = 1$ mA. These values yield $r_\pi = 2600\ \Omega$. The total collector current is $\beta$ times the total base current:

$$i_C(t) = \beta i_B(t) \qquad (12.36)$$

But, the total currents are the sum of the $Q$-point values and the signal components, so we have

$$I_{CQ} + i_c(t) = \beta I_{BQ} + \beta i_b(t) \qquad (12.37)$$

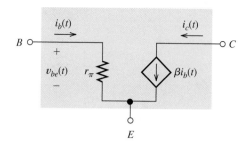

**Figure 12.26** Small-signal equivalent circuit for the BJT.

Thus, the signal components are related by

$$i_c(t) = \beta i_b(t) \tag{12.38}$$

## Small-Signal Equivalent Circuit for the BJT

Equations 12.33 and 12.38 relate the small-signal currents and voltages in a BJT. It is convenient to represent the BJT by the **small-signal equivalent circuit** shown in Figure 12.26. Notice that the circuit embodies the relationships of Equations 12.33 and 12.38.

It turns out that the *pnp* transistor has exactly the same small-signal equivalent circuit as the *npn* –even the reference directions for the *signal* currents are the same. The resistance $r_\pi$ is given by Equation 12.35 for both types of transistors. (We assume that $I_{CQ}$ is referenced out of the collector of the *pnp*, so it has a positive value.) In the next several sections, we will find that the small-signal equivalent circuit is very useful in analysis of BJT amplifier circuits.

The small-signal BJT equivalent circuit shown in Figure 12.26 is very useful in amplifier analysis.

## 12.8 COMMON-EMITTER AMPLIFIERS

In a BJT amplifier circuit, the power supply biases the transistor at an operating point in the active region for which amplification can take place. For example, we can use the four-resistor bias circuit discussed in Section 12.6. Coupling capacitors can be used to connect the load and the signal source without affecting the dc bias point.

We can analyze amplifier circuits to find gain, input resistance, and output resistance by use of the small-signal equivalent circuit. In this section and the next, we illustrate this procedure for two important BJT amplifier circuits.

Figure 12.27(a) shows the circuit diagram of a **common-emitter amplifier**. The resistors $R_1$, $R_2$, $R_E$, and $R_C$ form the four-resistor biasing network. The capacitor $C_1$ couples the signal source to the base of the transistor, and $C_2$ couples the amplified signal at the collector to the load $R_L$. The capacitor $C_E$ is called a **bypass capacitor**. It provides a low-impedance path for the ac emitter current to ground.

The coupling and bypass capacitors are chosen large enough so they have very low ac impedances at the signal frequencies. Thus, for simplicity in our small-signal ac analysis, we treat the capacitors as short circuits. However, at sufficiently low frequencies, the capacitors reduce the gain of the amplifier.

Because the bypass capacitor grounds the emitter for ac signals, the emitter terminal is common to the input source and to the load. This is the origin of the name *common-emitter amplifier*.

The analysis we give here is valid for the **midband region** of frequency. In the **low-frequency region**, the effects of the coupling and bypass capacitors must be

Coupling capacitors can be used to connect the signal source and the load to a BJT amplifier without affecting the dc bias point.

In small-signal ac analysis, we treat the coupling and bypass capacitors as short circuits.

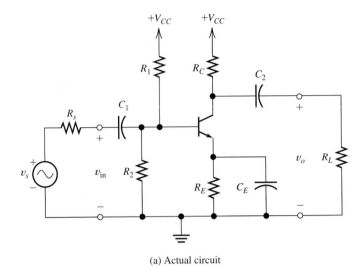

(a) Actual circuit

(b) Small-signal ac equivalent circuit

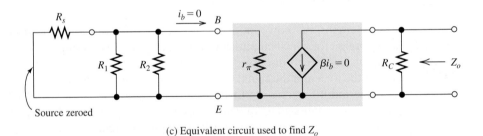

(c) Equivalent circuit used to find $Z_o$

**Figure 12.27** Common-emitter amplifier.

considered. In the **high-frequency region**, a more complex transistor model would be needed that includes its frequency limitations. Analysis for the low- and high-frequency regions is beyond the scope of this book.

## The Small-Signal Equivalent Circuit

Before we analyze the amplifier, it is very helpful to draw its small-signal ac equivalent circuit. This is shown in Figure 12.27(b). The coupling capacitors have been replaced by short circuits, and the transistor has been replaced by its small-signal equivalent, which was discussed in the preceding section.

*The dc power supply is replaced by a short circuit.* This is appropriate because no ac voltage can appear across an ideal dc voltage source that is assumed to have zero internal impedance.

Dc voltage sources act as short circuits for ac signals.

Carefully compare the actual circuit of Figure 12.27(a) with the small-signal ac equivalent shown in Figure 12.27(b). Notice that the signal source is connected directly to the base terminal because $C_1$ has been treated as a short circuit. Similarly, the emitter is connected directly to ground, and the load is connected to the collector.

Notice also that the top end of $R_1$ connects to the supply in the original circuit, but $R_1$ is connected from base to ground in the equivalent circuit, because the power-supply voltage is treated as a short circuit to ground for ac signals.

Notice that $R_1$ ends up in parallel with $R_2$. Similarly, $R_C$ and $R_L$ are in parallel. We find it convenient to define $R_B$ as the parallel combination of $R_1$ and $R_2$:

$$R_B = R_1 \| R_2 = \frac{1}{1/R_1 + 1/R_2} \tag{12.39}$$

Similarly, $R_L'$ is the parallel combination of $R_C$ and $R_L$:

$$R_L' = R_L \| R_C = \frac{1}{1/R_L + 1/R_C} \tag{12.40}$$

These parallel combinations are indicated in Figure 12.27(b).

## Voltage Gain

Now, we analyze the equivalent circuit to find an expression for the voltage gain of the amplifier. First, the input voltage is equal to the voltage across $r_\pi$, given by

$$v_{in} = v_{be} = r_\pi i_b \tag{12.41}$$

The output voltage is produced by the collector current flowing through $R_L'$, given by

$$v_o = -R_L' \beta i_b \tag{12.42}$$

The minus sign is necessary because of the reference directions for the current and voltage–the current flows out of the positive voltage reference. Dividing the respective sides of Equation 12.42 by those of 12.41 gives the voltage gain:

$$A_v = \frac{v_o}{v_{in}} = -\frac{R_L' \beta}{r_\pi} \tag{12.43}$$

Notice that the gain is negative showing that the common-emitter amplifier is inverting. The gain magnitude can be quite large–several hundred is not unusual.

The expression for gain given in Equation 12.43 is the gain with the load connected. We found the open-circuit voltage gain useful to characterize amplifiers in Chapter 11. With $R_L$ replaced by an open circuit, the voltage gain becomes

$$A_{voc} = \frac{v_o}{v_{in}} = -\frac{R_C \beta}{r_\pi} \tag{12.44}$$

## Input Impedance

Another important amplifier specification is the input impedance, which in this case can be obtained by inspection of the equivalent circuit. The input impedance is the impedance *seen* looking into the input terminals. For the equivalent circuit of Figure 12.27(b), it is the parallel combination of $R_B$ and $r_\pi$, given by

$$Z_{\text{in}} = \frac{v_{\text{in}}}{i_{\text{in}}} = \frac{1}{1/R_B + 1/r_\pi} \tag{12.45}$$

(In this case, the input impedance is a pure resistance. Therefore, we can find the input impedance by dividing the *instantaneous* voltage $v_{\text{in}}$ by the *instantaneous* current $i_{\text{in}}$. Of course, if there were capacitances or inductances in the equivalent circuit, it would be necessary to obtain the impedance as the ratio of the *phasor* voltage and the *phasor* current.)

## Current Gain and Power Gain

The current gain $A_i$ can be found by use of Equation 10.3. With some changes in notation, the equation is

$$A_i = \frac{i_o}{i_{\text{in}}} = A_v \frac{Z_{\text{in}}}{R_L} \tag{12.46}$$

The power gain $G$ of the amplifier is the product of the current gain and the voltage gain (assuming that the input and load impedances are pure resistive).

$$G = A_i A_v \tag{12.47}$$

## Output Impedance

The output impedance is the impedance seen looking back from the load terminals with the source voltage $v_s$ set to zero. This situation is shown in Figure 12.27(c). With $v_s$ set to zero, there is no driving source for the base circuit, so $i_b$ is zero. Therefore, the controlled source $\beta i_b$ produces zero current and appears as an open circuit. Thus, the impedance seen from the output terminals is simply $R_C$.

$$Z_o = R_C \tag{12.48}$$

---

**Example 12.8    Common-Emitter Amplifier**

Find $A_v$, $A_{voc}$, $Z_{\text{in}}$, $A_i$, $G$, and $Z_o$ for the amplifier shown in Figure 12.28. If $v_s(t) = 0.001 \sin(\omega t)$, find and sketch $v_o(t)$ versus time. Assume that the circuit operates at a temperature for which $V_T = 26$ mV.

**Solution**    First, we need to know $I_{CQ}$ to be able to find the value of $r_\pi$. Hence, we start by analyzing the dc conditions in the circuit. Only the dc supply, the transistor, and the resistors $R_1$, $R_2$, $R_C$, and $R_E$ need to be considered in the bias-point analysis. The capacitors, the signal source, and the load resistance have no effect on the $Q$ point (because the capacitors behave as open circuits for dc currents).

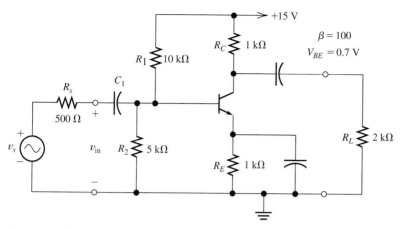

**Figure 12.28** Common-emitter amplifier of Example 12.8.

The dc circuit was shown earlier in Figure 12.23 and was analyzed in Example 12.7. For $\beta = 100$, the resulting $Q$ point was found to be $I_{CQ} = 4.12$ mA and $V_{CE} = 6.72$ V. Substituting values into Equation 12.35, we have

$$r_\pi = \frac{\beta V_T}{I_{CQ}} = 631 \ \Omega$$

Using Equations 12.39 and 12.40, we find that

$$R_B = R_1 \| R_2 = \frac{1}{1/R_1 + 1/R_2} = 3.33 \ \text{k}\Omega$$

$$R_L' = R_L \| R_C = \frac{1}{1/R_L + 1/R_C} = 667 \ \Omega$$

Equations 12.43 through 12.48 yield

$$A_v = \frac{v_o}{v_{\text{in}}} = -\frac{R_L'\beta}{r_\pi} = -106$$

$$A_{voc} = \frac{v_o}{v_{\text{in}}} = -\frac{R_C\beta}{r_\pi} = -158$$

$$Z_{\text{in}} = \frac{v_{\text{in}}}{i_{\text{in}}} = \frac{1}{1/R_B + 1/r_\pi} = 531 \ \Omega$$

$$A_i = \frac{i_o}{i_{\text{in}}} = A_v \frac{Z_{\text{in}}}{R_L} = -28.1$$

$$G = A_i A_v = 2980$$

$$Z_o = R_C = 1 \ \text{k}\Omega$$

Notice that $A_v$ is somewhat smaller in magnitude than $A_{voc}$. This is due to loading of the amplifier by $R_L$ as discussed in Chapter 10. Power gain is quite large for the common-emitter amplifier, and primarily for this reason, it is a commonly used configuration.

The common-emitter amplifier is inverting and has large voltage gain magnitude, large current gain, and large power gain.

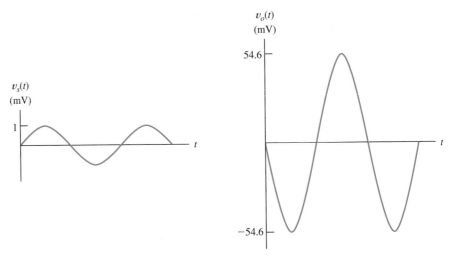

Figure 12.29 Source and output voltages for Example 12.8.

The source voltage divides between the internal source resistance and the input impedance of the amplifier. Thus, we can write

$$v_{in} = v_s \frac{Z_{in}}{Z_{in} + R_s} = 0.515v_s$$

Now, with the load connected, we have

$$v_o = A_v v_{in} = -54.6v_s$$

But, we are given that $v_s(t) = \sin(\omega t)$ mV, so we have

$$v_o(t) = -54.6 \sin(\omega t) \text{ mV}$$

The source voltage $v_s(t)$ and the output voltage are shown in Figure 12.29. Notice the phase inversion.    ∎

**Exercise 12.14**    Repeat Example 12.8 if $\beta = 300$. [*Hint:* Do not forget that the $Q$ point changes (slightly) when $\beta$ changes.]
**Answer**    $A_v = -109$, $A_{voc} = -163$, $Z_{in} = 1186 \ \Omega$, $A_i = -64.4$, $G = 7004$, $Z_o = 1 \text{ k}\Omega$, $v_o(t) = -76.5 \sin(\omega t)$ mV.    □

## 12.9 EMITTER FOLLOWERS

The circuit diagram of another type of BJT amplifier called an **emitter follower** is shown in Figure 12.30(a). The resistors $R_1$, $R_2$, and $R_E$ form the bias circuit. The collector resistor $R_C$ (used in the common-emitter amplifier) is not needed in this circuit. Thus, we have a version of the four-resistor bias circuit with $R_C = 0$. Analysis of this bias circuit is very similar to the examples that we considered in Section 12.6.

The input signal is applied to the base through the coupling capacitor $C_1$. The output signal is coupled from the emitter to the load by the coupling capacitor $C_2$.

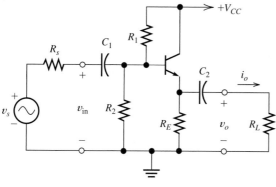

(a) Actual circuit

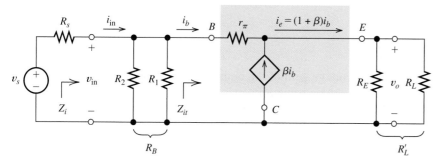

(b) Small-signal equivalent circuit

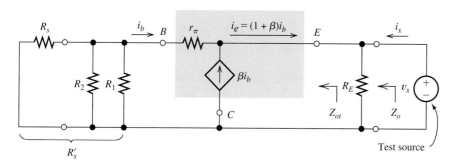

(c) Equivalent circuit used to find output impedance $Z_o$

**Figure 12.30** Emitter follower.

## Small-Signal Equivalent Circuit

The ac small-signal equivalent circuit is shown in Figure 12.30(b). As before, we replace the capacitors and power supply with short circuits. The transistor is replaced by its small-signal equivalent.

Notice that as a result, the collector terminal is connected directly to ground in the equivalent circuit. The transistor equivalent circuit is oriented with the collector at the bottom in Figure 12.30(b), but it is electrically the same as the transistor equivalent circuit we have used before. Because the collector is connected to ground, this circuit is sometimes called a **common-collector amplifier**.

Try to draw the small-signal
equivalent on your own,
referring to only Figure
12.30(a).

The ability to draw small-signal equivalent circuits is an important skill for analyzing electronic circuits. Carefully compare the small-signal equivalent in Figure 12.30(b) to the original circuit. Better still, try to draw the small-signal equivalent circuit on your own, starting from the original circuit.

Notice that $R_1$ and $R_2$ are in parallel in the equivalent circuit. We denote the combination by $R_B$. Also, $R_E$ and $R_L$ are in parallel, and we denote the combination by $R'_L$. In equation form, we have

$$R_B = R_1 \| R_2 = \frac{1}{1/R_1 + 1/R_2} \tag{12.49}$$

and

$$R'_L = R_L \| R_E = \frac{1}{1/R_L + 1/R_E} \tag{12.50}$$

### Voltage Gain

Next, we find the voltage gain of the emitter follower. The current flowing through $R'_L$ is $i_b + \beta i_b$. Thus, the output voltage is given by

$$v_o = R'_L (1 + \beta) i_b \tag{12.51}$$

Writing a voltage equation from the input terminals through $r_\pi$ and then through the load to ground, we have

$$v_{in} = r_\pi i_b + (1 + \beta) i_b R'_L \tag{12.52}$$

Division of each side of Equation 12.51 by the respective side of Equation 12.52 results in

$$A_v = \frac{(1 + \beta) R'_L}{r_\pi + (1 + \beta) R'_L} \tag{12.53}$$

Emitter followers have
voltage gains that are less
than unity.

*The voltage gain of the emitter follower is less than unity* because the denominator of the expression is larger than the numerator. However, the voltage gain is usually only slightly less than unity. An amplifier with voltage gain less than unity can sometimes be useful because it can have a large current gain.

Also, notice that the voltage gain is positive. In other words, the emitter follower is noninverting. Thus, if the input voltage changes, the output at the emitter changes by almost the same amount and in the same direction as the input. Thus, the output voltage *follows* the input voltage. This is the reason for the name *emitter follower*.

### Input Impedance

The input impedance of the
emitter follower is relatively
large compared with that of
other BJT amplifiers.

The input impedance $Z_i$ can be found as the parallel combination of $R_B$ and the input impedance seen looking into the base of the transistor, which is indicated as $Z_{it}$ in Figure 12.30(b). Thus, we can write

$$Z_i = \frac{1}{1/R_B + 1/Z_{it}} = R_B \| Z_{it} \tag{12.54}$$

The input impedance looking into the base can be found by dividing both sides of Equation 12.52 by $i_b$:

$$Z_{it} = \frac{v_{in}}{i_b} = r_\pi + (1 + \beta)R'_L \qquad (12.55)$$

The input impedance of the emitter follower is relatively high compared to other BJT amplifier configurations. (However, in Chapter 11, we saw that field-effect transistors are capable of providing much higher input impedance than BJTs.) Once we have found the voltage gain and input impedance of the emitter follower, the current gain and power gain can be found by use of Equations 10.3 and 10.5.

## Output Impedance

The output impedance of an amplifier is the Thévenin impedance seen from the output terminals. To find the output impedance of the emitter follower, we remove the load resistance, zero the signal source, and look back into the output terminals of the equivalent circuit. This is shown in Figure 12.30(c). We have attached a test source $v_x$, which delivers a current $i_x$ to the impedance that we want to find. The output impedance is given by

$$Z_o = \frac{v_x}{i_x} \qquad (12.56)$$

(Here again, the impedance can be found as the ratio of instantaneous time-varying quantities because the circuit is purely resistive. Otherwise, we should use phasors.)

To find this ratio, we write equations involving $v_x$ and $i_x$. For example, summing currents at the top end of $R_E$, we have

$$i_b + \beta i_b + i_x = \frac{v_x}{R_E} \qquad (12.57)$$

We must eliminate $i_b$ from this equation before we can find the desired expression for the output impedance. *We do not want any circuit variables such as $i_b$ in the result— only transistor parameters and resistance values.* Thus, we need to write another circuit equation.

First, we denote the parallel combination of $R_s$, $R_1$, and $R_2$ as

$$R'_s = \frac{1}{1/R_s + 1/R_1 + 1/R_2} \qquad (12.58)$$

The additional equation needed can now be obtained by applying Kirchhoff's voltage law to the loop consisting of $v_x$, $r_\pi$, and $R'_s$:

$$v_x + r_\pi i_b + R'_s i_b = 0 \qquad (12.59)$$

If we solve Equation 12.59 for $i_b$, substitute into Equation 12.57, and rearrange the result, we obtain the output impedance:

$$Z_o = \frac{v_x}{i_x} = \frac{1}{(1 + \beta)(R'_s + r_\pi) + 1/R_E} \qquad (12.60)$$

This can be recognized as the parallel combination of $R_E$ and the impedance

$$Z_{ot} = \frac{R_s' + r_\pi}{1 + \beta} \qquad (12.61)$$

[It can be shown that $Z_{ot}$ is the impedance seen looking into the emitter of the transistor, as indicated in Figure 12.30(c).]

The output impedance of the emitter follower tends to be smaller than that of other BJT amplifier configurations.

The output impedance of the emitter follower is relatively small compared with that of other BJT amplifiers.

---

### Example 12.9    Emitter-Follower Performance

Compute the voltage gain, input impedance, current gain, power gain, and output impedance for the emitter-follower amplifier shown in Figure 12.31. Assume that the circuit operates at a temperature for which $V_T = 26$ mV.

**Solution**    First, we must find the bias point so that the value of $r_\pi$ can be calculated. The dc circuit is shown in Figure 12.31(b). Because the coupling capacitors act as open circuits for dc, $R_s$ and $R_L$ do not appear in the dc bias circuit.

First, we must determine the bias point so the value of $r_\pi$ can be calculated.

Replacing the base bias circuit by its Thévenin equivalent, we obtain the equivalent circuit shown in Figure 12.31(c). Now, if we assume operation in the active

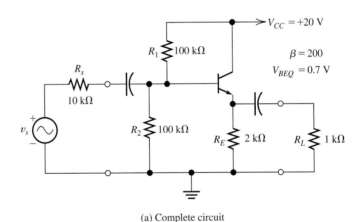

(a) Complete circuit

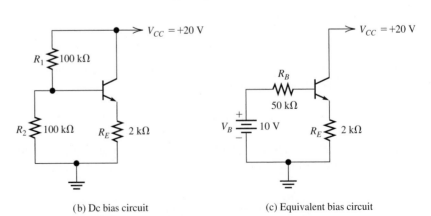

(b) Dc bias circuit

(c) Equivalent bias circuit

Figure 12.31  Emitter follower of Example 12.9.

region, we can write the following voltage equation around the base loop:

$$V_B = R_B I_{BQ} + V_{BEQ} + R_E(1 + \beta)I_{BQ}$$

Substituting values, we find $I_{BQ} = 20.6\ \mu\text{A}$. Then, we have

$$I_{CQ} = \beta I_{BQ} = 4.12\ \text{mA}$$

$$V_{CEQ} = V_{CC} - I_{EQ}R_E = 11.7\ \text{V}$$

Since $V_{CEQ}$ is greater than 0.2 V and $I_{BQ}$ is positive, the transistor is indeed operating in the active region.

Equation 12.35 yields

$$r_\pi = \frac{\beta V_T}{I_{CQ}} = 1260\ \Omega$$

Now that we have established that the transistor operates in the active region and found the value of $r_\pi$, we can proceed in finding the amplifier gains and impedances. Substituting values into Equations 12.49 and 12.50, we discover that

$$R_B = R_1 \| R_2 = \frac{1}{1/R_1 + 1/R_2} = 50\ \text{k}\Omega$$

$$R'_L = R_L \| R_E = \frac{1}{1/R_L + 1/R_E} = 667\ \Omega$$

Equation 12.53 gives the voltage gain:

$$A_v = \frac{(1 + \beta)R'_L}{r_\pi + (1 + \beta)R'_L} = 0.991$$

Equations 12.54 and 12.55 give the input impedance:

$$Z_{it} = r_\pi + (1 + \beta)R'_L = 135\ \text{k}\Omega$$

$$Z_i = \frac{1}{1/R_B + 1/Z_{it}} = 36.5\ \text{k}\Omega$$

Equations 12.58 and 12.60 yield

$$R'_s = \frac{1}{1/R_s + 1/R_1 + 1/R_2} = 8.33\ \text{k}\Omega$$

$$Z_o = \frac{1}{(1 + \beta)/(R'_s + r_\pi) + 1/R_E} = 46.6\ \Omega$$

From Equation 10.3, we can find the current gain:

$$A_i = A_v \frac{Z_i}{R_L} = 36.2$$

Using Equation 10.5, we find that the power gain is

$$G = A_v A_i = 35.8$$

Even though the voltage
gain of the emitter follower
is less than unity, the current
gain and power gain can be
large.

Notice that even though the voltage gain is less than unity, the current gain is large (compared with unity). Thus, the output power is larger than the input power, and the circuit is effective as an amplifier. ■

In general, the output impedance of the emitter follower is much lower and the input impedance is higher than those of other single-stage BJT amplifiers. *Thus, we use an emitter follower if high input impedance or low output impedance is needed.*

If the emitter follower is cascaded with common-emitter stages, amplifiers with many useful combinations of parameters are possible. Furthermore, there are several other useful amplifier configurations using the BJT.

**Exercise 12.15**   Repeat Example 12.9 for $\beta = 300$. Compare the results to those of the example.
**Answer**   $A_v = 0.991$, $Z_i = 40.1 \text{ k}\Omega$, $Z_o = 33.2 \text{ }\Omega$, $A_i = 39.7$, $G = 39.4$.    ☐

## Summary

1. An *npn* BJT consists of a layer of *p*-type material (the base) between two layers of *n*-type material (the collector and emitter, respectively).

2. In the active region, the collector current is an amplified version of the base current. In equation form, we have $i_C = \beta i_B$. A typical value of $\beta$ is 100.

3. Typical common-emitter characteristics for an npn BJT are shown in Figure 12.5 on page 603.

4. Load-line analysis provides a basic understanding of amplifier circuits.

5. Besides being amplified, signals are often distorted by BJT amplifiers. In the active region, distortion is due to curvature of the input characteristic or unequal spacing of the output characteristics. For large signals, a more severe type of distortion, known as clipping, occurs when the device operation swings either into saturation or cutoff.

6. The *pnp* BJT is similar to the *npn* except that the current directions and voltage polarities are reversed.

7. BJTs can operate in three regions: the active region, the saturation region, and the cutoff region. Circuit models for BJTs in each of the three regions of operation are shown in Figure 12.16 on page 613.

8. In large-signal analysis of BJT circuits, we assume operation in a given region, use the

corresponding circuit model to solve for currents and voltages, and then check to see if the results are consistent with the assumed operating region. The process is repeated until a self-consistent solution is found.

9. To be effective as an amplifier, the BJT must be biased in the active region. The four-resistor bias circuit [Figure 12.22(a) on page 620] is commonly used for simple amplifiers.

10. The small-signal equivalent circuit for the BJT is shown in Figure 12.26 on page 625.

11. To analyze an amplifier in the midrange of frequencies, we first draw the small-signal equivalent circuit. The BJT is replaced by its equivalent circuit, the coupling and bypass capacitors are replaced by short circuits, and the dc power voltage sources are replaced by short circuits. Then, circuit equations are written, and algebra is used to find expressions for the gains and impedances of interest.

12. Two important BJT amplifier circuits are the common-emitter amplifier [Figure 12.27(a) on page 626] and the emitter follower [Figure 12.30(a) on page 631]. The common-emitter amplifier is inverting, has relatively large voltage and current gain magnitudes, and has moderate input impedance. The emitter follower is noninverting and has nearly unity voltage gain, large current gain, and moderate to high input impedance.

## Problems

### Section 12.1: Current and Voltage Relationships

**P12.1.** Write the Shockley equation for the emitter current of an *npn* transistor.

**P12.2.** Which side of a *pn* junction should be connected to the positive voltage for forward bias? In normal operation, which type of bias (forward or reverse) is applied to the emitter–base junction of a BJT? To the collector–base junction?

**P12.3.** Sketch the construction of an *npn* BJT and label the three regions. In normal operation, does current flow into or out of the base terminal? The collector terminal? The emitter terminal?

**P12.4.** Give the definitions of $\alpha$ and $\beta$ for a BJT. What bias conditions for each junction are assumed in these definitions?

**P12.5.** Draw the circuit symbol for an *npn* BJT. Label the terminals and the currents. Choose reference directions that agree with the true current directions for operation in the active region.

**\*P12.6.** An *npn* transistor is operating with the base–emitter junction forward biased and the base–collector junction reverse biased. Given $i_C = 9$ mA for $i_B = 0.3$ mA, compute $i_E$, $\alpha$, and $\beta$.

**\*P12.7.** Suppose that we have an *npn* transistor at room temperature, with $I_{ES} = 10^{-13}$ A, $\beta = 100$, $v_{CE} = 10$ V, and $i_E = 10$ mA. Determine $v_{BE}$, $v_{BC}$, $i_B$, $i_C$, and $\alpha$. (Assume that $V_T = 26$ mV at room temperature.)

**P12.8.** Consider the circuit shown in Figure P12.8. The transistors $Q_1$ and $Q_2$ are identical, both having $I_{ES} = 10^{-14}$ A and $\beta = 100$. Calculate $V_{BE}$ and $I_{C2}$. Assume that $V_T = 26$ mV for both transistors. [*Hint:* Both transistors are operating in the active region. Because the transistors are identical and have identical values of $V_{BE}$, their collector currents are equal.]

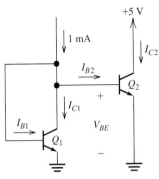

**Figure P12.8**

**P12.9.** Consider an *npn* transistor that is operating with the base–emitter junction forward biased and the base–collector junction reverse biased. Given that the transistor has $i_C = 10$ mA and $i_E = 10.5$ mA, determine the values of $i_B$, $\alpha$, and $\beta$.

**P12.10.** If a transistor has $\beta = 200$, what is its value for $\alpha$?

**P12.11.** Suppose that an *npn* transistor is operating with the base–emitter junction forward biased and the base–collector junction reverse biased. Given $\beta = 200$ and $i_B = 10\ \mu$A, determine the values of $i_C$ and $i_E$.

**P12.12.** Consider the circuit of Figure P12.12 in which $Q_1$ has $I_{ES1} = 10^{-14}$ A and $\beta_1 = 100$, while $Q_2$ has $I_{ES2} = 10^{-13}$ A and $\beta_2 = 100$. Determine the values of $V_{BE}$ and $I_{C2}$. Assume that both transistors have $V_T = 26$ mV and are operating in the active region.

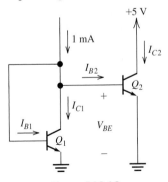

**Figure P12.12**

---

\* Denotes that answers are contained in the Student Solutions files. See Appendix E for more information about accessing the Student Solutions.

**P12.13.** Suppose that a certain *npn* transistor has $V_{BE} = 0.7$ V for $I_E = 10$ mA. Compute $V_{BE}$ for $I_E = 1$ mA. Repeat for $I_E = 1$ $\mu$A. Assume that $V_T = 26$ mV.

**P12.14.** At an absolute temperature of $T = 300$ K, a certain transistor has $i_E = 10$ mA and $v_{BE} = 0.600$ V. Determine the value of $I_{ES}$. At 310 K, the transistor has $i_E = 10$ mA and $v_{BE} = 0.580$ V. Determine the new value of $I_{ES}$. By what factor did $I_{ES}$ change for this 10 K increase in temperature? (Recall that Equation 9.2 on page 462 gives $V_T = kT/q$, in which $k = 1.38 \times 10^{-23}$ J/K is Boltzmann's constant and $q = 1.60 \times 10^{-19}$ C is the magnitude of the charge on an electron.)

**P12.15.** Determine the value of $\beta$ for the transistor of Figure P12.15.

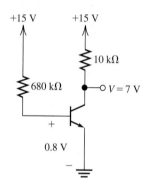

**Figure P12.15**

**\*P12.16.** Two transistors $Q_1$ and $Q_2$ connected in parallel are equivalent to a single transistor, as indicated in Figure P12.16. If the individual transistors have $I_{ES1} = I_{ES2} = 10^{-13}$ A and $\beta_1 = \beta_2 = 100$, determine $I_{ES}$ and $\beta_{eq}$ for the equivalent transistor. Assume that all transistors have the same temperature.

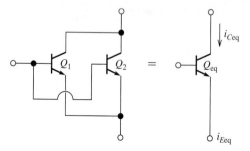

**Figure P12.16**

**P12.17.** The transistors $Q_1$ and $Q_2$ shown in Figure P12.17 are said to be **Darlington connected**

and can be considered to be equivalent to a single transistor, as indicated. Find an expression for $\beta_{eq}$ of the equivalent transistor in terms of $\beta_1$ and $\beta_2$.

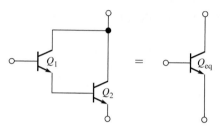

**Figure P12.17** Darlington pair.

## Section 12.2: Common-Emitter Characteristics

**\*P12.18.** Determine the values of $\alpha$ and $\beta$ for the transistor whose characteristics are shown in Figure P12.18.

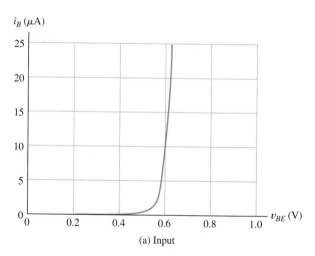

(a) Input

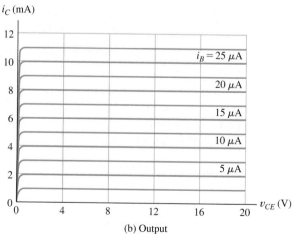

(b) Output

**Figure P12.18**

**\*P12.19.** A certain *npn* silicon transistor has $v_{BE} = 0.7$ V for $i_B = 0.1$ mA at a temperature of 30°C. Sketch the input characteristic to scale at 30°C. What is the approximate value of $v_{BE}$ for $i_B = 0.1$ mA at 180°C? (Use the rule of thumb that $v_{BE}$ is reduced in magnitude by 2 mV per degree increase in temperature.) Sketch the input characteristic to scale at 180°C.

**P12.20.** Consider an *npn* silicon transistor that has $\beta = 100$ and $i_B = 0.1$ mA. Sketch $i_C$ versus $v_{CE}$, for $v_{CE}$ ranging from 0 to 5 V. Repeat for $\beta = 300$.

**P12.21.** The transistor having the characteristics shown in Figure P12.18 is operating with $i_C = 8$ mA and $v_{CE} = 12$ V. Locate the point of operation on both the input and output characteristics.

**Section 12.3:** Load-Line Analysis of a Common-Emitter Amplifier

**\*P12.22.** Consider the circuit of Figure 12.7 on page 605. Assume that $V_{CC} = 20$ V, $V_{BB} = 0.8$ V, $R_B = 40$ k$\Omega$, and $R_C = 2$ k$\Omega$. The input signal is a 0.2-V-peak 1-kHz sinusoid given by $v_{in}(t) = 0.2 \sin(2000\pi t)$. The common-emitter characteristics for the transistor are shown in Figure P12.18. Determine the maximum, minimum, and $Q$-point values for $v_{CE}$. What is the approximate voltage gain for this circuit?

**P12.23.** List several reasons that distortion occurs in BJT amplifiers.

**P12.24.** Consider the circuit of Figure 12.7 on page 605. Assume that $V_{CC} = 20$ V, $V_{BB} = 0.8$ V, $R_B = 40$ k$\Omega$, and $R_C = 10$ k$\Omega$. The input signal is a 0.2-V-peak 1-kHz sinusoid given by $v_{in}(t) = 0.2 \sin(2000\pi t)$. The common-emitter characteristics for the transistor are shown in Figure P12.18. Determine the maximum, minimum, and $Q$-point values for $v_{CE}$. What can you say about the waveform for $v_{CE}(t)$? Why isn't voltage gain an appropriate concept in this case?

**P12.25.** Consider the circuit of Figure 12.7 on page 605. Assume that $V_{CC} = 20$ V, $V_{BB} = 0.3$ V, $R_B = 40$ k$\Omega$, and $R_C = 2$ k$\Omega$. The input signal is a 0.2-V-peak 1-kHz sinusoid given by $v_{in}(t) = 0.2 \sin(2000\pi t)$. The common-emitter characteristics for the transistor are shown in Figure P12.18. Determine the maximum, minimum, and $Q$-point values for $v_{CE}$. What is the approximate voltage gain for this circuit? Why is the gain so small in magnitude?

**P12.26.** Consider the circuit of Figure 12.7 on page 605. Given $V_{CC} = 10$ V and $R_C = 2$ k$\Omega$, construct the load line on the $i_C$ versus $v_{CE}$ axes. Repeat for $V_{CC} = 15$ V. How does the slope of the load line change when $V_{CC}$ changes?

**Section 12.4:** *pnp* Bipolar Junction Transistors

**P12.27.** Draw the circuit symbol for a *pnp* BJT. Label the terminals and the currents. Choose reference directions that agree with the true current direction for operation in the active region.

**P12.28.** Figure P12.28 shows an *npn* transistor and a *pnp* transistor connected as a **Sziklai pair**, which is equivalent to a single *npn* transistor, as indicated. Find an expression for $\beta_{eq}$ of the equivalent transistor in terms of $\beta_1$ and $\beta_2$.

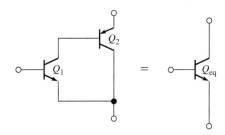

Figure P12.28  Sziklai pair.

**\*P12.29.** A certain *pnp* silicon transistor has $\beta = 100$ and $i_B = 50$ $\mu$A. Sketch $i_C$ versus $v_{CE}$, for $v_{CE}$ ranging from 0 to $-5$ V. Repeat for $\beta = 300$.

**\*P12.30.** The circuit shown in Figure P12.30 has $i_s(t) = 10 + 5 \sin(2000\pi t)$ $\mu$A. The transistor has $\beta = 100$.

  **a.** Sketch the output characteristics for $i_B = 0, 5, 10, 15, 20$, and 25 $\mu$A with $v_{CE}$ ranging from zero to $-20$ V.

  **b.** Draw the output load line on the characteristics sketched in part (a).

  **c.** Determine the values for $I_{Cmax}$, $I_{CQ}$, and $I_{Cmin}$.

  **d.** Sketch $v_{CE}(t)$ to scale versus time.

  **e.** Repeat parts (c) and (d) for $i_s(t) = 20 + 5 \sin(2000\pi t)$ $\mu$A.

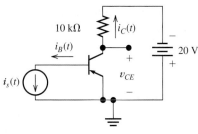

**Figure P12.30**

**P12.31.** Suppose we have a certain *pnp* BJT that has $V_{CE} = -5$ V, $I_C = 0.995$ mA, and $I_E = 1.000$ mA. Determine the values for $\alpha$ and $\beta$ for this transistor.

**P12.32.** At a temperature of 30°C, a particular *pnp* transistor has $V_{BE} = -0.7$ V for $I_E = 2$ mA. Estimate $V_{BE}$ for $I_E = 2$ mA at a temperature of 180°C.

**Section 12.5: Large-Signal DC Circuit Models**

**P12.33.** Sketch the input characteristic curve to scale for a typical small-signal silicon *npn* BJT at room temperature, allowing $i_B$ to range from 0 to 40 $\mu$A. Sketch the output characteristic curves to scale, with $i_B$ ranging from 0 to 40 $\mu$A in 10 $\mu$A steps, given that $\beta = 100$. Finally, label the cutoff, active, and saturation regions.

**P12.34.** Each of the transistors shown in Figure P12.34 has $\beta = 100$, $|V_{CE}| = 0.2$ V in saturation, and $|V_{BE}| = 0.6$ V in the active

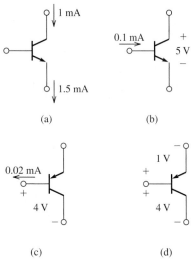

**Figure P12.34**

and saturation regions. For each transistor, determine the region of operation and the values of $V_{CE}$, $I_B$, $I_E$, and $I_C$.

**\*P12.35.** In the active region, how is the base–collector junction biased (forward or reverse)? How is the base–emitter junction biased? Repeat for the saturation region. Repeat for the cutoff region.

**P12.36.** Draw the large-signal dc circuit model for a silicon *npn* transistor in the active region at room temperature. Include the constraints of currents and voltages that guarantee operation in the active region. Repeat for the saturation region. Repeat for the cutoff region.

**P12.37.** Determine the region of operation for a room-temperature silicon *npn* transistor that has $\beta = 100$ if: **a.** $V_{CE} = 10$ V and $I_B = 20$ $\mu$A; **b.** $I_C = I_B = 0$; **c.** $V_{CE} = 3$ V and $V_{BE} = 0.4$ V; **d.** $I_C = 1$ mA and $I_B = 50$ $\mu$A.

**P12.38.** Shown in Figure P12.17 is the Darlington pair, which is equivalent to a single transistor, as indicated in the figure. Assuming that both $Q_1$ and $Q_2$ are operating in the active region, with $|V_{BE}| = 0.6$ V, determine the value for $V_{BE}$ for the equivalent transistor in the active region. Repeat for the Sziklai pair shown in Figure P12.28.

**P12.39.** Determine the region of operation for a room-temperature silicon *pnp* transistor that has $\beta = 100$ if: **a.** $V_{CE} = -5$ V and $V_{BE} = -0.3$ V; **b.** $I_C = 10$ mA and $I_B = 1$ mA; **c.** $I_B = 0.05$ mA and $V_{CE} = -5$ V.

**P12.40.** Draw the large-signal dc circuit model for a silicon *pnp* transistor in the active region at room temperature. Include the constraints of currents and voltages that guarantee operation in the active region. Repeat for the saturation region. Repeat for the cutoff region.

**Section 12.6: Large-Signal DC Analysis of BJT Circuits**

**\*P12.41.** List the step-by-step procedure for dc analysis of a BJT circuit, using the large-signal circuit models.

**\*P12.42.** Use the large-signal models shown in Figure 12.16 on page 613 for the transistors to determine $I_C$ and $V_{CE}$ for the circuits of Figure P12.42. Assume that $\beta = 100$

and $|V_{BE}| = 0.7$ V in both the active and saturation regions. Repeat for $\beta = 300$, and compare the results for both values.

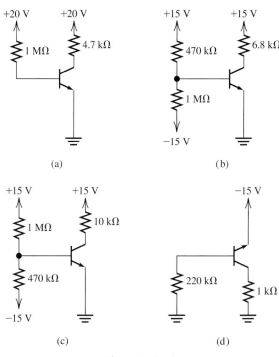

(a)

(b)

(c)

(d)

**Figure P12.42**

**P12.43.** The transistors shown in Figure P12.43 operate in the active region and have $\beta = 100$ and $V_{BE} = 0.7$ V. Determine $I_C$ and $V_{CE}$ for each transistor.

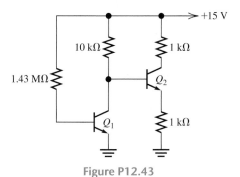

**Figure P12.43**

**P12.44.** Draw the four-resistor bias circuit for an *npn* BJT.

**P12.45.** Draw the fixed base bias circuit. What is the principal reason that this circuit is unsuitable for mass production of amplifier circuits?

**\*P12.46.** Consider the circuit shown in Figure P12.46. A $Q$-point value for $I_C$ between a minimum of 4 mA and a maximum of 5 mA is required. Assume that resistor values are constant and that $\beta$ ranges from 100 to 300. It is desired for $R_B$ to have the largest possible value while meeting the other constraints. Determine the values of $R_B$ and $R_E$.

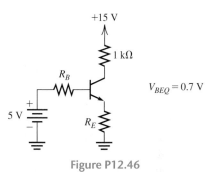

**Figure P12.46**

**P12.47.** Analyze the circuits shown in Figure P12.47 to determine $I$ and $V$. For all transistors, assume that $\beta = 100$ and $|V_{BE}| = 0.7$ V in both the active and saturation regions. Repeat for $\beta = 300$.

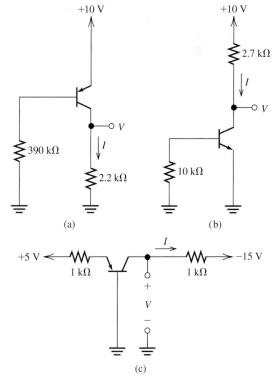

**Figure P12.47**

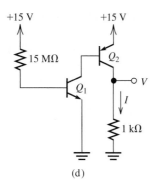

(d)

Figure P12.47 (Cont)

**\*P12.48.** The four-resistor bias network of Figure 12.22(a) on page 620 has $V_{CC} = 15$ V, $R_1 = 100$ kΩ, $R_2 = 47$ kΩ, $R_C = 4.7$ kΩ, and $R_E = 4.7$ kΩ. Suppose that $β$ ranges from 50 to 200, $V_{BE} = 0.7$ V, and the resistors have tolerances of ±5 percent. Calculate the maximum and minimum values for $I_C$.

**P12.49.** Consider the four-resistor bias network of Figure 12.22(a) on page 620, with $R_1 = 200$ kΩ, $R_2 = 100$ kΩ, $V_{CC} = 15$ V, $R_C = 10$ kΩ, $R_E = 10$ kΩ, and $β = 200$. Assume that $V_{BE} = 0.7$ V. Determine $I_{CQ}$ and $V_{CEQ}$.

**P12.50.** The four-resistor bias network of Figure 12.22(a) on page 620 has $R_1 = 100$ kΩ, $R_2 = 200$ kΩ, $V_{CC} = 15$ V, $R_C = 10$ kΩ, $R_E = 10$ kΩ, and $β = 200$. Assume that $V_{BE} = 0.7$ V in both the active and saturation regions. Determine $I_{CQ}$ and $V_{CEQ}$. [*Hint:* The transistor may *not* be operating in the active region.]

**P12.51.** Analyze the circuit of Figure P12.51 to determine $I_C$ and $V_{CE}$.

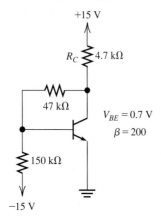

Figure P12.51

**P12.52.** Consider the circuit shown in Figure P12.52. Determine the values of $R_1$ and $R_C$ if a bias point of $V_{CE} = 5$ V and $I_C = 2$ mA is required.

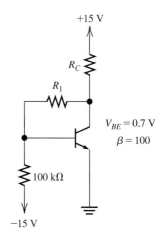

Figure P12.52

**Section 12.7: Small-Signal Equivalent Circuits**

**P12.53.** Give the formula for determination of $r_π$ assuming that $β$ and the $Q$ point are known.

**P12.54.** Draw the small-signal equivalent circuit for a BJT.

**\*P12.55.** Suppose that a new type of BJT has been invented for which $i_B = 10^{-5} v_{BE}^2$, where $i_B$ is in amperes and $v_{BE}$ is in volts. Also, this new transistor has $i_C = 100 i_B$. The small-signal equivalent circuit for the transistor is shown in Figure 12.26 on page 625. Find an equation relating $r_π$ of this new transistor to $I_{CQ}$. Evaluate $r_π$ for $I_{CQ} = 1$ mA.

**P12.56.** Shown in Figure P12.28 is the Sziklai pair, which is equivalent to a single transistor, as indicated in the figure. Draw the small-signal equivalent circuits for the pair and for the equivalent transistor. Find an expression for $r_{πeq}$ in terms of $r_{π1}, r_{π2}, β_1$, and $β_2$.

**P12.57.** Shown in Figure P12.17 is the Darlington pair, which is equivalent to a single transistor, as indicated in the figure. Draw the small-signal equivalent circuits for the pair and for the equivalent transistor. Find an expression for $r_{πeq}$ in terms of $r_{π1}, r_{π2}, β_1$, and $β_2$.

**P12.58.** A certain *npn* silicon transistor at room temperature has $β = 100$. Find the corresponding values of $r_π$ if $I_{CQ} = 1$ mA,

0.1 mA, and 1 μA. Assume operation in the active region and that $V_T = 26$ mV.

## Section 12.8: Common-Emitter Amplifiers

**P12.59.** Draw the circuit diagram of a common-emitter amplifier circuit that uses the four-resistor biasing network. Include a signal source and a load resistance.

**P12.60.** Why are coupling capacitors often used to connect the signal source and the load to amplifier circuits? Should coupling capacitors be used if it is necessary to amplify dc signals? Explain.

**P12.61.** Are common-emitter amplifiers inverting or noninverting? What can you say about the magnitudes of the voltage and current gains?

**\*P12.62.** Consider the common-emitter amplifier of Figure P12.62. Draw the dc circuit and find $I_{CQ}$. Find the value of $r_\pi$. Then, calculate values for $A_v$, $A_{voc}$, $Z_{in}$, $A_i$, $G$, and $Z_o$. Assume operation in the frequency range for which the coupling and bypass capacitors are short circuits.

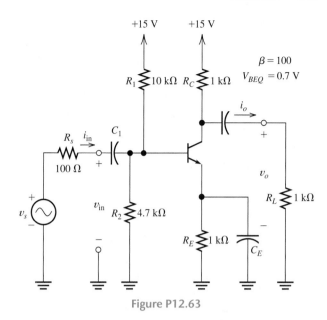

Figure P12.63

**P12.64.** Consider the common-emitter amplifier shown in Figure P12.64. **a.** Draw the small-signal equivalent circuit by assuming that the coupling capacitors are short circuits. **b.** Derive an expression for the voltage gain in terms of resistor values, $r_\pi$, and $\beta$. **c.** Derive an expression for the input resistance in terms of resistor values, $r_\pi$, and $\beta$. **d.** Suppose that $\beta = 100$, $V_{BEQ} = 0.7$ V, $R_C = 2$ kΩ, $R_L = 2$ kΩ, $R_E = 100$ Ω, and $V_{CC} = 20$ V. Determine the value required for $R_B$ so that $I_{CQ} = 5$ mA. **e.** Evaluate the expressions found in parts (c) and (d) for the values given in part (d).

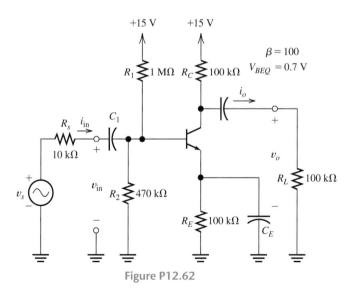

Figure P12.62

**P12.63.** Consider the common-emitter amplifier of Figure P12.63. Draw the dc circuit and find $I_{CQ}$. Find the value of $r_\pi$. Then, calculate values for $A_v$, $A_{voc}$, $Z_{in}$, $A_i$, $G$, and $Z_o$. Assume operation in the frequency range for which the coupling and bypass capacitors are short circuits.

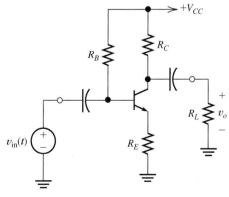

Figure P12.64

## Section 12.9: **Emitter Followers**

**P12.65.** Draw the circuit diagram of an emitter follower. Include a signal source and a load resistance.

**P12.66.** What can you say about the voltage gain of an emitter follower? The current gain? The power gain?

**\*P12.67.** Consider the emitter-follower amplifier of Figure P12.67. Draw the dc circuit and

find $I_{CQ}$. Next, determine the value of $r_\pi$. Then, calculate midband values for $A_v$, $A_{voc}$, $Z_{in}$, $A_i$, $G$, and $Z_o$.

**P12.68.** Consider the emitter-follower amplifier of Figure P12.68. Draw the dc circuit and find $I_{CQ}$. Next, determine the value of $r_\pi$. Then, calculate midband values for $A_v$, $A_{voc}$, $Z_{in}$, $A_i$, $G$, and $Z_o$.

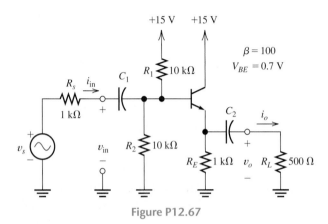

Figure P12.67

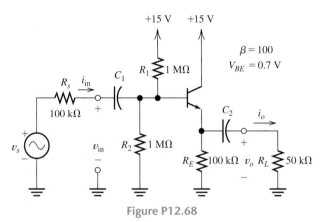

Figure P12.68

## Practice Test

Here is a practice test you can use to check your comprehension of the most important concepts in this chapter. Answers can be found in Appendix D and complete solutions are included in the Student Solutions files. See Appendix E for more information about the Student Solutions.

**T12.1.** Match each entry in Table T12.1(a) with the best choice from the list given in Table T12.1(b). [Items in Table T12.1(b) may be used more than once or not at all.]

**T12.2.** The simple amplifier of Figure 12.7 on page 605 has $R_B = 10\ k\Omega$, $V_{BB} = 0.8\ V$,

## Table T12.1

**(a)**

**a.** For a BJT in the active region, the collector–base junction is _____.

**b.** For a BJT in the active region, the emitter–base junction is _____.

**c.** For a BJT in the active region, the _____ current is usually much smaller than the current in the other two terminals.

**d.** Clipping occurs when the BJT in an amplifier circuit reaches _____ or _____.

**e.** In a BJT amplifier, nonuniform spacing of the collector characteristics can cause _____.

**f.** The large signal model for a BJT consists of two voltage sources in the _____ region.

**g.** The large signal model for a BJT has open circuits between all three terminals in the _____ region.

**h.** A BJT with $\beta = 50$, $I_C = 1\ mA$, and $I_E = 1.5\ mA$ is operating in the _____ region.

**i.** The voltage gain of an emitter follower is typically _____.

**j.** The voltage gain of a common-emitter amplifier is typically _____.

**k.** In a midband small-signal equivalent circuit, a coupling capacitance appears as _____.

**(b)**

1. cutoff
2. forward biased
3. reverse biased
4. emitter
5. base
6. collector
7. saturation
8. large gain
9. small gain
10. distortion
11. active
12. large in magnitude compared to unity and negative
13. small in magnitude compared to unity and negative
14. inverting
15. slightly less than unity
16. noninverting
17. large in magnitude compared to unity and positive
18. an open circuit
19. a short circuit

$v_{in}(t) = 0.2 \sin(2000\pi t)$ V, $R_C = 2.5$ kΩ, and $V_{CC} = 10$ V. The characteristics are shown in Figure 12.5 on page 603. Use load-line analysis to determine the values for $V_{CE\,min}$, $V_{CEQ}$, and $V_{CE\,max}$.

**T12.3.** An *npn* BJT is operating in the active region with $I_{CQ} = 1.0$ mA and $I_{EQ} = 1.04$ mA. Determine the values for $\alpha$, $\beta$, and $r_\pi$. Draw the small-signal equivalent circuit for the BJT. Assume that $V_T = 26$ mV.

**T12.4.** The BJT shown in Figure T12.4 has $\beta = 50$ and $V_{BE} = 0.7$ V. **a.** Determine the values of $I_C$ and $V_{CE}$; **b.** Repeat for $\beta = 250$.

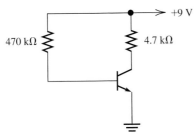

Figure T12.4

**T12.5.** Draw the midband small-signal equivalent circuit for the amplifier shown in Figure T12.5.

Be sure to label each element in your equivalent circuit.

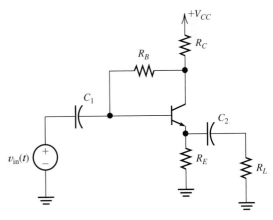

Figure T12.5

**T12.6.** The common-emitter amplifier shown in Figure 12.27(a) on page 626 has $R_1 = 100$ kΩ, $R_2 = 47$ kΩ, $R_C = 2.2$ kΩ, $R_L = 5.6$ kΩ, $\beta = 120$, $V_T = 26$ mV, and $I_{CQ} = 4$ mA. Determine the value of the voltage gain $A_v = v_o/v_{in}$ and the input impedance.

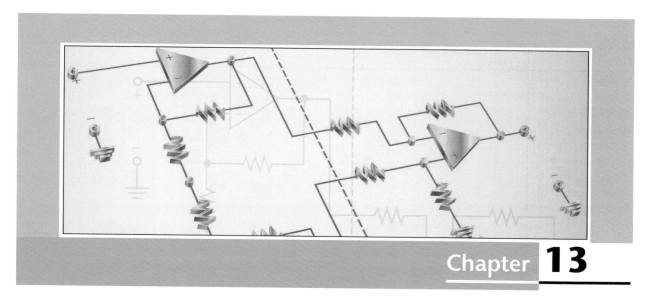

# Chapter 13

# Operational Amplifiers

## Study of this chapter will enable you to:

- List the characteristics of ideal op amps.
- Identify negative feedback in op-amp circuits.
- Use the summing-point constraint to analyze ideal-op-amp circuits that have negative feedback.
- Select op-amp circuit configurations suitable for various applications.
- Use op amps to design useful circuits.

- Identify practical op-amp limitations and recognize potential inaccuracies in instrumentation applications.
- Work with instrumentation amplifiers.
- Apply integrators, differentiators, and active filters.

## Introduction to this chapter:

In Chapter 11, we discussed the external characteristics of amplifiers in general. In Chapters 12 and 13, we saw how basic amplifiers can be built by using FETs or BJTs. In this chapter, we introduce an important device known as the **operational amplifier**, which finds application in a wide range of engineering instrumentation.

An operational amplifier is a circuit composed of perhaps 30 BJTs or FETs, 10 resistors, and several capacitors. These components are manufactured concurrently on a single piece of silicon crystal (called a chip) by a sequence of processing steps. Circuits manufactured in this way are called **integrated circuits** (ICs).

Because the manufacture of ICs is not much more complicated than the manufacture of individual transistors, operational amplifiers provide an economical and often better alternative to the discrete FET and BJT circuits that we studied Chapters 12 and 13.

Currently, the term *operational amplifier*, or less formally *op amp*, refers to ICs that are employed in a wide variety of general-purpose applications. However, this type of amplifier originated in analog-computer circuits in which it was used to perform such operations as integration or addition of signals—hence, the name *operational* amplifier.

We will see that inexpensive integrated-circuit op amps can be combined with resistors (and sometimes capacitors) to form many useful circuits. Furthermore, the characteristics of these circuits can be made to depend on the circuit configuration and the resistor values but only weakly on the op amp—which can have large unit-to-unit variations in some of its parameters.

## 13.1   IDEAL OPERATIONAL AMPLIFIERS

The circuit symbol for the operational amplifier is shown in Figure 13.1. The operational amplifier is a differential amplifier having both inverting and noninverting input terminals. (We discussed differential amplifiers in Section 10.11.) The input signals are denoted as $v_1(t)$ and $v_2(t)$. (As usual, we use lowercase letters to represent general time-varying voltages. Often, we will omit the time dependence and refer to the voltages simply as $v_1$, $v_2$, and so on.)

The input signal of a differential amplifier consists of a differential component and a common-mode component.

Recall that the average of the input voltages is called the **common-mode signal** and is given by

$$v_{icm} = \frac{1}{2}(v_1 + v_2)$$

Also, the difference between the input voltages is called the **differential signal**, given by

$$v_{id} = v_1 - v_2$$

An ideal operational amplifier has the following characteristics:

Characteristics of ideal op amps.

- Infinite input impedances
- Infinite gain for the differential input signal
- Zero gain for the common-mode input signal

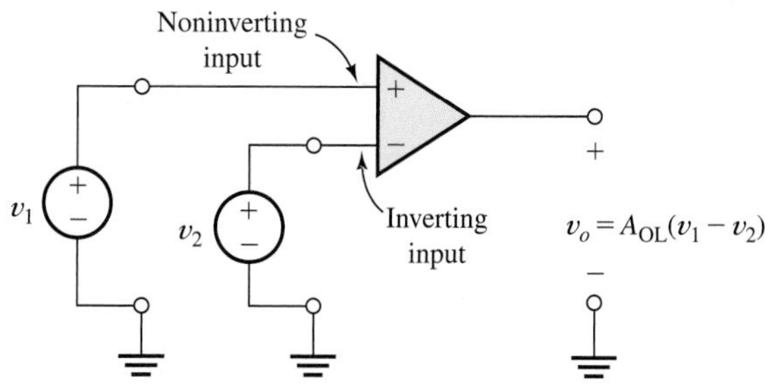

**Figure 13.1**  Circuit symbol for the op amp.

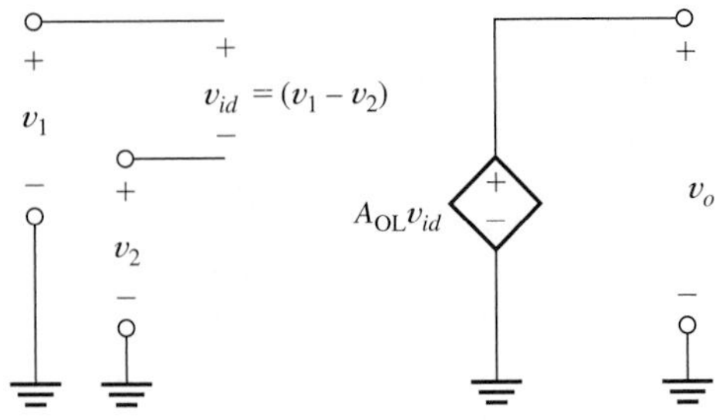

**Figure 13.2** Equivalent circuit for the ideal op amp. The open-loop gain $A_{\text{OL}}$ is very large (approaching infinity).

- Zero output impedance
- Infinite bandwidth

An equivalent circuit for the ideal operational amplifier consists simply of a controlled source as shown in Figure 13.2. The **open-loop gain** $A_{\text{OL}}$ is very large in magnitude—ideally, infinite.

As we will shortly see, op amps are generally used with feedback networks that return part of the output signal to the input. Thus, a *loop* is created in which signals flow through the amplifier to the output and back through the feedback network to the input. $A_{\text{OL}}$ is the gain of the op amp without a feedback network. That is why we call it the *open-loop gain*.

For now, we assume that the open-loop gain $A_{\text{OL}}$ is constant. Thus, there is no distortion, either linear or nonlinear, and the output voltage $v_o$ has a waveshape identical to that of the differential input $v_{id} = v_1 - v_2$. (Later, we will see that $A_{\text{OL}}$ is actually a function of frequency. Furthermore, we will learn that real op amps suffer from nonlinear imperfections.)

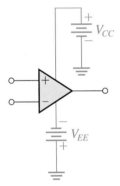

**Figure 13.3** Op-amp symbol showing the dc power supplies, $V_{CC}$ and $V_{EE}$.

## Power-Supply Connections

For a real op amp to function properly, one or more dc supply voltages must be applied, as shown in Figure 13.3. Often, however, we do not explicitly show the power-supply connections in circuit diagrams. (As indicated in the figure, it is standard practice to use uppercase symbols with repeated uppercase subscripts, such as $V_{CC}$ and $V_{EE}$, to represent dc power-supply voltages.)

## 13.2 INVERTING AMPLIFIERS

Operational amplifiers are almost always used with negative feedback, in which part of the output signal is returned to the input in opposition to the source signal.

Operational amplifiers are almost always used with **negative feedback**, in which part of the output signal is returned to the input in opposition to the source signal. (It is also possible to have *positive* feedback, in which the signal returned to the input *aids* the original source signal. However, negative feedback turns out to be more useful in amplifier circuits.) Frequently, we analyze op-amp circuits by assuming an ideal op amp and employing a concept that we call the summing-point constraint.

In a negative feedback system, the ideal-op-amp output voltage attains the value needed to force the differential input voltage and input current to zero. We call this fact the summing-point constraint.

For an ideal op amp, the open-loop differential gain is assumed to approach infinity, and even a very tiny input voltage results in a very large output voltage. In a negative-feedback circuit, a fraction of the output is returned to the inverting input terminal. This forces the differential input voltage toward zero. If we assume infinite gain, the differential input voltage is driven to zero exactly. Since the differential input voltage of the op amp is zero, the input current is also zero. The fact that the differential input voltage and the input current are forced to zero is called the **summing-point constraint**.

Ideal-op-amp circuits are analyzed by the following steps:

1. Verify that *negative* feedback is present.
2. Assume that the differential input voltage and the input current of the op amp are forced to zero. (This is the summing-point constraint.)
3. Apply standard circuit-analysis principles, such as Kirchhoff's laws and Ohm's law, to solve for the quantities of interest.

Next, we illustrate this type of analysis for some important circuits that are commonly used in engineering and scientific instrumentation.

## The Basic Inverter

An op-amp circuit known as the **inverting amplifier** is shown in Figure 13.4. We will determine the voltage gain $A_v = v_o/v_{in}$ by assuming an ideal op amp and employing the summing-point constraint. However, before starting analysis of an op-amp circuit, we should always check to make sure that negative feedback is present rather than positive feedback.

In Figure 13.4, the feedback is negative, as we shall demonstrate. For example, suppose that due to the input source $v_{in}$, a positive voltage $v_x$ appears at the inverting input. Then a negative output voltage of large (theoretically infinite) magnitude results at the output. Part of this output voltage is returned to the inverting input by the feedback path through $R_2$. Thus, the initially positive voltage at the inverting input is driven toward zero by the feedback action. A similar chain of events occurs for the appearance of a negative voltage at the inverting input terminal. Hence, the output voltage of the op amp takes precisely the value needed to oppose the source and produce (nearly) zero voltage at the op-amp input. Since we assume that the gain of the op amp is infinite, a negligible (theoretically zero) input voltage $v_x$ is needed to produce the required output.

Figure 13.5 shows the inverting amplifier, including the conditions of the summing-point constraint at the input of the op amp. Notice that the input voltage $v_{in}$ appears across $R_1$. Thus, the current through $R_1$ is

$$i_1 = \frac{v_{in}}{R_1} \tag{13.1}$$

Because the current flowing into the op-amp input terminals is zero, the current flowing through $R_2$ is

$$i_2 = i_1 \tag{13.2}$$

Thus, from Equations 13.1 and 13.2, we have

$$i_2 = \frac{v_{in}}{R_1} \tag{13.3}$$

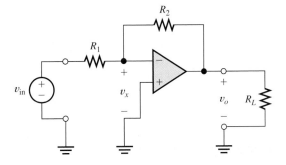

**Figure 13.4** The inverting amplifier.

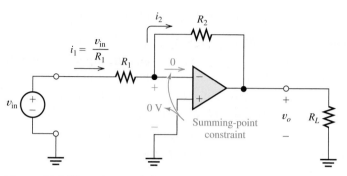

**Figure 13.5** We make use of the summing-point constraint in the analysis of the inverting amplifier.

Writing a voltage equation around the loop by including the output terminals, the resistor $R_2$, and the op-amp input, we obtain

$$v_o + R_2 i_2 = 0 \tag{13.4}$$

Using Equation 13.3 to substitute for $i_2$ in Equation 13.4 and solving for the circuit voltage gain, we have

$$A_v = \frac{v_o}{v_{in}} = -\frac{R_2}{R_1} \tag{13.5}$$

We refer to $A_v$ as the **closed-loop gain** because it is the gain of the circuit with the feedback network in place.

Under the ideal-op-amp assumption, the closed-loop voltage gain is determined solely by the ratio of the resistances. This is a very desirable situation because resistors are available with precise and stable values. Notice that the voltage gain is negative, indicating that the amplifier is inverting (i.e., the output voltage is out of phase with the input voltage).

The input impedance of the inverting amplifier is

> Under the ideal-op-amp assumption, the closed-loop voltage gain of the inverter is determined solely by the ratio of the resistances.

$$Z_{in} = \frac{v_{in}}{i_1} = R_1 \tag{13.6}$$

Thus, we can easily control the input impedance of the circuit by our choice of $R_1$.

Rearranging Equation 13.5, we have

$$v_o = -\frac{R_2}{R_1} v_{in} \tag{13.7}$$

Consequently, we see that the output voltage is independent of the load resistance $R_L$. We conclude that the output acts as an ideal voltage source (as far as $R_L$ is concerned). *In other words, the output impedance of the inverting amplifier is zero.*

Later, we will see that the characteristics of the inverting amplifier are influenced by nonideal properties of the op amp. Nevertheless, in many applications, the departure of actual performance from the ideal is insignificant.

> The inverter has a closed-loop voltage gain $A_v = -R_2/R_1$, an input impedance equal to $R_1$, and zero output impedance.

## Virtual-Short-Circuit Concept

Sometimes, the condition at the op-amp input terminals of Figure 13.5 is called a **virtual short circuit**. That terminology is used because even though the differential input voltage of the op amp is forced to zero (as if by a short circuit to ground), the op-amp input current is also zero.

This terminology can be confusing unless it is realized that it is the action at the output of the op amp acting through the feedback network that enforces zero differential input voltage. (Possibly, it would be just as valid to call the condition at the op-amp input terminals a "virtual open circuit" because no current flows.)

## Variations of the Inverter Circuit

Several useful versions of the inverter circuit exist. Analysis of these circuits follows the same pattern that we have used for the basic inverter: Verify that *negative* feedback is present, assume the summing-point constraint, and then apply basic circuit laws.

---

**Example 13.1**   **Analysis of an Inverting Amplifier**

Figure 13.6 shows a version of the inverting amplifier that can have high gain magnitude without resorting to as wide a range of resistor values as is needed in the standard inverter configuration. Derive an expression for the voltage gain under the ideal-op-amp assumption. Also, find the input impedance and output impedance. Evaluate the results for $R_1 = R_3 = 1\text{ k}\Omega$ and $R_2 = R_4 = 10\text{ k}\Omega$. Then, consider the standard inverter configuration of Figure 13.5 with $R_1 = 1\text{ k}\Omega$, and find the value of $R_2$ required to achieve the same gain.

**Solution**   First, we verify that negative feedback is present. Assume a positive value for $v_i$, which results in a negative output voltage of very large magnitude. Part of this negative voltage is returned through the resistor network and opposes the original input voltage. Thus, we conclude that negative feedback is present.

Next, we assume the conditions of the summing-point constraint:

$$v_i = 0 \quad \text{and} \quad i_i = 0$$

Generally, if a network of resistors is connected between the inverting input and the output, negative feedback exists.

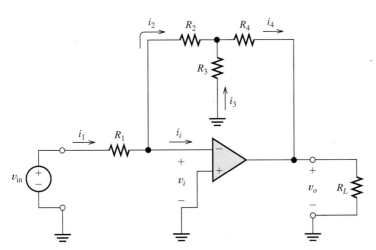

**Figure 13.6**  An inverting amplifier that achieves high gain magnitude with a smaller range of resistance values than required for the basic inverter. See Examples 13.1.

Then, we apply Kirchhoff's current law, Kirchhoff's voltage law, and Ohm's law to analyze the circuit. To begin, we notice that $v_{in}$ appears across $R_1$ (because $v_i = 0$). Hence, we can write

$$i_1 = \frac{v_{in}}{R_1} \tag{13.8}$$

Next, we apply Kirchhoff's current law to the node at the right-hand end of $R_1$, to obtain

$$i_2 = i_1 \tag{13.9}$$

(Here, we have used the fact that $i_i = 0$.)

Writing a voltage equation around the loop through $v_i$, $R_2$, and $R_3$, we obtain

$$R_2 i_2 = R_3 i_3 \tag{13.10}$$

(In writing this equation, we have used the fact that $v_i = 0$.) Applying Kirchhoff's current law at the top end of $R_3$ yields

$$i_4 = i_2 + i_3 \tag{13.11}$$

Writing a voltage equation for the loop containing $v_o$, $R_4$, and $R_3$ gives

$$v_o = -R_4 i_4 - R_3 i_3 \tag{13.12}$$

Next, we use substitution to eliminate the current variables ($i_1$, $i_2$, $i_3$, and $i_4$) and obtain an equation relating the output voltage to the input voltage. First, from Equations 13.8 and 13.9, we obtain

$$i_2 = \frac{v_{in}}{R_1} \tag{13.13}$$

Then, we use Equation 13.13 to substitute for $i_2$ in Equation 13.10 and rearrange terms to obtain

$$i_3 = v_{in} \frac{R_2}{R_1 R_3} \tag{13.14}$$

Using Equations 13.13 and 13.14 to substitute for $i_2$ and $i_3$ in Equation 13.11, we find that

$$i_4 = v_{in} \left( \frac{1}{R_1} + \frac{R_2}{R_1 R_3} \right) \tag{13.15}$$

Finally, using Equations 13.14 and 13.15 to substitute into 13.12, we obtain

$$v_o = -v_{in} \left( \frac{R_2}{R_1} + \frac{R_4}{R_1} + \frac{R_2 R_4}{R_1 R_3} \right) \tag{13.16}$$

Therefore, the voltage gain of the circuit is

$$A_v = \frac{v_o}{v_{in}} = -\left( \frac{R_2}{R_1} + \frac{R_4}{R_1} + \frac{R_2 R_4}{R_1 R_3} \right) \tag{13.17}$$

The input impedance is obtained from Equation 13.8:

$$R_{in} = \frac{v_{in}}{i_1} = R_1 \tag{13.18}$$

Inspection of Equation 13.16 shows that the output voltage is independent of the load resistance. Thus, the output appears as an ideal voltage source to the load. In other words, the output impedance of the amplifier is zero.

Evaluating the voltage gain for the resistor values given ($R_1 = R_3 = 1\,k\Omega$ and $R_2 = R_4 = 10\,k\Omega$) yields

$$A_v = -120$$

In the basic inverter circuit of Figure 13.5, the voltage gain is given by Equation 13.5, which states that

$$A_v = -\frac{R_2}{R_1}$$

Therefore, to achieve a voltage gain of $-120$ with $R_1 = 1\,k\Omega$, we need $R_2 = 120\,k\Omega$. Notice that a resistance ratio of 120:1 is required for the basic inverter, whereas the circuit of Figure 13.6 has a ratio of only 10:1. Sometimes, there are significant practical advantages in keeping the ratio of resistances in a circuit as close to unity as possible. Then, the circuit of Figure 13.6 is preferable to the basic inverter shown in Figure 13.5. ∎

Now that we have demonstrated how to make use of the summing-point constraint in analysis of ideal-op-amp circuits having negative feedback, we provide some exercises for you to practice applying the technique. Each of these circuits has negative feedback, and if we assume ideal op amps, the summing-point constraint can be used in analysis.

**Exercise 13.1**   A circuit known as a summer is shown in Figure 13.7. **a.** Use the ideal-op-amp assumption to solve for the output voltage in terms of the input voltages and resistor values. **b.** What is the input resistance seen by $v_A$? **c.** By $v_B$? **d.** What is the output resistance seen by $R_L$?
**Answer**   **a.** $v_o = -(R_f/R_A)v_A - (R_f/R_B)v_B$; **b.** the input resistance for $v_A$ is equal to $R_A$; **c.** the input resistance for $v_B$ is equal to $R_B$; **d.** the output resistance is zero. □

**Exercise 13.2**   Solve for the currents and voltages labeled in the circuits of Figure 13.8.
**Answer**   **a.** $i_1 = 1\,mA$, $i_2 = 1\,mA$, $i_o = -10\,mA$, $i_x = -11\,mA$, $v_o = -10\,V$; **b.** $i_1 = 5\,mA$, $i_2 = 5\,mA$, $i_3 = 5\,mA$, $i_4 = 10\,mA$, $v_o = -15\,V$. □

**Exercise 13.3**   Find an expression for the output voltage of the circuit shown in Figure 13.9.
**Answer**   $v_o = 4v_1 - 2v_2$. □

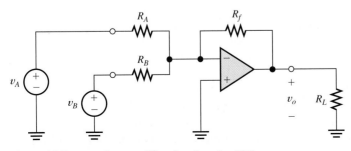

**Figure 13.7** Summing amplifier. See Exercise 13.1.

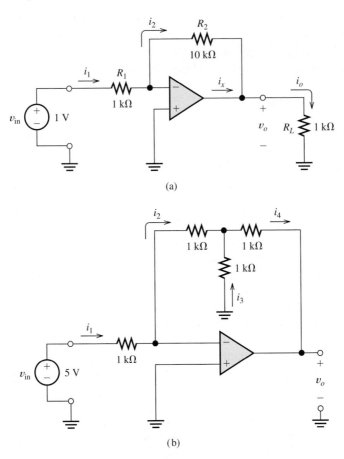

(a)

(b)

**Figure 13.8** Circuit for Exercise 13.2.

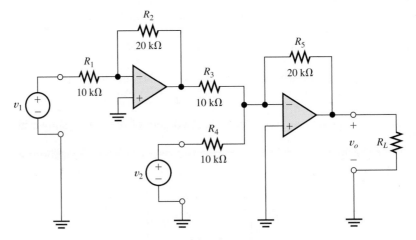

**Figure 13.9** Circuit for Exercise 13.3.

## Positive Feedback

It is interesting to consider the inverting amplifier configuration with the input terminals of the op amp interchanged as shown in Figure 13.10. In this case, the feedback is positive—in other words, the feedback signal *aids* the original input

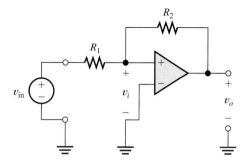

**Figure 13.10** Circuit with positive feedback.

signal. For example, if the input voltage $v_i$ is positive, a very large positive output voltage results. Part of the output voltage is returned to the op-amp input by the feedback network. Thus, the input voltage becomes larger, causing an even larger output voltage. The output quickly becomes saturated at the maximum possible voltage that the op amp can produce.

If an initial negative input voltage is present, the output saturates at its negative extreme. Hence, the circuit does not function as an amplifier—the output voltage is stuck at one extreme or the other and does not respond to the input voltage $v_{\text{in}}$. (However, if the input voltage $v_{\text{in}}$ becomes sufficiently large in magnitude, the output can be forced to switch from one extreme to the other. We saw in Chapter 7 that positive-feedback circuits [i.e., flip-flops] are useful for memory in digital systems.)

If we were to ignore the fact that the circuit of Figure 13.10 has positive rather than negative feedback and to apply the summing-point constraint erroneously, we would obtain $v_o = -(R_2/R_1)v_{\text{in}}$, just as we did for the circuit with negative feedback. This illustrates the importance of verifying that negative feedback is present before using the summing-point constraint.

With positive feedback, the op-amp's input and output voltages increase in magnitude until the output voltage reaches one of its extremes.

## 13.3  NONINVERTING AMPLIFIERS

The circuit configuration for a noninverting amplifier is shown in Figure 13.11. We assume an ideal op amp to analyze the circuit. First, we check to see whether the feedback is negative or positive. In this case, it is negative. To see this, assume that $v_i$ becomes positive and notice that it produces a very large positive output voltage. Part of the output voltage appears across $R_1$. Since $v_i = v_{\text{in}} - v_1$, the voltage $v_i$ becomes smaller as $v_o$ and $v_1$ become larger. Thus, the amplifier and feedback network act to

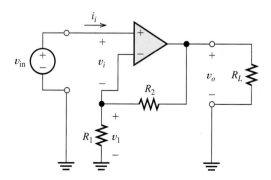

**Figure 13.11** Noninverting amplifier.

drive $v_i$ toward zero. This is negative feedback because the feedback signal opposes the original input.

Having verified that negative feedback is present, we utilize the summing-point constraint: $v_i = 0$ and $i_i = 0$. Applying Kirchhoff's voltage law and the fact that $v_i = 0$, we can write

$$v_{in} = v_1 \qquad (13.19)$$

Since $i_i$ is zero, the voltage across $R_1$ is given by the voltage-division principle:

$$v_1 = \frac{R_1}{R_1 + R_2} v_o \qquad (13.20)$$

Using Equation 13.20 to substitute into 13.19 and rearranging, we find that the closed-loop voltage gain is:

> Under the ideal-op-amp assumption, the noninverting amplifier is an ideal voltage amplifier having infinite input resistance and zero output resistance.

$$A_v = \frac{v_o}{v_{in}} = 1 + \frac{R_2}{R_1} \qquad (13.21)$$

Notice that the circuit is a noninverting amplifier ($A_v$ is positive), and the gain is set by the ratio of the feedback resistors.

The input impedance of the circuit is theoretically infinite because the input current $i_i$ is zero. Since the voltage gain is independent of the load resistance, the output voltage is independent of the load resistance. Thus, the output impedance is zero. *Therefore, under the ideal-op-amp assumption, the noninverting amplifier is an ideal voltage amplifier.* (Ideal amplifiers are discussed in Section 10.6.)

### Voltage Follower

Notice from Equation 13.21 that the minimum gain magnitude is unity, which is obtained with $R_2 = 0$. Usually, we choose $R_1$ to be an open circuit for unity gain. The resulting circuit, called a **voltage follower**, is shown in Figure 13.12.

**Exercise 13.4** Find the voltage gain $A_v = v_o/v_{in}$ and input impedance of the circuit shown in Figure 13.13: **a.** with the switch open; **b.** with the switch closed.
**Answer**   **a.** $A_v = +1$, $R_{in} = \infty$; **b.** $A_v = -1$, $R_{in} = R/2$. □

**Exercise 13.5** Assume an ideal op amp and use the summing-point constraint to find an expression for the output current $i_o$ in the circuit of Figure 13.14. Also find the input and output resistances of the circuit.
**Answer**   $i_o = v_{in}/R_F$, $R_{in} = \infty$, $R_o = \infty$ (because the output current is independent of the load resistance). □

**Exercise 13.6**  **a.** Derive an expression for the voltage gain $v_o/v_{in}$ of the circuit shown in Figure 13.15. **b.** Evaluate for $R_1 = 10\,\text{k}\Omega$ and $R_2 = 100\,\text{k}\Omega$. **c.** Find the input resistance of this circuit. **d.** Find the output resistance.
**Answer**   **a.** $A_v = 1 + 3(R_2/R_1) + (R_2/R_1)^2$; **b.** $A_v = 131$; **c.** $R_{in} = \infty$; **d.** $R_o = 0$. □

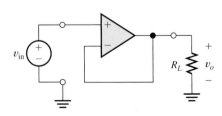

**Figure 13.12** The voltage follower which has $A_v = 1$.

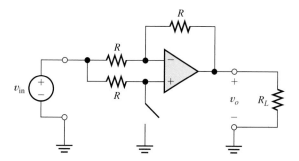

**Figure 13.13** Inverting or noninverting amplifier. See Exercise 13.4.

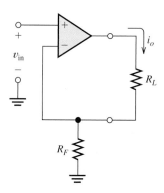

**Figure 13.14** Voltage-to-current converter (also known as a transconductance amplifier). See Exercise 13.5.

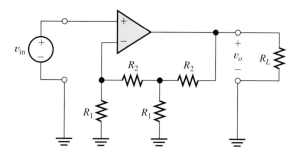

**Figure 13.15** Circuit for Exercise 13.6.

PRACTICAL APPLICATION   13.1

**Mechanical Application of Negative Feedback: Power Steering**

Besides its use in op-amp circuits, negative feedback has many other applications in engineering—automotive power-assisted steering systems provide one example. A simplified diagram of a typical system is shown in Figure PA13.1. A hydraulic pump driven by the engine continuously supplies pressure to a control valve that routes the fluid to two sides of the booster cylinder. For straight-ahead steering, pressure is applied equally to both sides of the cylinder and no turning force results.

As the steering wheel is moved by the driver, more pressure is applied to one side of the cylinder or the other to help turn the wheels in the desired direction. (The fluid path for one direction is illustrated in the figure.) A mechanical feedback arm from the steering linkage causes the valve to return to its neutral position as the wheels turn. Thus, there is a negative feedback path from the booster cylinder through the mechanical linkage back to the control valve.

Negative feedback is an important aspect of the power-steering system. Consider what would happen if the mechanical linkage between the output of the booster cylinder and the control valve is removed. Then, if the steering wheel were moved slightly off center, pressure would be applied continuously to the booster cylinder and the wheels would eventually move all the way to their extreme position. It would be very difficult for the driver to make a gradual turn.

On the other hand, with the feedback linkage in place, the wheels move only far enough to return the valve (nearly) to its neutral position. As the steering wheel is turned, the wheels move a proportional amount.

Notice that the control valve responds to the difference between the input from the steering wheel and the position of the steering linkage. This is similar to the way that the op amp responds to its differential input signal. The pump is analogous to the power supply in an op-amp circuit. Furthermore, the booster cylinder position is analogous to the op-amp output signal, and the mechanical linkage back to the control valve is analogous to a feedback circuit.

Steer-by-wire systems are a more modern alternative for mechanical/hydraulic steering systems and are under intense development. Here, the mechanical/hydraulic steering components are replaced by electrical sensors, microcontrollers, software, and electrical motors. Just as in the mechanical system, negative feedback is used in steer-by-wire systems. Actual and desired wheel positions are derived from sensors connected to the wheels and to the steering wheel. The software compares the actual wheel position to the desired position and uses the difference to control motors that turn the wheels. There is no steering column or actual mechanical path from the steering wheel to the vehicle wheels. These new electronic steering systems have the potential to significantly reduce weight and increase fuel economy of a vehicle. However, they also have some new safety issues. After all, loss of steering due to a software glitch could be very serious.

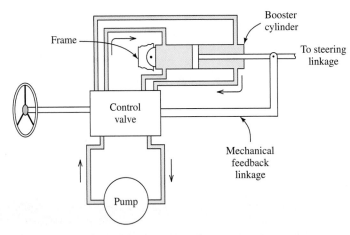

**Figure PA13.1** A simplified diagram of an automotive power-assisted steering system illustrating the importance of negative feedback.

## 13.4 DESIGN OF SIMPLE AMPLIFIERS

Many useful amplifiers can be designed by using resistive feedback networks with op amps.

Many useful amplifiers can be designed by using resistive feedback networks with op amps. For now, we consider the op amps to be ideal. Later, we consider the effects of the nonideal properties of real op amps. Often in practice, the performance requirements of the circuits to be designed are not extreme, and design can be carried out by assuming ideal op amps.

We illustrate design by using the op-amp circuits that we have considered in previous sections (including the exercises). For these circuits, design consists primarily of selecting a suitable circuit configuration and values for the feedback resistors.

### Example 13.2 | Design of a Noninverting Amplifier

Design a noninverting amplifier that has a voltage gain of 10, using an ideal op amp. The input signals lie in the range −1 to +1 V. Use standard 5-percent-tolerance resistors in the design. (See Appendix B for a list of standard 5-percent-tolerance resistor values.)

**Solution** We use the noninverting amplifier configuration of Figure 13.11. The gain is given by Equation 13.21. Thus, we have

$$A_v = 10 = 1 + \frac{R_2}{R_1}$$

Theoretically, any resistor values would provide the proper gain, provided that $R_2 = 9R_1$. However, very small resistances are not practical because the current through the resistors must be supplied by the output of the op amp, and ultimately, by the power supply. For example, if $R_1 = 1\ \Omega$ and $R_2 = 9\ \Omega$, for an output voltage of 10 V, the op amp must supply 1 A of current. This is illustrated in Figure 13.16. Most integrated-circuit op amps are not capable of such a large output current, and even if they were, the load on the power supply would be unwarranted. In the circuit at hand, we would want to keep $R_1 + R_2$ large enough so the current that must be supplied to them is reasonable. For general design with power supplies operated from the ac power line, currents up to several milliamperes are usually acceptable. (In battery-operated equipment, we would try harder to reduce the current and avoid having to replace batteries frequently.)

On the other hand, very large resistances, such as $R_1 = 10\ \text{M}\Omega$ and $R_2 = 90\ \text{M}\Omega$, also present problems. Such large resistances are unstable in value, particularly in a humid environment. Later, we will see that large resistances lead to problems due to an op-amp imperfection known as bias current. Furthermore, high-impedance circuits are prone to injection of unwanted signals from nearby circuits through stray capacitive coupling. This is illustrated in Figure 13.17.

Generally, resistance values between about 100 Ω and 1 MΩ are suitable for use in op-amp circuits. Since the problem statement calls for standard 5-percent-tolerance resistors (see Appendix B), we look for a pair of resistor values such that the ratio $R_2/R_1$ is 9. One possibility is $R_2 = 180\ \text{k}\Omega$ and $R_1 = 20\ \text{k}\Omega$. However, for many

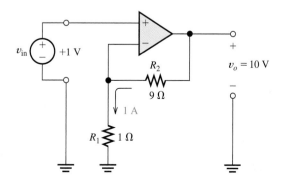

**Figure 13.16** If low resistances are used, an excessively large current is required.

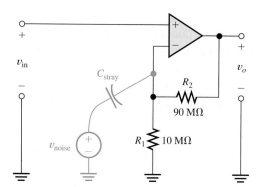

**Figure 13.17** If very high resistances are used, stray capacitance can couple unwanted signals into the circuit.

applications, we would find that $R_2 = 18\text{ k}\Omega$ and $R_1 = 2\text{ k}\Omega$ would work just as well. Of course, if 5-percent-tolerance resistors are used, we can expect unit-to-unit variations of about $\pm 10$ percent in the ratio $R_2/R_1$. This is because $R_2$ could be 5 percent low while $R_1$ is 5 percent high, or vice versa. Thus, the gain of the amplifier (which is $A_v = 1 + R_2/R_1$) varies by about $\pm 9$ percent.

If more precision is needed, 1-percent-tolerance resistors can be used. Another possibility is an adjustable resistor to set the gain to the desired value. ∎

---

### Example 13.3    Amplifier Design

A transducer for instrumentation of vibrations in a forge hammer has an internal impedance that is always less than 500 $\Omega$ but is variable over time. An amplifier that produces an amplified version of the internal source voltage $v_s$ is required. The voltage gain should be $-10 \pm 5$ percent. Design an amplifier for this application.

**Solution**   Since an inverting gain is specified, we choose to use the inverting amplifier of Figure 13.4. The proposed amplifier and the signal source are shown in Figure 13.18.

Using the summing-point constraint and conventional circuit analysis, we can show that

$$v_o = -\frac{R_2}{R_1 + R_s} v_s$$

Hence, we must select resistance values so that

$$\frac{R_2}{R_1 + R_s} = 10 \pm 5\%$$

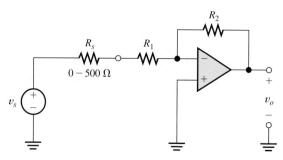

**Figure 13.18** Circuit of Example 13.3.

Because the value of $R_s$ is variable, we must choose $R_1$ much greater than the maximum value of $R_s$. Thus, we are led to choose $R_1 \cong 100$, $R_{smax} = 50\,k\Omega$. (Then, as $R_s$ ranges from zero to 500 $\Omega$, the sum $R_1 + R_s$ varies by only 1 percent.) To achieve the desired gain, we require that $R_2 \cong 500\,k\Omega$.

Since a gain tolerance of $\pm 5$ percent is specified, we resort to the use of 1 percent resistors. This is necessary because gain variations occur due to variations in $R_s$, variations in $R_1$, and variations in $R_2$. If each of these causes a $\pm 1$ percent gain variation, the gain varies by about $\pm 3$ percent, which is within the allowed range.

Consulting a table of standard values for 1 percent resistors (see Appendix B), we choose $R_1 = 49.9\,k\Omega$ and $R_2 = 499\,k\Omega$. As well as ensuring that the gain does not vary outside the specified limits, these values are not so small that large currents occur or so large that undue coupling of unwanted signals into the circuit is likely to be a problem.

Another solution would be to use 5-percent-tolerance resistors, but choose $R_1 = 51\,k\Omega$ and $R_2$ as the series combination of a 430-$k\Omega$ fixed resistor and a 200-$k\Omega$ adjustable resistor. Then the gain could be set initially to the desired value. Some gain fluctuation would occur in operation due to variation of $R_s$ and drift of the other resistance values due to aging, temperature changes, and so on.  ∎

**Exercise 13.7**   Find the maximum and minimum values of the gain $A_{vs} = v_o/v_s$ for the circuit designed in Example 13.3. The nominal resistor values are $R_1 = 49.9\,k\Omega$ and $R_2 = 499\,k\Omega$. Assume that the resistors $R_1$ and $R_2$ range as far as $\pm 1$ percent from their nominal values and that $R_s$ ranges from 0 to 500 $\Omega$.
**Answer**   The extreme gain values are $-9.71$ and $-10.20$. ☐

## Close-Tolerance Designs

When designing amplifiers with tight gain tolerances (1 percent or better), it is necessary to employ adjustable resistors. We might be tempted to use lower cost 5-percent-tolerance resistors rather than 1-percent-tolerance resistors and use the adjustable resistor to offset the larger variations. However, this is not good practice because 5-percent-tolerance resistors tend to be less stable than 1-percent-tolerance resistors. Furthermore, fixed resistors tend to be more stable than adjustable resistors. The best approach from the standpoint of long-term precision is to use 1-percent-tolerance fixed resistors and design for only enough adjustment to overcome the resulting gain variations.

Often, we combine various types of op-amp circuits in the design of a desired function. These points are illustrated in the next example.

---

**Example 13.4**   **Summing Amplifier Design**

Two signal sources have internal voltages $v_1(t)$ and $v_2(t)$, respectively. The internal resistances of the sources are known always to be less than 1 $k\Omega$, but the exact values are not known and are likely to change over time. Design an amplifier for which the output voltage is $v_o(t) = A_1 v_1(t) + A_2 v_2(t)$. The gains are to be $A_1 = 5 \pm 1$ percent and $A_2 = -2 \pm 1$ percent. Assume that ideal op amps are available.

**Solution**   The summer circuit of Figure 13.7 can be used to form the weighted sum of the input voltages given by

$$v_o = -\frac{R_f}{R_A}v_A - \frac{R_f}{R_B}v_B$$

in which the gains for both input signals are negative. However, the problem statement calls for a positive gain for $v_1$ and a negative gain for $v_2$. Thus, we first pass $v_1$ through an inverting amplifier. The output of this inverter and $v_2$ are then applied to the summer. The proposed circuit diagram is shown in Figure 13.19.

It can be shown that the output voltage of this circuit is given by

$$v_o = \frac{R_2}{R_{s1} + R_1} \frac{R_f}{R_A} v_1 - \frac{R_f}{R_{s2} + R_B} v_2 \tag{13.22}$$

We must select values for the resistances so that the gain for the $v_1$ input is $+5$ and the gain for the $v_2$ input is $-2$. Many combinations of resistances can be used to meet these specifications. However, we should keep the input impedances seen by the sources much larger than the internal source impedances, to avoid gain variations due to loading. This implies that we should choose large values for $R_1$ and $R_B$. (However, keep in mind that extremely large values are not practical.) Since we want the gain values to remain within $\pm1$ percent of the design values, we choose $R_1 = R_B \cong 500\,\text{k}\Omega$. Then, as the source impedances change, the gains change by only about 0.2 percent (because the input impedances are approximately 500 times larger than the highest value of the source impedances).

Even if we choose to use 1-percent-tolerance resistors, we must use adjustable resistors to trim the gain. For example, with 1 percent resistors, the gain

$$A_1 = \frac{R_2}{R_{s1} + R_1} \frac{R_f}{R_A}$$

varies by about $\pm4$ percent, due to the resistance tolerances. Thus, to provide for adjustment of $A_1$, we use the combination of a fixed resistance in series with a variable resistance for $R_1$. Similarly, to provide for adjustment of $A_2$, we use a second variable resistance in series with a fixed resistance for $R_B$.

Suppose that we select $R_1$ as a 453-k$\Omega$ (this is a standard nominal value for 1-percent-tolerance resistors) fixed resistor in series with a 100-k$\Omega$ trimmer (i.e., an adjustable resistor having a maximum value of 100 k$\Omega$). We use the same combination for $R_B$. (Recall that we plan to design for nominal values of $R_1$ and $R_B$ of 500 k$\Omega$ each.) The trimmers allow for approximately a $\pm10$ percent adjustment, which is more than adequate to allow for variations of the fixed resistors.

The gain for the $v_2$ input is

$$A_2 = -\frac{R_f}{R_{s2} + R_B}$$

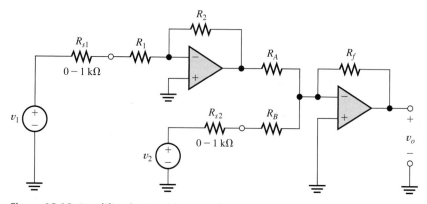

Figure 13.19 Amplifier designed in Example 13.4.

Because $R_{s2} + R_B$ has a nominal value 500 kΩ and we want to have $A_2 = -2$, $R_f$ is selected to be a 1-MΩ 1-percent-tolerance resistor. Now, since we want to achieve

$$A_1 = \frac{R_2}{R_{s1} + R_1} \frac{R_f}{R_A} = 5$$

and the values we have already selected result in $R_f/(R_{s1} + R_1) = 2$, we must choose values of $R_2$ and $R_A$ such that $R_2/R_A \cong 2.5$. Thus, we choose $R_2$ as a 1-MΩ resistor and $R_A$ as a 402-kΩ resistor. This completes the design. The following values are selected:

$R_1$ = a 453-kΩ fixed resistor in series with a 100-kΩ trimmer (500-kΩ nominal design value).

$R_B$ is the same as $R_1$.

$R_2 = 1\ \text{MΩ}$.

$R_A = 402\ \text{kΩ}$.

$R_f = 1\ \text{MΩ}$.

These are by no means the only values that can be used to meet the specifications. Usually, design problems have many "right" answers.  ■

Usually, design problems have many "right" answers.

**Exercise 13.8**   Derive Equation 13.22.  ☐

**Exercise 13.9**   A certain source has an internal impedance of 600 Ω ± 20 percent. Design an amplifier whose output voltage is $v_o = A_{vs}v_s$, where $v_s$ is the internal voltage of the source. Assume ideal op amps and design for $A_{vs} = 20 \pm 5$ percent.
**Answer**   Many answers are possible. A good solution is the circuit of Figure 13.11 with $R_2 \cong 19 \times R_1$. For example, we could use 1-percent-tolerance resistors having nominal values of $R_1 = 1\ \text{kΩ}$ and $R_2 = 19.1\ \text{kΩ}$.  ☐

**Exercise 13.10**   Repeat Exercise 13.9 for $A_{vs} = -25 \pm 3$ percent.
**Answer**   Many answers are possible. A good solution is the circuit of Figure 13.18 with $R_1 \geq 20R_s$ and with $R_2 \cong 25(R_1 + R_s)$. For example, we could use 1-percent-tolerance resistors having nominal values of $R_1 = 20\ \text{kΩ}$ and $R_2 = 515\ \text{kΩ}$.  ☐

**Exercise 13.11**   Repeat Example 13.4 if $A1 = +1 \pm 1$ percent and $A_2 = -3 \pm 1$ percent.
**Answer**   Many answers can be found by following the approach taken in Example 13.4.  ☐

## 13.5   OP-AMP IMPERFECTIONS IN THE LINEAR RANGE OF OPERATION

In Sections 13.1 through 13.4, we introduced the op amp, learned how to use the summing-point constraint to analyze negative-feedback amplifier circuits, and learned how to design simple amplifiers. So far, we have assumed ideal op amps. This assumption is appropriate for learning the basic principles of op-amp circuits, but not for high-performance circuits using real op amps. Therefore, in this and the next few sections, we consider the imperfections of real op amps and how to allow for these imperfections in circuit design.

The nonideal characteristics of real op amps fall into three categories: (1) nonideal properties in linear operation, (2) nonlinear characteristics, and (3) dc

Real op amps have several categories of imperfections compared with ideal op amps.

offsets. We discuss the imperfections for the linear range of operation in this section. In the next several sections, we consider nonlinear operation and dc offsets.

## Input and Output Impedances

Real op amps have finite input impedance and nonzero output impedance.

An ideal op amp has infinite input impedance and zero output impedance. However, a real op amp has finite input impedance and nonzero output impedance. The input impedances of IC op amps having BJT input stages are usually about 1 M$\Omega$. Op amps having FET input stages have much higher input impedances, as much as $10^{12}$ $\Omega$. Output impedance is ordinarily between 1 and 100 $\Omega$ for an IC op amp, although it can be as high as several thousand ohms for a low-power op amp.

In circuits with negative feedback, the impedances are drastically altered by the feedback action, and the input or output impedances of the op amps rarely place serious limits on closed-loop circuit performance.

## Gain and Bandwidth Limitations

Real op amps have finite open-loop gain and finite bandwidth.

Ideal op amps have infinite gain magnitude and unlimited bandwidth. Real op amps have finite open-loop gain magnitude, typically between $10^4$ and $10^6$. (We are referring to the open-circuit voltage gain of the op amp without feedback resistors.) Furthermore, the bandwidth of real op amps is limited. The gain of a real op amp is a function of frequency, becoming smaller in magnitude at higher frequencies.

Usually, the bandwidth of an IC op amp is intentionally limited by the op-amp designer. This is called *frequency compensation* and is necessary to avoid oscillation in feedback amplifiers. An in-depth discussion of frequency compensation is beyond the scope of this book. However, it turns out that the open-loop gain of most integrated-circuit op amps is of the form

$$A_{\text{OL}}(f) = \frac{A_{0\text{OL}}}{1 + j(f/f_{BOL})} \tag{13.23}$$

in which $A_{0\text{OL}}$ is the dc open-loop gain of the amplifier, and $f_{BOL}$ is the open-loop break frequency.

In a Bode plot for $A_{\text{OL}}(f)$, the gain magnitude is approximately constant up to $f_{BOL}$. Above $f_{BOL}$, the gain magnitude falls at 20 dB per decade. This is illustrated in Figure 13.20. (Bode plots are discussed in Section 6.4.)

## Closed-Loop Bandwidth

We will show that negative feedback reduces the dc gain of an op amp and extends its bandwidth. Consider the noninverting amplifier circuit shown in Figure 13.21. The phasor output voltage $\mathbf{V}_o$ is the open-loop gain times the phasor differential input voltage $\mathbf{V}_{id}$:

$$\mathbf{V}_o = A_{\text{OL}}(f)\mathbf{V}_{id} \tag{13.24}$$

We assume that the input impedance of the op amp is infinite, so the input current is zero. Then, the voltage across $R_1$ can be found by applying the voltage-division principle to the feedback network (which is composed of $R_1$ in series with $R_2$). The voltage across $R_1$ is shown as $\beta\mathbf{V}_o$, in which $\beta$ is the voltage-division ratio of $R_1$ and $R_2$:

$$\beta = \frac{R_1}{R_1 + R_2} \tag{13.25}$$

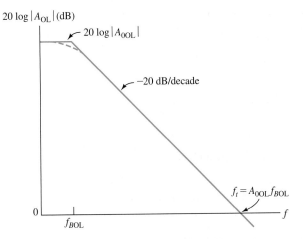

**Figure 13.20** Bode plot of open-loop gain for a typical op amp.

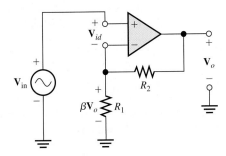

**Figure 13.21** Noninverting amplifier circuit used for analysis of closed-loop bandwidth.

Applying Kirchhoff's voltage law to Figure 13.21, we have

$$\mathbf{V}_{\text{in}} = \mathbf{V}_{id} + \beta \mathbf{V}_o$$

Solving for $\mathbf{V}_{id}$ and substituting into Equation 13.24, we have

$$\mathbf{V}_o = A_{\text{OL}}(\mathbf{V}_{\text{in}} - \beta \mathbf{V}_o) \qquad (13.26)$$

Now, we can solve for the gain of the circuit including the feedback resistors, which is called the **closed-loop gain**. This is given by

$$A_{\text{CL}} = \frac{\mathbf{V}_o}{\mathbf{V}_{\text{in}}} = \frac{A_{\text{OL}}}{1 + \beta A_{\text{OL}}} \qquad (13.27)$$

Using Equation 13.23 to substitute for the open-loop gain, we get

$$A_{\text{CL}}(f) = \frac{A_{\text{0OL}}/[1 + j(f/f_{\text{BOL}})]}{1 + \{\beta A_{\text{0OL}}/[1 + j(f/f_{\text{BOL}})]\}} \qquad (13.28)$$

This can be put into the form

$$A_{\text{CL}}(f) = \frac{A_{\text{0OL}}/(1 + \beta A_{\text{0OL}})}{1 + \{jf/[f_{\text{BOL}}(1 + \beta A_{\text{0OL}})]\}} \qquad (13.29)$$

Now, we define the closed-loop dc gain as

$$A_{0CL} = \frac{A_{0OL}}{1 + \beta A_{0OL}}$$

(13.30)

and the closed-loop bandwidth as

$$f_{BCL} = f_{BOL}(1 + \beta A_{0OL})$$

(13.31)

Using these definitions in Equation 13.29, we obtain

$$A_{CL}(f) = \frac{A_{0OL}}{1 + j(f/f_{BCL})}$$

(13.32)

Comparing this with Equation 13.23, we see that the closed-loop gain takes exactly the same form as the open-loop gain. The dc open-loop gain $A_{0OL}$ is very large, and we usually have $(1 + \beta A_{0OL}) \gg 1$. Thus, from Equation 13.30, we see that the closed-loop gain is much smaller than the open-loop gain. Furthermore, Equation 13.31 shows that the closed-loop bandwidth is much greater than the open-loop bandwidth. *In sum, we see that negative feedback reduces the gain magnitude and increases bandwidth.*

Negative feedback reduces gain magnitude and increases bandwidth.

### Gain–Bandwidth Product

Now, consider the product of the closed-loop gain and closed-loop bandwidth. From Equations 13.30 and 13.31, we have

$$A_{0CL}f_{BCL} = \frac{A_{0OL}}{1 + \beta A_{0OL}} \times f_{BOL}(1 + \beta A_{0OL}) = A_{0OL}f_{BOL}$$

(13.33)

Hence, we see that the product of dc gain and bandwidth is independent of the feedback ratio $\beta$. We denote the **gain–bandwidth** as $f_t$. Thus, we have

$$f_t = A_{0CL}f_{BCL} = A_{0OL}f_{BOL}$$

(13.34)

The gain–bandwidth product is constant for the noninverting amplifier. As we reduce the gain (by choosing a lower value for $1R_2/R_1$), the bandwidth becomes greater.

As indicated in Figure 13.20, it turns out that $f_t$ is the frequency at which the Bode plot of the open-loop gain crosses 0 dB. Recall that 0 dB corresponds to unity-gain magnitude. Consequently, $f_t$ is also called the **unity-gain–bandwidth**. General-purpose IC op amps have gain–bandwidth products of several megahertz.

> **Example 13.5**    Open-Loop and Closed-Loop Bode Plots

A certain op amp has a dc open-loop gain of $A_{0OL} = 10^5$ and $f_{BOL} = 40$ Hz. Find the closed-loop bandwidth if this op amp is used with feedback resistors to form a noninverting amplifier having a closed-loop dc gain of 10. Then, construct a Bode plot of the open-loop gain and a Bode plot of the closed-loop gain.

**Solution**    The gain–bandwidth product is

$$f_t = A_{0OL}f_{BOL} = 10^5 \times 40 \text{ Hz} = 4 \text{ MHz} = A_{0CL}f_{BCL}$$

Thus, if feedback is used to reduce the gain to $A_{0CL} = 10$, the bandwidth is $f_{BCL} = 400$ kHz.

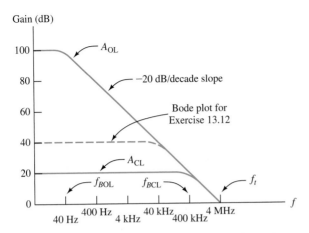

Figure 13.22 Bode plots for Example 13.5 and Exercise 13.12.

In decibels, the dc open-loop gain becomes

$$A_{0OL} = 20 \log(10^5) = 100 \text{ dB}$$

The break frequency is $f_{BOL} = 40$ Hz. Recall from our discussion of Bode plots in Section 6.4 that this gain function is approximated as being constant below the break frequency and falls at 20 dB/decade above the break frequency. The Bode plot is shown in Figure 13.22.

Converting $A_{0CL} = 10$ to decibels, we have

$$A_{0CL} = 10 \log(10) = 20 \text{ dB}$$

and the break frequency is $f_{BCL} = 400$ kHz. The resulting Bode plot is shown in Figure 13.22. Notice that the closed-loop gain plot is constant until it reaches the open-loop plot and then the closed-loop plot rolls off. ∎

**Exercise 13.12** Repeat Example 13.5 for $A_{0CL} = 100$.
**Answer** $f_{BCL} = 40$ kHz. The Bode plot is shown in Figure 13.22. ☐

## 13.6 NONLINEAR LIMITATIONS

### Output Voltage Swing

There are several nonlinear limitations of the outputs of real op amps. First, the output voltage range is limited. If an input signal is sufficiently large that the output voltage would be driven beyond these limits, clipping occurs.

The range of allowed output voltage depends on the type of op amp in use, on the load resistance value, and on the values of the power-supply voltages. For example, with supply voltages of +15 V and −15 V, the LM741 op amp (LM741 is the manufacturer's type number for a popular op amp) is capable of producing output voltages in the range from approximately −14 to +14 V. If smaller power-supply voltages are used, the linear range is reduced. (These are *typical* limits for load

The output voltage of a real op amp is limited to the range between certain limits that depend on the internal design of the op amp. When the output voltage tries to go beyond these limits, clipping occurs.

resistances greater than 10 kΩ. The *guaranteed* output range for the LM741 is only from −12 to +12 V. Smaller load resistances further restrict the range.)

Consider the noninverting amplifier with a sinusoidal input signal shown in Figure 13.23. Assuming an ideal op amp, the voltage gain is given by Equation 13.21, which is repeated here for convenience:

$$A_v = 1 + \frac{R_2}{R_1}$$

Substituting the values shown in Figure 13.23 ($R_1 = 1$ kΩ and $R_2 = 3$ kΩ), we find $A_v = 4$. The output waveform for $R_L = 10$ kΩ and $V_{im} = 1$ V is shown in Figure 13.24. The output waveform is sinusoidal because none of the nonlinear limits of the op amp have been exceeded. On the other hand, for $V_{im} = 5$ V, the output reaches the maximum output voltage limits and the output waveform is clipped. This is shown in Figure 13.25.

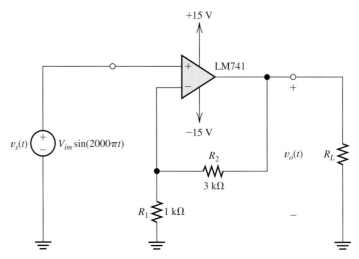

**Figure 13.23** Noninverting amplifier used to demonstrate various nonlinear limitations of op amps.

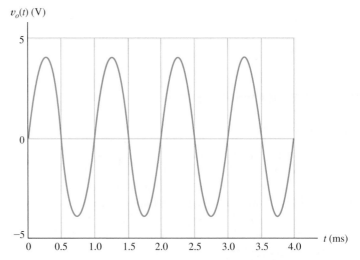

**Figure 13.24** Output of the circuit of Figure 13.23 for $R_L = 10$ kΩ and $V_{im} = 1$ V. None of the limitations are exceeded, and $v_o(t) = 4v_s(t)$.

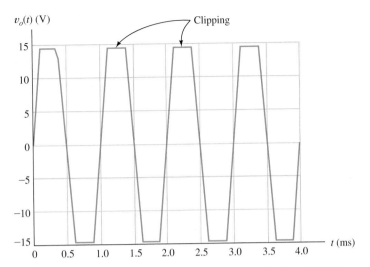

**Figure 13.25** Output of the circuit of Figure 13.23 for $R_L = 10\ k\Omega$ and $V_{im} = 5$ V. Clipping occurs because the maximum possible output voltage magnitude is reached.

## Output Current Limits

A second limitation is the maximum current that an op amp can supply to a load. For the LM741, the limits are $\pm 40$ mA. If a small-value load resistance would draw a current outside these limits, the output waveform becomes clipped.

The output current range of a real op amp is limited. If an input signal is sufficiently large that the output current would be driven beyond these limits, clipping occurs.

For example, suppose that we set the peak input voltage to $V_{im} = 1$ V and adjust the load resistance to $R_L = 50\ \Omega$ for the circuit of Figure 13.23. For an ideal op amp, we would expect a peak output voltage of $V_{om} = 4$ V and a peak load current of $V_{om}/R_L = 80$ mA. However, output current magnitude of the LM741 is limited to 40 mA. Therefore, clipping occurs due to current limiting. The output voltage waveform of the circuit is shown in Figure 13.26. Notice that the peak output voltage is 40 mA $\times R_L = 2$ V.

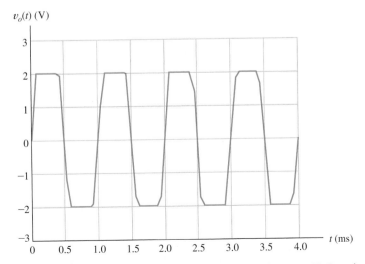

**Figure 13.26** Output of the circuit of Figure 13.23 for $R_L = 50\ \Omega$ and $V_{im} = 1$ V. Clipping occurs because the maximum output current limit is reached.

## Slew-Rate Limitation

Another nonlinear limitation of actual op amps is that the magnitude of the rate of change of the output voltage is limited. This is called the **slew-rate limitation**. The output voltage cannot increase (or decrease) in magnitude at a rate exceeding this limit. In equation form, the slew-rate limit is

$$\left| \frac{dv_o}{dt} \right| \leq \text{SR} \tag{13.35}$$

For various types of IC op amps, the slew-rate limit ranges from $\text{SR} = 10^5$ V/s to $\text{SR} = 10^8$ V/s. For the LM741 with $\pm 15$-V supplies and $R_L > 2$ k$\Omega$, the typical value is $5 \times 10^5$ V/s (which is often stated as 0.5 V/$\mu$s).

For example, consider the circuit of Figure 13.23, except that the input source voltage is changed to a 2.5-V-peak 50-kHz sine wave given by

$$v_s(t) = 2.5 \sin(10^5 \pi t)$$

starting at $t = 0$. [$v_s(t)$ is assumed to be zero prior to $t = 0$.] The output waveform is shown in Figure 13.27. Also plotted in the figure is four times the input voltage, which is the output assuming an ideal op amp. At $t = 0$, the output voltage is zero. The ideal output increases at a rate exceeding the slew-rate limit of the LM741, so the LM741 output increases at its maximum rate, which is approximately 0.5 V/$\mu$s. At point $A$, the actual output finally "catches up" with the ideal output, but by then, the ideal output is decreasing at a rate that exceeds the slew-rate limit. Thus, at point $A$, the output of the LM741 begins to decrease at its maximum possible rate. Notice that because of slew-rate limiting, the actual op-amp output is a triangular waveform rather than a sinusoid.

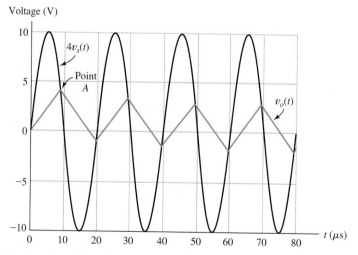

**Figure 13.27** Output of the circuit of Figure 13.23 for $R_L = 10$ k$\Omega$ and $v_s(t) = 2.5 \sin(10^5 \pi t)$. The output waveform is a triangular waveform because the slew-rate limit is exceeded. The output for an ideal op amp, which is equal to $4v_s(t)$, is shown for comparison.

## Full-Power Bandwidth

The **full-power bandwidth** of an op amp is the range of frequencies for which the op amp can produce an undistorted sinusoidal output with peak amplitude equal to the guaranteed maximum output voltage.

The full-power bandwidth of an op amp is the range of frequencies for which the op amp can produce an undistorted sinusoidal output with peak amplitude equal to the guaranteed maximum output voltage.

Next, we derive an expression for the full-power bandwidth in terms of the slew rate and peak amplitude. The output voltage is given by

$$v_o(t) = V_{om} \sin(\omega t)$$

Taking the derivative with respect to time, we have

$$\frac{dv_o(t)}{dt} = \omega V_{om} \cos(\omega t)$$

The maximum magnitude of the rate of change is $\omega V_{om} = 2\pi f V_{om}$. Setting this equal to the slew-rate limit, we get

$$2\pi f V_{om} = SR$$

Solving for frequency, we obtain

$$f_{FP} = \frac{SR}{2\pi V_{om}} \tag{13.36}$$

where we have denoted the full-power bandwidth as $f_{FP}$. An undistorted full-amplitude sinusoidal output waveform is possible only for frequencies less than $f_{FP}$.

---

### Example 13.6   Full-Power Bandwidth

Find the full-power bandwidth of the LM741 op amp given that the slew rate is $SR = 0.5 \text{ V}/\mu s$ and the guaranteed maximum output amplitude is $V_{om} = 12$ V.

**Solution**   We substitute the given data into Equation 13.36 to obtain

$$f_{FP} = \frac{SR}{2\pi V_{om}} \cong 6.63 \text{ kHz}$$

Thus, we can obtain an undistorted 12-V-peak sinusoidal output from the LM741 only for frequencies less than 6.63 kHz.   ■

**Exercise 13.13**   A certain op amp has a maximum output voltage range from $-4$ to $+4$ V. The maximum current magnitude is 10 mA. The slew-rate limit is $SR = 5 \text{ V}/\mu s$. This op amp is used in the circuit of Figure 13.28. Assume a sinusoidal input signal for all parts of this exercise. **a.** Find the full-power bandwidth of the op amp. **b.** For a frequency of 1 kHz and $R_L = 1 \text{ k}\Omega$, what peak output voltage is possible without distortion (i.e., clipping or slew-rate limiting)? **c.** For a frequency of 1 kHz and $R_L = 100 \ \Omega$, what peak output voltage is possible without distortion? **d.** For a frequency of 1 MHz and $R_L = 1 \text{ k}\Omega$, what peak output voltage is possible without distortion? **e.** If $R_L = 1 \text{ k}\Omega$ and $v_s(t) = 5 \sin(2\pi 10^6 t)$, sketch the steady-state output waveform to scale versus time.
**Answer   a.** $f_{FP} = 199$ kHz; **b.** 4 V; **c.** 1 V; **d.** 0.796 V; **e.** the output waveform is a triangular wave with a peak amplitude of 1.25 V.   □

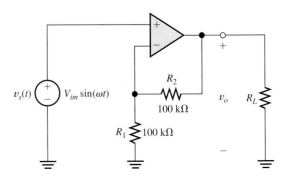

**Figure 13.28** Circuit for Exercise 13.13.

## 13.7 DC IMPERFECTIONS

Op amps have direct-coupled input circuits. Thus, dc bias currents that flow into (or from) the input devices of the op amp must flow through the elements that are connected to the input terminals, such as the signal source or feedback resistors.

The dc current flowing into the noninverting input is denoted as $I_{B+}$, and the dc current flowing into the inverting input is $I_{B-}$. The average of the dc currents is called **bias current** and is denoted as $I_B$. Thus, we have

$$I_B = \frac{I_{B+} + I_{B-}}{2} \tag{13.37}$$

Nominally, the input circuit of the op amp is symmetrical, and the bias currents flowing into the inverting and noninverting inputs are equal. However, in practice, the devices are not perfectly matched, and the bias currents are not equal. The difference between the bias currents, called the **offset current**, is denoted as

$$I_{\text{off}} = I_{B+} - I_{B-} \tag{13.38}$$

Another dc imperfection of op amps is that the output voltage may not be zero for zero input voltage. The op amp behaves as if a small dc source known as the **offset voltage** is in series with one of the input terminals.

The three dc imperfections (bias current, offset current, and offset voltage) can be modeled by placing dc sources at the input of the op amp as shown in Figure 13.29. The $I_B$ current sources model the bias current. The $I_{\text{off}}/2$ current source models the

The three dc imperfections (bias current, offset current, and offset voltage) can be modeled by placing dc sources at the input of the op amp as shown in Figure 13.29.

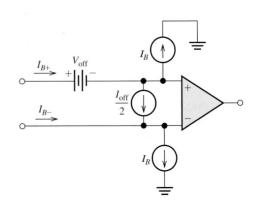

**Figure 13.29** Three current sources and a voltage source model the dc imperfections of an op amp.

offset current, and the $V_{\text{off}}$ voltage source models the offset voltage. (These sources were discussed in Section 11.12, and the discussion given there applies to op-amp circuits as well as to amplifiers in general.)

The bias-current sources are equal in magnitude and are referenced in the same direction (which is away from the input terminals in Figure 13.29). In some op amps, the bias current can have a negative value, so the currents flow toward the input terminals. Usually, the direction of the bias current is predictable for a given type of op amp. For example, if the input terminals of an op amp are the base terminals of *npn* BJTs, the bias current $I_B$ is positive (assuming the reference directions shown in Figure 13.29). On the other hand, *pnp* BJTs would result in a negative value for $I_B$.

Since the bias-current sources are matched in magnitude and direction, it is possible to design circuits in such a way that their effects cancel. On the other hand, the polarity of the offset voltage and the direction of the offset current are unpredictable—varying from unit to unit. For example, if the offset voltage of a given type of op amp is specified as a maximum of 2 mV, the value of $V_{\text{off}}$ ranges from $-2$ to $+2$ mV from unit to unit. Usually, most units have offset values close to zero, and only a few have values close to the maximum specification. A typical specification for the maximum offset-voltage magnitude for IC op amps is several millivolts.

Bias currents are usually on the order of 100 nA for op amps with BJT input devices. Bias currents are much lower for op amps with FET inputs—a typical specification is 100 pA at 25°C for a device with JFET input devices. Usually, offset-current specifications range from 20 to 50 percent of the bias current.

*The effect of bias current, offset current, and offset voltage on inverting or noninverting amplifiers is to add a (usually undesirable) dc voltage to the intended output signal.* We can analyze these effects by including the sources shown in Figure 13.29 and assuming an otherwise ideal op amp.

> The effect of bias current, offset current, and offset voltage on inverting or noninverting amplifiers is to add a (usually undesirable) dc voltage to the intended output signal.

---

### Example 13.7   Determining Worst-Case DC Output

Find the worst-case dc output voltage of the inverting amplifier shown in Figure 13.30(a), assuming that $v_{\text{in}} = 0$. The maximum bias current of the op amp is 100 nA, the maximum offset-current magnitude is 40 nA, and the maximum offset-voltage magnitude is 2 mV.

**Solution**   Our approach is to calculate the output voltage due to each of the dc sources acting individually. Then by using superposition, the worst-case output can be found by adding the outputs due to the various sources.

> First, we calculate the output voltage resulting from each of the dc sources acting individually. Then, we use superposition.

First, we consider the offset voltage. The circuit, including the offset-voltage source, is shown in Figure 13.30(b). The offset-voltage source can be placed in series with either input. We have elected to place it in series with the noninverting input. Then, the circuit takes the form of a noninverting amplifier. [Notice that although it is drawn differently, the circuit of Figure 13.30(b) is electrically equivalent to the noninverting amplifier of Figure 13.11.] Thus, the output voltage is the gain of the noninverting amplifier, given by Equation 13.21, times the offset voltage:

$$V_{o,v\text{off}} = -\left(1 + \frac{R_2}{R_1}\right) V_{\text{off}}$$

Substituting values, we find that

$$V_{o,v\text{off}} = -11 V_{\text{off}}$$

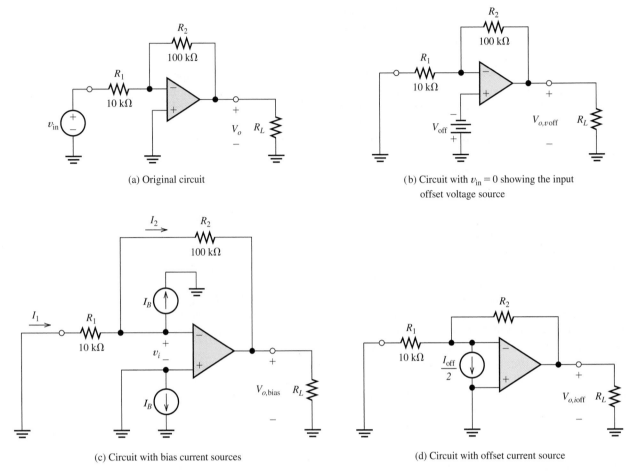

(a) Original circuit

(b) Circuit with $v_{\text{in}} = 0$ showing the input
offset voltage source

(c) Circuit with bias current sources

(d) Circuit with offset current source

**Figure 13.30** Circuits of Example 13.7.

Since the offset voltage $V_{\text{off}}$ is specified to have a maximum value of 2 mV, the value of $V_{o,v\text{off}}$ ranges between extremes of $-22$ and $+22$ mV. However, most units would have $V_{o,v\text{off}}$ closer to zero.

Next, we consider the bias-current sources. The circuit, including the bias-current sources, is shown in Figure 13.30(c). Because the noninverting input is connected directly to ground, one of the bias-current sources is short circuited and has no effect. Since we assume an ideal op amp (aside from the dc sources), the summing-point constraint applies, and $v_i = 0$. Thus, the current $I_1$ is zero. Applying Kirchhoff's current law, we have $I_2 = -I_B$. Writing a voltage equation from the output through $R_2$ and $R_1$, we have

$$V_{o,\text{bias}} = -R_2 I_2 - R_1 I_1$$

Substituting $I_1 = 0$ and $I_2 = -I_B$, we obtain

$$V_{o,\text{bias}} = R_2 I_B$$

Because the maximum value of $I_B$ is 100 nA, the maximum value of $V_{o,\text{bias}}$ is 10 mV. As is often the case, the maximum value of $I_B$ is specified, but the minimum is not. Thus, $V_{o,\text{bias}}$ ranges from some small indeterminate voltage (perhaps a few millivolts) up to 10 mV. (We will conservatively assume that the minimum value of $V_{o,\text{bias}}$ is zero.)

Next, we consider the offset-current source. The circuit is shown in Figure 13.30(d). By an analysis similar to that for the bias current, we can show that

$$V_{o,\text{ioff}} = R_2 \left( \frac{I_{\text{off}}}{2} \right)$$

The specification for the maximum magnitude of $I_{\text{off}}$ is 40 nA. Therefore, the value of $V_{o,\text{ioff}}$ ranges between extremes of $-2$ and $+2$ mV.

By superposition, the dc output voltage is the sum of the contributions of the various sources acting individually, yielding

$$V_o = V_{o,\text{voff}} + V_{o,\text{bias}} + V_{o,\text{ioff}}$$

Hence, the extreme values of the output voltage are

$$V_o = 22 + 10 + 2 = 34 \text{ mV}$$

and

$$V_o = -22 + 0 - 2 = -24 \text{ mV}$$

Thus, the output voltage ranges from $-24$ to $+34$ mV from unit to unit. (We have assumed a minimum contribution of zero for the bias current.) Typical units would have total output voltages closer to zero than to these extreme values. ∎

By superposition, the dc output voltage is the sum of the contributions of the various sources acting individually.

## Cancellation of the Effects of Bias Currents

As mentioned earlier, it is possible to design circuits in which the effects of the two bias-current sources cancel. For example, consider the inverting amplifier configuration. Adding a resistor $R_{\text{bias}}$ in series with the noninverting op-amp input, as shown in Figure 13.31, does not affect the gain of the amplifier, but results in cancellation of the effects of the $I_B$ sources. Notice that the value of $R_{\text{bias}}$ is equal to the parallel combination of $R_1$ and $R_2$.

**Exercise 13.14**   Consider the amplifier shown in Figure 13.31. **a.** Assume an ideal op amp, and derive an expression for the voltage gain $v_o/v_{\text{in}}$. Notice that the result is the same as Equation 13.5, which was derived for the inverting amplifier without the bias-current-compensating resistor $R_{\text{bias}}$. **b.** Redraw the circuit with $v_{\text{in}} = 0$, but include the bias-current sources. Show that the output voltage is zero. **c.** Assume that $R_1 = 10 \text{ k}\Omega$, $R_2 = 100 \text{ k}\Omega$, and a specification of 3 mV for the maximum magnitude of $V_{\text{off}}$. Find the range of output voltages resulting from

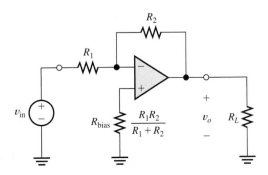

Figure 13.31 Adding the resistor $R_{\text{bias}}$ to the inverting amplifier circuit causes the effects of bias currents to cancel.

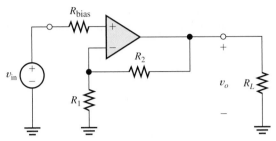

**Figure 13.32** Noninverting amplifier, including resistor $R_{bias}$ to balance the effects of the bias currents. See Exercise 13.15.

the offset-voltage source $V_{off}$. **d.** Assume that $R_1 = 10\,k\Omega$, $R_2 = 100\,k\Omega$, and a specification of 40 nA for the maximum magnitude of $I_{off}$. Find the range of output voltages resulting from the offset current. **e.** Assuming the values given in parts (c) and (d), what range of output voltages results from the combined action of the bias current, offset voltage, and offset current?

**Answer   a.** $v_o/v_{in} = -R_2/R_1$; **b.** $\pm 33\,mV$; **c.** $\pm 4\,mV$; **d.** $\pm 37\,mV$.         □

**Exercise 13.15**   Consider the noninverting amplifier shown in Figure 13.32. **a.** Derive an expression for the voltage gain $v_o/v_{in}$. Does the gain depend on the value of $R_{bias}$? Explain. **b.** Derive an expression for $R_{bias}$ in terms of the other resistance values so that the output voltage due to the bias currents is zero.

**Answer   a.** $v_o/v_{in} = 1 + R_2/R_1$. The gain is independent of $R_{bias}$ because the current through $R_{bias}$ is zero (assuming an ideal op amp). **b.** $R_{bias} = R_1 \| R_2 = 1/(1/R_1 + 1/R_2)$.         □

## 13.8  DIFFERENTIAL AND INSTRUMENTATION AMPLIFIERS

Figure 13.33 shows a differential amplifier. Assuming an ideal op amp and that $R_4/R_3 = R_2/R_1$, the output voltage is a constant times the differential input signal $(v_1 - v_2)$. The gain for the common-mode signal is zero. (See Section 11.11 for a discussion of common-mode signals.) To minimize the effects of bias current, we should choose $R_2 = R_4$ and $R_1 = R_3$.

Differential amplifiers are widely used in engineering instrumentation.

The output impedance of the circuit is zero. The input impedance for the $v_1$ source is $R_3 + R_4$.

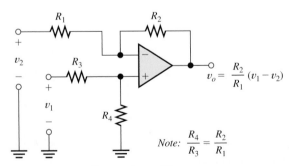

**Figure 13.33** Differential amplifier.

A current that depends on $v_1$ flows back through the feedback network ($R_1$ and $R_2$) into the input source $v_2$. Thus, as seen by the $v_2$ source, the circuit does not appear to be passive. Hence, the concept of input impedance does not apply for the $v_2$ source (unless $v_1$ is zero).

In some applications, the signal sources contain internal impedances, and the desired signal is the difference between the internal source voltages. Then, we could design the circuit by including the internal source resistances of $v_2$ and $v_1$ as part of $R_1$ and $R_3$, respectively. However, to obtain very high common-mode rejection, it is necessary to match the ratios of the resistances closely. This can be troublesome if the source impedances are not small enough to be neglected and are not predictable.

## Instrumentation-Quality Differential Amplifier

Figure 13.34 shows an improved differential amplifier circuit for which the common-mode rejection ratio is not dependent on the internal resistances of the sources. Because of the summing-point constraint at the inputs of $X_1$ and $X_2$, the currents drawn from the signal sources are zero. Hence, the input impedances seen by both sources are infinite, and the output voltage is unaffected by the internal source impedances. This is an important advantage of this circuit compared to the simpler differential amplifier of Figure 13.33. Notice that the second stage of the instrumentation amplifier is a unity-gain version of the differential amplifier.

A subtle point concerning this circuit is that the differential-mode signal experiences a higher gain in the first stage ($X_1$ and $X_2$) than the common-mode signal does. To illustrate this point, first consider a pure differential input (i.e., $v_1 = -v_2$). Then, because the circuit is symmetrical, point $A$ remains at zero voltage. Hence, in the analysis for a purely differential input signal, point $A$ can be considered to be grounded. In this case, the input amplifiers $X_1$ and $X_2$ are configured as noninverting

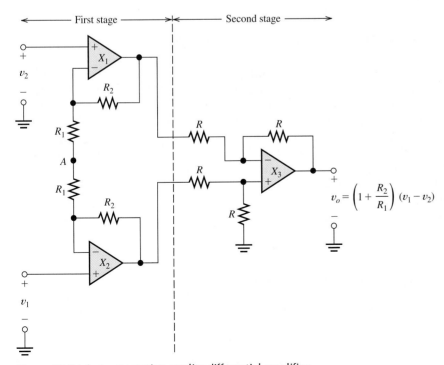

**Figure 13.34** Instrumentation-quality differential amplifier.

amplifiers having gains of $(1 + R_2/R_1)$. The differential gain of the second stage is unity. Thus, the overall gain for the differential signal is $(1 + R_2/R_1)$.

Now, consider a pure common-mode signal (i.e., $v_1 = v_2 = v_{cm}$). Because of the summing-point constraint, the voltage between the input terminals of $X_1$ (or $X_2$) is zero. Thus, the voltages at the inverting input terminals of $X_1$ and $X_2$ are both equal to $v_{cm}$. Hence, the voltage across the series-connected $R_1$ resistors is zero, and no current flows through the $R_1$ resistors. Therefore, no current flows through the $R_2$ resistors. Thus, the output voltages of $X_1$ and $X_2$ are equal to $v_{cm}$, and we have shown that the first-stage gain is unity for the common-mode signal. On the other hand, the differential gain of the first stage is $(1 + R_2/R_1)$, which can be much larger than unity, thereby achieving a reduction of the common-mode signal amplitude relative to the differential signal. [Notice that if point $A$ were actually grounded, the gain for the common-mode signal would be the same as for the differential signal, namely $(1 + R_2/R_1)$.]

In practice, the series combination of the two $R_1$ resistors is implemented by a single resistor (equal in value to $2R_1$) because it is not necessary to have access to point $A$. Thus, matching of component values for $R_1$ is not required. Furthermore, it can be shown that close matching of the $R_2$ resistors is not required to achieve a higher differential gain than common-mode gain in the first stage. Since the first stage reduces the relative amplitude of the common-mode signal, matching of the resistors in the second stage is not as critical.

Thus, although it is more complex, the differential amplifier of Figure 13.34 has better performance than that of Figure 13.33. Specifically, the common-mode rejection ratio is independent of the internal source resistances, the input impedance seen by both sources is infinite, and resistor matching is not as critical.

**Exercise 13.16**   Assume an ideal op amp, and derive the expression shown for the output voltage of the differential amplifier of Figure 13.33. Assume that $R_4/R_3 = R_2/R_1$.                                                                      ☐

## 13.9   INTEGRATORS AND DIFFERENTIATORS

Integrators produce output voltages that are proportional to the running-time integral of the input voltages. In a running-time integral, the upper limit of integration is $t$.

Figure 13.35 shows the diagram of an **integrator**, which is a circuit that produces an output voltage proportional to the running-time integral of the input voltage. (By the term *running-time integral*, we mean that the upper limit of integration is $t$.)

The integrator circuit is often useful in instrumentation applications. For example, consider a signal from an accelerometer that is proportional to acceleration. By integrating the acceleration signal, we obtain a signal proportional to velocity. Another integration yields a signal proportional to position.

In Figure 13.35, negative feedback occurs through the capacitor. Thus, assuming an ideal op amp, the voltage at the inverting op-amp input is zero. The input current is given by

$$i_{in}(t) = \frac{v_{in}(t)}{R} \tag{13.39}$$

The current flowing into the input terminal of the (ideal) op amp is zero. Therefore, the input current $i_{in}$ flows through the capacitor. We assume that the reset switch is opened at $t = 0$. Therefore, the capacitor voltage is zero at $t = 0$. The voltage across the capacitor is given by

$$v_c(t) = \frac{1}{C} \int_0^t i_{in}(t)dt \tag{13.40}$$

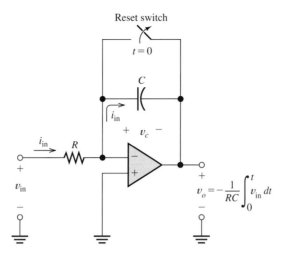

**Figure 13.35** Integrator.

Writing a voltage equation from the output terminal through the capacitor and then to ground through the op-amp input terminals, we obtain

$$v_o(t) = -v_c(t) \tag{13.41}$$

Using Equation 13.39 to substitute into 13.40 and the result into 13.41, we obtain

$$v_o(t) = -\frac{1}{RC} \int_0^t v_{in}(t)dt \tag{13.42}$$

Thus, the output voltage is $-1/RC$ times the running integral of the input voltage. If an integrator having positive gain is desired, we can cascade the integrator with an inverting amplifier.

The magnitude of the gain can be adjusted by the choice of $R$ and $C$. Of course, in selecting a capacitor, we usually want to use as small a value as possible to minimize cost, volume, and mass. However, for a given gain constant ($1/RC$), smaller $C$ leads to larger $R$ and smaller values of $i_{in}$. Therefore, the bias current of the op amp becomes more significant as the capacitance becomes smaller. As usual, we try to design for the best compromise.

**Exercise 13.17** Consider the integrator of Figure 13.35 with the square-wave input signal shown in Figure 13.36. **a.** If $R = 10 \text{ k}\Omega$, $C = 0.1 \mu\text{F}$, and the op amp is ideal, sketch the output waveform to scale. **b.** If $R = 10 \text{ k}\Omega$, what value of $C$ is required for the peak-to-peak output amplitude to be 2 V?
**Answer   a.** See Figure 13.37; **b.** $C = 0.5 \mu\text{F}$. ☐

**Exercise 13.18** Consider the circuit of Figure 13.35 with $v_{in} = 0$, $R = 10 \text{ k}\Omega$, and $C = 0.01 \mu\text{F}$. As indicated in the figure, the reset switch opens at $t = 0$. The op amp is ideal except for a bias current of $I_B = 100 \text{ nA}$. **a.** Find an expression for the output voltage of the circuit as a function of time. **b.** Repeat for $C = 1 \mu\text{F}$.
**Answer   a.** $v_o(t) = 10t$; **b.** $v_o(t) = 0.1t$. ☐

**Exercise 13.19** Add a resistance $R$ in series with the noninverting input of the op amp in Figure 13.35 and repeat Exercise 13.18.
**Answer   a.** $v_o(t) = -1 \text{ mV}$; **b.** $v_o(t) = -1 \text{ mV}$. ☐

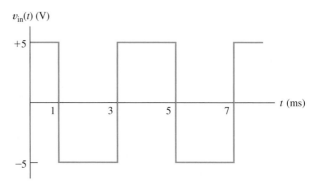

Figure 13.36  Square-wave input signal for Exercise 13.17.

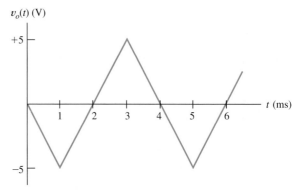

Figure 13.37  Answer for Exercise 13.17.

## Differentiator Circuit

Figure 13.38 shows a **differentiator** that produces an output voltage proportional to the time derivative of the input voltage. By an analysis similar to that used for the integrator, we can show that the circuit produces an output voltage given by

$$v_o(t) = -RC\frac{dv_{in}}{dt} \tag{13.43}$$

**Exercise 13.20**    Derive Equation 13.43. ☐

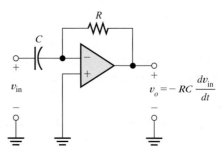

Figure 13.38  Differentiator.

## 13.10   ACTIVE FILTERS

Filters are circuits designed to pass input components with frequencies in one range to the output and prevent input components with frequencies in other ranges from reaching the output. For example, a lowpass filter passes low-frequency input components to the output but not high-frequency components. A common application for filters is to separate a signal of interest from other signals and noise. For example, in an electrocardiograph, we need a filter that passes the heart signals, which have frequencies below about 100 Hz, and rejects higher frequency noise that can be created by contraction of other muscles. We might use a lowpass filter to remove noise from historical phonograph recordings. In radio receivers, filters separate one station from the others. In digital instrumentation systems, a lowpass filter is often needed to remove noise and signal components that have frequencies higher than half of the sampling frequency in order to avoid a type of distortion, known as aliasing, during sampling and analog-to-digital conversion.

In Sections 6.2 and 6.8, we considered a few examples of passive-filter design. In this section, we show how to design lowpass filters composed of resistors, capacitors, and op amps. Filters composed of op amps, resistors, and capacitors are said to be **active filters**. In many respects, active filters have improved performance compared to passive circuits.

> Filters can be very useful in separating desired signals from noise.

Active filters have been studied extensively and many useful circuits have been found. Ideally, an active filter circuit should:

1. Contain few components.
2. Have a transfer function that is insensitive to component tolerances.
3. Place modest demands on the op amp's gain–bandwidth product, output impedance, slew rate, and other specifications.
4. Be easily adjusted.
5. Require a small spread of component values.
6. Allow a wide range of useful transfer functions to be realized.

Various circuits have been described in the literature that meet these goals to varying degrees. Many complete books have been written that deal exclusively with active filters. In this section, we confine our attention to a particular (but practical) means for implementing lowpass filters.

### Butterworth Transfer Function

The magnitude of the **Butterworth transfer function** is given by

$$|H(f)| = \frac{H_0}{\sqrt{1 + (f/f_B)^{2n}}} \qquad (13.44)$$

in which the integer $n$ is the *order* of the filter and $f_B$ is the 3-dB cutoff frequency. Substituting $f = 0$ yields $|H(0)| = H_0$; thus, $H_0$ is the dc gain magnitude. Plots of this transfer function are shown in Figure 13.39. Notice that as the order of the filter increases, the transfer function approaches that of an ideal lowpass filter.

An active lowpass Butterworth filter can be implemented by cascading modified **Sallen–Key circuits**, one of which is shown in Figure 13.40. In this version of the Sallen–Key circuit, the resistors labeled $R$ have equal values. Similarly, the capacitors

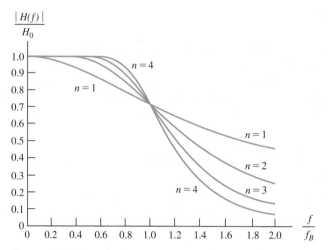

**Figure 13.39** Transfer-function magnitude versus frequency for lowpass Butterworth filters.

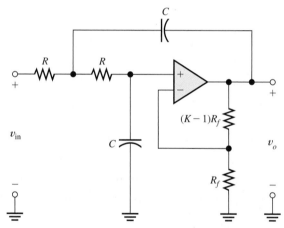

**Figure 13.40** Equal-component Sallen–Key lowpass active-filter section.

labeled $C$ have equal values. Useful circuits having unequal components are possible, but equal components are convenient.

The Sallen–Key circuit shown in Figure 13.40 is a second-order lowpass filter. To obtain an $n$th-order filter, $n/2$ circuits must be cascaded. (We assume that $n$ is even.)

The 3-dB cutoff frequency of the overall filter is related to $R$ and $C$ by

$$f_B = \frac{1}{2\pi RC} \tag{13.45}$$

Usually, we wish to design for a given cutoff frequency. We try to select small capacitance values because this leads to small physical size and low cost. However, Equation 13.45 shows that as the capacitances become small, the resistance values become larger (for a given cutoff frequency). If the capacitance is selected too small, the resistance becomes unrealistically large. Furthermore, stray wiring capacitance can easily affect a high-impedance circuit. Thus, we select a capacitance value that is small, but not too small (say not smaller than 1000 pF).

**Table 13.1** *K* Values for Lowpass
or Highpass Butterworth Filters
of Various Orders

| Order | K |
|-------|-------|
| 2 | 1.586 |
| 4 | 1.152 |
|  | 2.235 |
| 6 | 1.068 |
|  | 1.586 |
|  | 2.483 |
| 8 | 1.038 |
|  | 1.337 |
|  | 1.889 |
|  | 2.610 |

In selecting the capacitor, we should select a value that is readily available in the tolerance required. Then, we use Equation 13.45 to compute the resistance. It is helpful to select the capacitance first and then compute the resistance, because resistors are commonly available in more finely spaced values than capacitors. Possibly, we cannot find nominal values of $R$ and $C$ that yield exactly the desired break frequency; however, it is a rare situation for which the break frequency must be controlled to an accuracy less than a few percent. Thus, 1-percent-tolerance resistors usually result in a break frequency sufficiently close to the value desired.

Notice in the circuit of Figure 13.40 that the op amp and the feedback resistors $R_f$ and $(K - 1)R_f$ form a noninverting amplifier having a gain of $K$. At dc, the capacitors act as open circuits. Then, the resistors labeled $R$ are in series with the input terminals of the noninverting amplifier and have no effect on gain. Thus, the dc gain of the circuit is $K$. As $K$ is varied from zero to three, the transfer function displays more and more peaking (i.e., the gain magnitude increases with frequency and reaches a peak before falling off). For $K = 3$, infinite peaking occurs. It turns out that for $K$ greater than three, the circuit is unstable—it oscillates.

The most critical issue in selection of the feedback resistors $R_f$ and $(K - 1)R_f$ is their ratio. If desired, a precise ratio can be achieved by including a potentiometer, which is adjusted to yield the required dc gain for each section. To minimize the effects of bias current, we should select values such that the parallel combination of $R_f$ and $(K - 1)R_f$ is equal to $2R$. However, with FET input op amps, input bias current is often so small that this is not necessary.

An $n$th-order Butterworth lowpass filter is obtained by cascading $n/2$ stages having proper values for $K$. (Here again, we assume that $n$ is even.) Table 13.1 shows the required $K$ values for filters of various orders. The dc gain $H_0$ of the overall filter is the product of the $K$ values of the individual stages.

---

**Example 13.8**   Lowpass Active Filter Design

Design a fourth-order lowpass Butterworth filter having a cutoff frequency of 100 Hz.

**Solution**   We arbitrarily choose capacitor values of $C = 0.1\ \mu F$. This is a standard value and not prohibitively large. (Perhaps we could achieve an equally good design by using smaller capacitances, say $0.01\ \mu F$. However, as we have mentioned earlier, there is a practical limit to how small the capacitances can be.)

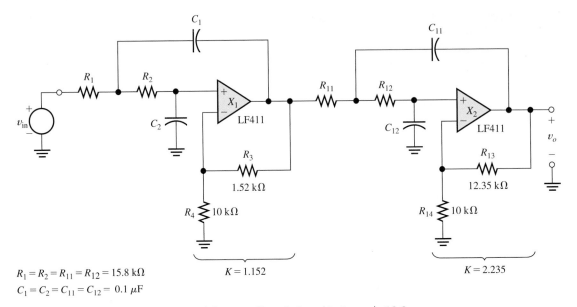

$R_1 = R_2 = R_{11} = R_{12} = 15.8 \ k\Omega$
$C_1 = C_2 = C_{11} = C_{12} = 0.1 \ \mu F$

**Figure 13.41** Fourth-order Butterworth lowpass filter designed in Example 13.8.

Active lowpass filters such as this are useful as antialias filters in computer-based instrumentation systems.

Next, we solve Equation 13.45 for $R$. Substituting $f_B = 100$ Hz and $C = 0.1 \ \mu F$ results in $R = 15.92 \ k\Omega$. In practice, we would select a 15.8-k$\Omega$ 1-percent-tolerance resistor. This results in a nominal cutoff frequency slightly higher than the design objective.

Consulting Table 13.1, we find that a fourth-order filter requires two sections having gains of $K = 1.152$ and $2.235$. This results in an overall dc gain of $H_0 = 1.152 \times 2.235 \cong 2.575$. We arbitrarily select $R_f = 10 \ k\Omega$ for both sections. The complete circuit diagram is shown in Figure 13.41. The resistors $R_3$ and $R_{13}$ consist of fixed resistors in series with small trimmers that can be adjusted to obtain the required gain for each stage. ∎

A Bode plot of the overall gain magnitude for the filter designed in Example 13.8 is shown in Figure 13.42. It can be verified that the dc gain in decibels is $20 \log H_0 \cong 8.2$ dB. As desired, the 3-dB frequency is very nearly 100 Hz.

Figure 13.43 shows the gain of each section normalized by its dc gain. The figure also shows the normalized overall gain. Of course, the overall normalized gain is the product of the normalized gains of the individual stages. (Notice that the gains are plotted as ratios rather than in decibels.) The transfer function of the first stage—which is the low-gain stage—rolls off without peaking. However, considerable peaking occurs in the second stage. It is this peaking that squares up the shoulder of the overall transfer characteristic.

**Exercise 13.21**   Show that for frequencies much greater than $f_B$, the magnitude of the lowpass Butterworth transfer function given in Equation 13.44 rolls off at $20 \times n$ decibels per decade. ☐

**Exercise 13.22**   Design a sixth-order Butterworth lowpass filter having a cutoff frequency of 5 kHz.
**Answer**   Many answers are possible. For a sixth-order filter, three stages like Figure 13.40 need to be cascaded. A good choice is to use capacitors in the range from

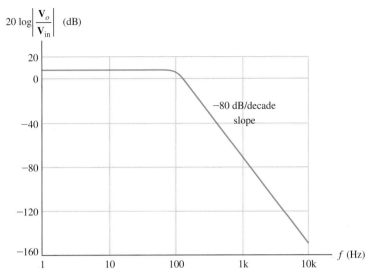

**Figure 13.42** Bode magnitude plot of the gain for the fourth-order lowpass filter of Example 13.8.

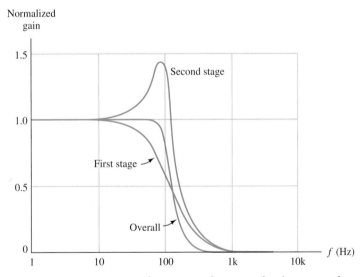

**Figure 13.43** Comparison of gain versus frequency for the stages of the fourth-order lowpass filter of Example 13.8.

1000 pF to 0.01 $\mu$F. With $C = 0.01\ \mu$F, we need $R = 3.183$ k$\Omega$. $R_f = 10$ k$\Omega$ is a good choice. From Table 13.1, we find the gain values to be 1.068, 1.586, and 2.483.    □

## Summary

1. If a differential amplifier has input voltages $v_1$ and $v_2$, the common-mode input is $v_{icm} = \frac{1}{2}(v_1 + v_2)$ and the differential input signal is $v_{id} = v_1 - v_2$.

2. An ideal operational amplifier has infinite input impedance, infinite gain for the differential input

signal, zero gain for the common-mode input signal, zero output impedance, and infinite bandwidth.

3. In an amplifier circuit with negative feedback, part of the output is returned to the input. The feedback signal opposes the input source.

4. To analyze ideal op-amp circuits with negative feedback, we assume that the differential input voltage and the input current of the op amp are driven to zero (this is the summing-point constraint), and then we use basic circuit principles to analyze the circuit.

5. The basic inverting amplifier configuration is shown in Figure 13.4 on page 649. Its closed-loop voltage gain is $A_v = -R_2/R_1$.

6. The basic noninverting amplifier configuration is shown in Figure 13.11 on page 655. Its closed-loop voltage gain is $A_v = 1 + R_2/R_1$.

7. Many useful amplifier circuits can be designed with the use of op amps. First, we select a suitable circuit configuration, and then we determine the resistor values that achieve the desired gain values.

8. In the design of op-amp circuits, very large resistances are unsuitable because their values are unstable and because high-impedance circuits are vulnerable to capacitive coupling of noise. Very low resistances are unsuitable because large currents flow in them for the voltages typically encountered in op-amp circuits.

9. In the linear range of operation, the imperfections of real op amps include finite input impedance, nonzero output impedance, and finite open-loop gain magnitude, which falls off with increasing frequency.

10. Negative feedback reduces gain magnitude and extends bandwidth. For the noninverting amplifier, the product of dc gain magnitude and bandwidth is constant for a given op-amp type.

11. The output voltage range and the output current range of any op amp are limited. If the output waveform reaches (and tries to exceed) either of these limits, clipping occurs.

12. The rate of change of the output voltage of any op amp is limited in magnitude. This is called the slew-rate limitation. The full-power bandwidth is the highest frequency for which the op amp can produce a full-amplitude sinusoidal output signal.

13. Dc imperfections of op amps are bias current, offset current, and offset voltage. These effects can be modeled by the sources shown in Figure 13.29 on page 672. The effect of dc imperfections is a (usually undesirable) dc component added to the intended output signal.

14. A single op amp can be used as a differential amplifier as shown in Figure 13.33 on page 676. However, the instrumentation amplifier shown in Figure 13.34 on page 677 has better performance.

15. The integrator circuit shown in Figure 13.35 on page 679 produces an output voltage that is proportional to the running time integral of the input voltage. A differentiator circuit is shown in Figure 13.38 on page 680.

16. Active filters often have better performance than passive filters. Active Butterworth lowpass filters can be obtained by cascading several Sallen–Key circuits having the proper gains.

## Problems

### Section 13.1: Ideal Operational Amplifiers

**P13.1.** What are the characteristics of an ideal op amp?

**P13.2.** A real op amp has five terminals. Name the probable function for each of the terminals.

**P13.3.** A differential amplifier has input voltages $v_1$ and $v_2$. Give the definitions of the differential input voltage and of the common-mode input voltage.

*__P13.4.__ The input voltages of a differential amplifier are

$$v_1(t) = 0.5 \cos(2000\pi t) + 20 \cos(120\pi t)$$
$$v_2(t) = -0.5 \cos(2000\pi t) + 20 \cos(120\pi t)$$

Find expressions for the common-mode and differential components of the input signal.

**P13.5.** Discuss the distinction between *open-loop gain* and *closed-loop gain*.

---

*Denotes that answers are contained in the Student Solutions files. See Appendix E for more information about accessing the Student Solutions.

### Section 13.2: Inverting Amplifiers

*P13.6.** What are the steps in analyzing an amplifier containing an ideal op amp?

**P13.7.** What do we mean by the term *summing-point constraint*? Does it apply when positive feedback is present?

**P13.8.** Draw the circuit diagram of the basic inverting amplifier configuration. Give an expression for the closed-loop voltage gain of the circuit in terms of the resistances, assuming an ideal op amp. Give expressions for the input impedance and output impedance of the circuit.

**P13.9.** Consider the circuit shown in Figure P13.9. Sketch $v_{in}(t)$ and $v_o(t)$ to scale versus time. The op amp is ideal.

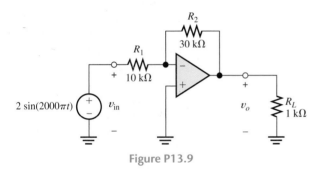

Figure P13.9

*P13.10.** Determine the closed-loop voltage gain of the circuit shown in Figure P13.10, assuming an ideal op amp.

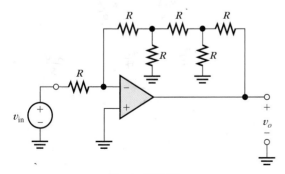

Figure P13.10

**P13.11.** Determine the closed-loop voltage gain of the circuit shown in Figure P13.11, assuming an ideal op amp.

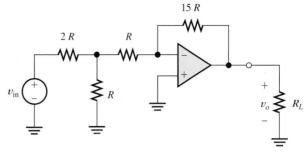

Figure P13.11

**P13.12.** Consider the inverting amplifier shown in Figure P13.12, in which one of the resistors has been replaced with a diode. Assume an ideal op amp, $v_{in}$ positive, and a diode current given by Equation 9.4, which states that $i_D = I_s \exp(v_D/nV_T)$. Derive an expression for $v_o$ in terms of $v_{in}$, $R$, $I_s$, $n$, and $V_T$.

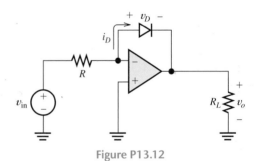

Figure P13.12

**P13.13.** Repeat Problem P13.12 by interchanging the resistance $R$ and the diode. Keep the diode pointing toward the right-hand side.

**P13.14.** Consider the circuit shown in Figure P13.12, with an unusual diode that has $i_D = Kv_D^3$. Derive an expression for $v_o$ in terms of $v_{in}$, $R$, and $K$.

**P13.15.** The op amp shown in Figure P13.15 is ideal, except that the extreme output voltages that it can produce are $\pm 10$ V. Determine two possible values for each of the voltages shown. [*Hint:* Notice that this circuit has *positive* feedback.]

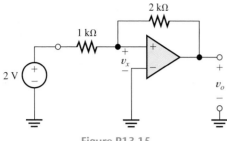

Figure P13.15

**P13.16.** Consider the inverting amplifier shown in Figure P13.16. Assuming an ideal op amp, solve for the currents and voltages shown. According to Kirchhoff's current law, the sum of the currents entering a closed surface must equal the sum of the currents leaving. Explain how the law is satisfied for the closed surface shown when we use a real op amp in this circuit.

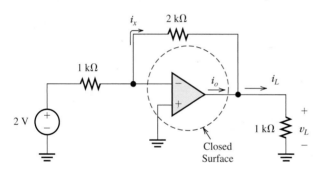

Figure P13.16

### Section 13.3: Noninverting Amplifiers

*P13.17. Draw the circuit diagram of an op-amp voltage follower. What value is its voltage gain? Input impedance? Output impedance?

*P13.18. The voltage follower of Figure 13.12 on page 656 has unity voltage gain so that $v_o = v_{in}$. Why not simply connect the load directly to the source, thus eliminating the op amp? Give an example of a situation in which the voltage follower is particularly good compared with the direct connection.

**P13.19.** Draw the circuit diagram of the basic noninverting amplifier configuration. Give an expression for the closed-loop voltage gain of the circuit in terms of the resistances, assuming an ideal op amp. Give expressions for the input impedance and output impedance of the circuit.

**P13.20.** For each of the circuits shown in Figure P13.20, assume that the op amp is ideal and find the value of $v_o$. Each of the circuits has negative feedback, so the summing-point constraint applies.

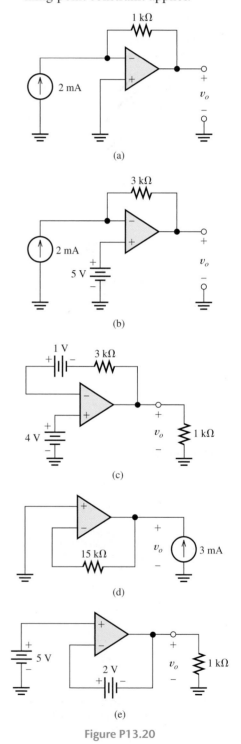

Figure P13.20

*P13.21. Analyze the ideal-op-amp circuit shown in Figure P13.21 to find an expression for $v_o$ in terms of $v_A$, $v_B$, and the resistance values.

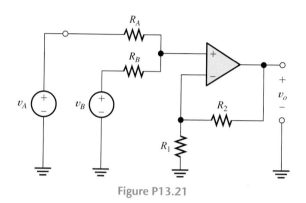

Figure P13.21

P13.22. The circuit shown in Figure P13.22 has

$$v_{in}(t) = 2 + 3\cos(2000\pi t)$$

Determine the value required for $R_2$ so that the dc component of the output $v_o(t)$ is zero. What is the resulting output voltage?

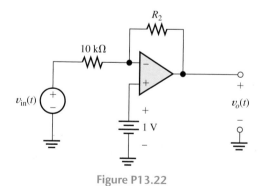

Figure P13.22

P13.23. Analyze each of the ideal-op-amp circuits shown in Figure P13.23 to find expressions for $i_o$. What is the value of the output impedance for each of these circuits? Why? [*Note:* The bottom end of the input voltage source is *not* grounded in part (b) of the figure. Thus, we say that this source is *floating.*]

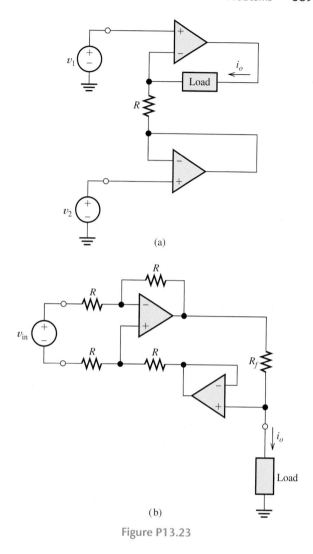

(a)

(b)

Figure P13.23

*P13.24. Consider the circuit shown in Figure P13.24. **a.** Find an expression for the output voltage in terms of the source current and resistance values. **b.** What value is the output impedance of this circuit? **c.** What value is the input impedance of this circuit? **d.** This circuit can

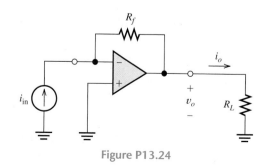

Figure P13.24

be classified as an ideal amplifier. What is the amplifier type? (See Section 11.6 for a discussion of various ideal-amplifier types.)

**P13.25.** Repeat Problem P13.24 for the circuit shown in Figure P13.25.

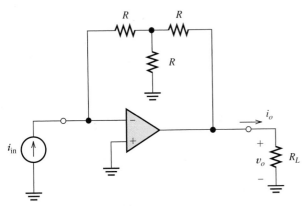

Figure P13.25

**P13.26.** Consider the circuit shown in Figure P13.26. **a.** Find an expression for the output current $i_o$ in terms of the source voltage and resistance values. **b.** What value is the output impedance of this circuit? **c.** What value is the input impedance of this circuit? **d.** This circuit can be classified as an ideal amplifier. What is the amplifier type? (See Section 11.6 for a discussion of various ideal-amplifier types.)

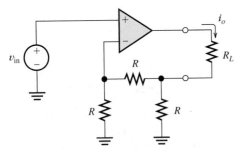

Figure P13.26

**P13.27.** The power gain $G$ of an amplifier is defined to be the power delivered to the load $R_L$ divided by the power delivered by the signal source $v_s$. Find an expression for the power gain of each of the amplifiers shown in Figure P13.27. Assume ideal op amps. Which circuit has the larger power gain?

*P13.28.** Consider the circuits shown in Figure P13.28(a) and (b). One of the circuits

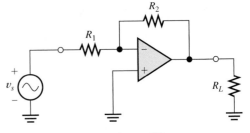

(a) Inverting amplifier

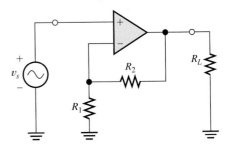

(b) Noninverting amplifier

Figure P13.27

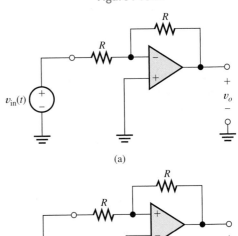

(a)

(b)

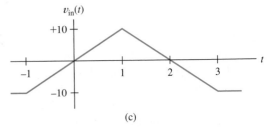

$v_{in}(t)$

(c)

Figure P13.28

has negative feedback, and the other circuit has positive feedback. Assume that the op amps are ideal, except that the output voltage is limited to extremes of $\pm 5$ V. For the input voltage waveform shown in Figure P13.28(c), sketch the output voltage $v_o(t)$ to scale versus time for each circuit.

**P13.29.** Repeat Problem P13.28 for the circuits of Figure P13.29(a) and (b). [The input voltage waveform is shown in Figure P13.28(c).]

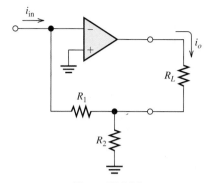

Figure P13.32

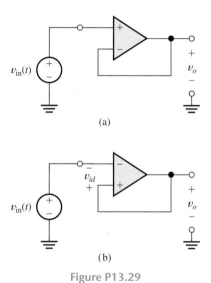

(a)

(b)

Figure P13.29

**P13.30.** Suppose that we design an inverting amplifier using 5-percent-tolerance resistors and an ideal op amp. The nominal amplifier gain is $-2$. What are the minimum and maximum gains possible, assuming that the resistances are within the stated tolerance? What is the percentage tolerance of the gain?

**P13.31.** Repeat Problem P13.30 for a noninverting amplifier having a nominal voltage gain of $+2$.

*****P13.32.** Consider the amplifier shown in Figure P13.32. Find an expression for the output current $i_o$. What is the input impedance? What is the output impedance seen by $R_L$?

**P13.33.** Derive an expression for the voltage gain of the circuit shown in Figure P13.33 as a

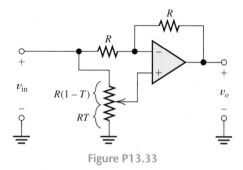

Figure P13.33

function of $T$, assuming an ideal op amp. ($T$ varies from 0 to unity, depending on the position of the wiper of the potentiometer.)

**P13.34.** The circuit shown in Figure P13.34 employs negative feedback. Use the summing-point

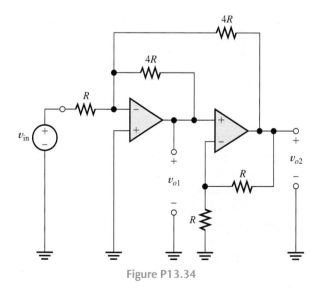

Figure P13.34

constraint (for both op amps) to derive expressions for the voltage gains $A_1 = v_{o1}/v_{in}$ and $A_2 = v_{o2}/v_{in}$.

### Section 13.4: Design of Simple Amplifiers

**P13.35.** Suppose that we are designing an amplifier, using an op amp. What problems are associated with using very small feedback resistances? With very large feedback resistances?

*__P13.36.__ Using the components listed in Table P13.36, design an amplifier having a voltage gain of $-10 \pm 20$ percent. The input impedance is required to be as large as possible (ideally, an open circuit). Remember to use practical resistance values. (*Hint:* Cascade a noninverting stage with an inverting stage.)

**Table P13.36** Available Parts for Design Problems

Standard 5%-tolerance resistors. (See Appendix B.)
Standard 1%-tolerance resistors. (Don't use these if a 5%-tolerance resistor will do, because 1%-tolerance resistors are more expensive.)
Ideal op amps.
Adjustable resistors (trimmers) having maximum values ranging from 100 $\Omega$ to 1 M$\Omega$ in a 1–2–5 sequence (i.e., 100 $\Omega$, 200 $\Omega$, 500 $\Omega$, 1 k$\Omega$, etc.). Don't use trimmers if fixed resistors will suffice.

*__P13.37.__ For Example 13.4 on page 661, it is possible to achieve a design by using only one op amp. Find a suitable circuit configuration and resistance values. For this problem, the gain tolerances are relaxed to $\pm 5$ percent.

**P13.38.** Using the components listed in Table P13.36, design an amplifier having an input impedance of at least 10 k$\Omega$ and a voltage gain of **a.** $-10 \pm 20$ percent; **b.** $-10 \pm 5$ percent; **c.** $-10 \pm 0.5$ percent.

**P13.39.** Design an amplifier having a voltage gain of $+10 \pm 3$ percent and an input impedance of 1 k$\Omega$ $\pm$ 1 percent, using the components listed in Table P13.36.

**P13.40.** Using the components listed in Table P13.36, design a circuit for which the output voltage is $v_o = A_1 v_1 + A_2 v_2$. The voltages $v_1$ and $v_2$ are input voltages. Design to achieve $A_1 = 5 \pm 5$ percent and $A_2 = -10 \pm 5$ percent. There is no restriction on input impedances.

*__P13.41.__ Repeat Problem P13.40 if the input impedances are required to be as large as possible (ideally, open circuits).

**P13.42.** Two signal sources have internal voltages $v_1(t)$ and $v_2(t)$, respectively. The internal (i.e., Thévenin) resistances of the sources are known always to be less than 2 k$\Omega$, but the exact values are not known and are likely to change over time. Using the components listed in Table P13.36, design an amplifier for which the output voltage is $v_o(t) = A_1 v_1(t) + A_2 v_2(t)$. The gains are to be $A_1 = -10 \pm 1$ percent and $A_2 = 3 \pm 1$ percent.

**P13.43.** Suppose we have a signal source with an internal (i.e., Thévenin) impedance that is always less than 1000 $\Omega$, but is variable over time. Using the components listed in Table P13.36, design an amplifier that produces an amplified version of the internal source voltage. The voltage gain should be $-20 \pm 5$ percent.

### Section 13.5: Op-Amp Imperfections in the Linear Range of Operation

**P13.44.** List the imperfections of real op amps in their linear range of operation.

*__P13.45.__ A certain op amp has a unity-gain–bandwidth of $f_t = 15$ MHz. If this op amp is used in a noninverting amplifier having a closed-loop dc gain of $A_{0CL} = 10$, determine the closed-loop break frequency $f_{BCL}$. Repeat for a dc gain of 100.

**P13.46.** A certain op amp has an open-loop dc gain of $A_{0OL} = 200{,}000$ and an open-loop 3-dB bandwidth of $f_{BOL} = 5$ Hz. Find the open-loop gain magnitude at a frequency of **a.** 100 Hz; **b.** 1000 Hz; **c.** 1 MHz.

**P13.47.** The objective of this problem is to investigate the effects of finite open-loop gain, finite input impedance, and nonzero output impedance of the op amp on the voltage follower. The circuit, including the op-amp model, is shown in Figure P13.47. **a.** Derive an expression for the circuit voltage gain $v_o/v_s$. Evaluate for $A_{OL} = 10^5$, $R_{in} = 1$ M$\Omega$, and

$R_o = 25\ \Omega$. Compare this result to the gain with an ideal op amp. **b.** Derive an expression for the circuit input impedance $Z_{in} = v_s/i_s$. Evaluate for $A_{OL} = 10^5$, $R_{in} = 1\ M\Omega$, and $R_o = 25\ \Omega$. Compare this result to the input impedance with an ideal op amp. **c.** Derive an expression for the circuit output impedance $Z_o$. Evaluate for $A_{OL} = 10^5$, $R_{in} = 1\ M\Omega$, and $R_o = 25\ \Omega$. Compare this result to the output impedance with an ideal op amp.

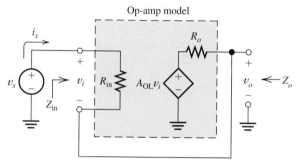

**Figure P13.47**

**P13.48.** The objective of this problem is to investigate the effects of finite gain, finite input impedance, and nonzero output impedance of the op amp on the inverting amplifier. The circuit, including the op-amp model, is shown in Figure P13.48. **a.** Derive an expression for the circuit voltage gain $v_o/v_s$. Evaluate for $A_{OL} = 10^5$, $R_{in} = 1\ M\Omega$, $R_o = 25\ \Omega$, $R_1 = 1\ k\Omega$, and $R_2 = 10\ k\Omega$. Compare this result to the gain with an ideal op amp. **b.** Derive an expression for the circuit input impedance

$Z_{in} = v_s/i_s$. Evaluate for $A_{OL} = 10^5$, $R_{in} = 1\ M\Omega$, $R_o = 25\Omega$, $R_1 = 1\ k\Omega$, and $R_2 = 10\ k\Omega$. Compare this result to the input impedance with an ideal op amp. **c.** Derive an expression for the circuit output impedance $Z_o$. Evaluate for $A_{OL} = 10^5$, $R_{in} = 1\ M\Omega$, $R_o = 25\ \Omega$, $R_1 = 1\ k\Omega$, and $R_2 = 10\ k\Omega$. Compare this result to the output impedance with an ideal op amp.

**P13.49.** We need a noninverting amplifier that has a dc gain of 10, and the gain magnitude at 10 kHz must be not less than 9. Determine the minimum gain–bandwidth specification required for the op amp.

**P13.50.** We need a noninverting amplifier that has a dc gain of 10, and the phase shift for the 200-kHz component must not exceed 10° magnitude. Determine the minimum gain–bandwidth specification required for the op amp.

**P13.51.** Consider two alternatives for designing an amplifier having a dc gain of 100. The first alternative is to use a single noninverting stage, having a gain of 100. The second alternative is to cascade two noninverting stages, each having a gain of 10. Op amps having a gain–bandwidth product of $10^6$ are to be used. Write an expression for the gain as a function of frequency for each alternative. Find the 3-dB bandwidth for each alternative.

**\*P13.52.** A certain op amp has an open-loop dc gain of $A_{0OL} = 200{,}000$ and an open-loop 3-dB bandwidth of $f_{BOL} = 5$ Hz. Sketch the Bode plot of the open-loop gain magnitude to scale. If this op amp is used in a noninverting amplifier having a closed-loop dc gain of 100, sketch the Bode plot of the closed-loop gain magnitude to scale. Repeat for a closed-loop dc gain of 10.

**Section 13.6: Nonlinear Limitations**

**P13.53.** List the nonlinear limitations of real op amps.

**P13.54.** Define *full-power bandwidth*.

**P13.55.** If the ideal output, with a sinusoidal input signal, greatly exceeds the full-power bandwidth, what is the waveform of the

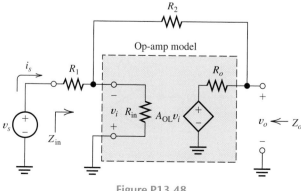

**Figure P13.48**

output signal? Under these conditions, if the slew rate of the op amp is 10 V/$\mu$s and the frequency of the input is 1 MHz, what is the peak-to-peak amplitude of the output signal?

**P13.56.** Suppose that we want to design an amplifier that can produce a 100-kHz sine-wave output having a peak amplitude of 5 V. What is the minimum slew-rate specification allowed for the op amp?

**\*P13.57.** Suppose that we have an op amp with a maximum output voltage range from $-10$ to $+10$ V. The maximum output current magnitude is 20 mA. The slew-rate limit is $SR = 10$ V/$\mu$s. This op amp is used in the circuit of Figure 13.28 on page 672. **a.** Find the full-power bandwidth of the op amp. **b.** For a frequency of 1 kHz and $R_L = 1$ k$\Omega$, what peak output voltage is possible without distortion? **c.** For a frequency of 1 kHz and $R_L = 100$ $\Omega$, what peak output voltage is possible without distortion? **d.** For a frequency of 1 MHz and $R_L = 1$ k$\Omega$, what peak output voltage is possible without distortion? **e.** If $R_L = 1$ k$\Omega$ and $v_s(t) = 5\sin(2\pi 10^6 t)$, sketch the steady-state output waveform to scale versus time.

**P13.58.** We need a noninverting amplifier with a dc gain of 10 to amplify an input signal given by

$$v_{in}(t) = 0 \qquad t \le 0$$
$$= t\exp(-t) \qquad t \ge 0$$

in which $t$ is in $\mu$s. Determine the minimum slew-rate specification required for the op amp if distortion must be avoided.

**P13.59.** We need a voltage follower to amplify an input signal given by

$$v_{in}(t) = 0 \qquad t \le 0$$
$$= t^2 \qquad 0 \le t \le 3$$
$$= 9 \qquad 3 \le t$$

in which $t$ is in $\mu$s. Determine the minimum slew-rate specification required for the op amp if distortion must be avoided.

**\*P13.60.** One way to measure the slew-rate limitation of an op amp is to apply a sine wave (or square wave) as the input to an amplifier and then increase the frequency until the output waveform becomes triangular. Suppose that

a 1-MHz input signal produces a triangular output waveform having a peak-to-peak amplitude of 4 V. Determine the slew rate of the op amp.

**P13.61.** An op amp has a maximum output voltage range from $-10$ to $+10$ V. The maximum output current magnitude is 25 mA. The slew-rate limit is 1 V/$\mu$s. The op amp is used in the amplifier shown in Figure P13.61. **a.** Find the full-power bandwidth of the op amp. **b.** For a frequency of 5 kHz and $R_L = 100$ $\Omega$, what peak output voltage is possible without distortion? **c.** For a frequency of 5 kHz and $R_L = 10$ k$\Omega$, what peak output voltage is possible without distortion? **d.** For a frequency of 100 kHz and $R_L = 10$ k$\Omega$, what peak output voltage is possible without distortion?

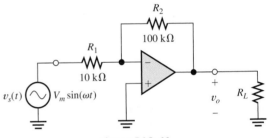

**Figure P13.61**

**P13.62.** Consider the **bridge amplifier** shown in Figure P13.62. **a.** Assuming ideal op amps, derive an expression for the voltage gain $v_o/v_s$. **b.** If $v_s(t) = 3\sin(\omega t)$, sketch $v_1(t)$, $v_2(t)$, and $v_o(t)$ to scale versus time. **c.** If the op amps are supplied from $\pm 15$ V

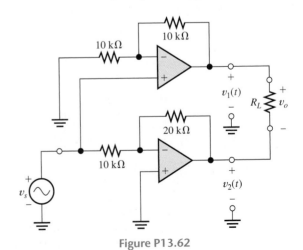

**Figure P13.62**

and clip at output voltages of $\pm14$ V, what is the peak value of $v_o(t)$ just at the threshold of clipping? (*Comment:* This circuit can be useful if a peak output voltage greater than the magnitude of the supply voltages is required.)

## Section 13.7: DC Imperfections

**\*P13.63.** Draw the circuit symbol for an op amp, adding sources to account for dc imperfections.

**P13.64.** Name the dc imperfections of real op amps. What is the net effect of these dc imperfections?

**P13.65.** What is an advantage of a FET-input op amp compared with a BJT-input op amp?

**\*P13.66.** Find the worst-case dc output voltages of the inverting amplifier shown in Figure 13.30(a) on page 674 for $v_{in} = 0$. The bias current ranges from 100 to 200 nA, the maximum offset-current magnitude is 50 nA, and the maximum offset-voltage magnitude is 4 mV.

**P13.67.** Sometimes, an ac-coupled amplifier is needed. The circuit shown in Figure P13.67 is a poor way to accomplish ac coupling. Explain why. [*Hint:* Consider the effect of bias current.] Show how to add a component (including its value) so that bias current has no effect on the output voltage of this circuit.

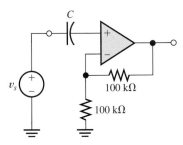

**Figure P13.67**

**P13.68.** Consider the amplifier shown in Figure P13.61. With zero dc input voltage from the signal source, it is desired that the dc output voltage be no greater than 100 mV. **a.** Ignoring other dc imperfections, what is the maximum offset voltage allowed for the op amp? **b.** Ignoring other dc imperfections, what is the maximum bias current allowed for the op amp? **c.** Show how to add a resistor to the circuit (including its value) so that

the effects of the bias currents cancel. **d.** Assuming that the resistor of part (c) is in place, and ignoring offset voltage, what is the maximum offset current allowed for the op amp?

## Section 13.8: Differential and Instrumentation Amplifiers

**P13.69.** In terms of the differential and common-mode components of a signal, what is the function of a differential amplifier?

**\*P13.70.** Using the parts listed in Table P13.36, design a single-op-amp differential amplifier having a nominal differential gain of 10.

**P13.71.** Repeat Problem P13.70, using the instrumentation-quality circuit shown in Figure 13.34 on page 677.

**P13.72.** Consider the instrumentation-quality differential amplifier shown in Figure 13.34 on page 677, with $R_1 = 1\,k\Omega$, $R_2 = 9\,k\Omega$, and $R = 10\,k\Omega$. The input signals are given by

$$v_1(t) = 0.5\cos(2000\pi t) + 2\cos(120\pi t)$$
$$v_2(t) = -0.5\cos(2000\pi t) + 2\cos(120\pi t)$$

**a.** Find expressions for the differential and common-mode components of the input signal.
**b.** Assuming ideal op amps, find expressions for the voltages at the output terminals of $X_1$ and $X_2$.
**c.** Again assuming ideal op amps, find an expression for the output voltage $v_o(t)$.

## Section 13.9: Integrators and Differentiators

**P13.73.** What do we mean by the term *running-time integral*?

**\*P13.74.** Sketch the output voltage of the circuit shown in Figure P13.74 to scale versus time. Sometimes, an integrator circuit is used as a (approximate) pulse counter. Suppose that the output voltage is $-10$ V. How many input pulses have been applied (assuming that the pulses have an amplitude of 5 V and a duration of 2 ms, as shown in the figure)?

**P13.75.** Sketch the output voltage of the ideal-op-amp circuit shown in Figure P13.75 to scale versus time.

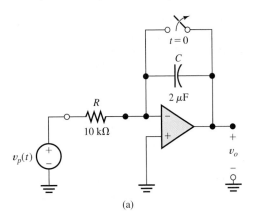

(a)

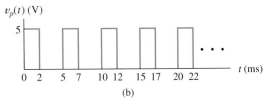

(b)

**Figure P13.74**

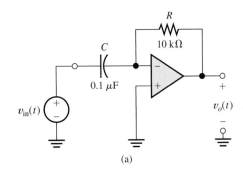

(a)

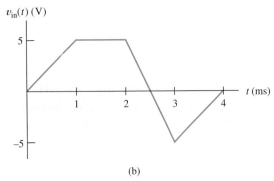

(b)

**Figure P13.75**

**P13.76.** The displacement of a robot arm in a given direction is represented by a voltage signal $v_{in}(t)$. The voltage is proportional to displacement, and 1 V corresponds to a displacement of 10 mm from the reference position. Design a circuit that produces a voltage $v_1(t)$ that is proportional to the velocity of the robot arm such that 1 m/s corresponds to 1 V. Design an

additional circuit that produces a voltage $v_2$ that is proportional to the acceleration of the robot arm such that 1 m/s² corresponds to 1 V. Use the components listed in Table P13.36, plus as many capacitors as needed.

### Section 13.10: Active Filters

**P13.77.** What is the function of a filter? What is a typical application? What is an active filter?

*P13.78.** Derive an expression for the voltage transfer ratio of each of the circuits shown in Figure P13.78. Also, sketch the magnitude

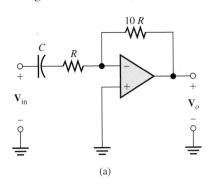

(a)

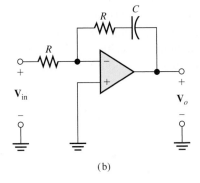

(b)

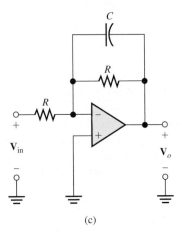

(c)

**Figure P13.78**

Bode plots to scale. Assume that the op amps are ideal.

**P13.79.** It is illuminating to look at the integrator circuit as a filter. Derive the transfer function for the integrator of Figure P13.74, and sketch the magnitude Bode plot to scale.

**P13.80.** Repeat Problem P13.79 for the differentiator circuit shown in Figure P13.75.

## Practice Test

Here is a practice test you can use to check your comprehension of the most important concepts in this chapter. Answers can be found in Appendix D and complete solutions are included in the Student Solutions files. See Appendix E for more information about the Student Solutions.

**T13.1.** Draw the circuit diagram for each of the following amplifiers. Clearly label the op amp input terminals and any resistances needed. Also, give the equation for the voltage gain of the amplifier in terms of the resistances on your diagram. **a.** The basic inverter. **b.** The noninverting amplifier. **c.** The voltage follower.

**T13.2.** Derive an expression for voltage gain $A_v = v_o/v_{in}$ of the circuit shown in Figure T13.2 assuming that the summing-point constraint applies. (There are both negative and positive feedback paths, but the resistances have been carefully selected so the circuit has net negative feedback and the summing-point constraint does apply.)

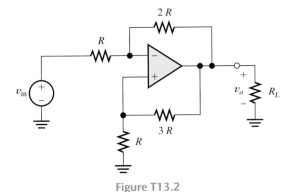

Figure T13.2

**T13.3.** A certain op amp has an open-loop dc gain of $A_{0OL} = 200,000$ and an open-loop 3-dB bandwidth of $f_{BOL} = 5$ Hz. This op amp is used in a noninverting amplifier having a closed-loop dc gain of $A_{0CL} = 100$. **a.** Determine the closed-loop break frequency $f_{BCL}$. **b.** Given that the input voltage to the noninverting amplifier is $v_{in}(t) = 0.05 \cos(2\pi \times 10^5 t)$ V, find the expression for the output voltage.

**T13.4.** We have an op amp with a maximum output voltage range from $-4.5$ to $+4.5$ V. The maximum output current magnitude is 5 mA. The slew-rate limit is SR $= 20$ V/$\mu$s. This op amp is used in the circuit of Figure 13.28 on page 672. **a.** Find the full-power bandwidth of the op amp. **b.** For a frequency of 1 kHz and $R_L = 200$ $\Omega$, what peak output voltage, $V_{om}$, is possible without distortion? **c.** For a frequency of 1 kHz and $R_L = 10$ k$\Omega$, what peak output voltage, $V_{om}$, is possible without distortion? **d.** For a frequency of 5 MHz and $R_L = 10$ k$\Omega$, what peak output voltage, $V_{om}$, is possible without distortion?

**T13.5.** Draw an op amp symbol including sources to account for offset voltage, offset current, and bias current. What is the principal effect of these sources in an amplifier circuit?

**T13.6.** Draw the circuit diagram of a differential amplifier using one op amp and resistances as needed. Give the output voltage in terms of the input voltages and resistances.

**T13.7.** Draw the circuit diagrams for an integrator and for a differentiator. Also give an expression for the output voltage in terms of the input voltage and component values.

**T13.8.** What is a filter? An active filter? Give one application for an active filter.

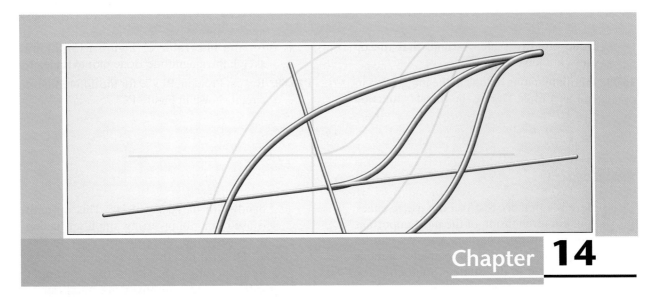

# Magnetic Circuits and Transformers

## Study of this chapter will enable you to:

- Understand magnetic fields and their interactions with moving charges.

- Use the right-hand rule to determine the direction of the magnetic field around a current-carrying wire or coil.

- Calculate forces on moving charges and current-carrying wires due to magnetic fields.

- Calculate the voltage induced in a coil by a changing magnetic flux or in a conductor cutting through a magnetic field.

- Use Lenz's law to determine the polarities of induced voltages.

- Apply magnetic-circuit concepts to determine the magnetic fields in practical devices.

- Determine the inductance and mutual inductance of coils, given their physical parameters.

- Understand hysteresis, saturation, core loss, and eddy currents in cores composed of magnetic materials such as iron.

- Understand ideal transformers and solve circuits that include transformers.

- Use the equivalent circuits of real transformers to determine their regulations and power efficiencies.

## Introduction to this chapter:

In describing interactions of matter, we often employ field concepts. For example, masses are attracted by gravitation. We envision gravitational fields produced by masses and explain the forces on other masses in terms of their interaction with these fields. Another example is stationary electrical charges. Charges of like sign repel one another, and unlike charges attract one another. Conceptually, each charge creates an electric field, and the other charge interacts with the field, resulting in a force.

In this and the next two chapters, we study some important engineering applications of magnetic fields, which are created by electrical charges in motion. Charges moving through magnetic fields experience forces. Furthermore, changing magnetic fields induce voltages in nearby conductors.

In this chapter, we start by reviewing basic magnetic-field concepts. Then, we consider the relationships between magnetic fields and inductance, including mutual inductance. Next, we study **transformers**, which greatly facilitate the distribution of electrical power.

Magnetic fields also form the basis of most practical devices for converting energy between electrical and mechanical forms. In subsequent chapters, we study the basic operating principles of several types of rotating energy conversion devices, collectively known as motors and generators.

## 14.1   MAGNETIC FIELDS

Magnetic fields exist in the space around permanent magnets and around wires that carry current. In both cases, the basic source of the magnetic field is electrical charge in motion. In an iron permanent magnet, fields are created by the spin of electrons in atoms. These fields aid one another, producing the net external field that we observe. (In most other materials, the magnetic fields of the electrons tend to cancel one another.) If a current-carrying wire is formed into a multiturn coil, the magnetic field is greatly intensified, particularly if the coil is wound around an iron core.

We can visualize a magnetic field as **lines of magnetic flux** that form closed paths. The lines are close together where the magnetic field is strong and farther apart where the field is weaker. This is illustrated in Figure 14.1. The units of magnetic flux are webers (Wb).

*Magnetic flux lines form closed paths that are close together where the field is strong and farther apart where the field is weak.*

The earth has a natural magnetic field that is relatively weak compared with those in typical transformers, motors, or generators. Due to interactions of the fields, magnets tend to align with the earth's field. Thus, a magnet has a north-seeking end (*N*) and a south-seeking end (*S*). Unlike ends of magnets are attracted. By convention, flux lines leave the north-seeking end (*N*) of a magnet and enter its south-seeking end (*S*). A compass can be used to investigate the direction of the lines of flux. The compass needle indicates north in the direction of the flux [i.e., the compass points toward the south-seeking (*S*) end of the magnet]. (Note that the earth's field lines are directed from south to north. Thus, if we were to place *N* and *S* marks on the earth as we do on a magnet, *S* would appear near the *north* geographic pole, because that is where the field lines enter the earth.)

*Flux lines leave the north-seeking end of a magnet and enter the south-seeking end.*

In equations, we represent the **magnetic flux density** as the vector quantity **B**. (Throughout our discussion, we use boldface for vector quantities. The corresponding lightface italic symbols represent the magnitudes of the vectors. Thus, $B$ represents the magnitude of the vector **B**. We also use boldface for phasors. However, it will be clear from the context which quantities are spatial vectors and which are phasors.) Furthermore, we use the International System of Units (SI), in which the units of **B** are webers/meter$^2$ (Wb/m$^2$) or, equivalently, teslas (T). The flux density vector **B** has a direction tangent to the flux lines, as illustrated in Figure 14.1.

*When placed in a magnetic field, a compass indicates north in the direction of the flux lines.*

### Right-Hand Rule

The direction of the magnetic field produced by a current can be determined by the right-hand rule. There are several interpretations of this rule. For example, as

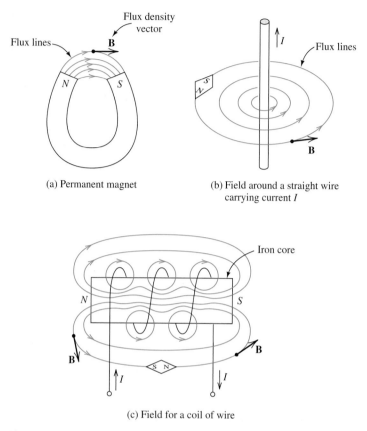

**Figure 14.1** Magnetic fields can be visualized as lines of flux that form closed paths. Using a compass, we can determine the direction of the flux lines at any point. Note that the flux density vector **B** is tangent to the lines of flux.

illustrated in Figure 14.2(a), if a wire is grasped with the thumb pointing in the direction of the current, the fingers encircle the wire, pointing in the direction of the magnetic field. Moreover, as illustrated in Figure 14.2(b), if the fingers are wrapped around a coil in the direction of current flow, the thumb points in the direction of the magnetic field that is produced inside the coil.

*The right-hand rule is used to determine the directions of magnetic fields.*

**Exercise 14.1**   A wire horizontal to the ground carries current toward the north. (Neglect the earth's field.) **a.** Directly underneath the wire, what is the direction of **B**? **b.** Directly above the wire, what is the direction of **B**?
**Answer    a.** west; **b.** east.    ▫

**Exercise 14.2**   A coil is wound around the periphery of a clock. If current flows clockwise, what is the direction of **B** in the center of the clock face?
**Answer**   Into the clock face.    ▫

### Forces on Charges Moving in Magnetic Fields

An electrical charge $q$ moving with velocity vector **u** through a magnetic field **B** experiences a force **f** as illustrated in Figure 14.3. The force vector is given by

$$\mathbf{f} = q\mathbf{u} \times \mathbf{B} \tag{14.1}$$

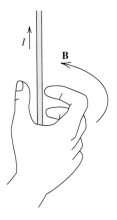

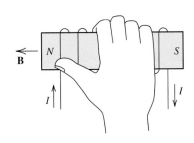

(a) If a wire is grasped with the thumb pointing in the current direction, the fingers encircle the wire in the direction of the magnetic field

(b) If a coil is grasped with the fingers pointing in the current direction, the thumb points in the direction of the magnetic field inside the coil

**Figure 14.2** Illustrations of the right-hand rule.

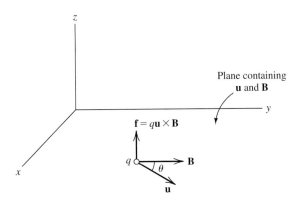

**Figure 14.3** A charge moving through a magnetic field experiences a force **f** perpendicular to both the velocity **u** and flux density **B**.

in which $\times$ represents the vector cross product. Note that due to the definition of the cross product, the force is perpendicular to the plane containing the magnetic flux density **B** and the velocity **u**. Furthermore, the magnitude of the force is given by

Force is exerted on a charge as it moves through a magnetic field.

$$f = quB \sin(\theta) \tag{14.2}$$

in which $\theta$ is the angle between **u** and **B**, as illustrated in the figure.

In the SI system, the force vector **f** has units of newtons (N), the charge is in coulombs (C), and the velocity vector **u** is in meters/second (m/s). Thus, for dimensional consistency in Equations 14.1 and 14.2, the magnetic field vector **B** must have units of newton seconds per coulomb meter (Ns/Cm), which is the dimensional equivalent of the tesla (T).

**Exercise 14.3** An electron ($q = -1.602 \times 10^{-19}$ C) travels at $10^5$ m/s in the positive $x$ direction. The magnetic flux density is 1 T in the positive $y$ direction. Find the magnitude and direction of the force on the electron. (Assume a right-hand coordinate system such as that shown in Figure 14.3.)
**Answer** $f = 1.602 \times 10^{-14}$ N in the negative $z$ direction. □

### Forces on Current-Carrying Wires

*Force is exerted on a current-carrying conductor when it is immersed in a magnetic field.*

Current flowing in a conductor consists of charge (usually, electrons) in motion. Thus, forces appear on a current-carrying wire immersed in a magnetic field. The force on an incremental length of the wire is given by

$$d\mathbf{f} = i \, d\mathbf{l} \times \mathbf{B} \tag{14.3}$$

in which the direction of $d\mathbf{l}$ and the reference direction for the current are the same. For a straight wire of length $l$ and a constant magnetic field, we have

$$f = ilB \sin(\theta) \tag{14.4}$$

in which $\theta$ is the angle between the wire and the field. Notice that the force is maximized if the direction of the field is perpendicular to the wire.

**Exercise 14.4** A wire of length $l = 1$ m carries a current of 10 A perpendicular to a field of $B = 0.5$ T. Compute the magnitude of the force on the wire.
**Answer** $f = 5$ N. □

### Flux Linkages and Faraday's Law

The magnetic flux passing through a surface area $A$ is given by the surface integral

$$\phi = \int_A \mathbf{B} \cdot d\mathbf{A} \tag{14.5}$$

*The magnetic flux passing through a surface is determined by integrating the dot product of B and incremental area over the surface.*

in which $d\mathbf{A}$ is an increment of area on the surface. The direction of the vector $d\mathbf{A}$ is perpendicular to the surface. If the magnetic flux density is constant and perpendicular to the surface, Equation 14.5 reduces to

$$\phi = BA \tag{14.6}$$

We say that the flux passing through the surface bounded by a coil **links** the coil. If the coil has $N$ turns, then the total flux linkages are given by

$$\lambda = N\phi \tag{14.7}$$

Here, we have assumed that the same flux links each turn of the coil. This is a good approximation when the turns are close together on an iron form, which is often the case in transformers and electrical machines.

According to **Faraday's law of magnetic induction**, a voltage

$$e = \frac{d\lambda}{dt} \tag{14.8}$$

is induced in a coil whenever its flux linkages are changing. This can occur either because the magnetic field is changing with time or because the coil is moving relative to a magnetic field.

   **Lenz's law** states that the polarity of the induced voltage is such that the voltage would produce a current (through an external resistance) that opposes the original change in flux linkages. (Think of the induced voltage as a voltage source.) For example, suppose that the magnetic field linking the coil shown in Figure 14.4 is pointing into the page and increasing in magnitude. (This field is the result of a coil or moving permanent magnet not shown in the figure.) Then, the voltage induced in the coil produces a counterclockwise current. According to the right-hand rule, this current produces a magnetic field directed out of the page, opposing the initial field change.

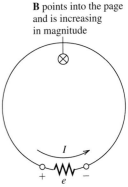

B points into the page and is increasing in magnitude

Induced voltage

**Figure 14.4** When the flux linking a coil changes, a voltage is induced in the coil. The polarity of the voltage is such that if a circuit is formed by placing a resistance across the coil terminals, the resulting current produces a field that tends to oppose the original change in the field.

## Voltages Induced in Field-Cutting Conductors

Voltage is also induced in a conductor moving through a magnetic field in a direction such that the conductor cuts through magnetic lines of flux. For example, consider Figure 14.5. A uniform magnetic field is directed into the page. The sliding conductor and the stationary rails form a loop having an area of $A = lx$. The flux linkages of the coil are

$$\lambda = BA = Blx$$

According to Faraday's law, the voltage induced in the coil is given by

$$e = \frac{d\lambda}{dt} = Bl\frac{dx}{dt}$$

However, $u = dx/dt$ is the velocity of the sliding conductor, so we have

$$e = Blu \tag{14.9}$$

Equation 14.9 can be used to compute the voltage induced across the ends of a straight conductor moving in uniform magnetic field, provided that the velocity, the conductor, and the magnetic-field vector are mutually perpendicular.

   For example, a conductor in a typical dc generator rated for 1 kW has a length of 0.2 m, a velocity of 12 m/s, and cuts through a field of 0.5 T. This results in an

Voltage is induced across the terminals of a coil if the flux linkages are changing with time. Moreover, voltage is induced between the ends of a conductor moving so as to cut through flux lines.

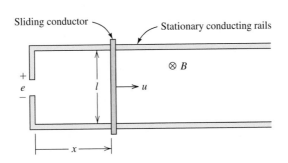

**Figure 14.5** A voltage is induced in a conductor moving so as to cut through magnetic flux lines.

Sliding conductor

Stationary conducting rails

⊗ B

induced voltage of 1.2 V. (Higher output voltages are produced by connecting many such conductors in series.)

**Exercise 14.5    a.** A 10-turn circular coil has a radius of 5 cm. A flux density of 0.5 T is directed perpendicular to the plane of the coil. Evaluate the flux linking the coil and the flux linkages. **b.** Suppose that the flux is reduced to zero at a uniform rate during an interval of 1 ms. Determine the voltage induced in the coil.
**Answer    a.** $\phi = 3.927$ mWb, $\lambda = 39.27$ mWb turns; **b.** $e = 39.27$ V. □

## Magnetic Field Intensity and Ampère's Law

So far, we have considered the magnetic flux density **B** and its effects. To summarize, **B** produces forces on moving charges and current-carrying conductors. It also induces voltage in a coil if the flux linkages are changing with time. Furthermore, voltage is induced across a moving conductor when it cuts through flux lines.

Now, we introduce another field vector, known as the **magnetic field intensity H**, and consider how magnetic fields are established. In general, magnetic fields are set up by charges in motion. In most of the applications that we consider, the magnetic fields are established by currents flowing in coils. We will see that **H** is determined by the currents and the configuration of the coils. Furthermore, we will see that the resulting flux density **B** depends on **H**, as well as the properties of the material filling the space around the coils.

The magnetic field intensity **H** and magnetic flux density **B** are related by

**B** is magnetic flux density with units of webers per square meter (Wb/m²) or teslas (T), and H is magnetic field intensity with units of amperes per meter (A/m).

$$\mathbf{B} = \mu\mathbf{H} \tag{14.10}$$

in which $\mu$ is the magnetic permeability of the material. The units of **H** are amperes/meter (A/m), and the units of $\mu$ are webers/ampere-meter (Wb/Am).

For free space, we have

$$\mu = \mu_0 = 4\pi \times 10^{-7} \text{ Wb/Am} \tag{14.11}$$

Some materials, most notably iron and certain rare-earth alloys, have a much higher magnetic permeability than free space. The relative permeability of a material is the ratio of its permeability to that of free space:

$$\mu_r = \frac{\mu}{\mu_0} \tag{14.12}$$

The value of $\mu_r$ ranges from several hundred to 1 million for various iron and rare-earth alloys. The iron used in typical transformers, motors, and generators has a relative permeability of several thousand.

**Ampère's law** states that the line integral of the magnetic field intensity around a closed path is equal to the algebraic sum of the currents flowing through the area enclosed by the path. In equation form, we have

$$\oint \mathbf{H} \cdot d\mathbf{l} = \sum i \tag{14.13}$$

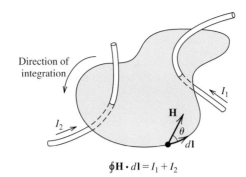

Figure 14.6 Ampère's law states that the line integral of magnetic field intensity around a closed path is equal to the sum of the currents flowing through the surface bounded by the path.

$$\oint \mathbf{H} \cdot d\mathbf{l} = I_1 + I_2$$

in which $d\mathbf{l}$ is a vector element of length having its direction tangent to the path of integration. Recall that the vector dot product is given by

$$\mathbf{H} \cdot d\mathbf{l} = Hdl\cos(\theta) \qquad (14.14)$$

in which $\theta$ is the angle between $\mathbf{H}$ and $d\mathbf{l}$.

Depending on its reference direction, a given current carries either a plus sign or a minus sign in the summation of Equation 14.13. If the reference direction for a current is related to the direction of integration by the right-hand rule, it carries a plus sign. (According to the right-hand rule, if you place the thumb of your right hand on the wire pointing in the reference direction, your fingers encircle the wires in the direction of integration.) Currents that are referenced in the opposite direction carry a negative sign in Equation 14.13. Ampère's law is illustrated by example in Figure 14.6, in which case the reference directions of both currents are related to the direction of integration by the right-hand rule.

If the magnetic intensity has constant magnitude and points in the same direction as the incremental length $d\mathbf{l}$ everywhere along the path, Ampère's law reduces to

$$Hl = \sum i \qquad (14.15)$$

in which $l$ is the length of the path.

In some cases, we can use Ampère's law to find formulas for the magnetic field in the space around a current-carrying wire or coil.

---

**Example 14.1    Magnetic Field around a Long Straight Wire**

Consider a long straight wire carrying current $I$ out of the page as shown in Figure 14.7. Find expressions for the magnetic field intensity and magnetic flux density in the space around the wire. Assume that the material surrounding the wire has permeability $\mu$.

**Solution**   By symmetry and the right-hand rule, we conclude that $\mathbf{B}$ and $\mathbf{H}$ fall in a plane perpendicular to the wire (i.e., in the plane of the paper) and are tangent to circles having their centers at the wire. This is illustrated in Figure 14.7. Furthermore, the magnitude of $H$ is constant for a given radius $r$. Applying Ampère's law (Equation 14.15) to the circular path shown in the figure, we have

$$Hl = H2\pi r = I$$

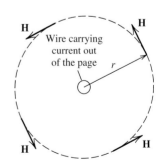

Figure 14.7 The magnetic field around a long straight wire carrying a current can be determined with Ampère's law aided by considerations of symmetry.

Solving for the magnetic intensity, we obtain

$$H = \frac{I}{2\pi r}$$

Then by Equation 14.10, we find the magnetic flux density as

$$B = \mu H = \frac{\mu I}{2\pi r}$$    ∎

---

### Example 14.2    Flux Density in a Toroidal Core

Consider the toroidal coil shown in Figure 14.8. Find an expression for the magnetic flux density $B$ on the center line of the core in terms of the number of coil turns $N$, the current $I$, the permeability $\mu$ of the core, and the physical dimensions. Then, assuming that the flux density is constant throughout the core (this is approximately true if $R \gg r$), find expressions for the total flux and the flux linkages.

**Solution**    By symmetry, the field intensity is constant in magnitude along the dashed circular center line shown in the figure. (We assume that the coil is wound in a symmetrical manner all the way around the toroidal core. For clarity, only part of the coil is shown in the figure.) Applying Ampère's law to the dashed path, we obtain

$$Hl = H2\pi R = NI$$

Solving for $H$ and using Equation 14.10 to determine $B$, we have

$$H = \frac{NI}{2\pi R} \tag{14.16}$$

and

$$B = \frac{\mu NI}{2\pi R} \tag{14.17}$$

Assuming that $R$ is much greater than $r$, the flux density is nearly constant over the cross section of the core. Then, according to Equation 14.6, the flux is equal to the product of the flux density and the area of the cross section:

$$\phi = BA = \frac{\mu NI}{2\pi R}\pi r^2 = \frac{\mu NIr^2}{2R} \tag{14.18}$$

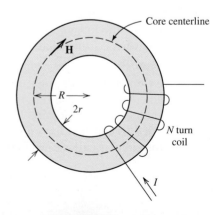

Figure 14.8 Toroidal coil analyzed in Examples 14.2, 14.3, and 14.4.

Finally, we note that all of the flux links all of the turns, and we have

$$\lambda = N\phi = \frac{\mu N^2 I r^2}{2R} \qquad (14.19)$$

■

### Example 14.3    Flux and Flux Linkages for a Toroidal Core

Suppose that we have a toroidal core with $\mu_r = 5000$, $R = 10$ cm, $r = 2$ cm, and $N = 100$. The current is

$$i(t) = 2 \sin(200\pi t)$$

Compute the flux and the flux linkages. Then, use Faraday's law of induction to determine the voltage induced in the coil.

**Solution**    First, the permeability of the core material is

$$\mu = \mu_r \mu_0 = 5000 \times 4\pi \times 10^{-7}$$

Using Equation 14.18, we compute the flux:

$$\phi = \frac{\mu N I r^2}{2R} = \frac{5000 \times 4\pi \times 10^{-7} \times 100 \times 2 \sin(200\pi t) \times (2 \times 10^{-2})^2}{2 \times 10 \times 10^{-2}}$$

$$= (2.513 \times 10^{-3}) \sin(200\pi t) \text{ Wb}$$

The flux linkages are

$$\lambda = N\phi$$

$$= 100 \times (2.513 \times 10^{-3}) \sin(200\pi t)$$

$$= 0.2513 \sin(200\pi t) \text{ weber turns}$$

Finally, using Faraday's law (Equation 14.8), we can find the voltage induced in the coil by the changing field:

$$e = \frac{d\lambda}{dt} = 0.2513 \times 200\pi \cos(200\pi t)$$

$$= 157.9 \cos(200\pi t) \text{ V}$$

■

**Exercise 14.6**    A long straight wire surrounded by air ($\mu_r \cong 1$) carries a current of 20 A. Compute the magnetic flux density at a point 1 cm from the wire.
**Answer**    $4 \times 10^{-4}$ T.                                                       □

**Exercise 14.7**    Figure 14.9 shows two wires carrying equal currents in opposite directions. Find the value of

$$\oint \mathbf{H} \cdot d\mathbf{l}$$

for each path shown in the direction indicated.
**Answer**    Path 1, 10 A; path 2, 0; path 3, −10 A.                              □

**Exercise 14.8**    Find the force between a 1-m length of the wires shown in Figure 14.9 if the distance between the wires is 10 cm. Is this a force of attraction or of repulsion?
**Answer**    $f = 2 \times 10^{-4}$ N; repulsion.                                        □

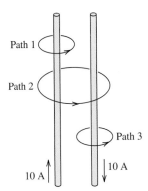

Figure 14.9  See Exercises 14.7 and 14.8.

## 14.2 MAGNETIC CIRCUITS

We will see that many useful devices (such as transformers, motors, and generators) contain coils wound on iron cores. In this section, we learn how to calculate the magnetic fields in these devices. A simple example, discussed in the preceding section, is the toroidal coil shown in Figure 14.8 and analyzed in Example 14.2. The toroid possesses sufficient symmetry that we readily applied Ampère's law to find an expression for the field intensity. However, in many applications, we need to analyze more complex configurations (such as cores that lack symmetry and those with multiple coils) for which the direct application of Ampère's law is not feasible. Instead, we use **magnetic circuit concepts**, which are analogous to those used to analyze electrical circuits.

The **magnetomotive force** (mmf) of an $N$-turn current-carrying coil is given by

$$\mathcal{F} = Ni \qquad (14.20)$$

A current-carrying coil is the magnetic-circuit analog of a voltage source in an electrical circuit. Magnetomotive force is analogous to source voltage. Usually, we give the units of magnetomotive force as A · turns; however, the number of turns is actually a pure number without physical units.

The **reluctance** of a path for magnetic flux, such as the bar of iron shown in Figure 14.10, is given by

$$\mathcal{R} = \frac{l}{\mu A} \qquad (14.21)$$

in which $l$ is the length of the path (in the direction of the magnetic flux), $A$ is the cross-sectional area, and $\mu$ is the permeability of the material. Reluctance is analogous to resistance in an electrical circuit. When the bar is not straight, the length of the path is somewhat ambiguous, and then we estimate its value as the length of the centerline. Thus, $l$ is sometimes called the **mean length** of the path.

Magnetic flux $\phi$ in a magnetic circuit is analogous to current in an electrical circuit. Magnetic flux, reluctance, and magnetomotive force are related by

$$\mathcal{F} = \mathcal{R}\phi \qquad (14.22)$$

which is the counterpart of Ohm's law ($V = Ri$). The units of reluctance are A · turns/Wb.

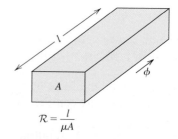

**Figure 14.10** The reluctance $\mathcal{R}$ of a magnetic path depends on the mean length $l$, the area $A$, and the permeability $\mu$ of the material.

### Example 14.4   The Toroidal Coil as a Magnetic Circuit

Using magnetic circuit concepts, analyze the toroidal coil shown in Figure 14.8 to find an expression for the flux.

**Solution**   As indicated in Figure 14.11, the magnetic circuit of the toroidal coil is analogous to a simple electrical circuit with a resistance connected across a voltage source.

The mean length of the magnetic path is

$$l = 2\pi R$$

The cross section of the core is circular with radius $r$. Thus, the area of the cross section is

$$A = \pi r^2$$

Substituting into Equation 14.21, we find the reluctance to be

$$\mathcal{R} = \frac{l}{\mu A} = \frac{2\pi R}{\mu \pi r^2} = \frac{2R}{\mu r^2}$$

The magnetomotive force is

$$\mathcal{F} = NI$$

Solving Equation 14.22 for the flux, we have

$$\phi = \frac{\mathcal{F}}{\mathcal{R}}$$

Substituting the expressions for $\mathcal{F}$ and $\mathcal{R}$ found earlier, we get

$$\phi = \frac{\mu N r^2 I}{2R}$$

This is the same expression for the flux that we obtained in Examples 14.2 and 14.3 by applying Ampère's law.   ∎

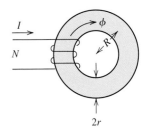

(a) Coil on a toroidal iron core

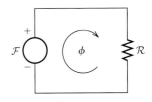

(b) Magnetic circuit

**Figure 14.11  The magnetic circuit for the toroidal coil.**

### Advantage of the Magnetic-Circuit Approach

The advantage of the magnetic-circuit approach is that it can be applied to unsymmetrical magnetic cores with multiple coils. Coils are sources of magnetomotive forces that can be manipulated as source voltages are in an electrical circuit. Reluctances in series or parallel are combined as resistances are. Fluxes are analogous to currents. The magnetic-circuit approach is not an exact method for determining magnetic fields, but it is sufficiently accurate for many engineering applications. We illustrate these methods with a few examples.

The advantage of the magnetic-circuit approach is that it can be applied to unsymmetrical magnetic cores with multiple coils.

### Example 14.5   A Magnetic Circuit with an Air Gap

Consider the magnetic core with an air gap as shown in Figure 14.12(a). The core material has a relative permeability of 6000 and a rectangular cross section 2 cm by 3 cm. The coil has 500 turns. Determine the current required to establish a flux density of $B_{\text{gap}} = 0.25$ T in the air gap.

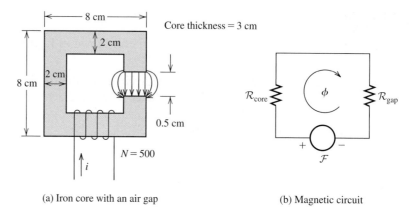

(a) Iron core with an air gap                    (b) Magnetic circuit

**Figure 14.12** Magnetic circuit of Example 14.5.

**Solution**    As shown in Figure 14.12(b), this magnetic circuit is analogous to an electrical circuit with one voltage source and two resistances in series. First, we compute the reluctance of the core. Notice that the centerline of the flux path is a square 6 cm by 6 cm. Thus, the mean length of the iron core is

$$l_{core} = 4 \times 6 - 0.5 = 23.5 \text{ cm}$$

The cross-sectional area of the core is

$$A_{core} = 2 \text{ cm} \times 3 \text{ cm} = 6 \times 10^{-4} \text{ m}^2$$

The permeability of the core is

$$\mu_{core} = \mu_r \mu_0 = 6000 \times 4\pi \times 10^{-7} = 7.540 \times 10^{-3}$$

Finally, the reluctance of the core is

$$\mathcal{R}_{core} = \frac{l_{core}}{\mu_{core} A_{core}} = \frac{23.5 \times 10^{-2}}{7.540 \times 10^{-3} \times 6 \times 10^{-4}}$$

$$= 5.195 \times 10^4 \text{ A} \cdot \text{turns/Wb}$$

Now, we compute the reluctance of the air gap. The flux lines tend to bow out in the air gap as shown in Figure 14.12(a). This is called **fringing**. *Thus, the effective area of the air gap is larger than that of the iron core. Customarily, we take this into account by adding the length of the gap to each of the dimensions of the air-gap cross section.* Thus, the effective area of the gap is

We approximately account for fringing by adding the length of the gap to the depth and width in computing effective gap area.

$$A_{gap} = (2 \text{ cm} + 0.5 \text{ cm}) \times (3 \text{ cm} + 0.5 \text{ cm}) = 8.75 \times 10^{-4} \text{ m}^2$$

The permeability of air is approximately the same as that of free space:

$$\mu_{gap} \cong \mu_0 = 4\pi \times 10^{-7}$$

Thus, the reluctance of the gap is

$$\mathcal{R}_{gap} = \frac{l_{gap}}{\mu_{gap} A_{gap}} = \frac{0.5 \times 10^{-2}}{4\pi \times 10^{-7} \times 8.75 \times 10^{-4}}$$

$$= 4.547 \times 10^6 \text{ A} \cdot \text{turns/Wb}$$

The total reluctance is the sum of the reluctance of the core and that of the gap:

$$\mathcal{R} = \mathcal{R}_{\text{gap}} + \mathcal{R}_{\text{core}} = 4.547 \times 10^6 + 5.195 \times 10^4 = 4.600 \times 10^6$$

Even though the gap is much shorter than the iron core, the reluctance of the gap is higher than that of the core because of the much higher permeability of the iron. Most of the magnetomotive force is dropped across the air gap. (This is analogous to the fact that the largest fraction of the applied voltage is dropped across the largest resistance in a series electrical circuit.)

Now, we can compute the flux:

$$\phi = B_{\text{gap}}A_{\text{gap}} = 0.25 \times 8.75 \times 10^{-4} = 2.188 \times 10^{-4} \text{ Wb}$$

The flux in the core is the same as that in the gap. However, the flux density is higher in the core, because the area is smaller. The magnetomotive force is given by

$$\mathcal{F} = \phi\mathcal{R} = 4.600 \times 10^6 \times 2.188 \times 10^{-4} = 1006 \text{ A} \cdot \text{turns}$$

According to Equation 14.20, we have

$$\mathcal{F} = Ni$$

Solving for the current and substituting values, we get

$$i = \frac{\mathcal{F}}{N} = \frac{1006}{500} = 2.012 \text{ A} \qquad \blacksquare$$

---

### Example 14.6   A Magnetic Circuit with Reluctances in Series and Parallel

The iron core shown in Figure 14.13(a) has a cross section of 2 cm by 2 cm and a relative permeability of 1000. The coil has 500 turns and carries a current of $i = 2$ A. Find the flux density in each air gap.

**Solution**   The magnetic circuit is depicted in Figure 14.13(b). First, we compute the reluctances of the three paths. For the center path, we have

$$\mathcal{R}_c = \frac{l_c}{\mu_r\mu_0 A_{\text{core}}} = \frac{10 \times 10^{-2}}{1000 \times 4\pi \times 10^{-7} \times 4 \times 10^{-4}}$$

$$= 1.989 \times 10^5 \text{ A} \cdot \text{turns/Wb}$$

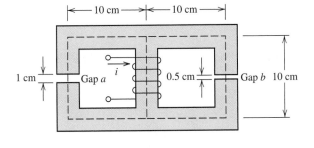

(a) Core

(b) Magnetic circuit

**Figure 14.13** Magnetic circuit of Example 14.6.

For the left-hand path, the total reluctance is the sum of the reluctance of the iron core plus the reluctance of gap $a$. We take fringing into account by adding the gap length to its width and depth in computing area of the gap. Thus, the area of gap $a$ is $A_a = 3\,\text{cm} \times 3\,\text{cm} = 9 \times 10^{-4}\,\text{m}^2$. Then, the total reluctance of the left-hand path is

$$\mathcal{R}_a = \mathcal{R}_{\text{gap}} + \mathcal{R}_{\text{core}}$$

$$= \frac{l_{\text{gap}}}{\mu_0 A_a} + \frac{l_{\text{core}}}{\mu_r \mu_0 A_{\text{core}}}$$

$$= \frac{1 \times 10^{-2}}{4\pi \times 10^{-7} \times 9 \times 10^{-4}} + \frac{29 \times 10^{-2}}{1000 \times 4\pi \times 10^{-7} \times 4 \times 10^{-4}}$$

$$= 8.842 \times 10^6 + 5.769 \times 10^5$$

$$= 9.420 \times 10^6 \,\text{A} \cdot \text{turns/Wb}$$

Similarly, the reluctance of the right-hand path is

$$\mathcal{R}_b = \mathcal{R}_{\text{gap}} + \mathcal{R}_{\text{core}}$$

$$= \frac{l_{\text{gap}}}{\mu_0 A_b} + \frac{l_{\text{core}}}{\mu_r \mu_0 A_{\text{core}}}$$

$$= \frac{0.5 \times 10^{-2}}{4\pi \times 10^{-7} \times 6.25 \times 10^{-4}} + \frac{29.5 \times 10^{-2}}{1000 \times 4\pi \times 10^{-7} \times 4 \times 10^{-4}}$$

$$= 6.366 \times 10^6 + 5.869 \times 10^5$$

$$= 6.953 \times 10^6 \,\text{A} \cdot \text{turns/Wb}$$

Next, we can combine the reluctances $\mathcal{R}_a$ and $\mathcal{R}_b$ in parallel. Then, the total reluctance is the sum of $\mathcal{R}_c$ and this parallel combination:

$$\mathcal{R}_{\text{total}} = \mathcal{R}_c + \frac{1}{1/\mathcal{R}_a + 1/\mathcal{R}_b}$$

$$= 1.989 \times 10^5 + \frac{1}{1/(9.420 \times 10^6) + 1/(6.953 \times 10^6)}$$

$$= 4.199 \times 10^6 \,\text{A} \cdot \text{turns/Wb}$$

Now, the flux in the center leg of the coil can be found by dividing the magnetomotive force by the total reluctance:

$$\phi_c = \frac{Ni}{\mathcal{R}_{\text{total}}} = \frac{500 \times 2}{4.199 \times 10^6} = 238.1 \,\mu\text{Wb}$$

Fluxes are analogous to currents. Thus, we use the current-division principle to determine the flux in the left-hand and right-hand paths, resulting in

$$\phi_a = \phi_c \frac{\mathcal{R}_b}{\mathcal{R}_a + \mathcal{R}_b}$$

$$= 238.1 \times 10^{-6} \times \frac{6.953 \times 10^6}{6.953 \times 10^6 + 9.420 \times 10^6}$$

$$= 101.1 \,\mu\text{Wb}$$

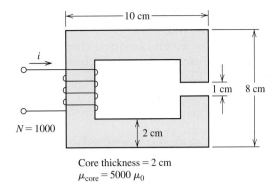

**Figure 14.14** Magnetic circuit of Exercise 14.9.

Core thickness = 2 cm
$\mu_{\text{core}} = 5000\,\mu_0$

Similarly, for gap $b$ we have

$$\phi_b = \phi_c \frac{\mathcal{R}_a}{\mathcal{R}_a + \mathcal{R}_b}$$

$$= 238.1 \times 10^{-6} \frac{9.420 \times 10^6}{6.953 \times 10^6 + 9.420 \times 10^6}$$

$$= 137.0\ \mu\text{Wb}$$

As a check on these calculations, we note that $\phi_c = \phi_a + \phi_b$.

Now, we find the flux densities in the gaps by dividing the fluxes by the areas:

$$B_a = \frac{\phi_a}{A_a} = \frac{101.1\ \mu\text{Wb}}{9 \times 10^{-4}\ \text{m}^2} = 0.1123\ \text{T}$$

$$B_b = \frac{\phi_b}{A_b} = \frac{137.0\ \mu\text{Wb}}{6.25 \times 10^{-4}\ \text{m}^2} = 0.2192\ \text{T} \qquad \blacksquare$$

Typically, we find that in magnetic circuits consisting of iron cores with air gaps, the reluctance of the iron has a negligible effect on the results. Furthermore, we usually do not have a precise value of the permeability for the iron. Thus, it is often sufficiently accurate to assume zero reluctance for the iron cores. This is the counterpart of assuming zero resistance for the wires in an electrical circuit.

**Exercise 14.9**  Consider the magnetic circuit shown in Figure 14.14. Determine the current required to establish a flux density of 0.5 T in the air gap.
**Answer**  $i = 4.03$ A.  ☐

**Exercise 14.10**  Repeat Example 14.6, taking the reluctance of the iron paths to be zero. Determine the error as a percentage of the flux densities computed in the example.
**Answer**  $\phi_a = 113.1\ \mu\text{Wb}$, $B_a = 0.1257$ T, 11.9 percent error; $\phi_b = 157.1\ \mu\text{Wb}$, $B_b = 0.2513$ T, 14.66 percent error.  ☐

## 14.3  INDUCTANCE AND MUTUAL INDUCTANCE

We have seen that when a coil carries current, a magnetic flux is produced that links the coil. If the current changes with time, the flux also changes, inducing a voltage in the coil. This is the physical basis of inductance that we introduced in Section 3.4.

Now, we relate inductance to the physical parameters of the coil and the core upon which it is wound.

Consider a coil carrying a current $i$ that sets up a flux $\phi$ linking the coil. The inductance of the coil can be defined as flux linkages divided by current:

$$L = \frac{\lambda}{i} \tag{14.23}$$

*Assuming that the flux is confined to the core so that all of the flux links all of the turns,* we can write $\lambda = N\phi$. Then, we have

$$L = \frac{N\phi}{i} \tag{14.24}$$

**Equation 14.25 is valid only if all of the flux links all of the turns.**

Substituting $\phi = Ni/\mathcal{R}$, we obtain

$$L = \frac{N^2}{\mathcal{R}} \tag{14.25}$$

Thus, we see that the inductance depends on the number of turns, the core dimensions, and the core material. Notice that inductance is proportional to the square of the number of turns.

According to Faraday's law, voltage is induced in a coil when its flux linkages change:

$$e = \frac{d\lambda}{dt} \tag{14.26}$$

Rearranging Equation 14.23, we have $\lambda = Li$. Substituting this for $\lambda$ in Equation 14.26, we get

$$e = \frac{d(Li)}{dt} \tag{14.27}$$

For a coil wound on a stationary core, the inductance is constant with time, and Equation 14.27 reduces to

$$e = L\frac{di}{dt} \tag{14.28}$$

Of course, this is the equation relating voltage and current that we used to analyze circuits containing inductance in Chapters 3 through 6.

---

**Example 14.7**    Calculation of Inductance

Determine the inductance of the 500-turn coil shown in Figure 14.12 and analyzed in Example 14.5.

**Solution**    In Example 14.5, we found that the reluctance of the magnetic path is

$$\mathcal{R} = 4.600 \times 10^6 \text{ A} \cdot \text{turns/Wb}$$

Substituting into Equation 14.25, we obtain

$$L = \frac{N^2}{\mathcal{R}} = \frac{500^2}{4.6 \times 10^6} = 54.35 \text{ mH} \qquad \blacksquare$$

## Mutual Inductance

When two coils are wound on the same core, some of the flux produced by one coil links the other coil. We denote the flux linkages of coil 2 caused by the current in coil 1 as $\lambda_{21}$. Correspondingly, the flux linkages of coil 1 produced by its own current are denoted as $\lambda_{11}$. Similarly, the current in coil 2 produces flux linkages $\lambda_{22}$ in coil 2 and $\lambda_{12}$ in coil 1.

The **self inductances** of the coils are defined as

$$L_1 = \frac{\lambda_{11}}{i_1} \qquad (14.29)$$

and

$$L_2 = \frac{\lambda_{22}}{i_2} \qquad (14.30)$$

The **mutual inductance** between the coils is

$$M = \frac{\lambda_{21}}{i_1} = \frac{\lambda_{12}}{i_2} \qquad (14.31)$$

The total fluxes linking the coils are

$$\lambda_1 = \lambda_{11} \pm \lambda_{12} \qquad (14.32)$$

and

$$\lambda_2 = \pm\lambda_{21} + \lambda_{22} \qquad (14.33)$$

where the $+$ sign applies if the fluxes are aiding and the $-$ sign applies if the fluxes are opposing.

## Dot Convention

It is standard practice to place a dot on one end of each coil in a circuit diagram to indicate how the fluxes interact. An example of this is shown in Figure 14.15. The dots are placed such that currents entering the dotted terminals produce aiding magnetic flux. Notice that (according to the right-hand rule) a current entering either of the dotted terminals in Figure 14.15 produces flux in a clockwise direction in the core. Thus, if both currents enter (or if both leave) the dotted terminals, the mutual flux linkages add to the self flux linkages. On the other hand, if one current enters a dotted terminal and the other leaves, the mutual flux linkages carry a minus sign.

Aiding fluxes are produced by currents entering like-marked terminals.

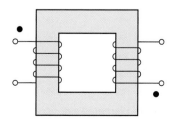

**Figure 14.15** According to convention, currents entering the dotted terminals produce aiding fluxes.

### Circuit Equations for Mutual Inductance

Solving Equations 14.29 through 14.31 for the flux linkages and substituting into Equations 14.32 and 14.33, we have

$$\lambda_1 = L_1 i_1 - M i_2 \tag{14.34}$$

and

$$\lambda_2 = \pm M i_1 + L_2 i_2 \tag{14.35}$$

Applying Faraday's law to find the voltages induced in the coils, we get

$$e_1 = \frac{d\lambda_1}{dt} = L_1 \frac{di_1}{dt} \pm M \frac{di_2}{dt} \tag{14.36}$$

and

$$e_2 = \frac{d\lambda_2}{dt} = \pm M \frac{di_1}{dt} + L_2 \frac{di_2}{dt} \tag{14.37}$$

Here again, we have assumed that the coils and core are stationary, so the inductances are constant with respect to time. These are the basic equations used to analyze circuits having mutual inductance.

---

**Example 14.8    Calculation of Inductance and Mutual Inductance**

Two coils are wound on a toroidal core as illustrated in Figure 14.16. The reluctance of the core is $10^7$ (ampere-turns)/Wb. Determine the self inductances and mutual inductance of the coils. Assume that the flux is confined to the core so that all of the flux links both coils.

**Solution**    The self inductances can be computed using Equation 14.25. For coil 1, we have

$$L_1 = \frac{N_1^2}{\mathcal{R}} = \frac{100^2}{10^7} = 1 \text{ mH}$$

Similarly, for coil 2 we get

$$L_2 = \frac{N_2^2}{\mathcal{R}} = \frac{200^2}{10^7} = 4 \text{ mH}$$

To compute the mutual inductance, we find the flux produced by $i_1$:

$$\phi_1 = \frac{N_1 i_1}{\mathcal{R}} = \frac{100 i_1}{10^7} = 10^{-5} i_1$$

The flux linkages of coil 2 resulting from the current in coil 1 are given by

$$\lambda_{21} = N_2 \phi_1 = 200 \times 10^{-5} i_1$$

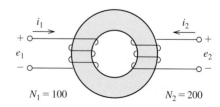

**Figure 14.16** Coils of Example 14.8.

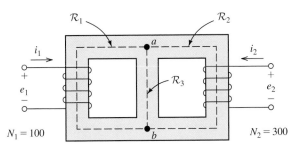

**Figure 14.17**  Magnetic circuit of Exercise 14.13.

Finally, the mutual inductance is

$$M = \frac{\lambda_{21}}{i_1} = 2 \text{ mH}$$

**Exercise 14.11**  Compute the mutual inductance in Example 14.8 by use of the formula $M = \lambda_{12}/i_2$.
**Answer**   $M = 2$ mH. Notice that we get the same value from $M = \lambda_{21}/i_1$ and from $M = \lambda_{12}/i_2$.  □

**Exercise 14.12**  Does the flux produced by $i_2$ aid or oppose the flux produced by $i_1$ for the coils shown in Figure 14.16? If a dot is placed on the top terminal of coil 1, which end of coil 2 should have a dot? Write the expressions for $e_1$ and $e_2$, taking care to select the proper sign for the mutual term.
**Answer**   The fluxes oppose one another. The dot should be on the bottom terminal of coil 2, so the correct expressions are

$$e_1 = L_1 \frac{di_1}{dt} - M \frac{di_2}{dt} \quad \text{and} \quad e_2 = -M \frac{di_1}{dt} + L_2 \frac{di_2}{dt}$$  □

**Exercise 14.13**  For the core shown in Figure 14.17, the reluctances of all three paths between points $a$ and $b$ are equal.

$$\mathcal{R}_1 = \mathcal{R}_2 = \mathcal{R}_3 = 10^6 \quad (\text{A} \cdot \text{turns})/\text{Wb}$$

Assume that all of the flux is confined to the core. **a.** Do the fluxes produced by $i_1$ and $i_2$ aid or oppose one another in path 1? In path 2? In path 3? If a dot is placed on the top end of coil 1, which end of coil 2 should carry a dot? **b.** Determine the values of $L_1$, $L_2$, and $M$. **c.** Should the mutual term for the voltages (in Equations 14.36 and 14.37) carry a plus sign or a minus sign?
**Answer**   **a.** Aid in paths 1 and 2, oppose in path 3; the dot should be on the top end of coil 2; **b.** $L_1 = 6.667$ mH, $L_2 = 60$ mH, $M = 10$ mH; **c.** a plus sign.  □

## 14.4  MAGNETIC MATERIALS

So far, we have assumed that the relationship between $B$ and $H$ is linear (i.e., $B = \mu H$). Actually, for the iron alloys used in motors, permanent magnets, and transformers, the relationship between $B$ and $H$ is not linear.

The relationship between $B$ and $H$ is not linear for the types of iron used in motors and transformers.

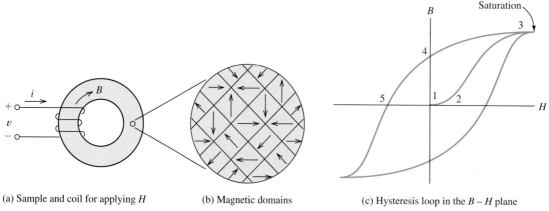

(a) Sample and coil for applying $H$          (b) Magnetic domains          (c) Hysteresis loop in the $B–H$ plane

**Figure 14.18** Materials such as iron display a $B–H$ relationship with hysteresis and saturation.

Figure 14.18(a) shows a coil used to apply magnetic field intensity $H$ to a sample of iron. Suppose that we start with a sample that is not magnetized. If we look at the material on a microscopic scale, we see that the magnetic fields of the atoms in small **domains** are aligned. However, the field directions are random for the various domains, and the external macroscopic field is zero. This is illustrated in Figure 14.18(b).

Figure 14.18(c) shows a plot of $B$ versus $H$. At point 1, both $B$ and $H$ are zero. As $H$ is increased by applying current to the coil, the magnetic fields of the domains tend to align with the applied field. At first (point 1 to point 2), this is a reversible process, so that if the applied field is reduced to zero, the domains return to their original random orientations. However, for greater applied field intensities, the domains align with the applied field such that they tend to maintain their alignment even if the applied field is reduced to zero (point 2 to point 3). Eventually, for sufficiently high fields, all of the domains are aligned with the applied field and the slope of the $B–H$ curve approaches $\mu_0$. We say that the material is **saturated**. For typical iron core materials, saturation occurs for $B$ in the range of 1 to 2 T.

If starting from point 3, the applied field $H$ is reduced to zero, a residual flux density $B$ remains in the core (point 4). This occurs because the magnetic domains continue to point in the direction imposed earlier by the applied field. If $H$ is increased in the reverse direction, $B$ is reduced to zero (point 5). Eventually, saturation occurs in the reverse direction. If an ac current is applied to the coil, a **hysteresis loop** is traced in the $B–H$ plane.

> For typical iron cores, saturation occurs for $B$ in the range from 1 to 2 T.

### Energy Considerations

Let us consider the energy flow to and from the coil shown in Figure 14.18(a). We assume that the coil has zero resistance. As the current is increasing, the increasing flux density induces a voltage, resulting in energy flow into the coil. The energy $W$ delivered to the coil is the integral of power. Thus, we get

$$ W = \int_0^t vi\, dt = \int_0^t N\frac{d\phi}{dt} i\, dt = \int_0^\phi Ni\, d\phi \tag{14.38}$$

Now $Ni = Hl$ and $d\phi = A\,dB$, where $l$ is the mean path length and $A$ is the cross-sectional area. Making these substitutions in the expression on the right-hand side of Equation 14.38, we have

$$W = \int_0^B AlH\,dB \tag{14.39}$$

However, the product of the cross-sectional area $A$ and the length of the core $l$ is the volume of the core. Dividing both sides of Equation 14.39 by the volume results in

$$W_v = \frac{W}{Al} = \int_0^B H\,dB \tag{14.40}$$

in which $W_v$ represents energy per unit volume of the core. As illustrated in Figure 14.19, the volumetric energy delivered to the coil is the area between the $B$–$H$ curve and the $B$-axis. Part of this energy is returned to the circuit when $H$ is reduced to zero, part of it remains stored in the residual field, and part of it is converted to heat in the process of magnetizing the core.

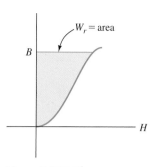

**Figure 14.19** The area between the $B$–$H$ curve and the $B$ axis represents the volumetric energy supplied to the core.

### Core Loss

When an ac current is applied to a coil having an iron core, more energy is put into the coil on each cycle than is returned to the circuit. Part of the energy is converted to heat in reversing the directions of the magnetic domains. This is similar to the heat produced when we repeatedly bend a piece of metal. The volumetric energy converted to heat per cycle is equal to the area of the hysteresis loop as illustrated in Figure 14.20. This energy loss is called **core loss**. Since a fixed amount of energy is converted to heat for each cycle, the power loss due to hysteresis is proportional to frequency.

In motors, generators, and transformers, conversion of energy into heat is undesirable. Therefore, we would choose an alloy having a thin hysteresis loop as in Figure 14.21(a). On the other hand, for a permanent magnet, we would choose a material having a large residual field, such as in Figure 14.21(b).

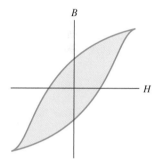

**Figure 14.20** The area of the hysteresis loop is the volumetric energy converted to heat per cycle.

### Eddy-Current Loss

Besides hysteresis, there is another effect that leads to core loss for ac operation. First, let us consider a solid iron core. Of course, the core itself is an electrical conductor, acting much like shorted turns. As the magnetic fields change, voltages are induced in

Power loss due to hysteresis is proportional to frequency, assuming constant peak flux.

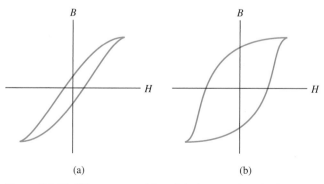

(a)                              (b)

**Figure 14.21** When we want to minimize core loss (as in a transformer or motor), we choose a material having a thin hysteresis loop. On the other hand, for a permanent magnet, we should choose a material with a wide loop.

the core, causing currents, known as **eddy currents**, to circulate in the core material. As a result, power is dissipated in the core according to $P = v^2/R$.

A partial solution to eddy-current loss is to laminate the core with thin sheets of iron that are electrically insulated from one another. The orientation of the sheets is selected to interrupt the flow of current. Thus, the resistance is higher for eddy currents, and the loss is greatly reduced. Another approach is to make the core with powdered iron held together by an insulating binder.

For operation with a given peak flux density, the voltages induced in the core are proportional to frequency (because of Faraday's law). Therefore, power loss due to eddy currents increases with the square of frequency (because $P = v^2/R$).

Power loss due to eddy currents is proportional to the square of frequency, assuming constant peak flux.

## Energy Stored in the Magnetic Field

Even though many core materials do not have a linear $B–H$ characteristic, we often perform initial design calculations assuming that $B = \mu H$. The properties of the core material are usually not accurately known, so the calculations for motor or transformer design are approximate. The linear approximation is convenient and sufficiently accurate as long as the cores are operated below the saturation level.

Substituting $H = B/\mu$ into Equation 14.40 and integrating, we obtain

$$W_v = \int_0^B \frac{B}{\mu}\, dB = \frac{B^2}{2\mu} \tag{14.41}$$

Notice that for a given flux density, the volumetric energy stored in the field is inversely proportional to the permeability.

In a magnetic circuit having an air gap, the flux density is roughly the same in the iron core as in the air gap. (It is usually a little less in the air gap, due to fringing.) The permeability of an iron core is much greater (by a factor of several thousand or more) than that of air. Thus, the volumetric energy of the gap is much higher than that of the core. *In a magnetic circuit consisting of an iron core with a substantial air gap, nearly all of the stored energy resides in the gap.*

**Exercise 14.14**　Consider a coil wound on an iron core. For 60-Hz ac operation with a given applied current, the hysteresis loop of the core material has an area of 40 J/m³. The core volume is 200 cm³. Find the power converted to heat because of hysteresis.
**Answer**　0.48 W. ☐

**Exercise 14.15**　A certain iron core has an air gap with an effective area of 2 cm by 3 cm and a length of 0.5 cm. The applied magnetomotive force is 1000 ampere turns and the reluctance of the iron is negligible. Find the flux density and the energy stored in the air gap.
**Answer**　$B = 0.2513$ T, $W = 0.0754$ J. ☐

## 14.5 IDEAL TRANSFORMERS

A transformer consists of several coils wound on a common core that usually consists of laminated iron (to reduce eddy-current loss). We will see that transformers can be used to adjust the values of ac voltages. A voltage can be *stepped up* by using a transformer. For example, 2400 V can be stepped up to 48 kV. Transformers can also be used to *step a voltage down*, such as 2400 V to 240 V.

Transformers find many applications in electric power distribution. In transporting power over long distances (from a hydroelectric power-generating station to a distant city, for example), it is desirable to use relatively large voltages, typically hundreds of kilovolts. Recall that the power delivered by an ac source is given by

$$P = V_{rms}I_{rms} \cos(\theta) \tag{14.42}$$

For a fixed power factor ($\cos \theta$), many combinations of voltage and current can be used in transferring a given amount of power. The wires that carry the current have nonzero resistances. Thus, some power is lost in the transmission lines, given by

$$P_{loss} = R_{line}I_{rms}^2 \tag{14.43}$$

in which $R_{line}$ is the resistance of the transmission line. By designing the power distribution system with a large voltage value and a small current value, the line loss can be made to be a small fraction of the power transported. Thus, larger voltage yields higher efficiency in power distribution.

For safety and other reasons, relatively small voltages must be employed where the power is consumed. For example, in U.S. residences, the nominal voltages are either 110 or 220 V rms. Thus, transformers are useful in stepping voltage levels up or down as needed in a power distribution system.

Transformers greatly facilitate power distribution by stepping voltage up and down at various points in the distribution system.

## Voltage Ratio

A transformer is illustrated in Figure 14.22. An ac voltage source is connected to the primary coil, which consists of $N_1$ turns of wire. Current flows into the primary side and causes an ac magnetic flux $\phi(t)$ to appear in the core. This flux induces a voltage in the $N_2$-turn secondary coil, which delivers power to the load. Depending on the turns ratio $N_2/N_1$, the rms secondary voltage can be greater or less than the rms primary voltage.

For now, we neglect the resistances of the coils and the core loss. Furthermore, we assume that the reluctance of the core is very small and that all of the flux links all of the turns of both coils.

The primary voltage is given by

$$v_1(t) = V_{1m} \cos(\omega t) \tag{14.44}$$

According to Faraday's law, we have

$$v_1(t) = V_{im} \cos(\omega t) = N_1 \frac{d\phi}{dt} \tag{14.45}$$

which can be rearranged and integrated to yield

$$\phi(t) = \frac{V_{1m}}{N_1 \omega} \sin(\omega t) \tag{14.46}$$

Assuming that all of the flux links all of the turns, the secondary voltage is given by

$$v_2(t) = N_2 \frac{d\phi}{dt} \tag{14.47}$$

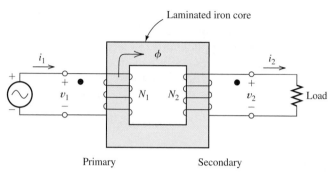

**Figure 14.22** A transformer consists of several coils wound on a common core.

Using Equation 14.46 to substitute for $\phi(t)$, we have

$$v_2(t) = N_2 \frac{V_{1m}}{N_1\omega} \frac{d}{dt}[\sin(\omega t)] \tag{14.48}$$

$$v_2(t) = \frac{N_2}{N_1} V_{1m} \cos(\omega t) \tag{14.49}$$

$$v_2(t) = \frac{N_2}{N_1} v_1(t) \tag{14.50}$$

In an ideal transformer, all of the flux links all of the turns, and the voltage across each coil is proportional to its number of turns.

Notice that the voltage across each coil is proportional to its number of turns. This is an important relationship to know when working with transformers.

Also, notice that we have included a dot on the end of each winding in Figure 14.22. As usual, the dots are placed so that if the currents entered the dotted terminals, they would produce aiding magnetic fields. Furthermore, application of Lenz's law shows that the induced voltages have positive polarity at both dotted terminals when $\phi$ is increasing, and negative polarity at both dotted terminals when $\phi$ is decreasing. *Thus in a transformer, the polarities of the voltages at the dotted terminals agree. When a voltage with positive polarity at the dotted terminal appears across coil 1, the voltage across coil 2 is also positive at the dotted terminal.*

Voltage polarities are the same at like-dotted terminals.

Hence, we have established the fact that the voltage across each winding is proportional to the number of turns. Clearly, the peak and rms values of the voltages are also related by the turns ratio:

$$V_{2\mathrm{rms}} = \frac{N_2}{N_1} V_{1\mathrm{rms}} \tag{14.51}$$

---

**Example 14.9    Determination of Required Turns Ratio**

Suppose that we have a 4700-V-rms ac source and we need to deliver 220 V rms to a load. Determine the turns ratio $N_1/N_2$ of the transformer needed.

**Solution**   Rearranging Equation 14.51, we have

$$\frac{N_1}{N_2} = \frac{V_{1\mathrm{rms}}}{V_{2\mathrm{rms}}} = \frac{4700}{220} = 21.36$$

■

## Current Ratio

Again let us consider the transformer shown in Figure 14.22. Notice that the currents $i_1$ and $i_2$ produce opposing magnetic fields (because $i_1$ enters a dotted terminal and $i_2$ leaves a dotted terminal). Thus, the total mmf applied to the core is

$$\mathcal{F} = N_1 i_1(t) - N_2 i_2(t) \tag{14.52}$$

Furthermore, the mmf is related to the flux and the reluctance of the core by

$$\mathcal{F} = \mathcal{R}\phi \tag{14.53}$$

In a well-designed transformer, the core reluctance is very small. Ideally, the reluctance is zero, and the mmf required to establish the flux in the core is zero. Then, Equation 14.52 becomes

The net mmf is zero for an ideal transformer.

$$\mathcal{F} = N_1 i_1(t) - N_2 i_2(t) = 0 \tag{14.54}$$

Rearranging this equation, we obtain

$$i_2(t) = \frac{N_1}{N_2} i_1(t) \tag{14.55}$$

This relationship also applies for the rms values of the currents:

$$I_{2\text{rms}} = \frac{N_1}{N_2} I_{1\text{rms}} \tag{14.56}$$

Compare Equation 14.51 for the voltages to Equation 14.56 for the currents. Notice that if the voltage is stepped up (i.e., $N_2/N_1 > 1$), the current is stepped down, and vice versa.

## Power in an Ideal Transformer

Again consider Figure 14.22. The power delivered to the load by the secondary winding is

$$p_2(t) = v_2(t)i_2(t) \tag{14.57}$$

Using Equations 14.50 and 14.55 to substitute for $v_2(t)$ and $i_2(t)$, respectively, we have

$$p_2(t) = \frac{N_2}{N_1} v_1(t) \frac{N_1}{N_2} i_1(t) = v_1(t)i_1(t) \tag{14.58}$$

However, the power delivered to the primary winding by the source is $p_1(t) = v_1(t)i_1(t)$, and we get

$$p_2(t) = p_1(t) \tag{14.59}$$

Thus, we have established the fact that the power delivered to the primary winding by the source is delivered in turn to the load by the secondary winding. *Net power is neither generated nor consumed by an ideal transformer.*

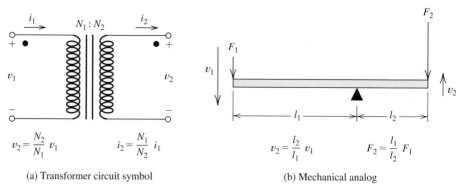

(a) Transformer circuit symbol                    (b) Mechanical analog

**Figure 14.23** The circuit symbol for a transformer and its mechanical analog.

Power is neither generated nor consumed by an ideal transformer.

**Summary.**    Let us summarize the idealizing assumptions and their consequences for the transformer.

1. We assumed that all of the flux links all of the windings of both coils and that the resistance of the coils is zero. Thus, the voltage across each coil is proportional to the number of turns on the coil. This led to the voltage relationship

$$v_2(t) = \frac{N_2}{N_1}v_1(t)$$

2. We assumed that the reluctance of the core is negligible, so the total mmf of both coils is zero. This led to the current relationship

$$i_2(t) = \frac{N_1}{N_2}i_1(t)$$

3. A consequence of the voltage and current relationships is that all of the power delivered to an ideal transformer by the source is transferred to the load. Thus, an ideal transformer has a power efficiency of 100 percent.

4. The circuit symbol for the transformer is shown in Figure 14.23(a).

## Mechanical Analog of the Transformer: The Lever

The lever illustrated in Figure 14.23(b) is a mechanical analog of the electrical transformer. The velocities of the ends of the lever are related by the length ratio of the lever, $v_2 = v_1(l_2/l_1)$, just as transformer voltages are related by the turns ratio. Similarly, the forces are related by $F_2 = F_1(l_1/l_2)$, which is analogous to the relationship between currents in the transformer. As in a transformer, the frictionless lever neither generates nor consumes energy. On one end of the lever, we have small force and large velocity, while on the other end the force is large and the velocity is small. This mirrors the effects of the transformer on current and voltage.

**Example 14.10**    Analysis of a Circuit Containing an Ideal Transformer

Consider the source, transformer, and load shown in Figure 14.24. Determine the rms values of the currents and voltages: **a.** with the switch open; **b.** with the switch closed.

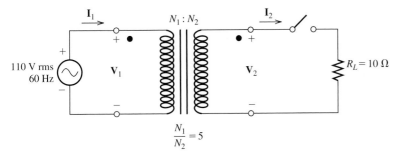

**Figure 14.24** Circuit of Example 14.10.

**Solution**   Because of the applied source, the primary voltage is $V_{1\text{rms}} = 110$ V. The primary and secondary voltages are related by Equation 14.51:

$$V_{2\text{rms}} = \frac{N_2}{N_1} V_{1\text{rms}} = \frac{1}{5} \times 110 = 22 \text{ V}$$

Rearranging Equation 14.56, we have

$$I_{1\text{rms}} = \frac{N_2}{N_1} I_{2\text{rms}}$$

**a.** With the switch open, the secondary current is zero. Therefore, the primary current $I_{1\text{rms}}$ is also zero, and no power is taken from the source.

**b.** With the switch closed, the secondary current is

$$I_{2\text{rms}} = \frac{V_{2\text{rms}}}{R_L} = \frac{22}{10} = 2.2 \text{ A}$$

Then, the primary current is

$$I_{1\text{rms}} = \frac{N_2}{N_1} I_{2\text{rms}} = \frac{1}{5} \times 2.2 = 0.44 \text{ A}$$

Let us consider the sequence of events in this example. When the source voltage is applied to the primary winding, a very small primary current (ideally zero) flows, setting up the flux in the core. The flux induces the voltage in the secondary winding. Before the switch is closed, no current flows in the secondary. After the switch is closed, current flows in the secondary opposing the flux in the core. However, because of the voltage applied to the primary, the flux must be maintained in the core. (Otherwise, Kirchhoff's voltage law would not be satisfied in the primary circuit.) Thus, current must begin to flow into the primary to offset the magnetomotive force of the secondary winding.                                              ∎

## Impedance Transformations

Consider the circuit shown in Figure 14.25. The phasor current and voltage in the secondary are related to the load impedance by

$$\frac{\mathbf{V}_2}{\mathbf{I}_2} = Z_L \qquad\qquad (14.60)$$

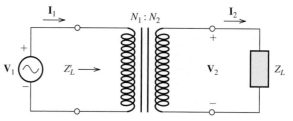

**Figure 14.25** The impedance seen looking into the primary is $Z'_L = (N_1/N_2)^2 \times Z_L$.

Using Equations 14.51 and 14.56 to substitute for $\mathbf{I}_2$ and $\mathbf{V}_2$, we have

$$\frac{(N_2/N_1)\mathbf{V}_1}{(N_1/N_2)\mathbf{I}_1} = Z_L \qquad (14.61)$$

Rearranging this, we get

$$Z'_L = \frac{\mathbf{V}_1}{\mathbf{I}_1} = \left(\frac{N_1}{N_2}\right)^2 Z_L \qquad (14.62)$$

in which $Z'_L$ is the impedance *seen* by the source. We say that the load impedance is *reflected* to the primary side by the square of the turns ratio.

---

### Example 14.11    Using Impedance Transformations

Consider the circuit shown in Figure 14.26(a). Find the phasor currents and voltages. Also, find the power delivered to the load.

**Solution**    First, we reflect the load impedance $Z_L$ to the primary side of the transformer as shown in Figure 14.26(b). The impedance seen from the primary side is

$$Z'_L = \left(\frac{N_1}{N_2}\right)^2 Z_L = (10)^2(10 + j20) = 1000 + j2000$$

The total impedance seen by the source is

$$Z_s = R_1 + Z'_L = 1000 + 1000 + j2000 = 2000 + j2000$$

Converting to polar form, we have

$$Z_s = 2828\underline{/45°}$$

Now, we can compute the primary current and voltage:

$$\mathbf{I}_1 = \frac{\mathbf{V}_s}{Z_s} = \frac{1000\underline{/0°}}{2828\underline{/45°}} = 0.3536\underline{/-45°} \text{ A peak}$$

$$\mathbf{V}_1 = \mathbf{I}_1 Z'_L = 0.3536\underline{/-45°} \times (1000 + j2000)$$

$$= 0.3536\underline{/-45°} \times (2236\underline{/63.43°}) = 790.6\underline{/18.43°} \text{ V peak}$$

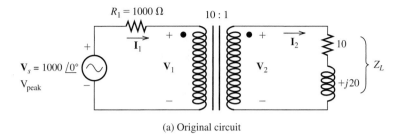

(a) Original circuit

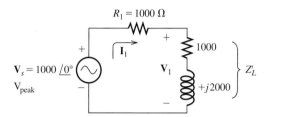

(b) Circuit with $Z_L$ reflected to the primary side

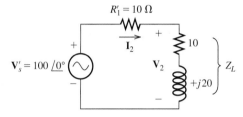

(c) Circuit with $V_s$ and $R_1$ reflected to the secondary side

**Figure 14.26**  The circuit of Examples 14.11 and 14.12.

Next, we can use the turns ratio to compute the secondary current and voltage:

$$\mathbf{I}_2 = \frac{N_1}{N_2}\mathbf{I}_1 = \frac{10}{1}0.3536\underline{/-45°} = 3.536\underline{/-45°} \text{ A peak}$$

$$\mathbf{V}_2 = \frac{N_2}{N_1}\mathbf{V}_1 = \frac{1}{10}790.6\underline{/18.43°} = 79.06\underline{/18.43°} \text{ V peak}$$

Finally, we compute the power delivered to the load:

$$P_L = I_{2\text{rms}}^2 R_L = \left(\frac{3.536}{\sqrt{2}}\right)^2 (10) = 62.51 \text{ W} \qquad \blacksquare$$

Besides transferring impedances from one side of a transformer to the other by using the square of the turns ratio, we can also reflect voltage sources or current sources by using the turns ratio.

---

**Example 14.12   Reflecting the Source to the Secondary**

Consider Figure 14.26(a). Reflect $V_s$ and $R_1$ to the secondary side.

**Solution**   The voltage is reflected by using the turns ratio. Thus, we have

$$\mathbf{V}_s' = \frac{N_2}{N_1}\mathbf{V}_s = \frac{1}{10}1000\underline{/0°} = 100\underline{/0°}$$

On the other hand, the resistance is reflected using the square of the turns ratio. This yields

$$R_1' = \left(\frac{N_2}{N_1}\right)^2 R_1 = \left(\frac{1}{10}\right)^2 (1000) = 10 \ \Omega$$

The circuit with $\mathbf{V}_s$ and $R_1$ transferred to the secondary side is shown in Figure 14.26(c). $\qquad \blacksquare$

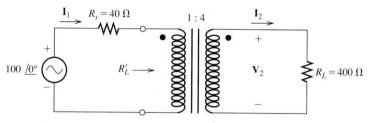

**Figure 14.27** Circuit of Exercises 14.16 and 14.17.

**Exercise 14.16** Working from the circuit of Figure 14.26(c), find the values of $\mathbf{V}_2$ and the power delivered to the load. (Of course, the answers should be the same as the values found in Example 14.11.)
**Answer** $\mathbf{V}_2 = 79.06\underline{/18.43°}$, $P_L = 62.51$ W.

**Exercise 14.17** Consider the circuit shown in Figure 14.27. Compute the values of $\mathbf{I}_1$, $\mathbf{I}_2$, $\mathbf{V}_2$, the power delivered to $R_L$, and $R_L'$.
**Answer** $\mathbf{I}_1 = 1.538\underline{/0°}$, $\mathbf{I}_2 = 0.3846\underline{/0°}$, $\mathbf{V}_2 = 153.8\underline{/0°}$, $P_L = 29.60$ W, $R_L' = 25$ Ω.

**Exercise 14.18** Recall that to achieve maximum power transfer from a source with an internal resistance of $R_s$, we want the effective load resistance $R_L'$ to equal $R_s$. Find the turns ratio that would result in maximum power delivered to the load in Figure 14.27.
**Answer** $N_1/N_2 = 1/\sqrt{10}$.

## 14.6 REAL TRANSFORMERS

Well-designed transformers approximately meet the conditions that we assumed in our discussion of the ideal transformer. Often for initial design calculations, we can assume that a transformer is ideal. However, a better model is needed for accurate calculations in the final stages of design. Moreover, a better understanding of transformers and their limitations is gained by considering a refined model.

The equivalent circuit of a real transformer is shown in Figure 14.28. The resistances $R_1$ and $R_2$ account for the resistance of the wires used to wind the coils of the transformer.

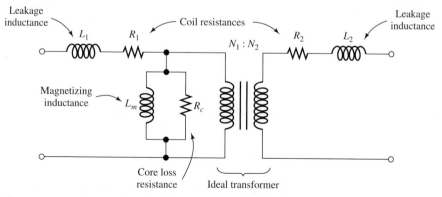

**Figure 14.28** The equivalent circuit of a real transformer.

**Table 14.1** Circuit Values of a 60-Hz 20-kVA 2400/240-V Transformer Compared with Those of an Ideal Transformer

| Element Name | Symbol | Ideal | Real |
|---|---|---|---|
| Primary resistance | $R_1$ | 0 | 3.0 Ω |
| Secondary resistance | $R_2$ | 0 | 0.03 Ω |
| Primary leakage reactance | $X_1 = \omega L_1$ | 0 | 6.5 Ω |
| Secondary leakage reactance | $X_2 = \omega L_2$ | 0 | 0.07 Ω |
| Magnetizing reactance | $X_m = \omega L_m$ | ∞ | 15 kΩ |
| Core-loss resistance | $R_c$ | ∞ | 100 kΩ |

For the ideal transformer, we assumed that all of the flux links all of the turns of both coils. In fact, some of the flux produced by each coil leaves the core and does not link the other coil. We account for this **leakage flux** by adding the inductances $L_1$ and $L_2$ to the ideal transformer, as shown in Figure 14.28.

In discussing the ideal transformer, we assumed that the core reluctance was zero and ignored core loss. This meant that zero magnetomotive force was required to establish the flux in the core. Neither of these assumptions is exactly true. The **magnetizing inductance** $L_m$ shown in Figure 14.28 accounts for the nonzero core reluctance. The current needed to establish the flux flows through $L_m$. Finally, the resistance $R_c$ accounts for power dissipated in the core due to hysteresis and eddy currents.

Table 14.1 compares the values of the circuit elements of a real transformer with those of an ideal transformer.

## Variations of the Transformer Model

Figure 14.29 shows several variations of the transformer equivalent circuit. In Figure 14.29(a), the secondary inductance and resistance have been referred to the primary side. In Figure 14.29(b), the magnetizing inductance and loss resistance have been moved to the input side of the circuit. [Actually, the circuit in Figure 14.29(b) is not precisely equivalent to that in Figure 14.29(a). However, in normal operation, the voltage drop across $L_1$ and $R_1$ is very small compared to either the input voltage or the voltage across $L_m$ and $R_m$. Thus, for normal operating conditions, virtually

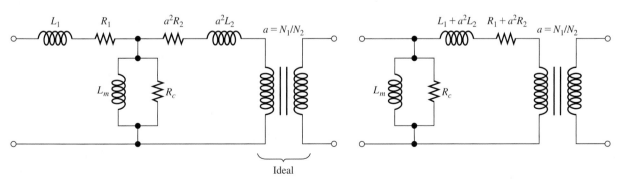

(a) All elements referred to the primary side

(b) Approximate equivalent circuit that is sometimes more convenient to use than that of part (a)

**Figure 14.29** Variations of the transformer equivalent circuit. The circuit of (b) is not exactly equivalent to that of (a), but is sufficiently accurate for practical applications.

identical results are obtained from either circuit.] Other equivalent circuits can be obtained by moving the circuit elements to the secondary side and by moving $L_m$ and $R_c$ to the right-hand side. Usually, we select the equivalent circuit configuration that is most convenient for the problem at hand.

## Regulation and Efficiency

Because of the elements $L_1$, $L_2$, $R_1$, and $R_2$, the voltage delivered to the load side of a transformer varies with the load current. Usually, this is an undesirable effect. The regulation of a transformer is defined as

$$\text{percent regulation} = \frac{V_{\text{no-load}} - V_{\text{load}}}{V_{\text{load}}} \times 100\%$$

in which $V_{\text{no-load}}$ is the rms voltage across the load terminals for an open-circuit load and $V_{\text{load}}$ is the rms voltage across the actual load.

Ideally, we usually want the percentage regulation to be zero. For instance, poor regulation in a residence would mean that the lights dim when an electric clothes dryer is started. Clearly, this is not a desirable situation.

Because of the resistances in the transformer equivalent circuit, not all of the power input to the transformer is delivered to the load. We define the power efficiency as

$$\text{power efficiency} = \frac{P_{\text{load}}}{P_{\text{in}}} \times 100\% = \left(1 - \frac{P_{\text{loss}}}{P_{\text{in}}}\right) \times 100\%$$

in which $P_{\text{load}}$ is the power delivered to the load, $P_{\text{loss}}$ is the power dissipated in the transformer, and $P_{\text{in}}$ is the power delivered by the source to the transformer primary terminals.

In this example, we are taking the rms values of currents and voltages (rather than peak values) as the phasor magnitudes. This is often done by power-distribution engineers. We will clearly indicate when phasors represent rms values rather than peak values.

### Example 14.13    Regulation and Efficiency Calculations

Find the percentage regulation and power efficiency for the transformer of Table 14.1 for a rated load having a lagging power factor of 0.8.

**Solution**    First, we draw the circuit as shown in Figure 14.30. Notice that we have placed the magnetizing reactance $X_m$ and core loss resistance $R_c$ on the left-hand

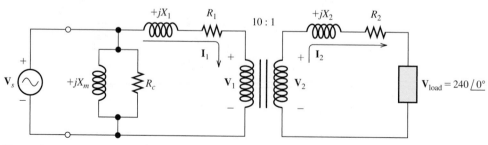

**Figure 14.30** Circuit of Example 14.13.

side of $R_1$ and $X_1$, because this makes the calculations a bit simpler and is sufficiently accurate. We assume a zero phase reference for the load voltage. It is customary in power-system engineering to take the values of phasors as the rms values (rather than the peak values) of the currents and voltages. Thus, as a phasor, we have

$$\mathbf{V}_{\text{load}} = 240 \underline{/0^\circ} \text{ V rms}$$

For rated load (20 kVA), the load current is

$$I_2 = \frac{20 \text{ kVA}}{240 \text{ V}} = 83.33 \text{ A rms}$$

The load power factor is

$$\text{power factor} = \cos(\theta) = 0.8$$

Solving, we find that

$$\theta = 36.87^\circ$$

Thus, the phasor load current is

$$\mathbf{I}_2 = 83.33 \underline{/-36.87^\circ} \text{ A rms}$$

where the phase angle is negative because the load was stated to have a lagging power factor.

The primary current is related to the secondary current by the turns ratio:

$$\mathbf{I}_1 = \frac{N_2}{N_1} \mathbf{I}_2 = \frac{1}{10} \times 83.33 \underline{/-36.87^\circ} = 8.333 \underline{/-36.87^\circ} \text{ A rms}$$

Next, we can compute the voltages:

$$\mathbf{V}_2 = \mathbf{V}_{\text{load}} + (R_2 + jX_2)\mathbf{I}_2$$
$$= 240 + (0.03 + j0.07)83.33 \underline{/-36.87^\circ}$$
$$= 240 + 6.346 \underline{/29.93^\circ}$$
$$= 245.50 + j3.166 \text{ V rms}$$

The primary voltage is related to the secondary voltage by the turns ratio:

$$\mathbf{V}_1 = \frac{N_1}{N_2} \mathbf{V}_2 = 10 \times (245.50 + j3.166)$$
$$= 2455.0 + j31.66 \text{ V rms}$$

Now, we can compute the source voltage:

$$\mathbf{V}_s = \mathbf{V}_1 + (R_1 + jX_1)\,\mathbf{I}_1$$
$$= 2455.0 + j31.66 + (3 + j6.5) \times (8.333\underline{/-36.87°})$$
$$= 2508.2\underline{/1.37°} \text{ V rms}$$

Next, we compute the power loss in the transformer:

$$P_{\text{loss}} = \frac{V_s^2}{R_c} + I_1^2 R_1 + I_2^2 R_2$$
$$= 62.91 + 208.3 + 208.3$$
$$= 479.5 \text{ W}$$

The power delivered to the load is given by

$$P_{\text{load}} = V_{\text{load}} I_2 \times \text{power factor}$$
$$= 20 \text{ kVA} \times 0.8 = 16{,}000 \text{ W}$$

The input power is given by

$$P_{\text{in}} = P_{\text{load}} + P_{\text{loss}}$$
$$= 16{,}000 + 479.5 = 16{,}479.5 \text{ W}$$

At this point, we can compute the power efficiency:

$$\text{efficiency} = \left(1 - \frac{P_{\text{loss}}}{P_{\text{in}}}\right) \times 100\%$$
$$= \left(1 - \frac{479.5}{16{,}479.5}\right) \times 100\% = 97.09\%$$

Next, we can determine the no-load voltages. Under no-load conditions, we have

$$I_1 = I_2 = 0$$
$$V_1 = V_s = 2508.2$$
$$V_{\text{no-load}} = V_2 = V_1 \frac{N_2}{N_1} = 250.82 \text{ V rms}$$

Finally, the percentage regulation is

$$\text{percent regulation} = \frac{V_{\text{no-load}} - V_{\text{load}}}{V_{\text{load}}} \times 100\%$$
$$= \frac{250.82 - 240}{240} \times 100\%$$
$$= 4.51\%$$

## Summary

1. The right-hand rule can be used to determine the direction of the magnetic field produced by a current. This is illustrated in Figure 14.2 on page 701.

2. Force is exerted on a charge moving through a magnetic field according to the equation

$$\mathbf{f} = q\mathbf{u} \times \mathbf{B}$$

Similarly, forces appear on a current-carrying wire immersed in a magnetic field. The force on an incremental length of wire is given by

$$d\mathbf{f} = id\mathbf{l} \times \mathbf{B}$$

3. According to Faraday's law of induction, voltage is induced in a coil when its magnetic flux linkages change with time. Similarly, voltages are induced in conductors that cut through magnetic flux lines. We can determine the polarity of the induced voltage by using Lenz's law.

4. Magnetic flux density B and the magnetic field intensity H are related by

$$\mathbf{B} = \mu\mathbf{H}$$

where $\mu$ is the magnetic permeability of the material. For air or vacuum, $\mu = \mu_0 = 4\pi \times 10^{-7}$.

5. According to Ampère's law, the line integral of H around a closed path is equal to the algebraic sum of the currents flowing through the area bounded by the path. We can use this law to find the field around a long straight wire or inside a toroidal coil.

6. Practical magnetic devices can be approximately analyzed by using circuit concepts. Magnetomotive forces are analogous to voltage sources, reluctance is analogous to resistance, and flux is analogous to current.

7. The inductance and mutual inductance of coils can be computed from knowledge of the physical parameters of the coils and the core on which they are wound.

8. The B–H relationship for iron takes the form of a hysteresis loop, which displays saturation in the neighborhood of 1 to 2T. The area of the loop represents energy converted to heat per cycle. Eddy currents are another cause of core loss. Energy can be stored in magnetic fields. In a magnetic circuit consisting of an iron core with an air gap, most of the energy is stored in the gap.

9. In an ideal transformer, the voltage across each coil is proportional to its number of turns, the net mmf is zero, and the power efficiency is 100 percent.

10. Equivalent circuits for real transformers are shown in Figures 14.28 and 14.29 on pages 728 and 729, respectively.

11. Efficiency and regulation are important aspects of transformer operation.

## Problems

### Section 14.1: Magnetic Fields

**P14.1.** What is the fundamental cause of magnetic fields?

**P14.2.** State Faraday's law of magnetic induction and Lenz's law.

**P14.3.** State Ampère's law, including the reference directions for the currents.

**P14.4.** State the right-hand rule as it applies to: **a.** a current-carrying conductor; **b.** a current-carrying coil.

*__P14.5.__ A bar magnet is inserted into a single-turn coil as illustrated in Figure P14.5. Is the voltage $v_{ab}$ positive or negative as the bar approaches the coil?

---

*Denotes that answers are contained in the Student Solutions files. See Appendix E for more information about accessing the Student Solutions.

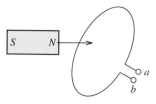

**Figure P14.5**

**\*P14.6.** The magnetic field of the earth is approximately $3 \times 10^{-5}$ T. At what distance from a long straight wire carrying a steady current of 10 A is the field equal to 10 percent of the earth's field? Suggest at least two ways to help reduce the effect of electrical circuits on the navigation compass in a boat or airplane.

**P14.7.** An irregular loop of wire carries an electrical current as illustrated in Figure P14.7. Is there a *net* force on the loop due to the magnetic fields created? Justify your answer. [*Hint:* Consider Newton's third law of motion.]

**Figure P14.7**

**\*P14.8.** A 0.5-m length of wire carries a 10-A current perpendicular to a magnetic field. Determine the magnetic flux density needed so that the force on the wire is 3 N.

**P14.9.** A long copper pipe carries a dc current. Is there a magnetic field inside the pipe due to the current? Outside? Justify your answers.

**\*P14.10.** Suppose that we test a material and find that $B = 0.1$ Wb/m$^2$ for an applied $H$ of 50 A/m. Compute the relative permeability of the material.

**P14.11.** Consider two coils which are wound on nonmagnetic forms such that part of the flux produced by each coil links the other, as shown in Figure P14.11. Assume that the inductance of the left-hand coil is small enough so that $i_1(t)$ is equal to the voltage induced by the magnetic field of the right-hand coil divided by the resistance. At $t = 1$ s, is the force between the coils

attraction, repulsion, or zero? Explain your reasoning. Repeat for $t = 2, 3$, and 5 s.

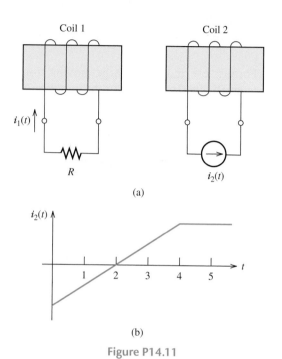

**Figure P14.11**

**\*P14.12.** A uniform flux density of 1 T is perpendicular to the plane of a five-turn circular coil of radius 10 cm. Find the flux linking the coil and the flux linkages. Suppose that the field is decreased to zero at a uniform rate in 1 ms. Find the magnitude of the voltage induced in the coil.

**P14.13.** Two very long parallel wires are 1 cm apart and carry currents of 10A in the same direction. The material surrounding the wires has $\mu_r = 1$. Determine the force on a 0.5-m section of one of the wires. Do the wires attract or repel one another?

**P14.14.** Suppose that the flux $\phi$ linking the coils shown in Figure P14.14 is increasing in magnitude. Find the polarity of the voltage across each coil.

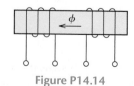

**Figure P14.14**

**P14.15.** Using equations given in Section 14.1, perform dimensional analyses to determine the units of $\mu$, **B**, and **H** in terms of meters, kilograms, seconds, and coulombs.

**P14.16.** Use the right-hand rule to find the direction of the magnetic flux for each coil shown in Figure P14.16. Mark the $N$ and $S$ ends of each coil. Do the coils attract or repel one another?

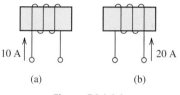

(a)                    (b)

Figure P14.16

**P14.17.** A very long, straight wire carrying current $i(t)$ and a rectangular single-turn coil lie in the same plane, as illustrated in Figure P14.17. The wire and the coil are surrounded by air. **a.** Derive an expression for the flux linking the coil. **b.** Derive an expression for the voltage $v_{ab}(t)$ induced in the coil. **c.** Determine the rms value of $v_{ab}$, given that $i(t)$ is a 10-A rms 60-Hz sinusoid, $l = 10$ cm, $r_1 = 1$ cm, and $r_2 = 10$ cm.

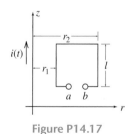

Figure P14.17

**P14.18.** A 120-V-rms 60-Hz sinusoidal voltage appears across a 500-turn coil. Determine the peak and rms values of the flux linking the coil.

**P14.19.** A uniform flux density given by $B = 0.3 \sin(377t)$ T is perpendicular to the plane of a 1000-turn circular coil of radius 20 cm. Find the flux linkages and the voltage as functions of time.

**P14.20.** A 200-turn toroidal coil (see Figure 14.8 on page 706) has $r = 1$ cm and $R = 10$ cm.

When a current given by $0.05 \sin(200t)$ A flows in the coil, the voltage is $0.5 \cos(200t)$ A. Determine the flux $\phi$ as a function of time and the relative permeability of the core material.

**P14.21.** Suppose that, in designing an electrical generator, we need to produce a voltage of 120 V by moving a straight conductor through a uniform magnetic field of 0.5 T at a speed of 30 m/s. The conductor, its motion, and the field are mutually perpendicular. What is the required length of the conductor? It turns out that in generator design, a conductor of this length is impractical, and we must use $N$ conductors of length 0.1 m. However, by connecting the conductors in series, we can obtain the required 120 V. What is the number $N$ of conductors needed?

**P14.22.** A very long, straight wire carrying a constant current $i(t) = I_1$ and a rectangular single-turn coil lie in the same plane, as illustrated in Figure P14.17. The wire and the coil are surrounded by air. A source is applied to the coil causing a constant clockwise current $I_2$ to flow in the loop. **a.** Derive an expression for the net force exerted on the coil due to the magnetic field of the wire. **b.** Evaluate the force given that $I_1 = I_2 = 10$ A, $l = 10$ cm, $r_1 = 1$ cm, and $r_2 = 10$ cm. **c.** Is the loop attracted or repelled by the wire?

**P14.23.** Two infinitely long, very thin wires lie on the $x$ and $y$ axes and carry currents as shown in Figure P14.23. **a.** Show the direction of the forces on the wires due to the magnetic fields on the positive and negative part of each axis, assuming that $I_x$ and $I_y$ are both positive. **b.** Compute the torque on the wire that lies on the $y$ axis.

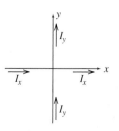

Figure P14.23

## Section 14.2: Magnetic Circuits

*P14.24. An air gap has a length of 0.1 cm. What length of iron core has the same reluctance as the air gap? The relative permeability of the iron is 5000. Assume that the cross-sectional areas of the gap and the core are the same.

*P14.25. Consider the magnetic circuit of Figure 14.13 on page 711 that was analyzed in Example 14.6. Suppose that the length of gap $a$ is reduced to zero. Compute the flux in gap $b$. Why is the result less than that found in Example 14.6?

P14.26. What happens to the reluctance of a magnetic path if its length is doubled? If the cross-sectional area is doubled? If the relative permeability is doubled?

*P14.27. What quantity in a magnetic circuit is analogous to a voltage source in an electrical circuit? To resistance? To current?

P14.28. What are the physical units of reluctance in terms of kilograms, coulombs, meters, and seconds?

P14.29. Consider the magnetic circuit shown in Figure P14.29. Assume that the reluctance of the iron is small enough so it can be neglected. The lengths of the air gaps are 0.1 cm, and the effective area of each gap is 20 cm². Determine the total number of turns needed to produce a flux density of 0.5 T in the gaps.

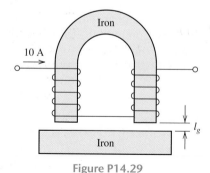

Figure P14.29

P14.30. Compute the flux in each leg of the magnetic core shown in Figure P14.30.

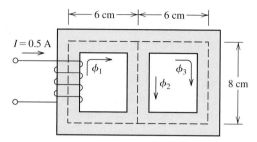

$N = 1000$ turns
Core cross-section: 2 cm × 2 cm
$\mu_r = 5000$

Figure P14.30

*P14.31. Consider the *solenoid* shown in Figure P14.31, which is typical of those commonly used as actuators for mechanisms and for operating valves in chemical processes. Neglect fringing and the reluctance of the core. Derive an expression for the flux as a function of the physical dimensions, $\mu_0$, the number of turns $N$, and the current.

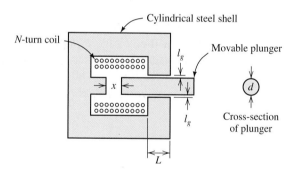

Figure P14.31

P14.32. Consider the core shown in Figure P14.32, which has two coils of $N$ turns, each connected so that the fluxes aid in the center leg. Determine the value of $N$ so that

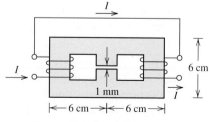

Cross-section: 2 cm × 2 cm square
$\mu_r = 2500$

Figure P14.32

$I = 2$ A produces a flux density of 0.25 T in the gap. The gap and the core have square cross sections 2 cm on each side. Account for fringing by adding the gap length to the length of each side of the gap.

**P14.33.** Compute the flux in each leg of the core shown in Figure P14.33. Account for fringing by adding the gap length to each of the cross-sectional dimensions of the gap.

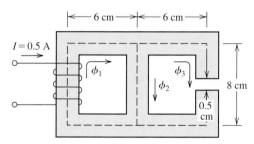

$N = 1000$ turns
Core cross-section: 2 cm $\times$ 2 cm
$\mu_r = 5000$

Figure P14.33

**P14.34.** Draw the electrical circuit analog for the magnetic circuit shown in Figure P14.34. Pay special attention to the polarities of the voltage sources. Determine the flux density in the core.

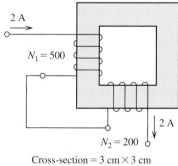

Cross-section = 3 cm $\times$ 3 cm
Path length = 36 cm
$\mu_r = 1000$

Figure P14.34

**P14.35.** Consider a toroidal core, as shown in Figure 14.11 on page 709, that has a relative permeability of 1000, $R = 5$ cm, and $r = 2$ cm. Two windings are wound on the core, one with 200 turns and the other with

400 turns. A voltage source described by $v(t) = 10 \cos(10^5 t)$ is applied to the 200-turn coil, and the 400-turn coil is open circuited. Determine the current in the 200-turn coil and the voltage across the 400-turn coil. Assume that all of the magnetic flux is confined to the core and that the current is a pure sine wave.

**Section 14.3:** **Inductance and Mutual Inductance**

*$\ast$**P14.36.** Consider the circuit shown in Figure P14.36. The two coils have $L_1 = 0.1$ H, $L_2 = 10$ H, and $M = 0.5$ H. Prior to $t = 0$, the currents in the coils are zero. At $t = 0$, the switch closes. Determine and sketch $i_1(t)$ and $i_2(t)$ to scale versus time.

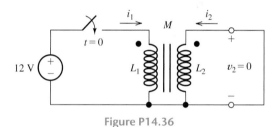

Figure P14.36

*$\ast$**P14.37.** A 500-turn coil is wound on an iron core. When a 120-V-rms 60-Hz voltage is applied to the coil, the current is 1 A rms. Neglect the resistance of the coil. Determine the reluctance of the core. Given that the cross-sectional area of the core is 5 cm$^2$ and the length is 20 cm, determine the relative permeability of the core material.

*$\ast$**P14.38.** A 100-turn coil wound on a magnetic core is found to have an inductance of 200 mH. What inductance will be obtained if the number of turns is increased to 200, assuming that all of the flux links all of the turns?

*$\ast$**P14.39.** Two coils wound on a common core have $L_1 = 1$ H, $L_2 = 2$ H, and $M = 0.5$ H. The currents are $i_1 = \cos(377t)$ A and $i_2 = 0.5 \cos(377t)$ A. Both of the currents enter dotted terminals. Find expressions for the voltages across the coils.

**P14.40.** Write one or two paragraphs that explain the voltage–current relationship of an inductor in terms of basic principles of magnetic fields.

**P14.41.** Two coils having inductances $L_1$ and $L_2$ are wound on a common core. The fraction of the flux produced by one coil that links the other coil is called the *coefficient of coupling* and is denoted by $k$. Derive an expression for the mutual inductance $M$ in terms of $L_1$, $L_2$, and $k$.

**P14.42.** Two coils wound on a common core have $L_1 = 0.2$ H, $L_2 = 0.5$ H, and $M = 0.1$ H. The currents are $i_1 = \exp(-1000t)$ A and $i_2 = 2 \exp(-1000t)$ A. Both of the currents enter dotted terminals. Find expressions for the voltages across the coils.

**P14.43.** A 200-turn coil is wound on a core having a reluctance of $5 \times 10^5$ A · turns/Wb. Determine the inductance of the coil.

**P14.44.** Consider the circuit shown in Figure P14.44. The two coils have $L_1 = 0.1$ H, $L_2 = 10$ H, and $M = 1$ H. Prior to $t = 0$, the currents in the coils are zero. At $t = 0$, the switch closes. Determine and sketch $i_1(t)$ and $v_2(t)$ to scale versus time.

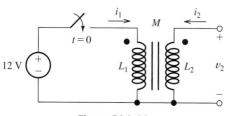

Figure P14.44

**P14.45.** Two coils wound on a common core have $L_1 = 1$ H, $L_2 = 2$ H, and $M = 0.5$ H. The currents are $i_1 = 1$ A and $i_2 = 0.5$ A. If both currents enter dotted terminals, find the flux linkages of both coils. Repeat if $i_1$ enters a dotted terminal and $i_2$ leaves a dotted terminal.

**P14.46.** Consider the coils shown in Figure P14.14. Suppose that a dot is placed on the leftmost terminal. Place a dot on the appropriate terminal of the right-hand coil to indicate the sense of the coupling.

**P14.47.** A 100-turn coil is wound on a toroidal core and has an inductance of 100 mH. Suppose

that a 200-turn coil is wound on a second toroidal core having dimensions ($r$ and $R$ as shown in Figure 14.11 on page 709) that are double those of the first core.. Both cores are made of the same material, which has very high permeability. Determine the inductance of the second coil.

**P14.48.** A symmetrical toroidal coil is wound on a plastic core ($\mu_r \cong 1$) and is found to have an inductance of 1 mH. What inductance will result if the core material is changed to a ferrite having $\mu_r = 200$? Assume that the entire magnetic path is composed of ferrite.

**P14.49.** A relay has a 500-turn coil that draws 50 mA rms when a 60-Hz voltage of 24 V rms is applied. Assume that the resistance of the coil is negligible. Determine the peak flux linking the coil, the reluctance of the core, and the inductance of the coil.

**Section 14.4: Magnetic Materials**

*P14.50. What are two causes of core loss for a coil with an iron core excited by an ac current? What considerations are important in minimizing loss due to each of these causes? What happens to the power loss in each case if the frequency of operation is doubled while maintaining constant peak flux density?

*P14.51. For operation at 60 Hz and a given peak flux density, the core loss of a given core is 1 W due to hysteresis and 0.5 W due to eddy currents. Estimate the core loss for 400-Hz operation with the same peak flux density.

**P14.52.** Sketch the $B$–$H$ curve for a magnetic material such as iron. Show hysteresis and saturation.

**P14.53.** What characteristic is desirable in the $B$–$H$ curve for a prospective material to be used in a permanent magnet? In a motor or transformer? Explain.

**P14.54.** Consider a coil wound on an iron core. Suppose that for operation with a given 60-Hz ac current, the hysteresis loop of the core material takes the form of a rectangle, as shown in Figure P14.54. The core volume is 1000 cm$^3$. Find the power converted to heat because of hysteresis.

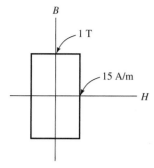

Figure P14.54

**P14.55.** A magnetic core has a mean length of 20 cm, a cross-sectional area of 4 cm², and a relative permeability of 2000. A 500-turn coil wound on the core carries a dc current of 0.1 A. **a.** Determine the reluctance of the core, the flux density in the core, and the inductance. **b.** Compute the energy stored in the magnetic field as $W = (1/2)LI^2$. **c.** Use Equation 14.41 on page 720 to compute the energy density in the core. Then, compute the energy by taking the product of energy density and volume. Compare to the value found in part (b).

**P14.56.** At a frequency of 60 Hz, the core loss of a certain coil with an iron core is 1.8 W, and at a frequency of 120 Hz, it is 5.6 W. The peak flux density is the same for both cases. Determine the power loss due to hysteresis and that due to eddy currents for 60-Hz operation.

**P14.57.** A certain iron core has an air gap with an effective area of 2 cm × 3 cm and a length $l_g$. The applied magnetomotive force is 1000 A · turns, and the reluctance of the iron is negligible. Find the flux density and the energy stored in the air gap as a function of $l_g$.

**Section 14.5: Ideal Transformers**

**P14.58.** What assumptions did we make in deriving the relationships between the voltages and currents in an ideal transformer?

*__P14.59.__ Consider the transformer having three windings, as shown in Figure P14.59. **a.** Place dots on windings to indicate the sense of the coupling between coil 1 and coil 2; between

coil 1 and coil 3. **b.** Assuming that all of the flux links all of the turns, determine the voltages $V_2$ and $V_3$. **c.** Assuming that the net mmf required to establish the core flux is zero, find an expression for $I_1$ in terms of $I_2$, $I_3$, and the turns ratios. Then, compute the value of $I_1$.

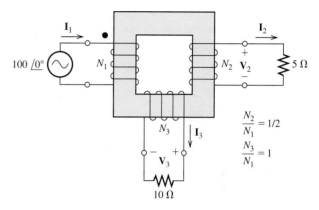

Figure P14.59

*__P14.60.__ Suppose that we need to cause a 25-Ω load resistance to appear as a 100-Ω resistance to the source. Instead of using a transformer, we could place a 75-Ω resistance in series with the 25-Ω resistance. From the standpoint of power efficiency, which approach is better? Explain.

*__P14.61.__ In U.S. residences, electrical power is generally utilized at a nominal voltage of 120 V rms. What problems would become pronounced if the power distribution system and household appliances had been designed for a lower voltage (say, 12 V rms)? For a higher voltage (say, 12 kV)?

**P14.62.** A voltage source $V_s$ is to be connected to a resistive load $R_L = 10$ Ω by a transmission line having a resistance $R_{line} = 10$ Ω, as shown in Figure P14.62. In part (a) of the figure, no transformers are used. In part (b) of the figure, one transformer is used to step up the source voltage at the sending end of the line, and another transformer is used to step the voltage back down at the load. For each case, determine the power delivered by the source; the power dissipated in the line resistance; the power delivered to the load; and the efficiency, defined as the power delivered to the load as a percentage of the source power.

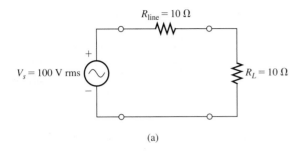

(a)

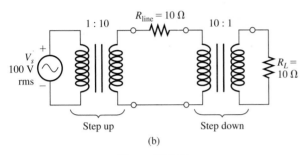

Step up                    Step down

(b)

Figure P14.62

**P14.63.** A transformer is needed that will cause an actual load resistance of 25 Ω to appear as 100 Ω to an ac voltage source of 240 V rms. Draw the diagram of the circuit required. What turns ratio is required for the transformer? Find the current taken from the source, the current flowing through the load, and the load voltage.

**P14.64.** Consider the circuit shown in Figure P14.64. Find the secondary voltage $V_{2rms}$, the secondary current $I_{2rms}$, and the power delivered to the load if the turns ratio is $N_1/N_2 = 10$. Repeat for $N_1/N_2 = 1$ and for $N_1/N_2 = 0.1$.

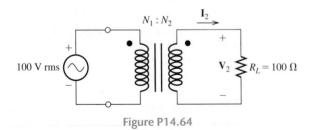

Figure P14.64

**P14.65.** A type of transformer known as an *autotransformer* is shown in Figure P14.65. **a.** Assuming that all of the flux links all of the turns, determine the relationship between

$v_1$, $v_2$, and the number of turns. **b.** Assuming that the total mmf required to establish the flux is zero, find the relationship between the currents $i_1$ and $i_2$.

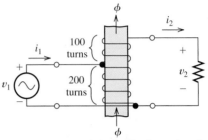

(*Note*: The return path for the flux is not shown)

(a) Auto transformer

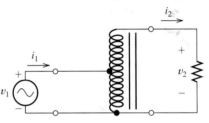

(b) Circuit symbol for the auto transformer

Figure P14.65

**P14.66. a.** Reflect the resistances and voltage sources to the left-hand side of the circuit shown in Figure P14.66 and solve for the current $\mathbf{I}_1$. **b.** Repeat with the dot moved to the top of the right-hand coil.

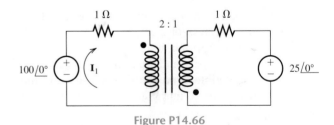

Figure P14.66

**P14.67.** Consider the circuit shown in Figure P14.67. **a.** Determine the values of $\mathbf{I}_1$ and $\mathbf{V}_2$. **b.** For each of the sources, determine the average power and state whether power is delivered by or absorbed by the source. **c.** Move the dot on the secondary to the bottom end of the coil and repeat parts (a) and (b).

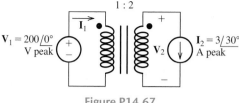

Figure P14.67

**P14.68.** An autotransformer is shown in Figure P14.68. Assume that all of the flux links all of the turns and that negligible mmf is needed to establish the flux. Determine the values of $I_1$, $I_2$, $I_3$, and $V_2$.

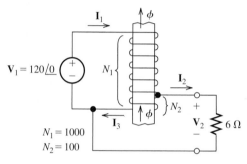

Figure P14.68

**P14.69.** Find the equivalent resistance $R'_L$ and capacitance $C'_L$ seen looking into the transformers in Figure P14.69. [*Hint:* Keep in mind that it is the *impedance* that is reflected by the square of the turns ratio.]

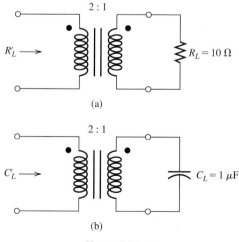

(a)

(b)

Figure P14.69

## Section 14.6: Real Transformers

*P14.70. Draw an equivalent circuit for a real transformer. Briefly discuss the reason that each element appears in the equivalent circuit.

*P14.71. A 60-Hz 20-kVA 8000/240-V-rms transformer has the following equivalent-circuit parameters:

| | | |
|---|---|---|
| Primary resistance | $R_1$ | 15 Ω |
| Secondary resistance | $R_2$ | 0.02 Ω |
| Primary leakage reactance | $X_1 = \omega L_1$ | 120 Ω |
| Secondary leakage reactance | $X_2 = \omega L_2$ | 0.15 Ω |
| Magnetizing reactance | $X_m = \omega L_m$ | 30 kΩ |
| Core-loss resistance | $R_c$ | 200 kΩ |

Find the percentage regulation and power efficiency for the transformer for a 2-kVA load (i.e., 10 percent of rated capacity) having a lagging power factor of 0.8.

*P14.72. Usually, transformers are designed to operate with peak flux densities just below saturation of the core material. Why would we not want to design them to operate far below the saturation point? Far above the saturation point? Assume that the voltage and current ratings are to remain constant.

**P14.73.** A certain residence is supplied with electrical power by the transformer of Table 14.1 on page 729. The residence uses 400 kWh of electrical energy per month. From the standpoint of energy efficiency, which of the equivalent circuit elements listed in the table are most significant? You will need to use good judgment and make some assumptions in obtaining an answer.

**P14.74.** When operating with an open-circuit load and with rated primary voltage, a certain 60-Hz 20-kVA 8000/240-V-rms transformer has a primary current of 0.315 A rms and absorbs 360 W. Which of the elements of the equivalent circuit of Figure 14.29(b) on page 729 can be determined from this data? Find the numerical values of these elements.

**P14.75.** Under the assumptions that we made for the ideal transformer, we determined that $v_2 = (N_2/N_1)v_1$. Theoretically, if a dc voltage is applied to the primary of an ideal transformer, a dc voltage should appear across the secondary winding. However, real transformers are ineffective for dc. Use the equivalent circuit of Figure 14.28 on page 728 to explain.

**P14.76.** A certain 60-Hz, 20-kVA, 8000/240 V-rms transformer is operated with a short-circuited secondary and reduced primary voltage. It is found that, for an applied primary voltage of 500 V rms, the primary current is 2.5 A rms (i.e., this is the rated primary current) and the transformer absorbs 270 W. Consider the equivalent circuit of Figure 14.29(b) on page 729. Under the stated conditions, the current and power for $L_m$ and $R_c$ can be neglected. Explain why. Determine the values of the total leakage inductance referred to the primary $(L_1 + a^2 L_2)$ and the total resistance referred to the primary $(R_1 + a^2 R_2)$.

**P14.77.** A 60-Hz 20-kVA 8000/240-V-rms transformer has the following equivalent-circuit parameters:

| | | |
|---|---|---|
| Primary resistance | $R_1$ | 15 Ω |
| Secondary resistance | $R_2$ | 0.02 Ω |
| Primary leakage reactance | $X_1 = \omega L_1$ | 120 Ω |
| Secondary leakage reactance | $X_2 = \omega L_2$ | 0.15 Ω |
| Magnetizing reactance | $X_m = \omega L_m$ | 30 kΩ |
| Core-loss resistance | $R_c$ | 200 kΩ |

Find the percentage regulation and power efficiency for the transformer for a rated load having a lagging power factor of 0.8.

**P14.78.** We have a transformer designed to operate at 60 Hz. The voltage ratings are 4800 V rms and 240 V rms for the primary and secondary windings, respectively. The transformer is rated for 10 kVA. Now, we want to use this transformer at 120 Hz. Discuss the factors that must be considered in setting ratings appropriate for operation at the new frequency. (Keep in mind that for best utilization of the material in the transformer, we want the peak flux density to be nearly at saturation for both frequencies.)

## Practice Test

Here is a practice test you can use to check your comprehension of the most important concepts in this chapter. Answers can be found in Appendix D and complete solutions are included in the Student Solutions files. See Appendix E for more information about the Student Solutions.

**T14.1.** Consider a right-hand Cartesian coordinate system as shown in Figure 14.3 on page 701. We have a wire along the $x$-axis carrying 12 A in the positive $x$ direction and a constant flux density of 0.3 T directed in the positive $z$ direction. **a.** Determine the force and its direction on a 0.2 m length of the wire. **b.** Repeat if the field is directed in the positive $x$ direction.

**T14.2.** Suppose we have a ten-turn square coil 25 cm on each side lying in the $x$-$y$ plane. A magnetic flux density is directed in the positive $z$ direction and is given by $0.7 \sin(120\pi t)$ T. The flux is constant with respect to $x$, $y$, and $z$. Determine the voltage induced in the coil.

**T14.3.** A 20-cm length of wire moves at 15 m/s in a constant flux density of 0.4 T. The wire, direction of motion, and direction of the flux are mutually perpendicular. Determine the voltage induced in the wire segment.

**T14.4.** Consider the magnetic circuit shown in Figure T14.4. The core has a relative permeability of

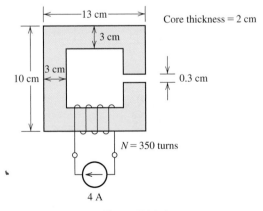

**Figure T14.4**

1500. **a.** Carefully estimate the flux density in the air gap. **b.** Determine the inductance of the coil.

**T14.5.** Suppose we have an ac current flowing through a coil wound on an iron core. Name two mechanisms by which energy is converted to heat in the core material. For each, how is the core material selected to minimize the power loss? How does each power loss depend on the frequency of the ac current?

**T14.6.** Consider the circuit shown in Figure T14.6 which has $R_s = 0.5\ \Omega$, $R_L = 1000\ \Omega$, and $N_1/N_2 = 0.1$. **a.** Determine the rms values

of the currents and voltages with the switch open. **b.** Repeat with the switch closed.

**T14.7.** You have been assigned to select a transformer to supply a peak power of 100 kW to a load that draws peak power only a very small percentage of the time and draws very little power the rest of the time. Two transformers, $A$ and $B$, are both suitable. While both transformers have the same efficiency at peak load, most of the loss in $A$ is due to core loss, and most of the loss in $B$ is due to the resistances of the coils. From the standpoint of operating costs, which transformer is better? Why?

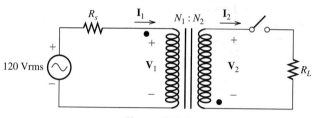

**Figure T14.6**

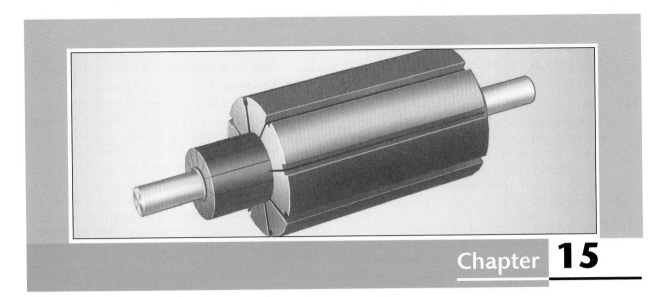

# Chapter 15

# DC Machines

## Study of this chapter will enable you to:

- Select the proper motor type for various applications.
- State how torque varies with speed for various motors.
- Use the equivalent circuit for dc motors to compute electrical and mechanical quantities.

- Use motor nameplate data.
- Understand the operation and characteristics of shunt-connected dc motors, series-connected dc motors, and universal motors.

## Introduction to this chapter:

In this chapter and the next, we consider machines that convert mechanical energy to and from electrical energy. **Motors** convert electrical energy into rotational mechanical energy. Conversely, **generators** convert mechanical energy into electrical energy. Most electrical machines can be used either as motors or as generators.

Electrical motors are used to power hundreds of the devices that we use in everyday life, such as computer disks, refrigerators, garage-door openers, washing machines, food mixers, vacuum cleaners, DVD players, ventilation fans, automotive power windows, windshield wipers, elevators, and so on. Industrial applications include bulk material handling, machining operations, pumps, rock crushers, fans, compressors, and hoists. Electrical motors are the best choice in the vast majority of stationary applications where mechanical energy is needed, whether it is a tiny fraction of a horsepower or thousands of horsepower. It is important for designers of mechanical systems that employ motors to have a good understanding of the external characteristics of various motors so they can choose the proper types to power their systems.

## 15.1   OVERVIEW OF MOTORS

We will see that there are many kinds of electrical motors. In this section, we give a brief overview of electrical motors, their specifications, and operating characteristics. Then, in the remainder of this chapter, we discuss dc machines in detail. In Chapter 16, we treat ac machines. We cite the three-phase ac induction motor as an example frequently in this section because it is the type in most widespread use. However, many of the concepts discussed in this section apply to other types of electrical motors as well.

You will find this section useful both as a preview of motor characteristics and as a convenient summary after you finish studying this chapter and the next.

### Basic Construction

An electrical motor consists of a stationary part, or **stator**, and a **rotor**, which is the rotating part connected to a shaft that couples the machine to its mechanical load. The shaft and rotor are supported by bearings so that they can rotate freely. This is illustrated in Figure 15.1.

Depending on the type of machine, either the stator or the rotor (or both) contains current-carrying conductors configured into coils. Slots are cut into the stator and rotor to contain the windings and their insulation. Currents in the windings set up magnetic fields and interact with fields to produce torque.

Usually, the stator and the rotor are made of iron to intensify the magnetic field. As in transformers, if the magnetic field alternates in direction through the iron with time, the iron must be laminated to avoid large power losses due to eddy currents. (In certain parts of some machines, the field is steady and lamination is not necessary.)

The characteristics of several common types of machines are summarized in Table 15.1. At this point, many of the entries in the table will probably not be very meaningful to you, particularly if this is the first time that you have studied rotating

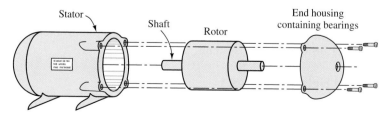

**Figure 15.1**  An electrical motor consists of a cylindrical rotor that spins inside a stator.

**Table 15.1.** Characteristics of Electrical Motors

| | | Type | Power Range (hp) | Rotor | Stator | Comments and Applications |
|---|---|---|---|---|---|---|
| Ac motors | Three phase | Induction | 1–5000 | Squirrel cage | Three-phase armature windings | Simple rugged construction; very common; fans, pumps |
| | | | | Wound field | | Adjustable speed using rotor resistance; cranes, hoists |
| | | Synchronous | 1–5 | Permanent magnet | | Precise speed; transport sheet materials |
| | | | 1000–50,000 | Dc field winding | | Large constant loads; potential for power-factor correction |
| | Single phase | Induction | $\frac{1}{3}$–5 | Squirrel cage | Main and auxiliary windings | Several types: split phase, capacitor start, capacitor run; simple and rugged; many household applications: fans, water pumps, refrigerators |
| | | Synchronous | $\frac{1}{10}$ or less | Reluctance or hysteresis | Armature winding | Low torque, fixed speed; timing applications |
| Dc motors | Wound field | Shunt connected | 10–200 | Armature winding | Field winding | Industrial applications, grinding, machine tools, hoists |
| | | Series connected | | | | High torque at low speed; dangerous if not loaded; drills, automotive starting motors, (universal motor used for single-phase ac has high power/weight ratio) |
| | | Compound connected | | | | Can be designed to tailor torque–speed characteristic; traction motors |
| | Permanent-magnet field | | $\frac{1}{20}$–10 | Armature winding | Permanent magnets | Servo applications, machine tools, computer peripherals, automotive fans, window motors |

electrical machinery. However, as we progress through this chapter and the next, the table will become a useful tool for comparing the various types of motors. Also, it will provide a convenient starting point for you when you face the problem of selecting the proper motor for one of your systems.

## Armature and Field Windings

The purpose of the field winding is to set up the magnetic field required to produce torque.

As we have mentioned, a machine may contain several sets of windings. In most types of machines, a given winding can be classified either as a **field winding** or as an **armature winding**. (We avoid classification of armature and field windings for induction motors and simply refer to stator windings and rotor conductors.) The primary purpose of a field winding is to set up the magnetic field in the machine. The current in the field winding is independent of the mechanical load imposed on the motor (except in series-connected motors). On the other hand, the armature winding carries a current that depends on the mechanical power produced. Typically, the armature current amplitude is small when the load is light and larger for heavier loads. If the machine acts as a generator, the electrical output is taken from the armature. In some machines, the field is produced by permanent magnets (PMs), and a field winding is not needed.

Table 15.1 shows the location (stator or rotor) of the field and armature windings for some common machine types. For example, in three-phase synchronous ac machines, the field winding is on the rotor, and the armature is on the stator. In other machines, such as the wound-field dc machine, the locations are reversed. You may find it convenient to refer to Table 15.1 from time to time throughout this chapter and the next to help avoid confusion between the different types of machines.

> The armature windings carry currents that vary with mechanical load. When the machine is used as a generator, the output is taken from the armature windings.

## AC Motors

Motors can be powered from either ac or dc sources. Ac power can be either single phase or three phase. (Three-phase ac sources and circuits are discussed in Section 5.7.) Ac motors include several types:

1. Induction motors, which are the most common type because they have relatively simple rugged construction and good operating characteristics.

2. Synchronous motors, which run at constant speed regardless of load torque, assuming that the frequency of the electrical source is constant, which is usually the case. Three-phase synchronous machines generate most of the electrical energy used in the world.

3. A variety of special-purpose types.

About two-thirds of the electrical energy generated in the United States is consumed by motors. Of this, well over half is used by induction motors. Thus, you are likely to encounter ac induction motors very frequently. We discuss the various types of ac motors in Chapter 16.

## DC Motors

Dc motors are those that are powered from dc sources. One of the difficulties with dc motors is that nearly all electrical energy is distributed as ac. If only ac power is available and we need to use a dc motor, a rectifier or some other converter must be used to convert ac to dc. This adds to the expense of the system. Thus, ac machines are usually preferable if they meet the needs of the application.

Exceptions are automotive applications in which dc is readily available from the battery. Dc motors are employed for starting, windshield wipers, fans, and power windows.

> Dc motors are common in automotive applications.

In the most common types of dc motors, the direction of the current in the armature conductors on the rotor is reversed periodically during rotation. This is accomplished with a mechanical switch composed of **brushes** mounted on the stator and a **commutator** mounted on the shaft. The commutator consists of conducting segments insulated from one another. Each commutator segment is connected to some of the armature conductors (on the rotor). The brushes are in sliding contact with the commutator. As the commutator rotates, switching action caused by the

A significant disadvantage of dc motors is the need for relatively frequent maintenance of brushes and commutators.

brushes moving from one segment to another changes the direction of current in the armature conductors. We explain this in more detail later; the important point here is that the brushes and commutator are subject to wear, and a significant disadvantage of dc motors is their relatively frequent need for maintenance.

Until recently, an important advantage of dc motors was that their speed and direction could be controlled more readily than those of ac motors. However, this advantage is rapidly disappearing because electronic systems that can vary the frequency of an ac source have become economically advantageous. These variable-frequency sources can be used with simple rugged ac induction motors to achieve speed control.

Nevertheless, dc motors are still useful in some control applications and wherever dc power is readily available, such as in vehicles. Later in this chapter, we examine the various types of dc motors in more detail.

### Losses, Power Ratings, and Efficiency

Figure 15.2 depicts the flow of power from a three-phase electrical source through an induction motor to a mechanical load such as a pump. Part of the electrical power is lost (converted to heat) due to resistance of the windings, hysteresis, and eddy currents. Similarly, some of the power that is converted to mechanical form is lost due to friction and windage (i.e., moving the air surrounding the rotor and shaft). Part of the power loss to windage is sometimes intentional, because fan blades to promote cooling are fabricated as an integral part of the rotor.

The electrical input power $P_{\text{in}}$, in watts, supplied by the three-phase source is given by

$$P_{\text{in}} = \sqrt{3} V_{\text{rms}} I_{\text{rms}} \cos(\theta) \tag{15.1}$$

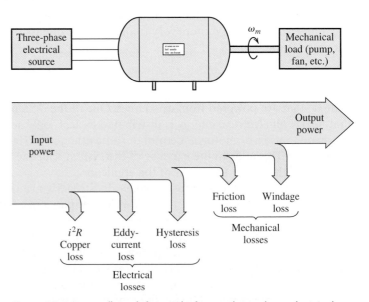

**Figure 15.2** Power flows left to right from a three-phase electrical source into an induction motor and then to a mechanical load. Some of the power is lost along the way due to various causes.

where $V_{rms}$ is the rms value of the line-to-line voltage, $I_{rms}$ is the rms value of the line current, and $\cos(\theta)$ is the power factor.

The mechanical output power is

$$P_{out} = T_{out}\omega_m \qquad (15.2)$$

in which $P_{out}$ is the output power in watts, $T_{out}$ is the output torque in newton-meters, and $\omega_m$ is the angular speed of the load in radians per second.

Rotational speed may be given in revolutions per minute denoted by $n_m$, or in radians per second denoted by $\omega_m$. These quantities are related by

$$\omega_m = n_m \times \frac{2\pi}{60} \qquad (15.3)$$

Also, torque may be given in foot-pounds instead of in newton-meters. The conversion relationship is

$$T_{foot\text{-}pounds} = T_{newton\text{-}meters} \times 0.7376 \qquad (15.4)$$

In the United States, the mechanical output power for a given electric motor is frequently stated in horsepower (hp). To convert from watts to horsepower, we have

$$P_{horsepower} = \frac{P_{watts}}{746} \qquad (15.5)$$

The **power rating** of a motor is the output power that the motor can safely produce on a continuous basis. For example, we can safely operate a 5-hp motor with a load that absorbs 5 hp of mechanical power. If the power required by the load is reduced, the motor draws less input power from the electrical source, and in the case of an induction motor, speeds up slightly. It is important to realize that most motors can supply output power varying from zero to several times their rated power, depending on the mechanical load. It is up to the system designer to ensure that the motor is not overloaded.

The chief output power limitation of motors is their temperature rise due to losses. Thus, a brief overload that does not cause significant rise in temperature is often acceptable.

The power **efficiency** of a motor is given by

$$\eta = \frac{P_{out}}{P_{in}} \times 100\% \qquad (15.6)$$

Well-designed electrical motors operating close to their rated capacity have efficiencies in the range of 85 to 95 percent. On the other hand, if the motor is called upon to produce only a small fraction of its rated power, its efficiency is generally much lower.

## Torque–Speed Characteristics

Consider a system in which a three-phase induction motor drives a load such as a pump. Figure 15.3 shows the torque produced by the motor versus speed. (In Chapter 16, we will see why the torque–speed characteristic of the induction motor has this shape.)

*It is up to the system designer to ensure that the motor is not overloaded.*

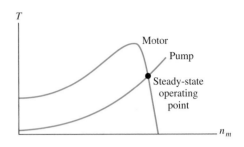

**Figure 15.3** The torque–speed characteristics of an induction motor and a load consisting of a pump. In steady state, the system operates at the point for which the torque produced by the motor equals the torque required by the load.

The torque required to drive the load is also shown. Suppose that the system is at a standstill and then a switch is closed connecting the electrical source to the motor. At low speeds, the torque produced by the motor is larger than that needed to drive the load. The excess torque causes the system to accelerate. Eventually, the speed stabilizes at the point for which the torque produced by the motor equals the torque needed to drive the load.

Now consider the torque–speed characteristics for a three-phase induction motor and a load consisting of a hoist shown in Figure 15.4. Here, the starting torque of the motor is less than that demanded by the load. Thus, if power is applied from a standing start, the system does not move. In this case, excessive currents are drawn by the motor, and unless fuses or other protection equipment disconnect the source, the motor could overheat and be destroyed.

Even though the motor cannot start the load shown in Figure 15.4, notice that the motor is capable of keeping the load moving once the speed exceeds $n_1$. Perhaps this could be accomplished with a mechanical clutch.

The various types of motors have different torque–speed characteristics. Some examples are shown in Figure 15.5. It is important for the system designer to choose a motor suitable for the load requirements.

> Designers must be able to choose motors having torque–speed characteristics appropriate for various loads.

## Speed Regulation

Depending on the torque–speed characteristics, a motor may slow down as the torque demanded by the load increases. Speed regulation is defined as the difference between the no-load speed and the full-load speed, expressed as a percentage of the full-load speed:

$$\text{speed regulation} = \frac{n_{\text{no-load}} - n_{\text{full-load}}}{n_{\text{full-load}}} \times 100\% \tag{15.7}$$

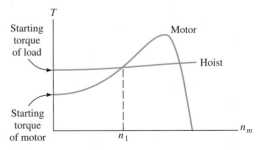

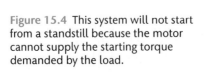

**Figure 15.4** This system will not start from a standstill because the motor cannot supply the starting torque demanded by the load.

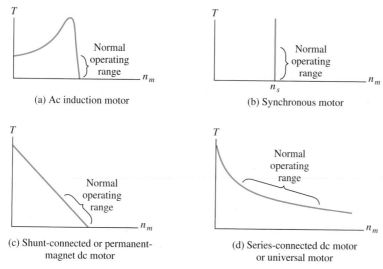

Figure 15.5 Torque versus speed characteristics for the most common types of electrical motors.

## Synchronous-Motor Operating Characteristics

The torque–speed characteristic for ac synchronous motors is shown in Figure 15.5(b). The operating speed of a synchronous motor is constant and is given by

$$\omega_s = \frac{2\omega}{P} \tag{15.8}$$

in which $\omega$ is the angular frequency of the ac source and $P$ is the number of magnetic poles possessed by the machine. (We examine the internal structure of these machines in Section 16.3.) In rpm, the synchronous speed is

$$n_s = \frac{120f}{P} \tag{15.9}$$

in which $f$ is the frequency of the ac source in hertz.

We will see that the number of magnetic poles $P$ is always an even integer. Substituting various values for $P$ into Equation 15.9 and assuming 60-Hz operation, the available speeds are 3600 rpm, 1800 rpm, 1200 rpm, 900 rpm, and so on. If some other speed is required, a synchronous machine is usually not a good choice. (Electronic systems known as cycloconverters can be used to convert 60-Hz power into any desired frequency. Thus, this speed limitation can be circumvented at additional cost.)

As shown in Figure 15.5(b), the starting torque of a synchronous motor is zero. Therefore, special provisions must be made for starting. We will see that one approach is to operate the motor as an induction motor with reduced load until the speed approaches synchronous speed, and then to switch to synchronous operation.

Synchronous motors operate at $(60 \times f)$ rpm or one of its submultiples. If the frequency $f$ is fixed and if none of the available speeds is suitable for the load, some other type of motor is needed.

The starting torque of a synchronous motor is zero.

## Induction-Motor Operating Characteristics

The torque–speed characteristic typical of an induction motor is shown in Figure 15.5(a). The motor has good starting torque. In normal operation, the speed of an induction motor is only slightly less than synchronous speed, which is given by

Induction motors have starting torques that are comparable to their rated full-load torques.

Induction motors operate
in narrow ranges of speed
that are slightly less than
$(60 \times f)$ rpm or one of
its submultiples. If the
frequency is fixed, the
speeds available may not be
suitable for the load.

Equations 15.8 and 15.9. For example, at full load, a typical four-pole ($P = 4$) induction motor runs at 1750 rpm, and at no load, its speed approaches 1800 rpm. The comments given earlier about speed limitations of synchronous motors also apply to induction motors.

During startup, the current drawn by an induction motor can be many times larger than its rated full-load current. To avoid excessive currents, large induction motors are usually started with reduced voltage. As you might expect, the torque produced by a motor depends on the applied voltage. At a given speed, the torque of an induction motor is proportional to the square of the magnitude of the voltage applied to the armature. When starting a motor at, say, half of its rated voltage, its torque is one-fourth of its value at rated voltage.

## Shunt-Connected DC Motor Operating Characteristics

Dc motors contain field windings on the stator and armature windings located on the rotor. Depending on whether the field windings are connected in shunt (i.e., parallel) or in series with the armature windings, the torque–speed characteristics are quite different. We examine why this is true later in this chapter.

The torque–speed characteristic of the shunt-connected dc motor is shown in Figure 15.5(c). The shunt-connected motor has very high starting torque and draws very large starting currents. Usually, resistance is inserted in series with the armature during starting to limit the current to reasonable levels.

For fixed supply voltage and fixed field current, the shunt dc machine shows only a small variation in speed within its normal operating range. However, we will see that several methods can be used to shift the torque–speed characteristic of the shunt motor to achieve excellent speed control. Unlike ac induction and synchronous motors, the speeds of dc motors are not limited to specific values.

Dc motors can be designed
to operate over a wide range
of speeds.

## Series-Connected DC Motor Operating Characteristics

The torque–speed characteristic of the series dc motor is shown in Figure 15.5(d). The series-connected dc motor has moderate starting torque and starting current. Its speed automatically adjusts over a large range as the load torque varies. Because it slows down for heavier loads, its output power is more nearly constant than for other motor types. This is advantageous because the motor can operate within its maximum power rating for a wide range of load torque. The starter motor in automobiles is a series dc motor. When the engine is cold and stiff, the starter motor operates at a lower speed. On the other hand, when the engine is warm, the starter spins faster. In either case, the current drawn from the battery remains within acceptable limits. (On the other hand, without sophisticated controls, a shunt motor would attempt to turn the load at a constant speed and would draw too much current in starting a cold engine.)

In some cases, the no-load speed of a series dc motor can be excessive—to the point of being dangerous. A control system that disconnects the motor from the electrical source is needed if the possibility of losing the mechanical load exists. We will see that a very useful type of ac motor known as a universal motor is essentially identical to the series-connected dc motor.

---

**Example 15.1**    Motor Performance Calculations

A certain 5-hp three-phase induction motor operates from a 440-V-rms (line-to-line) three-phase source and draws a line current of 6.8 A rms at a power factor of 78 percent lagging [i.e., $\cos(\theta) = 0.78$] under rated full-load conditions. The full-load speed

Section 15.1   Overview of Motors    **753**

is 1150 rpm. Under no-load conditions, the speed is 1195 rpm, and the line current is 1.2 A rms at a power factor of 30 percent lagging. Find the power loss and efficiency with full load, the input power with no load, and the speed regulation.

**Solution**   The rated output power is 5 hp. Converting to watts, we have

$$P_{out} = 5 \times 746 = 3730 \text{ W}$$

Substituting into Equation 15.1, we find the input power under full load:

$$P_{in} = \sqrt{3} V_{rms} I_{rms} \cos(\theta)$$
$$= \sqrt{3}(440)(6.8)(0.78) = 4042 \text{ W}$$

The power loss is given by

$$P_{loss} = P_{in} - P_{out} = 4042 - 3730 = 312 \text{ W}$$

The full-load efficiency is

$$\eta = \frac{P_{out}}{P_{in}} \times 100\% = \frac{3730}{4042} \times 100\% = 92.28\%$$

Under no-load conditions, we have

$$P_{in} = \sqrt{3}(440)(1.2)(0.30) = 274.4 \text{ W}$$
$$P_{out} = 0$$
$$P_{loss} = P_{in} = 274.4 \text{ W}$$

and the efficiency is

$$\eta = 0\%$$

Speed regulation for the motor is given by Equation 15.7. Substituting values, we get

$$\text{speed regulation} = \frac{n_{\text{no-load}} - n_{\text{full-load}}}{n_{\text{full-load}}} \times 100\%$$

$$= \frac{1195 - 1150}{1150} \times 100\% = 3.91\% \qquad \blacksquare$$

Now that we have presented an overall view of electrical motors, we will make a more detailed examination of the most common and useful types. In the remainder of this chapter, we consider dc machines, and in Chapter 16, we discuss ac machines.

---

**Exercise 15.1**   A certain 50-hp dc motor operates from a 220-V dc source with losses of 3350 W under rated full-load conditions. The full-load speed is 1150 rpm. Under no-load conditions, the speed is 1200 rpm. Find the source current, the efficiency with full load, and the speed regulation.
**Answer**   $I_{source} = 184.8$ A, $\eta = 91.76$ percent, speed regulation $= 4.35$ percent. ◻

**Exercise 15.2**    Consider the torque–speed characteristics shown in Figure 15.5. **a.** Which type of motor would have the most difficulty in starting a high-inertia load from a standing start? **b.** Which type of motor would have the poorest (i.e., largest) speed regulation in its normal operating range? **c.** Which would have the best (i.e., smallest) speed regulation? **d.** Which has the best combination of high starting torque and good speed regulation? **e.** Which should not be operated without a load? **Answer**    **a.** The synchronous motor, because its starting torque is zero; **b.** the series-connected dc motor; **c.** the synchronous motor; **d.** the ac induction motor; **e.** the series-connected dc motor because the speed can become excessive for zero load torque.

□

## 15.2 PRINCIPLES OF DC MACHINES

Study of the idealized linear dc machine clearly demonstrates how the principles of electromagnetism apply to dc machines in general.

In this section, we introduce the basic principles of dc machines by considering the idealized linear machine shown in Figure 15.6. Later, we will see that the operation of rotating dc machines is very similar to that of this simple linear machine. In Figure 15.6, a dc voltage source $V_T$ is connected through a resistance $R_A$ and a switch that closes at $t = 0$ to a pair of conducting rails. A conducting bar slides without friction on the rails. We assume that the rails and the bar have zero resistance. A magnetic field is directed into the page, perpendicular to the plane of the rails and the bar.

Suppose that the bar is stationary when the switch is closed at $t = 0$. Then, just after the switch is closed, an initial current given by $i_A(0+) = V_T/R_A$ flows clockwise around the circuit. A force given by

$$\mathbf{f} = i_A \mathbf{l} \times \mathbf{B} \tag{15.10}$$

is exerted on the bar. The direction of the current (and $\mathbf{l}$) is toward the bottom of the page. Thus, the force is directed to the right. Because the current and the field are mutually perpendicular, the force magnitude is given by

$$f = i_A l B \tag{15.11}$$

This force causes the bar to be accelerated toward the right. As the bar gains velocity $u$ and cuts through the magnetic field lines, a voltage is induced across the bar. The voltage is positive at the top end of the bar and is given by Equation 15.9 (with a change in notation):

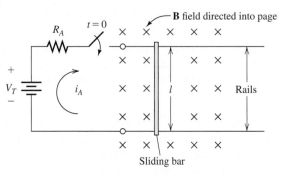

**Figure 15.6** A simple dc machine consisting of a conducting bar sliding on conducting rails.

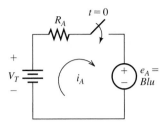

Figure 15.7 Equivalent circuit for the linear machine operating as a motor.

$$e_A = Blu \qquad\qquad (15.12)$$

An equivalent circuit for the system is shown in Figure 15.7. Notice that the induced voltage $e_A$ opposes the source $V_T$. The current is

$$i_A = \frac{V_T - e_A}{R_A} \qquad\qquad (15.13)$$

As the velocity of the bar builds up, energy is absorbed by the induced voltage $e_A$, and this energy shows up as the kinetic energy of the bar. Eventually, the bar speed becomes high enough that $e_A = V_T$. Then, the current and the force become zero, and the bar coasts at constant velocity.

## Operation as a Motor

Now, suppose that a mechanical load exerting a force to the left is connected to the moving bar. Then, the bar slows down slightly, resulting in a reduction in the induced voltage $e_A$. Current flows clockwise in the circuit, resulting in a magnetically induced force directed to the right. Eventually, the bar slows just enough so that the force created by the magnetic field ($f = i_A lB$) equals the load force. Then, the system moves at constant velocity.

In this situation, power delivered by the source $V_T$ is converted partly to heat in the resistance $R_A$ and partly to mechanical power. It is the power $p = e_A i_A$ delivered to the induced voltage that shows up as mechanical power $p = fu$.

## Operation as a Generator

Again suppose that the bar is moving at constant velocity such that $e_A = V_T$ and the current is zero. Then, if a force is applied pulling the bar even faster toward the right, the bar speeds up, the induced voltage $e_A$ exceeds the source voltage $V_T$, and current circulates counterclockwise as illustrated in Figure 15.8. Because the current has reversed direction, the force induced in the bar by the field also reverses and points to the left. Eventually, the bar speed stabilizes with the pulling force equal to the induced force. Then, the induced voltage delivers power $p = e_A i_A$, partly to the resistance ($p_R = R_A i_A^2$) and partly to the battery ($p_t = V_T i_A$). Thus, mechanical energy is converted into electrical energy that eventually shows up as loss (i.e., heat) in the resistance or as stored chemical energy in the battery.

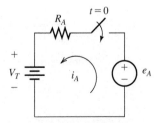

Figure 15.8 Equivalent circuit for the linear machine operating as a generator.

---

**Example 15.2**   **Idealized Linear Machine**

Suppose that for the linear machine shown in Figure 15.6, we have $B = 1$ T, $l = 0.3$ m, $V_T = 2$ V, and $R = 0.05$ $\Omega$. **a.** Assuming that the bar is stationary at

$t = 0$, compute the initial current and the initial force on the bar. Also, determine the final (i.e., steady-state) speed assuming that no mechanical load is applied to the bar. **b.** Now, suppose that a mechanical load of 4 N directed to the left is applied to the moving bar. In steady state, determine the speed, the power delivered by $V_T$, the power delivered to the mechanical load, the power lost to heat in the resistance $R_A$, and the efficiency. **c.** Now, suppose that a mechanical pulling force of 2 N directed to the right is applied to the moving bar. In steady state, determine the speed, the power taken from the mechanical source, the power delivered to the battery, the power lost to heat in the resistance $R_A$, and the efficiency.

**Solution**

**a.** Initially, for $u = 0$, we have $e_A = 0$, and the initial current is given by

$$i_A(0+) = \frac{V_T}{R_A} = \frac{2}{0.05} = 40 \text{ A}$$

The resulting initial force on the bar is

$$f(0+) = Bli_A(0+) = 1(0.3)40 = 12 \text{ N}$$

In steady state with no load, the induced voltage equals the battery voltage. Thus, we have

$$e_A = Blu = V_T$$

Solving for the velocity and substituting values, we get

$$u = \frac{V_T}{Bl} = \frac{2}{1(0.3)} = 6.667 \text{ m/s}$$

**b.** Because the mechanical force opposes the motion of the bar, we have motor action. In steady state, the net force on the bar is zero—the force created by the magnetic field equals the load force. Thus, we obtain

$$f = Bli_A = f_{load}$$

Solving for the current and substituting values, we find that

$$i_A = \frac{f_{load}}{Bl} = \frac{4}{1(0.3)} = 13.33 \text{ A}$$

From the circuit shown in Figure 15.7, we have

$$e_A = V_T - R_A i_A = 2 - 0.05(13.33) = 1.333 \text{ V}$$

Now, we can find the steady-state speed:

$$u = \frac{e_A}{Bl} = \frac{1.333}{1(0.3)} = 4.444 \text{ m/s}$$

The mechanical power delivered to the load is

$$P_m = f_{load}u = 4(4.444) = 17.77 \text{ W}$$

The power taken from the battery is

$$p_t = V_T i_A = 2(13.33) = 26.67 \text{ W}$$

The power dissipated in the resistance is

$$p_R = i_A^2 R = (13.33)^2 \times 0.05 = 8.889 \text{ W}$$

As a check, we note that $p_t = p_m + p_R$ to within rounding error. Finally, the efficiency of converting electrical power from the battery into mechanical power is

$$\eta = \frac{p_m}{p_t} \times 100\% = \frac{17.77}{26.67} \times 100\% = 66.67\%$$

c. With a pulling force applied to the bar to the right, the bar speeds up, the induced voltage exceeds $V_T$, and current circulates counterclockwise, as illustrated in Figure 15.8. Thus, the machine operates as a generator. In steady state, the force induced by the field is directed to the left and equals the pulling force. Thus, we have

$$f = Bli_A = f_{\text{pull}}$$

Solving for $i_A$ and substituting values, we find that

$$i_A = \frac{f_{\text{pull}}}{Bl} = \frac{2}{1(0.3)} = 6.667 \text{ A}$$

From the circuit shown in Figure 15.8, we obtain

$$e_A = V_T + R_A i_A = 2 + 0.05(6.67) = 2.333 \text{ V}$$

Now, we can find the steady-state speed:

$$u = \frac{e_A}{Bl} = \frac{2.333}{1(0.3)} = 7.778 \text{ m/s}$$

The mechanical power delivered by the pulling force is

$$p_m = f_{\text{pull}} u = 2(7.778) = 15.56 \text{ W}$$

The power absorbed by the battery is

$$p_t = V_T i_A = 2(6.667) = 13.33 \text{ W}$$

The power dissipated in the resistance is

$$p_R = i_A^2 R = (6.667)^2 \times 0.05 = 2.222 \text{ W}$$

a. As a check, we note that $p_m = p_t + p_R$ to within rounding error. Finally, the efficiency of converting mechanical power into electrical power charging the battery is

$$\eta = \frac{p_t}{p_m} \times 100\% = \frac{13.33}{15.56} \times 100\% = 85.67\%$$

In Example 15.2, we have seen that only modest forces (12 N) were produced on a conductor carrying a fairly large current (40 A). The force could be increased by using a longer conductor, but this increases the size of the machine. Another option would be to increase the field strength. However, because of the fact that magnetic materials used in motors saturate in the neighborhood of 1 T, it is not practical to increase the forces on conductors greatly by increasing the field.

On the other hand, a cylindrical rotor containing many conductors is a practical way to obtain large forces in a compact design. Furthermore, rotary motion is more useful than translation in many applications. Thus, most (but not all) practical motors are based on rotational motion. We study rotating dc machines in the remaining sections of this chapter.

---

**Exercise 15.3**  Repeat the calculations of Example 15.2 if the field strength is doubled to 2 T.

**Answer**  **a.** $i_A(0+) = 40$ A, $f(0+) = 24$ N, $u = 3.333$ m/s; **b.** $i_A = 6.667$, $e_A = 1.667$, $u = 2.778$ m/s,    $p_m = 11.11$ W,    $p_t = 13.33$ W,    $p_R = 2.22$ W,    $\eta = 83.33\%$; **c.** $i_A = 3.333$ A,  $e_A = 2.167$ V,  $u = 3.612$ m/s,  $p_m = 7.222$ W,  $p_t = 6.667$ W, $p_R = 0.555$ W, $\eta = 92.3\%$.    □

---

## PRACTICAL APPLICATION    15.1

### Magnetic Flowmeters, Faraday, and The Hunt for Red October

Flowmeters measure the flow rate of liquids through pipes and are very important sensors in chemical-process-control systems. A commonly used type is the magnetic flowmeter (also called a magflow), which operates on the same principles as the linear machine discussed in Section 15.2.

The basic operation of a magflow is illustrated in Figure PA15.1. Coils set up a vertical magnetic field in the fluid, and electrodes are located at opposite sides of the pipe, which is lined with an electrical insulating material such as ceramic or epoxy resin. Thus, the magnetic field, the direction of flow, and the line between the electrodes are mutually perpendicular. As the conductive fluid moves through the magnetic field, a voltage proportional to velocity is induced between the electrodes. The flow rate can be determined by multiplying the cross-sectional area of the pipe times the velocity. Hence, the meter measures the induced voltage, but can be calibrated in units of volumetric flow.

Faraday realized the potential for using his law of electromagnetic induction to measure water flow, and attempted to measure the flow rate of the Thames River with a device suspended from a bridge. However, lacking the advantages of modern electronics, he was not successful.

In modern meters, an electronic amplifier is used to amplify the induced voltage. In many units, this voltage is converted to digital form by an analog-to-digital converter and processed by a microcomputer that drives a display or sends data to a central plant-control computer.

If the fluid has low electrical conductivity, the Thévenin resistance seen looking into the electrode terminals is very high. Then, it is important for the amplifier input impedance to be very high; otherwise, the observed voltage would vary with changes in electrical conductivity of the fluid, leading to gross inaccuracies in flow rate. Of course, it is important for chemical engineers employing magflows to understand this limitation. Even with their limitations, well-designed meters are available for a wide range of applications.

As a meter, the magflow acts as a generator. However, it can also act as a motor if an electrical current is passed through the fluid between the electrodes. Then force is exerted directly on the fluid by the interaction of the magnetic field with the electrical current. Of course, if we wanted to build a pump based on this approach, we would want a fluid with high conductivity, such as seawater. We would also need a strong magnetic field and high

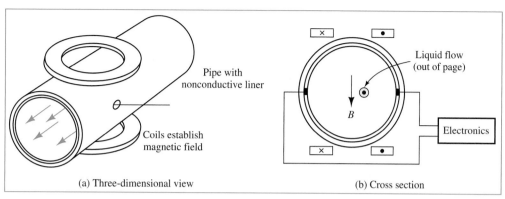

**Figure PA15.1** A magnetic flowmeter.

current. By making these modifications, a powerful pump can be constructed. This is the operating principle of the ultraquiet submarine propulsion system mentioned by Tom Clancy in *The Hunt for Red October*. Such a system can be very quiet because force is applied smoothly and directly to the

seawater without rotating parts or valves that could cause vibration.

*Sources:* Ian Robertson, "Magnetic flowmeters: the whole story," *The Chemical Engineer*, February 24, 1994, pp. 17–18; The Magmeter Flowmeter Home-page, http://www.magmeter.com.

## 15.3 ROTATING DC MACHINES

We have gained some familiarity with basic principles of dc machines from our analysis of the linear machine in the preceding section. In this section, we will see that the same principles apply to rotating dc machines.

> The basic principles of rotating dc machines are the same as those of the linear dc machine.

### Structure of the Rotor and Stator

The most common type of dc machine contains a cylindrical stator with an even number $P$ of magnetic poles that are established by field windings or by PMs. The poles alternate between north and south around the periphery of the stator.

Inside the stator is a rotor consisting of a laminated iron cylinder mounted on a shaft that is supported by bearings so that it can rotate. Slots cut lengthwise into the surface of the rotor contain the armature windings. A rotor with armature conductors (and other features to be discussed shortly) is illustrated in Figure 15.9.

The cross section of a two-pole machine showing the flux lines in the air gap is illustrated in Figure 15.10. Magnetic flux tends to take the path of least reluctance. Because the reluctance of air is much higher than that of iron, the flux takes the shortest path from the stator into the rotor. Thus, the flux in the air gap is perpendicular to the surface of the rotor and to the armature conductors. Furthermore, the flux density is nearly constant in magnitude over the surface of each pole face. Between poles, the gap flux density is small in magnitude.

In a motor, external electrical sources provide the currents in the field windings and in the armature conductors. The current directions shown in Figure 15.10 result in a counterclockwise torque. This can be verified by applying the equation $\mathbf{f} = i\mathbf{l} \times \mathbf{B}$ that gives the force on a current-carrying conductor.

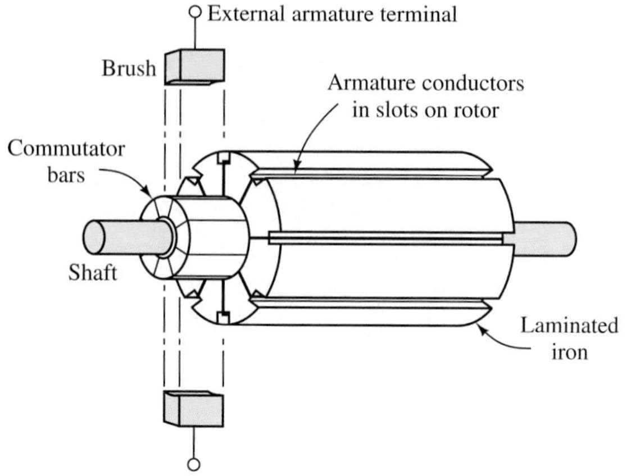

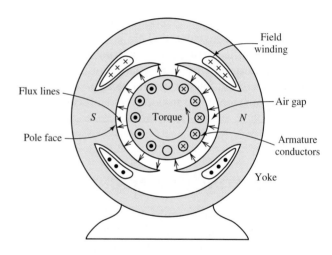

**Figure 15.9** Rotor assembly of a dc machine.

**Figure 15.10** Cross section of a two-pole dc machine.

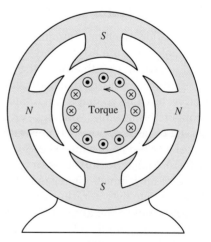

**Figure 15.11** Cross section of a four-pole dc machine.

The cross section of a four-pole machine is shown in Figure 15.11. Notice that the directions of the currents in the armature must be reversed under south poles relative to the direction under the north poles to achieve aiding contributions to total torque.

## Induced EMF and Commutation

As the rotor turns, the conductors move through the magnetic field produced by the stator. Under the pole faces, the conductors, the field, and the direction of motion are mutually perpendicular, just as in the linear machine discussed in Section 15.2. Thus, a nearly constant voltage is induced in each conductor as it moves under a pole. However, as the conductors move between poles, the field direction reverses. Therefore, the induced voltages fall to zero and build up with the opposite polarity. A mechanical switch known as a **commutator** reverses the connections to the conductors as they move between poles so that the polarity of the induced voltage seen from the external machine terminals is constant.

Let us illustrate these points with a two-pole machine containing one armature coil, as shown in Figure 15.12. In this case, the ends of the coil are attached to a two-segment commutator mounted on the shaft. The segments are insulated from one another and from the shaft. Brushes mounted to the stator make electrical contact with the commutator segments. (For clarity, we have shown the brushes inside the commutator, but in a real machine, they ride on the outside surface of the commutator. A more realistic version of the commutator and brushes was shown in Figure 15.9.)

Notice that as the rotor turns in Figure 15.12, the left-hand brush is connected to the conductor under the south stator pole, and the right-hand brush is connected to the conductor under the north stator pole.

The voltage $v_{ad}$ induced across the terminals of the coil is an ac voltage, as shown in the figure. As mentioned earlier, this voltage passes through zero when the conductors are between poles where the flux density goes to zero. While the conductors are under the pole faces where the flux density is constant, the induced voltage has nearly constant magnitude. Because the commutator reverses the

*The commutator and brushes form a mechanical switch that reverses the external connections to conductors as they move from pole to pole.*

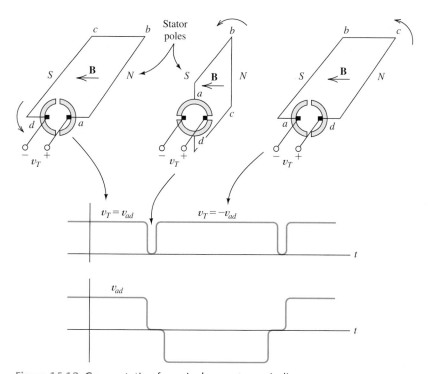

**Figure 15.12** Commutation for a single armature winding.

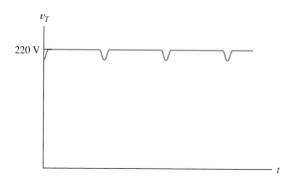

**Figure 15.13** Voltage produced by a practical dc machine. Because only a few (out of many) conductors are commutated (switched) at a time, the voltage fluctuations are less pronounced than in the single-loop case illustrated in Figure 15.12.

external connections to the coil as it rotates, the voltage $v_T$ seen at the external terminals is of constant polarity.

Notice that the brushes short the armature winding briefly during the switching process. This occurs because the brushes are wider than the insulation between commutator segments. This shorting is not a problem, provided that the voltage is small when it occurs. (Actual machines have various provisions to ensure that the coil voltage is close to zero during commutation for all operating conditions.)

Commutators in typical machines contain 20 to 50 segments. Because only part of the coils are commutated at a time, the terminal voltage of a real machine shows relatively little fluctuation compared to the two-segment example that we used for the illustration of concepts. The terminal voltage of an actual dc machine is shown in Figure 15.13.

Generally, the commutator segments are copper bars insulated from one another and from the shaft. The brushes contain graphite that lubricates the sliding contact. Even so, a significant disadvantage of dc machines is the need to replace brushes and redress the commutator surface because of mechanical wear.

Actual armatures consist of a large number of conductors placed around the circumference of the rotor. To attain high terminal voltages, many conductors are placed in series, forming coils. Furthermore, there are usually several parallel current paths through the armature. The armature conductors and their connections to the commutator are configured so that the currents flow in the opposite direction under south stator poles than they do under north stator poles. As mentioned earlier, this is necessary so that the forces on the conductors produce aiding torques. The construction details needed to produce these conditions are beyond the scope of our discussion. As a user of electrical motors, you will find the external behavior of machines more helpful than the details of their internal design.

### Equivalent Circuit of the DC Motor

The equivalent circuit of the dc motor is shown in Figure 15.14. The field circuit is represented by a resistance $R_F$ and an inductance $L_F$ in series. We consider steady-state operation in which the currents are constant, and we can neglect the

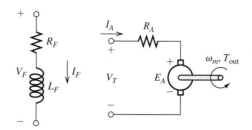

**Figure 15.14** Equivalent circuit for the rotating dc machine.

inductance because it behaves as a short circuit for dc currents. Thus, for dc field currents, we have

$$V_F = R_F I_F \qquad (15.14)$$

The voltage $E_A$ shown in the equivalent circuit represents the average voltage induced in the armature due to the motion of the conductors relative to the magnetic field. In a motor, $E_A$ is sometimes called a **back emf** (electromotive force) because it opposes the applied external electrical source. The resistance $R_A$ is the resistance of the armature windings plus the brush resistance. (Sometimes, the drop across the brushes is estimated as a constant voltage of about 2 V rather than as a resistance. However, in this book, we lump the brush drop with the armature resistance.)

The induced armature voltage is given by

$$E_A = K\phi\omega_m \qquad (15.15)$$

in which $K$ is a **machine constant** that depends on the design parameters of the machine, $\phi$ is the magnetic flux produced by each stator pole, and $\omega_m$ is the angular velocity of the rotor.

The torque developed in the machine is given by

$$T_{dev} = K\phi I_A \qquad (15.16)$$

in which $I_A$ is the armature current. (We will see that the *output* torque of a dc motor is less than the *developed* torque because of friction and other rotational losses.)

The **developed power** is the power converted to mechanical form, which is given by the product of developed torque and angular velocity:

$$P_{dev} = \omega_m T_{dev} \qquad (15.17)$$

This is the power delivered to the induced armature voltage, and therefore, is also given by

$$P_{dev} = E_A I_A \qquad (15.18)$$

## Magnetization Curve

The **magnetization curve** of a dc machine is a plot of $E_A$ versus the field current $I_F$ with the machine being driven at a constant speed. ($E_A$ can be found by measuring

The magnetization curve of a dc machine is a plot of $E_A$ versus the field current $I_F$ with the machine being driven at a constant speed.

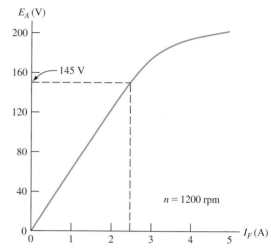

**Figure 15.15** Magnetization curve for a 200-V 10-hp dc motor.

the open-circuit voltage at the armature terminals.) A typical magnetization curve is shown in Figure 15.15.

Because $E_A$ is proportional to the flux $\phi$, the magnetization curve has the same shape as a $\phi$ versus $I_F$ plot, which depends on the parameters of the magnetic circuit for the field. The magnetization curve flattens out for high field currents due to magnetic saturation of the iron. Of course, different machines usually have differently shaped magnetization curves.

As Equation 15.15 shows, the induced armature voltage $E_A$ is directly proportional to speed. If $E_{A1}$ represents the voltage at speed $n_1$, and $E_{A2}$ is the voltage at a second speed $n_2$, we have

> The equivalent circuit, Equations 15.14 through 15.19, and the magnetization curve are the key elements needed to solve dc machine problems.

$$\frac{E_{A1}}{E_{A2}} = \frac{n_1}{n_2} = \frac{\omega_1}{\omega_2} \qquad (15.19)$$

Equations 15.14 through 15.19, in combination with the equivalent circuit shown in Figure 15.14 and the magnetization curve, provide the basis for analyzing a dc machine.

### Example 15.3    DC Machine Performance Calculations

The machine having the magnetization curve shown in Figure 15.15 is operating as a motor at a speed of 800 rpm with $I_A = 30$ A and $I_F = 2.5$ A. The armature resistance is $0.3\ \Omega$ and the field resistance is $R_F = 50\ \Omega$. Find the voltage $V_F$ applied to the field circuit, the voltage $V_T$ applied to the armature, the developed torque, and the developed power.

**Solution**    Equation 15.14 allows us to find the voltage for the field coil:

$$V_F = R_F I_F = 50 \times 2.5 = 125 \text{ V}$$

From the magnetization curve, we see that the induced voltage is $E_{A1} = 145$ V at $I_F = 2.5$ A and $n_1 = 1200$ rpm. Rearranging Equation 15.19, we can find the induced voltage $E_{A2}$ for $n_2 = 800$ rpm:

$$E_{A2} = \frac{n_2}{n_1} \times E_{A1} = \frac{800}{1200} \times 145 = 96.67 \text{ V}$$

The machine speed in radians per second is

$$\omega_m = n_2 \times \frac{2\pi}{60} = 800 \times \frac{2\pi}{60} = 83.78 \text{ rad/s}$$

Rearranging Equation 15.15, we have

$$K\phi = \frac{E_A}{\omega_m} = \frac{96.67}{83.78} = 1.154$$

From Equation 15.16, the developed torque is

$$T_{dev} = K\phi I_A = 1.154 \times 30 = 34.62 \text{ Nm}$$

The developed power is

$$P_{dev} = \omega_m T_{dev} = 2900 \text{ W}$$

As a check, we can also compute the developed power by using Equation 15.18:

$$P_{dev} = I_A E_A = 30 \times 96.67 = 2900 \text{ W}$$

Applying Kirchhoff's voltage law to the armature circuit in Figure 15.14, we have

$$V_T = R_A I_A + E_A = 0.3(30) + 96.67 = 105.67 \qquad \blacksquare$$

**Exercise 15.4**   Find the voltage $E_A$ for the machine having the magnetization curve shown in Figure 15.15 for $I_F = 2$ A and a speed of 1500 rpm.
**Answer**   $E_A \cong 156$ V.   □

**Exercise 15.5**   The machine having the magnetization curve shown in Figure 15.15 is operating as a motor at a speed of 1500 rpm with a developed power of 10 hp and $I_F = 2.5$ A. The armature resistance is 0.3 Ω and the field resistance is $R_F = 50$ Ω. Find the developed torque, the armature current $I_A$, and the voltage $V_T$ applied to the armature circuit.
**Answer**   $T_{dev} = 47.49$ Nm, $I_A = 41.16$ A, $V_T = 193.6$ V.   □

In the next several sections, we will see that different torque–speed characteristics can result, depending on how the field windings and the armature are connected to the dc source.

## 15.4   SHUNT-CONNECTED AND SEPARATELY EXCITED DC MOTORS

In a shunt-connected dc machine, the field circuit is in parallel with the armature, as shown in Figure 15.16. The field circuit consists of a rheostat having a variable resistance, denoted as $R_{adj}$, in series with the field coil. Later, we will see that the rheostat can be used to adjust the torque–speed characteristic of the machine.

We assume that the machine is supplied by a constant voltage source $V_T$. The resistance of the armature circuit is $R_A$, and the induced voltage is $E_A$. We denote the mechanical shaft speed as $\omega_m$ and the developed torque as $T_{dev}$.

By designing the field windings to be connected either in parallel or in series with the armature, we can obtain drastically different torque–speed characteristics.

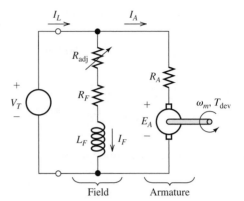

**Figure 15.16** Equivalent circuit of a shunt-connected dc motor. $R_{\text{adj}}$ is a rheostat that can be used to adjust motor speed.

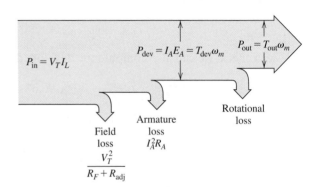

**Figure 15.17** Power flow in a shunt-connected dc motor.

## Power Flow

Figure 15.17 shows the flow of power in the shunt-connected dc machine. The electrical source supplies an input power given by the product of the supply voltage and the line current $I_L$:

$$P_{\text{in}} = V_T I_L \tag{15.20}$$

Some of this power is used to establish the field. The power absorbed by the field circuit is converted to heat. The **field loss** is given by

$$P_{\text{field-loss}} = \frac{V_T^2}{R_F + R_{\text{adj}}} = V_T I_F \tag{15.21}$$

Furthermore, **armature loss** occurs due to heating of the armature resistance:

$$P_{\text{arm-loss}} = I_A^2 R_A \tag{15.22}$$

Sometimes, the sum of the field loss and armature loss is called **copper loss**.

The power delivered to the induced armature voltage is converted to mechanical form and is called the **developed power**, given by

$$P_{\text{dev}} = I_A E_A = \omega_m T_{\text{dev}} \tag{15.23}$$

in which $T_{\text{dev}}$ is the developed torque.

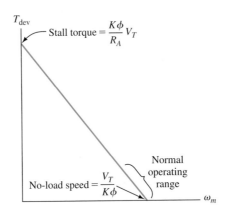

Figure 15.18 Torque–speed characteristic of the shunt dc motor.

The output power $P_{\text{out}}$ and output torque $T_{\text{out}}$ are less than the developed values because of **rotational losses**, which include friction, windage, eddy-current loss, and hysteresis loss. Rotational power loss is approximately proportional to speed.

## Torque–Speed Characteristic

Next, we derive the torque–speed relationship for the shunt-connected machine. Applying Kirchhoff's voltage law to the equivalent circuit shown in Figure 15.16, we obtain

$$V_T = R_A I_A + E_A \tag{15.24}$$

Next, rearranging Equation 15.16 yields

$$I_A = \frac{T_{\text{dev}}}{K\phi} \tag{15.25}$$

Then, using Equation 15.15 to substitute for $E_A$ and Equation 15.25 to substitute for $I_A$ in Equation 15.24, we obtain

$$V_T = \frac{R_A T_{\text{dev}}}{K\phi} + K\phi\omega_m \tag{15.26}$$

Finally, solving for the developed torque, we get

$$T_{\text{dev}} = \frac{K\phi}{R_A}(V_T - K\phi\omega_m) \tag{15.27}$$

which is the torque–speed relationship that we desire. Notice that this torque–speed relationship plots as a straight line, as illustrated in Figure 15.18. The speed for no load (i.e., $T_{\text{dev}} = 0$) and the stall torque are labeled in the figure. The normal operating range for most motors is on the lower portion of the torque–speed characteristic, as illustrated in the figure.

The starting or stall torque of a shunt-connected machine is usually many times higher than the rated full-load torque.

---

| Example 15.4 | Shunt-Connected DC Motor |

A 50-hp shunt-connected dc motor has the magnetization curve shown in Figure 15.19. The dc supply voltage is $V_T = 240$ V, the armature resistance is $R_A = 0.065$ Ω, the field resistance is $R_F = 10$ Ω, and the adjustable resistance is $R_{\text{adj}} = 14$ Ω. At a

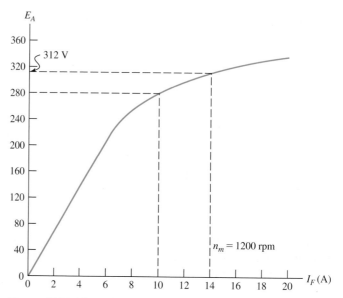

**Figure 15.19** Magnetization curve for the motor of Example 15.4.

speed of 1200 rpm, the rotational loss is $P_{rot} = 1450$ W. If this motor drives a hoist that demands a torque of $T_{out} = 250$ Nm independent of speed, determine the motor speed and efficiency.

**Solution**    The equivalent circuit is shown in Figure 15.20. The field current is given by

$$I_F = \frac{V_T}{R_F + R_{adj}} = \frac{240}{10 + 14} = 10 \text{ A}$$

Next, we use the magnetization curve to find the machine constant $K\phi$ for this value of field current. From the curve shown in Figure 15.19, we see that the induced armature voltage is $E_A = 280$ V at $I_F = 10$ A and $n_m = 1200$ rpm. Thus, rearranging Equation 15.15 and substituting values, we find that the machine constant is

$$K\phi = \frac{E_A}{\omega_m} = \frac{280}{1200(2\pi/60)} = 2.228$$

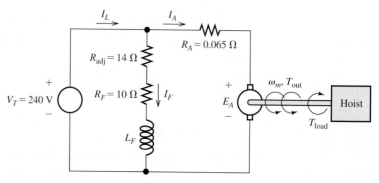

**Figure 15.20** Equivalent circuit for Example 15.4.

We assume that the rotational power loss is proportional to speed. This is equivalent to assuming constant torque for the rotational loss. The rotational-loss torque is

$$T_{rot} = \frac{P_{rot}}{\omega_m} = \frac{1450}{1200(2\pi/60)} = 11.54 \text{ Nm}$$

Thus, the developed torque is

$$T_{dev} = T_{out} + T_{rot} = 250 + 11.54 = 261.5 \text{ Nm}$$

Now, we use Equation 15.16 to find the armature current:

$$I_A = \frac{T_{dev}}{K\phi} = \frac{261.5}{2.228} = 117.4 \text{ A}$$

Then, applying Kirchhoff's voltage law to the armature circuit, we have

$$E_A = V_T - R_A I_A = 240 - 0.065(117.4) = 232.4 \text{ V}$$

Solving Equation 15.15 for speed and substituting values, we get

$$\omega_m = \frac{E_A}{K\phi} = \frac{232.4}{2.228} = 104.3 \text{ rad/s}$$

or

$$n_m = \omega_m \left(\frac{60}{2\pi}\right) = 996.0 \text{ rpm}$$

To find efficiency, we first compute the output power and the input power, given by

$$P_{out} = T_{out}\omega_m = 250(104.3) = 26.08 \text{ kW}$$

$$P_{in} = V_T I_L = V_T(I_F + I_A) = 240(10 + 117.4) = 30.58 \text{ kW}$$

$$\eta = \frac{P_{out}}{P_{in}} \times 100\% = \frac{26.08}{30.58} \times 100\% = 85.3\% \qquad \blacksquare$$

**Exercise 15.6**   Repeat Example 15.4 with the supply voltage increased to 300 V while holding the field current constant by increasing $R_{adj}$. What is the new value of $R_{adj}$? What happens to the speed?
**Answer**   $R_{adj} = 20 \ \Omega$; the speed increases to $\omega_m = 131.2$ rad/s or $n_m = 1253$ rpm. ☐

**Exercise 15.7**   Repeat Example 15.4 if the adjustable resistance $R_{adj}$ is increased in value to 30 $\Omega$ while holding $V_T$ constant at 240 V. What happens to the speed?
**Answer**   $I_F = 6$ A, $E_A = 229.3$ V, $I_A = 164.3$ A, $\omega_m = 144.0$, $\eta = 88.08$ percent, $n_m = 1376$ rpm; thus, speed increases as $R_{adj}$ is increased. ☐

## Separately Excited DC Motors

A separately excited dc motor is similar to a shunt-connected motor except that different sources are used for the armature and field circuits. The equivalent circuit for a separately excited dc machine is shown in Figure 15.21. Analysis of a separately excited machine is very similar to that of a shunt-connected machine. The chief

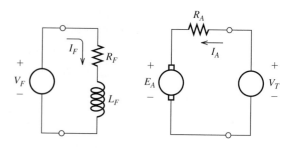

**Figure 15.21** Equivalent circuit for a separately excited dc motor. Speed can be controlled by varying either source voltage ($V_F$ or $V_T$).

reason for using two separate sources for the armature and field is to be able to control speed by varying one of the two sources.

### Permanent-Magnet Motors

Separately excited and permanent-magnet motors have similar characteristics to those of shunt-connected motors.

In a PM dc motor, the field is supplied by magnets mounted on the stator rather than by field coils. Its characteristics are similar to those of the separately excited machine except that the field cannot be adjusted. PM motors have several advantages compared to motors with field windings. First, no power is required to establish the field—leading to better efficiency. Second, PM motors can be smaller than equivalent machines with field windings. PM motors are common in applications calling for fractional- or subfractional-horsepower sizes. Typical applications include fan and power-window motors in automobiles.

PM motors also have some disadvantages. The magnets can become demagnetized by overheating or because of excessive armature currents. Also, the flux density magnitude is smaller in PM motors than in wound-field machines. Consequently, the torque produced per ampere of armature current is smaller in PM motors than in wound-field motors with equal power ratings. PM motors are confined to operation at lower torque and higher speed than wound-field motors with the same power rating.

## 15.5 SERIES-CONNECTED DC MOTORS

The equivalent circuit of a **series-connected dc motor** is shown in Figure 15.22. Notice that the field winding is in series with the armature. In this section, we will see that the series connection leads to a torque–speed characteristic that is useful in many applications.

Field windings are designed differently for series-connected machines than they are for shunt-connected machines.

In series dc motors, the field windings are made of larger diameter wire and the field resistances are much smaller than those of shunt machines of comparable size. This is necessary to avoid dropping too much of the source voltage across the field winding.

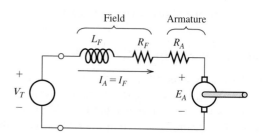

**Figure 15.22** Equivalent circuit of the series-connected dc motor.

Next, we derive the relationship between torque and speed for the series motor. We use a linear equation to approximate the relationship between magnetic flux and field current. In equation form, we have

$$\phi = K_F I_F \tag{15.28}$$

in which $K_F$ is a constant that depends on the number of field windings, the geometry of the magnetic circuit, and the $B$–$H$ characteristics of the iron. Of course, the actual relationship between $\phi$ and $I_F$ is nonlinear, due to magnetic saturation of the iron. (A plot of $\phi$ versus $I_F$ has exactly the same shape as the magnetization curve of the machine.) However, Equation 15.28 will give us insight into the behavior of the series dc motor. Later, we consider saturation effects.

Because $I_A = I_F$ in the series machine, we have

$$\phi = K_F I_A \tag{15.29}$$

Using Equation 15.29 to substitute for $\phi$ in Equations 15.15 and 15.16, we obtain

$$E_A = K K_F \omega_m I_A \tag{15.30}$$

and

$$T_{\text{dev}} = K K_F I_A^2 \tag{15.31}$$

If we apply Kirchhoff's voltage law to the equivalent circuit shown in Figure 15.22, we get

$$V_T = R_F I_A + R_A I_A + E_A \tag{15.32}$$

As usual, we are assuming steady-state conditions so that the voltage across the inductance is zero.

Then using Equation 15.30 to substitute for $E_A$ in Equation 15.32 and solving for $I_A$, we have

$$I_A = \frac{V_T}{R_A + R_F + K K_F \omega_m} \tag{15.33}$$

Finally, using Equation 15.33 to substitute for $I_A$ in Equation 15.31, we obtain the desired relationship between torque and speed:

$$T_{\text{dev}} = \frac{K K_F V_T^2}{(R_A + R_F + K K_F \omega_m)^2} \tag{15.34}$$

Torque–speed relationship for series-connected dc machines.

A plot of torque versus speed for the series dc motor is shown in Figure 15.23. The figure shows a plot of Equation 15.34 as well as an actual curve of torque versus speed, illustrating the effects of rotational loss and magnetic saturation. Equation 15.34 predicts infinite no-load speed. (In other words, for $T_{\text{dev}} = 0$, the speed must be infinite.) Yet, at high speeds, rotational losses due to windage and eddy currents become large, and the motor speed is limited.

However, in some cases, *the no-load speed can become large enough to be dangerous. It is important to have protection devices that remove electrical power to a series machine when the load becomes disconnected.*

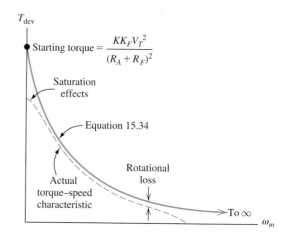

**Figure 15.23** Torque–speed characteristic of the series-connected dc motor.

At very low speeds, Equation 15.33 shows that the current $I_F = I_A$ becomes large. Then, magnetic saturation occurs. Therefore, the starting torque is not as large as predicted by Equation 15.34.

---

**Example 15.5    Series-Connected DC Motor**

A series-connected dc motor runs at $n_{m1} = 1200$ rpm while driving a load that demands a torque of 12 Nm. Neglect the resistances, rotational loss, and saturation effects. Find the power output. Then, find the new speed and output power if the load torque increases to 24 Nm.

**Solution**    Since we are neglecting losses, the output torque and power are equal to the developed torque and power, respectively. First, the angular speed is

$$\omega_{m1} = n_{m1} \times \frac{2\pi}{60} = 125.7 \text{ rad/s}$$

and the output power is

$$P_{dev1} = P_{out1} = \omega_{m1} T_{out1} = 1508 \text{ W}$$

Setting $R_A = R_F = 0$ in Equation 15.34 gives

$$T_{dev} = \frac{KK_F V_T^2}{(R_A + R_F + KK_F \omega_m)^2} = \frac{V_T^2}{KK_F \omega_m^2}$$

Thus, for a fixed supply voltage $V_T$, torque is inversely proportional to speed squared, and we can write

$$\frac{T_{dev1}}{T_{dev2}} = \frac{\omega_{m2}^2}{\omega_{m1}^2}$$

Solving for $\omega_{m2}$ and substituting values, we have

$$\omega_{m2} = \omega_{m1} \sqrt{\frac{T_{dev1}}{T_{dev2}}} = 125.7 \sqrt{\frac{12}{24}} = 88.88 \text{ rad/s}$$

which corresponds to

$$n_{m2} = 848.5 \text{ rpm}$$

Finally, the output power with the heavier load is

$$P_{out2} = T_{dev2}\omega_{m2} = 2133 \text{ W}$$

**Exercise 15.8**   Find the speed and output power in Example 15.5 for a load torque of $T_{dev3} = 6$ Nm.
**Answer**   $P_{out3} = 1066$ W, $\omega_{m3} = 177.8$ rad/s, $n_{m3} = 1697$ rpm.

**Exercise 15.9**   Repeat Example 15.5 for a shunt-connected motor. (In an ideal shunt-connected motor, the field resistance would be very large rather than zero.)
**Answer**   With $R_A = 0$ and fixed $V_T$, the shunt motor runs at constant speed, independent of load. Thus, $n_{m1} = n_{m2} = 1200$ rpm, $P_{out1} = 1508$ W, and $P_{out2} = 3016$ W.

Notice that by comparing the results of Exercise 15.9 with those of Example 15.5 we find that the output power variation is larger for the shunt-connected motor than for the series-connected motor.

### Universal Motors

Equation 15.34 shows that the torque produced by the series dc motor is proportional to the square of the source voltage. Thus, the direction of the torque is independent of the polarity of the applied voltage. The series-connected machine can be operated from a single-phase ac source, provided that the stator is laminated to avoid excessive losses due to eddy currents. Since the field and armature inductances have nonzero impedances for ac currents, the current is not as large for an ac source as it would be for a dc source of the same average magnitude.

The universal motor is an important type of ac motor.

Series motors that are intended for use with ac sources are called **universal motors** because in principle they can operate from either ac or dc. Any time that you examine an ac motor and find brushes and a commutator, you have a universal motor. Compared with other types of single-phase ac motors, the universal motor has several advantages:

1. For a given weight, universal motors produce more power than other types. This is a large advantage for hand-held tools and small appliances, such as drills, saws, mixers, and blenders.

2. The universal motor produces large starting torque without excessive current.

3. When load torque increases, the universal motor slows down. Hence, the power produced is relatively constant, and the current magnitude remains within reasonable bounds. (In contrast, shunt dc motors or ac induction motors tend to run at constant speed and are more prone to drawing excessive currents for high-torque loads.) Thus, the universal motor is more suitable for loads that demand a wide range of torque, such as drills and food mixers. (For the same reason, series dc motors are used as starter motors in automobiles.)

4. Universal motors can be designed to operate at very high speeds, whereas we will see that other types of ac motors are limited to 3600 rpm, assuming a 60-Hz source.

One disadvantage of universal motors (as well as dc machines in general) is that the brushes and commutators wear out relatively quickly. Thus, the service life is much less than for ac induction motors. Induction motors are better choices than universal motors in applications that need to run often over a long life, such as refrigerator compressors, water pumps, or furnace fans.

## 15.6  SPEED CONTROL OF DC MOTORS

Speed control is an important consideration in the design of many systems.

Several methods can be used to control the speed of dc motors:

**1.** Vary the voltage supplied to the armature circuit while holding the field constant.

**2.** Vary the field current while holding the armature supply voltage constant.

**3.** Insert resistance in series with the armature circuit.

In this section, we discuss briefly each of these approaches to speed control.

### Variation of the Supply Voltage

This method is applicable to separately excited motors and PM motors. For the shunt motor, varying the supply voltage is not an appropriate method of speed control, because the field current and flux vary with $V_T$. The effects of increasing both armature supply voltage and the field current tend to offset one another, resulting in little change in speed.

In normal operation, the drop across the armature resistance is small compared to $E_A$, and we have

$$E_A \cong V_T$$

Since we also have

$$E_A = K\phi\omega_m$$

we can write

$$\omega_m \cong \frac{V_T}{K\phi} \tag{15.35}$$

Thus, the speed of a separately excited motor with constant field current or of a PM motor is approximately proportional to the source voltage.

Variation of the supply voltage is also appropriate for control of a series-connected dc motor; however, the flux does not remain constant in this case. Equation 15.34 shows that the torque of a series machine is proportional to the square of the source voltage at any given speed. Thus, depending on the torque–speed characteristic of the load, the speed varies with applied voltage. Generally, higher voltage produces higher speed.

### Variable DC Voltage Sources

Historically, variable dc voltages were obtained from dc generators. For example, one popular approach was the **Ward Leonard system**, in which a three-phase induction motor drives a dc generator that in turn supplies a variable dc voltage to the motor to be controlled. The magnitude and polarity of the dc supply voltage are controlled by using a rheostat or switches to vary the field current of the dc generator. A disadvantage of this scheme is that three machines are needed to drive one load.

Since the advent of high-power electronics, a more economical approach is to use a rectifier to convert three-phase ac into dc, as illustrated in Figure 15.24. The resulting dc voltage $v_L$ has some ripple, but a smoother voltage can be obtained with a full-wave version of the rectifier using six diodes. In any case, it is not necessary for the dc source supplying motors to be absolutely free of ripple, because the inductances and inertia tend to smooth the response.

Once a constant dc source has been created, an electronic switching circuit can be used to control the average voltage delivered to a load, as illustrated in Figure 15.25.

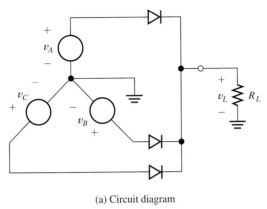

(a) Circuit diagram

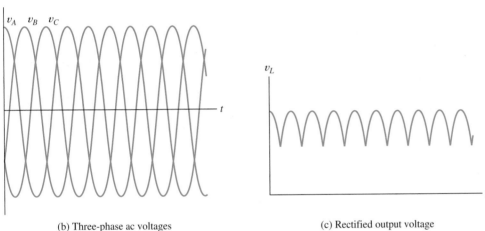

(b) Three-phase ac voltages

(c) Rectified output voltage

**Figure 15.24** Three-phase half-wave rectifier circuit used to convert ac power to dc.

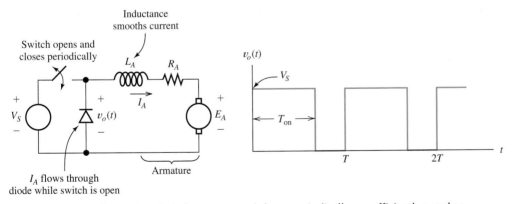

**Figure 15.25** An electronic switch that opens and closes periodically can efficiently supply a variable dc voltage to a motor from a fixed dc supply voltage.

(In Chapter 11 and Chapter 12, we showed how electronic devices such as BJTs and FETs can be used as switches. In high-power electronics, similar devices such as silicon-controlled rectifiers perform the switching function.)

The switch periodically opens and closes with period $T$, spending $T_{\text{on}}$ in the closed state and the remainder of the period open. The inductance $L_A$ tends to cause the armature current to continue flowing when the switch is open. Thus, the armature current $I_A$ is nearly constant even though the voltage $v_o(t)$ switches rapidly between zero and $V_s$. The diode provides a path for the armature current while the switch is open. The average value of the voltage applied to the motor is given by

$$V_T = V_s \frac{T_{on}}{T} \tag{15.36}$$

Thus, the average voltage, and therefore the speed of the motor, can be controlled by varying the fraction of the period for which the switch is closed.

## Speed Control by Varying the Field Current

The speed of either a shunt-connected or a separately excited motor can be controlled by varying the field current. The circuit for the shunt-connected machine was shown in Figure 15.16, in which the rheostat $R_{adj}$ provides the means to control field current.

On the other hand, PM motors have constant flux. In series-connected motors, the field current is the same as the armature current and cannot be independently controlled. Thus, using field current to control speed is not appropriate for either of these types of motors.

To understand the effect of field current on motor torque and speed, let us review the following equations for the shunt-connected or separately excited motor:

$$E_A = K\phi\omega_m$$

$$I_A = \frac{V_T - E_A}{R_A}$$

$$T_{dev} = K\phi I_A$$

Now, consider what happens when $I_F$ is reduced (by increasing $R_{adj}$). Reducing $I_F$ reduces the flux $\phi$. Immediately, the induced voltage $E_A$ is reduced. This in turn causes $I_A$ to increase. In fact, the percentage increase in $I_A$ is much greater than the percentage reduction in $\phi$, because $V_T$ and $E_A$ are nearly equal. Thus, $I_A = (V_T - E_A)/R_A$ increases rapidly when $E_A$ is reduced. Two of the terms in the equation for torque $T_{dev} = K\phi I_A$ change in opposite directions; specifically, $\phi$ falls and $I_A$ rises. However, the change in $I_A$ is much greater and the torque rises rapidly when $I_F$ falls. (You can verify this by comparing your solution to Exercise 15.7 to the values in Example 15.4.)

## Danger of an Open Field Circuit

What happens in a shunt or separately excited motor if the field circuit becomes open circuited and $\phi$ falls to nearly zero? (Because of residual magnetization, the field is not zero for zero field current.) The answer is that $I_A$ becomes very large and the machine speeds up very rapidly. In fact, it is possible for excessive speeds to cause the armature to fly apart. Then, in a matter of seconds, the machine is reduced to a pile of useless scrap consisting of loose windings and commutator bars. Thus, it is important to operate shunt machines with well-designed protection circuits that open the armature circuit automatically when the field current vanishes.

It is important to operate shunt machines with well-designed protection circuits that automatically open the armature circuit when the field current vanishes.

## Speed Control by Inserting Resistance in Series with the Armature

Another method for controlling the speed of a dc motor is to insert additional resistance in series with the armature circuit. This approach can be applied to all

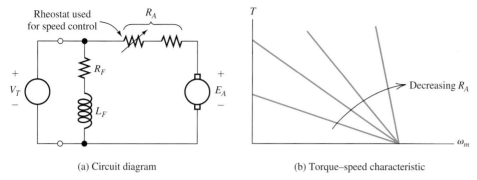

(a) Circuit diagram                     (b) Torque–speed characteristic

**Figure 15.26** Speed can be adjusted by varying a rheostat that is in series with the armature.

types of dc motors: shunt, separately excited, series, or PM. For example, a shunt-connected motor with added armature resistance is illustrated in Figure 15.26(a). We denote the total resistance as $R_A$, which consists of the control resistance plus the resistance of the armature winding. The torque–speed relationship for a shunt-connected motor is given by Equation 15.27, repeated here for convenience:

$$T_{dev} = \frac{K\phi}{R_A}(V_T - K\phi\omega_m)$$

Plots of the torque–speed characteristic for various resistances are shown in Figure 15.26(b). Similar results apply to separately excited and PM motors.

Starting controls for shunt or separately excited dc motors usually place resistance in series with the armature to limit armature current to reasonable values while the machine comes up to speed.

A disadvantage of inserting resistance in series with the armature to control speed is that it is wasteful of energy. When running at low speeds, much of the energy taken from the source is converted directly into heat in the series resistance.

Equation 15.34 gives the torque for series-connected machines. For convenience, the equation is repeated here:

$$T_{dev} = \frac{KK_FV_T^2}{(R_A + R_F + KK_F\omega_m)^2}$$

Notice that if $R_A$ is made larger by adding resistance in series with the armature, the torque is reduced for any given speed.

Look at Figures 15.26, 15.27, and 15.28 to see how the torque–speed characteristics of shunt-connected and separately excited dc motors can be changed by varying armature resistance, armature supply voltage, or field current to achieve speed control.

**Exercise 15.10**   Why is variation of the supply voltage $V_T$ in the shunt machine (see Figure 15.16) an ineffective way to control speed?
**Answer**   Decreasing $V_T$ decreases the field current, and therefore the flux. In the linear portion of the magnetization curve, the flux is proportional to the field current. Thus, reduction of $V_T$ leads to reduction of $\phi$, and according to Equation 15.35, the speed remains constant. (Actually, some variation of speed may occur due to saturation effects.)                                                                              □

**Exercise 15.11**   Figure 15.26(b) shows a family of torque–speed curves for various values of $R_A$. Sketch a similar family of torque–speed characteristics for a separately excited machine (see Figure 15.21) with various values of $V_T$ and constant field current.
**Answer**   The torque–speed relationship is given by Equation 15.27. With constant field current, $\phi$ is constant. The family of characteristics is shown in Figure 15.27.   □

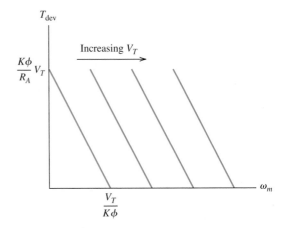

**Figure 15.27** Torque versus speed for the separately excited dc motor for various values of armature supply voltage $V_T$.

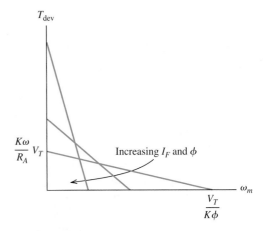

**Figure 15.28** Effect of varying $I_F$ on the torque–speed characteristics of the shunt-connected or separately excited dc motor.

**Exercise 15.12**  Sketch a family of torque–speed characteristics for a shunt-connected or separately excited machine with constant $V_T$ and variable $I_F$.

**Answer**  The torque–speed relationship is given in Equation 15.27. As field current is increased, the flux $\phi$ increases. The family of characteristics is shown in Figure 15.28.    □

## 15.7  DC GENERATORS

Generators convert kinetic energy from a **prime mover**, such as a steam turbine or a diesel engine, into electrical energy. When dc power is needed, we can use a dc generator or an ac source combined with a rectifier. The trend is toward ac sources and rectifiers; however, many dc generators are in use, and for some applications they are still a good choice.

Several connections, illustrated in Figure 15.29, are useful for dc generators. We will discuss each type of connection briefly and conclude with an example illustrating performance calculations for the separately excited generator.

### Separately Excited DC Generators

The equivalent circuit for a separately excited dc generator is shown in Figure 15.29(a). A prime mover drives the armature shaft at an angular speed $\omega_m$, and the external dc

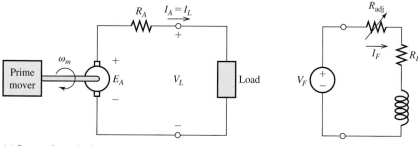

(a) Separately excited

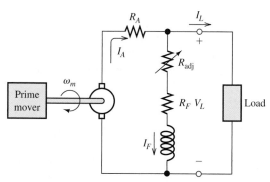

(b) Shunt connected

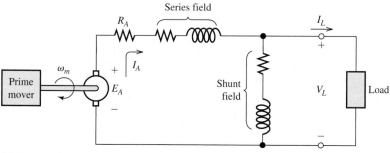

(c) Compound connected

**Figure 15.29** DC generator equivalent circuits.

source $V_F$ supplies current $I_F$ to the field coils. The induced armature voltage causes current to flow through the load. Because of the drop across the armature resistance, the load voltage $V_L$ decreases as the load current $I_L$ increases, assuming constant speed and field current. This is illustrated in Figure 15.30(a).

For some applications, it is desirable for the load voltage to be nearly independent of the load current. A measure of the amount of decrease in load voltage with current is the percentage load voltage regulation given by

$$\text{voltage regulation} = \frac{V_{NL} - V_{FL}}{V_{FL}} \times 100\% \qquad (15.37)$$

in which $V_{NL}$ is the no-load voltage (i.e., $I_L = 0$) and $V_{FL}$ is the full-load voltage (i.e., with full-rated load current).

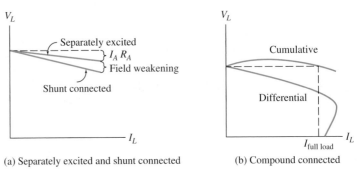

**Figure 15.30** Load voltage versus load current for various dc generators.

One of the advantages of the separately excited dc generator is that the load voltage can be adjusted over a wide range by varying the field current either by changing $V_F$ or by changing $R_{adj}$. Also, the load voltage is proportional to speed.

### Shunt-Connected DC Generators

One of the disadvantages of the separately excited dc generator is the need for a separate dc source to supply the field windings. This disadvantage is overcome in the shunt-connected machine, for which the field circuit is in parallel with the armature and the load, as shown in Figure 15.29(b). The output voltage can be adjusted by changing the resistance $R_{adj}$ that is in series with the field winding.

Initial buildup of voltage in the shunt-connected machine usually occurs because of the residual magnetic field of the iron. (Adjusting $R_{adj}$ to its minimum value and reversing the connections to the field winding may be needed to ensure voltage buildup.) However, if the machine becomes demagnetized, the induced armature voltage is zero, resulting in zero field current, and no output is produced. This can be remedied by briefly applying a dc source of the correct polarity to the field winding to create a residual field in the machine. Also, depending on the machine history, it is possible for the polarity of the output voltage to build up with the opposite polarity to that desired. This can be corrected by applying an external source of the correct polarity to the field winding or by reversing the connections to the machine.

The load regulation of the shunt-connected generator is poorer (i.e., larger) than that of the separately excited machine because the field current falls as the load current increases due to the drop across the armature resistance. This increased drop due to field weakening is illustrated in Figure 15.30(a).

### Compound-Connected DC Generators

It is possible to design a dc generator with both series and shunt windings. When both are connected, we have a compound-connected machine. Several variations of the connections are possible. Figure 15.29(c) illustrates a **long-shunt compound connection**. Another possibility is the **short-shunt compound connection**, in which the shunt field is directly in parallel with the armature and the series field is in series with the load. Furthermore, in either the short-shunt or the long-shunt, the field of the series field coil can either aid or oppose the field of the shunt field coil. If the fields aid, we have a **cumulative shunt connection**. On the other hand, if the fields oppose, we have a **differential shunt connection**. Thus, we have four types of connections in all.

It is possible to design a **fully compensated** cumulative-connected machine for which the full-load voltage is equal to the no-load voltage, as illustrated in Figure 15.30(b). Curvature of the voltage versus current characteristic is due to saturation effects. If the full-load voltage is less than the no-load voltage, the machine is said to be **undercompensated**. In an **overcompensated** machine, the full-load voltage is greater than the no-load voltage.

In a differential shunt connection, the output voltage falls rapidly with load current because the field of the series winding opposes that of the shunt winding. Considerable load current may flow even after the load voltage drops to zero. This is illustrated in Figure 15.30(b).

## Performance Calculations

Next, we illustrate performance calculations for the separately excited generator. Analysis of the other connections is left for the problems.

As for dc motors, the following equations apply to dc generators:

$$E_A = K\phi\omega_m \tag{15.38}$$

$$T_{\text{dev}} = K\phi I_A \tag{15.39}$$

Referring to Figure 15.29(a), we can write:

$$E_A = R_A I_A + V_L \tag{15.40}$$

$$V_F = (R_F + R_{\text{adj}})I_F \tag{15.41}$$

Figure 15.31 illustrates the power flow of a dc generator. The efficiency is given by

$$\text{efficiency} = \frac{P_{\text{out}}}{P_{\text{in}}} \times 100\% \tag{15.42}$$

Equations 15.37 through 15.42, the magnetization curve of the machine, and Figure 15.31 are the tools for analysis of the separately excited dc generator. (Recall that the magnetization curve is a plot of $E_A$ versus $I_F$ for a given speed.)

---

**Example 15.6**   **Separately Excited DC Generator**

A separately excited dc generator has $V_F = 140$, $R_F = 10 \, \Omega$, $R_{\text{adj}} = 4 \, \Omega$, $R_A = 0.065 \, \Omega$, the prime mover rotates the armature at a speed of 1000 rpm, and the magnetization curve is shown in Figure 15.19 on page 768. Determine the field current, the no-load voltage, the full-load voltage, and the percentage voltage regulation for a full-load current of 200 A. Assuming that the overall efficiency (not including the power supplied to the field circuit) of the machine is 85 percent, determine the input torque, the developed torque, and the losses associated with friction, windage, eddy currents, and hysteresis.

**Solution**   The field current is

$$I_F = \frac{V_F}{R_{\text{adj}} + R_F} = \frac{140}{4 + 10} = 10 \text{ A}$$

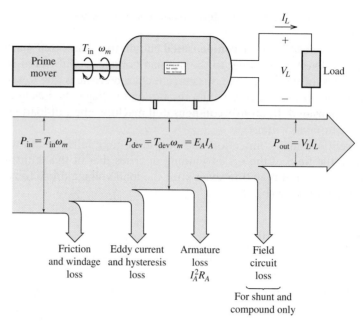

**Figure 15.31** Power flow in dc generators.

Then, referring to the magnetization curve on page 768, we find that $E_A = 280$ V for a speed of 1200 rpm. Equation 15.38 shows that $E_A$ is proportional to speed. So, for a speed of 1000 rpm, we have

$$E_A = 280 \frac{1000}{1200} = 233.3 \text{ V}$$

which is also the no-load voltage of the machine. For a load current of 200 A, we get

$$V_{FL} = E_A - R_A I_A = 233.3 - 200 \times 0.065 = 220.3 \text{ V}$$

Finally, we obtain

$$\text{voltage regulation} = \frac{V_{NL} - V_{FL}}{V_{FL}} \times 100\% = \frac{233.3 - 220.3}{220.3} \times 100\% = 5.900\%$$

The output power is

$$P_{\text{out}} = I_L V_{FL} = 200 \times 220.3 = 44.06 \text{ kW}$$

The developed power is the sum of the output power and the armature loss:

$$P_{\text{dev}} = P_{\text{out}} + R_A I_A^2 = 44060 + 0.065(200)^2 = 46.66 \text{ kW}$$

The angular speed is

$$\omega_m = n_m \frac{2\pi}{60} = 104.7 \text{ rad/s}$$

The input power is

$$P_{in} = \frac{P_{out}}{0.85} = \frac{44.06}{0.85} = 51.84 \text{ kW}$$

The power losses associated with friction, windage, eddy currents, and hysteresis are

$$P_{losses} = P_{in} - P_{dev} = 51.84 - 46.66 = 5.18 \text{ kW}$$

The input and developed torques are

$$T_{in} = \frac{P_{in}}{\omega_m} = \frac{51,840}{104.7} = 495.1 \text{ N} \cdot \text{m}$$

$$T_{dev} = \frac{P_{dev}}{\omega_m} = \frac{46,660}{104.7} = 445.7 \text{ N} \cdot \text{m} \qquad \blacksquare$$

**Exercise 15.13**   Repeat Example 15.6 for $R_{adj} = 0$.
**Answer**   $I_F = 14$ A, $V_{NL} = 260$, $V_{FL} = 247$,  voltage  regulation $= 5.263\%$, $P_{losses} = 6.1$ kW, $T_{in} = 555$ N $\cdot$ m, $T_{dev} = 497$ N $\cdot$ m. □

## Summary

1. An electrical motor consists of a rotor and a stator, both of which may contain windings. A given winding may be classified as either an armature winding or as a field winding.

2. The electrical source for a motor can be dc, single-phase ac, or three-phase ac.

3. Factors to consider in selecting an electrical motor for a given application include the electrical sources available, the output power required, load torque versus speed, the service-life requirements, efficiency, speed regulation, starting current, the desired operating speed, ambient temperature, and the acceptable frequency of maintenance.

4. Dc motors in common use contain brushes and commutators that reverse the connections to the armature conductors with rotation. The need for frequent maintenance of these parts is a significant disadvantage of dc motors.

5. An advantage of dc motors is the ease with which their speed can be controlled. However, ac motors used with modern electronic systems that can change the frequencies of ac sources are taking over in many speed-control applications.

6. The speed of a motor load combination self-adjusts to the point at which the output torque of the motor equals the torque demanded by the load. Thus, if the torque–speed characteristics of the motor and of the load are plotted on the same axes, the steady-state operating point is at the intersection of the characteristics.

7. The linear dc machine illustrates the basic principles of dc machines in an uncomplicated manner. However, it is not practical for the vast majority of applications.

8. Commonly used dc machines have field windings on the stator that establish an even number of magnetic poles. The armature windings on the rotor carry currents, resulting in forces induced by the field. To achieve aiding contributions to torque (and induced armature voltage), a commutator and a set of brushes reverse the connections to the armature conductors as they move between poles.

9. The magnetization curve of a dc machine is a plot of the induced armature voltage versus field current with the machine being driven at a constant speed. The equivalent circuit shown

in Figure 15.14 on page 763, the magnetization curve, and the following equations provide the basis for analyzing dc machines:The equivalent circuit shown in Figure 15.14, the magnetization curve, and the following equations provide the basis for analyzing dc machines:

$$E_A = K\phi\omega_m$$
$$T_{dev} = K\phi I_A$$
$$P_{dev} = E_A I_A = \omega_m T_{dev}$$

10. The equivalent circuit of a shunt-connected dc motor is shown in Figure 15.16 on page 766.The torque–speed characteristic is shown in Figure 15.18 on page 767. Speed can be controlled either by using $R_{adj}$ to vary the field current or by inserting a variable resistance in series with the armature. Variation of $V_T$ is not an effective means of speed control.

11. The equivalent circuit of a separately excited dc motor is shown in Figure 15.21 on page 770. The torque–speed characteristic is identical to that of the shunt-connected machine shown in Figure 15.18 on page 767. Speed can be controlled by varying the field current, by varying the armature source voltage $V_T$, or by inserting additional resistance in series with the armature.

12. The characteristics of a permanent-magnet dc motor are similar to those of a separately excited motor, except that no means is available to vary the field. Speed control can be achieved by varying the armature supply voltage or by inserting additional resistance in series with the armature.

13. The equivalent circuit of the series-connected dc motor is shown in Figure 15.22 on page 770. In the normal range of operation, its torque is almost inversely proportional to the square of its speed. The series-connected dc motor is suitable for starting heavy loads. It can reach dangerous speeds if the load is totally removed.

14. The universal motor is basically a series dc motor designed for operation from an ac source. It is to be found in applications where a large power-to-weight ratio is needed. However, due to commutator wear, its service life is limited. Its speed can be controlled by varying the applied voltage or by inserting series resistance.

15. Historically, variable dc voltages for speed control of dc motors were obtained by employing dc generators. Presently, the approach of choice is to use rectifiers that first convert ac into dc. Then, electronic chopper circuits with adjustable duty factors provide variable average voltages.

16. Figures 15.26, 15.27, and 15.28 on pages 777, 778 and 778 show how the torque–speed characteristics of shunt-connected and separately excited dc motors can be changed by varying armature resistance, armature supply voltage, or field current to achieve speed control.

17. Dc generators can be separately excited, shunt connected, or compound connected. Analysis of dc generators is similar to that for dc motors.

## Problems

**Section 15.1: Overview of Motors**

**P15.1.** List the two principal types of three-phase ac motors. Which is in more common use?

**P15.2.** What types of motors contain brushes and a commutator? What is the function of these parts?

**P15.3.** What two types of windings are used in electrical machines? Which type is not used in permanent-magnet machines? Why not?

**P15.4.** In what application would dc motors be more advantageous than ac motors?

**P15.5.** Name the principal parts of a rotating electrical machine.

**\*P15.6.** List two practical disadvantages of dc motors compared to single-phase ac induction motors for supplying power to a ventilation fan in a home or small business.

**\*P15.7.** A three-phase induction motor is rated at 5 hp, 1760 rpm, with a line-to-line voltage of

---

\*Denotes that answers are contained in the Student Solutions files. See Appendix E for more information about accessing the Student Solutions.

220 V rms. The no-load speed of the motor is 1800 rpm. Determine the percentage speed regulation.

**\*P15.8.** A 440-V-rms (line-to-line voltage) three-phase induction motor runs at 1150 rpm driving a load requiring 15 Nm of torque. The line current is 3.4 A rms at a power factor of 80 percent lagging. Find the output power, the power loss, and the efficiency.

**\*P15.9.** A certain 25-hp three-phase induction motor operates from a 440-V-rms (line-to-line) three-phase source. The full-load speed is 1750 rpm. The motor has a starting torque equal to 200 percent of its full-load torque when started at rated voltage. For an engineering estimate, assume that the starting torque of an induction motor is proportional to the square of the applied voltage. To reduce the starting current of the motor, we decide to start it with a line-to-line voltage of 220 V. Estimate the starting torque with this reduced line voltage.

**P15.10.** A three-phase induction motor is rated at 5 hp, 1760 rpm, with a line-to-line voltage of 220 V rms. Find the output torque and angular velocity of the motor under full-load conditions.

**P15.11.** Operating from a line-to-line voltage of 440 V rms with a line current of 14 A rms and a power factor of 85 percent, a three-phase induction motor produces an output power of 6.5 hp. Determine the losses in watts and the efficiency of the motor.

**P15.12.** A motor drives a load for which the torque required is given by

$$T_{\text{load}} = \frac{800}{20 + \omega_m} \text{ Nm}$$

The torque–speed characteristic of the motor is shown in Figure P15.12. **a.** Will this system start from zero speed? Why or why not? **b.** Suppose that the system is brought up to speed by an auxiliary driver, which is then disengaged. In principle, at what two constant speeds can the system rotate? **c.** If the system is operating at the lower of the two speeds and power to the motor is briefly interrupted so the system slows by a few radians per second, what will happen after power is restored? Why? **d.** Repeat

(c) if the system is initially operating at the higher speed.

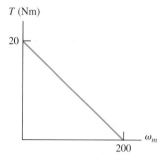

**Figure P15.12**

**P15.13.** A motor has output torque given by

$$T_{\text{out}} = 10^{-2} (60\pi - \omega_m)\omega_m$$

where $\omega_m$ is angular velocity in rad/s and $T_{\text{out}}$ is the output torque in newton meters. **a.** Find the no-load speed of the motor. **b.** At what speed between zero and the no-load speed is the output torque maximum? What is the maximum output torque? **c.** At what speed between zero and the no-load speed is the output power maximum? What is the maximum output power? **d.** Find the starting torque of the motor. How could this motor be started?

**P15.14.** A three-phase induction motor is rated at 5 hp, 1760 rpm, with a line-to-line voltage of 220 V rms. The motor has a power factor of 80 percent lagging and an efficiency of 75 percent under full-load conditions. Find the electrical input power absorbed by the motor under full-load conditions. Also, find the rms line current.

**P15.15.** We want a four-pole synchronous motor that operates at a speed of 1000 rpm. Determine the frequency of the ac source. List several other speeds that can be achieved by using synchronous motors operating from this ac source. What is the highest speed achievable?

**P15.16.** A 220-V-rms (line-to-line), 60-Hz, three-phase induction motor operates at 3500 rpm while delivering its rated output power of 3 hp. The line current is 8 A rms and the losses are 300 W. Find the input power, power factor, and efficiency.

**P15.17.** A certain 25-hp three-phase induction motor operates from a 440-V-rms (line-to-line) three-phase source and draws a line current of 35 A rms at a power factor of 83 percent lagging under rated full-load conditions. The full-load speed is 1750 rpm. Under no-load conditions, the speed is 1797 rpm, and the line current is 6.5 A rms at a power factor of 30 percent lagging. Find the power loss and efficiency with full load, the input power with no load, and the speed regulation.

**P15.18.** Consider a system consisting of a motor driving a load. The motor has the torque–speed characteristic shown in Figure P15.12. The load is a fan requiring a torque given by

$$T_{\text{load}} = K\omega_m^2$$

At a speed of $n = 1000$ rpm, the power absorbed by the load is 0.75 hp. Determine the speed of the system in radians per second and the power delivered to the fan in watts. Convert your answers to rpm and horsepower.

**P15.19.** A 220-V-rms (line-to-line), 60-Hz, three-phase induction motor operates at 3500 rpm while delivering its rated full-load output power. Estimate the no-load speed and speed regulation for the motor.

### Section 15.2: Principles of DC Machines

**\*P15.20.** Consider the linear dc machine of Figure P15.20. When the switch closes, is the force on the bar toward the top of the page or toward the bottom? Determine the magnitude of the initial (starting) force. Also, determine the final velocity of the bar neglecting friction.

**\*P15.21.** Consider the linear dc machine shown in Figure 15.6 on page 754 with no load force applied. What happens to the steady-state velocity of the bar if: **a.** the source voltage $V_T$ is doubled in magnitude; **b.** the resistance $R_A$ is doubled; **c.** the magnetic flux density $B$ is doubled in magnitude?

**P15.22.** Consider the linear dc machine shown in Figure 15.6 on page 754. What happens to the initial force (i.e., starting force) induced in the bar if: **a.** the source voltage $V_T$ is doubled in magnitude; **b.** the resistance $R_A$ is doubled; **c.** the magnetic flux density $B$ is doubled in magnitude?

**P15.23.** Suppose that an external force of 10 N directed toward the top of the page is applied to the bar as shown in Figure P15.23. In steady state, is the machine acting as a motor or as a generator? Find the power supplied by or absorbed by: **a.** the electrical voltage source $V_T$; **b.** $R_A$; **c.** the external force.

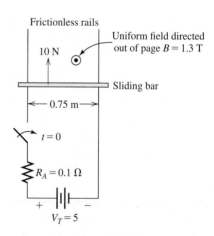

**Figure P15.23**

**P15.24.** Suppose that we wish to design a linear motor based on Figure 15.6 on page 754 that can deliver 1 hp at a steady bar velocity of 20 m/s. The flux density is limited to 1 T by the magnetic properties of the materials to be used. The length of the bar is to be 0.5 m, and the resistance is 0.05 $\Omega$. Find the current

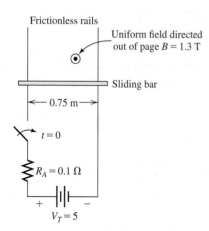

**Figure P15.20**

$i_A$, the source voltage $V_T$, and the efficiency of the machine. Assume that the only loss is due to the resistance $R_A$.

**P15.25.** Suppose that an external force of 10 N directed toward the bottom of the page is applied to the bar as shown in Figure P15.25. In steady state, is the machine acting as a motor or as a generator? Find the power supplied by or absorbed by: **a.** the electrical voltage source $V_T$; **b.** $R_A$; **c.** the external force.

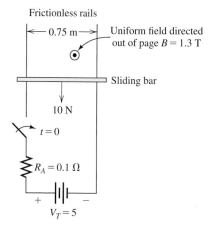

Figure P15.25

**P15.26.** We have presented the linear dc machine with an applied magnetic field, such as we might have in a dc motor. However, forces are exerted on the slider by the magnetic field produced by the currents in the rails. This is the principle of the **electromagnetic rail gun** for which you can find many references on the web including many practical construction tips. A version of such a rail gun is shown in Figure P15.26. **a.** When the switch closes, in which direction is force exerted on the projectile? Use physical principles discussed in this text to explain how you arrived at your answer. **b.** Suppose that the projectile mass is 3 g (which is about that of a penny). Assuming that all of the energy stored in the capacitor is transferred to kinetic energy in the slider, determine its final velocity. (*Note:* The highest velocity achievable by ordinary rifle bullets is about 1200 m/s.) **c.** List as many

effects as you can that cause the velocity to be lower than the value calculated in part (b).

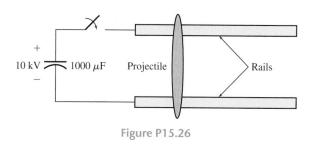

Figure P15.26

## Section 15.3: Rotating DC Machines

*\*P15.27.* Suppose that we are designing a 1200-rpm dc motor to run from a 240-V source. We have determined that the flux density will be 1 T because smaller fluxes make inefficient use of the iron and larger fluxes result in saturation. The radius of the rotor (and thus, the torque arm for the armature conductors) is 0.1 m. The lengths of the armature conductors are 0.3 m. Approximately how many armature conductors must be placed in series in this machine?

*\*P15.28.* An alternative way to determine the direction of the torque produced by a dc machine is to consider the interaction of the magnetic poles produced by the armature current with the stator poles. Consider the cross section of a two-pole dc motor shown in Figure 15.10 on page 760. The magnetic poles established on the stator by the field windings are shown. For the armature current directions shown in the figure, find the locations and label the rotor poles. Of course, the rotor poles try to align with unlike stator poles. Which direction is the resulting torque, clockwise or counterclockwise? Repeat for the four-pole machine shown in Figure 15.11 on page 760.

*\*P15.29.* A dc motor operates with a load that demands constant developed torque. With $V_T = 200$ V, the motor operates at 1200 rpm and has $I_A = 10$ A. The armature resistance is 5 Ω and the field current remains constant. Determine the speed if $V_T$ is increased to 250 V.

**\*P15.30.** A certain dc motor has $R_A = 1.3\ \Omega$, $I_A = 10$ A, and produces a back emf $E_A = 240$ V, while operating at a speed of 1200 rpm. Determine the voltage applied to the armature, the developed torque, and the developed power.

**P15.31.** Consider a motor having the model shown in Figure 15.14 on page 763. The motor runs at a speed of 1200 rpm and has $R_F = 150\ \Omega$, $V_F = V_T = 180$ V, $I_A = 10$ A, and $R_A = 1.2\ \Omega$. Find $E_A$, $T_{dev}$, $P_{dev}$, and the power converted to heat in the resistances.

**P15.32.** A permanent-magnet dc motor has $R_A = 7\ \Omega$, $V_T = 240$ V, and operates under no-load conditions at a speed of 1500 rpm with $I_A = 1$ A. A load is connected and the speed drops to 1300 rpm. Determine the efficiency of the motor under loaded conditions. Assume that the losses consist solely of heating of $R_A$ and frictional loss torque that is independent of speed.

**P15.33.** A certain dc motor produces a back emf of $E_A = 240$ V at a speed of 1200 rpm. Assume that the field current remains constant. Find the back emf for a speed of 600 rpm and for a speed of 1500 rpm.

**P15.34.** Under no-load conditions, a certain motor operates at 1200 rpm with an armature current of 0.5 A and a terminal voltage of 480 V. The armature resistance is $2\ \Omega$. Determine the speed and speed regulation if a load demanding a torque of 50 Nm is connected to the motor. Assume that the losses consist solely of heating of $R_A$ and frictional loss torque that is independent of speed.

**P15.35.** Consider the two-pole motor shown in Figure 15.10 on page 760. The gap between the rotor and stator is 1.5 mm. Each of the two field coils has 250 turns and carries a current of 3 A. Assume that the permeability of the iron is infinite. **a.** Determine the flux density in the air gap. **b.** Each armature conductor carries a current of 30 A and has a length of 0.5 m. Find the force induced in each conductor.

**P15.36.** Sometimes the stator, particularly the yoke (shown in Figure 15.10 on page 760 shown in Figure 15.10), of a dc machine is not laminated. However, it is always necessary to laminate the rotor. Explain.

**P15.37.** A certain motor has an induced armature voltage of 200 V at $n_{m1} = 1200$ rpm. Suppose that this motor is operating at a speed of $n_{m2} = 1500$ rpm with a developed power of 5 hp. Find the armature current and the developed torque.

### Section 15.4: Shunt-Connected and Separately Excited DC Motors

**\*P15.38.** A certain shunt-connected dc motor has $R_A = 1\ \Omega$, $R_F + R_{adj} = 200\ \Omega$, and $V_T = 200$ V. At a speed of 1200 rpm, the rotational losses are 50 W and $E_A = 175$ V. **a.** Find the no-load speed in rpm. **b.** Plot $T_{dev}$, $I_A$, and $P_{dev}$ versus speed for speed ranging from zero to the no-load speed.

**\*P15.39.** Is a magnetization curve needed for a permanent-magnet dc motor? Explain.

**\*P15.40.** A shunt dc motor has $R_A = 0.1\ \Omega$ and $V_T = 440$ V. For an output power of 50 hp, we have $n_m = 1500$ rpm and $I_A = 103$ A. The field current remains constant for all parts of this problem. **a.** Find the developed power, power lost in $R_A$, and the rotational losses. **b.** Assuming that the rotational power loss is proportional to speed, find the no-load speed of the motor.

**\*P15.41.** A permanent-magnet dc motor has $R_A = 0.5\ \Omega$. With no load, it operates at 1070 rpm and draws 0.5 A from a 12.6-V source. Assume that rotational power loss is proportional to speed. Find the output power and efficiency for a load that drops the speed to 950 rpm.

**\*P15.42.** A shunt-connected dc motor has $K\phi = 1$ V/(rad/s), $R_A = 1.2\ \Omega$, and $V_T = 200$ V. Find the two speeds for which the developed power is 5 hp. Neglect field loss and rotational loss. Find the value of $I_A$ and efficiency for each speed. Which answer is most likely to be in the normal operating range of the machine?

**P15.43.** A permanent-magnet automotive fan motor draws 20 A from a 12-V source when the rotor is locked (i.e., held motionless). The motor has a speed of 800 rpm and draws 3.5

A when operating the fan with a terminal voltage of $V_T = 12$ V. Assume that the load (including the rotational losses) requires a developed torque that is proportional to the square of the speed. Find the speed for operation at 10 V. Repeat for 14 V.

**\*P15.44.** A shunt-connected motor has the magnetization curve shown in Figure P15.44. Ignore rotational losses in this problem. The motor is supplied from a source of $V_T = 240$ V and has $R_A = 1.5$ $\Omega$. The total field resistance is $R_F + R_{adj} = 240$ $\Omega$. **a.** Find the no-load speed. **b.** A load is connected and the speed drops by 6 percent. Find the load torque, output power, armature current, field loss, and armature loss.

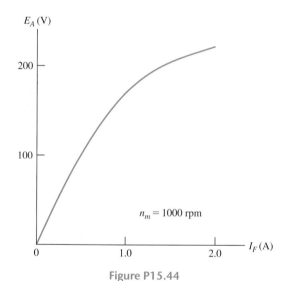

**Figure P15.44**

**P15.45.** A shunt-connected motor delivers an output power of 24 hp at 1200 rpm while operating from a source voltage of 440 V and drawing $I_L = 50$ A. The resistances are $R_A = 0.05$ $\Omega$ and $R_F + R_{adj} = 100$ $\Omega$. Find the developed torque and efficiency of the motor.

**P15.46.** A separately excited dc motor (see the equivalent circuit shown in Figure 15.21 on page 770) has $R_A = 1.3$ $\Omega$ and $V_T = 220$ V. For an output power of 3 hp, $n_m = 950$ rpm and $I_A = 12.2$ A. The field current remains constant for all parts of this problem. **a.** Find the developed power, developed torque, power lost in $R_A$, and the rotational losses.

**b.** Assuming that the rotational power loss is proportional to speed, find the no-load speed of the motor.

**P15.47.** A shunt-connected 5-hp dc motor is rated for operation at $V_T = 200$ V, $I_L = 23.3$ A, and $n_m = 1500$ rpm. Furthermore, $I_F = 1.5$ A and $R_A = 0.4$ $\Omega$. Under rated conditions find: **a.** the input power; **b.** the power supplied to the field circuit; **c.** the power lost in the armature resistance; **d.** the rotational loss; **e.** the efficiency.

**P15.48.** A permanent-magnet automotive fan motor draws 20 A from a 12-V source when the rotor is locked (i.e., held motionless). **a.** Find the armature resistance. **b.** Find the maximum developed power that this motor can produce when operated from a 12-V source. **c.** Repeat part (b) for operation at 10 V (this represents a nearly dead battery) and at 14 V (this is the terminal voltage during charging of the battery after the engine is started).

**P15.49.** A shunt-connected dc motor has zero rotational losses and $R_A = 0$. Assume that $R_F + R_{adj}$ is constant [except in part (d)] and that $\phi$ is directly proportional to field current. For $V_T = 200$ V and $P_{out} = 2$ hp, the speed is 1200 rpm. What is the effect on $I_A$ and speed if: **a.** the load torque doubles; **b.** the load power doubles; **c.** $V_T$ is changed to 100 V and $P_{out}$ remains constant; **d.** $R_F + R_{adj}$ is doubled in value and $P_{out}$ remains constant?

**P15.50.** A shunt-connected motor has the magnetization curve shown in Figure P15.50. Ignore

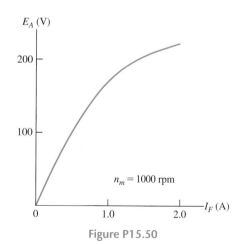

**Figure P15.50**

rotational losses in this problem. The motor is supplied from a source of $V_T = 240$ V and has $R_A = 1.5\ \Omega$. The total field resistance is $R_F + R_{adj} = 160\ \Omega$. **a.** Find the no-load speed. **b.** A load is connected and the speed drops by 6 percent. Find the load torque, output power, armature current, field loss, and armature loss.

**P15.51.** A shunt-connected dc motor has $R_A = 0.5\ \Omega$ and $R_F + R_{adj} = 400\ \Omega$. For $V_T = 200$ V and no load, the motor runs at 1150 rpm and the line current is $I_L = 1.2$ A. Find the rotational loss at this speed.

**P15.52.** A certain shunt-connected dc motor has $R_A = 4\ \Omega$ and $V_T = 240$ V. At a speed of 1000 rpm, the induced armature voltage is 120 V. Plot the torque–speed characteristic ($T_{dev}$ versus $n_m$) to scale.

**P15.53.** Suppose that a dc machine is designed such that the voltage applied to the field $V_F$ is equal to the voltage $V_T$ applied to the armature. (Refer to Figure 15.14 on page 763.) In a well-designed machine, which current, $I_F$ or $I_A$, is larger under full-load operating conditions? Why? What do you estimate as an acceptable value for the ratio $I_A/I_F$ under full load?

**P15.54.** Consider a shunt-connected dc motor that has the magnetization curve shown in Figure 15.19 on page 768. The dc supply voltage is $V_T = 200$ V, the armature resistance is $R_A = 0.085\ \Omega$, the field resistance is $R_F = 10\ \Omega$, and the adjustable resistance is $R_{adj} = 2.5\ \Omega$. At a speed of 1200 rpm, the rotational loss is $P_{rot} = 1000$ W. If this motor drives a load that demands a torque of $T_{out} = 200$ Nm independent of speed, determine the motor speed and efficiency.

**Section 15.5: Series-Connected DC Motors**

**\*P15.55.** A series-connected dc motor has $R_F + R_A = 0.6\ \Omega$ and draws $I_A = 40$ A from the dc source voltage $V_T = 220$ V, while running at 900 rpm. What is the speed for $I_A = 20$ A? Assume a linear relationship between $I_A$ and $\phi$.

**P15.56.** List four potential advantages of universal motors compared to other types of ac motors.

**P15.57.** Would a universal motor be a good choice for a clock? For a furnace fan motor? For a home coffee grinder? Give reasons for your answers.

**P15.58.** In examining a single-phase ac motor, what features could we look for to identify it as a universal motor?

**P15.59.** Running at 1200 rpm from a 280-V source, a series-connected dc motor draws an armature current of 25 A. The field resistance is 0.2 $\Omega$, and the armature resistance is 0.3 $\Omega$. Assuming that the flux is proportional to the field current, determine the speed at which the armature current is 10 A.

**P15.60.** Running at 1200 rpm from a 280-V source, a series-connected dc motor draws an armature current of 25 A. The field resistance is 0.2 $\Omega$, and the armature resistance is 0.3 $\Omega$. Rotational losses are 350 W and can be assumed to be proportional to speed. Determine the output power and developed torque. Determine the new armature current and speed if the load torque is increased by a factor of two.

**P15.61.** A series-connected dc motor has $R_F + R_A = 0.6\ \Omega$ and draws $I_A = 40$ A from the dc source voltage $V_T = 220$ V, while running at 900 rpm. The rotational losses are 400 W. Find the output power and developed torque. Suppose that the load torque is reduced by a factor of two and that the rotational power loss is proportional to speed. Find the new values of $I_A$ and speed.

**P15.62.** A series-connected dc motor has $R_A = 0.5\ \Omega$ and $R_F = 1.5\ \Omega$. In driving a certain load at 1200 rpm, the current is $I_A = 20$ A from a source voltage of $V_T = 220$ V. The rotational loss is 150 W. Find the output power and efficiency.

**P15.63.** A series-connected dc motor is designed to operate with a variable load. The resistances $R_A$ and $R_F$ are negligible. To attain high efficiency, the motor has been designed to have very small rotational losses. For a load torque of 100 Nm, the machine runs at its maximum rated speed of 1200 rpm. **a.** Find the speed for a load torque of 300 Nm. **b.** What is the no-load speed? What are the potential consequences of having the load become disconnected from the motor without disconnecting the dc source?

### Section 15.6: Speed Control of DC Motors

**\*P15.64.** A series-connected dc motor operates at 1400 rpm from a source voltage of $V_T = 75$ V. The developed torque (load torque plus loss torque) is constant at 25 Nm. The resistance is $R_A + R_F = 0.1$ $\Omega$. Determine the value of resistance that must be placed in series with the motor to reduce the speed to 1000 rpm.

**\*P15.65.** Sketch the family of torque–speed characteristics for a separately excited dc motor obtained by:

**a.** varying the field current;

**b.** varying the voltage applied to the armature;

**c.** varying the resistance in series with the armature.

**\*P15.66.** A series-connected motor drives a constant torque load from a 50-V source at a speed of 1500 rpm. The resistances $R_F = R_A = 0$. Neglect rotational losses. What average source voltage is needed to achieve a speed of 1000 rpm? If this is achieved by chopping the 50-V source as illustrated in Figure 15.25 on page 775, find the duty ratio $T_{on}/T$.

**P15.67.** Suppose that a shunt-connected machine operates at 800 rpm on the linear portion of its magnetization characteristic. The motor drives a load that requires constant torque. Assume that $R_A = 0$. The resistances in the field circuit are $R_F = 50$ $\Omega$ and $R_{adj} = 25$ $\Omega$. Find a new value for $R_{adj}$ so that the speed becomes 1200 rpm. What is the slowest speed that can be achieved by varying $R_{adj}$?

**P15.68.** Consider a PM motor that operates from a 12-V source with a no-load speed of 1700 rpm. Neglect rotational losses. What average source voltage is needed to achieve a no-load speed of 1000 rpm? If this is achieved by chopping the 12-V source as illustrated in Figure 15.25 on page 775, find the duty ratio $T_{on}/T$.

**P15.69.** List three methods that can be used to control the speed of dc motors. Which of these apply to shunt-connected motors? To separately excited motors? To permanent-magnet motors? To series-connected motors?

**P15.70.** Consider a shunt-connected dc motor that has the magnetization curve shown in Figure 15.19 on page 768. The dc supply voltage is $V_T = 200$ V, the armature resistance is $R_A = 0.085$ $\Omega$, the field resistance is $R_F = 10$ $\Omega$, and the adjustable resistance is $R_{adj} = 2.5$ $\Omega$. At a speed of 1200 rpm, the rotational loss is $P_{rot} = 1000$ W. Assume that the rotational power loss is proportional to speed. **a.** With a load that demands a torque of $T_{load} = 200$ Nm independent of speed, determine the steady-state armature current. **b.** Suppose that in starting this machine, the field circuit has reached steady state and the motor is not moving when power is applied to the armature circuit. What is the initial value of $I_A$? Determine the starting value of the developed torque. Compare these values to the steady-state values from part (a). **c.** What additional resistance must be inserted in series with the armature to limit the starting current to 200 A? Find the starting torque with this resistance in place.

**P15.71.** A series-connected motor drives a load from a 50-V source at a speed of 1500 rpm. The load torque is proportional to speed. The resistances $R_F = R_A = 0$. Neglect rotational losses. What average source voltage is needed to achieve a speed of 1000 rpm? If this is achieved by chopping the 50-V source as illustrated in Figure 15.25 on page 775, find the duty ratio $T_{on}/T$.

### Section 15.7: DC Generators

**\*P15.72.** A separately excited dc generator is rated for a load voltage of 150 V for a full load current of 20 A at 1500 rpm. With the load disconnected, the output voltage is 160 V. **a.** Determine the voltage regulation, the load resistance, the armature resistance, and the developed torque at full load. **b.** The speed of the generator is decreased to 1200 rpm, and the load resistance is unchanged. Determine the load current, the load voltage, and the developed power.

**P15.73.** Using Figure 15.30 on page 780 as a guide, list the types of connections for dc generators considered in Section 15.7, in order of percentage voltage regulation from highest to lowest.

**P15.74.** Name the four types of compound connections for dc generators.

**P15.75.** What methods can be used to increase the load voltage of: **a.** a separately excited dc generator? **b.** a shunt-connected dc generator?

**P15.76.** What is the value of the voltage regulation for a fully compensated compound dc generator?

**P15.77.** A separately excited dc generator has the magnetization curve shown in Figure P15.77, $V_F = 150$, $R_F = 40 \ \Omega$, $R_{\text{adj}} = 60 \ \Omega$, and $R_A = 1.5 \ \Omega$. The prime mover rotates the armature at a speed of 1300 rpm. Determine the field current, the no-load voltage,

the full-load voltage, and the percentage voltage regulation for a full-load current of 10 A. Assuming that the overall efficiency (not including the power supplied to the field circuit) of the machine is 80 percent, determine the input torque, the developed torque, and the losses associated with friction, windage, eddy currents, and hysteresis.

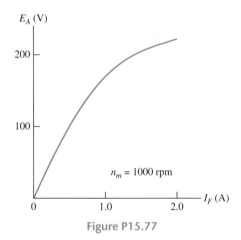

**Figure P15.77**

## Practice Test

Here is a practice test you can use to check your comprehension of the most important concepts in this chapter. Answers can be found in Appendix D and complete solutions are included in the Student Solutions files. See Appendix E for more information about the Student Solutions.

**T15.1.** Consider a shunt-connected dc motor. What are the names of the two windings? Which is on the stator? On the rotor? For which does the current vary with mechanical load?

**T15.2.** Sketch the torque versus speed characteristic for a shunt-connected dc motor. What happens to the speed if the machine is lightly loaded and the field winding becomes disconnected?

**T15.3.** Sketch the torque versus speed characteristic for a series-connected dc motor.

**T15.4.** Give the definition of percentage speed regulation.

**T15.5.** Explain how the magnetization curve is measured for a dc machine.

**T15.6.** Name and briefly discuss the types of power loss in a shunt-connected dc motor.

**T15.7.** What is a universal motor? What are its advantages and disadvantages compared to other types of ac motors?

**T15.8.** List three methods for controlling the speed of a dc motor.

**T15.9.** Suppose we have a dc motor that produces a back emf of $E_A = 240$ V at a speed of 1500 rpm.

    **a.** If the field current remains constant, what is the back emf for a speed of 500 rpm?

    **b.** For a speed of 2000 rpm?

**T15.10.** We have a dc motor that has an induced armature voltage of 120 V at $n_{m1} = 1200$ rpm. If the field remains constant and this motor is operating at a speed of $n_{m2} = 900$ rpm with a developed power of 4 hp, what are the values for the armature current and the developed torque?

**T15.11.** Consider a separately excited dc motor (see the equivalent circuit shown in Figure 15.21 on page 770) that has $R_A = 0.5$ Ω and $V_T = 240$ V. For a full-load output power of 6 hp, we have $n_m = 1200$ rpm and $I_A = 20$ A. The field current remains constant for all parts of this problem. **a.** Find the developed power, developed torque, power lost in $R_A$, and the rotational losses. **b.** Assuming that the rotational power loss is proportional to speed, find the speed regulation of the motor.

**T15.12.** We have a series-connected dc motor that draws an armature current of 20 A while running at 1000 rpm from a 240-V source. The field resistance is 0.3 Ω, and the armature resistance is 0.4 Ω. Assuming that the flux is proportional to the field current, determine the speed at which the armature current is 10 A.

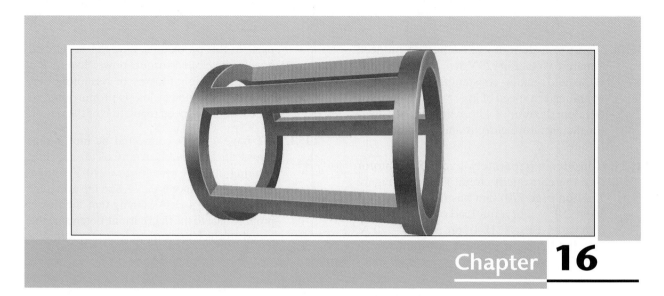

# Chapter 16

# AC Machines

## Study of this chapter will enable you to:

- Select the proper ac motor type for various applications.
- State how torque varies with speed for various ac motors.
- Compute electrical and mechanical quantities for ac motors.

- Use motor nameplate data.
- Understand the operation and characteristics of three-phase induction motors, three-phase synchronous machines, various types of single-phase ac motors, stepper motors, and brushless dc motors.

## Introduction to this chapter:

In this chapter, we continue our discussion of electrical machines. It is not necessary to have studied all of Chapter 15 before reading this chapter. However, you should read Section 15.1 to become familiar with the general concepts related to electrical machines before starting this chapter.

## 16.1   THREE-PHASE INDUCTION MOTORS

Three-phase induction machines account for the great majority of applications that call for motors with power ratings over 5 hp. They are used to power pumps, fans, compressors, and grinders, and in other industrial applications. In this section, we describe the construction and principles of these important devices.

### Rotating Stator Field

The stator of a three-phase induction machine contains a set of windings to which three-phase electrical power is applied. In the first part of this section, we show that these windings establish a rotating magnetic field in the gap between the stator and rotor. The stator field can be visualized as a set of north and south poles rotating around the circumference of the stator. (North stator poles are where magnetic flux lines leave the stator, and south stator poles are where magnetic flux lines enter the stator.) Because north and south poles occur in pairs, the total number of poles $P$ is always even. The field is illustrated for two-pole and four-pole machines in Figure 16.1. Similarly, it is possible for a three-phase induction motor to have six, eight, or more poles.

In this section, we see that the stator windings of three-phase induction machines set up magnetic poles that rotate around the circumference of the stator.

Next, we examine the stator windings and how the rotating field is established in a two-pole machine. The stator of the two-pole machine contains three windings embedded in slots cut lengthwise on the inside of the stator. One of the three stator windings is illustrated in Figure 16.2.

For simplicity, we represent each winding by only two conductors on opposite sides of the stator. However, each winding actually consists of a large number of conductors distributed in various slots in such a manner that the resulting air-gap flux varies approximately sinusoidally with the angle $\theta$ (which is defined in Figure 16.2). Thus, the field in the air gap due to the current $i_a(t)$ in winding $a$ is given by

$$B_a = Ki_a(t)\cos(\theta) \tag{16.1}$$

where $K$ is a constant that depends on the geometry and materials of the stator and rotor as well as the number of turns in winding $a$. $B_a$ is taken as positive when directed from the stator toward the rotor and negative when directed in the opposite direction.

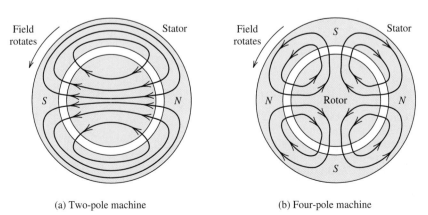

(a) Two-pole machine          (b) Four-pole machine

**Figure 16.1** The field established by the stator windings of a three-phase induction machine consists of an even number of magnetic poles. The field rotates at a speed known as synchronous speed.

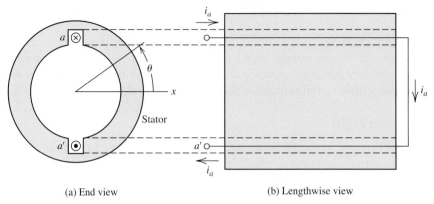

(a) End view                    (b) Lengthwise view

**Figure 16.2** Two views of a two-pole stator showing one of the three windings. For simplicity, we represent the winding with a single turn, but in a real machine, each winding has many turns distributed around the circumference of the stator such that the air-gap flux varies sinusoidally with $\theta$.

The field in the gap due to winding $a$ is shown in Figure 16.3. Notice that the field is strongest at $\theta = 0$ and at $\theta = 180°$. Although it fluctuates in strength and polarity as the current changes with time, the field produced by winding $a$ alone does not rotate. However, we are about to demonstrate that the combined field produced by all three windings does rotate.

The other two windings ($b$ and $c$) are identical to winding $a$, except that they are rotated in space by $120°$ and $240°$, respectively. This is illustrated in Figure 16.4. Thus, the fields in the air gap due to windings $b$ and $c$ are given by

> Each winding sets up a field that varies sinusoidally around the circumference of the gap and varies sinusoidally with time. These fields are displaced from one another by 120° in both time and space.

$$B_b = Ki_b(t) \cos(\theta - 120°) \tag{16.2}$$

$$B_c = Ki_c(t) \cos(\theta - 240°) \tag{16.3}$$

The total field in the gap is the sum of the individual fields produced by the three coils. Thus, the total field is

$$B_{\text{gap}} = B_a + B_b + B_c \tag{16.4}$$

Using Equations 16.1 through 16.3 to substitute into Equation 16.4, we have

$$B_{\text{gap}} = Ki_a(t) \cos(\theta) + Ki_b(t) \cos(\theta - 120°) + Ki_c(t) \cos(\theta - 240°) \tag{16.5}$$

**Figure 16.3** The field produced by the current in winding $a$ varies sinusoidally in space around the circumference of the gap. The field is shown here for the positive maximum of the current $i_a(t)$. As illustrated, the field is strongest in magnitude at $\theta = 0$ and at $\theta = 180°$. Furthermore, the current and the field vary sinusoidally with time. Over time, the field dies to zero and then builds up in the opposite direction.

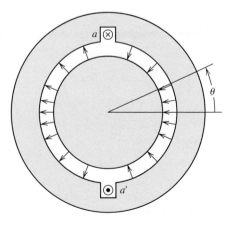

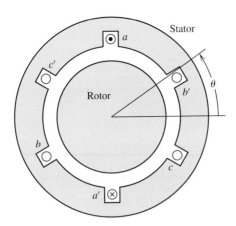

**Figure 16.4** The stator of a two-pole machine contains three identical windings spaced 120° apart.

Application of a balanced three-phase source to the windings results in currents given by

$$i_a(t) = I_m \cos(\omega t) \tag{16.6}$$

$$i_b(t) = I_m \cos(\omega t - 120°) \tag{16.7}$$

$$i_c(t) = I_m \cos(\omega t - 240°) \tag{16.8}$$

Now, using Equations 16.6, 16.7, and 16.8 to substitute for the currents in Equation 16.5 yields

$$B_{\text{gap}} = KI_m \cos(\omega t) \cos(\theta) + KI_m \cos(\omega t - 120°) \cos(\theta - 120°)$$
$$+ KI_m \cos(\omega t - 240°) \cos(\theta - 240°) \tag{16.9}$$

The trigonometric identity $\cos(x)\cos(y) = (1/2)[\cos(x - y) + \cos(x + y)]$ can be used to write Equation 16.9 as

$$B_{\text{gap}} = \frac{3}{2}KI_m \cos(\omega t - \theta) + \frac{1}{2}KI_m[\cos(\omega t + \theta)$$
$$+ \cos(\omega t + \theta - 240°) + \cos(\omega t + \theta - 480°)] \tag{16.10}$$

Furthermore, we can write

$$[\cos(\omega t + \theta) + \cos(\omega t + \theta - 240°) + \cos(\omega t + \theta - 480°)] = 0 \tag{16.11}$$

because the three terms form a balanced three-phase set. A phasor diagram for these terms is shown in Figure 16.5. (Notice that −240° is equivalent to +120°, and that −480° is equivalent to −120°.) Thus, Equation 16.10 reduces to

$$B_{\text{gap}} = B_m \cos(\omega t - \theta) \tag{16.12}$$

where we have defined $B_m = (3/2)KI_m$. An important conclusion can be drawn from Equation 16.12: *The field in the gap rotates counterclockwise with an angular speed of $\omega$.* To verify this fact, notice that the maximum flux density occurs for

$$\theta = \omega t$$

Thus, in the two-pole machine, the point of maximum flux rotates counterclockwise with an angular velocity of $d\theta/dt = \omega$.

The field in the gap rotates counterclockwise with an angular speed $\omega$.

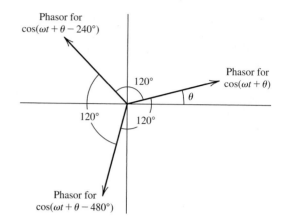

**Figure 16.5** Phasor diagram for the three terms on the left-hand side of Equation 16.11. Regardless of the value of $\theta$, the phasors add to zero.

### Synchronous Speed

In a $P$-pole machine, the field rotates at the synchronous speed $\omega_s$.

By a similar analysis, it can be shown that for a $P$-pole machine, the field rotates at an angular velocity of

$$\omega_s = \frac{\omega}{P/2} \tag{16.13}$$

which is called the **synchronous angular velocity**. In rpm, the synchronous speed is

$$n_s = \frac{120f}{P} \tag{16.14}$$

Table 16.1 gives synchronous speeds versus the number of poles assuming a frequency of 60 Hz.

In summary, we have shown that the field set up by the stator windings consists of a system of $P$ magnetic poles that rotate at synchronous speed. These fields were illustrated in Figure 16.1 for two- and four-pole machines.

The direction of rotation of a three-phase induction motor can be reversed by interchanging any two of the three line connections to the three-phase source.

**Exercise 16.1**   If the connections of $b$ and $c$ to the three-phase source are interchanged, the currents become

$$i_a(t) = I_m \cos(\omega t)$$
$$i_b(t) = I_m \cos(\omega t - 240°)$$
$$i_c(t) = I_m \cos(\omega t - 120°)$$

Show that, in this case, the field rotates clockwise rather than counterclockwise.  □

**Table 16.1 Synchronous Speed Versus Number of Poles for $f = 60$ Hz**

| $P$ | $n_s$ |
|---|---|
| 2 | 3600 |
| 4 | 1800 |
| 6 | 1200 |
| 8 | 900 |
| 10 | 720 |
| 12 | 600 |

The result of Exercise 16.1 shows that the direction of rotation of the field in a three-phase induction machine can be reversed by interchanging any two of the line connections to the electrical source. We will see that this reverses the direction of mechanical rotation. You may find the fact that interchanging two of the electrical connections to the source reverses the direction of rotation to be useful in working with three-phase motors.

## Squirrel-Cage Induction Machines

The rotor windings of a three-phase induction machine can take two forms. The simplest, least expensive, and most rugged is known as a **squirrel-cage rotor**. It consists simply of bars of aluminum with shorting rings at the ends, as illustrated in Figure 16.6. The squirrel cage is embedded in the laminated iron rotor by casting molten aluminum into slots cut into the rotor. In the squirrel-cage induction machine, there are no external electrical connections to the rotor. The other type of rotor construction, which we discuss later, is known as a **wound rotor**.

Next, we consider how torque is produced in squirrel-cage induction machines. We have seen earlier in this section that the stator sets up a system of $P$ magnetic poles that rotate at synchronous speed. As this magnetic field moves past, voltages are induced in the squirrel-cage conductors. Since the field, the direction of relative motion, and the length of the conductors are mutually perpendicular, the induced voltage $v_c$ is given by Equation 15.9, which is repeated here for convenience:

$$v_c = Blu \qquad (16.15)$$

in which $B$ is the flux density, $l$ is the length of the conductor, and $u$ is the relative velocity between the conductor and the field.

This voltage causes currents to flow in the conductors as illustrated in Figure 16.7. Of course, the largest voltages are induced in the conductors that are directly under the stator poles because that is where the flux density $B$ is largest in magnitude. Furthermore, for conductors under south poles, the voltage polarity and current direction are opposite to those for conductors under north poles. Currents flow through the bars under the north pole, around the shorting ring, and back in the opposite direction through the bars under the south pole.

The rotor currents establish magnetic poles on the rotor. It is the interaction of the rotor poles with the stator poles that produces torque. The north rotor pole $N_r$ attempts to align itself with the south stator pole $S_s$.

If the impedances of the rotor conductors were purely resistive, the largest currents would occur directly under the stator poles $S_s$ and $N_s$, as shown in Figure 16.7. Consequently, the rotor poles would be displaced by $\delta_{rs} = 90°$ with respect to the stator poles, as illustrated for a two-pole machine in Figure 16.7. This is exactly the angular displacement between the sets of magnetic poles that produces maximum torque.

*The rotating stator field induces voltages in the rotor conductors resulting in currents that produce magnetic poles on the rotor. The interaction of the rotor poles and stator poles produces torque.*

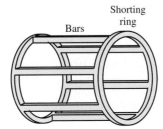

**Figure 16.6** The rotor conductors of a squirrel-cage induction machine are aluminum bars connected to rings that short the ends together. These conductors are formed by casting molten aluminum into slots in the laminated iron rotor.

Bars

Shorting ring

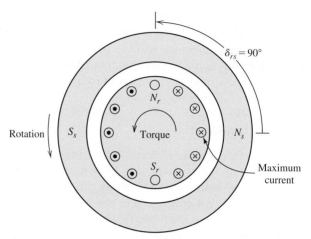

**Figure 16.7** Cross section of a squirrel-cage induction motor. The rotating stator field induces currents in the conducting bars which in turn set up magnetic poles on the rotor. Torque is produced because the rotor poles are attracted to the stator poles.

### Slip and Slip Frequency

The frequency of the voltages induced in the rotor conductors depends on the rotational speed of the stator field relative to the rotor and on the number of poles. We have seen that the stator field rotates at synchronous speed denoted as either $\omega_s$ or $n_s$. We denote the mechanical speed of the rotor as $\omega_m$ (or $n_m$). In an induction motor, the mechanical speed $\omega_m$ varies from zero to almost synchronous speed. Thus, the speed of the stator field relative to the rotor is $\omega_s - \omega_m$ (or $n_s - n_m$).

The **slip** $s$ is defined to be the relative speed as a fraction of synchronous speed:

$$s = \frac{\omega_s - \omega_m}{\omega_s} = \frac{n_s - n_m}{n_s} \tag{16.16}$$

Slip $s$ varies from 1 when the rotor is stationary to 0 when the rotor turns at synchronous speed.

The angular frequency of the voltages induced in the squirrel cage, called the **slip frequency**, is given by

$$\omega_{\text{slip}} = s\omega \tag{16.17}$$

Notice that when the mechanical speed approaches the speed of the stator field (which is the synchronous speed), the frequency of the induced voltages approaches zero.

The frequency of the rotor currents is called the slip frequency.

### Effect of Rotor Inductance on Torque

In Figure 16.7, we saw how torque is produced in an induction motor assuming purely resistive impedances for the rotor conductors. However, the impedances of the conductors are not purely resistive. Because the conductors are embedded in iron, there is significant series inductance associated with each conductor. The equivalent circuit for a given conductor is shown in Figure 16.8, in which $\mathbf{V}_c$ is the phasor for the induced voltage, $R_c$ is the resistance of the conductor, and $L_c$ is its inductance. Both the frequency and the amplitude of the induced voltage are proportional to slip.

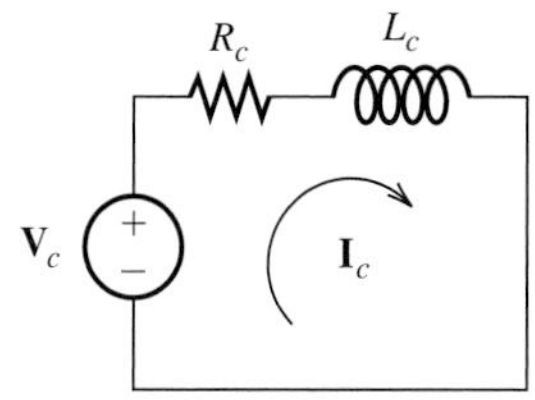

Since the frequency of the induced voltage is $\omega_{slip} = s\omega$, the impedance is

$$Z_c = R_c + js\omega L_c \qquad (16.18)$$

The current is

$$\mathbf{I}_c = \frac{\mathbf{V}_c}{R_c + js\omega L_c} \qquad (16.19)$$

Because of the inductance, the current lags the induced voltage. As the slip $s$ increases, the amount of phase lag approaches 90°. Consequently, the peak current in a given rotor conductor occurs somewhat after the stator pole passes by. Furthermore, the rotor poles are displaced from the stator poles by less than 90°. This is illustrated in Figure 16.9. Because of this, the torque is reduced. (If the stator and rotor poles were aligned, no torque would be produced.)

## Torque–Speed Characteristic

Now we are in a position to explain qualitatively the torque–speed characteristic shown in Figure 16.10 for the squirrel-cage induction motor. First, assume that the rotor speed $n_m$ equals the synchronous speed $n_s$ (i.e., the slip $s$ equals zero). In this case, the relative velocity between the conductors and the field is zero (i.e., $u = 0$). Then according to Equation 16.15, the induced voltage $v_c$ is zero. Consequently, the rotor currents are zero and the torque is zero.

As the rotor slows down from synchronous speed, the stator field moves past the rotor conductors. The magnitudes of the voltages induced in the rotor conductors increase linearly with slip. For small slips, the inductive reactances of the conductors, given by $s\omega L_c$, are negligible, and maximum rotor current is aligned with maximum stator field, which is the optimum situation for producing torque. Because the

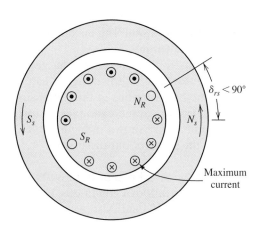

**Figure 16.9** As slip $s$ increases, the conductor currents lag the induced voltages. Consequently, the angular displacement $\delta_{rs}$ between the rotor poles and the stator poles approaches 0°.

**Figure 16.8** Equivalent circuit for a rotor conductor. $\mathbf{V}_c$ is the phasor for the induced voltage, $R_c$ is the resistance of the conductor, and $L_c$ is the inductance.

Because of the inductances of the rotor conductors, the rotor currents lag the induced voltages.

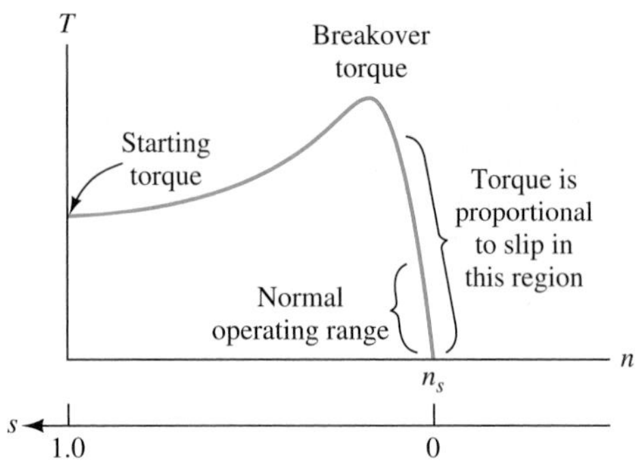

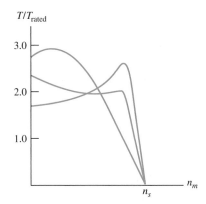

**Figure 16.10** Torque-versus-speed characteristic for a typical three-phase induction motor.

**Figure 16.11** Depending on various design features, the torque–speed characteristic of the three-phase induction motor can be modified to better suit particular applications.

**For small values of slip, developed torque is proportional to slip.**

induced voltage is proportional to slip and the impedance is independent of slip, the currents are proportional to slip. Torque is proportional to the product of the field and the current. Hence, we conclude that *torque is proportional to slip, assuming small slip*. This fact is illustrated in Figure 16.10.

As the motor slows further, the inductive reactance eventually dominates the denominator of Equation 16.19. Then, the magnitude of the current is nearly independent of slip. Thus, the torque tends to level out as the motor slows. Because the poles on the rotor tend to become aligned with the stator poles, the torque decreases as the motor slows to a stop. The torque for zero speed is called either the **starting torque** or the **stall torque**. The maximum torque is called either the **pull-out torque** or the **breakover torque**.

**Motor designers can modify the shape of the torque–speed curve by changing various aspects of the machine design, such as the cross section and depth of the rotor conductors.**

Our discussion has revealed the general characteristics of the three-phase induction motor. The motor designer can modify the shape of the torque–speed characteristic by variations in the dimensions and geometry of the motor and by materials selection. Some examples of torque–speed characteristics available in induction motors are shown in Figure 16.11. However, the details of motor design are beyond the scope of our discussion.

**Exercise 16.2**   A 5-hp four-pole 60-Hz three-phase induction motor runs at 1750 rpm under full-load conditions. Determine the slip and the frequency of the rotor currents at full load. Also, estimate the speed if the load torque drops in half.
**Answer**    $s = 50/1800 = 0.02778, f_{slip} = 1.667$ Hz, $n = 1775$ rpm.     ☐

## 16.2  EQUIVALENT-CIRCUIT AND PERFORMANCE CALCULATIONS FOR INDUCTION MOTORS

In Section 16.1, we described the induction motor and its operation in qualitative terms. In this section, we develop an equivalent circuit and show how to calculate the performance of induction motors.

Consider the induction motor with the rotor locked so that it cannot turn. Then, the magnetic field of the stator links the rotor windings and causes current to flow in them. Basically, the locked-rotor induction motor is the same as a three-phase transformer with the stator windings acting as the primary. The rotor conductors act as short-circuited secondary windings. Thus, we can expect the equivalent circuit for each phase of the motor to be very similar to the transformer equivalent circuit shown in Figure 14.28 on page 728. Of course, modifications to the transformer equivalent circuit are necessary before it can be applied to the induction motor because of rotation and the conversion of electrical energy into mechanical form.

*The equivalent circuit for each phase of an induction motor is similar to that of a transformer with the secondary winding shorted.*

### Rotor Equivalent Circuit

An equivalent circuit for one phase of the rotor windings is shown in Figure 16.12(a). (Equivalent circuits for the other two phases are identical except for the phase angles of the current and voltage.) $\mathbf{E}_r$ represents the induced voltage in phase $a$ of the rotor under locked conditions. As discussed in the preceding section, the voltage induced in the rotor is proportional to slip $s$. Thus, the induced voltage is represented by the voltage source $s\mathbf{E}_r$. (Recall that for a stationary rotor $s = 1$.)

We have seen that the frequency of the rotor currents is $s\omega$. The rotor inductance (per phase) is denoted by $L_r$ and has a reactance of $js\omega L_r = jsX_r$, where $X_r = \omega L_r$ is the reactance under locked-rotor conditions. The resistance per phase is denoted by $R_r$, and the current in one phase of the rotor is $\mathbf{I}_r$, which is given by

$$\mathbf{I}_r = \frac{s\mathbf{E}_r}{R_r + jsX_r} \qquad (16.20)$$

Dividing the numerator and denominator of the right-hand side of Equation 16.20 by $s$, we have

$$\mathbf{I}_r = \frac{\mathbf{E}_r}{R_r/s + jX_r} \qquad (16.21)$$

This can be represented by the circuit shown in Figure 16.12(b).

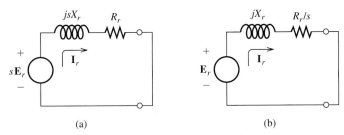

(a)                             (b)

**Figure 16.12** Two equivalent circuits for one phase of the rotor windings.

Figure 16.13 provides a convenient reference for the information needed to analyze induction machines.

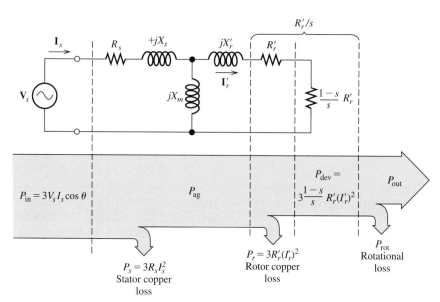

**Figure 16.13** Equivalent circuit for one phase of an induction motor and the associated power-flow diagram. $V_s$ is the rms phase voltage and $I_s$ is the rms phase current.

### Complete Induction-Motor Equivalent Circuit

As in a transformer, the induced rotor voltage $E_r$ under locked conditions is related to the stator voltage by the turns ratio. Thus, we can reflect the rotor impedances of Figure 16.12(b) to the primary (stator) side of the equivalent circuit. We denote the reflected values of $X_r$ and $R_r/s$ by $X_r'$ and $R_r'/s$, respectively.

The completed (per phase) induction-motor equivalent circuit is shown in Figure 16.13. The resistance of the stator winding is $R_s$, and the stator leakage reactance is $X_s$. The magnetizing reactance $X_m$ accounts for the current needed to set up the rotating stator field. Except for changes in notation, these parts of the equivalent circuit are the same as the transformer equivalent circuit.

### Phase versus Line Quantities

The voltage $V_s$ across each winding and current $I_s$ through each winding shown in Figure 16.13 are called the **phase voltage** and **phase current**, respectively.

The windings of an induction motor may be connected in either a delta or a wye. In the case of a delta connection, the phase voltage $V_s$ is the same as the line-to-line voltage $V_{\text{line}}$. The line current $I_{\text{line}}$ is $\sqrt{3}$ times the phase current $I_s$. (This is explained in Section 5.7.) In equation form for the delta connection, we have

Relationships between line and phase quantities for a delta-connected motor.

$$V_s = V_{\text{line}}$$

$$I_{\text{line}} = I_s \sqrt{3}$$

On the other hand, for a wye connection, we get

Relationships between line and phase quantities for a wye-connected motor.

$$V_s = \frac{V_{\text{line}}}{\sqrt{3}}$$

$$I_{\text{line}} = I_s$$

The voltage rating stated for a machine is invariably the line-to-line voltage. For a given three-phase source, the voltage across the windings is smaller by a factor of $\sqrt{3}$ for the wye connection.

We will see that the starting current for induction motors can be very large compared to the full-load running current. Sometimes, motors are started in the wye configuration and then switched to delta as the motor approaches its running speed to lessen starting currents.

The voltage rating stated for a machine is invariably the line-to-line voltage.

### Power and Torque Calculations

In Figure 16.13, notice that we have split the reflected resistance $R'_r/s$ into two parts as follows:

$$\frac{R'_r}{s} = R'_r + \frac{1 - s}{s}R'_r \qquad (16.22)$$

A power-flow diagram for induction motors is also shown in Figure 16.13. The power delivered to the resistance $[(1 - s)/s]R'_r$ is the part that is converted to mechanical form. This portion of the power, called the **developed power**, is denoted by $P_{\text{dev}}$. The equivalent circuit shown in Figure 16.13 represents one of three phases, so the total developed power is

$$P_{\text{dev}} = 3 \times \frac{1 - s}{s}R'_r(I'_r)^2 \qquad (16.23)$$

On the other hand, the power delivered to the rotor resistance $R'_r$ is converted to heat. Generally, we refer to $I^2R$ losses as **copper losses** (even though the conductors are sometimes aluminum). The total copper loss in the rotor is

$$P_r = 3R'_r(I_r)^2 \qquad (16.24)$$

and the stator copper loss is

$$P_s = 3R_s I_s^2 \qquad (16.25)$$

The input power from the three-phase source is

$$P_{\text{in}} = 3I_s V_s \cos(\theta) \qquad (16.26)$$

in which $\cos(\theta)$ is the power factor.

Part of the developed power is lost to friction and windage. Another loss is core loss due to hysteresis and eddy currents. Sometimes, a resistance is included in parallel with the magnetization reactance $jX_m$ to account for core loss. However, we will include the core loss with the rotational losses. Unless stated otherwise, we assume that the rotational power loss is proportional to speed. The output power is the developed power minus the rotational loss:

$$P_{\text{out}} = P_{\text{dev}} - P_{\text{rot}} \qquad (16.27)$$

As usual, the efficiency of the machine is given by

$$\eta = \frac{P_{\text{out}}}{P_{\text{in}}} \times 100\%$$

The developed torque is

$$T_{dev} = \frac{P_{dev}}{\omega_m} \qquad (16.28)$$

The power $P_{ag}$ that crosses the air gap into the rotor is delivered to the rotor resistances. Thus, we can find the air-gap power by adding the respective sides of Equations 16.23 and 16.24:

$$P_{ag} = P_r + P_{dev} \qquad (16.29)$$

$$P_{ag} = 3R'_r(I'_r)^2 + 3 \times \frac{1-s}{s}R'_r(I'_r)^2 \qquad (16.30)$$

$$P_{ag} = 3 \times \frac{1}{s}R'_r(I'_r)^2 \qquad (16.31)$$

Comparing Equations 16.23 and 16.31, we have

$$P_{dev} = (1 - s)P_{ag} \qquad (16.32)$$

Using Equation 16.32 to substitute for $P_{dev}$ in Equation 16.28, we get

$$T_{dev} = \frac{(1-s)P_{ag}}{\omega_m} \qquad (16.33)$$

However, we also have $\omega_m = (1 - s)\omega_s$. Using this to substitute into Equation 16.33, we obtain

$$T_{dev} = \frac{P_{ag}}{\omega_s} \qquad (16.34)$$

Equation 16.34 can be used to compute starting torque.

For speed to increase from a standing start, the initial torque or starting torque produced by the motor must be larger than the torque required by the load. We can find starting torque as follows. Under starting conditions (i.e., $\omega_m = 0$), we have $s = 1$ and $P_{ag} = 3R'_r(I'_r)^2$. Then, the starting torque can be computed by using Equation 16.34.

### Example 16.1   Induction-Motor Performance

A certain 30-hp four-pole 440-V-rms 60-Hz three-phase delta-connected induction motor has

$$R_s = 1.2 \ \Omega \qquad\qquad R'_r = 0.6 \ \Omega$$
$$X_s = 2.0 \ \Omega \qquad\qquad X'_r = 0.8 \ \Omega$$
$$X_m = 50 \ \Omega$$

Under load, the machine operates at 1746 rpm and has rotational losses of 900 W. Find the power factor, the line current, the output power, copper losses, output torque, and efficiency.

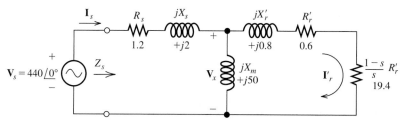

**Figure 16.14** Equivalent circuit for one phase of the motor of Example 16.1.

**Solution**   From Table 16.1, we find that synchronous speed for a four-pole motor is $n_s = 1800$ rpm. Then, we utilize Equation 16.16 to compute the slip:

$$s = \frac{n_s - n_m}{n_s} = \frac{1800 - 1746}{1800} = 0.03$$

We can use the data given to draw the equivalent circuit shown in Figure 16.14 for one phase of the motor. The impedance seen by the source is

$$Z_s = 1.2 + j2 + \frac{j50(0.6 + 19.4 + j0.8)}{j50 + 0.6 + 19.4 + j0.8}$$

$$= 1.2 + j2 + 16.77 + j7.392$$

$$= 17.97 + j9.392$$

$$= 20.28\underline{/27.59°}\ \Omega$$

The power factor is the cosine of the impedance angle. Because the impedance is inductive, we know that the power factor is lagging:

$$\text{power factor} = \cos(27.59°) = 88.63\% \text{ lagging}$$

For a delta-connected machine, the phase voltage is equal to the line voltage, which is specified to be 440 V rms. The phase current is

$$I_s = \frac{V_s}{Z_s} = \frac{440\underline{/0°}}{20.28\underline{/27.59°}} = 21.70\underline{/-27.59°} \text{ A rms}$$

Thus, the magnitude of the line current is

$$I_{\text{line}} = I_s\sqrt{3} = 21.70\sqrt{3} = 37.59 \text{ A rms}$$

The input power is

$$P_{\text{in}} = 3I_sV_s \cos\theta$$

$$= 3(21.70)440 \cos(27.59°)$$

$$= 25.38 \text{ kW}$$

In ac machine calculations, we take the rms values of currents and voltages for the phasor magnitudes (instead of peak values as we have done previously).

Next, we compute $\mathbf{V}_x$ and $\mathbf{I}'_r$:

$$\mathbf{V}_x = \mathbf{I}_s \frac{j50(0.6 + 19.4 + j0.8)}{j50 + 0.6 + 19.4 + j0.8}$$

$$= 21.70\underline{/-27.59°} \times 18.33\underline{/23.78°}$$

$$= 397.8\underline{/-3.807°} \text{ V rms}$$

$$\mathbf{I}'_r = \frac{\mathbf{V}_x}{j0.8 + 0.6 + 19.4}$$

$$= \frac{397.8\underline{/-3.807°}}{20.01\underline{/1.718°}}$$

$$= 19.88\underline{/-5.52°} \text{ A rms}$$

The copper losses in the stator and rotor are

$$P_s = 3R_s I_s^2$$

$$= 3(1.2)(21.70)^2$$

$$= 1695 \text{ W}$$

and

$$P_r = 3R'_r(I'_r)^2$$

$$= 3(0.6)(19.88)^2$$

$$= 711.4 \text{ W}$$

Finally, the developed power is

$$P_{\text{dev}} = 3 \times \frac{1-s}{s} R'_r(I'_r)^2$$

$$= 3(19.4)(19.88)^2$$

$$= 23.00 \text{ kW}$$

As a check, we note that

$$P_{\text{in}} = P_{\text{dev}} + P_s + P_r$$

to within rounding error.

The output power is the developed power minus the rotational loss, given by

$$P_{\text{out}} = P_{\text{dev}} - P_{\text{rot}}$$

$$= 23.00 - 0.900$$

$$= 22.1 \text{ kW}$$

This corresponds to 29.62 hp, so the motor is operating at nearly its rated load. The output torque is

$$T_{\text{out}} = \frac{P_{\text{out}}}{\omega_m}$$

$$= \frac{22,100}{1746(2\pi/60)}$$

$$= 120.9 \text{ Nm}$$

The efficiency is

$$\eta = \frac{P_{out}}{P_{in}} \times 100\%$$

$$= \frac{22,100}{25,380} \times 100\%$$

$$= 87.0\%$$   ∎

---

**Example 16.2**   **Starting Current and Torque**

Calculate the starting line current and torque for the motor of Example 16.1.

**Solution**   For starting from a standstill, we have $s = 1$. The equivalent circuit is shown in Figure 16.15(a). Combining the impedances to the right of the dashed line, we have

$$Z_{eq} = R_{eq} + jX_{eq} = \frac{j50(0.6 + j0.8)}{j50 + 0.6 + j0.8} = 0.5812 + j0.7943 \ \Omega$$

The circuit with the combined impedances is shown in Figure 16.15(b).
   The impedance seen by the source is

$$Z_s = 1.2 + j2 + Z_{eq}$$

$$= 1.2 + j2 + 0.5812 + j0.7943$$

$$= 1.7812 + j2.7943$$

$$= 3.314\underline{/57.48°} \ \Omega$$

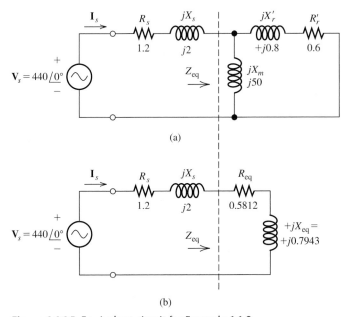

(a)

(b)

**Figure 16.15**  Equivalent circuit for Example 16.2.

Thus, the starting phase current is

$$\mathbf{I}_{s,\,starting} = \frac{\mathbf{V}_s}{Z_s} = \frac{440\underline{/0°}}{3.314\underline{/57.48°}}$$

$$= 132.8\underline{/-57.48°} \text{ A rms}$$

and, because the motor is delta connected, the starting-line-current magnitude is

$$I_{line,\,starting} = \sqrt{3}I_{s,\,starting} = 230.0 \text{ A rms}$$

In Example 16.1, with the motor running under nearly a full load, the line current is $I_{line} = 37.59$ A. Thus, the starting current is approximately six times larger than the full-load running current. This is typical of induction motors.

The power crossing the air gap is three times the power delivered to the right of the dashed line in Figure 16.15, given by

$$P_{ag} = 3R_{eq}(I_{s,\,starting})^2$$

$$= 30.75 \text{ kW}$$

Finally, Equation 16.34 gives us the starting torque:

$$T_{dev,\,starting} = \frac{P_{ag}}{\omega_s}$$

$$= \frac{30{,}750}{2\pi(60)/2}$$

$$= 163.1 \text{ Nm}$$

Notice that the starting torque is larger than the torque while running under full-load conditions. This is also typical of induction motors.    ■

---

### Example 16.3    Induction-Motor Performance

A 220-V-rms 60-Hz three-phase wye-connected induction motor draws 31.87 A at a power factor of 75 percent lagging. For all three phases, the total stator copper losses are 400 W, and the total rotor copper losses are 150 W. The rotational losses are 500 W. Find the power crossing the air gap $P_{ag}$, the developed power $P_{dev}$, the output power $P_{out}$, and the efficiency.

**Solution**    The phase voltage is $V_s = V_{line}/\sqrt{3} = 127.0$ V rms. Next, we find the input power:

$$P_{in} = 3V_sI_s\cos(\theta)$$

$$= 3(127)(31.87)(0.75)$$

$$= 9107 \text{ W}$$

The air-gap power is the input power minus the stator copper loss:

$$P_{ag} = P_{in} - P_s$$

$$= 9107 - 400$$

$$= 8707 \text{ W}$$

The developed power is the input power minus the copper losses:

$$P_{dev} = 9107 - 400 - 150 = 8557 \text{ W}$$

Next, by subtracting the rotational losses, we find that the output power is

$$\begin{aligned} P_{out} &= P_{dev} - P_{rot} \\ &= 8557 - 500 \\ &= 8057 \text{ W} \end{aligned}$$

Finally, the efficiency is

$$\eta = \frac{P_{out}}{P_{in}} \times 100\%$$

$$= 94.0\%$$

**Exercise 16.3**   Repeat Example 16.1 for a running speed of 1764 rpm.
**Answer**   $s = 0.02$; power factor $= 82.62\%$; $P_{in} = 17.43$ kW; $P_{out} = 15.27$ kW; $P_s = 919$ W; $P_r = 330$ W; $T_{out} = 82.66$ Nm; $\eta = 87.61\%$.

**Exercise 16.4**   Repeat Example 16.2 if the rotor resistance is increased to 1.2 $\Omega$. Compare the starting torque with the value found in the example.
**Answer**   $\mathbf{I}_{s, \text{starting}} = 119.7\underline{/-50°}$; $T_{dev, \text{starting}} = 265.0$ Nm.

## Wound-Rotor Induction Machine

A variation of the induction motor is the **wound-rotor machine**. The stator is identical to that of a squirrel-cage motor. Instead of a cast aluminum cage, the rotor contains a set of three-phase coils that are placed in slots. The windings are configured to produce the same number of poles on the rotor as on the stator. The windings are usually wye connected and the three terminals are brought out to external terminals through slip rings.

The results of Exercise 16.4 show that the starting torque of an induction motor can be increased by increasing the rotor resistance. By using a set of variable resistances connected to the rotor terminals, the torque–speed characteristic of the machine can be modified as illustrated in Figure 16.16. A degree of speed control can be achieved by varying the resistances. However, efficiency becomes poorer as the resistance is increased.

A disadvantage of the wound-rotor machine is that it is more expensive and less rugged than the cage machine.

## Selection of Induction Motors

Some of the most important considerations in selecting an induction motor are

1. Efficiency
2. Starting torque
3. Pull-out torque
4. Power factor
5. Starting current

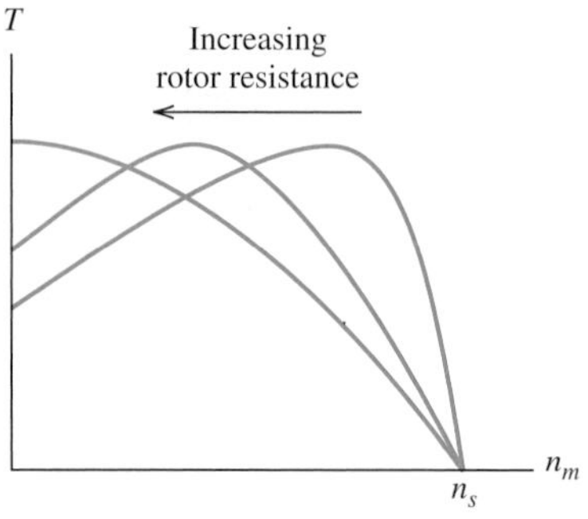

Figure 16.16 Variation of resistance in series with the rotor windings changes the torque–speed characteristic of the wound-rotor machine.

High values for the first four factors and low starting current are generally most desirable. Unfortunately, it is not possible to design a motor having the most desirable values for all of these criteria. It turns out that in the design of a motor, various trade-offs between these criteria must be made. For example, higher rotor resistance leads to lower efficiency and higher starting torque. Larger leakage reactance $X_s$ leads to lower starting current but poorer power factor. The design engineer must consider the various motors available and select the one that best meets the needs of the application at hand.

## 16.3   SYNCHRONOUS MACHINES

Generation of electrical energy by utility companies is done almost exclusively with synchronous machines.

In this section, we discuss synchronous ac machines. These machines are used for nearly all electrical-energy generation by utility companies. As motors, they tend to be used in higher-power, lower-speed applications than those for which induction motors are used. Unlike other types of ac and dc motors that we have studied to this point, the speed of a synchronous motor does not vary with mechanical load (assuming a constant-frequency ac source). Instead, we will see that they run at synchronous speed $\omega_s$, which is given by Equation 16.13, repeated here for convenience:

Assuming a constant frequency source, the speed of a synchronous motor does not vary with load.

$$\omega_s = \frac{\omega}{P/2}$$

(Recall that $\omega$ is the angular frequency of the ac source and $P$ is the number of magnetic poles of the stator or rotor.) Unless stated otherwise, we assume that the rotor is turning at synchronous speed throughout our discussion of synchronous machines.

The stator of a synchronous machine has the same construction as the stator of a three-phase induction motor, which was described in Section 15.1. In review, the stator contains a set of three-phase windings that establish the stator field. This field consists of $P$ magnetic poles, alternating between north and south around the circumference of the stator and rotating at synchronous speed. In a synchronous machine, the set of stator windings is called the **armature**.

The stator windings of a synchronous machine are basically the same as those of an induction machine.

The rotor of a synchronous machine is usually a $P$-pole electromagnet with **field windings** that carry dc currents. (In smaller machines, the rotor can be a permanent magnet, but we will concentrate on machines with field windings.) The field current can be supplied from an external dc source through stationary brushes to **slip rings** mounted on the shaft. The slip rings are insulated from one another and from the shaft. Another method is to place a small ac generator, known as an **exciter**, on the

The rotor of a synchronous machine is a $P$-pole electromagnet or (in low-power machines) a permanent magnet.

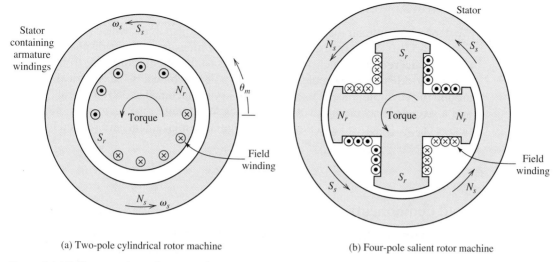

(a) Two-pole cylindrical rotor machine          (b) Four-pole salient rotor machine

**Figure 16.17** Cross-sections of two synchronous machines. The relative positions of the stator and rotor poles are shown for motor action. Torque is developed in the direction of rotation because the rotor poles try to align themselves with the opposite stator poles.

same shaft and use diodes mounted on the shaft to rectify the ac. This avoids the maintenance associated with brushes and slip rings.

Two- and four-pole synchronous machines are illustrated in Figure 16.17. The rotor can either be **cylindrical**, as shown for the two-pole machine, or it can have **salient poles** as illustrated for the four-pole machine. Generally, salient-pole construction is less costly but is limited to low-speed machines having many poles. High-speed machines usually have cylindrical rotors. Salient-pole machines are common in hydroelectric power generation, whereas cylindrical-rotor machines are common in thermal (coal, nuclear, etc.) power plants.

## Automobile Alternator

The alternators found in most automobiles are basically synchronous machines, except that the armature is not connected to an independent ac source. Therefore, the speed of the alternator is not fixed. As the rotor spins, the rotating field cuts the armature conductors, inducing a set of ac voltages. The ac armature voltages are rectified, and the resulting dc is used to power the headlights, charge the battery, and so on. The frequency and amplitude of the ac voltages increase with speed. The amplitude of the induced ac armature voltages is proportional to the flux density, which in turn depends on the field current. An electronic control circuit (or regulator) varies the field current to maintain approximately 14 V dc at the output of the rectifier.

## Motor Action

In using the machine as a motor, the armature is connected to a three-phase ac source. We have seen that the resulting three-phase currents in the armature windings set up a rotating stator field. The rotor turns at synchronous speed with the rotor poles lagging behind the stator poles. Torque is produced because the rotor poles attempt to align with the stator poles. This is illustrated in Figure 16.17.

### Electrical Angles

Angles can be measured in electrical degrees for which the angle between adjacent north and south poles is 180°.

We denote the angular displacement around the air gap as $\theta_m$, which is illustrated in Figure 16.17(a). Sometimes, it is convenient to measure angular displacements in **electrical degrees**, for which 180° corresponds to the angular distance from a north pole to the adjacent south pole. Thus, a four-pole machine has 720 electrical degrees around the circumference of its air gap, a two-pole machine has 360 electrical degrees, and a six-pole machine has $3 \times 360$ degrees. We denote displacement in electrical degrees as $\theta_e$. Electrical and mechanical angular displacements are related by

$$\theta_e = \theta_m \frac{P}{2} \tag{16.35}$$

### Field Components

The total field rotating in the air gap is partly due to the dc currents in the rotor windings and partly due to the ac currents in the stator (armature) windings.

Synchronous machines are designed so the flux varies sinusoidally around the air gap.

The total rotating field in the air gap is due partly to dc current in the field windings on the rotor and partly to ac currents flowing in the armature windings. The air-gap flux components are functions of both time and angular displacement. The field lines cross perpendicular to the gap, because that is the path of least reluctance. Thus, at any given point, the field is directed perpendicular to the armature conductors, which lie in slots cut lengthwise into the inside face of the stator.

Most synchronous machines are designed so that the flux density varies sinusoidally with $\theta_m$. Because the field rotates at a uniform rate, the flux density at any point in the gap varies sinusoidally with time. Thus, we can represent the field components at $\theta_m = 0$ by phasors denoted by $\mathbf{B}_s$, $\mathbf{B}_r$, and $\mathbf{B}_{total}$, which correspond to the stator flux component, rotor flux component, and total flux, respectively. Then, we can write

$$\mathbf{B}_{total} = \mathbf{B}_s + \mathbf{B}_r \tag{16.36}$$

The torque developed in the rotor is given by

$$T_{dev} = KB_r B_{total} \sin(\delta) \tag{16.37}$$

in which $K$ is a constant that depends on the dimensions and other features of the machine. $B_{total}$ and $B_r$ are the magnitudes of the phasors $\mathbf{B}_{total}$ and $\mathbf{B}_r$, respectively. $\delta$ is the electrical angle, called the **torque angle**, by which the rotor field lags the total field.

### Equivalent Circuit

The rotating field components induce corresponding voltage components in the armature windings. We concentrate on the $a$ phase of the armature winding. The voltages and currents in the other two armature windings are identical except for phase shifts of $\pm 120°$.

The voltage component induced by the rotor flux can be represented as a phasor that is given by

$$\mathbf{E}_r = k\mathbf{B}_r \tag{16.38}$$

in which $k$ is a constant that depends on the machine construction features.

A second voltage component is induced in each winding by the rotating stator field. This voltage component is given by

$$\mathbf{E}_s = k\mathbf{B}_s \tag{16.39}$$

As we have seen, the stator field is established by the armature currents. The stator is a mutually coupled three-phase inductor, and the voltage due to the stator field can be written as

$$\mathbf{E}_s = jX_s\mathbf{I}_a \qquad (16.40)$$

where $X_s$ is an inductive reactance known as the **synchronous reactance**, and $\mathbf{I}_a$ is the phasor for the armature current. [Actually, the stator windings also have resistance, and more precisely, we have $\mathbf{E}_s = (R_a + jX_s)\mathbf{I}_a$. However, the resistance $R_a$ is usually very small compared to the synchronous reactance, so Equation 16.40 is sufficiently accurate.]

The voltage observed at the terminals of the armature winding is the sum of these two components. Thus, we can write

$$\mathbf{V}_a = \mathbf{E}_r + \mathbf{E}_s \qquad (16.41)$$

where $\mathbf{V}_a$ is the phasor for the terminal voltage for the $a$-phase winding. Using Equation 16.40 to substitute for $\mathbf{E}_s$, we have

$$\mathbf{V}_a = \mathbf{E}_r + jX_s\mathbf{I}_a \qquad (16.42)$$

Also, we can write

$$\mathbf{V}_a = k\mathbf{B}_{\text{total}} \qquad (16.43)$$

because the total voltage is proportional to the total flux.

The equivalent circuit of the synchronous motor is shown in Figure 16.18. Only the $a$ phase of the armature is shown. The three-phase source $\mathbf{V}_a$ supplies current $\mathbf{I}_a$ to the armature. The ac voltage induced in the armature by the rotor field is represented by the voltage source $\mathbf{E}_r$. The dc voltage source $V_f$ supplies the field current $I_f$ to the rotor. An adjustable resistance $R_{\text{adj}}$ is included in the field circuit so that the field current can be varied. This in turn adjusts the magnitudes of the rotor field $\mathbf{B}_r$ and the resulting induced voltage $\mathbf{E}_r$.

The armature windings can be connected either in a wye or in a delta configuration. In our discussion, we do not specify the way in which the windings are connected. For either connection, $\mathbf{V}_a$ represents the voltage across the $a$ winding. In a wye connection, $\mathbf{V}_a$ corresponds to the line-to-neutral voltage, whereas in a delta connection, $\mathbf{V}_a$ corresponds to the line-to-line voltage. Similarly, $\mathbf{I}_a$ is the current through the $a$ winding, which corresponds to the line current in a wye connection but not in a delta connection. The important things to remember are that $\mathbf{V}_a$ is the voltage across the $a$ winding and that $\mathbf{I}_a$ is the current through the a winding, regardless of the manner in which the machine is connected.

The phasor diagram for the current and voltages is shown in Figure 16.19(a). The corresponding phasor diagram for the fields is shown in Figure 16.19(b). Because the

> $\mathbf{V}_a$ and $\mathbf{I}_a$ represent the rms phase voltage and phase current, respectively. The relationship to line voltage and line current depends on whether the machine is wye or delta connected.

> Throughout our discussion, we assume that the phase angle of $\mathbf{V}_a$ is 0.

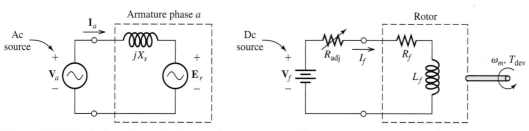

**Figure 16.18** Equivalent circuit for the synchronous motor. The armature circuit is based on Equation 16.42.

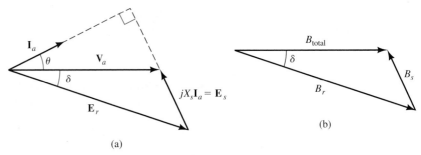

**Figure 16.19** Phasor diagrams for the synchronous motor. Notice that the stator component of voltage $\mathbf{E}_s = jX_s\mathbf{I}_a$ is at right angles to the current $\mathbf{I}_a$. The developed torque is given by $T_{\text{dev}} = KB_rB_{\text{total}} \sin \delta$, and the power factor is $\cos(\theta)$.

rotor field is lagging the total field, positive torque (given by Equation 16.37) and output power are being developed. In other words, the machine is acting as a motor.

The input power taken from the three-phase ac source is given by

$$P_{\text{dev}} = P_{\text{in}} = 3V_aI_a \cos(\theta) \qquad (16.44)$$

in which the factor of three accounts for the fact that there are three sets of windings. Since the equivalent circuit does not include any losses, the input power and the developed mechanical power are equal.

### Potential for Power-Factor Correction

The total reactive power absorbed by the three windings is given by

$$Q = 3V_aI_a \sin(\theta) \qquad (16.45)$$

> The synchronous motor can act as a source of reactive power.

in which $\theta$ is defined to be the angle by which the phase current $\mathbf{I}_a$ *lags* the phase voltage $\mathbf{V}_a$.

Notice in Figure 16.19(a) that $\theta$ takes a negative value because the phase current $\mathbf{I}_a$ *leads* the phase voltage $\mathbf{V}_a$. Therefore, the reactive power for the machine is negative, indicating that the synchronous motor can supply reactive power. This is a significant advantage because most industrial plants have an overall lagging power factor (due largely to the widespread employment of induction motors). Poor power factor leads to larger currents in the transmission lines and transformers supplying the plant. Thus, utility companies invariably charge their industrial customers more for energy supplied while the power factor is low. By using some synchronous motors in an industrial plant, part of the reactive power taken by inductive loads can be supplied locally, thereby lowering energy costs. Should you someday be employed as a plant engineer, you will need to have a good understanding of these issues.

> Proper use of synchronous motors can lower energy costs of an industrial plant by increasing the power factor.

Unloaded synchronous machines have sometimes been installed solely for the purpose of power-factor correction. With zero load (and neglecting losses), the rotor field and the total field align so that the torque angle $\delta$ is zero, and according to Equation 16.37, the developed torque is zero. Phasor diagrams for unloaded synchronous machines are shown in Figure 16.20.

If we have

$$V_a > E_r \cos(\delta) \qquad (16.46)$$

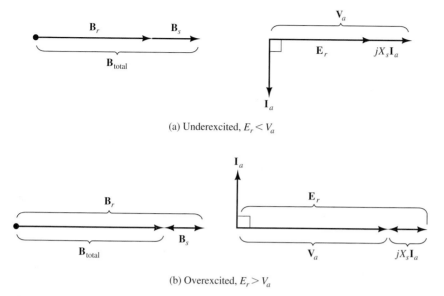

(a) Underexcited, $E_r < V_a$

(b) Overexcited, $E_r > V_a$

**Figure 16.20** Phasor diagrams for unloaded synchronous machines. When a machine has $E_r > V_a$, the current $\mathbf{I}_a$ leads the applied voltage $\mathbf{V}_a$ by 90°, and each phase of the machine is electrically equivalent to a capacitor. Thus, the machine supplies reactive power.

we say that the machine is **underexcited**. For an unloaded machine with $\delta = 0$, the machine is underexcited if the magnitude of $\mathbf{E}_r$ is less than the magnitude of the applied phase voltage $\mathbf{V}_a$. Then, the current $\mathbf{I}_a$ lags $\mathbf{V}_a$ by $\theta = 90°$. Consequently, the real power supplied (which is given by Equation 16.44) is zero, as we expect for an unloaded machine (neglecting losses). The underexcited machine absorbs reactive power. This is the opposite of the desired result for most applications.

However, if the field current is increased such that

$$V_a < E_r \cos(\delta) \tag{16.47}$$

we say that the machine is **overexcited**. The phasor diagram is shown in Figure 16.20(b) for an unloaded overexcited machine. In this case, the current leads the voltage by 90°, and the machine supplies reactive power. In the overexcited state, an unloaded synchronous machine appears as a pure capacitive reactance to the ac source. Machines used in this manner are called **synchronous capacitors**.

## Operation with Variable Load and Constant Field Current

Motors are usually operated from ac voltage sources of constant magnitude and phase. This fact in combination with Equation 16.43 shows that the total flux phasor $\mathbf{B}_{\text{total}}$ is constant in magnitude and phase. Because speed is constant in a synchronous machine, power is proportional to torque, which in turn is proportional to $B_r \sin(\delta)$, as shown by Equation 16.37. Thus, we can write

$$P_{\text{dev}} \propto B_r \sin(\delta) \tag{16.48}$$

This fact is illustrated in Figure 16.21(a).

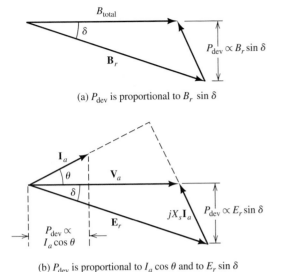

(a) $P_{\text{dev}}$ is proportional to $B_r \sin \delta$

(b) $P_{\text{dev}}$ is proportional to $I_a \cos \theta$ and to $E_r \sin \delta$

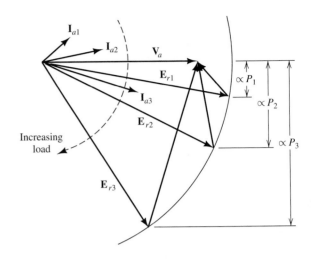

(c) Phasor diagram with increasing load and constant field current

**Figure 16.21** Phasor diagrams for a synchronous motor.

Furthermore, $E_r$ is proportional to $B_r$. Hence, we have established that

$$P_{\text{dev}} \propto E_r \sin(\delta) \qquad (16.49)$$

Since $P_{\text{dev}} = P_{\text{in}} = 3V_a I_a \cos(\theta)$ (neglecting stator copper loss) and since $V_a$ is constant, we also have

$$P_{\text{dev}} \propto I_a \cos(\theta) \qquad (16.50)$$

Equations 16.49 and 16.50 are illustrated in the phasor diagram shown in Figure 16.21(b).

Now suppose that we have a synchronous motor operating with a variable load and constant field current. Because the field current is constant, $\mathbf{E}_r$ is constant in magnitude. As the load changes, $\mathbf{E}_r$ can change in phase, but not in magnitude. Therefore, the locus formed by $\mathbf{E}_r$ is a circle. The phasor diagram for a machine with a variable load is shown in Figure 16.21(c). Notice that the power factor tends to become lagging as the load is increased.

As the load varies with constant field current, the locus of $\mathbf{E}_r$ is a circle. Notice that if $jX_s\mathbf{I}_a$ and $\mathbf{I}_a$ are extended, they meet at right angles. Careful study of Figure 16.21 shows that the power factor tends to become lagging as load is increased.

---

**Example 16.4    Synchronous-Motor Performance**

A 480-V-rms 200-hp 60-Hz eight-pole delta-connected synchronous motor operates with a developed power (including losses) of 50 hp and a power factor of 90 percent leading. The synchronous reactance is $X_s = 1.4\ \Omega$.

**a.** Find the speed and developed torque.

**b.** Determine the values of $\mathbf{I}_a$, $\mathbf{E}_r$, and the torque angle.

**c.** Suppose that the excitation remains constant and the load torque increases until the developed power is 100 hp. Determine the new values of $\mathbf{I}_a$, $\mathbf{E}_r$, the torque angle, and the power factor.

**Solution**

**a.** The speed of the machine is given by Equation 16.14:

$$n_s = \frac{120f}{P} = \frac{120(60)}{8} = 900 \text{ rpm}$$

$$\omega_s = n_s \frac{2\pi}{60} = 30\pi = 94.25 \text{ rad/s}$$

For the first operating condition, the developed power is

$$P_{dev1} = 50 \times 746 = 37.3 \text{ kW}$$

and the developed torque is

$$T_{dev1} = \frac{P_{dev1}}{\omega_s} = \frac{37{,}300}{94.25} = 396 \text{ Nm}$$

**b.** The voltage rating refers to the rms line-to-line voltage. Because the windings are delta connected, we have $V_a = V_{line} = 480$ V rms. Solving Equation 16.44 for $I_a$ and substituting values, we have

$$I_{a1} = \frac{P_{dev1}}{3V_a \cos(\theta_1)} = \frac{37{,}300}{3(480)(0.9)} = 28.78 \text{ A rms}$$

Next, the power factor is $\cos(\theta_1) = 0.9$, which yields

$$\theta_1 = 25.84°$$

Because the power factor was given as leading, we know that the phase of $\mathbf{I}_{a1}$ is positive. Thus, we have

$$\mathbf{I}_{a1} = 28.78\underline{/25.84°} \text{ A rms}$$

Then from Equation 16.42, we have

$$\mathbf{E}_{r1} = \mathbf{V}_{a1} - jX_s\mathbf{I}_a = 480 - j1.4(28.78\underline{/25.84°})$$

$$= 497.6 - j36.3$$

$$= 498.9\underline{/-4.168°} \text{ V rms}$$

Consequently, the torque angle is $\delta_1 = 4.168°$.

**c.** When the load torque is increased while holding excitation constant (i.e., the values of $I_f$, $B_r$, and $E_r$ are constant), the torque angle must increase. In Figure 16.21(b), we see that the developed power is proportional to $\sin(\delta)$. Hence, we can write

$$\frac{\sin(\delta_2)}{\sin(\delta_1)} = \frac{P_2}{P_1}$$

Solving for $\sin(\delta_2)$ and substituting values, we find that

$$\sin(\delta_2) = \frac{P_2}{P_1}\sin(\delta_1) = \frac{100 \text{ hp}}{50 \text{ hp}}\sin(4.168°)$$

$$\delta_2 = 8.360°$$

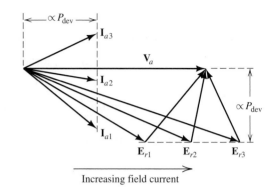

**Figure 16.22** Phasor diagram for constant developed power and increasing field current.

Because $E_r$ is constant in magnitude, we get

$$\mathbf{E}_{r2} = 498.9\underline{/-8.360°} \text{ V rms}$$

(We know that $\mathbf{E}_{r2}$ lags $\mathbf{V}_a = 480\underline{/0°}$ because the machine is acting as a motor.) Next, we can find the new current:

$$\mathbf{I}_{a2} = \frac{\mathbf{V}_a - \mathbf{E}_{r2}}{jX_s} = 52.70\underline{/10.61°} \text{ A rms}$$

Finally, the new power factor is

$$\cos(\theta_2) = \cos(10.61°) = 98.3\% \text{ leading} \qquad \blacksquare$$

**Exercise 16.5**    For the motor of Example 16.4, suppose that the excitation remains constant and the load torque increases until the developed power is $P_{dev3} = 200$ hp. Determine the new values of $\mathbf{I}_a$, $\mathbf{E}_r$, the torque angle, and the power factor.
**Answer**    $\mathbf{I}_{a3} = 103.6\underline{/-1.05°}$; $\mathbf{E}_{r3} = 498.9\underline{/-16.90°}$; $\delta_3 = 16.90°$; power factor = 99.98% lagging.    □

## Operation with Constant Load and Variable Field Current

When operating with constant developed power $P_{dev}$, Figure 16.21(b) shows that the values of $I_a \cos(\theta)$ and $E_r \sin(\delta)$ are constant. Then, if the field current increases, the magnitude of $E_r$ increases. The resulting phasor diagram for several values of field current is shown in Figure 16.22. Notice that as the field current increases, the armature current decreases in magnitude, reaching a minimum for $\theta = 0°$ (or unity power factor) and then increases with a leading power factor. The current magnitude reaches a minimum when $\mathbf{I}_a$ is in phase with $\mathbf{V}_a$ (i.e., when $\theta = 0$ and the power factor is unity). Plots of $I_a$ versus field current are shown in Figure 16.23. These plots are called **V curves** because of their shape.

Power factor tends to become leading as field current, and consequently, $\mathbf{E}_r$ increase in magnitude.

### Example 16.5    Power-Factor Control

A 480-V-rms 200-hp 60-Hz eight-pole delta-connected synchronous motor operates with a developed power (including losses) of 200 hp and a power factor of 85 percent lagging. The synchronous reactance is $X_s = 1.4 \, \Omega$. The field current is $I_f = 10$ A. What must the new field current be to produce 100 percent power factor? Assume that magnetic saturation does not occur, so that $B_r$ is proportional to $I_f$.

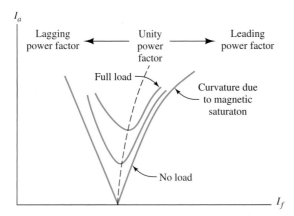

**Figure 16.23** V curves for a synchronous motor with variable excitation.

**Solution**    First, we determine the initial value of $E_r$. Because the initial power factor is $\cos(\theta_1) = 0.85$, we can determine that

$$\theta_1 = 31.79°$$

Then, the phase current is

$$I_{a1} = \frac{P_{\text{dev}}}{3V_a \cos(\theta_1)} = \frac{200(746)}{3(480)0.85} = 121.9 \text{ A rms}$$

Thus, the phasor current is

$$\mathbf{I}_{a1} = 121.9 \underline{/-31.79°} \text{ A rms}$$

The induced voltage is

$$\mathbf{E}_{r1} = \mathbf{V}_{a1} - jX_s\mathbf{I}_{a1} = 480 - j1.4(121.9\underline{/-31.79°})$$
$$= 390.1 - j145.0$$
$$= 416.2\underline{/-20.39°} \text{ V rms}$$

The phasor diagram for the initial excitation is shown in Figure 16.24(a).

To achieve 100 percent power factor, we need to increase the field current and the magnitude of $\mathbf{E}_r$ until $\mathbf{I}_a$ is in phase with $\mathbf{V}_a$, as shown in Figure 16.24(b). The new value of the phase current is

$$I_{a2} = \frac{P_{\text{dev}}}{3V_a \cos(\theta_2)} = \frac{200(746)}{3(480)} = 103.6 \text{ A rms}$$

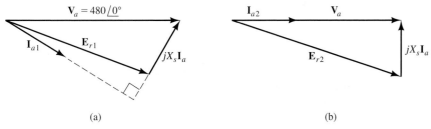

(a)                                                                 (b)

**Figure 16.24** Phasor diagrams for Example 16.5.

Then, we have

$$\mathbf{E}_{r2} = \mathbf{V}_{a2} - jX_s\mathbf{I}_{a2} = 480 - j1.4(103.6)$$
$$= 480 - j145.0$$
$$= 501.4\underline{/-16.81°} \text{ V rms}$$

Now the magnitude of $\mathbf{E}_r$ is proportional to the field current, so we can write

$$I_{f2} = I_{f1}\frac{E_{r2}}{E_{r1}} = 10\frac{501.4}{416.2} = 12.05 \text{ A dc} \qquad \blacksquare$$

**Exercise 16.6**    Find the field current needed to achieve a power factor of 90 percent leading for the motor of Example 16.5.
**Answer**    $I_f = 13.67$ A.    □

## Pull-Out Torque

The developed torque of a synchronous motor is given by Equation 16.37, which states that

$$T_{dev} = KB_rB_{total}\sin(\delta)$$

The pull-out torque is the maximum torque that the synchronous motor can produce, which occurs for a torque angle of 90°.

This is plotted in Figure 16.25. The maximum or **pull-out torque** $T_{max}$ occurs for a torque angle of $\delta = 90°$:

$$T_{max} = KB_rB_{total} \qquad (16.51)$$

Typically, the rated torque is about 30 percent of the maximum torque.

Suppose that a synchronous motor is initially unloaded. Then, it runs at synchronous speed with $\delta = 0$. As the load increases, the motor slows momentarily and $\delta$ increases just enough so that the developed torque meets the demands of the load plus losses. Then, the machine again runs at synchronous speed.

However, if the load on a synchronous machine was to exceed the pull-out torque, it would no longer be possible for the machine to drive the load at synchronous speed and $\delta$ would keep on increasing. Then, the machine would produce enormous surges in torque back and forth, resulting in great vibration. Once the rotor pulls out of synchronism with the rotating armature field, the average torque falls to zero, and the system slows to a stop.

Generally, it is desirable to operate synchronous motors in the overexcited state to obtain large pull-out torque and to generate reactive power.

The torque–speed characteristic of a synchronous motor is shown in Figure 16.26. Generally, it is desirable to operate synchronous motors in an overexcited state

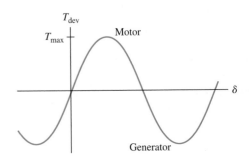

Figure 16.25 Torque versus torque angle. $T_{max}$ is the maximum or pull-out torque.

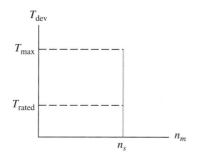

**Figure 16.26** Torque–speed characteristic of synchronous motors.

(i.e., large values of $I_f$, $B_r$, and $E_r$), for several reasons. First, the machine produces reactive power. Second, as shown by Equation 16.51, the pull-out torque is higher with higher $B_r$.

## Starting Methods

Because a synchronous motor develops zero starting torque, we need to make special provisions for starting. Several methods can be used:

1. Vary the frequency of the ac source starting very low (a fraction of a hertz) and gradually increasing to the operating speed desired. This can be accomplished with power electronic circuits known as *cycloconverters* that can convert 60-Hz ac power into three-phase power of any desired frequency. Such a system can also be used for very accurate speed control.

2. Use a prime mover to bring the synchronous motor up to speed. Then, the motor is connected to the ac source and the load is connected. Before the ac source is connected, it is important to wait until the phases of the voltages induced in the armature closely match those of the line voltages. In other words, we want the torque angle $\delta$ to be close to zero before closing the switches to the ac source. Otherwise, excessive currents and torques occur as the rotor tries to rapidly align itself with the stator field.

3. The rotors of many synchronous motors contain **amortisseur** or **damper conductors**, which are similar in structure to the squirrel-cage conductors used in induction motors. Then, the motor can be started as an induction motor with the field windings shorted and without load. After the motor has approached synchronous speed, the dc source is connected to the field and the motor pulls into synchronism. Then, the load is connected.

Damper conductors have another purpose besides use in starting. It is possible for the speed of a synchronous motor to oscillate above and below synchronous speed so that the torque angle $\delta$ swings back and forth. This action is similar to that of a pendulum. By including the damper bars, the oscillation is damped out. When running at synchronous speed, no voltage is induced in the damper bars and they have no effect.

> Because the starting torque of the synchronous motor is zero, special starting provisions are needed.

**Exercise 16.7**   A synchronous motor produces maximum torque and maximum power for $\delta = 90°$. Draw the phasor diagram for this case and show that $P_{max} = 3(V_aE_r/X_s)$ and $T_{max} = 3(V_aE_r/\omega_mX_s)$.
**Answer**   The phasor diagram is shown in Figure 16.27. ☐

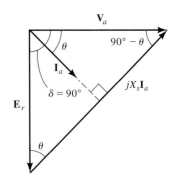

**Figure 16.27** Phasor diagram for maximum developed torque and maximum power conditions. See Exercise 16.7.

## 16.4  SINGLE-PHASE MOTORS

In Chapter 15, we examined the universal motor, which can be operated from single-phase ac. In this section, we discuss briefly several additional types of single-phase ac motors. Single-phase motors are important because three-phase power is not available for homes, most offices, and many small businesses.

Compared with induction motors, universal motors have a higher power/weight ratio, but they do not have as long a service life, due to wear of the brushes. Assuming constant source frequency, induction motors are essentially constant-speed devices. On the other hand, the speed of a universal motor can be varied by changing the amplitude of the applied voltage.

Universal motors have relatively large power-to-weight ratios but short service lives.

### Basic Single-Phase Induction Motor

Let us begin by considering the basic single-phase induction motor shown in Figure 16.28. The stator of this motor has a **main winding** that is connected to an ac source. (Later, we will see that an auxiliary winding is needed for starting.) It has a squirrel-cage rotor that is identical to the rotor of the three-phase induction motor shown in Figure 16.6.

Ideally, the air-gap flux varies sinusoidally in space around the circumference of the gap. Thus, the flux is given by

$$B = Ki(t)\cos(\theta) \tag{16.52}$$

which is the same, except for changes in notation, as Equation 16.1 for the flux due to winding $a$ of a three-phase induction motor. The stator current is given by

$$i(t) = I_m \cos(\omega t) \tag{16.53}$$

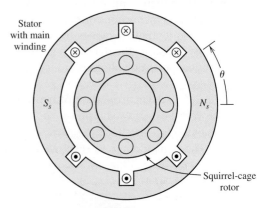

**Figure 16.28** Cross-section of the basic single-phase induction motor.

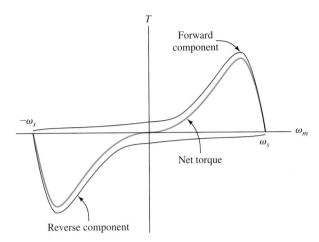

**Figure 16.29** The main winding produces two counter-rotating flux components each of which induces torque in the rotor. The main winding alone induces no net starting torque.

Substituting this expression for the current into Equation 16.52, we have

$$B = KI_m \cos(\omega t) \cos(\theta) \qquad (16.54)$$

Instead of rotating, this flux pulsates, switching direction twice per cycle.

However, by applying the trigonometric identity for the product of cosines, we can write Equation 16.54 as

$$B = \frac{1}{2}KI_m \cos(\omega t - \theta) + \frac{1}{2}KI_m \cos(\omega t + \theta) \qquad (16.55)$$

The first term on the right-hand side of Equation 16.55 represents a flux that rotates counterclockwise (i.e., in the positive $\theta$ direction), while the second term rotates clockwise. Thus, the pulsating flux in the basic single-phase induction motor can be resolved into two counter rotating components. On the other hand, the three-phase motor has flux rotating in one direction only.

*The pulsating flux produced by the main winding can be resolved into two counter rotating components.*

We assume that the rotor spins counterclockwise with speed $\omega_m$. The field component that rotates in the same direction as the rotor is called the **forward component**. The other component is called the **reverse component**. Each of these components produces torque, but in opposite directions. The torque versus speed characteristic for each component is similar to that of a three-phase induction motor. The torques produced by the forward component, the reverse component, and the total torque are shown in Figure 16.29.

Notice that the net starting torque is zero, and therefore the main winding will not start a load from a standing start. Once started, however, the motor develops torque and accelerates loads within its ratings to nearly synchronous speed. Its running characteristics (in the vicinity of synchronous speed) are similar to those of the three-phase induction motor. Because of the symmetry of its torque–speed characteristic, the basic single-phase motor is capable of running equally well in either direction.

## Auxiliary Windings

Lack of starting torque is a serious flaw for a motor in most applications. However, the basic single-phase induction motor can be modified to provide starting torque and improve its running characteristics. It can be shown (see Problem P16.8) that equal-amplitude currents having a 90° phase relationship and flowing in windings

*Two windings that are 90° apart physically and carry currents 90° apart in phase produce a rotating magnetic field.*

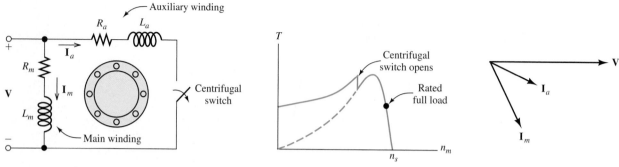

**Figure 16.30** The split-phase induction motor.

that are at right angles produce only a forward rotating component of flux. (This is similar to the rotating flux produced by balanced three-phase currents flowing in windings displaced by 120° from one another.) If the two currents differ in phase by less than 90° (but by more than 0°), the forward flux component is larger than the reverse flux component and net starting torque results. Thus, nearly all single-phase induction motors have an **auxiliary winding** rotated in space by 90 electrical degrees from the main winding. Various provisions can be made to achieve the requisite phase shift between the current in the main winding and the current in the auxiliary winding.

*Single-phase induction motors contain an auxiliary winding displaced by 90 electrical degrees from the main winding.*

One possibility is to wind the auxiliary winding with smaller wire that has a higher ratio of resistance to inductive reactance than the main winding. Then, the current in the auxiliary winding has a different phase angle than that of the main current. Motors using this approach are called **split-phase motors** (Figure 16.30). Usually, the auxiliary winding is designed to be used only briefly during starting, and a centrifugal switch disconnects it when the motor approaches rated speed. (A common failure in this type of motor is for the switch to fail to open, and then the auxiliary winding overheats and burns out.)

When running on the main winding, the torque of a single-phase motor pulsates at twice the frequency of the ac source, because no torque is produced when the stator current passes through zero. On the other hand, torque is constant in a three-phase motor because the current is nonzero in at least two of the three windings at all instants. Thus, single-phase induction motors display more noise and vibration than do three-phase motors. Furthermore, single-phase induction motors are larger and heavier than three-phase motors of the same ratings.

*Single-phase induction motors produce more noise and vibration and are larger than three-phase motors with equal power ratings.*

In a **capacitor-start motor**, a capacitor is placed in series with the auxiliary winding, resulting in much higher starting torque than that of the split-phase motor, because the phase relationship between $\mathbf{I}_a$ and $\mathbf{I}_m$ is closer to 90°. In a **capacitor-run motor**, the auxiliary winding is a permanent part of the circuit, resulting in smoother torque and less vibration. Another variation is the **capacitor-start, capacitor-run motor** shown in Figure 16.31.

## Shaded-Pole Motors

*Shaded-pole motors are used for inexpensive low-power applications.*

The least expensive approach to providing self-starting for single-phase induction motors is the **shaded-pole motor**, shown in Figure 16.32. A shorted copper band is placed around part of each pole face. As the field builds up, current is induced in this *shading ring*. The current retards changes in the field for that part of the pole face encircled by the ring. As the current in the ring decays, the center of the magnetic

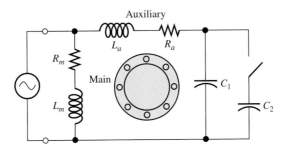

**Figure 16.31** The capacitor-start, capacitor-run motor.

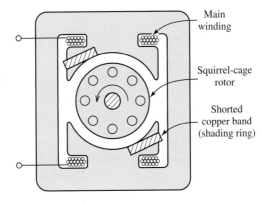

**Figure 16.32** The shaded-pole motor.

pole moves in the direction of the ring. This favors rotation in one direction over the other, resulting in starting torque. This approach is used only for very small motors (1/20 hp or less).

## 16.5 STEPPER MOTORS AND BRUSHLESS DC MOTORS

Stepper motors are used for accurate, repeatable positioning such as machine tool applications or for moving the head in an ink-jet printer. By using an electronic controller that applies electrical pulses to the motor windings, the motor shaft can be rotated in either direction in multiples of the step angle, which can range from 0.72° (500 steps per revolution) to 15° (24 steps per revolution). Stepper motors are available with rotational accuracies on the order of 3 percent of a step, which is noncumulative as the motor is stepped back and forth. By controlling the rate at which pulses are applied to the windings of the stepper motor, speed can be varied continuously from a standing stop to a maximum that depends on the motor and load.

There are several types of stepper motors. Figure 16.33(a) shows the cross-section of the simplest, which is known as a **variable-reluctance stepper motor**. Notice that the stator has eight salient poles that are 45° apart. On the other hand, the rotor has six salient poles 60° apart. Thus, when 1 is aligned with $A$ as shown, 2 is 15° counterclockwise from $B$, and 3 is 15° clockwise from $D$.

The stator contains four windings (which are not shown in the cross-section). A controller applies power to one of the coils at a time as shown in Figure 16.33(b). Coil $A$ is wound partly around pole $A$ and partly around $A'$, such that, when current is applied, $A$ becomes a north magnetic pole and $A'$ becomes a south pole. Then, the rotor moves to shorten the air gaps between $A$ (and $A'$) and the rotor. As long as power is applied to coil $A$, the rotor is held in the position shown

Anytime you need accurate repeatable positioning, consider using a stepper motor.

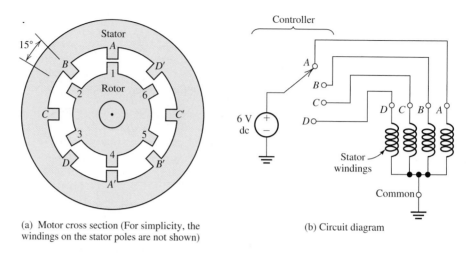

(a)  Motor cross section (For simplicity, the windings on the stator poles are not shown)

(b) Circuit diagram

Figure 16.33 Variable-reluctance stepper motor.

in the figure. However, if power is removed from $A$ and applied to $B$ by moving the controller switch, the rotor moves 15° clockwise so that 2 is aligned with $B$. Then if power is removed from $B$ and applied to $C$, the shaft rotates another 15° clockwise. Thus, applying power to the coils in the sequence $ABCDABC$... results in clockwise rotation of the shaft in 15° increments. By changing the switching rate, the motor speed can be varied upward from a standing stop. Furthermore, by reversing the switching sequence to $ADCBADCB$ ..., the direction of rotation can be reversed.

Another type is the **permanent-magnet stepper motor**, which has a cylindrical rotor (called a "tin-can rotor") that is permanently magnetized with north and south poles alternating around its circumference. The stator of the permanent magnet motor is similar to that of the reluctance motor. As in the reluctance type, the rotor position is stepped by applying a sequence of pulses to the stator windings. **Hybrid stepper motors** that combine variable reluctance with permanent magnets are also available. Of course, detailed specifications for stepper motors can be found from manufacturers' websites.

## Brushless DC Motors

Conventional dc motors are particularly useful in applications that require high speeds and in those for which dc power is available, such as those in aircraft and automobiles. However, because they contain commutators and brushes, conventional dc motors have several disadvantages. These include relatively short service lives due to brush and commutator wear, particularly at very high speeds. Also, arcing as the brushes move between commutator segments can pose a hazard in explosive environments and can create severe radio interference. A relatively new development, the **brushless dc motor**, provides an excellent alternative when the disadvantages of conventional dc motors are prohibitive.

Brushless dc motors are essentially permanent-magnet stepping motors equipped with position sensors (either Hall effect or optical) and enhanced control units. As in the stepper motor, power is applied to one stator winding at a time.

When the position sensor indicates that the rotor has approached alignment with the stator field, the controller electronically switches power to the next stator winding so that smooth motion continues. By varying the amplitude and duration of the pulses applied to the stator windings, speed can be readily controlled. The result is a motor that can operate from a dc source with characteristics similar to those of a conventional shunt dc motor.

Brushless dc motors are used primarily in low-power applications. Their advantages include relatively high efficiency, long service life with little maintenance, freedom from radio interference, ability to operate in explosive chemical environments, and capability for very high speeds (50,000 rpm or more).

## Summary

1. Application of a three-phase source to the stator windings of an induction motor produces a magnetic field in the air gap that rotates at synchronous speed. Interchanging any two of the connections to the three-phase source reverses the direction of rotation.

2. A squirrel-cage rotor contains aluminum conductors embedded in the rotor. As the stator field moves past, currents are induced in the rotor resulting in torque. The torque–speed characteristic takes the form shown in Figure 16.10 on page 802. In normal steady-state operation, typical motors operate with 0 to 5 percent slip, and the output power and torque are approximately proportional to slip.

3. The per-phase equivalent circuit shown in Figure 16.13 on page 804 is useful in performance calculations for induction motors.

4. Some of the most important considerations in selecting an induction motor are efficiency, starting torque, pull-out torque, power factor, and starting current.

5. Typically, the starting torque of an induction motor is 150 percent or more of the rated full-load running torque. Thus, induction motors can start all constant-torque loads that are within their full-load ratings. Starting current with rated voltage is typically five to six times the full-load running current.

6. A three-phase synchronous machine has stator windings that produce a magnetic field consisting of $P$ poles that rotate at synchronous speed. The rotor is an electromagnet. The machine runs at synchronous speed, and the torque–speed characteristic is shown in Figure 16.26 on page 823.

7. When operated in the overexcited state, synchronous machines produce reactive power and can help to correct the power factor of industrial plants, saving on energy costs.

8. The equivalent circuit for three-phase synchronous machines shown in Figure 16.18 on page 815 can be used in performance calculations.

9. Synchronous motors have zero starting torque, and special provisions must be made for starting.

10. Single-phase induction motors have a main winding and an auxiliary winding displaced by 90 electrical degrees. With power applied to only the main winding, the motor can run but has zero starting torque. Because of differences in the resistance/reactance ratio or because a capacitor is in the circuit, the currents in the two windings have different phases and this produces starting torque. Frequently, the starting winding is disconnected from the ac source when the motor approaches rated speed.

11. Single-phase induction motors are heavier and produce more vibration than do three-phase motors of the same power rating.

12. Stepper motors are useful in applications that require accurate repeatable positioning.

13. Brushless dc motors are a good alternative to conventional dc motors for low-power applications that require long life with little maintenance, operation in explosive environments, freedom from radio interference, or very high speed.

## Problems

### Section 16.1: Three-Phase Induction Motors

*P16.1. It is necessary to reduce the voltage applied to an induction motor as the frequency is reduced from the rated value. Explain why this is so.

*P16.2. The air-gap flux density of a two-pole induction motor is given by

$$B = B_m \cos(\omega t - \theta)$$

where $B_m$ is the peak flux density, $\theta$ is the angular displacement around the air gap, and we have assumed clockwise rotation. Give the corresponding expression for the flux density of a four-pole induction motor; of a six-pole induction motor.

*P16.3. A four-pole induction motor drives a load at 2500 rpm. This is to be accomplished by using an electronic converter to convert a 400-V dc source into a set of three-phase ac voltages. Find the frequency required for the ac voltages assuming that the slip is 4 percent. The load requires 2 hp. If the dc-to-ac converter has a power efficiency of 88 percent and the motor has a power efficiency of 80 percent, estimate the current taken from the dc source.

*P16.4. A 60-Hz induction motor is needed to drive a load at approximately 850 rpm. How many poles should the motor have? What is the slip of this motor for a speed of 850 rpm?

P16.5. The magnetic field produced in the air gap of an induction motor by the stator windings is given by $B = B_m \cos(\omega t - 2\theta)$, in which $\theta$ is angular displacement in the counterclockwise direction as illustrated in Figure 16.4 on page 797. How many poles does this machine have? Given that the frequency of the source is 50 Hz, determine the speed of rotation of the field. Does the field rotate clockwise or counterclockwise? Repeat for a field given by $B = B_m \cos(\omega t + 3\theta)$.

P16.6. Prepare a table that shows synchronous speeds for three-phase induction motors operating at 50 Hz. Consider motors having eight or fewer poles. Repeat for 400-Hz motors.

P16.7. Explain why induction motors develop zero torque at synchronous speed.

P16.8. Consider the two-pole two-phase induction motor having two windings displaced 90° in space shown in Figure P16.8. The fields produced by the windings are given by $B_a = Ki_a(t) \cos(\theta)$ and $B_b = Ki_b(t) \cos(\theta - 90°)$. The two-phase source produces currents given by $i_a(t) = I_m \cos(\omega t)$ and $i_b(t) = I_m \cos(\omega t - 90°)$. Show that the total field rotates. Determine the speed and direction of rotation. Also find the maximum flux density of the rotating field in terms of $K$ and $I_m$.

P16.9. In a proposed design for an electric automobile, the shaft of a four-pole three-phase induction motor is connected directly to the drive axle; in other words, there is no gear train. The outside diameter of the tires is 20 inches. Instead of a transmission, an electronic converter produces variable-frequency three-phase ac from a 48-V battery. Assuming negligible slip, find the range of frequencies needed for speeds ranging from 5 to 70 mph. The vehicle, including batteries and occupants, has a mass of 1000 kg. The power efficiency of the dc-to-ac converter is 85 percent, and the power efficiency of the motor is 89 percent. **a.** Find the current taken from the battery as a function of time while accelerating from 0 to 40 mph uniformly (i.e., acceleration is constant) in 10 seconds. Neglect wind load and road friction. **b.** Repeat assuming that the vehicle is accelerated with constant power.

P16.10. A 10-hp six-pole 60-Hz three-phase induction motor runs at 1160 rpm under full-load conditions. Determine the slip and the frequency of the rotor currents at full load.

---

*Denotes that answers are contained in the Student Solutions files. See Appendix E for more information about accessing the Student Solutions.

Also estimate the speed if the load torque drops in half.

**P16.11.** Consider the induction motor shown in Figure 16.7 on page 800. Redraw the figure showing the current directions in the rotor conductors, the magnetic rotor poles, and the direction of the developed torque if a prime mover drives the rotor at a speed higher than synchronous speed. In this case, does the machine operate as a motor or as a generator?

**P16.12.** Consider the two-pole two-phase induction motor having two windings displaced $90°$ in space shown in Figure P16.8. The fields produced by the windings are given by $B_a = Ki_a(t) \cos(\theta)$ and $B_b = Ki_b(t) \cos(\theta - 90°)$. The two-phase source produces currents given by $i_a(t) = I_m \cos(2\omega t)$ and $i_b(t) = I_m \cos(2\omega t + 90°)$. Show that the total field rotates. Determine the speed and direction of rotation. Also find the maximum flux density of the rotating field in terms of $K$ and $I_m$.

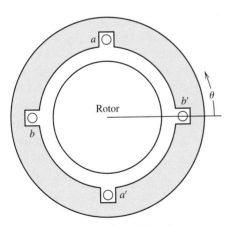

**Figure P16.8** Two-phase induction motor.

**P16.13.** Suppose that we could use superconducting material for the rotor conductors (i.e., rotor conductors with inductance and zero resistance) of an induction motor. Would this improve the performance of the motor? Explain by considering the torque–speed characteristic.

**Section 16.2: Equivalent-Circuit and Performance Calculations for Induction Motors**

*P16.14.** A two-pole 60-Hz induction motor produces an output power of 5 hp at a speed of 3500 rpm. With no load, the speed is 3598 rpm. Assume that the rotational torque loss is independent of speed. Find the rotational power loss at 3500 rpm.

*P16.15.** A certain four-pole 240-V-rms 60-Hz delta-connected three-phase induction motor has

$$R_s = 1 \ \Omega \qquad R'_r = 0.5 \ \Omega$$
$$X_s = 1.5 \ \Omega \qquad X'_r = 0.8 \ \Omega$$
$$X_m = 40 \ \Omega$$

Under load, the machine operates at 1728 rpm and has rotational losses of 200 W. Neglecting the rotational losses, find the no-load speed, line current, and power factor for the motor.

*P16.16.** Sketch the torque–speed characteristic of a delta-connected 220-V-rms 5-hp four-pole 60-Hz three-phase induction motor. Estimate values and label key features for things such as the full-load running speed, the full-load torque, the pull-out torque, and the starting torque. Estimate the full-load line current and the starting line current.

*P16.17.** Sometimes, to reduce starting current to reasonable values, induction motors are started with reduced source voltage. When the motor approaches its operating speed, the voltage is increased to full rated value. Compute the starting line current and torque for the motor of Example 16.2 if it is started with a source voltage of 220 V. Compare results with the values found in the example, and comment.

*P16.18.** A certain six-pole 440-V-rms 60-Hz three-phase delta-connected induction motor has

$$R_s = 0.08 \ \Omega \qquad R'_r = 0.06 \ \Omega$$
$$X_s = 0.20 \ \Omega \qquad X'_r = 0.15 \ \Omega$$
$$X_m = 7.5 \ \Omega$$

Neglecting the rotational losses, find the no-load speed, line current, and power factor for the motor.

**\*P16.19.** A 440-V-rms (line-to-line) 60-Hz three-phase wye-connected induction motor draws 16.8 A at a power factor of 80 percent lagging. The stator copper losses are 350 W, the rotor copper loss is 120 W, and the rotational losses are 400 W. Find the power crossing the air gap $P_{ag}$, the developed power $P_{dev}$, the output power $P_{out}$, and the efficiency.

**P16.20.** A certain four-pole 240-V-rms 60-Hz delta-connected three-phase induction motor has

$$R_s = 1 \ \Omega \qquad R_r' = 0.5 \ \Omega$$
$$X_s = 1.5 \ \Omega \qquad X_r' = 0.8 \ \Omega$$
$$X_m = 40 \ \Omega$$

Neglecting losses, find the starting torque and starting line current for the motor.

**P16.21.** A 2-hp six-pole 60-Hz delta-connected three-phase induction motor is rated for 1140 rpm, 220 V rms, and 5.72 A rms (line current) at an 80 percent lagging power factor. Find the full-load efficiency.

**P16.22.** Another method that is used to limit starting current is to place additional resistance in series with the stator windings during starting. The resistance is switched out of the circuit when the motor approaches full speed. Compute the resistance that must be placed in series with each phase of the motor of Examples 16.1 and 16.2 to limit the starting line current to $50\sqrt{3}$ A rms. Determine the starting torque with this resistance in place. Compare the starting torque with the value found in Example 16.2 and comment.

**P16.23.** A certain six-pole 440-V-rms 60-Hz three-phase delta-connected induction motor has

$$R_s = 0.08 \ \Omega \qquad R_r' = 0.06 \ \Omega$$
$$X_s = 0.20 \ \Omega \qquad X_r' = 0.15 \ \Omega$$
$$X_m = 7.5 \ \Omega$$

Neglecting losses, find the starting torque and starting line current for the motor.

**P16.24.** What are the two basic types of construction used for the rotors of induction motors? Which is the most rugged?

**P16.25.** The torque–speed characteristics of a 60-Hz induction motor and a load are shown in Figure P16.25. How many poles does the motor have? In steady-state operation, find the speed, the slip, the output power, and the rotor copper loss. Neglect rotational losses.

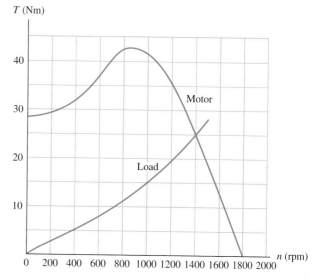

**Figure P16.25**

**P16.26.** A 60-Hz wound-rotor induction motor operates at 40 percent slip with added rotor resistance to achieve speed control. The stator resistance is negligible compared with the rotor resistance. Neglect rotational losses. Find the efficiency of this motor.

**P16.27.** List five important specifications to be considered (besides cost) in selecting induction motors. Indicate the optimum value or whether a high or low value is desirable for each specification.

**P16.28.** The torque–speed characteristics of a 60-Hz induction motor and a load are shown in Figure P16.25. The rotational inertia of the motor and load is 5 kgm². Estimate the time required for the motor to accelerate the load from a standing start to 1000 rpm. [*Hint:* The difference between the motor output torque and the load torque is approximately 25 Nm in the range of speeds under consideration.]

**P16.29.** A four-pole 60-Hz 240-V-rms induction motor operates at 1750 rpm and produces 2 hp of output power. The load is a hoist that requires constant torque versus speed.

Assume that the motor is operating in the range for which torque is proportional to slip and write an equation for motor torque in terms of slip when operating at rated voltage. Then modify the equation for operation from a 220-V-rms source. Estimate the speed when operating the hoist from a 220-V-rms source.

**P16.30.** An eight-pole 60-Hz ac induction motor produces an output power of 2 hp and has rotational losses of 100 W at a speed of 850 rpm. Determine the slip, the frequency of the stator currents, the frequency of the rotor currents, and the rotor copper loss.

**P16.31.** A certain six-pole 440-V-rms 60-Hz three-phase delta-connected induction motor has

$$R_s = 0.08 \ \Omega \qquad R_r' = 0.06 \ \Omega$$
$$X_s = 0.20 \ \Omega \qquad X_r' = 0.15 \ \Omega$$
$$X_m = 7.5 \ \Omega$$

Under load, the machine operates with a slip of 4 percent and has rotational losses of 2 kW. Determine the power factor, output power, copper losses, output torque, and efficiency.

**P16.32.** A certain four-pole 240-V-rms 60-Hz delta-connected three-phase induction motor has

$$R_s = 1 \ \Omega \qquad R_r' = 0.5 \ \Omega$$
$$X_s = 1.5 \ \Omega \qquad X_r' = 0.8 \ \Omega$$
$$X_m = 40 \ \Omega$$

Under load, the machine operates at 1728 rpm and has rotational losses of 200 W. Find the power factor, output power, copper losses, output torque, and efficiency.

### Section 16.3: Synchronous Machines

*\*P16.33.* List several methods for starting synchronous motors.

*\*P16.34.* **a.** A 12-pole 60-Hz synchronous motor drives a 10-pole synchronous machine that acts as a generator. What is the frequency of the voltages induced in the armature windings of the generator? **b.** Suppose that we need to drive a load at exactly 1000 rpm. The power available is 60-Hz three-phase.

Diagram a system of synchronous machines to drive the load, specifying the number of poles and frequency of operation for each. (Multiple correct answers exist.)

*\*P16.35.* A certain 480-V-rms delta-connected synchronous motor operates with zero developed power and draws a phase current of 15 A rms, which lags the voltage. The synchronous reactance is 5 Ω. The field current is 5 A. Assuming that the rotor field magnitude is proportional to field current, what field current is needed to reduce the armature current to zero?

*\*P16.36.* A synchronous motor is running at 75 percent of rated load with unity power factor. If the load increases to the rated output power, how do the following quantities change? **a.** field current; **b.** mechanical speed; **c.** output torque; **d.** armature current; **e.** power factor; **f.** torque angle.

*\*P16.37.* A 10-pole 60-Hz synchronous motor operates with a developed power of 100 hp, which is its rated full load. The torque angle is 20°. Plot the torque–speed characteristic to scale, showing the values for rated torque and for the pull-out torque.

**P16.38.** A 60-Hz 480-V-rms 200-hp delta-connected synchronous motor runs under no-load conditions. The field current is adjusted for minimum line current, which turns out to be 16.45 A rms. The per-phase armature impedance is $R_s + jX_s = 0.05 + j1.4$. (Until now in this chapter, we have neglected $R_s$. However, it is significant in efficiency calculations.) Estimate the efficiency of the machine under full-load conditions operating with 90 percent leading power factor.

**P16.39.** What is a synchronous capacitor? What is the practical benefit of using one?

**P16.40.** Sketch the V curve for a synchronous motor. Label the axes. Indicate where the power factor is lagging and where it is leading. Draw the phasor diagram corresponding to the minimum point on the V curve.

**P16.41.** A synchronous motor is running at 100 percent of rated load with unity power factor. If the field current is increased, how do the following quantities change? **a.** output power; **b.** mechanical speed;

**c.** output torque; **d.** armature current; **e.** power factor; **f.** torque angle.

**P16.42.** A six-pole 60-Hz synchronous motor is operating with a developed power of 5 hp and a torque angle of 5°. Find the speed and developed torque. Suppose that the load increases such that the developed torque doubles. Find the new torque angle. Find the pull-out torque and maximum developed power for this machine.

**P16.43.** An eight-pole 240-V-rms 60-Hz delta-connected synchronous motor operates with a constant developed power of 50 hp, unity power factor, and a torque angle of 15°. Then, the field current is increased such that $B_r$ increases in magnitude by 20 percent. Find the new torque angle and power factor. Is the new power factor leading or lagging?

**P16.44.** A 240-V-rms delta-connected 100-hp 60-Hz six-pole synchronous motor operates with a developed power (including losses) of 50 hp and a power factor of 90 percent leading. The synchronous reactance is $X_s = 0.5\ \Omega$. **a.** Find the speed and developed torque. **b.** Determine the values of $\mathbf{I}_a$, $\mathbf{E}_r$, and the torque angle. **c.** Suppose that the excitation remains constant and the load torque increases until the developed power is 100 hp. Determine the new values of $\mathbf{I}_a$, $\mathbf{E}_r$, the torque angle, and the power factor.

**P16.45.** A 240-V-rms 100-hp 60-Hz six-pole delta-connected synchronous motor operates with a developed power (including losses) of 100 hp and a power factor of 85 percent lagging. The synchronous reactance is $X_s = 0.5\ \Omega$. The field current is $I_f = 10$ A. What must the new field current be to produce 100 percent power factor? Assume that magnetic saturation does not occur so that $B_r$ is proportional to $I_f$.

**P16.46.** A six-pole 240-V-rms 60-Hz delta-connected synchronous motor operates with a developed power of 50 hp, unity power factor, and a torque angle of 15°. Find the phase current. Suppose that the load is removed so that the developed power is zero. Find the new values of the current, power factor, and torque angle.

**P16.47.** Suppose that a synchronous motor is instrumented to measure its armature current, armature voltage, and field current. The field circuit contains a rheostat so that the field current can be adjusted. Discuss how to adjust the field current to obtain unity power factor.

**P16.48.** Give two situations for which a synchronous motor would be a better choice than an induction motor in an industrial application.

## Section 16.4: Single-Phase Motors

**\*P16.49.** A farm house is located at the end of a country road in northern Michigan. The Thévenin impedance seen looking back into the power line from the electrical distribution panel is $0.2 + j0.2\ \Omega$. The Thévenin voltage is 240 V rms 60 Hz ac. A 2-hp 240-V-rms capacitor-start motor is used for pumping water. We want to estimate the voltage drop observed in the house when the motor starts. Typically, such a motor has a power factor of 75 percent and an efficiency of 80 percent at full load. Also, the starting current can be estimated as six times the full-load current. Estimate the worst-case percentage voltage drop observed when the motor starts.

**\*P16.50.** A 1-hp 120-V-rms 1740-rpm 60-Hz capacitor-start induction-run motor draws a current of 10.2 A rms at full load and has an efficiency of 80 percent. Find the values of **a.** the power factor and **b.** the impedance of the motor at full load. **c.** Determine the number of poles that the motor has.

**P16.51.** Which would be more suitable for use in a portable vacuum cleaner, an induction motor or a universal motor? For the fan in a home heating system? For the compressor motor in a refrigerator? For a variable-speed hand-held drill? Give the reasons for your answer in each case.

**P16.52.** Assuming small slip, the output power of a single-phase induction motor can be written as $P_{out} = K_1 s - K_2$, where $K_1$ and $K_2$ are constants and $s$ is the slip. A 0.5-hp motor has a full-load speed of 3500 rpm and a no-load speed of 3595 rpm. Determine the speed for 0.2-hp output.

**P16.53.** How could the direction of rotation of a single-phase capacitor-start induction motor be reversed?

**P16.54.** The winding impedances under starting conditions for a 60-Hz 0.5-hp motor are shown in Figure P16.54. Determine the capacitance $C$ needed so that the phase angle between the currents $\mathbf{I}_a$ and $\mathbf{I}_m$ is 90°.

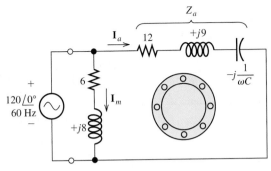

Figure P16.54

**Section 16.5: Stepper Motors and Brushless DC Motors**

**P16.55.** Sketch the cross-section of a reluctance stepper motor that has six stator poles and eight rotor poles. Your sketch should be similar to Figure 16.33(a) on page 828. Label the windings and specify the sequence of activation for clockwise rotation. What is the rotation angle per step?

**P16.56.** Use the Web to find two or more sources for stepper motors.

**P16.57.** List several advantages of brushless dc motors compared to conventional dc motors.

## Practice Test

Here is a practice test you can use to check your comprehension of the most important concepts in this chapter. Answers can be found in Appendix D and complete solutions are included in the Student Solutions files. See Appendix E for more information about the Student Solutions.

**T16.1. a.** Qualitatively describe the magnetic field set up in the air gap of a four-pole three-phase induction motor by the currents in the stator windings. **b.** Give an equation for the field intensity as a function of angular position around the gap and time, defining the terms in your equation.

**T16.2.** Besides low cost, what are five desirable characteristics for an induction motor?

**T16.3.** A certain eight-pole, 240-V-rms, 60-Hz, wye-connected, three-phase induction motor has

$$R_s = 0.5\ \Omega \qquad R'_r = 0.5\ \Omega$$
$$X_s = 2\ \Omega \qquad X'_r = 0.8\ \Omega$$
$$X_m = 40\ \Omega$$

Under load, the machine operates at 864 rpm and has rotational losses of 150 W. Find the power factor, output power, line current, copper losses, output torque, and efficiency.

**T16.4.** We have a 20-hp, eight-pole, 60-Hz, three-phase induction motor that runs at 850 rpm under full-load conditions. What are the values of the slip and the frequency of the rotor currents at full load? Also, estimate the speed if the load torque drops by 20 percent.

**T16.5.** In one or two paragraphs, describe the construction and principles of operation for a six-pole 60-Hz three-phase synchronous motor.

**T16.6.** We have a six-pole, 440-V-rms, 60-Hz delta-connected synchronous motor operating with a constant developed power of 20 hp, unity power factor, and a torque angle of 10°. Then, the field current is reduced such that $B_r$ is reduced in magnitude by 25 percent. Find the new torque angle and power factor. Is the new power factor leading or lagging?

# APPENDIX

# Complex Numbers

In Chapter 5, we learnt that sinusoidal steady-state analysis is greatly facilitated if the currents and voltages are represented as complex numbers known as phasors. In this appendix, we review complex numbers.

## Basic Complex-Number Concepts

Complex numbers involve the imaginary number $j = \sqrt{-1}$. (Electrical engineers use $j$ to represent the square root of $-1$ rather than $i$, because $i$ is often used for currents.) Several examples of complex numbers are

$$3 + j4 \quad \text{and} \quad -2 + j5$$

We say that a complex number $Z = x + jy$ has a **real part** $x$ and an **imaginary part** $y$. We can represent complex numbers by points in the **complex plane**, in which the real part is the horizontal coordinate and the imaginary part is the vertical coordinate. We often show the complex number by an arrow directed from the origin of the complex plane to the point defined by the real and imaginary components. This is illustrated in Figure A.1.

A **pure imaginary number**, $j6$ for example, has a real part of zero. On the other hand, a **pure real number**, such as 5, has an imaginary part of zero.

We say that complex numbers of the form $x + jy$ are in **rectangular form**. The **complex conjugate** of a number in rectangular form is obtained by changing the sign of the imaginary part. For example, if

$$Z_2 = 3 - j4$$

then the complex conjugate of $Z_2$ is

$$Z_2^* = 3 + j4$$

(Notice that we denote the complex conjugate by the symbol *.)

We add, subtract, multiply, and divide complex numbers that are in rectangular form in much the same way as we do algebraic expressions, making the substitution $j^2 = -1$.

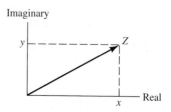

**Figure A.1** Complex plane.

---

**Example A.1**   Complex Arithmetic in Rectangular Form

Given that $Z_1 = 5 + j5$ and $Z_2 = 3 - j4$, reduce $Z_1 + Z_2, Z_1 - Z_2, Z_1 Z_2$, and $Z_1/Z_2$ to rectangular form.

**Solution**   For the sum, we have

$$Z_1 + Z_2 = (5 + j5) + (3 - j4) = 8 + j1$$

Notice that we add (algebraically) real part to real part and imaginary part to imaginary part.

The difference is

$$Z_1 - Z_2 = (5 + j5) - (3 - j4) = 2 + j9$$

In this case, we subtract each part of $Z_2$ from the corresponding part of $Z_1$.

For the product, we get

$$\begin{aligned}
Z_1 Z_2 &= (5 + j5)(3 - j4) \\
&= 15 - j20 + j15 - j^2 20 \\
&= 15 - j20 + j15 + 20 \\
&= 35 - j5
\end{aligned}$$

Notice that we expanded the product in the usual way for binomial expressions. Then, we used the fact that $j^2 = -1$.

To divide the numbers, we obtain

$$\frac{Z_1}{Z_2} = \frac{5 + j5}{3 - j4}$$

We can reduce this expression to rectangular form by multiplying the numerator and denominator by the complex conjugate of the denominator. This causes the denominator of the fraction to become pure real. Then, we divide each part of the numerator by the denominator. Thus, we find that

$$\begin{aligned}
\frac{Z_1}{Z_2} &= \frac{5 + j5}{3 - j4} \times \frac{Z_2^*}{Z_2^*} \\
&= \frac{5 + j5}{3 - j4} \times \frac{3 + j4}{3 + j4} \\
&= \frac{15 + j20 + j15 + j^2 20}{9 + j12 - j12 - j^2 16} \\
&= \frac{15 + j20 + j15 - 20}{9 + j12 - j12 + 16} \\
&= \frac{-5 + j35}{25} \\
&= -0.2 + j1.4 \qquad \blacksquare
\end{aligned}$$

**Exercise A.1**   Given that $Z_1 = 2 - j3$ and $Z_2 = 8 + j6$, reduce $Z_1 + Z_2, Z_1 - Z_2$, $Z_1 Z_2$, and $Z_1/Z_2$ to rectangular form.
**Answer**   $Z_1 + Z_2 = 10 + j3, Z_1 - Z_2 = -6 - j9, Z_1 Z_2 = 34 - j12, Z_1/Z_2 = -0.02 - j0.36$. ☐

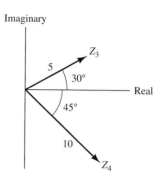

**Figure A.2** Complex numbers in polar form.

## Complex Numbers in Polar Form

Complex numbers can be expressed in **polar form** by giving the length of the arrow that represents the number and the angle between the arrow and the positive real axis. Examples of complex numbers in polar form are

$$Z_3 = 5\underline{/30°} \quad \text{and} \quad Z_4 = 10\underline{/-45°}$$

These numbers are shown in Figure A.2. The length of the arrow that represents a complex number $Z$ is denoted as $|Z|$ and is called the **magnitude** of the complex number.

Complex numbers can be converted from polar to rectangular form, or vice versa, by using the fact that the magnitude $|Z|$, the real part $x$, and the imaginary part $y$ form a right triangle. This is illustrated in Figure A.3. Using trigonometry, we can write the following relationships:

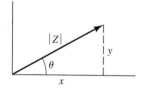

**Figure A.3** Complex number representation.

$$|Z|^2 = x^2 + y^2 \tag{A.1}$$

$$\tan(\theta) = \frac{y}{x} \tag{A.2}$$

$$x = |Z| \cos(\theta) \tag{A.3}$$

$$y = |Z| \sin(\theta) \tag{A.4}$$

These equations can be used to convert numbers from polar to rectangular form, or vice versa.

---

**Example A.2**    **Polar-to-Rectangular Conversion**

Convert $Z_3 = 5\underline{/30°}$ to rectangular form.

**Solution**    Using Equations A.3 and A.4, we have

$$x = |Z| \cos(\theta) = 5\cos(30°) = 4.33$$

and

$$y = |Z| \sin(\theta) = 5\sin(30°) = 2.5$$

Thus, we can write

$$Z_3 = 5\underline{/30°} = x + jy = 4.33 + j2.5$$

∎

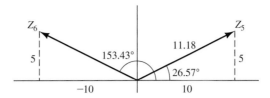

**Figure A.4** Complex numbers of Example A.3.

---

**Example A.3**   **Rectangular-to-Polar Conversion**

Convert $Z_5 = 10 + j5$ and $Z_6 = -10 + j5$ to polar form.

**Solution**   The complex numbers are illustrated in Figure A.4. First, we use Equation A.1 to find the magnitudes of each of the numbers. Thus,

$$|Z_5| = \sqrt{x_5^2 + y_5^2} = \sqrt{10^2 + 5^2} = 11.18$$

and

$$|Z_6| = \sqrt{x_6^2 + y_6^2} = \sqrt{(-10)^2 + 5^2} = 11.18$$

To find the angles, we use Equation A.2.

$$\tan(\theta_5) = \frac{y_5}{x_5} = \frac{5}{10} = 0.5$$

Taking the arctangent of both sides, we have

$$\theta_5 = \arctan(0.5) = 26.57°$$

Thus, we can write

$$Z_5 = 10 + j5 = 11.18\underline{/26.57°}$$

This is illustrated in Figure A.4.

Evaluating Equation A.2 for $Z_6$, we have

$$\tan(\theta_6) = \frac{y_6}{x_6} = \frac{5}{-10} = -0.5$$

Now if we take the arctan of both sides, we obtain

$$\theta_6 = -26.57°$$

However, $Z_6 = -10 + j5$ is shown in Figure A.4. Clearly, the value that we have found for $\theta_6$ is incorrect. The reason for this is that the arctangent function is multivalued. The value actually given by most calculators or computer programs is the principal value. *If the number falls to the left of the imaginary axis (i.e., if the real part is negative), we must add (or subtract) 180° to arctan (y/x) to obtain the correct angle.* Thus, the true angle for $Z_6$ is

$$\theta_6 = 180 + \arctan\left(\frac{y_6}{x_6}\right) = 180 - 26.57 = 153.43°$$

Finally, we can write

$$Z_6 = -10 + j5 = 11.18\underline{/153.43°}$$

∎

The procedures that we have illustrated in Examples A.2 and A.3 can be carried out with a relatively simple calculator. However, if we find the angle by taking the arctangent of $y/x$, we must consider the fact that the principal value of the arctangent is the true angle only if the real part $x$ is positive. If $x$ is negative, we have

$$\theta = \arctan\left(\frac{y}{x}\right) \pm 180° \tag{A.5}$$

Many scientific calculators are capable of converting complex numbers from polar to rectangular, and vice versa, in a single operation. Practice with your calculator to become proficient using this feature. *It is always a good idea to make a sketch of the number in the complex plane as a check on the conversion process.*

**Exercise A.2**   Convert the numbers $Z_1 = 15\underline{/45°}$, $Z_2 = 10\underline{/-150°}$, and $Z_3 = 5\underline{/90°}$ to rectangular form.
**Answer**   $Z_1 = 10.6 + j10.6$, $Z_2 = -8.66 - j5$, $Z_3 = j5$.    ☐

**Exercise A.3**   Convert the numbers $Z_1 = 3 + j4$, $Z_2 = -j10$, and $Z_3 = -5 - j5$ to polar form.
**Answer**   $Z_1 = 5\underline{/53.13°}$, $Z_2 = 10\underline{/-90°}$, $Z_3 = 7.07\underline{/-135°}$.    ☐

## Euler's Identities

You may have been wondering what complex numbers have to do with sinusoids. The connection is through Euler's identities, which state that

Equations A.6 through A.9 are the bridge between sinusoidal currents or voltages and complex numbers.

$$\cos(\theta) = \frac{e^{j\theta} + e^{-j\theta}}{2} \tag{A.6}$$

and

$$\sin(\theta) = \frac{e^{j\theta} - e^{-j\theta}}{2j} \tag{A.7}$$

Another form of these identities is

$$e^{j\theta} = \cos(\theta) + j\sin(\theta) \tag{A.8}$$

and

$$e^{-j\theta} = \cos(\theta) - j\sin(\theta) \tag{A.9}$$

Thus, $e^{j\theta}$ is a complex number having a real part of $\cos(\theta)$ and an imaginary part of $\sin(\theta)$. This is illustrated in Figure A.5. The magnitude is

$$|e^{j\theta}| = \sqrt{\cos^2(\theta) + \sin^2(\theta)}$$

By the well-known identity $\cos^2(\theta) + \sin^2(\theta) = 1$, this becomes

$$|e^{j\theta}| = 1 \tag{A.10}$$

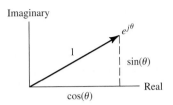

Figure A.5   Euler's identity.

Furthermore, the angle of $e^{j\theta}$ is $\theta$. Thus, we can write

$$e^{j\theta} = 1\underline{/\theta} = \cos(\theta) + j\sin(\theta) \qquad (A.11)$$

Similarly, we have

$$e^{-j\theta} = 1\underline{/-\theta} = \cos(\theta) - j\sin(\theta) \qquad (A.12)$$

Notice that $e^{-j\theta}$ is the complex conjugate of $e^{j\theta}$.

A complex number such as $A\underline{/\theta}$ can be written as

$$A\underline{/\theta} = A \times (1\underline{/\theta}) = Ae^{j\theta} \qquad (A.13)$$

We call $Ae^{j\theta}$ the **exponential form** of a complex number. Hence, a given complex number can be written in three forms: the rectangular form, the polar form, and the exponential form. Using Equation A.11 to substitute for $e^{j\theta}$ on the right-hand side of Equation A.13, we obtain the three forms of a complex number:

$$A\underline{/\theta} = Ae^{j\theta} = A\cos(\theta) + jA\sin(\theta) \qquad (A.14)$$

---

### Example A.4   Exponential Form of a Complex Number

Express the complex number $Z = 10\underline{/60°}$ in exponential and rectangular forms. Sketch the number in the complex plane.

**Solution**   Conversion from polar to exponential forms is based on Equation A.13. Thus, we have

$$Z = 10\underline{/60°} = 10e^{j60°}$$

The rectangular form can be found by using Equation A.8:

$$Z = 10 \times (e^{j60°})$$
$$= 10 \times [\cos(60°) + j\sin(60°)]$$
$$= 5 + j8.66$$

The graphical representation of $Z$ is shown in Figure A.6.   ■

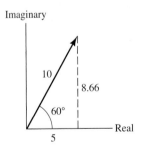

Figure A.6   See Example A.4.

---

**Exercise A.4**   Express $Z_1 = 10 + j10$ and $Z_2 = -10 + j10$ in polar and exponential forms.

**Answer**   $Z_1 = 14.14\underline{/45°} = 14.14e^{j45°}$, $Z_2 = 14.14\underline{/135°} = 14.14e^{j135°}$.   ☐

## Arithmetic Operations in Polar and Exponential Form

To add (or subtract) complex numbers, we must first convert them to rectangular form. Then, we add (or subtract) real part to real part and imaginary to imaginary.

Consider two complex numbers in exponential form given by

$$Z_1 = |Z_1|e^{j\theta_1} \quad \text{and} \quad Z_2 = |Z_2|e^{j\theta_2}$$

The polar forms of these numbers are

$$Z_1 = |Z_1|\underline{/\theta_1} \quad \text{and} \quad Z_2 = |Z_2|\underline{/\theta_2}$$

For multiplication of numbers in exponential form, we have

$$Z_1 \times Z_2 = |Z_1|e^{j\theta_1} \times |Z_2|e^{j\theta_2} = |Z_1||Z_2|e^{j(\theta_1+\theta_2)}$$

As usual, in multiplying exponentials, we add the exponents. In polar form, this is

$$Z_1 \times Z_2 = |Z_1|\underline{/\theta_1} \times |Z_2|\underline{/\theta_2} = |Z_1||Z_2|\underline{/\theta_1 + \theta_2}$$

*Thus, to multiply numbers in polar form, we multiply the magnitudes and add the angles.*

Now consider division:

$$\frac{Z_1}{Z_2} = \frac{|Z_1|e^{j\theta_1}}{|Z_2|e^{j\theta_2}} = \frac{|Z_1|}{|Z_2|}e^{j(\theta_1-\theta_2)}$$

As usual, in dividing exponentials, we subtract the exponents. In polar form, this is

$$\frac{Z_1}{Z_2} = \frac{|Z_1|\underline{/\theta_1}}{|Z_2|\underline{/\theta_2}} = \frac{|Z_1|}{|Z_2|}\underline{/\theta_1 - \theta_2}$$

*Thus, to divide numbers in polar form, we divide the magnitudes and subtract the angle of the divisor from the angle of the dividend.*

### Example A.5    Complex Arithmetic in Polar Form

Given $Z_1 = 10\underline{/60°}$ and $Z_2 = 5\underline{/45°}$, find $Z_1Z_2$, $Z_1/Z_2$, and $Z_1 + Z_2$ in polar form.

Solution    For the product, we have

$$Z_1 \times Z_2 = 10\underline{/60°} \times 5\underline{/45°} = 50\underline{/105°}$$

Dividing the numbers, we have

$$\frac{Z_1}{Z_2} = \frac{10\underline{/60°}}{5\underline{/45°}} = 2\underline{/15°}$$

Before we can add (or subtract) the numbers, we must convert them to rectangular form. Using Equation A.14 to convert the polar numbers to rectangular, we get

$$Z_1 = 10\underline{/60°} = 10\cos(60°) + j10\sin(60°)$$
$$= 5 + j8.66$$
$$Z_2 = 5\underline{/45°} = 5\cos(45°) + j5\sin(45°)$$
$$= 3.54 + j3.54$$

Now, we can add the numbers. We denote the sum as $Z_s$:

$$Z_s = Z_1 + Z_2 = 5 + j8.66 + 3.54 + j3.54$$
$$= 8.54 + j12.2$$

Next, we convert the sum to polar form:

$$|Z_s| = \sqrt{(8.54)^2 + (12.2)^2} = 14.9$$

$$\tan \theta_s = \frac{12.2}{8.54} = 1.43$$

Taking the arctangent of both sides, we have

$$\theta_s = \arctan(1.43) = 55°$$

Because the real part of $Z_s$ is positive, the correct angle is the principal value of the arctangent (i.e., $55°$ is the correct angle). Thus, we obtain

$$Z_s = Z_1 + Z_2 = 14.9 \underline{/55°}$$

∎

---

**Exercise A.5**   Given $Z_1 = 10\underline{/30°}$ and $Z_2 = 20\underline{/135°}$, find $Z_1 Z_2$, $Z_1/Z_2$, $Z_1 - Z_2$, and $Z_1 + Z_2$ in polar form.
**Answer**   $Z_1 Z_2 = 200\underline{/165°}$, $Z_1/Z_2 = 0.5\underline{/-105°}$, $Z_1 - Z_2 = 24.6\underline{/-21.8°}$, $Z_1 + Z_2 = 19.9\underline{/106°}$.
□

---

## Summary

1. Complex numbers can be expressed in rectangular, polar, or exponential forms. Addition, subtraction, multiplication, and division of complex numbers are necessary operations in solving steady-state ac circuits by the phasor method.

2. Sinusoids and complex numbers are related through Euler's identities.

## Problems*

**PA.1.** Give that $Z_1 = 2 + j3$ and $Z_2 = 4 - j3$, reduce $Z_1 + Z_2$, $Z_1 - Z_2$, $Z_1 Z_2$, and $Z_1/Z_2$ to rectangular form.

**PA.2.** Given that $Z_1 = 1 - j2$ and $Z_2 = 2 + j3$, reduce $Z_1 + Z_2$, $Z_1 - Z_2$, $Z_1 Z_2$, and $Z_1/Z_2$ to rectangular form.

**PA.3.** Given that $Z_1 = 10 + j5$ and $Z_2 = 20 - j20$, reduce $Z_1 + Z_2$, $Z_1 - Z_2$, $Z_1 Z_2$, and $Z_1/Z_2$ to rectangular form.

**PA.4.** Express each of these complex numbers in polar form and in exponential form: **a.** $Z_a = 5 - j5$; **b.** $Z_b = -10 + j5$; **c.** $Z_c = -3 - j4$; **d.** $Z_d = -j12$.

---

*Solutions for these problems are contained in the Student Solutions files. See Appendix E for more information about accessing the Student Solutions.

**PA.5.** Express each of these complex numbers in rectangular form and in exponential form: **a.** $Z_a = 5\underline{/45°}$; **b.** $Z_b = 10\underline{/120°}$; **c.** $Z_c = -15\underline{/-90°}$; **d.** $Z_d = -10\underline{/60°}$.

**PA.6.** Express each of these complex numbers in rectangular form and in polar form: **a.** $Z_a = 5e^{j30°}$; **b.** $Z_b = 10e^{-j45°}$; **c.** $Z_c = 100e^{j135°}$; **d.** $Z_d = 6e^{j90°}$.

**PA.7.** Reduce each of the following to rectangular form:

**a.** $Z_a = 5 + j5 + 10\underline{/30°}$

**b.** $Z_b = 5\underline{/45°} - j10$

**c.** $Z_c = \dfrac{10\underline{/45°}}{3 + j4}$

**d.** $Z_d = \dfrac{15}{5\underline{/90°}}$

# Nominal Values and the Color Code for Resistors

Several types of resistors are available for use in electronic circuits. Carbon-film and carbon-composition resistors with tolerances of 5 percent, 10 percent, or 20 percent are available with various power ratings (such as 1/8, 1/4, and 1/2 W). These resistors are used in noncritical applications such as biasing.

Metal-film 1-percent-tolerance resistors are used where greater precision is required. For example, we often choose metal-film resistors in applications such as the feedback resistors of an op amp.

Wire-wound resistors are available with high power-dissipation ratings. Wire-wound resistors often have significant series inductance because they consist of resistance wire that is wound on a form, such as ceramic. Thus, they are often not suitable for use as a resistance at high frequencies.

The value and tolerance are marked on 5-percent, 10-percent, and 20-percent-tolerance resistors by color bands as shown in Figure B.1. The first band is closest to one end of the resistor. The first and second bands give the significant digits of the resistance value. The third band gives the exponent of the multiplier. The fourth band indicates the tolerance. The fifth band is optional and indicates whether the resistors meet certain military reliability specifications.

Table B.1 shows the combinations of significant figures available as nominal values for 5-percent, 10-percent, and 20-percent-tolerance resistors. Table B.2 shows the standard nominal significant digits for 1-percent-tolerance resistors.

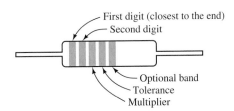

| Digit | Color | Tolerance | Color |
|-------|-------|-----------|-------|
| 0 | Black | 2% | Red |
| 1 | Brown | 5% | Gold |
| 2 | Red | 10% | Silver |
| 3 | Orange | 20% | No fourth band |
| 4 | Yellow | | |
| 5 | Green | | |
| 6 | Blue | | |
| 7 | Violet | | |
| 8 | Gray | | |
| 9 | White | | |

Examples

| Yellow | Violet | Black | $= 47 \times 10^0 = 47 \ \Omega$ |
|--------|--------|-------|-------|
| Yellow | Violet | Red | $= 47 \times 10^2 = 4700 \ \Omega$ |
| Brown | Black | Yellow | $= 10 \times 10^4 = 100 \ \text{k}\Omega$ |

**Figure B.1**  Resistor color code.

**Table B.1.** Standard Nominal Values for 5-Percent-Tolerance Resistors[a]

| | | | | |
|---|---|---|---|---|
| **10** | 16 | **27** | 43 | **68** |
| 11 | **18** | 30 | **47** | 75 |
| **12** | 20 | **33** | 51 | **82** |
| 13 | **22** | 36 | **56** | 91 |
| **15** | 24 | **39** | 62 | |

[a] Resistors having tolerances of 10 percent and 20 percent are available only for the values given in boldface.

**Table B.2.** Standard Values for 1-Percent-Tolerance Metal-Film Resistors

| | | | | | | |
|---|---|---|---|---|---|---|
| 100 | 140 | 196 | 274 | 383 | 536 | 750 |
| 102 | 143 | 200 | 280 | 392 | 549 | 768 |
| 105 | 147 | 205 | 287 | 402 | 562 | 787 |
| 107 | 150 | 210 | 294 | 412 | 576 | 806 |
| 110 | 154 | 215 | 301 | 422 | 590 | 825 |
| 113 | 158 | 221 | 309 | 432 | 604 | 845 |
| 115 | 162 | 226 | 316 | 442 | 619 | 866 |
| 118 | 165 | 232 | 324 | 453 | 634 | 887 |
| 121 | 169 | 237 | 332 | 464 | 649 | 909 |
| 124 | 174 | 243 | 340 | 475 | 665 | 931 |
| 127 | 178 | 249 | 348 | 487 | 681 | 953 |
| 130 | 182 | 255 | 357 | 499 | 698 | 976 |
| 133 | 187 | 261 | 365 | 511 | 715 | |
| 137 | 191 | 267 | 374 | 523 | 732 | |

# The Fundamentals of Engineering Examination

Becoming licensed as a Professional Engineer (PE) is a very important step toward success in your engineering career. In the United States, a PE license is required by all 50 states for engineers whose work may affect life, health, or property, or who offer their services to the public. Thus, a license is an absolute requirement for many types of work. Furthermore, licensed engineers have more opportunities and earn higher salaries (from 15 percent to 25 percent) than other engineers.

Licensure requirements are set by the various states, but are similar from state to state. Furthermore, by reciprocal agreements, many states recognize licenses granted in other states. Generally, a degree from an ABET-accredited engineering program, four years of relevant work experience, and successful completion of two state examinations are required. The National Council of Examiners for Engineering and Surveying (NCEES) prepares and scores the examinations. The examinations are the Fundamentals of Engineering (FE) Examination, which can be taken any time, and the Principles and Practice of Engineering (PE) Examination, which must be taken after at least four years of experience.

You should plan on taking the FE Examination before or shortly after you graduate, rather than later, because it contains questions on a wide variety of subjects. To pass the FE Examination, many engineers struggle to relearn topics that they were familiar with at graduation. Most likely, if you are a mechanical, civil, or chemical engineer, you will not routinely work with electrical circuits. After a number of years, you may not be able to answer questions regarding electrical circuits that you could have easily answered when you finished the courses for which you used this book. Thus, taking the examination in your senior year is best. Keep this book to refresh your knowledge of electrical circuits just before taking the FE Examination.

You should look carefully at the NCEES website www.ncees.org for current information about the FE exam as it applies to your discipline. Practice exams are available there.

Additional very useful information about professional licensure is provided by the National Society of Professional Engineers at their website www.nspe.org.

Up-to-date information about the PE license can be found at the following website: http://www.ncees.org/

# APPENDIX

# D

# Answers for the Practice Tests

Complete solutions for the practice tests are included in the Student Solutions files. See Appendix E for information on how to access these files.

**T1.1.** **a.** 4; **b.** 7; **c.** 16; **d.** 18; **e.** 1; **f.** 2; **g.** 8; **h.** 3; **i.** 5; **j.** 15; **k.** 6; **l.** 11; **m.** 13; **n.** 9; **o.** 14.

**T1.2.** **a.** $v_R = -6$ V. **b.** The voltage source is delivering 30 W. **c.** There are 3 nodes. **d.** The current source is absorbing 12 W.

**T1.3.** **a.** $v_{ab} = -8$ V. **b.** Source $I_1$ is supplying 24 W. Source $I_2$ is absorbing 8 W. **c.** $P_{R1} = 5.33$ W and $P_{R2} = 10.67$ W.

**T1.4.** **a.** $v_1 = 8$ V; **b.** $i = 2$ A; **c.** $R_2 = 2\ \Omega$.

**T1.5.** $i_{sc} = -3$ A.

**T1.6.** $v_4 = 80$ V, $i_3 = 5$ A, $i_2 = 4$ A, $i_1 = 11$ A, $v_1 = 110$ V, $v_s = 190$ V.

**T2.1.** **a.** 6; **b.** 10; **c.** 2; **d.** 7; **e.** 10 or 13; **f.** 1 or 4; **g.** 11; **h.** 3; **i.** 8; **j.** 15; **k.** 17; **l.** 14.

**T2.2.** $i_s = 6$ A; $i_4 = 1$ A.

**T2.3.**
```
G = [0.95 -0.20 -0.50; -0.20 0.30 0; -0.50 0 1.50]
I = [0; 2; -2]
V = G\ I % As an alternative, we could use V = inv(G)*I
```

**T2.4.** A proper set of equations consists of any two of the following three.

1. KVL mesh 1:

$$R_1 i_1 - V_s + R_3(i_1 - i_3) + R_2(i_1 - i_2) = 0$$

2. KVL for the supermesh obtained by combining meshes 2 and 3:

$$R_4 i_2 + R_2(i_2 - i_1) + R_3(i_3 - i_1) + R_5 i_3 = 0$$

3. KVL around the periphery of the circuit:

$$R_1 i_1 - V_s + R_4 i_2 + R_5 i_3 = 0$$

in combination with this equation for the current source:

$$i_2 - i_3 = I_s$$

**T2.5.** $V_t = 24$ V, $R_t = 24\ \Omega$, and $I_n = 1$ A. The reference direction for $I_n$ should point toward terminal $b$. The positive reference for $V_t$ should be on the side of the $b$ terminal.

**T2.6.** By superposition, 25 percent of the current through the 5-$\Omega$ resistance is due to the 5-V source. Superposition does not apply for power, but we can see from analysis of the complete circuit that all of the power is supplied by the 15-V source. Thus, 0 percent of the power in the 5-$\Omega$ resistance is due to the 5-V source.

**T2.7.** The equivalent resistance between terminals $a$ and $b$ is 26 $\Omega$.

**T2.8.** The final circuit consists of a 20-V voltage source in series with a 10-$\Omega$ resistance. The positive polarity of the voltage source is closest to terminal $a$.

**T3.1.** $v_{ab}(t) = 15 - 15 \exp(-2000t)$ V; $w_C(\infty) = 1.125$ mJ.

**T3.2.** $C_{eq} = 5\,\mu F$.

**T3.3.** $C = 4248$ pF.

**T3.4.** $v_{ab}(t) = 1.2\cos(2000t)$ V; $w_{peak} = 90\,\mu J$.

**T3.5.** $L_{eq} = 3.208$ H.

**T3.6.** $v_s(t) = 5\sin(1000t)$ V.

**T3.7.** $v_1(t) = -40\sin(500t) - 16\exp(-400t)$ V;
$\qquad v_2(t) = 20\sin(500t) - 24\exp(-400t)$ V.

**T3.8.** One set of commands and the result for $v_{ab}(t)$ are:

```
syms vab iab t
iab = 3*(10^5)*(t^2)*exp(-2000*t);
vab = (1/20e-6)*int(iab,t,0,t)
subplot(2,1,1)
ezplot(iab, [0 5e-3]), title('\ iti_a_b\rm (A) versus \itt\rm (s)')
subplot(2,1,2)
ezplot(vab, [0 5e-3]), title('\ itv_a_b\rm (V) versus \itt\rm (s)')
```

$$v_{ab} = \frac{15}{4} - \frac{15}{4}\exp(-2000t) - 7500t\exp(-2000t) - 7.5\times10^6 t^2 \exp(-2000t)$$

You can test your commands using MATLAB to see if they produce this result for $v_{ab}(t)$ and plots like those in the Student Solutions.

**T4.1.** $t_x = 4\ln(4) = 5.545$ s.

**T4.2.** **a.** $i_1(0-) = 10$ mA, $i_2(0-) = 5$ mA, $i_3(0-) = 0$, $i_L(0-) = 15$ mA,
$\qquad v_C(0-) = 10$ V;

**b.** $i_1(0+) = 15$ mA, $i_2(0+) = 2$ mA, and $i_3(0+) = -2$ mA, $i_L(0+) = $
$\quad$ 15 mA, $v_C(0+) = 10$ V;

**c.** $i_L(t) = 10 + 5\exp(-5\times10^5 t)$ mA;

**d.** $v_C(t) = 10\exp(-200t)$ V.

**T4.3.** **a.** $2\frac{di(t)}{dt} + i(t) = 5\exp(-3t)$;

**b.** $\tau = L/R = 2$ s, $i_c(t) = A\exp(-0.5t)$ A;

**c.** $i_p(t) = -\exp(-3t)$ A;

**d.** $i(t) = \exp(-0.5t) - \exp(-3t)$ A.

**T4.4.** **a.** $\frac{d^2 v_C(t)}{dt^2} + 2000\frac{dv_C(t)}{dt} + 25\times10^6 v_C(t) = 375\times10^6$;

**b.** $v_{Cp}(t) = 15$ V;

**c.** Underdamped; $v_{Cc}(t) = K_1\exp(-1000t)\cos(4899t) + K_2\exp(-1000t)$
$\quad \sin(4899t)$;

**d.** $v_C(t) = 15 - 15\exp(-1000t)\cos(4899t) - (3.062)\exp(-1000t)$
$\quad \sin(4899t)$ V.

**T4.5.** The commands are

```
syms vC t
S = dsolve('D2vC + 2000*DvC + (25e6)*vC = 375e6', 'vC(0) = 0, DvC(0) = 0');
simple(vpa(S,4))
```

The commands are stored in the m-file named T_4_4 that can be found in the Hambley MATLAB folder. See Appendix E for information about accessing this folder.

**T5.1.** $I_{rms} = \sqrt{8} = 2.828$ A; $P = 400$ W.

**T5.2.** $v(t) = 9.914 \cos(\omega t - 37.50°)$

**T5.3.** **a.** $V_{1rms} = 10.61$ V; **b.** $f = 200$ Hz; **c.** $\omega = 400\pi$ radians/s; **d.** $T = 5$ ms; **e.** $V_1$ lags $V_2$ by 15° or $V_2$ leads $V_1$ by 15°.

**T5.4.** $V_R = 7.071 \underline{/-45°}$ V; $V_L = 10.606 \underline{/45°}$ V; $V_C = 5.303 \underline{/-135°}$ V.

**T5.5.** $v_1(t) = 94.299 \cos(500t - 28.237°)$ V.

**T5.6.** $S = 5500 \underline{/40°} = 4213 + j3535$ VA;

$$P = 4213 \text{ W}; Q = 3535 \text{ VAR}; \text{apparent power} = 5500 \text{ VA};$$

Power factor $= 76.6$ percent lagging.

**T5.7.** $I_{aA} = 54.26\underline{/-23.13°}$ A.

**T5.8.** The commands are:

```
Z = [(15+i*10) -15; -15 (15-i*5)]
V = [pin(10,45); -15]
I = inv(Z)*V
pout(I(1))
pout(I(2))
```

**T6.1.** All real-world signals (which are usually time-varying currents or voltages) are sums of sinewaves of various frequencies, amplitudes, and phases. The transfer function of a filter is a function of frequency that shows how the amplitudes and phases of the input components are altered to produce the output components.

**T6.2.** $v_{out}(t) = 1.789 \cos(1000\pi t - 63.43°) + 3.535 \cos(2000\pi t + 15°)$.

**T6.3.** **a.** The slope of the low-frequency asymptote is +20 dB/decade. **b.** The slope of the high-frequency asymptote is zero. **c.** The coordinates at which the asymptotes meet are $20 \log(50) = 34$ dB and 200 Hz. **d.** This is a first-order highpass filter. **e.** The break frequency is 200 Hz.

**T6.4.** **a.** 1125 Hz; **b.** 28.28; **c.** 39.79 Hz; **d.** 5 $\Omega$; **e.** infinite impedance; **f.** infinite impedance.

**T6.5.** **a.** 159.2 kHz; **b.** 10.0; **c.** 15.92 kHz; **d.** 10 k$\Omega$; **e.** zero impedance; **f.** zero impedance.

**T6.6.** **a.** A band-reject filter with the transfer function shown in Figure 6.32(d) on page 321 is needed. The lower cutoff frequency $f_L$ should be slightly less than 800 Hz, and the upper cutoff frequency $f_H$ should be slightly more than 800 Hz. **b.** A bandpass filter with the transfer function shown in Figure 6.32(c) on page 321 is needed. The lower cutoff frequency $f_L$ should be slightly less than 800 Hz, and the upper cutoff frequency $f_H$ should be slightly more than 800 Hz.

**T6.7.** **a.** First-order lowpass filter; **b.** second-order lowpass filter; **c.** second-order band-reject (or notch) filter; **d.** first-order highpass filter.

**T6.8.** One set of commands is:

```
f = logspace(1,4,400);
H = 50*i*(f/200)./(1 + i*f/200);
```

```
semilogx(f,20*log10(abs(H)))
```
Other sets of commands will also work. Check to see if your commands produce a plot equivalent to the one produced by the set given above.

**T7.1.** **a.** 12; **b.** 19 (18 is incorrect because it omits the first step, inverting the variables); **c.** 20; **d.** 23; **e.** 21; **f.** 24; **g.** 16; **h.** 25; **i.** 7; **j.** 10; **k.** 8; **l.** 1 (the binary codes for hexadecimal symbols A through F do not occur in BCD).

**T7.2.** **a.** $101100001.111_2$; **b.** $541.7_8$; **c.** $161. \, E_{16}$; **d.** $001101010011.100001110101_{BCD}$.

**T7.3.** $FA. \, 7_{16} = 372.34_8$.

**T7.4.** **a.** $+97_{10}$; **b.** $-70_{10}$.

**T7.5.** **a.** $D = \overline{A}\,\overline{B} + \overline{(B + C)}$. **b.** Ones should appear only in the truth-table rows and map cells corresponding to $ABC = 000$, 001, and 100. **c.** $D = \overline{A}\,\overline{B} + \overline{B}\,\overline{C}$; **d.** $D = \overline{B}(\overline{A} + \overline{C})$.

**T7.6.** **a.** Ones should appear only in the cells corresponding to $B_8B_4B_2B_1 = 0001$, 0101, 1011, and 1111. **b.** $G = B_1\overline{B_2}\,\overline{B_8} + B_1B_2B_8$. **c.** $G = B_1(\overline{B_2} + B_8)(B_2 + \overline{B_8})$.

**T7.7.** The successive states are $Q_0Q_1Q_2 = 100$ (the initial state), $110, 111, 011, 001, 100, 111$. The state of the register returns to its initial state after 5 shifts.

**T8.1.** **a.** 11; **b.** 17; **c.** 21; **d.** 24; **e.** 27; **f.** 13; **g.** 26; **h.** 9; **i.** 20; **j.** 12; **k.** 15; **l.** 16; **m.** 8; **n.** 29; **o.** 23; **p.** 30.

**T8.2.** **a.** direct, 61; **b.** indexed, F3; **c.** inherent, FF; **d.** inherent, 01; **e.** immediate, 05; **f.** immediate, A1.

**T8.3.** After the four commands have been executed, the contents of the registers and memory locations are:

| | |
|---|---|
| A: 32 | 1034: 00 |
| B: 32 | 1035: 19 |
| SP: 1035 | 1036: 58 |
| X: 1958 | 1037: 19 |
| | 1038: 58 |
| | 1039: 00 |
| | 103A: 00 |
| | 103B: 00 |
| | 103C: 00 |

**T8.4.** The four main elements are sensors, a DAQ board, software, and a general-purpose computer.

**T8.5.** The four types of systematic (bias) errors are offset, scale error, nonlinearity, and hysteresis.

**T8.6.** Bias errors are the same for measurements repeated under identical conditions, while random errors are different for each measurement.

**T8.7.** Ground loops occur when the sensor and the input of the amplifier are connected to ground by separate connections. The effect is to add noise (often with frequencies equal to that of the power line and its harmonics) to the desired signal.

**T8.8.** If we are using a sensor that has one end grounded, we should choose an amplifier with a differential input to avoid a ground loop.

**T8.9.**  Coaxial cable or shielded twisted pair cable.

**T8.10.**  If we need to sense the open-circuit voltage, the input impedance of the amplifier should be very large compared to the internal impedance of the sensor.

**T8.11.**  The sampling rate should be more than twice the highest frequency of the components in the signal. Otherwise, higher frequency components can appear as lower frequency components known as aliases.

**T9.1.**  **a.** $i_D \cong 9.6$ mA; **b.** $i_D \cong 4.2$ mA.

**T9.2.**  The diode is on, $v_x = 2.286$ V and $i_x = 0.571$ mA.

**T9.3.**  The resistance is 1 k$\Omega$, and the voltage is 3 V.

**T9.4.**  Your diagram should be equivalent to Figure 9.28. It may be correct even if it is laid out differently. Check to see that you have four diodes and that current flows from the source through a diode in the forward direction, then through the load and finally through a second diode in the forward direction back to the opposite end of the source. On the opposite half cycle, the path should be through one of the other two diodes, through the load in the same direction as before, and back through the fourth diode to the opposite end of the source.

**T9.5.**  Your diagram should be equivalent to Figure 9.29(a) with the 6-V source changed to 5 V, the 9-V source changed to 4 V, and the peak ac voltage changed to 10 V. However, your diagram may be somewhat different in appearance. For example, the 4-V source and diode $B$ can be interchanged as long as the source polarity and direction of the diode don't change; similarly for the 5-V source and diode $A$. (In other words, the order of the elements doesn't matter in a series connection.) The parallel branches can be interchanged in position. The problem does not give enough information to properly select the value of the resistance, but any value from about 1 k$\Omega$ to 1 M$\Omega$ is acceptable.

**T9.6.**  Your diagram should be equivalent to Figure 9.33(a) with a 4-V source in place of the 5-V source. The time constant $RC$ should be much longer than the period of the source voltage. Thus, we should select component values so that $RC >> 0.1$ s.

**T9.7.**  The small-signal equivalent circuit for the diode is simply a 10.4-$\Omega$ resistance.

**T10.1.**  $A_{voc} = 1500$; $R_i = 60$ $\Omega$; $R_o = 40$ $\Omega$.

**T10.2.**  Your answer should be similar to Table 10.1 on page 521.

**T10.3.**  **a.** Transconductance amplifier; **b.** Current amplifier; **c.** Voltage amplifier; **d.** Transresistance amplifier.

**T10.4.**  $A_{voc} = 250$ (no units or V/V); $R_{moc} = 50$ k$\Omega$; $G_{msc} = 0.25$ S.

**T10.5.**  $P_d = 12$ W; $\eta = 60$ percent.

**T10.6.**  To avoid linear waveform distortion, the gain magnitude should be constant and the phase response should be a linear function of frequency over the frequency range from 1 to 10 kHz. Because the gain is 100 and the peak input amplitude is 100 mV, the peak output amplitude should be 10 V. The amplifier must not display clipping or unacceptable nonlinear distortion for output amplitudes of this value.

**T10.7.**  The principal effect of offset current, bias current, and offset voltage of an amplifier is to add a dc component to the signal being amplified.

**T10.8.**  Harmonic distortion can occur when a pure sinewave test signal is applied to the input of an amplifier. The distortion appears in the output as components whose frequencies are integer multiples of the input frequency. Harmonic distortion is caused by a nonlinear relationship between the input voltage and output voltage.

**T10.9.** Common mode rejection ratio (CMRR) is the ratio of the differential gain to the common mode gain of a differential amplifier. Ideally, the common mode gain is zero, and the amplifier produces an output only for the differential signal. CMRR is important when we have a differential signal of interest in the presence of a large common-mode signal not of interest. For example, in recording an electrocardiogram, two electrodes are connected to the patient; the differential signal is the heart signal of interest to the cardiologist; and the common mode signal is due to the 60-Hz power line.

**T11.1.** For $v_{GS} = 0.5$ V, the transistor is in cutoff, and the drain current is zero, because $v_{GS}$ is less than the threshold voltage $V_{to}$. Thus, the drain characteristic for $v_{GS} = 0.5$ V lies on the horizontal axis. The drain characteristic for $v_{GS} = 4$ V is identical to that of Figure 11.11 on page 567.

**T11.2.** The results of the load-line analysis are $V_{DS\,min} \cong 1.0$ V, $V_{DSQ} \cong 2.05$ V, and $V_{DS\,max} \cong 8.2$ V.

**T11.3.** $R_S = 2.586$ kΩ.

**T11.4.** We can determine that $g_m = 2.5$ mS. The $Q$-point values are $V_{DSQ} = 5$ V, $V_{GSQ} = 2$ V, and $I_{DQ} = 0.5$ mA.

**T11.5.** **a.** a short circuit; **b.** a short circuit; **c.** an open circuit.

**T11.6.** See Figure 11.31(b) and (c) on page 587. The NMOS is on and the PMOS is off.

**T12.1.** **a.** 3; **b.** 2; **c.** 5; **d.** 7 and 1 (either order); **e.** 10; **f.** 7; **g.** 1; **h.** 7; **i.** 15; **j.** 12; **k.** 19.

**T12.2.** $V_{CE\,min} \cong 0.2$ V, $V_{CEQ} \cong 5.0$ V, and $V_{CE\,max} \cong 9.2$ V.

**T12.3.** $\alpha = 0.9615$, $\beta = 25$, and $r_\pi = 650$ Ω. The small-signal equivalent circuit is shown in Figure 12.26 on page 625.

**T12.4.** **a.** $I_C = 0.8830$ mA and $V_{CE} = 4.850$ V; **b.** $I_C = 1.872$ mA and $V_{CE} = 0.2$ V.

**T12.5.** We need to replace $V_{CC}$ by a short circuit to ground, the coupling capacitances with short circuits, and the BJT with its equivalent circuit. The result is shown in Figure T12.5.

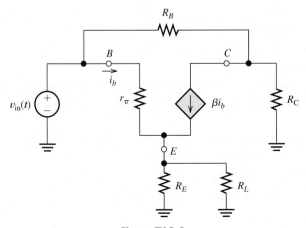

**Figure T12.5.**

**T12.6.** $A_v = -243.0$, $Z_{in} = 761.4\ \Omega$.

**T13.1. a.** The circuit diagram is shown in Figure 13.4 on page 649, and the voltage gain is $A_v = -R_2/R_1$. (Of course, you could use different resistance labels, such as $R_A$ and $R_B$, so long as your equation for the gain is modified accordingly.) **b.** The circuit diagram is shown in Figure 13.11 on page 655 and the voltage gain is $A_v = 1 + R_2/R_1$. The circuit diagram is shown in Figure 13.11 and the voltage gain is $A_v = 1 + R_2/R_1$. **c.** The circuit diagram is shown in Figure 13.12 on page 656 and the voltage gain is $A_v = 1$.

**T13.2.** $A_v = -8$.

**T13.3. a.** $f_{BCL} = 10$ kHz; **b.** $v_o(t) = 0.4975 \cos(2\pi \times 10^5 t - 84.29°)$.

**T13.4. a.** $f_{FP} = 707.4$ kHz; **b.** $V_{om} = 1$ V. **c.** $V_{om} = 4.5$ V; **d.** $V_{om} = 0.637$ V.

**T13.5.** See Figure 13.29 on page 672 for the circuit diagram. The principal effect of bias current, offset current, and offset voltage in amplifier circuits is to add a (usually undesirable) dc voltage to the intended output signal.

**T13.6.** See Figure 13.33 on page 676. Usually, we would have $R_1 = R_3$ and $R_2 = R_4$.

**T13.7.** See Figure 13.35 and 13.38 on pages 679 and 682, respectively.

**T13.8.** Filters are circuits designed to pass input components with frequencies in one range to the output and prevent input components with frequencies in other ranges from reaching the output. An active filter is a filter composed of op amps, resistors, and capacitors. Several filter applications are mentioned in the first paragraph of Section 13.10 starting on page 681.

**T14.1. a.** The force is 0.72 N pointing in the negative $y$ direction. **b.** The force is zero.

**T14.2.** $v = 164.9 \cos(120\pi t)$ V.

**T14.3.** 1.2 V.

**T14.4. a.** $B_{gap} = 0.5357$ T; **b.** $L = 35.58$ mH.

**T14.5.** The two mechanisms by which power is converted to heat in an iron core are hysteresis and eddy currents. To minimize loss due to hysteresis, we choose a material for which the plot of $B$ versus $H$ displays a thin hysteresis loop. To minimize loss due to eddy currents, we make the core from laminated sheets or from powdered iron held together by an insulating binder. Hysteresis loss is proportional to frequency and eddy-current loss is proportional to the square of frequency.

**T14.6. a.** $I_{1rms} = 0$, $I_{2rms} = 0$, $V_{1rms} = 120$ V, $V_{2rms} = 1200$ V. **b.** $I_{1rms} = 11.43$ A, $I_{2rms} = 1.143$ A, $V_{1rms} = 114.3$ V, $V_{2rms} = 1143$ V.

**T14.7.** Transformer $B$ is better from the standpoint of total energy loss and operating costs.

**T15.1.** The windings are the field winding, which is on the stator, and the armature winding, which is on the rotor. The armature current varies with mechanical load.

**T15.2.** See Figure 15.5(c) on page 751. If the field becomes disconnected, the speed becomes very high, and the machine can be destroyed.

**T15.3.** See Figure 15.5(d) on page 751.

**T15.4.** Speed regulation $= [(n_{no-load} - n_{full-load})/n_{full-load}] \times 100$ percent.

**T15.5.** To obtain the magnetization curve, we drive the machine at constant speed and plot the open-circuit armature voltage $E_A$ versus field current $I_F$.

**T15.6.** Power losses in a shunt-connected dc motor are: 1. Field loss, which is the power consumed in the resistances of the field circuit. 2. Armature loss, which is the power converted to heat in the armature resistance. 3. Rotational losses, which include friction, windage, eddy-current loss, and hysteresis loss.

**T15.7.** A universal motor is an ac motor that is similar in construction to a series-connected dc motor. In principle, it can be operated from either ac or dc sources. The stator of a universal motor is usually laminated to reduce eddy-current loss. Compared to other single-phase ac motors, the universal motor has a higher power-to-weight ratio, produces a larger starting torque without excessive current, slows down under heavy loads so the power is more nearly constant, and can be designed to operate at higher speeds. A disadvantage of the universal motor is that it contains brushes and a commutator resulting in shorter service life.

**T15.8.** 1. Vary the voltage supplied to the armature circuit while holding the field constant. 2. Vary the field current while holding the armature supply voltage constant. 3. Insert resistance in series with the armature circuit.

**T15.9.** **a.** 80 V; **b.** 320 V.

**T15.10.** $I_A = 33.16$ A; $T_{dev} = 31.66$ Nm.

**T15.11.** **a.** $P_{dev} = 4600$ W; $T_{dev} = 36.60$ Nm; $P_{RA} = 200$ W; $P_{rot} = 124$ W.
**b.** speed regulation $= 4.25$ percent.

**T15.12.** 2062 rpm.

**T16.1.** **a.** The magnetic field set up in the air gap of a four-pole three-phase induction motor consists of four magnetic poles spaced 90° from one another in alternating order (i.e., north-south-north-south). The field points from the stator toward the rotor under the north poles and in the opposite direction under the south poles. The poles rotate with time at synchronous speed around the axis of the motor. **b.** $B_{gap} = B_m \cos(\omega t - 2\theta)$ in which $B_m$ is the peak field intensity, $\omega$ is the angular frequency of the three-phase source, and $\theta$ denotes angular position around the gap.

**T16.2.** Five of the most important characteristics for an induction motor are:

  **1.** nearly unity power factor;

  **2.** high starting torque;

  **3.** close to 100 percent efficiency;

  **4.** low starting current;

  **5.** high pull-out torque.

**T16.3.** power factor $= 88.16$ percent lagging; $P_{out} = 3.429$ kW; $I_{line} = 10.64$ A rms; $P_s = 169.7$ W; $P_R = 149.1$ W; $T_{out} = 37.90$ Nm; $\eta = 87.97$ percent.

**T16.4.** $s = 5.556$ percent; $f_{slip} = 3.333$ Hz; 860 rpm.

**T16.5.** The stator of a six-pole synchronous motor contains a set of windings (collectively known as the armature) that are energized by a three-phase ac source. These windings produce six magnetic poles spaced 60° from one another in alternating order (i.e., north-south-north-south-north-south). The field points from the stator toward the rotor under the north stator poles and in the opposite direction under the south stator poles. The poles rotate with time at synchronous speed (1200 rpm) around the axis of the motor.

The rotor contains windings that carry dc currents and set up six north and south magnetic poles evenly spaced around the rotor. When driving a load, the rotor spins at synchronous speed with the north poles of the rotor lagging slightly behind and attracted by the south poles of the stator. (In some cases, the rotor may be composed of permanent magnets.)

**T16.6.** $\delta_2 = 13.39°$; power factor = 56.25 percent lagging.

# On-Line Student Resources

Users of the book can access the Student Solutions Manual in electronic form by following links starting from the website:

`www.pearsonhighered.com/hambley`

A pdf file for each chapter includes full solutions for the in-chapter exercises, answers for the end-of-chapter problems that are marked with asterisks, and full solutions for the Practice Tests.

The MATLAB folder contains the m-files discussed in the book.

# Index

# List of Examples (Cont.)